Basic Electronics:
Devices, Circuits and Systems

Basic Electronics:
Devices, Circuits and Systems

Second Edition

By Michael M. Cirovic
California Polytechnic State University
San Luis Obispo, California

Reston Publishing Company, Inc.
A Prentice-Hall Company
Reston, Virginia

Library of Congress Cataloging in Publication Data

Cirovic, Michael M
Basic electronics.

Includes index.
1. Electronics. I. Title.
TK7815.C53 1979 621.381 78-15014
ISBN 0-8359-0370-2

10 9 8 7 6

Printed in the United States of America

Dedicated to my mother
Smilja
to whom I am grateful
for much more than my existence

Contents

Preface

The age we live in has been termed the "Age of Technology." Electronics has played a major role in this technology: It has provided us with the hardware to improve communication, to enable us to explore the universe, to help physicians provide us with better health care, to process huge amounts of information and data (through digital computers), to liberate us from the more mundane tasks. The list of accomplishments, as well as of future possibilities, is quite lengthy. In many cases, the hardware of the technological revolution has created just as many problems as it has solved. With this in mind, the future technologist must be aware of, and concerned with, the consequences of this technology on people and their environment to a much greater extent than his predecessors seem to have been.

The purpose of this book is three-fold: first, to introduce a variety of electronic devices, their basic operation, and their characteristics; second, to illustrate how these devices are used in simple electronic circuits, as well as how these circuits are analyzed and designed; third, to present complex electronic systems as simple extensions and examples of the utilization of devices and simple circuits.

The prerequisites for the proper understanding of the material presented here are basic college mathematics and the first course in electric circuits.

The Introduction is presented as a review of basic circuit principles from a practical "how to use" standpoint. This is done in the hope of making the book easier to understand as well as showing unifying principles in all of electronics—i.e., the same basic laws and relationships that govern and describe the behavior of the simplest circuit apply to the most complex system. In addition, operational characteristics of resistors, capacitors, and inductors are presented in the Introduction.

Part 1 presents the basic physics and physical principles that make the understanding of the operation of electronic devices possible. This is a brief description, not a mathematical discussion, leading to the terminal characteristics of devices. The terminal characteristics directly lead to and suggest biasing schemes which follow. With the devices properly biased, terminal characteristics under signal conditions are presented, leading to the utilization of models and equivalent circuits in the systematic analysis of circuits containing devices.

Part 2 deals with the application of the devices introduced in Part 1 in simple circuits. Methods of analysis stressing approximations and practical considerations are used, and some design problems are illustrated.

In Part 3, more complex electronic circuits and systems are described. In some cases actual circuits are examined; in other cases a block diagram approach is used.

A building-block approach to electronics is used: Starting with simple, basic concepts, complex systems become understandable. The reader is able to see how even the most complex electronic system is a logical extension of very simple circuits.

I would be remiss if I did not acknowledge the contribution of Sherry Goldbecker in making the book more readable, and the help of my student, Harry Banks, with the pictures in the Introduction. I am also grateful to the manufacturers who provided much useful information.

Michael M. Cirovic

Introduction

In this introductory section we shall present the most basic concepts, theorems, and laws in a simple manner, so that you can understand why they exist and how they can be used. A rigorous theoretical approach is left to the mathematician. The technologist, however, must be thoroughly grounded in the basic principles governing all electronics.

The circuits and systems described throughout this book contain active electronic devices, which will be treated in Part 1. But even the most complicated electronic systems also contain passive elements, i.e., resistors, capacitors, and inductors. We shall, therefore, briefly discuss these three passive elements.

The Resistor. A resistor is a two-terminal element. Between its terminals it exhibits a voltage drop which is directly proportional to the current passing through it, as shown in Fig. I-1(a). We specify the size of the resistor in ohms (Ω), which is the ratio of voltage to current. Thus in Fig.I-1(b), resistor 1 has a smaller resistance than resistor 2. Although both resistors have the same voltage v, it causes a larger current i_1 in resistor 1 and a smaller current i_2 in resistor 2. This relationship between voltage and current, called Ohm's Law, can be stated in an equation

$$v = iR \qquad \text{(I-1)}$$

This description suffices for an *ideal* resistor. In the real world of electronic circuits, resistors are manufactured and only *approximate* the voltage-current relationships just indicated.

Resistors are made in a number of ways. One type is the *wirewound* resistor. In this type of resistor, the resistance to electrical current is provided by a length of special resistance wire wound inside an insulating tube, which is further coated by insulating enamel. The size and rating of the resistor is usually printed on the resistor. Examples of this type of resistor are shown in Fig. I-2.

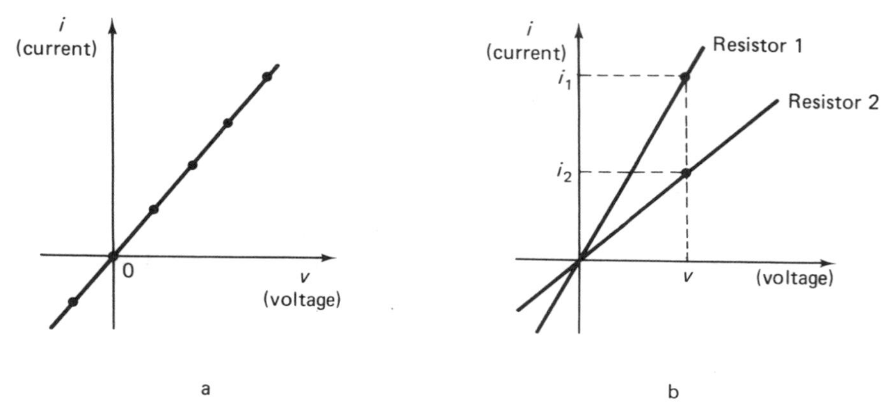

Figure I-1. Voltage-current relationship in a resistor.

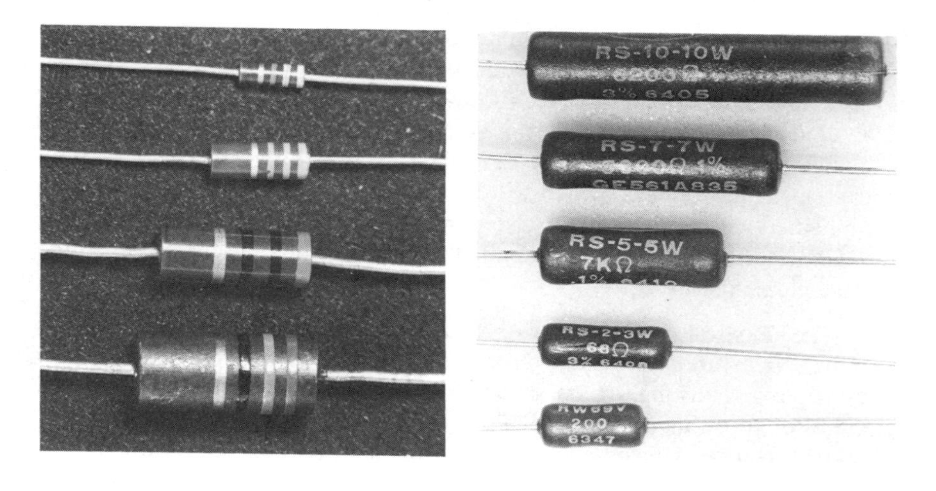

Figure I-2. Assorted resistors: (a) carbon composition—$\frac{1}{4}$W (watt), $\frac{1}{2}$W, 1W, 2W (*top to bottom*); (b) wirewound.

A second type of resistor, probably the most common, is the *carbon* or *carbon-composition* resistor. In this type of resistor, the resistance element is a carbon-composition rod.* Metal wire leads are attached to the two ends, and the rod is enclosed in a covering of bakelite or plastic. The size and rating of the resistor are indicated by color bands painted near one end of the resistor. The Electronic Industries Association (EIA) color code is shown in Fig. I-3. For example, a brown-green-orange-silver color code represents a 15,000 $\Omega \pm 10\%$ (or 15 $k\Omega \pm 10\%$) resistor.

---

*The exact nature of this "composition" is a trade secret, guarded by the manufacturers.

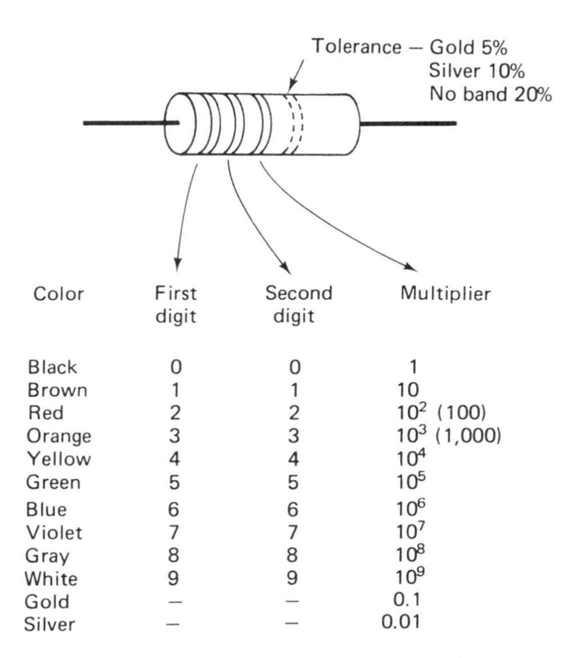

Color	First digit	Second digit	Multiplier
Black	0	0	1
Brown	1	1	10
Red	2	2	10^2 (100)
Orange	3	3	10^3 (1,000)
Yellow	4	4	10^4
Green	5	5	10^5
Blue	6	6	10^6
Violet	7	7	10^7
Gray	8	8	10^8
White	9	9	10^9
Gold	–	–	0.1
Silver	–	–	0.01

Figure I-3. EIA resistor color code.

To appreciate what at first must seem a strange choice of standard available (EIA) resistor values listed in Table I-1, you must know something about their manufacture. First, note, for example, that all $\frac{1}{2}$W (Watt–a unit of power) resistors of all assorted resistance values are of the same size as shown in Fig. I-2. The only way to gain this similarity is by slightly varying the composition in order to obtain a different resistance. The composition and the actual manufacturing process cannot be controlled exactly, so we can expect that we would not be able to produce a target valued resistor exactly. As a slight exaggeration, let us say that in trying to make many 50 Ω resistors, we can expect resistor values anywhere from 10 Ω to 100 Ω. Yet, we, as the manufacturer, would like to be able to sell all of the resistors. Therefore, we judiciously choose nominal resistor values to accompany tolerance limits. This choice is illustrated in Fig. I-4 for 20% tolerance. As can be seen, the nominal resistor values are such that the whole range is covered if the tolerance is taken into account; thus, a nominal 33 Ω 20% tolerance resistor can be anywhere from 26.4 Ω to 39.6 Ω. The lower limit overlaps with the upper limit of a 22 Ω 20% tolerance resistor, and the upper limit overlaps with the lower limit of a 47 Ω 20% tolerance resistor.

Precision resistors in tighter tolerances are also available but at an appropriate increase in cost.

Variable resistors, or *potentiometers*, are made in the same types as fixed resistors, i.e., wirewound and carbon. No standardization seems to

Table I-1. Standard EIA (Retma) Resistor Values by Tolerance

Tolerance	*5%*	*10%*	*20%*
N	10	10	10
O	11		
M	12	12	
I	13		
N	15	15	15
A	16		
L	18	18	
	20		
R	22	22	22
E	24		
S	27	27	
I	30		
S	33	33	33
T	36		
O	39	39	
R	43		
	47	47	47
V	51		
A	56	56	
L	62		
U	68	68	68
E	75		
S	82	82	
	91		
(in Ohms*)	100	100	100

*The range from 10 to 100 Ω is covered in this table. However, all decimal multipliers may be used to obtain the whole range, i.e., a 62 kΩ resistor is available with 5% tolerance, but not with 10% or 20%.

exist, but commonly available sizes are: 10 Ω, 20 Ω, 50 Ω, 100 Ω, 200 Ω, 500 Ω, 1 kΩ, 2 kΩ, 5 kΩ, 10 kΩ, 20 kΩ, 100 kΩ, 200 kΩ, 500 kΩ, and 1 MΩ (1,000,000 Ω). A potentiometer has a fixed resistance between two of its terminals (this is the nominal resistance) and has a slider or wiper third terminal. The resistance between the slider terminal and either of the other terminals may be varied almost continually, usually between zero and the nominal potentiometer resistance. A tolerance of usually ±10% or 20% is implied in the nominal value. The variation of the resistance provided by the slider is accomplished in from somewhat less than one turn of the control (about 320°) to some special-purpose potentiometers, which require 10 turns or more.

Besides these differences among potentiometers, the variation of resistance may be either linear for normal applications or logarithmic for special applications (e.g., audio circuits). The potentiometer is a very versatile circuit element, as attested by the vast variety of types and physical sizes. Some of these are shown in Fig. I-5.

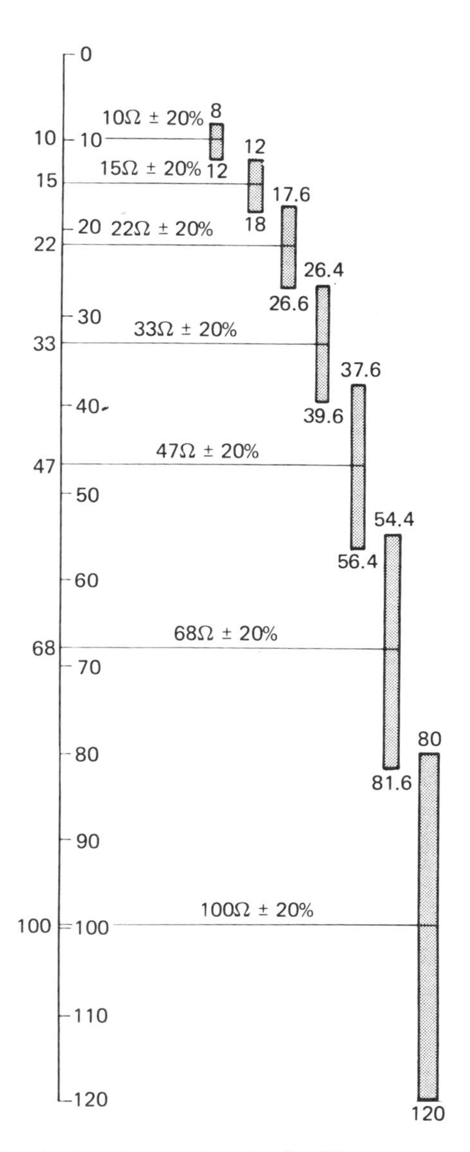

Figure I-4. Nominal resistor values in the 20 percent tolerance range.

So far, we have described only the makeup and types of ideal resistors and real resistors. What of the circuit properties of real resistors? First, a real resistor only approximates a linear V-I characteristic. Furthermore, in a real resistor, we cannot increase the current without bounds and expect an appropriately larger voltage drop across it. This power is dissipated in the form of heat given off to the surroundings. When the power (product of voltage and current) supplied to the resistor exceeds that

Figure I-5. Assorted potentiometers.

which the resistor can dissipate in terms of heat, a failure occurs (usually in the form of an open and occasionally in the form of a short) and the linearity of the *V-I* characteristics is permanently lost. In other words, the resistor is said to have burned out. So with real resistors, we need concern ourselves with the power rating.

Resistors dissipating upwards of a Watt tend to get hot (the higher the power, the hotter the resistor). Never touch a resistor in an operating circuit with your bare hands. Besides the dangers of electrical shock, you may burn your fingers.

An ideal resistor exhibits the same *V-I* characteristics at all voltage and current levels. A real resistor does not. An ideal resistor exhibits the same resistance at all temperatures. A real resistor changes its resistance (sometimes quite drastically) with temperature.

Another striking difference between ideal and real resistors becomes evident when high-frequency properties are examined. As might be expected, the real resistor may behave more as a capacitor or an inductor at very high frequencies, whereas the ideal resistor still remains a pure resistor.

The Capacitor. All types of capacitors are designed the same way. Two conducting surfaces are separated by a dielectric.* This basic scheme, however, cannot even begin to suggest the many different capacitor types in use.

Let us first describe the basic relationship between current and voltage in an ideal capacitor. Because of the presence of a dielectric, no net current can flow from one end of the capacitor to the other. If a *dc* voltage

---

*A dielectric is a substance that does not conduct but is capable of being electrically polarized.

is applied to a previously uncharged capacitor, a current is observed to flow for a time and then stop. During this time, electrons have accumulated on one of the conducting surfaces, not being able to pass through the dielectric. At the same time, an equal number of electrons have been repelled from the other conducting surface. This process is referred to as *charging* of a capacitor. The factors that determine when a capacitor ceases to charge (the charging and the current flow stop simultaneously) for a particular applied voltage are (1) the area of the conducting surfaces, (2) the type of dielectric used, and (3) the separation between the surfaces. A unified measure of these factors is called *capacitance (C)* and is measured in units of Farads (F). It has been determined experimentally that

$$C=\frac{eA}{d} \tag{I-2}$$

where e is the dielectric constant of the medium between the conducting plates; A, the area of the plates; and d, the separation between the plates.

It turns out that capacitance can also be determined experimentally as the ratio of the charge Q stored by the plates of the capacitor to the voltage V needed to set up the charge. Thus,

$$C=\frac{Q}{V} \tag{I-3}$$

In looking for the V-I relationship for a capacitor, we noted that when a fixed voltage was applied, a variable current (i.e., one that eventually stopped flowing) resulted. So, we see that the V-I relationship for a capacitor definitely is not one of proportionality as was the case with the resistor. Moreover, the V-I relationship for a capacitor depends on the instant of time at which it is evaluated; in other words, it is a function of time.

Note from Eq. (I-3) that if we allow the charge and voltage to be time-varying (and denote the time-varying quantities by lower-case symbols), we can write

$$q(t)=Cv(t) \tag{I-4}$$

Realizing that current is the flow of charge, i.e., $i=dQ/dt$, we finally have the V-I characteristic for a capacitor:

$$i(t)=\frac{dq(t)}{dt}=C\frac{dv(t)}{dt} \tag{I-5}$$

This equation tells us that, at any instant of time, the current supplied to a capacitor is directly proportional to the *rate* at which the capacitor voltage changes. In addition, a negative current supplied is equivalent to a charge taken away from the capacitor. So, if we multiply Eq. (I-5) by a minus sign, it can be read as: Charge removed from a capacitor (the negative current) results in a lowering of the voltage across the capacitor (the

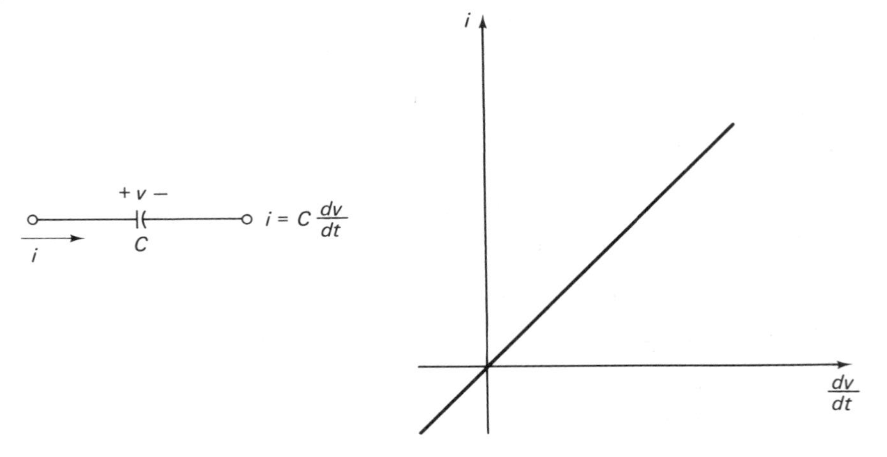

Figure I-6. Symbol and *V-I* characteristics of an ideal capacitor.

negative rate of change in voltage). These *V-I* characteristics are given in Fig. I-6.

The preceding discussion should also illustrate the one-to-one correspondence between the mathematical statement of fact—the equation—and our understanding and knowledge of the physical world. The mathematical formulation is nothing more than a shorthand method of stating and summarizing our observations of and conjectures about the physical world.

As a direct consequence of the relationship between voltage and current summarized in Eq. (I-5), *the voltage across a capacitor cannot change instantaneously*. The reasoning is as follows. If the voltage did change instantaneously (in zero time), dv/dt would tend towards infinity and would require an infinite current, a physical impossibility because charge can be neither created nor destroyed but is conserved. In all circuits containing capacitors, we shall make use of the fact that a capacitor prevents the voltage across its terminals from undergoing an instantaneous change.

We now turn our attention to the construction of capacitors. Figure I-7 shows some of the many varieties of capacitors. One of the most basic kinds is the paper capacitor in which the dielectric (paper) is sandwiched between two sheets of aluminum foil and tightly wound. The cylinder thus created is inserted into a cardboard tube filled with wax as a binder or encased in a plastic jacket. Various other dielectric materials are also used; mica and ceramic capacitors are quite common.

Similar in construction to the paper capacitor, but quite different in other respects, is the electrolytic capacitor. Here, one plate is made of aluminum foil, with a very thin file of insulating material (usually an oxide coating) acting as a dielectric. The second plate is formed by a conducting fluid termed the *electrolyte* (thus the name, *electrolytic*). Electrical contact

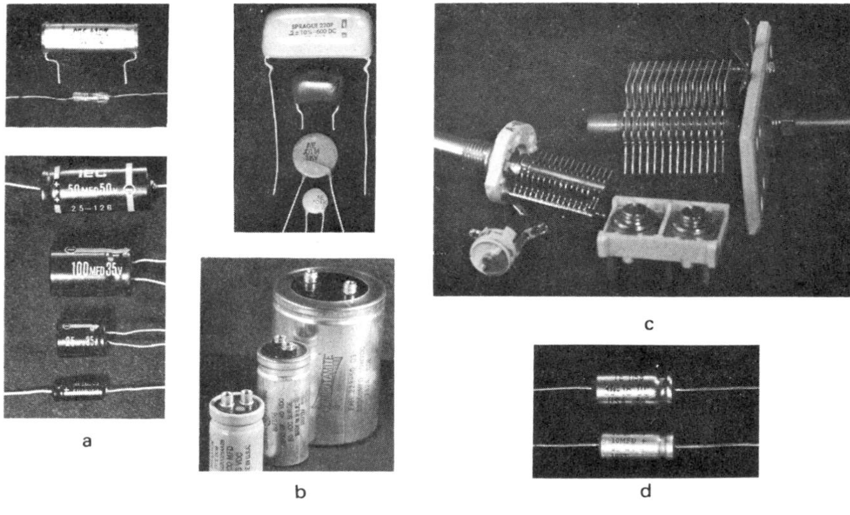

Figure I-7. Assorted capacitors: (a) polystyrene; (b) ceramic; (c) variable; (d) electrolytic.

to the second plate is provided by a strip of aluminum foil. The liquid electrolyte is kept between the two foils by soaking it into porous paper, which is then wound concentrically with the two plates. The cylinder is placed inside a metal case for protection. The rather close separation of the plates in the electrolytic capacitor contributes to a relatively high capacitance in a fairly small size. The major distinction of electrolytic capacitors is that they are polarized; that is, they can be safely charged only in the polarity (+ or −) indicated. A voltage of the wrong polarity can damage the thin insulating layer on the positive electrode and thus destroy the ability of the capacitor to store charge. Indeed, a voltage of the wrong polarity causes the capacitor to become a short circuit (essentially a zero resistance), and the rather high current that results causes the formation of gases in the capacitor, which may then burst or explode.

One type of electrolytic capacitor (called *tantalum* capacitor) uses tantalum electrodes instead of aluminum ones and offers greater capacitance than the aluminum capacitor of the same physical size. Many other types of capacitors employ a synthetic film of polystyrene, polyester, or Mylar.

Capacitors are made variable usually by varying the effective area of the plates. Two sets of semicircular metal plates are interleaved with air as the dielectric. One set is stationary. The other set is mounted on a shaft and may be rotated to cause a larger or smaller area of overlap and thus a larger or smaller capacitance.

Another method of varying the capacitance is suggested by Eq. (I-3), that of changing the separation between the plates. Sheets of foil separated by a sheet of mica (the dielectric) are mounted on a ceramic base with a screw passing through all three sheets. Tightening the screw brings the sheets of foil closer to each other and thus increases the capacitance. This type of capacitor is usually called a *trimmer* capacitor because of its small change in capacitance. It is used as a fine tuning (trimming) adjustment for larger capacitors.

There are profound differences between the real capacitors, whose construction we just examined, and ideal capacitors. An ideal capacitor stores all the charge supplied to its plates. The dielectric in a real capacitor is never a perfect insulator, and some (although very little) charge passes from one plate to the other. The amount of charge transmitted through a real capacitor, when compared to the charge stored in it, is extremely small and can be neglected in most cases. Therefore, a reasonable approximation is to represent a real capacitor by an ideal one. However, sometimes we must account for the small amount of charge that does leak through a real capacitor, and so we represent a real capacitor by the parallel combination of an ideal capacitor and an ideal resistor. This equivalent *leakage resistance* is usually very high (in the order of Megohms) and in most cases does not alter the results significantly.

An ideal capacitor is postulated to be capable of storing as much charge as it needs to, or of withstanding any voltage applied, in accordance with Eq. (I-3). The behavior of a real capacitor is given by Eq. (I-3) so long as the voltage does not exceed a certain maximum, which is specified on the capacitor in writing. Given a high enough voltage, any dielectric breaks down and conducts. When this conduction occurs, the capacitor is permanently damaged. Consequently, the maximum working voltage specified by the manufacturer should never be exceeded, even for the shortest amount of time.

Unlike resistors, capacitors do not dissipate the energy supplied to them. Instead, they store it in the form of an electric field, set up between the two plates. All of the energy supplied to a capacitor, with the exception of a relatively insignificant amount lost in the leakage process, can therefore be retrieved.

The Inductor. An inductor is formed by winding wire around a suitable mold to form a coil. When a current is caused to flow in the coil, a magnetic field, proportional to the current, is established in and around the coil. The motion of charged particles—in this case, electrons—is the cause of the magnetic field. It has been determined experimentally that if a magnetic field, varying with time, acts on a coil, a voltage drop across the coil is induced. This voltage drop is proportional to the rate at which the magnetic field varies. Let us now see what happens when a time-varying current is made to flow through a coil. The motion of electrons, which varies with time (i.e., the time-varying current), sets up a magnetic field in and around the coil, a field that is also time-varying. This variation of the magnetic field acts on the coil and induces a voltage drop across the coil.

We can conclude that, in an inductor, the voltage is directly proportional to the time rate of change in the current. The constant of propor-

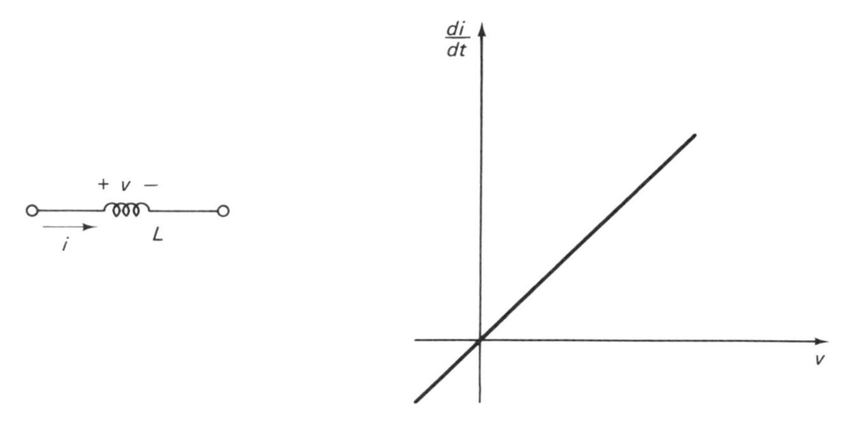

Figure I-8. Symbol and *V-I* characteristics of an ideal inductor.

tionality is called *inductance* (L) and measured in Henries (H). Thus,

$$v = L\frac{di}{dt} \tag{I-6}$$

Uniform current (i.e., not varying with time) sets up a uniform or stationary magnetic field. As a result, *no* voltage is induced in an inductor by a *dc* current. The *V-I* relationship of Eq. (I-6) is shown in Fig. I-8.

The inductance of an ideal coil is a function of the number of turns (N), the area (A) of the cross section of the coil, the length (l) of the coil, and the permeability of the medium inside the coil. In equation form:*

$$L = \mu\frac{N^2A}{l} \tag{I-7}$$

From this relationship we see that the inductance of a coil may be changed by inserting a material into the coil, which has a permeability (μ) other than air.The most common practice is to use a ferrite core or, if even larger inductance is needed, to wind the coil over a laminated-steel core. Examples of air and ferrite core inductors, as well as the larger laminated-steel core inductors, are shown in Fig. I-9.

A real inductor, sometimes called a *choke*, differs from the ideal inductor in many ways. First, the real inductor, containing many feet of wire wound over a suitable core, has a definite *dc* resistance, whereas the ideal inductor has none. This *dc* resistance is the result of ohmic resistance of the wire used in the windings and may be as small as a tenth of an Ohm or as large as a few hundred Ohms, depending on the coil. In a circuit representation of a real inductor, this *dc* resistance is accounted for by an ideal resistor in series with the ideal inductor. Another important difference is that in a real coil, we observe some very small capacitance to exist between the windings. This capacitance makes itself evident only at

*This is a relationship for a simple single layer coil. For other coil geometries, relationships are listed in a number of reference books.

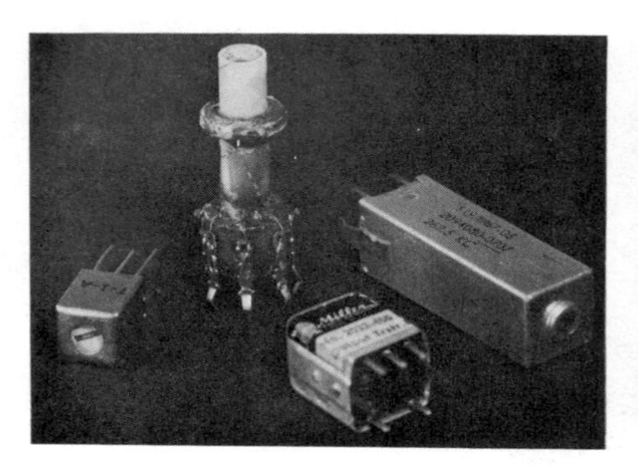

a

b

c

Figure I-9. Assorted inductors: (a) variable; (b) fixed; (c) tapped.

high frequencies; in most cases, it may be assumed to have negligible effect.

An inductor, like a capacitor, stores the energy supplied to it, but it stores energy in the form of a magnetic rather than an electric field. In a real inductor, some (usually very little) of this energy is lost in the series resistance.

In many circuit applications using inductors, it is important to protect the inductor with a metal shield of some sort. This shield protects against the effects of unwanted magnetic fields which may be present in the surroundings; it also shields other parts of the circuit from the effects of the magnetic fields set up by the coil.

To summarize, you must realize the limitations imposed by the physical world. No ideal resistors, capacitors, or inductors are to be found. Our calculations based on ideal elements are only approximations and at times may not be valid. We should, therefore, be ready to abandon the simple models and incorporate more of the properties of the real elements to make our calculations more closely conform to observed results.

Sources of Current and Voltage. When discussing sources of current and voltage, again we must discuss both ideal and real sources.

For convenience of analysis, we have idealized sources of current and voltage. Simply stated, an *ideal current source* is one that will cause a specified current i_s to flow out of its terminals, immaterial of the load it is required to drive or the resulting voltage. This situation is depicted in Fig. I-10(a).

An *ideal voltage source* is one that will maintain a specified voltage v_s across its terminals, immaterial of the load it is required to drive or the resulting current. This is depicted in Fig. I-10(b).

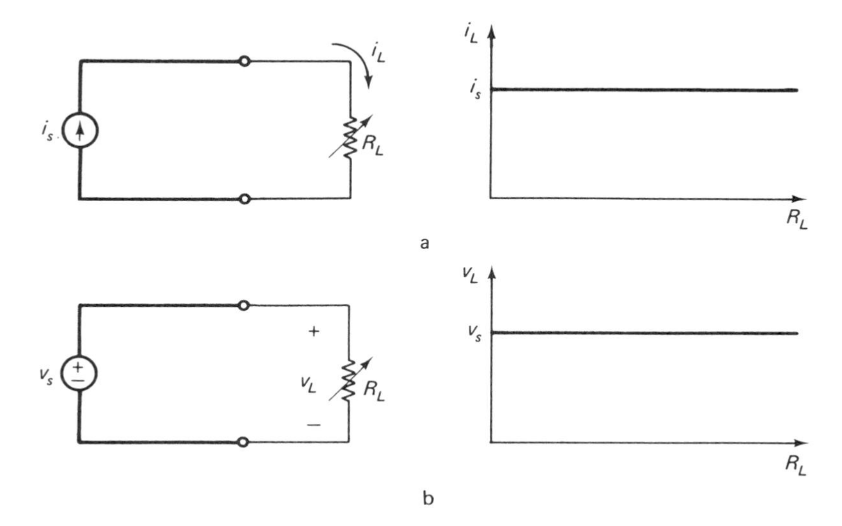

Figure I-10 Ideal sources: (a) current and (b) voltage.

Note that in the physical world as we know it, these ideal sources do not exist. Nevertheless, their importance cannot be stressed enough. Many examples in the real physical world approach either ideal current or ideal voltage generators. But the true importance of these ideal generators lies in the fact that they are extremely useful tools to be used in deriving models that correspond closely to the real world.

Any source, be it current or voltage, that can actually be constructed has some resistance, as shown in Fig. I-11. The important point here is that we can make a very useful and accurate model of a real source by employing ideal sources and other circuit elements (in this case, resistors) to represent the internal resistances of the real sources.

The current that a real current source is capable of delivering to a load decreases as the resistance of the load increases and falls to zero when the load becomes infinite. This decrease is shown in Fig. I-11(a). The current delivered to the load of resistance equal to the source resistance is exactly one-half of the nominal value of the current source i_s.

The voltage output of a real voltage source increases as the load resistance increases and approaches a maximum v_s when the load resistance becomes infinite. This increase is shown in Fig. I-11(b).

Both *ac* (alternating current) and *dc* (direct current) sources are considered here. However, if *ac* sources are depicted, we stipulate that the arrow on the current generator and the polarity on the voltage generator denote reference directions and polarities, those present at a particular instant of time.

When is a generator considered a source of current and when is it a source of voltage? As we shall see, there really is little difference. But a good procedure to follow is to use the current source representation if the

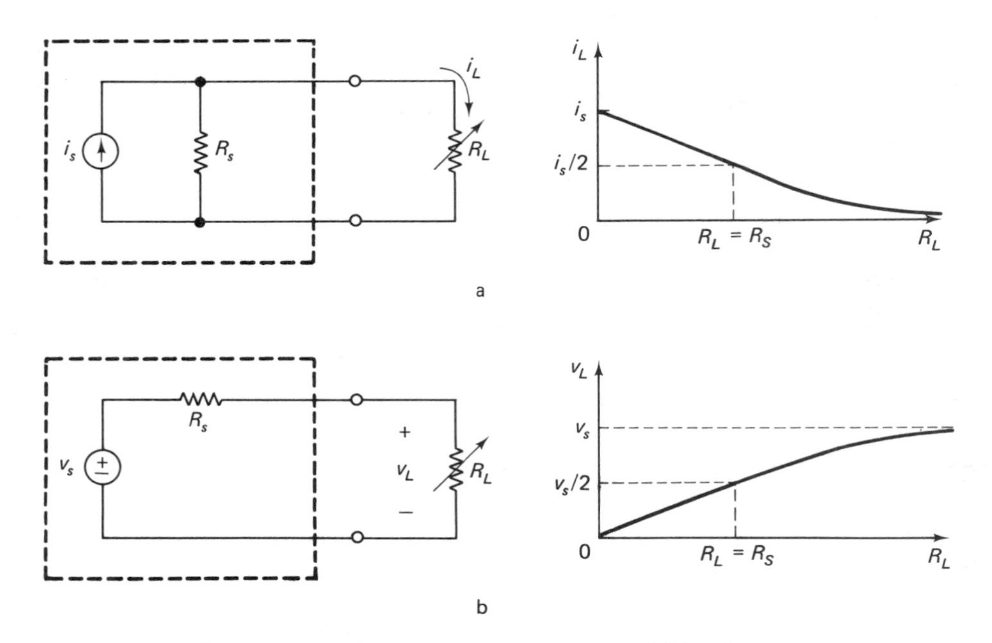

Figure I-11. Real sources: (a) current and (b) voltage.

internal resistance R_s is high and the voltage source representation if R_s is low.

In our discussion of electronic devices, we shall use sources of current and voltage whose nominal value depends on a current or voltage in another part of the circuit. Such generators are called *dependent generators*. Because of their nature, they must be treated with care. To emphasize the distinction between independent and dependent generators, a concentric circle will be added to the symbol to denote a dependent generator, as shown in Fig. I-12. We shall return to dependent generators in subsequent sections.

Equivalence. In a broad sense, the concept of equivalence helps us solve problems dealing with the physical world. We come up against many situations in nature that are just too complex for us to consider as they exist. We then idealize the situation and express the pertinent physical relationships in mathematical form. We say that the mathematical representation is the *equivalent* of the physical situation, with an obvious qualification as to the extent of approximation and idealization involved. For example, a ball rolls off the edge of a table and falls to the ground. We can write the pertinent equation relating the starting point of the ball, its mass, the acceleration due to gravity, and in a mathematical way describe the motion of the ball. Here, the equation is the mathematical equivalent of the physical situation.

Another way to look at the mathematical formulation is as a useful model of the physical situation. We use this model to obtain a mathemati-

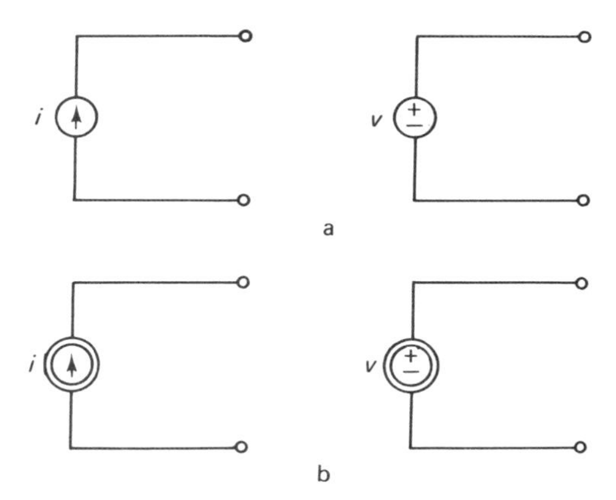

Figure I-12. Ideal generators: (a) independent and (b) dependent.

cal solution to the problem, which can then be interpreted in terms of the physical conditions it represents.

In electronics we make frequent use of the principle of equivalence. We have already noted one example in discussing the circuit representation of a generator. We saw that any generator could be represented by an ideal voltage generator in series with a resistance or by an ideal current generator in parallel with a resistance. Because both representations are valid and yield the same results, they are *equivalent*. There are three theorems–Thevenin, Norton, and superposition–that concern this principle and can be used advantageously.

Both the Thevenin and Norton theorems deal with equivalent representations of networks connected between two terminals. If the open-circuit voltage V_{oc} and short-circuit current I_{sc} can be measured (or otherwise determined) as shown in Fig. I-13, the equivalent resistance R_{AB} of the network between terminals A and B is given by

$$R_{AB} = \frac{V_{oc}}{I_{sc}} \tag{I-8}$$

The Thevenin theorem provides for one equivalent representation of a network appearing between terminals A and B characterized by the open-circuit voltage V_{oc} in series with resistance R_{AB}, as shown in Fig. I-14. An.examination of the equivalent circuit in Fig. I-14(b) shows the power of the Thevenin theorem. Let us imagine as complicated a circuit as we can and consider it to be in the box shown in Fig. I-14(a). The Thevenin theorem says that as far as observations at terminals A and $F$are concerned, we can replace the complicated circuit in the box by the simple Thevenin equivalent circuit containing one voltage source and one resistance (or impedance, as the case may be). One method of determining the values of Thevenin voltage and resistance is by direct measurement, as indicated in Fig. I-13 and using Eq. (I-8). However, the usefulness of Thevenin's theorem is even greater when dealing with circuits on paper as

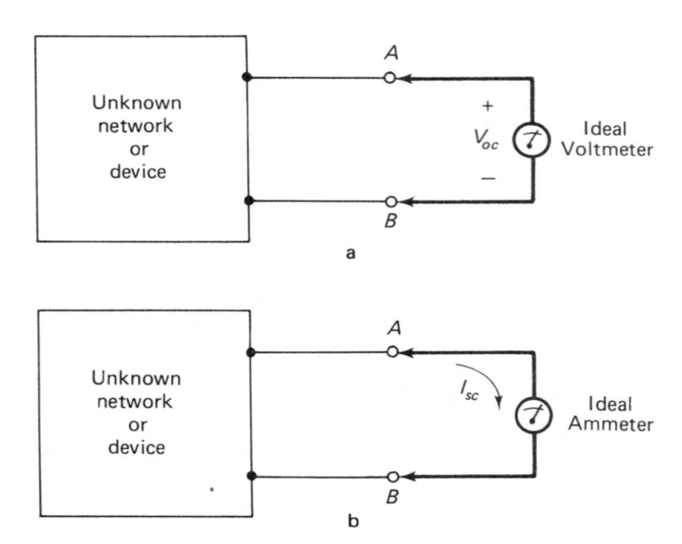

Figure I-13. Measuring: (a) the open-circuit voltage and (b) the short-circuit current.

opposed to circuits in the laboratory. In such cases, the open-circuit voltage at terminals A and B is determined by any suitable means.* To determine the equivalent resistance R_{AB}, we replace all voltage and current generators by their equivalent resistance. An independent ideal voltage generator has zero internal resistance and is replaced by a short circuit. An independent ideal current generator has infinite internal resistance and is replaced by an open circuit. The treatment of dependent generators, both voltage and current, is quite different. They cannot be replaced by a simple short or open circuit, and their dependence on values in other parts of the circuit must be included.

The Norton theorem offers an alternate but equivalent representation of a network connected between terminals A and B in Fig. I-14(a) by the short-circuit current I_{sc} in parallel with resistance R_{AB}, as shown in Fig. I-15. The value of the short-circuit current can be determined by placing a short between terminals A and B and calculating the resulting current through the short.* The resistance is calculated in the same way as in the application of Thevenin's theorem.

Thus, any network appearing between terminals A and B in Fig. I-14(a) can be represented either by the Thevenin equivalent in Fig. I-14(b) or by the Norton equivalent in Fig. I-15. The Thevenin and Norton representations are the simplest equivalent circuits for any network appearing between two terminals. The two representations themselves are equivalent; that is, both are just as valid and both eventually yield the same

*These means include any and all tools that circuit analysis provides.

*On paper, this procedure can be followed without damage to any components but should not be attempted in the laboratory.

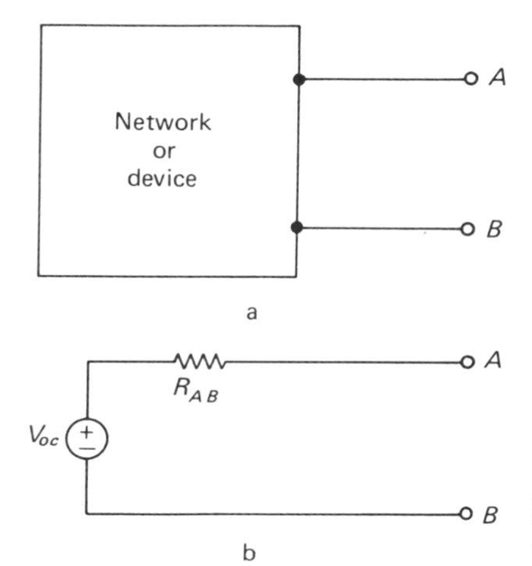

Figure I-14. The Thevenin equivalent circuit (b) of the network in (a).

results. In any case, the choice between Thevenin and Norton representations is one of convenience only, rather than any inherent superiority of one over the other.

We next consider the *superposition theorem*. In most circuits, the voltage across an element (or the current through it) results from the action of more than a single voltage or current generator. In these cases, the superposition theorem allows us to evaluate the voltage across an element (or the current through it) as the algebraic sum of the voltages across (or currents through) the element, because each source acts independently of the others. For example, suppose that we want to find the voltage across a certain resistor in a circuit containing three independent sources (either voltage or current, or a mixture of the two). To apply the superposition theorem, we replace all but one of the sources in a circuit by a short (if voltage sources) or by an open (if current sources). We then determine the voltage across the particular element as a result of the sole remaining source in the circuit. We thus proceed to determine the appropriate voltage resulting from each of the sources considered by itself, having replaced all other sources present in the circuit by a short (if a voltage source) or an open (if a current source). The algebraic sum of the voltages, due to each source acting alone, is totally equivalent to (i.e., exactly the same as) the voltage across the same element with all the sources acting simultaneously.

Application of the superposition theorem to a circuit containing independent voltage and current generators results in a systematic and usually easier method of obtaining desired results.

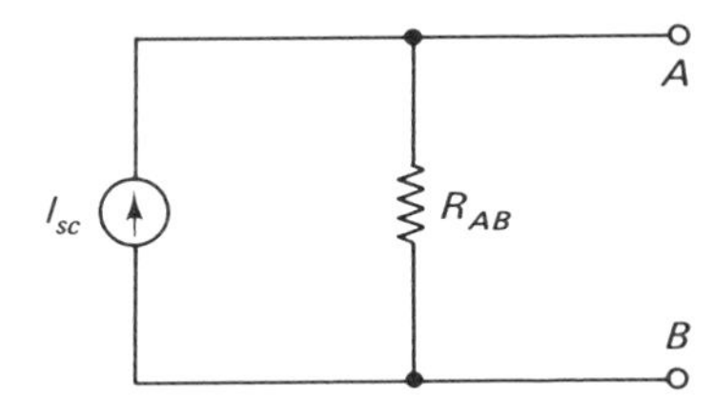

Figure I-15. Norton equivalent circuit of the network in Fig. I-14(a).

1

Devices

Electronic circuits utilize many different components. Besides resistors, capacitors, and inductors, there is a large group of active elements called *electronic devices*, or simply devices. In circuits, these devices exist in two different forms: either as separate components individually encased or as units with many components and devices in one package, called an *integrated circuit*. In either form, the device is usually the most important and, at the same time, the most complicated part of a circuit. You can understand and fully utilize electronic circuits only if you study each device separately at first and then as a part of the whole circuit. If you understand the working of the device, predicting the operation of a circuit containing the device becomes easy and straightforward.

Part 1 studies various semiconductor devices leading to their terminal characteristics and discusses the reasons behind these characteristics.

1

Semiconductor Physics

The rapid advances in semiconductor technology of the 1950's and 1960's revolutionized electronics. In replacing their vacuum-tube counterparts, semiconductor devices brought smaller size, increased lifetime, lower power consumption, lower operating temperatures, and eventually lower costs. To understand how these results were possible, we have to examine the physics of semiconductors. Our treatment will be descriptive rather than mathematical. The emphasis will be on the physical principles involved and not on their formal mathematical descriptions.

1.1 CLASSIFYING MATTER

Many ways of classifying matter have been devised. The one we shall use here is based on the ability of a material to conduct electricity or to sustain an electric current. We know that conduction takes place as a result of the motion of charged particles, usually electrons.* Therefore, we expect that the ability of any material to conduct should be directly proportional to how many charged particles inside the material can be set in motion. Materials (for example, metals) that have relatively large numbers of free electrons are very capable of sustaining an electric current and are termed *conductors*. Other materials having very few (or no) free electrons do not readily sustain an electric current under normal conditions and are called *insulators*. It should be realized that the terms are not absolute; that is, some conductors do not conduct as well as other conductors, while some insulators do not insulate as well as other insulators.

The reason for electrical conductivity, or the lack of it, in all materials becomes evident if the structure of the individual atoms, as well as the manner in which these atoms are arranged inside the material, is examined. All matter is made up of atoms in assorted configurations. Each atom has its electrons arranged in shells or orbits around the nucleus. The nucleus contains exactly the same amount of positive charge as the

*As we shall see, charged particles other than electrons also take part in conduction.

negative charge possessed by the electrons in orbit around it, thus causing each atom to be electrically neutral. The distinguishing factor among atoms of different materials is the size of the nucleus and the number of electrons in orbit around it. For example, an oxygen atom has a large nucleus with 16 electrons in orbit, whereas a hydrogen atom has a small nucleus with only one electron in orbit.

In atomic structure, nature has seen fit to establish a specific scheme for the arrangement of electrons into orbits called *shells*. Nature dictates that a maximum prescribed number of electrons can be sustained in any one shell or orbit at any given time. The shells are further divided into subshells (suborbits), each of which can sustain a maximum allowed number of electrons. This structure, together with the number of electrons allowed for the first three shells, is given in Table 1-1. Perhaps the easiest way to visualize this setup is to think of each electron as requiring its own little space, with only a limited number of spaces in each shell.

Table 1-1. Summary of Shell Structure

Shell Number (letter)	*Subshell Number (letter)*	*Maximum Number of Electrons Allowed*	*Total Possible in Shell*
1(L)	1(1s)	2	2
2(M)	1(2s)	2	8
	2(2p)	6	
3(N)	1(3s)	2	
	2(3p)	6	18
	3(3d)	10	

The shells and subshells closest to the nucleus are filled first, until all the electrons for that particular atom are accommodated. For example, in a hydrogen atom, the one and only electron is found in the L shell (the first shell). In an oxygen atom, its 16 electrons are distributed as follows: 2 in the first shell (the L shell); 8 in the second shell (2 in the 2s subshell, 6 in the 2p subshell); and 6 in the third shell (2 in the 3s subshell, 4 in the 3p subshell). Thus, the pattern is clear. The innermost shells and subshells are filled completely before any of the other shells. The outermost shell containing electrons is called the *valence shell*, and it plays the important role of determining the electrical as well as the chemical properties of elements.

Let us use aluminum, with a total of 13 electrons, for an example. The shell structure for aluminum is: $1s^2$, $2s^2$, $2p^6$, $3s^2$, $3p^1$ (where the superscript numbers indicate the number of electrons in that particular subshell). The valence shell is incomplete and contains three electrons. For the valence shell to be complete, it needs either to gain five electrons or to give up three electrons. In either case, the atom becomes *ionized*.* Aluminum is known to be trivalent; that is, it gives up three electrons

*An atom is electrically neutral. When it gains or loses electrons, it develops a net charge and is said to be *ionized*.

when reacting with other elements. The reason for giving up three and not acquiring five is that less energy is involved in liberating three electrons simply because of the lower number of electrons involved. We know that aluminum is a good conductor. This electrical property can also be traced to the valence shell of aluminum: The energy binding the three valence electrons to the nucleus of the atom is weak. Thus, only a small amount of energy is needed to liberate the three valence electrons of aluminum. At room temperature, enough energy is present in the form of thermal vibration. Therefore, in a bar of aluminum, made up of literally billions of atoms, there are countless free electrons that are available for conduction. Consequently, it is not in the least surprising that aluminum is a good conductor.

As an example of a good insulator, we can take any element which has a filled or complete valence shell. In these cases, no electrons are liberated at room temperature because of the strong binding forces between the electrons in the filled shells and the nucleus. A material made up of such elements is called *inert*; it does not provide any free electrons which could take part in conduction.

We can use this difference in the binding energy of valence electrons as the basis for another means of classifying materials. An electron in the valence shell is said to have energy corresponding to the *valence band*† of energy (or, simply, valence band). However, as a result of acquiring a specific amount of additional energy, an electron in the valence shell becomes free of the nucleus; with its new energy, it is characterized as being in the *conduction band* of energy (or, simply, conduction band). Differentiation among materials can be made on the basis of the amount of energy needed to liberate a single valence electron from the influence of the nucleus. The amount of energy between the highest energy in the valence band, labeled E_v, and the lowest energy in the conduction band, labeled E_c, is a characteristic of the material and is called the *energy gap*, labeled E_g. From the respective definitions, we can write:

$$E_G = E_C - E_V \tag{1-1}$$

In a metal or other good conductor at room temperature, there is an overlap between the conduction and valence bands, as shown in Fig. 1-1. Consequently, in a conductor, many electrons are free to take part in conduction and very little energy is needed to sustain fairly high electric currents.

In an insulator, the energy gap is large; that is, the conduction and valence bands are far apart, as shown in Fig. 1-1. As a result, a large amount of energy is required to liberate even a small number of electrons that could then contribute to conduction.

†Extremely large numbers of electrons are involved in even small samples, and each electron has a discrete amount of energy slightly different from any other electron. The range of energy possessed by all the electrons in all the valence shells constitutes a dense set of energy values called a *band*—in this case, a valence band.

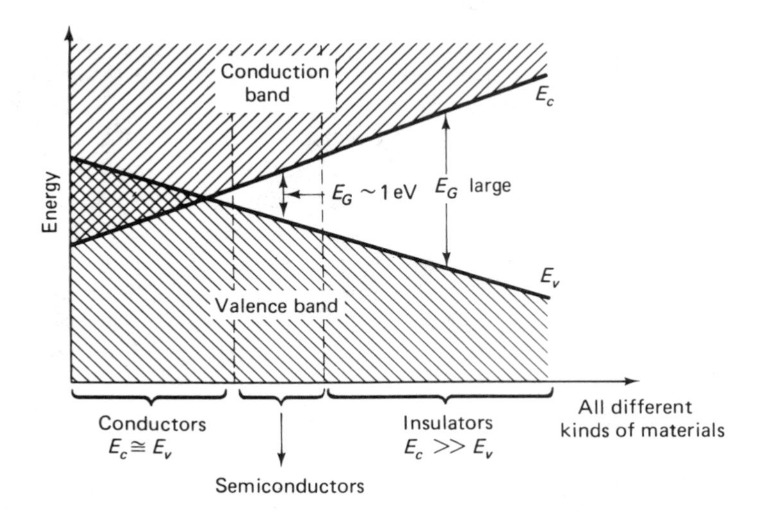

Figure 1-1. Classification of matter on the basis of conductivity.

1.2 SEMI-CONDUCTORS

As we can see, there is no sharp dividing line between conductors and insulators; there are materials which are neither good conductors nor good insulators. These materials are called *semiconductors* and are characterized by energy gaps on the order of 1 electron Volt (eV), as shown in Fig. 1-1. Semiconductor materials principally utilized in electronics are crystals of germanium and silicon.

An atom of silicon has a total number of 14 electrons, and its electron-shell configuration is: $1s^2$, $2s^2$, $2p^6$, $3s^2$, $3p^2$. The third shell is incomplete, and so it is the valence shell. In order to be complete, it needs either to acquire four electrons into the $3p$ subshell or to lose the four electrons which it already has in the $3s$ and $3p$ subshells. In either process, gaining or losing electrons, exactly the same number of electrons is involved; therefore, exactly the same energy is involved. Neither process is more likely; silicon neither acquires nor gives up electrons.

A silicon atom when close to other silicon atoms enters into a unique sharing of electrons called *covalent bonding*. In this scheme, each silicon atom shares two electrons with each of its four nearest neighbors. Two electrons shared by any two atoms are said to constitute a *covalent bond*, so that each silicon atom takes part in four covalent bonds. As a result of the covalent bonding scheme, atoms arrange themselves spatially in a characteristic structure, shown in Fig. 1-2. This basic structure is repeated millions of times in a crystal and is illustrated schematically in Fig. 1-3.

The situation is much the same in the case of germanium. Its valence shell has four electrons with space for four more. So germanium also forms covalent bonds, and a crystalline structure similar to that of silicon results.

The picture of a semiconductor that we have drawn thus far is somewhat oversimplified. In reality, atoms in a crystal are not stationary; they are three-dimensional, vibrating in essentially random fashion. The

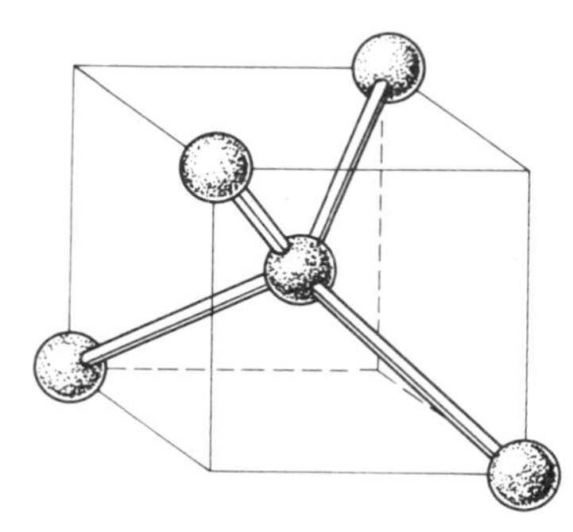

Figure 1-2. Diamond lattice construction.

temperature of the crystal is a quantitative measure of how rapid and violent this motion actually is. You should think of the positions of atoms depicted in Figs. 1-2 and 1-3 as average positions, and none of the atoms is ever completely stationary.

Electrons are also in constant motion, but their path cannot be predicted with certainty. As the temperature of the crystal increases, the vibration of the atoms becomes more violent; that is, excursions away from the average position become longer. As we shall see later, this increased vibration inside a crystal at elevated temperatures plays an important role in the limitations on semiconductor devices.

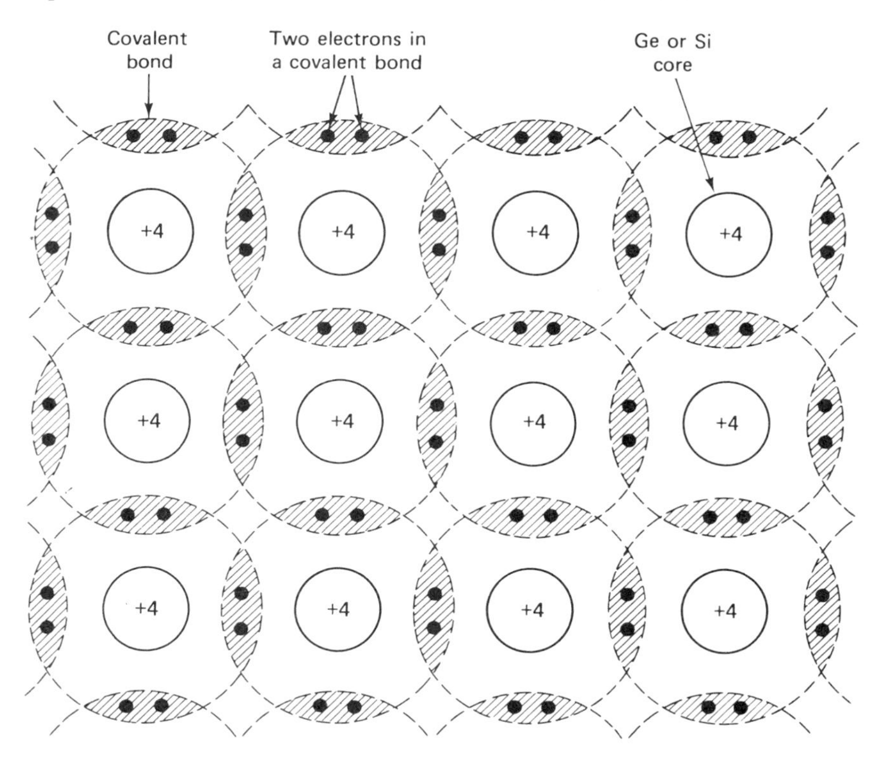

Figure 1-3. Schematic representation of a semiconductor crystal.

The binding forces on electrons exerted by the covalent bonds are quite strong, much stronger than the ionic binding forces on the electrons exerted by the nucleus. However, even at room temperature, the vibration of the atoms is enough to place a serious strain on the covalent bonds. Let us draw an analogy between the covalent bond and its nearest atoms with an elastic pod (the bond) containing two peas (the two electrons in the bond). The elastic pod is attached to two large balls (the two atoms between which the bond exists). The balls are in constant motion, vibrating about some average position. As the balls move apart, the elastic pod stretches and shrinks to accommodate their motion. Because the vibration is random, the two balls may move directly away from each other at some instant. The pod may not be able to stretch that much and may break, releasing one of its peas. Exactly the same process can happen in a semiconductor crystal. The vibration causes some of the covalent bonds to break, liberating free electrons.

Each atom before engaging in covalent bonding, started out being electrically neutral; that is, it had exactly the same positive charge in its nucleus as the negative charge carried by its electrons in orbit around the nucleus. The crystal, as a whole, is also electrically neutral because it contains only atoms that themselves are neutral. In the schematic representation shown in Fig. 1-3, we have conceptually combined the nucleus and the inner complete shells of each atom into one unit called the *core*. The core for both silicon and germanium has a net charge of +4 (the magnitude of the charge of one electron being used as a unit). The core together with four valence electrons, each one of which appears in each of the four surrounding covalent bonds, constitute a neutral atom.

Upon the breakup of a covalent bond, an electron has gained enough energy to become free. While in the bond, it had an amount of energy corresponding to the valence band; once the bond is broken, the electron energy corresponds to the conduction band. Each electron that is liberated by the breakup of a covalent bond has gained energy in the amount of the energy gap for that material.

Returning to our analogy, when the pod ruptures to release a pea, a hole is left behind in the pod where the electron used to be. In semiconductor terminology, the absence of an electron in a covalent bond is also termed a *hole*. Every time a covalent bond is broken, a hole-electron pair is formed. (From now on, "electron" denotes a free electron unless otherwise stated.) The liberated electron roams around in a random fashion and carries with it a negative charge. It therefore leaves behind a net positive charge in the proximity of the broken bond. The positive charge comes from the core of one of the atoms near the broken bond, but there is no way of identifying which core. We can associate the positive charge with the broken bond, specifically with the absence of an electron we have called a hole. Thus, the breakup of a covalent bond results in a free electron of negative charge and a hole of positive charge. In a pure (also called *intrinsic*) semiconductor, the number of free electrons is exactly

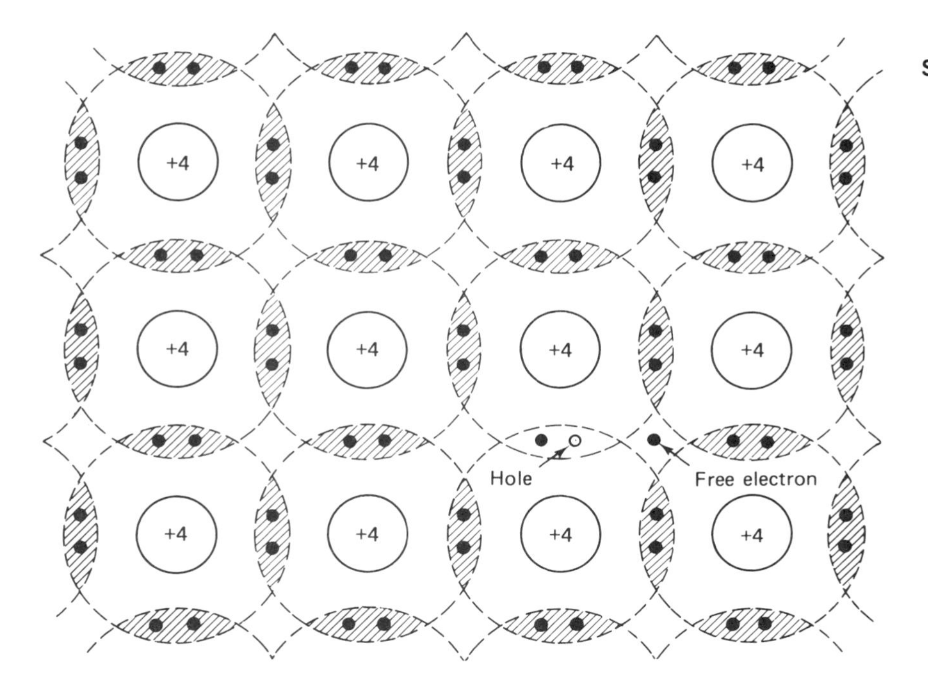

Figure 1-4. Crystal lattice showing a broken covalent bond.

equal to the number of holes, as illustrated schematically in Fig. 1-4. On the average, at room temperature, about 1 out of 10^{13} bonds are broken in silicon crystals, and about 1 out of 10^{10} bonds in germanium crystals. The fewer broken bonds in silicon are a reflection of its larger energy gap.

Let us examine what happens inside an intrinsic semiconductor if we apply a voltage to it. The negatively charged (free) electrons are attracted to the positive terminal, and an electric current results. Inside the semiconductor the electric field that is set up by the applied voltage also acts on the positively charged hole. As we noted earlier, the positive charge is actually in a nucleus of one of the atoms and cannot move as such. However, the electric field acting in the vicinity of the hole can cause a *bound* electron from an adjacent bond to move over and occupy the hole. This shift, in effect, accomplishes transport of a hole together with a positive charge, as shown in Fig. 1-5. If the electric field is directed from left to right, the hole will successively move to the right. The concept of the motion of a hole is at first hard to accept, for it may seem silly to say that "nothing" can move. We can again return to our analogy with pods and peas and imagine many large balls (cores) connected by many pods (bonds). Let us say that the pods are also interconnected in such a way that peas can be interchanged among them at will. When an attempted interchange occurs between one pod containing two peas and another pod

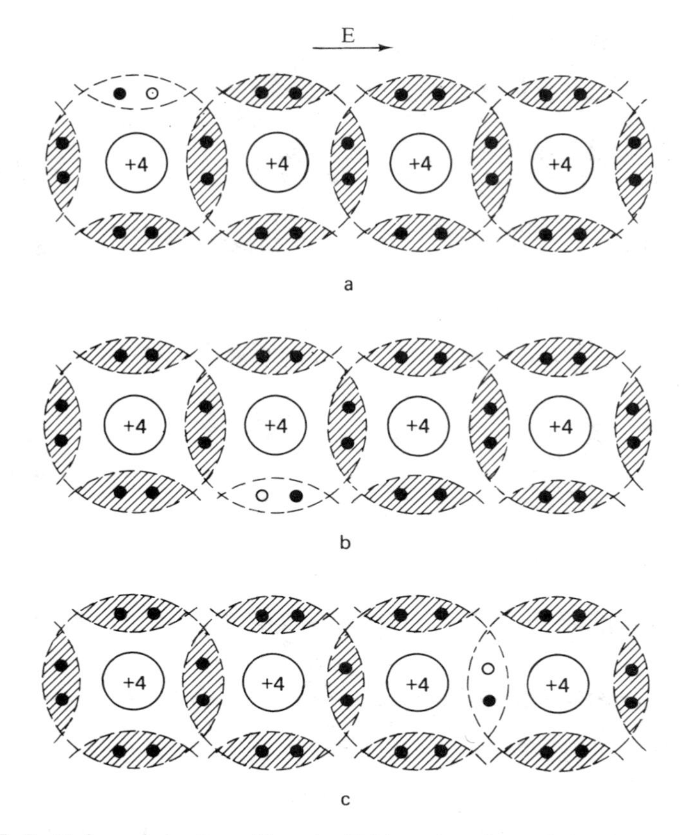

Figure 1-5. Hole conduction: Electric field acting from left to right and the successive motion of a hole (a), (b), and (c).

containing one pea and one hole, the pod that had two peas may wind up with only one pea and a hole. In other words, it traded a pea for a hole.

In reality, when we say that a hole has moved, it is actually a series of different bound electrons that has moved from one bond to another. The key here is that the electrons contributing to hole conduction are bound electrons; that is, they start in one bound state and wind up in another bound state. Note that it is specifically incorrect to say that the motion of holes is the result of free electron movement in the opposite direction. Experimental evidence for the motion of positively charged "particles" (holes) is contained in the Hall effect.*

We now see that current flow in a semiconductor results from the motion of two distinct charge carriers: negatively charged electrons and positively charged holes.

---

*See the discussion of this effect in M.M. Cirovic, *Semiconductors: Physics, Devices and Circuits* (Englewood Cliffs, N.J.: Prentice-Hall, Inc., 1971), pp. 38–41.

There are two inherent drawbacks to the use of intrinsic (i.e., pure) semiconductors. First, the number of charge carriers available to sustain an electric current is relatively low–hence, the name semiconductor. The second problem concerns recombination. Free electrons may be captured by vacancies in covalent bonds (holes), and in the process useful charge carriers are lost.

1.3 DOPED SEMICONDUCTORS

Conduction in semiconductor materials is greatly enhanced by the addition of carefully chosen impurities in precisely controlled amounts; this process is called *doping*.

Through the addition of appropriate impurities, semiconductors of two types can be formed. In one kind of semiconductor, called *N-type*, conduction results primarily from electron flow; whereas in the second type of semiconductor, called *P-type*, conduction by means of holes is the predominant effect.

N-type semiconductors are made by the addition of small amounts (typically, one part in a million) of phosphorus, arsenic, or antimony to the intrinsic semiconductor crystal. All of these elements have valence shells with five electrons. The impurity atoms take the place of silicon atoms in the crystal structure; they also form covalent bonds with neighboring

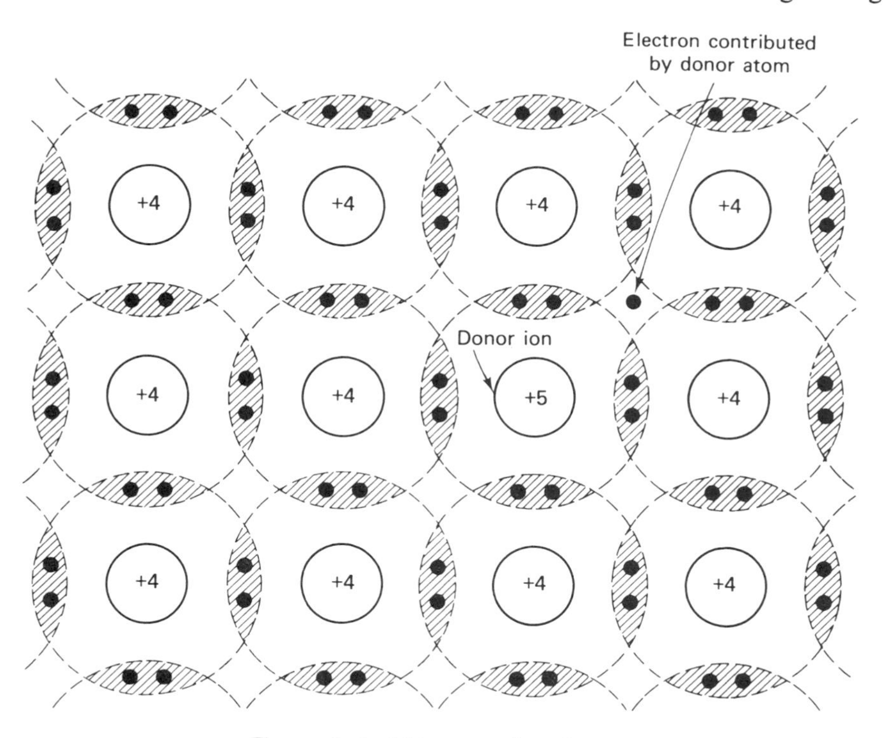

Figure 1-6. *N*-type semiconductor.

silicon atoms. Each impurity atom contributes one electron to each of four covalent bonds, and each has one *excess* electron that is not taking part in a covalent bond. This excess electron is only weakly bound to the core and, at room temperature, usually has enough energy to be considered as a free electron. This setup is shown schematically in Fig. 1-6. The addition of *pentavalent* atoms (those having five valence electrons), called *donor atoms* or just *donors*, increases the free electron population in a semiconductor without increasing the hole population. Therefore, the essential characteristic of an *N*-type semiconductor is that electrons are more (and usually much more) plentiful than holes.

Because current in an *N*-type semiconductor depends on electrons, they are called the *majority* (charge) *carriers*, and holes are referred to as the *minority* (charge) *carriers*.

P-type semiconductors are formed by the introduction of trivalent impurities into the intrinsic semiconductor. Typically, such impurities are boron, gallium, and indium. These impurity atoms all have three valence electrons. Upon replacing silicon atoms in the crystal, they cannot satisfy all four covalent bonds around them. One covalent bond near each of the *P*-type impurity atoms is incomplete; that is, it contains a hole. *P*-type impurity atoms are called *acceptors*, because each can accept one electron into its incomplete bond. A *P*-type semiconductor is illustrated schemati-

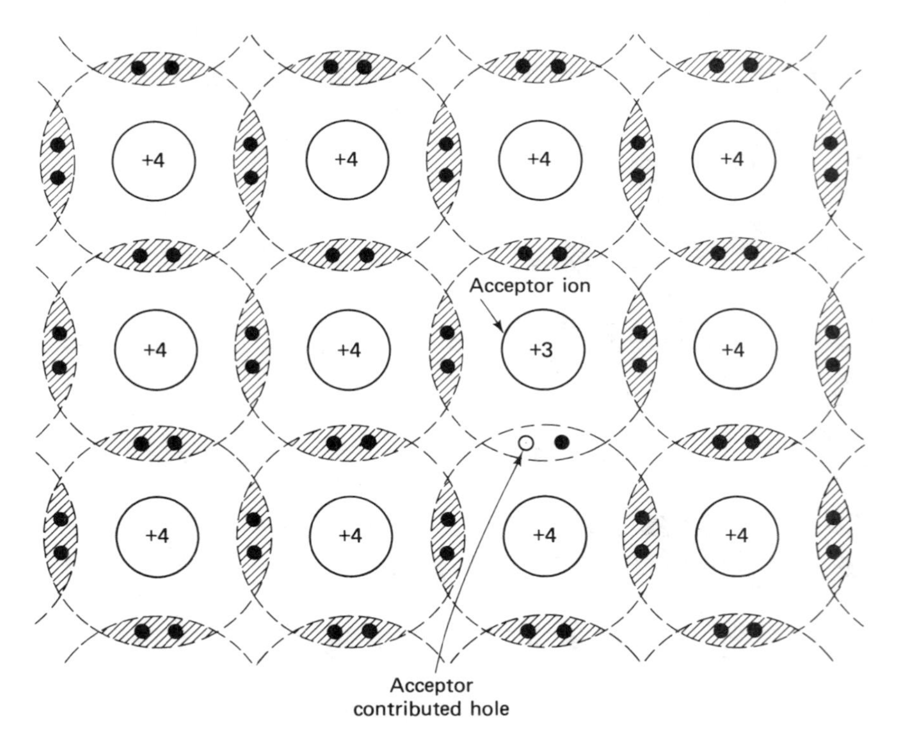

Figure 1-7. *P*-type semiconductor.

cally in Fig. 1-7. The effect of adding acceptor atoms to a semiconductor is to increase the number of holes without an increase in the number of electrons.

Because current in a P-type semiconductor is mainly the result of holes, they are termed *majority carriers*, and electrons in P-type semiconductors are called the *minority carriers*. We must emphasize that the terms "majority" and "minority" carriers only have significance and specific meaning when the type of semiconductor, N or P, is specified.

1.4 CONDUCTION IN SEMICONDUCTORS

There are two processes in semiconductor materials that account for the flow of current. The first process is called *drift*. For example, suppose that we apply a potential difference between the ends of an N-type sample of semiconductor. Because of the electric field that is established inside the semiconductor by the applied voltage, both majority carriers (in this case, electrons) and minority carriers (holes) will move in opposite directions to constitute a net current. We can now define drift as the motion of carriers under the influence of an electric field.

For any material, we can measure the current resulting from a known applied voltage. The ratio of the voltage to the current is the resistance, so

$$\frac{V}{I} = R = \rho \frac{L}{A} \tag{1-2}$$

where ρ is the *resistivity* of the material (in Ohm-cm), L is the length of the sample (in cm), and A is the cross-sectional area of the sample (in cm^2). Resistance is a function of the shape, size, and nature of the material; resistivity is a function of the nature of the material itself. For example, one ton of copper has the same *resistivity* as one ounce of copper. However, the *resistance* of the two quantities of copper is drastically different. The lower the resistivity of a material, the better it is able to conduct. In semiconductor materials, the higher the level of doping, the higher the number of free charge carriers, and the lower the resistivity. Thus, the effect of doping is to lower the resistivity. As you can see from Eq. (1-2), drift current is proportional to the voltage applied. For the same magnitude of voltage applied to two samples of silicon, one intrinsic and the other doped, we would expect a larger current in the doped semiconductor.

The second process in semiconductors contributing to conduction is known as *diffusion*. We frequently encounter it in nature. A drop of ink in a glass of water diffuses through the water until all the water is evenly discolored by the ink. In semiconductors the same process occurs. If, for example, we had a sample of P-type silicon and, in some manner, introduce an excess number of electrons into a small region of the sample, these electrons will diffuse through the sample so as to be evenly distributed throughout. This motion of minority carriers is strictly a statistical phenomenon and not the result of any electrostatic attraction or repulsion.

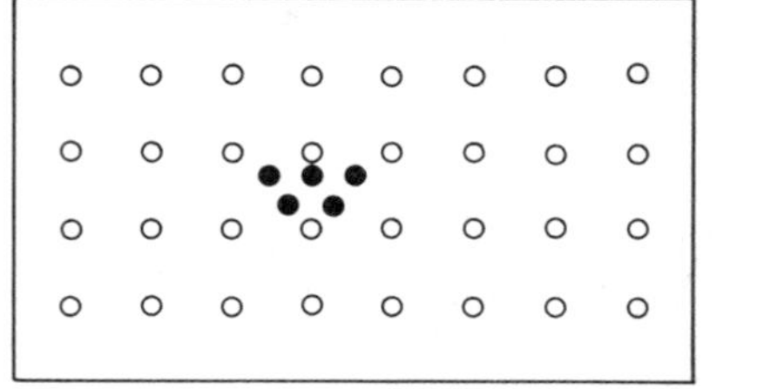

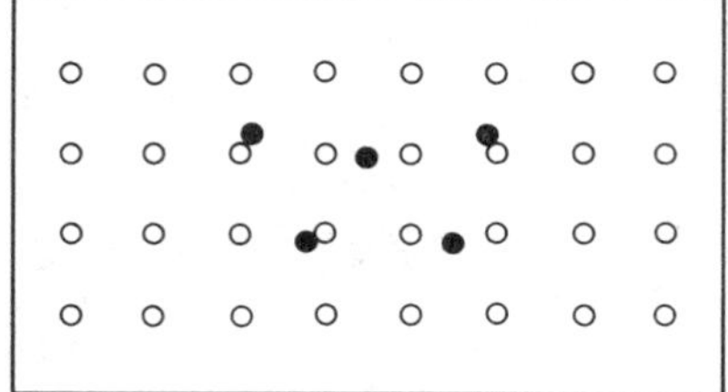

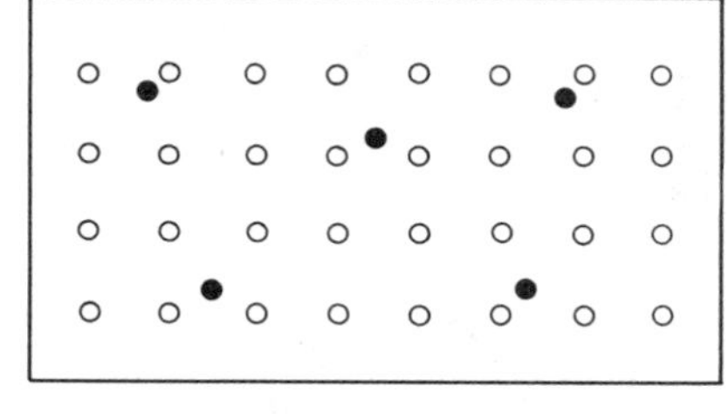

Figure 1-8. Diffusion: (a) large electron concentration in a localized area; (b) and (c) electrons diffuse *away* from the area of highest density.

Diffusion takes place in a direction away from the region of highest density of the minority carriers; that is, in our example, electrons diffuse away from the localized region into which they were introduced. This phenomenon is shown in Fig. 1-8. Diffusion is important because it is a process that describes the motion of excess minority carriers in a semiconductor: electrons in *P*-type and holes in *N*-type.

As a sidelight to diffusion, note that when a few minority carriers are in the process of diffusing, they encounter a tremendously large number of majority carriers. There is always danger that recombination of the two will occur; thus, majority carriers may be lost as far as conduction is concerned.

Review Questions

1. What are the pertinent properties of conductors?
2. Give examples of materials that are good conductors and state the reasons why they are good conductors.
3. What are the pertinent properties of insulators?
4. Give examples of materials that are good insulators and state why they are good insulators.
5. What is the mechanism by which conduction takes place inside a semiconductor?
6. What are some of the properties of an element that can be deduced from its electron-shell and subshell structure?

7. What are the ways in which atoms of unlike elements differ? In what ways are they similar?
8. What is the importance of the valence shell and valence electrons?
9. What is energy gap?
10. Characterize conductors, insulators, and semiconductors on the basis of energy gap.
11. Which material conducts electricity better, one with a small energy gap or one with a large energy gap? Why?
12. What is a covalent bond?
13. Under what conditions do atoms form covalent bonds?
14. What are some of the distinguishing characteristics of semiconductors as compared with conductors and insulators?
15. Give examples of semiconductors. Specifically, why are these materials classified as semiconductors?
16. What is the characteristic structure of a semiconductor?
17. What is an intrinsic semiconductor?
18. What is doping?
19. What is the name for an impurity atom that makes a semiconductor *N*-type?
20. Give examples of commonly used *N*-type impurities.
21. What is the characteristic that makes an impurity an *N*-type impurity?
22. What are the majority carriers in *N*-type semiconductors? Why?
23. What are the minority carriers in *N*-type semiconductors? Why?
24. What is the name for an impurity atom that makes a semiconductor *P*-type?
25. Give examples of commonly used *P*-type impurities.
26. What is the characteristic that makes an impurity *P*-type?
27. What are the majority carriers in *P*-type semiconductors? Why?
28. What are the minority carriers in *P*-type semiconductors? Why?
29. What are the two conduction processes in semiconductors? Give the conditions for each.
30. Describe conduction in an *N*-type semiconductor. Which carriers are responsible for most of the current? Why?
31. Describe conduction in a *P*-type semiconductor. Which carriers are responsible for most of the current? Why?
32. Describe the action as a result of the presence of an excess localized minority carrier concentration.

2 Semiconductor Diodes

With the introduction of the semiconductor diode, we start our study of semiconductor devices. If you can understand the basic physics of semiconductors, you can more easily understand and even predict the properties and behavior of these devices.

A diode results when a *P*-type semiconductor is brought into physical and electrical contact with an *N*-type semiconductor. A junction between the *N*-type and *P*-type semiconductors is formed, and, as we shall see, this *PN* junction is the essential building block for a variety of devices.

To manufacture a diode, a crystal of semiconductor is grown, usually 1 to 3 in. in diameter and 12 to 24 in. in length. The basic scheme for growing a crystal is depicted in Fig. 2-1. A "seed" crystal is inserted into the molten germanium or silicon and slowly drawn upwards. The molten semiconductor crystallizes and "grows" on the seed. A small, controlled amount of an impurity may be dissolved in the molten semiconductor if doped (either *N*- or *P*-type) crystals are desired.

Using a diamond saw, the crystal is then cut into thin slices called *wafers*. The wafers, once polished, are ready for further processing into diodes, transistors, or integrated circuits. From each wafer, literally hundreds of individual devices can be fabricated. The actual number depends on the type of device and the number of imperfections in the crystal (i.e., the wafer).

In actual practice, many different methods and processes are used in the manufacture of diodes, transistors, and integrated circuits. To illustrate some of the basic principles involved, we shall confine our discussion to one method. For additional processes, you should consult other books for discussion and explanation.*

Let us say that the wafers are cut from an *N*-type silicon crystal. The first step is to form a thin layer of silicon-oxide (SiO_2) on the surface of the wafer. We do this procedure by passing hot steam over the wafer. A thin

*For example, see: H. E. Thomas, *Handbook of Transistors, Semiconductors, Instruments and Microelectronics* (Englewood Cliffs, N.J.: Prentice-Hall, 1968).

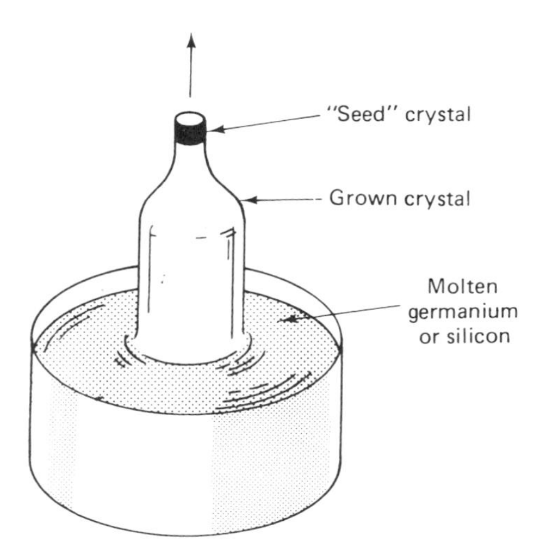

Figure 2-1. Crystal growing.

layer of oxide forms, as shown in Fig. 2-2(a). Next, the oxide is coated with a *photoresist*, a substance that reacts to light (usually fluorescent light). A *mask* is made, and the photoresist is exposed through the mask. With a suitable chemical agent, we then etch out the desired region of oxide. The agent is chosen so that it will react with only the oxide that has been exposed, i.e., the region where the photoresist has been exposed and removed. The resultant window in the oxide is shown in Fig. 2-2(b). A suitable acceptor impurity (for example, boron gas) is passed over the exposed region and *diffuses* into the lightly doped *N* region to form the *P* region, as shown in Fig. 2-2(c). Successive oxidation, masking, and etching processes expose a small part of the *N* and *P* regions onto which aluminum layers are deposited to form the metallic contacts for the connection of leads later on. Each of the many diodes thus formed on the single wafer is

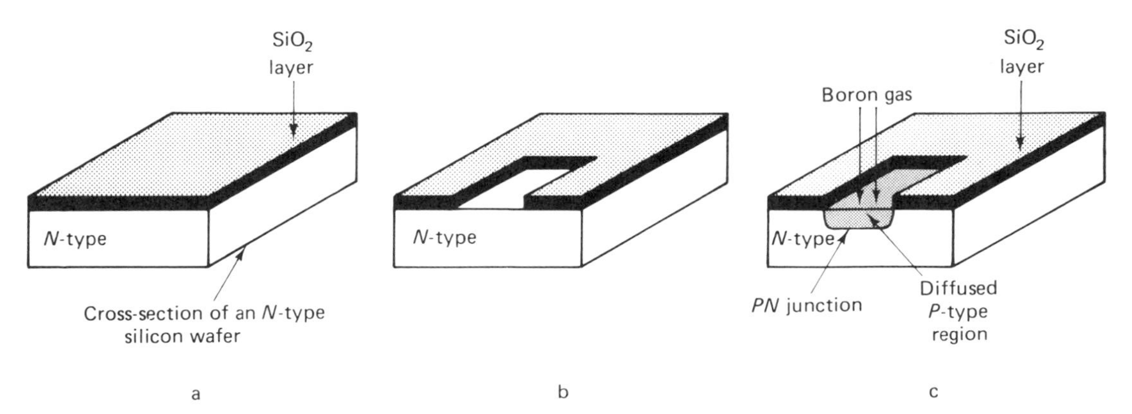

Figure 2-2. Formation of a *PN* diode: (a) oxide film is deposited, (b) the film is etched, and (c) the *P*-type region is diffused.

tested. Those which do not pass the test are inked and later discarded. The wafer is then cut into small pieces called *die*, each containing a single device, in this case a diode. Leads are attached. Finally the devices are packaged in protective jackets of epoxy or other suitable material.

2.1 DEPLETION REGION AND CONTACT POTENTIAL

Let us now consider the behavior of these devices, specifically *PN* diodes. Imagine that we have *P*-type and *N*-type samples separately and that we have just brought them into contact to form a diode. (It should be apparent that this method will not yield a diode. We can only use it in our mind to see what actually does happen when a *PN* junction is formed.) At the junction that is formed, one side (the *N* side) has a large number of electrons, whereas the other side (the *P* side) has a large number of holes. Conditions are quite favorable for diffusion to take place: Electrons from the *N* side diffuse across the junction into the *P* side, holes from the *P* side diffuse across the junction into the *N* side, as shown in Fig. 2-3(a). Before they were brought into contact, both sides were electrically neutral; that is, in the *N* region, for each electron with a negative charge, there is a donor ion in the crystal with a positive charge; in the *P* region, for each hole with its positive charge, we can associate an acceptor ion with a negative charge. As a result of the diffusion, the *N* side develops a net positive charge and the *P* side develops a net negative charge. This difference is due to the fact that the *N* side lost electrons and gained holes at the same time that the *P* side lost holes and gained electrons. The diode as a whole is still electrically neutral. It has not lost or gained any charge, but there is a localized imbalance of charge. Since the diode as a whole is still electrically neutral, the net positive charge in the *N* side must be exactly equal to the net negative charge in the *P* side.

The charges thus developed, negative in the *P* side and positive in the *N* side, are not mobile or free to move. They are *stationary* because they are caused by the ions in the crystal. The net negative charge in the *P* side

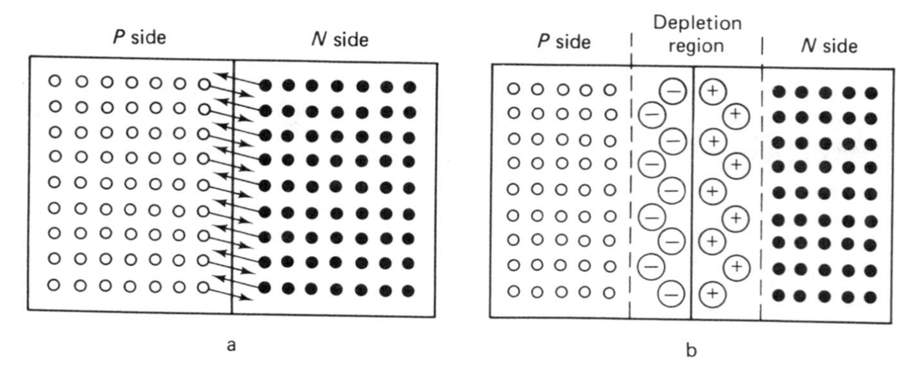

Figure 2-3. Formation of depletion region: (a) carriers diffuse across the junction until (b) equilibrium is reached.

is in the form of electrons that have fallen into holes near acceptor atoms; the net positive charge in the *N* side comes from the donor nuclei which have lost their fifth electrons.

Diffusion of carriers across the junction stops and equilibrium is reached when no additional carriers have enough energy to overcome the electric field which has built up at the junction as a result of the redistribution of charge. This situation is depicted in Fig. 2-3(b). The positively charged donor ions in the *N* side repel holes, while negatively charged acceptor ions in the *P* side repel electrons.

The region on either side of the junction where the stationary charges appear is called the *depletion region*. The name refers to the fact that this region has been depleted of mobile charge carriers (holes and free electrons). Other names often used for this region are transition region and space-charge region.

We stated that an electric field is set up across the depletion region. This field acts from + to −, or from the *N* side to the *P* side. A field acting over the width of the depletion region causes a potential difference between the *N* and *P* regions. This difference in potential is usually called the *contact potential* (or *barrier potential*) and is the equilibrium potential difference across a *PN* junction. By "equilibrium," we mean that neither holes nor electrons are crossing the junction, and the electric field across the depletion region is constant and not changing. The magnitude of the contact potential is in the order of a few tenths of a volt. It depends on the doping concentrations in the *N* and *P* regions as well as on the type of semiconductor (germanium or silicon) used.

2.2 DIODE WITH BIAS

In section 2.1, we described the diode under equilibrium conditions with no externally applied voltage, as shown in Fig. 2-4(a). (The diode circuit symbol is also shown.) Let us now consider the action of the diode when an external voltage is applied, or when the diode is *biased*.

When the positive end of a battery is connected to the *P* side (called the *anode*) and the negative end of the battery to the *N* side (called the *cathode*), a fairly high current is observed. This current is called the forward current, and the diode is said to be *forward biased*, as shown in Fig. 2-4(b). If we reverse the polarity of the battery and connect the negative end to the anode and the positive end to the cathode, a minute current is observed. This current is called the reverse current, and the diode is said to be *reverse biased*, as shown in Fig. 2-4(c).

Under conditions of forward bias, the externally applied voltage acts in opposition to the contact potential, thus lowering the effective or net potential at the junction. Physically, electrons enter the *N* side through the cathode lead, and as majority carriers in the *N* side, *drift* toward the junction. Because the potential barrier at the junction is lowered by the forward bias, electrons cross the junction and, as minority carriers in the *P* side, *diffuse* toward the anode lead. At the same time, holes, created at the

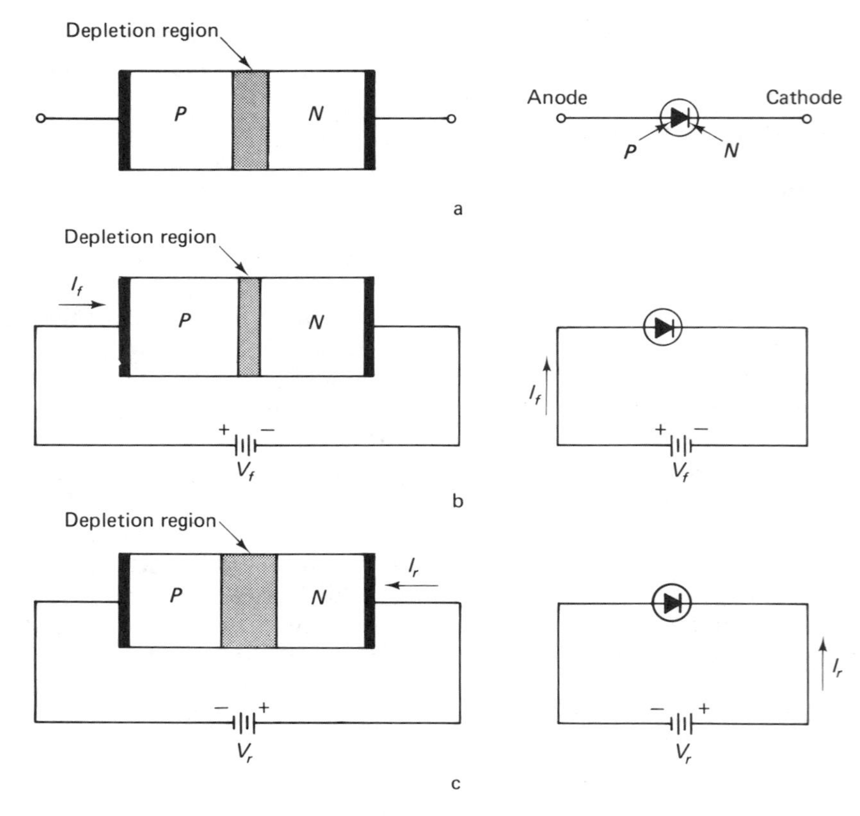

Figure 2-4. *PN* **junction diode: (a) no bias, (b) forward biased, (c) reverse biased.**

anode lead by the liberation of bound electrons, *drift* through the *P* side, cross the junction, and, as minority carriers in the *N* side, diffuse toward the cathode lead. At the cathode, these holes are filled by electrons from the external wiring. At any instant of time, the same number of electrons enter the *N* side as leave the *P* side.

In Fig. 2-4(b), the motion of electrons is from left to right, that of holes from right to left. However, remember that holes and electrons have opposite charges. So their motions in opposite directions cause conventional currents which add. The net current is in the direction of hole flow.

We can now explain the rather high current observed in the forward direction. It is due to the motion of majority carriers, which by definition are large in number, so that a large number of charge carriers should cause a high current.

Another consequence of applying a forward bias to the diode is the *narrowing* of the depletion region. This result is also illustrated in Fig. 2-4(b).

When the diode is reverse biased, the externally applied voltage acts in the same direction as the contact potential, thus *increasing* the effective potential barrier across the depletion region, while at the same time *widening* the depletion region itself. The external voltage causes holes in the N side and electrons in the P side to move in the direction of the junction. However, holes in N type and electrons in P type are both *minority* carriers. The resulting current is of necessity very low because minority carriers are by definition extremely few in numbers.

2.3 DIODE CHARACTERISTICS

In examining the terminal characteristics of other devices as well as diodes, we find an instrument called a *curve tracer* (depicted in Fig. 2-5) extremely useful. It gives us an instant display of the pertinent $V-I$ relationships. A camera may be attached to the display face of a curve tracer if a permanent record is desired.

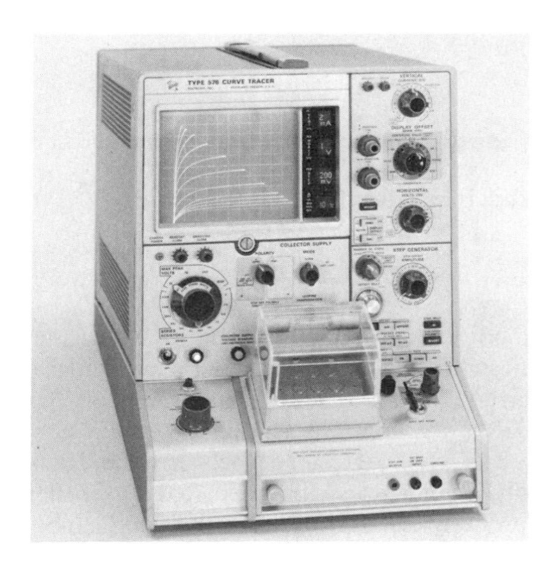

Figure 2-5. Commercially available transistor curve tracer. (Courtesy of Tektronix)

The forward characteristics of typical silicon and germanium diodes, obtained on a curve tracer, are shown in Fig. 2-6. We shall discuss the differences between silicon and germanium diodes later. Now we note only the general shape of the V-I characteristic and see that it is anything but linear.

The basic shape of any diode V-I characteristic is shown in Fig. 2-7 for reverse bias as well as forward bias. In the forward direction even small voltages result in appreciable currents; whereas in the reverse direction, the current is almost negligible until a certain voltage, labeled V_B on the characteristic, is reached. For voltages more positive than V_B, a good

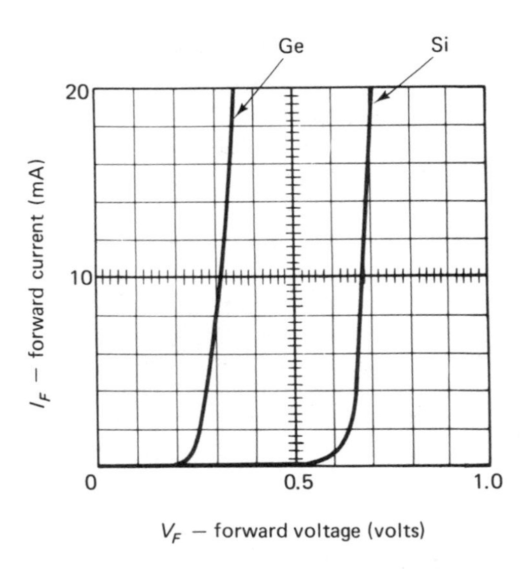

Figure 2-6. Comparison of silicon and germanium diode characteristics.

mathematical model for the diode characteristic is given by the *diode equation*:*

$$I = I_o(e^{V/V_t} - 1) \tag{2-1}$$

where I is the diode current resulting from an externally applied voltage V, I_o is a constant (whose value depends on the particular diode under test) with units of current, V_t is the voltage equivalent of temperature (whose value equals 0.025 V at room temperature), and e is the base of natural logarithms. There is good agreement between the actual diode current measured for a given voltage and that predicted by the diode equation with the constant I_o properly evaluated. To see the significance of this constant, we can evaluate the diode current for some value of applied voltage, say, for example, -0.25 V. At room temperature, this applied voltage in the diode equation yields a current of approximately $-I_o$. In fact, we get this answer for the current for any reverse voltage between V_B and about -0.2 V. In other words, the current *saturates* or levels off at a fixed value $(-I_o)$ and is not a function of the applied voltage. The negative sign indicates that the current actually flows in the reverse direction. The name attached to I_o now becomes evident; it is called the *reverse saturation current*. As we stated earlier, the reverse saturation current is different for different diodes. The factors that determine the value of I_o at room temperature for a particular diode are (1) the type of material (silicon or germanium), (2) the doping levels of the P and N regions, and (3) the geometry of the junction.

---

*For a derivation of the diode equation, see for example: Cirovic, *Semiconductors*, pp. 53–58.

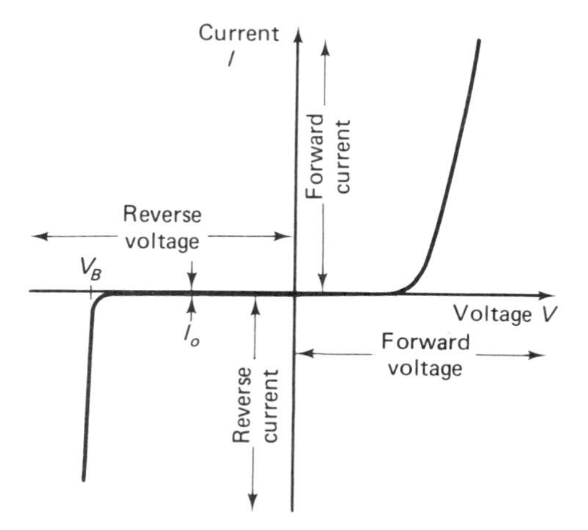

Figure 2-7. Typical diode characteristics.

Because of the differences in energy gap and the mobility of carriers in silicon and germanium semiconductors, diodes made of these materials exhibit different reverse saturation currents (as well as the difference in the forward characteristics already seen in Fig. 2-6). In general, germanium diodes have reverse saturation currents in the microAmpere (10^{-6}A) range, while silicon diodes show reverse saturation currents in the nanoAmpere (10^{-9}A) range.

Temperature plays an important part in the magnitude of the reverse saturation current. When temperature is increased, a larger number of covalent bonds are broken; thus, a larger number of charge carriers become available for conduction. For example, a 10°C rise in the temperature of the semiconductor causes the reverse saturation current to almost *double* in value. Obviously, a large number of devices are extremely temperature sensitive. A visual example of this phenomenon will be shown in the discussion of the bipolar junction transistor characteristics in Chap. 3.

We stated that the current quickly saturates as the voltage is advanced in the reverse bias direction. The current becomes essentially insensitive to additional changes in the reverse voltage. However, the voltage cannot be advanced in the reverse direction indefinitely. At some critical reverse voltage, labeled V_B in Fig. 2-7, we observe large changes in the reverse current for minute changes in the reverse voltage. Under these conditions, the diode is voltage saturated and is said to be in *breakdown*. The voltage at which this occurs, V_B, is called the *breakdown voltage*. There are two mechanisms that account for the sudden increase in reverse current.

The first mechanism is called *Zener breakdown*. It occurs when the applied reverse voltage sets up an electric field across the depletion region which is high enough to cause the breakup of covalent bonds. As a result, we have mobile carriers far in excess of those which had set up the reverse saturation current. We observe, therefore, a sudden increase in the current because of a sudden increase in the number of minority charge carriers.

The second mechanism is called *avalanche breakdown*. Here the highly accelerated carriers collide with the stationary ions inside the depletion region. A high enough field exists in the depletion region so that the carriers are moving fast enough (with enough energy) to break covalent bonds upon impact. This collision results in the liberation of many more mobile charge carriers, which are in turn accelerated in such a way that they now have the capability of producing other free charge carriers upon collision. The effect is the same as that of an avalanche. A few free carriers liberate many other free carriers, which in turn liberate others, and so on. The effect is cumulative. As a result, the abrupt creation of many extra free charge carriers causes an abrupt increase in the reverse current.

Diodes exhibiting reverse breakdowns anywhere from about 1 Volt to hundreds of Volts are available commercially. Breakdowns below 5 V are usually explained by the Zener mechanism, and those above 8 V, by the avalanche mechanism. However, there is nothing to preclude the possibility of both mechanisms occurring simultaneously.

Note that if protective circuitry is used to limit the current, a diode may be operated under breakdown conditions safely and without damage. In fact, as we shall see in later sections, special diodes are constructed for the purpose of exhibiting breakdown at a certain voltage and are used specifically for this property.

2.4 DIODE EQUIVALENT CIRCUIT

We have examined the essential property of diodes; that is, they pass current in the forward direction while in effect blocking it in the reverse direction. Now let us look at the ways by which we can replace the diode symbol in a circuit with an appropriate combination of elements and sources that are the *equivalent*, in terms of V–I behavior, to the diode. If we examine the diode characteristics in Fig. 2-7, we see that a single *equivalent circuit* for the diode would be extremely complicated and, therefore, cumbersome to use. To avoid this complexity, we break up the diode characteristics into three distinct regions, as shown in Fig. 2-8, and use a different *linear model* of the diode in each region.

In region 1, the diode is forward biased. We can fit a straight tangent line to the characteristic as shown in Fig. 2-8, intersecting the voltage axis at V_A, called the *cut-in voltage*. Thus, for forward voltages above V_A, the diode characteristic can be approximated by a straight line. For example, the point on the characteristic labeled P_1, where the current is I_1 and the voltage is V_1, can be represented by the cut-in voltage and a fixed resistance r, called the *dynamic resistance*. This dynamic resistance is characterized by the slope of the tangent line in Fig. 2-8:

$$r=\frac{\Delta V}{\Delta I}=\frac{\text{change in voltage}}{\text{change in current}}=\frac{1}{\text{slope of tangent}} \qquad (2\text{-}2)$$

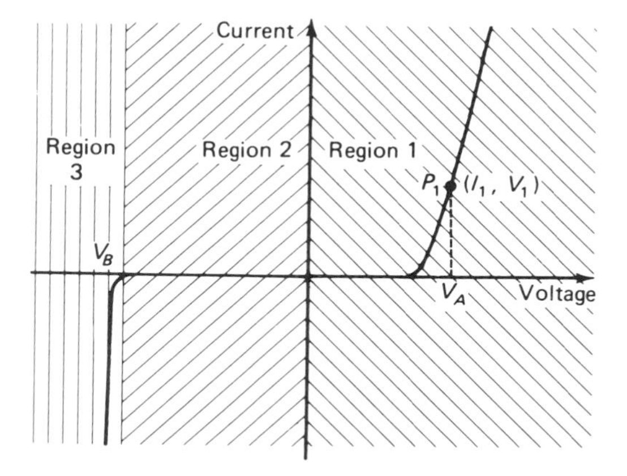

Figure 2-8. Definition of three regions of the diode characteristic for the purpose of obtaining piece-wise linear diode models.

This circuit representation, called an *equivalent circuit*, is shown in Fig. 2-9. We can evaluate r from the inverse of the slope of the tangent line in Fig. 2-8, noting that ΔV is $V_1 - V_A$ and ΔI is $I_1 - 0$ or just I_1. Thus,

$$r = \frac{V_1 - V_A}{I_1} \tag{2-3}$$

To check the equivalence between Figs. 2-9(a) and 2-9(b), we can calculate that the voltage drop between the terminals in Fig. 2-9(b) is $V_A + I_1 r$. Substituting for r from Eq. (2-3), we obtain this voltage drop of $V_A + V_1 - V_A$, or simply V_1. This result is identical to the voltage drop in Fig. 2-9(a). Therefore, the equivalent circuit in Fig. 2-9(b) can be used to replace the diode symbol if the cut-in voltage and dynamic resistance are evaluated properly. This procedure is best illustrated by an example.

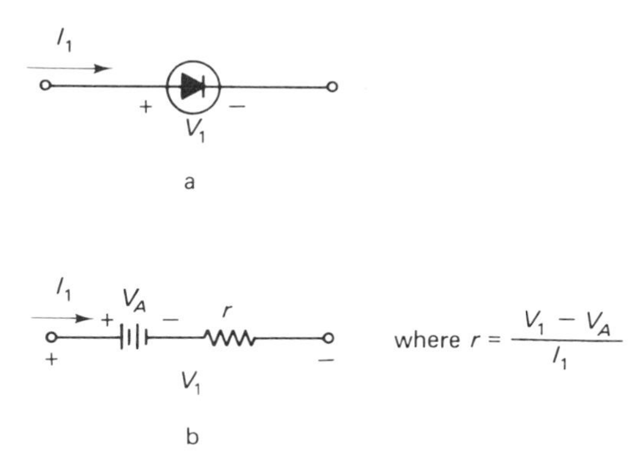

Figure 2-9. Diode under forward bias: (a) diode symbol and (b) the equivalent circuit to replace the symbol.

Example 2-1. We want to determine the cut-in voltage and dynamic resistance for the silicon diode whose forward characteristics are indicated in Fig. 2-6.

Solution: A straight line drawn tangent to the characteristics intercepts the voltage axis at about 0.6 V. Therefore, the cut-in voltage $V_A \cong 0.6$ V. To find the slope, we take ΔI from 0 to 8 mA, which yields a ΔV from 0.6 to 0.65 V. The dynamic resistance is then calculated

$$r \cong \frac{0.65-0.6}{8-0} \frac{V}{mA} \cong 6.3\Omega$$

We must make one important qualification: Both the cut-in voltage and the dynamic resistance values depend on the range of current (and voltage) where they are evaluated. If we take the same silicon diode as in Example 2-1 and display its characteristics in the range between 0 and 1 mA, we would calculate a higher dynamic resistance with a lower cut-in voltage. On the other hand, if we looked at the range from, say, 0 to 100 mA, we would obtain a lower dynamic resistance and a slightly higher cut-in voltage. Nevertheless, it in no way invalidates the use of the piece-wise linear equivalent circuit. It does, however, restrict the use of values for the equivalent circuit to a certain range of current and voltage.

In region 2, the current is essentially equal to the reverse saturation current I_o and is just about independent of the voltage applied. The characteristic curve does have a slight slope in region 2, but it is so small that it can be ignored without fear of a significant error being introduced. The equivalent circuit for the diode in region 2 is then just a current generator of magnitude I_o, as shown in Fig. 2-10(a). We are assuming that

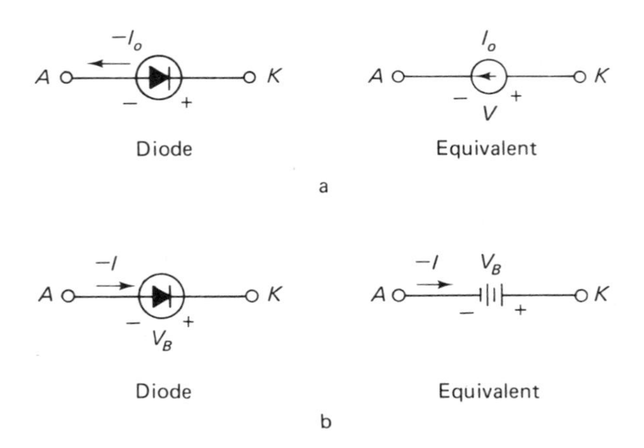

Figure 2-10. Diode under reverse bias: (a) reverse voltage less than the breakdown voltage, (b) at breakdown voltage.

the dynamic resistance in region 2 is essentially infinite or is an open circuit.

In region 3, the diode is in breakdown and maintains a fixed voltage, V_B, while the current varies over a wide range. The characteristic is almost perpendicular to the voltage axis in this region. Thus, the dynamic resistance is very small and so may be assumed to be zero. The diode equivalent circuit in breakdown (region 3) is just a battery of magnitude equal to the breakdown voltage, as shown in Fig. 2-10(b).

Let us now briefly examine the capacitive effects in a diode. The equivalent circuits developed previously are insufficient to predict the behavior of diodes at very high frequencies. At very high frequencies, these models may be modified by the addition of a capacitor in parallel. To appreciate why the capacitor is necessary, we must study the internal operation of the diode.

When the diode is forward biased, large numbers of holes are being injected into the *N* side and, simultaneously, large numbers of electrons are being injected into the *P* side. These processes cause *excess* concentrations of charge in the two regions. If the applied voltage is changing very rapidly, as is the case at very high frequencies, these excess concentrations of charge cannot increase or decrease instantaneously. Thus, there is a time lag between the change in voltage and the resulting change in current. This lag is a capacitive effect. The equivalent capacitance present in a diode under forward bias conditions is called *diffusion capacitance* and may be a few hundred picoFarads (pF) in magnitude.

Under reverse bias conditions, the depletion region is widened, so that there are more uncovered stationary charges on either side of the junction. These stored charges, positive in the *N* side and negative in the *P* side, give rise to a capacitance similar to a parallel-plate capacitance. The equivalent capacitance under reverse bias conditions is called *depletion-region capacitance* (also known as *space-charge capacitance, transition-region capacitance*, or *barrier capacitance*) and is typically a few picofarads in magnitude.

At high frequencies, we have two separate capacitive effects associated with a *PN* junction: depletion-region capacitance when the junction is reverse biased and diffusion capacitance when the junction is forward biased. Both effects can be neglected at low frequencies.

2.5 SPECIAL PURPOSE DIODES

Some of the general diode characteristics already mentioned are potentially so useful that special diodes are designed to emphasize a specific property. Such is the case of *Zener diodes*.

Zener, or breakdown, diodes are much like ordinary diodes except that they are manufactured to exhibit breakdown in the reverse direction at a specific voltage. These diodes are almost exclusively used because of their breakdown characteristics. The Zener diode circuit symbol is shown

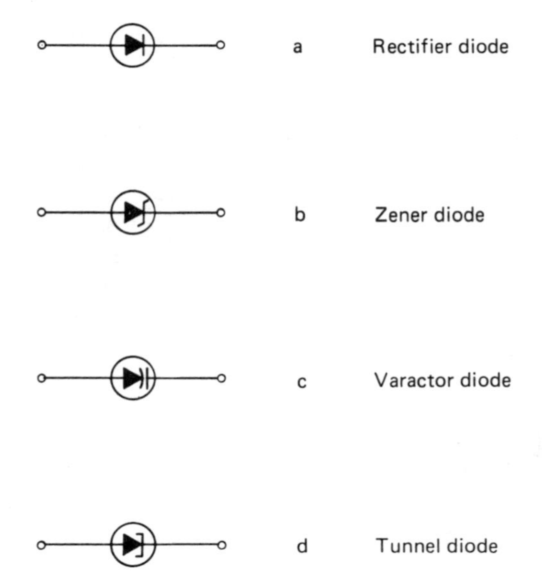

Figure 2-11. Various diode symbols.

in Fig. 2-11(b). Zener diodes can exhibit breakdowns at a very wide range of voltages, starting at about $\frac{1}{2}$ V. Figure 2-12 gives typical reverse characteristics of a Zener diode with a 6.8 V breakdown.

Another common type of diode that is also usually used in the reverse biased condition is the *varactor diode* (also called a *varicap* or *voltacap*). Remember that under reverse bias in a varactor the depletion

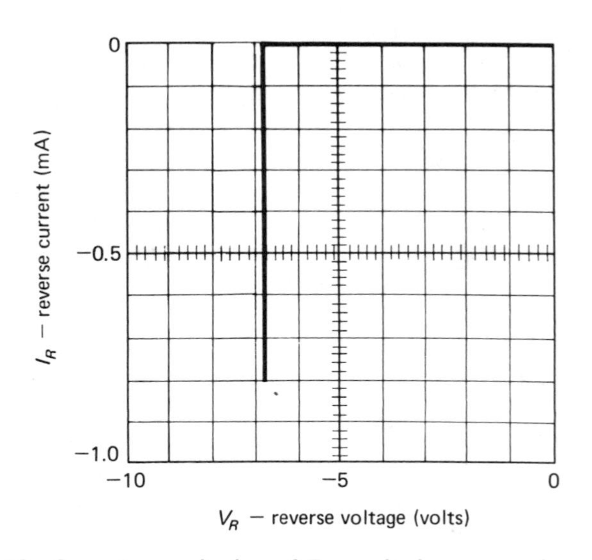

Figure 2-12. Curve-tracer display of Zener diode reverse characteristics.

region widens with an increase in negative voltage. Consider that the depletion region constitutes a parallel-plate capacitor. Then the varactor is effectively a voltage-controlled capacitor, because the voltage varies the depletion-region width which is analogous to the separation between the plates of a capacitor. The circuit symbol for a varactor is given in Fig. 2-11(c).

A diode with characteristics quite unlike those of conventional diodes also has found some limited applications. It is called a *tunnel diode* (or *Esaki diode*, after its discoverer). It is characterized by a *negative resistance* region in the forward bias direction. The circuit symbol for a tunnel diode is shown in Fig. 2-11(d), and a typical set of terminal characteristics, in Fig. 2-13. In this type of diode, both the N and P regions are heavily doped (typically 1 part impurity to 10^3 Ge or Si). As a result, the depletion region is extremely narrow. In both the reverse and the forward directions, the predominant mechanism by which carriers traverse the junction is called *tunneling*.*

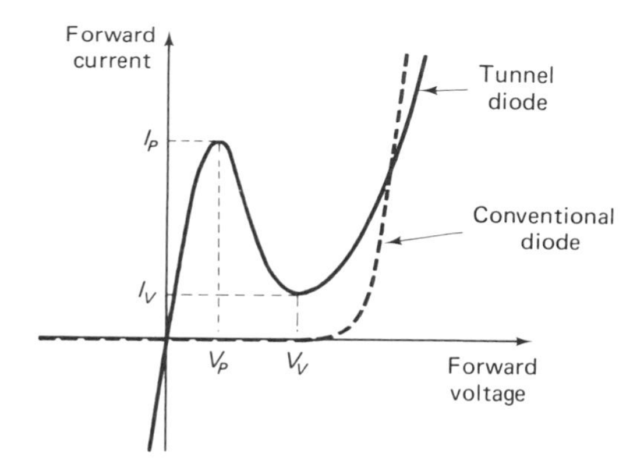

Figure 2-13. Tunnel diode characteristics.

The key quantities for the tunnel diode are the valley voltage and current, V_v and I_v, and the peak voltage and current, V_p and I_p. The negative resistance region is between the peak and valley points on the characteristics. Actual tunnel diode characteristics obtained on a curve tracer are shown in Fig. 2-14. For this particular tunnel diode: $I_p = 1$ mA; $V_p = 0.15$ V; $I_v = 0.1$ mA; $V_v = 0.6$ V; and the forward voltage $V_f = 0.9$ V.

Although the tunnel diode is not widely used, its negative resistance region makes it valuable in oscillator and other circuits.

*A full explanation of tunneling requires a quantum-mechanical treatment. If you are interested in a more in-depth discussion, consult for example: D. Le Croissette, *Transistors* (Englewood Cliffs, N.J.: Prentice-Hall, Inc., 1963).

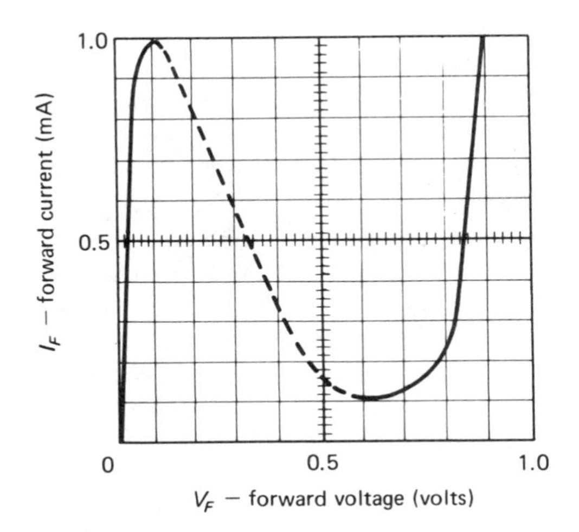

Figure 2-14. Tunnel diode (IN2939) forward characteristics.

Review Questions

1. What is the role of a *seed* in crystal growing?
2. What are some of the aspects of diode manufacture?
3. Enumerate and briefly describe the steps used in the diffusion process of diode manufacture.
4. Describe the dynamics of the formation of the *depletion region.*
5. In a diode under equilibrium, what causes the *contact potential*?
6. When a diode is *forward biased*, which terminal is positive and which is negative?
7. When a diode is *reverse biased*, which terminal is positive and which is negative?
8. Does a voltage in the forward bias direction add to or subtract from the contact potential? Why?
9. Describe the motion of charge carriers inside a forward-biased diode.
10. What is the effect of forward bias on the depletion region in a diode?
11. Does a reverse voltage add to or subtact from the contact potential? Why?
12. What is the effect of reverse bias on the depletion region in a diode?
13. Describe the motion of charge carriers inside a reverse-biased diode.
14. In the diode characteristics, what is the significance of the *reverse saturation current*?
15. What factors affect the reverse saturation current in a diode?
16. What order of magnitude is the reverse saturation current at room temperature in silicon? In germanium?
17. What effect does temperature have on the reverse saturation current? Why?

18. Describe the conditions inside a diode under *Zener breakdown*.
19. Describe the conditions inside a diode under *avalanche breakdown*.
20. What precautions must be taken to protect a diode from permanent damage when it is operating under reverse breakdown conditions?
21. What is *cut-in voltage*?
22. What is *dynamic resistance*?
23. What order of magnitude (ohms, kilohms, or megohms) is the diode dynamic resistance in region 1? (Refer to Fig. 2-8.)
24. What order of magnitude is the diode dynamic resistance in region 2? (Refer to Fig. 2-8.)
25. What order of magnitude is the diode dynamic resistance in region 3? (Refer to Fig. 2-8.)
26. What is meant by a *piece-wise linear model* in the case of the diode?
27. What is the capacitance in a forward-biased diode called? What is its cause?
    When must you account for it?
28. What is the capacitance in a reverse-biased diode called? What is its cause?
    When must you account for it?
29. Which gives rise to a larger capacitance in a diode, forward or reverse bias?
    Why?
30. What are the unique characteristics of a *Zener diode*?
31. What are the unique characteristics of a *varactor diode*?
32. What are the unique characteristics of a *tunnel diode*?
33. What is the meaning of *negative resistance* in the tunnel diode?
34. Draw the circuit symbols for: rectifier diode, Zener diode, varactor diode, and tunnel diode.

Problems

1. Make a plot of the diode equation [Eq. (2-1)] for a silicon diode with $I_o = 1$ nA. Use 0.02 V increments in the voltage between -0.1 and 0.7 V. Use 0.1 V increments between -2 and -0.1 V. (Assume the breakdown voltage to be below -2 V.)
2. Repeat Problem 1 for a germanium diode with $I_o = 1\ \mu$A, and go up to 0.3 V in the forward direction.
3. Plot the two diode curves obtained in Problems 1 and 2 on the same graph paper and compare them, stating pertinent differences.
4. From the characteristics in Fig. 2-6 and the diode equation, make a reasonable prediction of the saturation current for the germanium diode.
5. Repeat Problem 4 for the silicon diode.

6. Determine the cut-in voltage and dynamic resistance for the germanium diode (whose characteristics are given in Fig. 2-6) that would produce a good piece-wise linear model in the vicinity of 0.3 V.
7. Repeat Problem 6 in the vicinity of 0.5 V.
8. Repeat Problem 6 for the silicon diode in the vicinity of 0.55 V.
9. Draw the V–I characteristics for the circuit consisting of two diodes in series, as shown in Fig. 2-15, if the two diodes are identical and have reverse saturation currents of 10 μA.

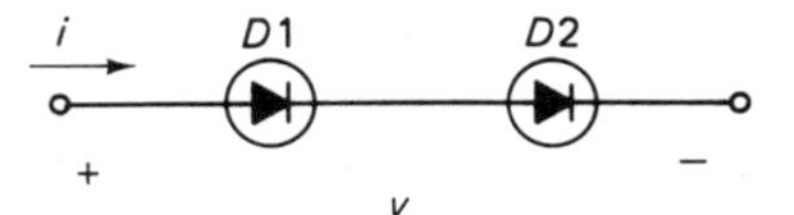

Figure 2-15.

10. Draw the V–I characteristics for the diode circuit shown in Fig. 2-16 if the two diodes are identical and have reverse saturation currents of 5 μA.

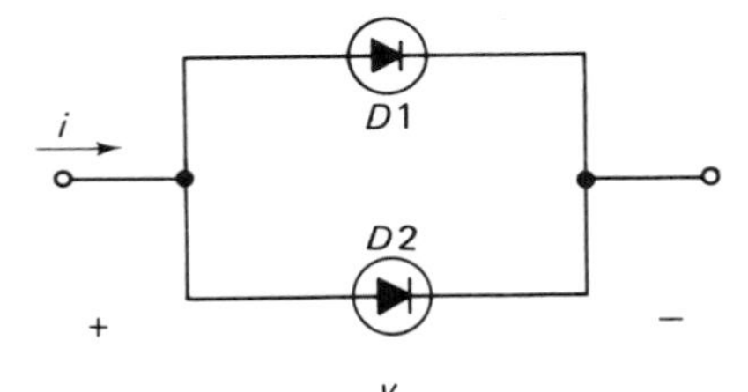

Figure 2-16.

11. The two Zener diodes in Fig. 2-17 are identical and have the reverse characteristics indicated in Fig. 2-12. Sketch the resulting V–I characteristics for the combination shown in Fig. 2-17.

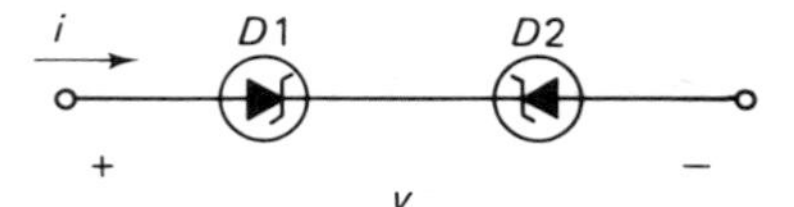

Figure 2-17.

12. The diodes in Fig. 2-18 have 5 V applied as shown. If the diodes are identical, with $I_o = 100$ nA, determine (a) the current, I; (b) the voltage, V_{D1}, across $D1$; (c) the voltage, V_{D2}, across $D2$. (Hint: The voltages across the two diodes are *not* equal, but the current is the same through both diodes.)

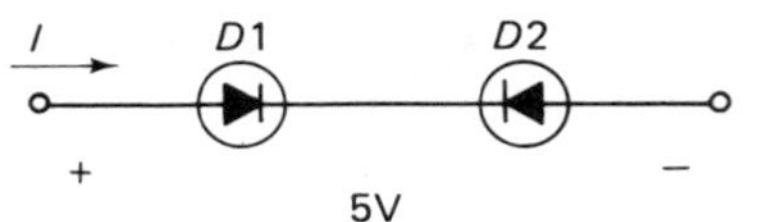

Figure 2-18.

3

Bipolar Junction Transistors

The bipolar junction transistor (BJT) is a three-terminal device and contains two *PN* diodes sandwiched back to back. The three regions corresponding to the three terminals have names suggestive of the function they perform. The *emitter* emits or dispatches charge carriers into the *base*, where control over the carriers is exercised; eventually the carriers are gathered in the *collector* region. The name "base" stems from a historical development of transistors. Originally a transistor was made by alloying the emitter and collector regions to a relatively large sample of doped semiconductor. This sample served as the base for manufacture, and the corresponding region took on that name.

The base region is situated between the emitter and collector regions. Two diodes are formed by the same type of doping in both the emitter and collector regions. The base always has a doping opposite to that of the emitter and collector. Consequently, there are two kinds of BJTs. The first kind has an *N*-type base and *P*-type emitter and collector. In the second kind, the base is *P*-type and both emitter and collector are *N*-type. These two types of transistors are labeled *PNP* and *NPN* respectively; the type of doping in the base is specified by the middle letter.

Many varied methods are used in manufacturing these two kinds of transistors. We have already discussed one method—*diffusion*—in connection with diodes. The same basic process (with some dimensions changed) is repeated in the manufacture of transistors. The first three steps that were illustrated in Fig. 2-2 are followed by additional oxidation, masking, etching, and diffusion processes, as shown in Fig. 3-1. In this case, they form an *NPN* structure of a BJT. This process is used in fabricating individual transistors, as well as transistors in integrated circuits (ICs) with many other components on the same chip. The geometry in Fig. 3-1(d), although not to scale, is important. The emitter region is surrounded by the base, which in turn is surrounded by the collector. As we shall see later, this arrangement allows the most efficient operation of the transistor.

Another common method of transistor fabrication is by *epitaxial growth*. In this process, the *PNP* or *NPN* structure is formed by growing

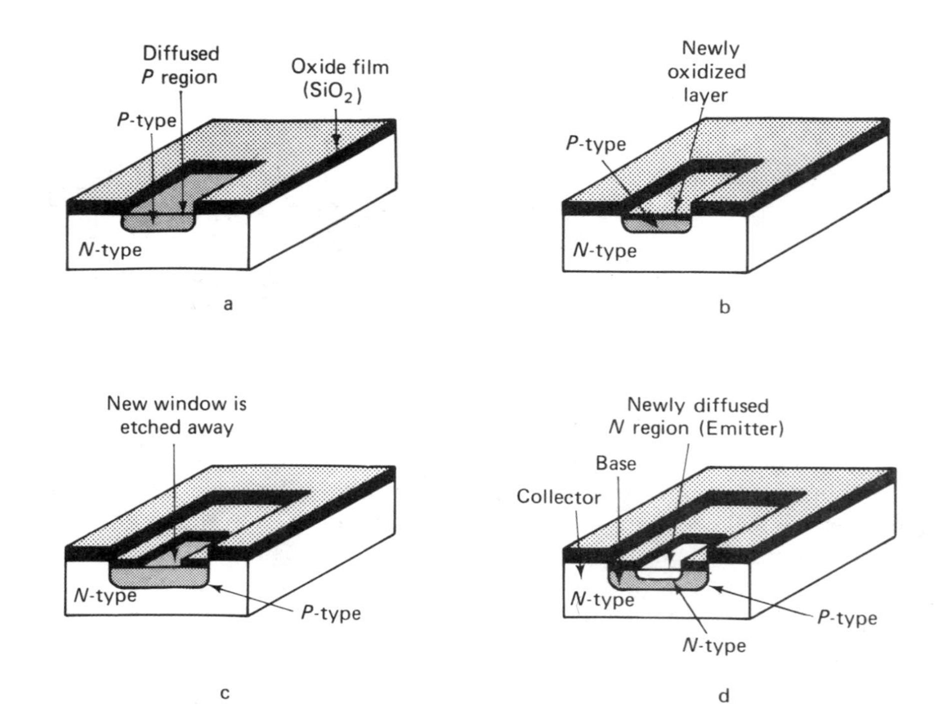

Figure 3-1. Formation of a planar *NPN* transistor by the diffusion process: (a) end result of processes shown in Fig. 2-2, (b) oxidation, (c) new window is masked and etched away, and (d) an *N* region is diffused to complete *NPN* structure.

regions one on top of another. The resulting transistor is called a *mesa* transistor because of its appearance when viewed through a microscope.

Processes utilizing a combination of epitaxial growth, diffusion, and alloying are also employed. In addition, we can use numerous different geometries. Nevertheless, irrespective of the actual geometry or fabrication technique, the operation of all BJTs is based on the same principles.

3.1 CURRENTS IN THE BJT

The two types of BJTs are shown schematically in Fig. 3-2, together with their circuit symbols. An easy way to remember the convention for the circuit symbols is to note the direction of the arrow on the emitter terminal. If the arrow points in, it indicates a *PNP* transistor; if it points out, it indicates an *NPN* transistor.

Let us now see how the BJT operates. A fairly high current is caused to flow through the BJT. Control of this current is exercised at the base, by either increasing or decreasing the current. We shall use an *NPN* transistor as an example first and then extend our discussion to include *PNP* transistors.

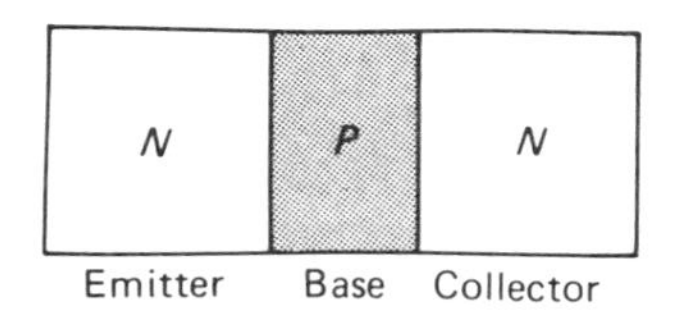

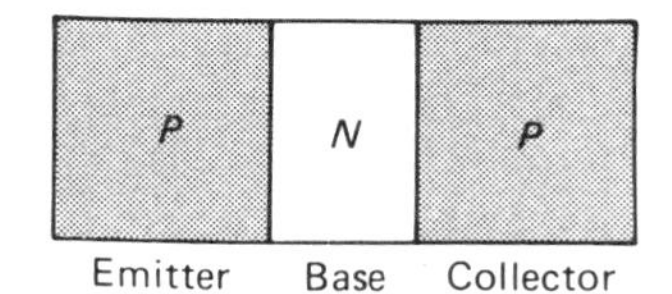

a

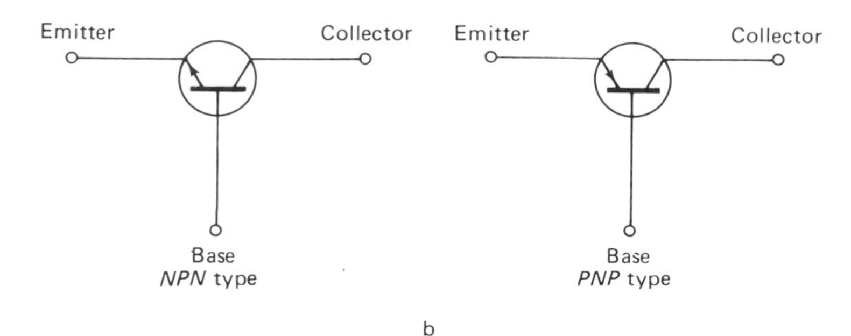

b

Figure 3-2. ***NPN*** **and** ***PNP*** **transistors: (a) block diagram and (b) circuit symbols.**

In an *NPN* transistor, electrons are the majority carriers inside the emitter. In order to cause these electrons to cross the base-emitter junction and enter the base region, an external voltage is applied between the base and emitter terminals. The base and emitter regions constitute a diode which must be *forward biased* in order to cause majority carriers—electrons in the *N*-type emitter—to cross the junction. Once electrons from the emitter enter the *P*-type base, they diffuse through the base. Some of them find their way to the base terminal and flow out; most of them reach the collector-base junction. In order for electrons to cross the junction from the *P*-type base into the *N*-type collector, a *reverse-biased* voltage must be applied between collector and base terminals. In this manner, the electrons that started in the emitter and wound up in the collector constitute the main current, whereas those electrons that flowed out of the base make up the small controlling current.

However, we have not accounted for all the currents in the BJT. As a result of the forward bias on the base-emitter junction, we would expect not only the injection of electrons into the base but also holes from the base to enter the emitter. Although holes do enter, their effect is minimal on the net current flow in the transistor, because the base is very lightly doped whereas the emitter is very heavily doped. The BJT is purposely made that way. Thus, when the junction is forward biased, the heavily doped *N*-type emitter offers extremely large numbers of electrons for conduction, but the lightly doped *P*-type base has only a small number of holes to offer. The current across the base-emitter junction in an *NPN* transistor, therefore, is essentially the result of electron flow.

At the collector-base junction, the reverse bias transports electrons, which started in the emitter, from the base into the collector. Besides this action, the reverse bias causes another current to flow across the collector base junction. This current is made up of the *minority* charge carriers in both the base and collector; that is, electrons from the *P*-type base and holes from the *N*-type collector. The current component resulting from this minority charge flow across the collector-base junction is quite small. It is called the *collector-base reverse saturation current*, or *collector cutoff current*, and is usually labeled I_{CBO} or simply I_{CO}. As a reverse saturation current, it is very sensitive to temperature.

Obviously, there is no way that we can actually measure the separate components of the currents inside the transistor. We simply note that electrons supplied to the emitter and flowing toward the base correspond to a conventional current flow out of the emitter. We label the net terminal current I_E for the emitter current. Similarly, electrons flowing out of the base and collector correspond to net terminal currents I_B and I_C, respectively. The three terminal currents, together with the internal motion of charge carriers, are indicated in Fig. 3-3. The circuit diagram for the normal operation of an *NPN* transistor, with the base-emitter diode forward biased and the collector-base diode reverse biased, is shown in Fig. 3-4.

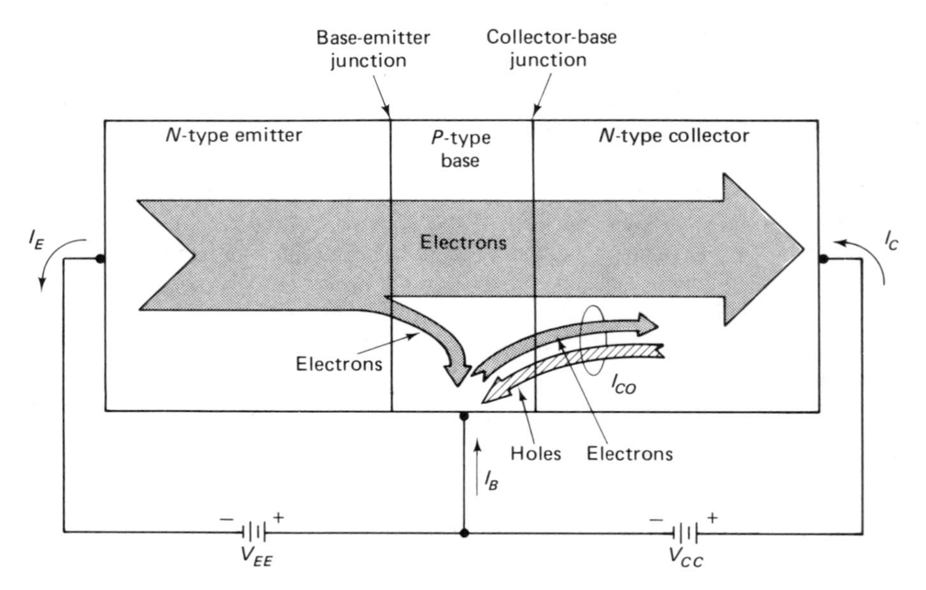

Figure 3-3 Motion of charge carriers in an *NPN* transistor, showing the directions of the resulting terminal currents.

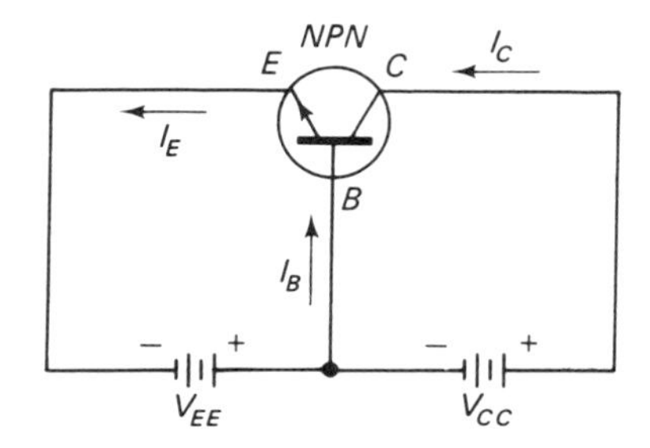

Figure 3-4. Bias for normal operation of an *NPN* transistor, showing actual directions of terminal currents.

As a result of the externally applied voltages, the depletion region at the base-emitter junction narrows and the depletion region at the collector-base junction widens, as shown in Fig. 3-5.

We can see that the net current entering the transistor in Fig. 3-4 is the sum of I_B and I_C and the net current leaving the transistor is I_E. Because charge is neither piling up nor being depleted inside the transistor, the net current into the transistor must be equal to the net current out of the transistor. Thus,

$$I_E = I_C + I_B \tag{3-1}$$

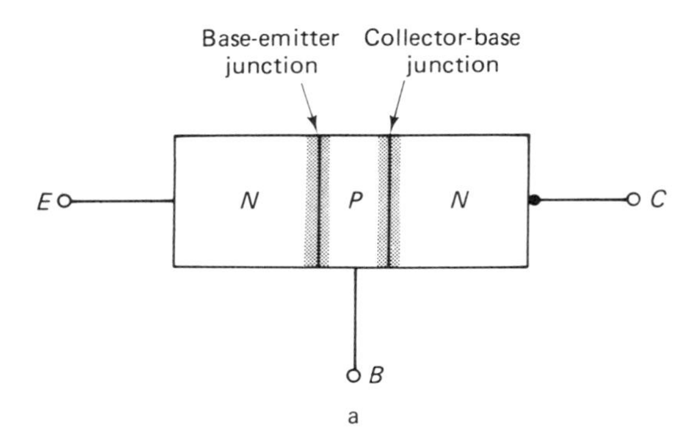

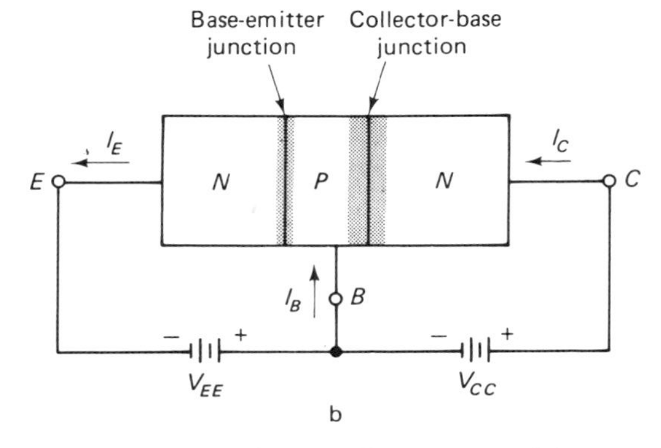

Figure 3-5. Depletion regions inside an *NPN* transistor: (a) no external bias, (b) with normal bias.

For a constant emitter current, the smaller the base current, the larger the collector current. In order to make the collector current as large as possible therefore, the base current must be minimized. We saw that, in an *NPN* transistor, electrons from the emitter diffuse through the *P*-type base and eventually reach the collector. Some of these electrons are lost in the base due to recombination. To minimize this recombination, the base is only a lightly doped *P*-type. Furthermore, the physical width of the base (distance from the edge of the emitter to the edge of the collector) is made small (typically 1/10,000 in.). Consequently, many electrons are assured of reaching the collector.

The collector current is made up of the current in the collector resulting from electrons that started in the emitter, labeled I_{NC}, together with the reverse saturation current I_{CO}. Thus,

$$I_C = I_{NC} + I_{CO} \tag{3-2}$$

The ratio of the electron current in the collector to the total electron current in the emitter, labeled α (alpha), is an important parameter for the BJT. In terms of an equation, α is defined as

$$\alpha = \frac{I_{NC}}{I_E} \tag{3-3}$$

Note that no matter how few of the electrons that started in the emitter are lost in the base, I_{NC} is always less than I_E. This is due to the simple fact that we can never collect a larger number of carriers in the collector than the number that started in the emitter; thus, α is always less than 1. It is, however, usually very close to 1, typically between 0.98 and 0.9995. The significance of α is that it is the *dc short-circuit current gain in the common-base configuration*. We shall discuss this gain shortly.

If we use Eq. (3-2) in the defining equation for α, we have

$$\alpha = \frac{I_C - I_{CO}}{I_E} \tag{3-4}$$

Solving this equation for I_C gives us an important current relationship for the BJT:

$$I_C = \alpha I_E + I_{CO} \tag{3-5}$$

We must digress briefly to consider the *PNP* transistor. Its operation is completely analogous to that of the *NPN*. The majority carriers (in this case, holes) from the *P*-type emitter are injected into the base by forward biasing the base-emitter diode. Holes are further transported through the base, and some of them are gathered by the collector with the aid of a reverse bias on the collector-base diode. In fact, all of the preceding

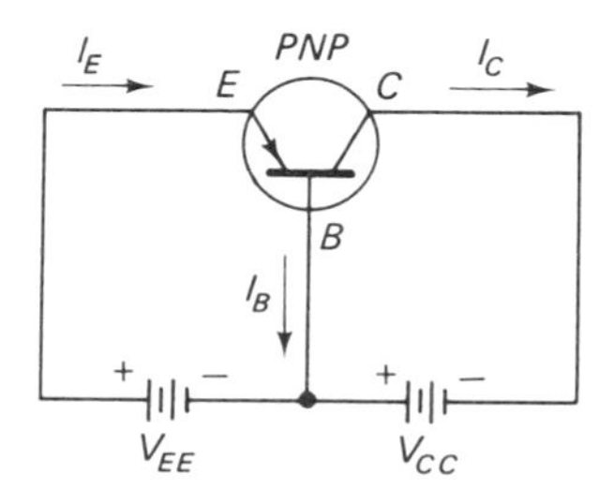

Figure 3-6. Bias for normal operation of a *PNP* transistor, showing actual directions of current flow.

discussion for the *NPN* transistor holds for the *PNP* transistor if the appropriate substitutions (*N* instead of *P*, *P* instead of *N*, and hole instead of electron) are made.

Some aspects are common to both *NPN* and *PNP* types. The base-emitter diode is forward biased and the collector-base diode reverse biased, irrespective of the type. In the case of the *PNP* transistor, the main current is caused by hole motion; therefore, the current is in the same direction as the flow of holes. Thus, the emitter current flows into a *PNP* transistor; the base and collector currents flow out, as shown in Fig. 3-6. (Note that the battery polarity is reversed from what it was in Fig. 3-4.) Taking the algebraic sum of the terminal currents in a *PNP* transistor yields the same result as for the *NPN* transistor, given in Eq. (3-1).

It is always good practice to indicate by arrows the directions of the terminal currents. We can be certain of indicating the *actual* directions under normal operating conditions if we use the arrow on the emitter of the transistor symbol to get the actual direction of the emitter current and remember from Eq. (3-1) that both I_C and I_B are in a direction opposite that of I_E. For example, the arrow on a *PNP* transistor points into the transistor. Therefore, I_E flows into the transistor, whereas I_C and I_B both flow out.

We can make a substitution in Eq. (3-5) for I_E, using Eq. (3-1), and obtain:

$$I_C = \alpha(I_C + I_B) + I_{CO} \tag{3-6}$$

When the above equation is solved for I_C, the following results:

$$I_C = \frac{\alpha}{1-\alpha} I_B + \frac{1}{1-\alpha} I_{CO} \tag{3-7}$$

We must now define β (beta), the *dc short-circuit current gain in the common-emitter configuration*:

$$\beta = \frac{\alpha}{1-\alpha} \tag{3-8}$$

Equation (3-7) can be rewritten in a slightly more useful form:

$$I_C = \beta I_B + (\beta + 1) I_{CO} \tag{3-9}$$

The most important and useful transistor equations are those given in Eqs. (3-1), (3-5), and (3-9).

3.2 BJT STATIC CHARACTERISTICS

When the transistor is operated as an amplifier, we apply the input between two terminals and obtain the (hopefully) amplified output at another pair of terminals. The transistor has only three terminals, so we have to designate one terminal as being common to both input and output. Because of the versatility of the BJT, it can operate with any one of its three terminals as the common terminal. The configuration is usually named by the common terminal, i.e., common base (CB), common emitter (CE), and common collector (CC).

The static characteristics for the BJT in any of the three possible configurations contain two sets of curves. One set, the *input characteristics*, gives the V-I relationship at the input terminals for different values of a parameter (either output current or voltage). The second set, the *output characteristics*, gives the V-I relationship at the output terminals, with either the input current or voltage as the parameter.

3.2.1 Common-Base

In the common-base (CB) configuration, the input is applied between emitter and base, while the output is taken between collector and base (as illustrated in Fig. 3-7 for an NPN transistor). The input characteristics, shown in Fig. 3-8(a), are a plot of the emitter current as a function of the emitter-base voltage, with the collector-base voltage as a parameter. For normal operation, the base-emitter junction is forward biased, so the general appearance of the CB input characteristics resembles that of a forward-biased diode. The effect of the output voltage V_{CB} is not great. As V_{CB} is increased (in the direction of reverse-biasing the collector-base junction), a larger emitter current is observed for the same V_{EB}.

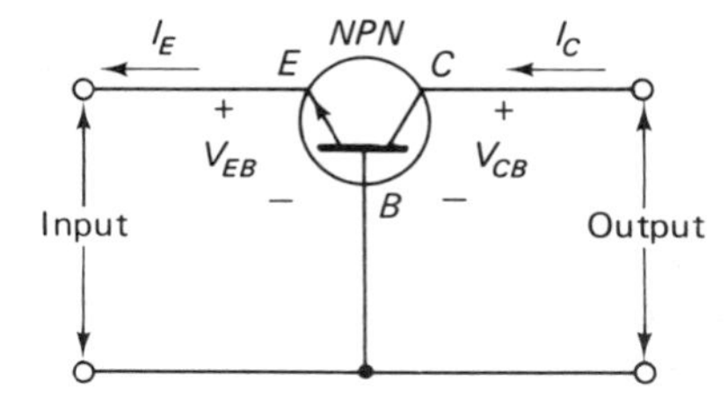

Figure 3-7. *CB* **configuration using an** *NPN* **transistor. For normal operation,** V_{EB} **is negative,** V_{CB} **positive, and currents in the directions shown.**

The CB output characteristics, shown in Fig. 3-8(b), are a plot of the collector current as a function of the collector-base voltage, with the emitter current as a parameter. For normal operation, the collector-base junction is always reverse biased. Under these conditions, the collector current saturates to a value determined by the emitter current and is essentially independent of the collector-base voltage. Note that collector

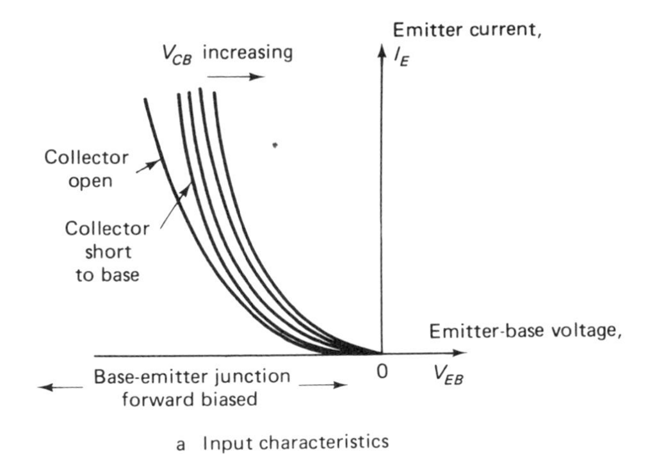

a Input characteristics

b Output characteristics

Figure 3-8. *CB* **characteristics for an** *NPN* **transistor.**

current flows even for some small forward bias at the collector-base junction, because carriers injected from the emitter into the base have enough energy to cross the collector junction. Some small forward bias at the collector is required to prevent the carriers from crossing into the collector. For silicon transistors, this voltage is typically $\frac{1}{2}$ V.

Actual *CB* output characteristics for a silicon *NPN* transistor are shown in Fig. 3-9. Using Eq. (3-5) and noting that I_{CO} is small enough to be negligible, we can write:

$$I_C \cong \alpha I_E \tag{3-10}$$

We can now evaluate α from Fig. 3-9. For example, at any positive value of V_{CB} and for $I_E = 0.3$ mA, I_C is approximately 0.27 mA. This calculation gives us $\alpha \cong 0.9$.

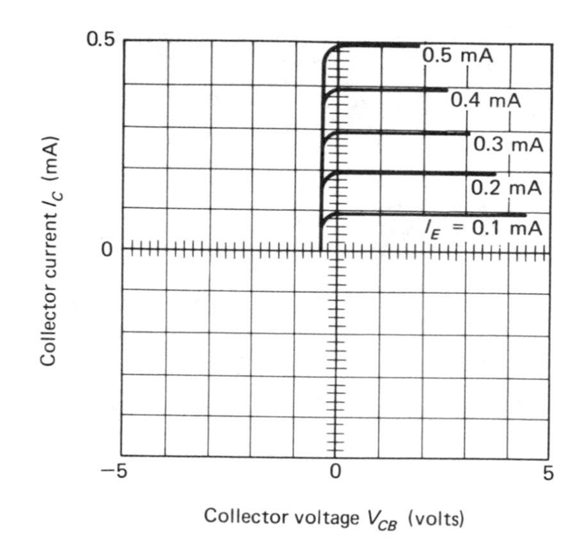

Figure 3-9. *CB* **output characteristics as displayed on a curve tracer.**

3.2.2 Common-Emitter

In the common-emitter (*CE*) configuration, the input is applied between base and emitter, while the output is taken between collector and emitter, as shown in Fig. 3-10 for an *NPN* transistor.

The *CE* input characteristics are a plot of the base current as a function of the base-emitter voltage, with V_{CE} as a parameter. As can be seen in Fig. 3-11(a), V_{CE} does not have much control on the input characteristics. For all V_{CE} larger than 0.5 V, all the input curves just about coincide with one another. The shape of the curves, as might be expected, is that of a forward-biased diode.

The *CE* output characteristics are a plot of the collector current as a function of collector-emitter voltage, with the input current, I_B, as a parameter. For all V_{CE} larger than about 0.2 to 0.5 V, I_C saturates to a value essentially independent of V_{CE} and only controlled by the magnitude of the base current. When the base is open-circuited (that is, $I_B = 0$), the collector current, labeled I_{CEO}, is given by:

$$I_C \cong I_{CEO} = (\beta + 1)I_{CO} \tag{3-11}$$

The collector leakage current in the *CE* configuration with the base disconnected, I_{CEO}, shown in Fig. 3-11(b), although appreciably larger

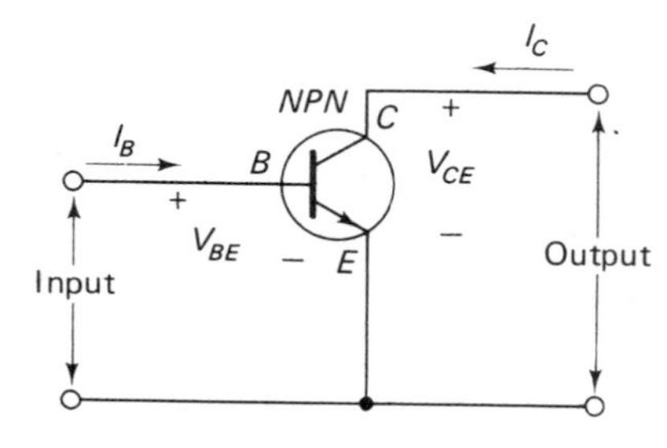

Figure 3-10. *CE* **configuration using an** *NPN* **transistor. For normal operation, both** V_{BE} **and** V_{CE} **are positive, with**

$$V_{CE} > V_{BE}$$

to give positive V_{CB} **(a reverse bias on the collector).**

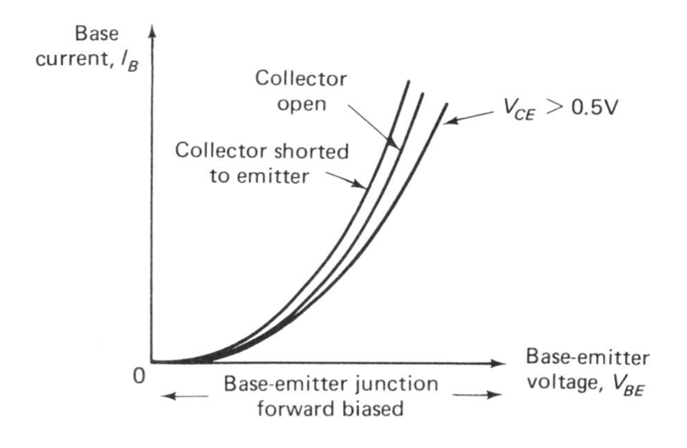

a Input characteristics

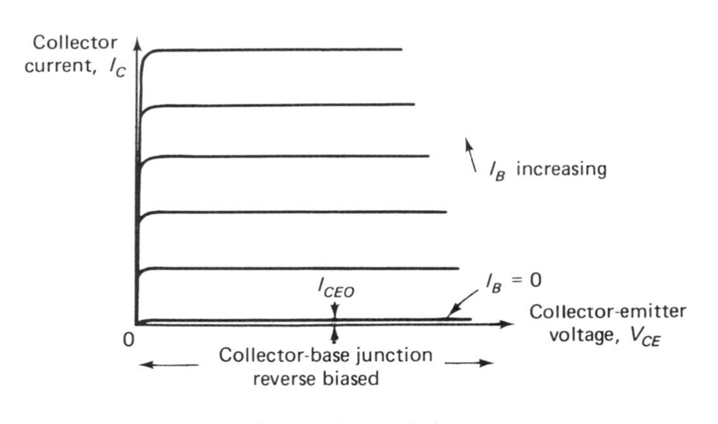

b Output characteristics

Figure 3-11. *CE* **characteristics for an** *NPN* **transistor.**

than I_{CO}, is usually very small and may be neglected in the determination of β. Actual *CE* characteristics for an *NPN* transistor, illustrated in Fig. 3-12, reveal I_{CEO} of less than 0.1 mA. Equation (3-9) can be rewritten as:

$$I_C = \beta I_B + I_{CEO} \cong \beta I_B \tag{3-12}$$

The value of β can be evaluated from the *CE* output characteristics, of the type shown in Fig. 3-12(b). For example, at $V_{CE} = 5V$, with $I_B = 100$ μA, we calculate I_C to be about 9 mA. This result corresponds to a β of 9/0.1 or 90. Note here that the symbol h_{FE} (or H_{FE}) is also used instead of β. However, because of the similarity between the dc and ac symbols, h_{FE} and h_{fe}, and the possibility of confusing the two, β will be used to denote the dc short-circuit current gain in the *CE* configuration.

As you might expect, β is a function of the operating point. In the *CB* configuration, as the number of carriers injected from the emitter into the base is increased, so is the number of carriers reaching the collector.

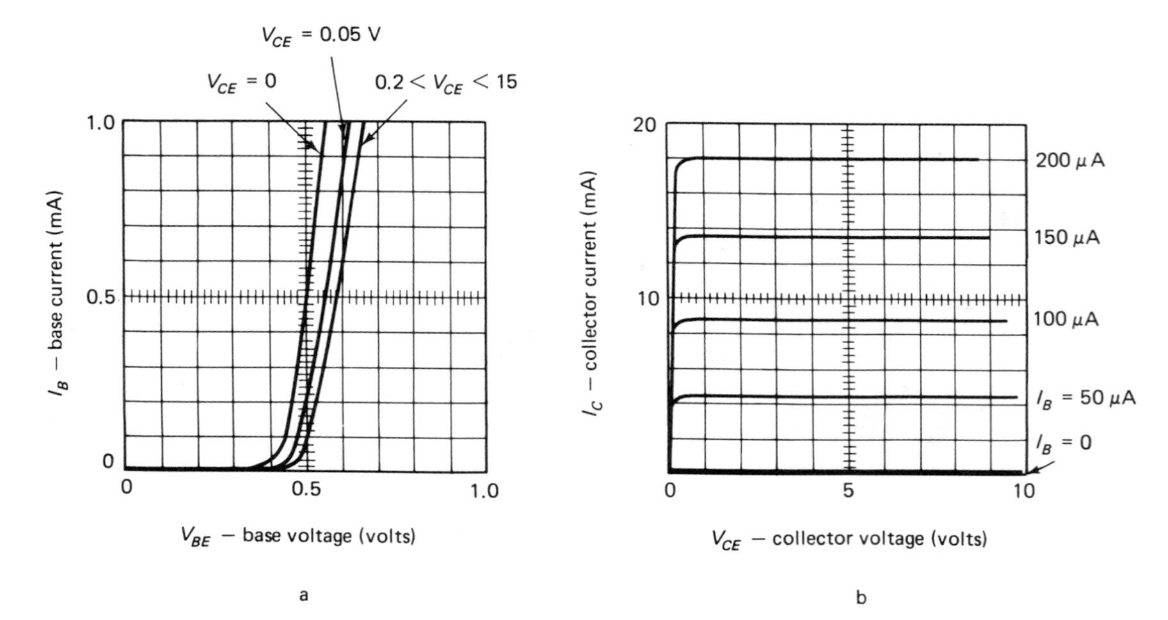

Figure 3-12. (a) *CE* input characteristics and (b) *CE* output characteristics for silicon *NPN* transistor (2N2369).

However, these two increases are not directly proportional; the number of carriers reaching the collector increases by a smaller amount than the increase in the carriers leaving the emitter. Consequently, there is a decrease in the transistor α, which is also a decrease in β. The same is true at very low injection levels. Not only does the β vary among transistors of the same type number, but it also varies for the same transistor under different operating conditions, i.e., different voltage and current. Figure 3-13 demonstrates the typical variation of β as a function of collector current.

Another property of the *CE* output characteristics that should not be overlooked is the extreme temperature sensitivity of these characteristics. Figure 3-14 shows a double exposure of the *CE* output characteristics as

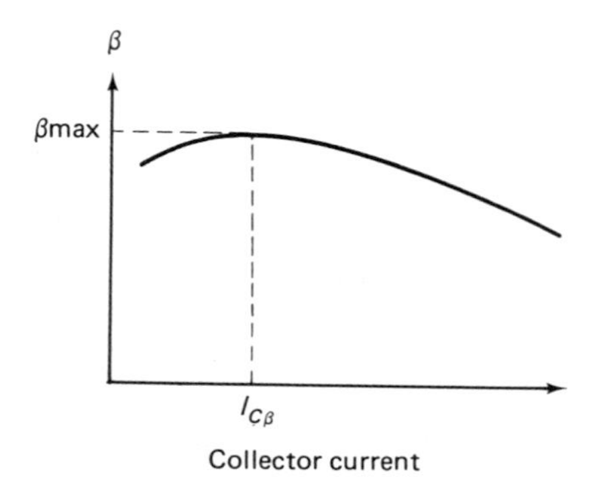

Figure 3-13. Variation of β as a function of quiescent collector current.

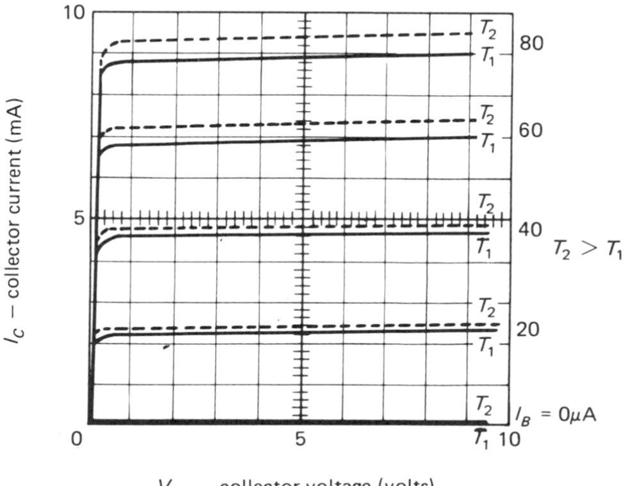

Figure 3-14. Double exposure of a curve-tracer display of *CE* output characteristics showing the effect of increased temperature.

displayed on a curve tracer. The first exposure was taken at room temperature (T_1); the second exposure was taken after the transistor case was heated by placing a match near it. For the same base current, at the higher temperature a larger collector current is observed. This difference is caused by the increase in the collector leakage current I_{CO}, as well as a slight increase in β. The collector leakage current I_{CO} approximately doubles in value for every 10°C rise in temperature. The *CE* output characteristics depend on $(\beta+1)I_{CO}$[see Eq. (3-9)]. Therefore, the most noticeable change as a result of temperature increase is in the *CE* characteristics.

In general, we can say that transistor characteristics and parameters are very sensitive to temperature. When we design a circuit, we must allow for heat dissipation and for the temperature sensitivity of transistors.

The *CC* transistor characteristics will not be discussed separately. We can show that the *CC* transistor configuration may be treated as a special case of the *CE* configuration.

3.3 TRANSISTOR RATINGS

Literally thousands of different BJTs are made, including both silicon and germanium *PNP* and *NPN* as well as many others with different types of construction. Some examples are given in Fig. 3-15.

There are so many various types of transistors mainly because there are such varied applications in which transistors are used. Readily available transistors range in price from a few cents to almost $100 for special-purpose ultra-high-frequency transistors.

In order to help customers select the proper transistor for a particular application as well as provide transistor specifications, manufacturers

Figure 3-15. Different semiconductor devices. (Courtesy of Texas Instruments)

publish data sheets, manuals, and application notes. Examples of typical data sheets for diodes, BJTs, FETs, and assorted other devices and circuits are given in Appendix 3. Among other quantities, the manufacturer usually specifies the maximum permissible V_{CE}, I_C, and device dissipation. Also specified is the maximum allowable junction temperature (referring to the collector junction, although both junctions are at just about the same temperature because of their proximity). In operation, exceeding any one of these maximum values may result in permanent damage to the transistor.

For example, a Motorola *NPN* silicon power transistor TIP29A (according to data sheets in Appendix 3) is listed by the manufacturer to have maximum continuous I_C of 1 A, maximum V_{CE} of 60 V, and maximum continuous device dissipation of 30 W. These three maximum values define the permissible region of operation for this particular transistor. The total continuous device dissipation can be approximated by the power dissipated at the collector junction, because the power dissipated in the forward-biased emitter junction is usually a few orders of magnitude smaller. A reasonable approximation for the power dissipated at the collector is the product of the collector voltage and current:

$$P_J = I_C V_{CE} \tag{3-13}$$

where P_J is the permissible power dissipation; I_C, the quiescent collector current; and V_{CE}, the operating point collector-emitter voltage. The permissible region of operation may be shown graphically on the *CE* output characteristics. It has five distinct boundaries. Two boundaries are provided by the transistor characteristics; one being the cutoff region, the

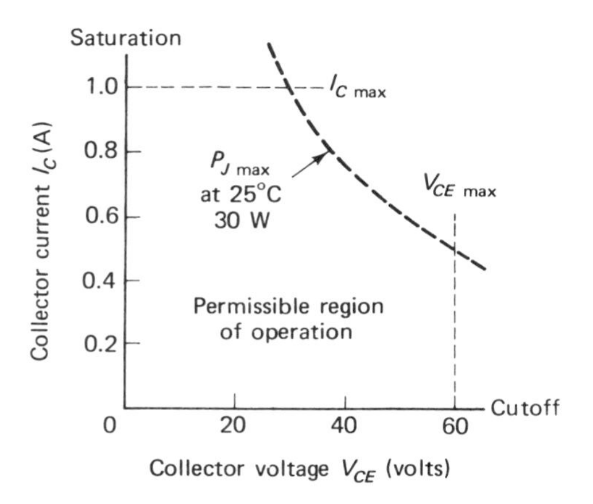

Figure 3-16. Permissible region of operation.

other the saturation region, as shown in Fig. 3-16. The other three boundaries are provided by the maximum values specified by the manufacturer for the collector current, collector-emitter voltage, and total power. The maximum power curve is obtained by choosing values for I_C lower than $I_{C\max}$ and by calculating from Eq. (3-13) values of V_{CE} to give the specified maximum permissible power.

In their data sheets, manufacturers also specify different breakdown voltages (usually denoted by a capital B before the symbol: e.g., BV_{CEO}, BV_{CBO}). These symbols use triple subscript notation. The first two subscripts denote the two terminals between which the voltage is to exist; the last subscript O denotes that the test is made with the unmentioned terminal held open. Thus, BV_{CBO} is the collector-base breakdown voltage with the emitter terminal open-circuited. The last subscript may also be R; in which case, a specific resistance is placed in the unmentioned terminal circuit.

We shall discuss other aspects of manufacturers' data sheets as the need arises in treating specific devices.

Review Questions

1. Discuss the role of the emitter region in the operation of a BJT.
2. Discuss the role of the base region in the operation of a BJT. Why is the width of the base usually very small?
3. Discuss the role of the collector region in the operation of a BJT.
4. Name some of the common methods used in the fabrication of transistors.
5. Briefly describe one method used for the manufacture of transistors.
6. Identify the doping type for each of the regions inside a *PNP* transistor. Repeat for an *NPN* transistor.
7. For normal operation, how is the base-emitter diode biased? How is the collector-base diode biased?

8. Draw a block diagram of an *NPN* transistor and include batteries with the proper polarity for normal operation.
9. Repeat Problem 8 for a *PNP* transistor.
10. What is α? How is it related to the transistor terminal currents?
11. What typical range of values can you expect for α?
12. What is β? How is it related to the transistor terminal currents?
13. What typical range of values can you expect for β?
14. How are the transistor α and β related?
15. What is the collector reverse-saturation current and what is it caused by?
16. What carrier makes up the largest component of the current inside the transistor: *PNP* and *NPN*?
17. What are the three possible transistor configurations?
18. In the *CB* input characteristics, what are the functions that are plotted?
    What is the parameter?
19. In the *CB* output characteristics, what are the functions that are plotted?
    What is the parameter?
20. Repeat Question 18 for the *CE* input characteristics.
21. Repeat Question 19 for the *CE* output characteristics.
22. What is the meaning of the symbol I_{CEO}?
23. What is the meaning of the symbol BV_{CBO}? Under what conditions is it measured?
24. What is the meaning of the symbol BV_{CEO}? Under what conditions is it measured?
25. What is the meaning of the symbol BV_{CER}? Under what conditions is it measured?
26. From the data sheets for a 2N5449 *NPN* transistor (see Appendix 3), determine the following: (a) $V_{CE\max}$, (b) $I_{C\max}$, (c) $P_{\max}$.
27. What are the five boundaries that determine the permissible region of operation for a BJT?
28. What are the consequences of exceeding the manufacturer's maximum specifications?
29. How does the transistor β vary as a function of the operating point (I_C)?
    Support your answer with an example from a manufacturer's data sheet (see Appendix 3).
30. What effect does an increase in temperature have on transistor characteristics?

Problems

1. In an *NPN* transistor, the collector and emitter currents are measured to be 2.0 and 2.01 mA, respectively. Determine the base current, transistor α and β.
2. The α of a certain transistor is determined to be 0.99 with an uncertainty in the last digit of ± 1. What is the β for this transistor and what is the uncertainty in β?
3. For the *CB* output characteristics in Fig. 3-9, determine the transistor α and β when $V_{CB}=2$ V and $I_E=0.1$ mA.
4. Repeat Problem 3 for I_E of 0.2, 0.3, 0.4, and 0.5 mA. Make a plot of α as a function of I_E. Also make a plot of β as a function of I_E.
5. On the same graph, sketch the *CB* output characteristics for two transistors, one of which has $\alpha \cong 1$ and the other $\alpha = 0.8$. State the difference between their characteristics in words.
6. Refer to Fig. 3-12. Determine the collector current when the transistor voltages are: $V_{BE}=0.6$ V and $V_{CE}=5$ V. [Hint: First determine I_B from Fig. 3-12(a).]
7. Determine β for the transistor with the *CE* output characteristics in Fig. 3-12(b) if $I_B=200\mu$A, and $V_{CE}=7$ V.
8. Determine the β for the transistor with the characteristics in Fig. 3-12 when $I_C=10$ mA and $V_{CE}=9$ V. (Hint: First determine β for the nearest I_B curve.)
9. Determine the base current for a 2N5450 transistor (see Appendix 3) when the collector current is 1, 10, and 100 mA. Make a sketch of I_B as a function of I_C.
10. Sketch the permissible region of operation for a 2N5449 transistor (see Question 26).

4 BJT Biasing, Bias Stability, and Small-Signal Models

The basic operation of a BJT was introduced in Chap. 3. Before we can consider the transistor as an amplifier, we must take up the problem of selecting and establishing the proper operating point for the transistor. A few of the basic biasing schemes will be discussed from both an anlytical and a graphical standpoint.

Once the transistor is properly biased, we shall consider it a two-port device and examine its characteristics as such. In this analysis we shall assume that operation is over a small portion of the active region where the transistor acts in a reasonably linear fashion. The analysis will lead to a discussion of ac small-signal models that can be used to determine the performance of transistors under small-signal operation.

4.1 BIASING

In the following analysis, we shall use approximations. You may be somewhat confused to see the same quantity neglected in one case and kept in the analysis in another case. The key to when and why something may be ignored lies in a comparison of the quantity in question with some other quantity. For example, in a series circuit, a voltage drop of 0.5 V cannot be neglected if the total voltage drop across the elements of interest is 2 or 3 V. However, the same voltage drop of 0.5 V may be ignored in another circuit where the total voltage drop under consideration is in excess of, say, 10 V. Follow the rule of thumb that when one quantity is at least one order of magnitude (one order of magnitude is considered a factor of 10) larger than another quantity, the smaller quantity may be justifiably neglected, with little loss in accuracy, since $\pm 10\%$ is typically the resistor tolerance.*

---

*It should go without saying that a comparison of unlike quantities (e.g., voltage and current, resistance and conductance) should not be attempted. Only like quantities may be compared (e.g., two voltages, two currents, two resistors,) with the aim of neglecting one of these.

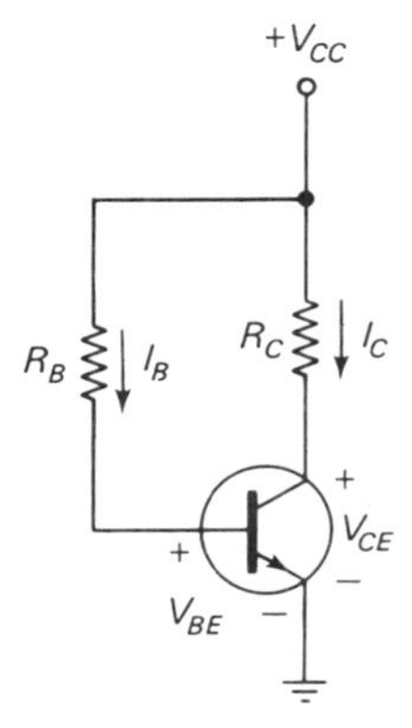

Figure 4-1. Fixed-current bias circuit.

4.1.1 Calculation of the Operating Point

To establish proper operation of the BJT, the base-emitter junction must be forward biased and the collector-base junction must be reverse biased. The circuit in Fig. 4-1, called the *fixed-bias circuit*, provides the transistor with proper bias for normal operation. The ac input is applied between the base and emitter terminals shown in Fig. 4-1, with the output taken between the collector and emitter terminals.

A fairly common notation for the indication of the connection of a dc power supply (or battery) is shown. We use the convention that the point labeled $+V_{CC}$ in the diagram has the dc source connected to it. The other terminal of the source is understood to be connected to ground.

The starting point in the analysis is to define voltage and current polarities and directions in the circuit. For the *NPN* transistor used, both the collector and base current flow *into* the transistor. Although we shall analyze *NPN* transistors here, the procedure applies equally well to *PNP* transistors with the appropriate reversal of the dc supplies, voltage, and current polarities.

In the analysis that follows we are interested in determining the operating, or *quiescent (Q)*, point. This point is specified by the values of the base and collector currents as well as the collector-emitter voltage set up by the specific values of the dc supply V_{CC} and bias resistors R_B and R_C. We shall denote the transistor currents and voltages at the operating point by adding a Q subscript to the symbol. Thus, I_{BQ} would denote the quiescent base current. We shall assume that the supply voltage V_{CC} and the bias resistors are all known. In addition, we shall assume that the necessary transistor parameters are also known.

Taking the summation of voltages from the battery, through R_B, across the base-emitter junction to ground, and setting it equal to zero, we obtain:

$$V_{CC} = I_B R_B + V_{BE} \tag{4-1}$$

We now come to the first approximation. The base-emitter voltage drop is a forward-biased diode drop. Typically for silicon transistors, this drop is 0.5 to 0.6 V; for germanium, between 0.2 and 0.3 V. In Eq. (4-1) when V_{BE} of this magnitude is compared to V_{CC}, which is typically larger than 9 or 10 V, V_{BE} may be justifiably neglected. Thus, we have:

$$V_{CC} \cong I_B R_B \tag{4-2}$$

This is the *bias curve equation*. Because both V_{CC} and R_B are fixed in value, Eq. (4-2) determines the quiescent base current. Therefore,

$$I_{BQ} \cong \frac{V_{CC}}{R_B} \tag{4-3}$$

In a similar manner, we take the summation of voltages in the collector circuit and set it equal to zero to obtain:

$$V_{CC} = I_C R_C + V_{CE} \tag{4-4}$$

This is the *load line equation*. We shall discuss the significance of this name when we examine graphical techniques for determining the operating point.

If the transistor β is known, the quiescent collector current is calculated from the relationship:

$$I_{CQ} = \beta I_{BQ} \tag{4-5}$$

where I_{BQ} has already been determined in Eq. (4-3). Using the value of I_{CQ} thus calculated in Eq. (4-4), we determine the quiescent collector-emitter voltage from the relationship:

$$V_{CEQ} = V_{CC} - I_{CQ} R_C \tag{4-6}$$

So we see that if the values of the dc supply voltage, the bias resistors, and the transistor β are known, the dc operating (quiescent) point for the transistor in Fig. 4-1 may be determined in the straightforward manner just indicated.

Example 4-1. The circuit values in Fig. 4-1 are: $R_B = 200$ kΩ, $R_C = 2$ kΩ and $V_{CC} = 20$ V. A silicon *NPN* transistor with $\beta = 50$ is used. We want to determine the *Q*-point.

Solution: We first check to make certain that V_{BE} may indeed be neglected. For silicon transistors, $V_{BE} \cong 0.6$ V, which is indeed negligible with the V_{CC} of 20 V. Thus,

$$I_{BQ} \cong \frac{20}{200} \text{ mA} \cong 0.1 \text{ mA}$$

The collector current is calculated next:

$$I_{CQ} \cong 50(0.1 \text{ mA}) \cong 5 \text{ mA}$$

The collector voltage is then:

$$V_{CEQ} = 20 - (5)(2) \cong 10 \text{ V}$$

The operating point, therefore, is seen to depend on the dc supply voltage, the bias resistors, the transistor used and, as we shall see subsequently, temperature.

We must now turn our attention to a practical problem. The manufacturer usually specifies a range for β rather than a single specific value. In Appendix 3, for a type 2N5451, the minimum β is specified as 30, while the maximum value specified is 600. In other words, a specific transistor of type 2N5451 may have a β any where within the specified range; it might be 30 or 600 or any value in between. Large β spreads in transistors of the same type number are by no means unusual; instead, they are the rule. Even if the α of a transistor is controlled within fairly tight tolerances, the corresponding β has a wide range of values.

From the preceding example, we can readily see that any variation in β causes the operating point in the circuit shown in Fig. 4-1 to vary drastically. To see this effect, note that if the transistor used in the example had a β of 100 or more, the transistor would be saturated, i.e., $V_{CEQ} \cong 0$, and could not be used to amplify any signal.

Another problem is related to the temperature sensitivity of transistor terminal currents. Recall that the relationship between the collector and base currents also contained a term involving the collector reverse-saturation current. The equation is repeated here for convenience:

$$I_C = \beta I_B + (\beta + 1) I_{CO} \tag{4-7}$$

The collector reverse-saturation current doubles for every 10°C rise in junction temperature. Moreover, this increase is multiplied by $\beta + 1$. Therefore, the collector current is also very much a function of the junction operating temperature.

A condition known as *thermal runaway* may occur with the increase in collector current. Such a condition can cause the transistor to saturate. Assume that the collector junction is at room temperature. As soon as any collector current is caused to flow, the junction temperature increases because of the power being dissipated at the junction. This increase in temperature is reflected as an increase in I_{CO}. From Eq. (4-7) we see that any increase in I_{CO} also causes I_C to increase. As a result of the increase in I_C, the power dissipated at the collector junction increases, causing the junction temperature to rise accordingly. This cycle may be repeated until the maximum junction temperature is exceeded, in which case the transistor may be damaged permanently, or until the collector current increases sufficiently to cause the transistor to be saturated. Either of these possibilities should be avoided.

A numerical measure, called the thermal stability factor S, has been defined to indicate how well the possibility of thermal runaway has been eradicated. The thermal stability factor is defined as:

$$S = \frac{\text{change in } I_C}{\text{change in } I_{CO}} \tag{4-8}$$

From the definition we see that the lower the numerical value of S, the less likely it is for thermal runaway to occur. The thermal stability factor is determined by the bias circuit configuration. For the circuit in Fig. 4-1, S is given by*

$$S=\beta+1 \tag{4-9}$$

The above expression for S indicates that for the fixed-bias circuit (Fig. 4-1), the thermal stability factor is a constant that depends only on the β of the transistor used. Moreover, because β is typically quite large, so is S. Consequently, the circuit is not very well protected from thermal runaway.

The self-bias arrangement, shown in Fig. 4-2, offers a number of advantages over the fixed-bias circuit. It provides us with the ability to design for specific operating point with almost complete immunity to the β spread. We can also design for a specified stability factor.

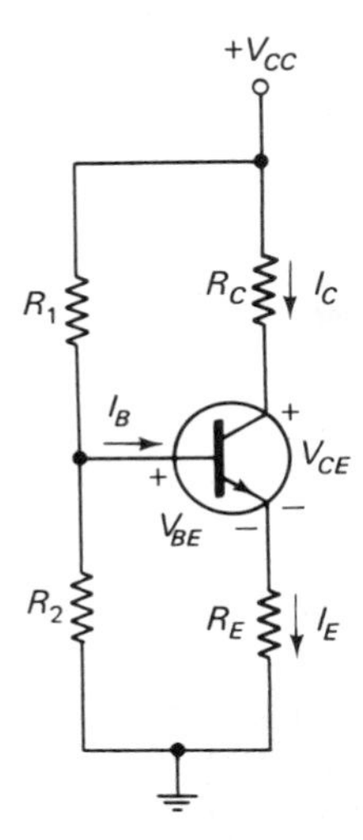

Figure 4-2. Self-bias circuit.

Analysis of the self-bias circuit is similar to that used in the fixed-bias circuit. First, we recognize that the two resistors (R_1 and R_2) in the base circuit provide a voltage divider. The circuit is redrawn for more convenience in Fig. 4-3(a). We may simplify it by finding the Thevenin equivalent circuit to the left of points A and N. If we label the equivalent resistance by R_B and the equivalent open-circuit voltage by V_{BB}, we have:

$$R_B=\frac{R_1R_2}{R_1+R_2} \tag{4-10}$$

and

$$V_{BB}=V_{CC}\frac{R_2}{R_1+R_2} \tag{4-11}$$

*For a derivation of the thermal stability factor for this and other bias configurations, see: Cirovic, *Semiconductors*, pp. 115–136.

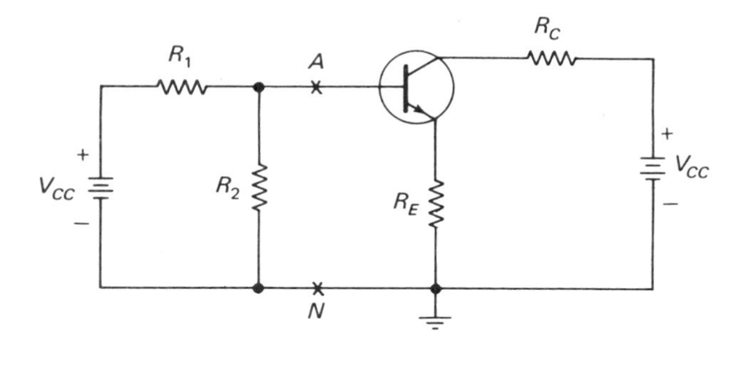

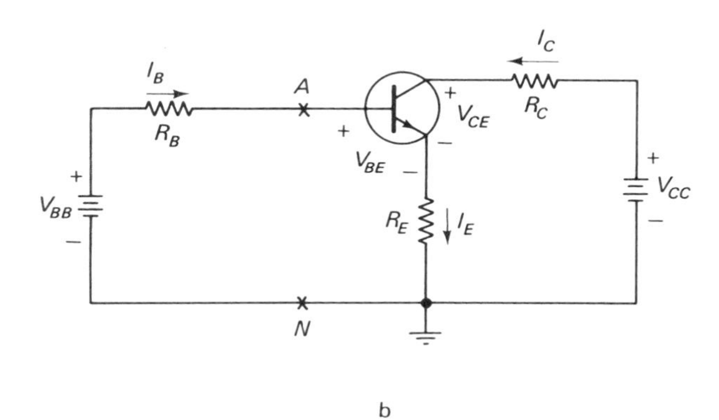

Figure 4-3. (a) The self-bias circuit redrawn for the purpose of finding the Thevenin equivalent at points *A* and *N*, (b) with the Thevenin equivalent circuit replacing the voltage divider.

Using these equivalent values, we redraw the circuit as indicated in Fig. 4-3(b). We can now proceed to determine the operating point by adding the voltages in the base and collector loops. In the base loop, we have:

$$V_{BB} = I_B R_B + V_{BE} + I_E R_E \tag{4-12}$$

Before we proceed, note that in this case the base-emitter voltage V_{BE} may not always be negligible, because it is compared to V_{BB}, which is always smaller than V_{CC}. The decision whether or not to neglect V_{BE} must be made in each case and depends on the relative size of V_{BB}.

We can make use of the fact that the summation of the transistor terminal currents is zero ($I_E = I_C + I_B$) in Eq. (4-12). At the same time, we substitute βI_B for I_C, neglecting the effect of I_{CO}.

$$V_{BB} = I_B R_B + V_{BE} + (\beta + 1) I_B R_E \tag{4-13}$$

Before we solve for the quiescent base current, note the result of the substitutions. Saying that the voltage drop from emitter to ground is

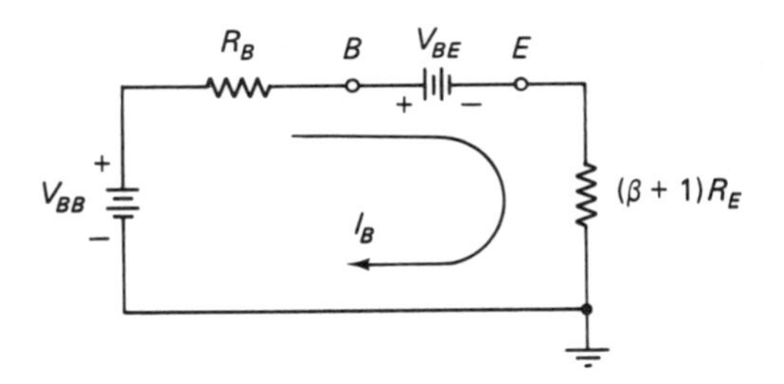

Figure 4-4. Equivalent circuit for the evaluation of base current.

caused by a current I_E flowing through R_E is equivalent to saying that it is the result of a current I_B flowing through an equivalent resistance $(\beta+1)R_E$. If we represent the base-emitter voltage drop by a fixed battery, the sum of the voltages in Eq. (4-13) corresponds to the equivalent loop shown in Fig. 4-4, where I_B is a loop instead of branch current. This equivalent circuit may be used to simplify calculations. Furthermore, it illustrates that a resistance R_E in the emitter circuit corresponds to a resistance $(\beta+1)R_E$ in the base circuit.

The quiescent base current is now calculated from Eq. (4-13) or Fig. 4-4:

$$I_{BQ}=\frac{V_{BB}-V_{BE}}{R_B+(\beta+1)R_E} \tag{4-14}$$

As was the case in the fixed-bias circuit, we may assume that V_{BE} is essentially constant to determine I_{BQ} from Eq. (4-14).

Setting the voltage rises (i.e., the supply voltage) equal to the voltage drops in the collector or output loop in Fig. 4-3, we obtain the load line equation:

$$V_{CC}=I_C(R_C+R_E)+V_{CE} \tag{4-15}$$

where we have used the approximation that $I_E \cong I_C$. With the value of β known, the quiescent collector current is determined:

$$I_{CQ} \cong \beta I_{BQ} \tag{4-16}$$

This value is then used in the load line equation to solve for the quiescent collector voltage:

$$V_{CEQ}=V_{CC}-I_{CQ}(R_C+R_E) \tag{4-17}$$

The procedure outlined is illustrated in the following example.

Example 4-2. The circuit values in Fig. 4-2 are: $R_1=33\text{ k}\Omega$, $R_2=10\text{ k}\Omega$, $R_C=2.2\text{ k}\Omega$, $R_E=1\text{ k}\Omega$, and $V_{CC}=18$ V. A silicon *NPN* transistor with $\beta=50$ is used. We want to determine the operating point.

Solution: We first calculate the base equivalent resistance and open-circuit voltage.

$$R_B = \frac{(33)(10)}{33+10}\,\text{k}\Omega = 7.7\text{k}\Omega$$

$$V_{BB} = (18)\frac{10}{33+10}\,\text{V} = 4.2\ \text{V}$$

The quiescent base current is now calculated from Eq. (4-14):

$$I_{BQ} = \frac{4.2-0.6}{7.7+(51)1}\,\text{mA} = 61\,\mu\text{A}$$

With $\beta=50$, the collector current is:

$$I_{CQ} = 50(0.061\ \text{mA}) = 3\ \text{mA}$$

The collector voltage is determined to complete the calculation of the operating point:

$$V_{CEQ} = 18-(3)(2.2+1) = 8.4\ V$$

The thermal stability factor for the self-bias circuit* is given by:

$$S = \frac{\beta+1}{1+\dfrac{\beta}{K}} \tag{4-18}$$

where K is a constant that is a function of the circuit values:

$$K = \frac{R_B+R_E}{R_E} \tag{4-19}$$

If in the limit $K=1$, note that S is also 1. This is the absolute minimum value, because for any real resistor values, the minimum value of K is 1. We achieve thermal stability in this circuit by the following manner. Should the collector current tend to increase because of an increase in I_{CO}, the voltage drop across R_E would also increase. Because V_{BB} is a constant for given resistor values, it is evident from Eq. (4-12) that I_B would decrease if the voltage drop $I_E R_E$ increased. Thus, a tendency for I_C to increase (because of increasing I_{CO}) is offset by the corresponding tendency for I_B to decrease. So I_C remains essentially fixed and relatively insensitive to I_{CO}. This method of achieving better thermal stability is based on sensing the output current (I_C) and causing a response in the input current (I_B). It is also an example of *negative feedback*. This principle will be discussed in detail in Chap. 13.

To see the improvement in thermal stability offered by the self-bias circuit, we determine S for the values in Example 4-2.

---

*See Cirovic, *Semiconductors*, pp. 115–136.

Example 4-3. For the circuit in Example 4-2 we want to determine the thermal stability factor S.

Solution: We first determine K:

$$K=\frac{7.7+1}{1}=8.7$$

S is now calculated from Eq. (4-18):

$$S=\frac{50+1}{1+\frac{50}{8.7}}=7.5$$

Note that this value is significantly lower than the corresponding value in the fixed-bias circuit. Moreover, by changing the values of the circuit resistors, you can obtain a different value of S. Remember that in the fixed-bias circuit the circuit resistor values did not have any effect on S.

4.1.2 Designing a Bias Circuit

In the simplest sense, designing a bias circuit involves specifying (1) resistor values to be used and (2) a certain dc supply voltage. Calculating the resistors to be used is quite similar to the analysis discussed in the previous section. The equations are just rearranged. In the design problem, the operating point quantities, voltages, and currents are known; it is the resistors that are to be determined.

For designing the self-bias circuit, we shall assume that the load resistor (R_C) and the supply voltage are given. In a case where they are not specified, they may simply be chosen. If we know the operating point, the thermal stability factor, and the transistor parameters, we have enough information to determine specific values for the remaining circuit components. The design equations are summarized here:

$$R_E=\frac{V_{CC}-V_{CEQ}-I_{CQ}R_C}{I_{CQ}} \qquad (4\text{-}20)$$

$$K=\frac{S\beta}{\beta+1-S} \qquad (4\text{-}21)$$

$$R_B=(K-1)R_E \qquad (4\text{-}22)$$

$$V_{BB}=I_{BQ}[R_B+(\beta+1)R_E]+V_{BE} \qquad (4\text{-}23)$$

$$R_1=R_B\frac{V_{CC}}{V_{BB}} \qquad (4\text{-}24)$$

$$R_2=\frac{R_1R_B}{R_1-R_B} \qquad (4\text{-}25)$$

The basic design procedure is illustrated in the following example.

Example 4-4. We want to design a self-bias circuit using a silicon transistor ($V_{BE}=0.6$ V) with a β of 80. The load resistor (R_C) is to be 1.2 kΩ.

With the supply voltage at 12 V, the operating point is to be: $V_{CEQ} = 6$ V and $I_{CQ} = 4$ mA. The circuit is to be designed with a thermal stability factor of 10.

Solution: We first calculate the operating-point base current corresponding to the given β:

$$I_{BQ} = \frac{I_{CQ}}{\beta} \cong \frac{4}{80}\,\text{mA} \cong 0.05\,\text{mA}$$

We are now ready to proceed with the steps outlined in Eqs. (4-20) to (4-25):

$$R_E = \frac{12-6-(4)(1.2)}{4}\,k\Omega \cong 300\Omega$$

$$K = \frac{(10)(80)}{80+1-10} \cong 11.3$$

$$R_B = (11.3-1)(0.3)\text{ k}\Omega \cong 3.1\text{ k}\Omega$$

$$V_{BB} = (0.05)[3.1+(80+1)(0.3)] + 0.6 \cong 2\text{ V}$$

$$R_1 = 3.1\frac{12}{2}\text{k}\Omega \cong 18.6\text{ k}\Omega$$

$$R_2 = \frac{(18.6)(3.1)}{18.6+3.1}\text{k}\Omega \cong 3.8\text{ k}\Omega$$

The design is implemented by specifying standard resistor values. In this case, we may choose:

$$R_E = 330\ \Omega, \qquad R_1 = 18\text{ k}\Omega, \qquad R_2 = 3.9\text{ k}\Omega$$

The choice of the thermal stability factor is governed by the following guidelines: For small-signal (low-power) stages, an S of 20 or lower is satisfactory; for power amplifiers, usually an S of 5 or less may be desirable.

The design procedure just outlined is quite satisfactory when the β of the transistor to be used can be measured. However, when units are mass produced, the β of each individual transistor cannot be measured. The problem, therefore, is to design the bias circuit so that any transistor with a β within a specified range (the β spread) can be used in the circuit and produce an operating point within a desired range. If we label the difference between the minimum operating-point collector current (I_{Cm}) and the maximum operating-point collector current (I_{CM}) by ΔI_C, then the needed stability factor S_M is calculated from:*

$$S_M = \frac{\Delta I_C}{I_{Cm}} \frac{\beta_m \beta_M}{\Delta\beta} \tag{4-26}$$

where β_m is the minimum value of β; β_M, the maximum value of β; and

*For a more complete discussion, see: Cirovic, *Semiconductors*, pp. 141–150.

$\Delta\beta$, the β spread, defined as $\beta_M - \beta_m$. This stability factor corresponds to the maximum value of β: β_M. The remainder of the design procedure follows that given in the previous example.

Example 4-5. We want to design the self-bias circuit of Fig. 4-2 using a 2N5450 silicon *NPN* transistor. The constraints are: $V_{CC}=15$ V, $R_C=330$ Ω. The nominal operating point is to be: $I_{CQ}=10$ mA and $V_{CEQ}=8$ V. The variation in the operating-point collector current, because of the uncertainty in β, cannot be lower than 9 mA or higher than 11 mA.

Solution: First, we determine that the β range for the 2N5450 transistor (from Appendix 3) is from 50 to 150. Thus, $\beta_M=150$, $\beta_m=50$, and $\Delta\beta=150-50=100$. Also, $I_{CM}=11$ mA, $I_{Cm}=9$ mA, and $\Delta I_C=11-9=2$ mA. We can now proceed to determine the thermal stability factor needed:

$$S_M \cong \frac{2}{9}\,\frac{(50)(150)}{100} \cong 16.7$$

We next use this value in Eq. (4-21) with appropriate subscripts:

$$K=\frac{S_M\beta_M}{\beta_M+1-S_M} \cong \frac{(16.7)(150)}{150+1-16.7} \cong 18.6$$

Using the nominal Q-point values, we determine R_E from Eq. (4-20):

$$R_E=\frac{15-7-(10)(0.33)}{10} \cong 470\ \Omega$$

From this point on, we can use one of two methods. Each gives the same results. Because we have no idea of a "nominal" β value, we may calculate either I_{BM} or I_{Bm}:

$$I_{BM}=\frac{I_{CM}}{\beta_M} \qquad \text{and} \qquad I_{Bm}=\frac{I_{Cm}}{\beta_m}$$

These values are: $I_{BM} \cong 0.073$ mA and $I_{Bm} \cong 0.18$ mA. Note that I_{Bm} is larger than I_{BM}. The interpretation of the symbols is important. The M subscript on I_B denotes the I_B value that results when the transistor in the circuit has β equal to β_M. It does not stand for the maximum I_B, and in fact it will not usually be larger than I_{Bm}. Similarly, the m subscript on I_B denotes the I_B value that results from the transistor being used having a β equal to β_m. Only one value of I_B is needed for the actual solution.

V_{BB} is calculated from Eq. (4-23), using the proper combination of I_B and β values; that is, if we use I_{BM}, then we must use β_M; if we use I_{Bm}, then we must use β_m. R_B is obtained from Eq. (4-22):

$$R_B \cong (18.6-1)(0.47)\ \text{k}\Omega = 8.3\ \text{k}\Omega$$

Using I_{Bm} and β_m in Eq. (4-23), we obtain:

$$V_{BB}=(0.18)[8.3+(51)(0.47)]+0.6 \cong 6.4\ \text{V}$$

A subsequent exercise will show that the same value of V_{BB} results if I_{BM} and β_M are used in Eq. (4-22).

The calculation of R_1 and R_2 proceeds as before:

$$R_1 = 8.3\frac{15}{6.4}\text{k}\Omega \cong 19.5\text{ k}\Omega$$

$$R_2 = \frac{(19.5)(8.3)}{19.5 - 8.3}\text{k}\Omega \cong 14.5\text{ k}\Omega$$

Again we round off the calculated values to standard available values to complete the design:

$$R_E = 470\ \Omega, \qquad R_1 = 20\text{ k}\Omega, \qquad R_2 = 15\text{ k}\Omega$$

4.1.3 Drawing the DC Load Line

Equation (4-6) is the dc load line equation for the fixed-bias circuit; Eq. (4-15) is the dc load line equation for the self-bias circuit. For either circuit, we can evaluate the Q-point graphically by using the same basic method.

In order to plot the dc load line on the CE output characteristics, note first that the line is a straight line, so we need plot only two points. Secondly, the most convenient points to calculate are the two axis intercepts. We can determine the I_C axis intercept by making $V_{CE} = 0$ in the load line equation. For the self-bias circuit we find:

$$\text{point 1} \qquad \text{when } V_{CE} = 0 \qquad I_C = \frac{V_{CC}}{(R_C + R_E)}$$

In a similar fashion, we can determine the V_{CE} axis intercept by forcing I_C to be zero in the load line equation. Again for the self-bias circuit, we find:

$$\text{point 2} \qquad \text{when } I_C = 0 \qquad V_{CE} = V_{CC}$$

These two points are indicated on Fig. 4-5. We then obtain the load line by

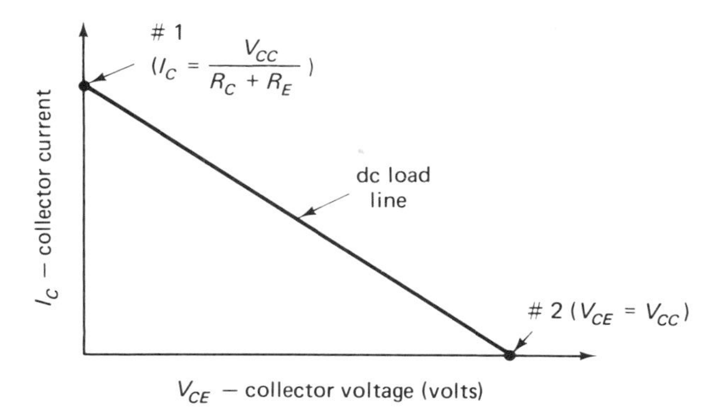

Figure 4-5. Plotting the load line on the output characteristics. (The curves are left out for simplicity.)

drawing a straight line between them. If we know the device output characteristics (they can be displayed on a curve tracer) and the base current, then we find that the Q-point is the intersection of the dc load line, as just plotted, with the bias curve corresponding to the proper I_B value. This method is especially useful when we use the curve tracer to determine the transistor small-signal parameters.

4.2 BJT LOW-FREQUENCY MODEL

Having described the operation and biasing of BJTs, we now examine the operation of a BJT, which has been properly biased, as an amplifier of time-varying signals.

To see how amplification takes place, consider the following example. We want to operate a silicon *NPN* transistor, whose characteristics are shown in Fig. 4-6, in the bias circuit of Fig. 4-1. The circuit values are: $V_{CC}=18$ V, $R_B=450$ kΩ, and $R_C=3$ kΩ. Using the methods discussed in the previous sections, we determine the operating point. Assuming V_{BE} negligible as compared with the supply voltage of 18 V, we find that the Q-point base current is: $I_{BQ}=18/450$ mA$=40$ μA. Figure 4-6(a) tells us that $V_{BEQ}=0.6$ V (when $I_{BQ}=40$ μA). This result justifies the earlier assumption that V_{BE} is negligible. The ac load line is obtained in the same manner as the dc load line. The ac resistance is exactly the same as the dc resistance in the circuit, so that the ac and dc load lines for this particular circuit are identical. The plot is shown in Fig. 4-6(b). The Q-point is the point where the $I_{BQ}=40$ μA curve intersects the load line, where $I_{CQ}=4$ mA and $V_{CEQ}=6$ V.

Now let us assume that, in some manner, the input (base-emitter) voltage is caused to vary about its quiescent value of 0.6 V. For example, let V_{BE} have a total excursion of 25 mV, i.e., 12.5 mV above V_{BEQ} and 12.5 mV below V_{BEQ}. By projecting this variation on the input characteristic, we can determine the corresponding variation in the base current. We see that $\Delta I_B=20$ μA, 10 μA above and below I_{BQ}. Using this variation in the base current on the output characteristics, we can determine the corresponding variation in the collector current and collector-emitter voltage, as shown in Fig. 4-6(b). Thus, $\Delta I_C=2$ mA and $\Delta V_{CE}=6$ V. We conclude from this graphical analysis that voltagc amplification has taken place. A variation of 25 mV in the input (base-emitter) voltage has caused a variation of 6 V in the output (collector-emitter) voltage. The amount of voltage amplification, called *gain*, is the ratio of the output variation to the input variation. In this case, it is approximately 240.

This graphical analysis method is an extremely useful technique in the analysis of amplifiers. We shall apply it later in our discussion of large-signal (power) amplifiers in Chap. 12. If the signals involved are small and the excursions that they cause are over an essentially linear portion of the transistor characteristics, the graphical techniques may be replaced by a systematic analysis utilizing an approximate model of the transistor.

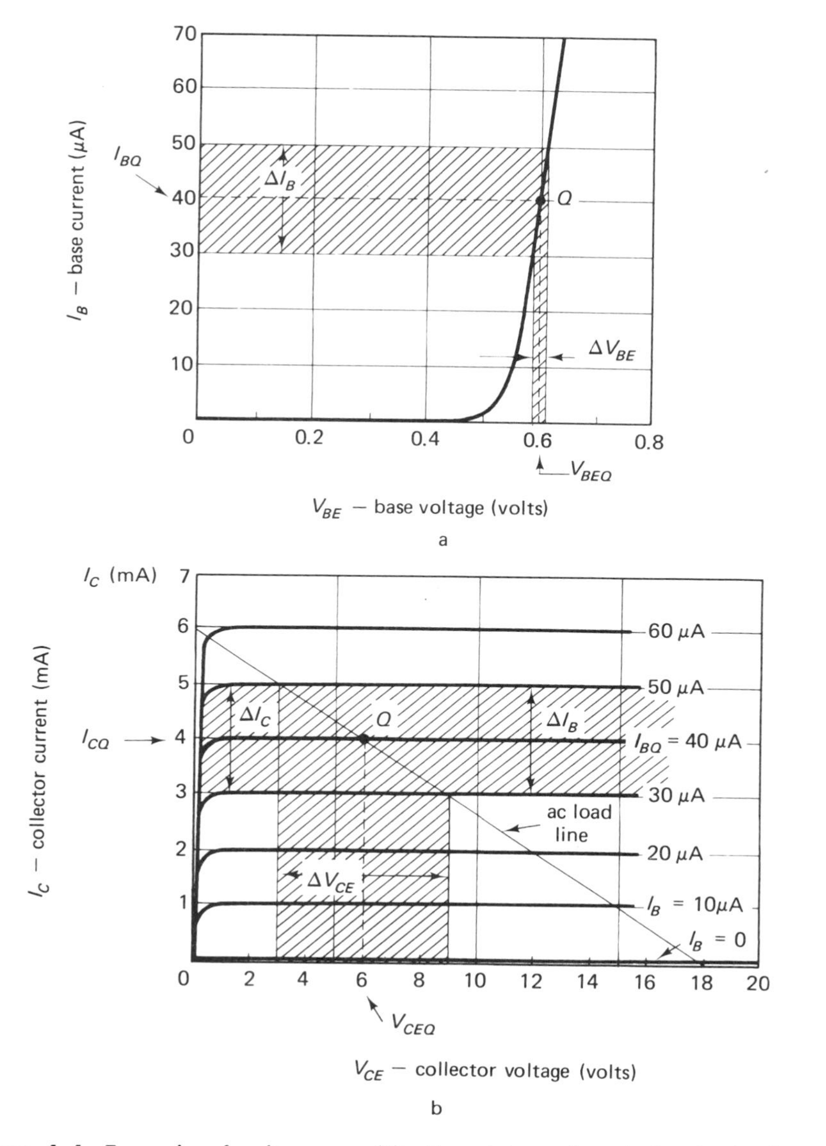

Figure 4-6. Example of voltage amplification: (a) small variation in the base-emitter voltage results in (b) a large variation in collector-emitter voltage.

The common-emitter transistor configuration is shown in two-port notation in Fig. 4-7, using the conventional directions for the terminal currents and polarity for the voltages. Lower-case symbols are used to denote ac (time-varying) currents and voltages. By ac signals we mean small excursions around the dc Q-point.

One word of caution is offered here. In dealing with BJT small-signal models, we shall consider the transistor by itself. However, remember that

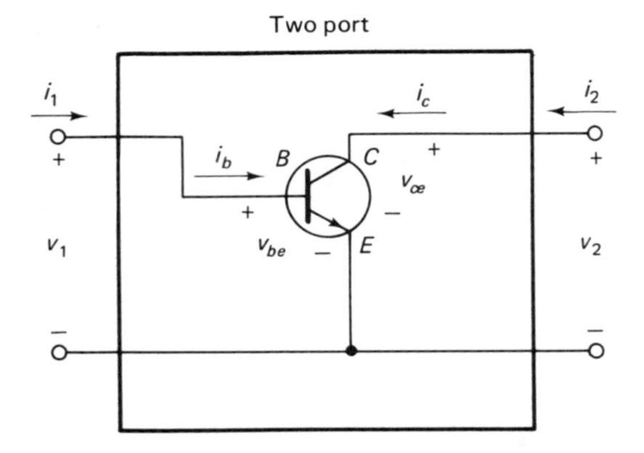

Figure 4-7. The *CE* configuration as a two-port network.

the transistor cannot operate without the proper bias being provided by one of the bias circuits, which we have already discussed. All these bias circuits contain resistors, which are deleted here for the sake of simplicity. At this time, we are concerned with the behavior of the transistor alone. We shall include the effects of the bias resistors in future analyses of amplifiers and other circuits.

We recall that a two-port, like the one shown in Fig. 4-7, may be completely represented by a set of four parameters. The most suitable set of parameters for the BJT is the set of h parameters. The defining equations for the h parameters are:

$$v_1 = h_{11} i_1 + h_{12} v_2$$
$$i_2 = h_{21} i_1 + h_{22} v_2 \tag{4-27}$$

Because of the importance of the BJT and the fact that it can be operated in any one of three possible configurations (*CE*, *CB*, or *CC*), the simple numeral subscript notation is replaced by one more suitable for transistors. The two numerals are replaced by two letters, the first of which denotes the specific h parameter, and the second of which denotes the transistor terminal common to both input and output. Thus,

$$h_{11} = h_i \quad h_{12} = h_r \quad h_{21} = h_f \quad h_{22} = h_o \tag{4-28}$$

The second subscript—either e, b, or c—denotes whether the emitter, base, or collector terminal is common to the input and the output. For example, in the *CE* configuration we have:

h_{ie} — input impedance (in Ohms)
h_{re} — reverse voltage ratio (no units)
h_{fe} — forward current ratio (no units)
h_{oe} — output admittance (in Mhos or Siemens)

In a similar manner, the parameters in the *CB* configuration are: h_{ib}, h_{rb}, h_{fb}, and h_{ob}. In the *CC* configuration, we have: h_{ic}, h_{rc}, h_{fc}, and h_{oc}.

Making the appropriate substitutions in Eq. (4-27) for the *CE* configuration, we obtain:

$$v_{be} = h_{ie} i_b + h_{re} v_{ce}$$
$$i_c = h_{fe} i_b + h_{oe} v_{ce} \tag{4-29}$$

Before we proceed, we must explain exactly what we mean by the voltage polarities and current directions. Because all these quantities (voltages and currents) are time-varying, their polarities and directions reverse periodically. The notation denotes the conditions at one instant of time. We would know, for example, that when the base is positive with respect to the emitter, the collector is also assumed positive with respect to the emitter.

Equation (4-29) is used to develop the small-signal model for the transistor. We identify the input and output voltages and currents from Fig. 4-7 and note that the emitter is common. Then we have the beginning of the model shown in Fig. 4-8(a). The first equation in Eq. (4-29) is nothing more than the sum of the voltages. The voltage applied (v_{be}) is equal to the sum of the voltage drops. This calculation is shown in Fig. 4-8(b). The second equation is the sum of the currents. The current applied (i_c) to the output node is equal to the sum of currents leaving the same node. This result is indicated in Fig. 4-8(c), completing the *CE* model for the transistor. Note that both the voltage generator in the input and the current generator in the output are controlled, or dependent, generators.

The determination of the four *h* parameters in the *CE* configuration may be made in two ways. One method involves making actual measurements in a circuit with the transistor biased and ac signals applied. The second method involves a graphical evaluation from the transistor static characteristics, which can be displayed on a curve tracer.

The conditions under which the parameters may be calculated are as follows: When the ac component of the output voltage (v_{ce}) is zero, h_{ie} and h_{fe} are calculated as:

$$h_{ie} = \frac{v_{be}}{i_b} \quad \text{at} \quad V_{CEQ} \tag{4-30}$$

$$h_{fe} = \frac{i_c}{i_b} \quad \text{at} \quad V_{CEQ} \tag{4-31}$$

Note here that the condition $v_{ce} = 0$ means that there is no ac collector-emitter voltage; that is, the collector-emitter voltage is kept constant at its *Q*-point value: V_{CEQ}.

In a similar fashion, the other two parameters may be calculated by causing the ac component of the base current to be zero:

$$h_{re} = \frac{v_{be}}{v_{ce}} \quad \text{at} \quad I_{BQ} \tag{4-32}$$

$$h_{oe} = \frac{i_c}{v_{ce}} \quad \text{at} \quad I_{BQ} \tag{4-33}$$

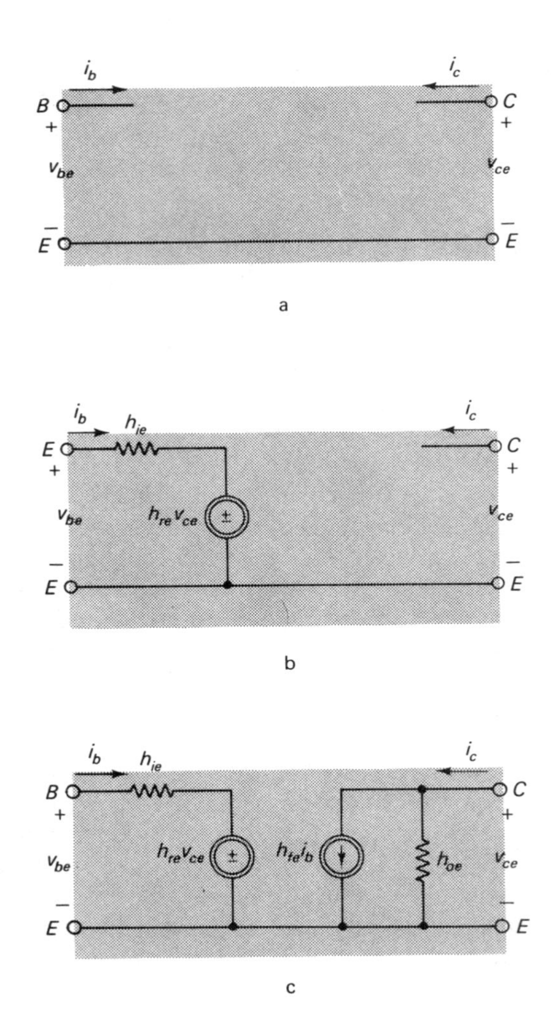

Figure 4-8. Development of the *CE h*-parameter equivalent circuit.

The simplest method for actually obtaining values for the four h parameters of a particular BJT is to display its input and output characteristics on a curve tracer. Another less direct technique is described in Appendix 1. The graphical technique is best illustrated through an example.

Example 4-6. The input and output static characteristics obtained on a curve tracer for a typical *NPN* transistor are given in Figs. 4-9 and 4-10. We want to evaluate the *CE h* parameters using these characteristics. (Using the Tektronix 576 curve tracer in the manner shown in Fig. 4-11 provides a display of the input characteristics.) The operating point is indicated.

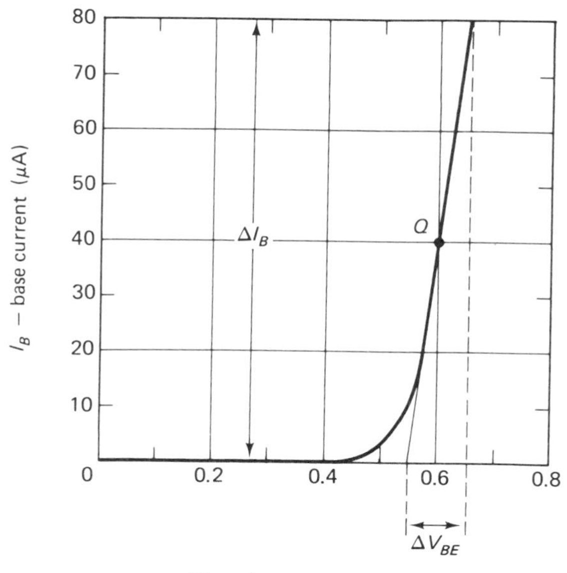

Figure 4-9. Graphical evaluation of h_{ie}.

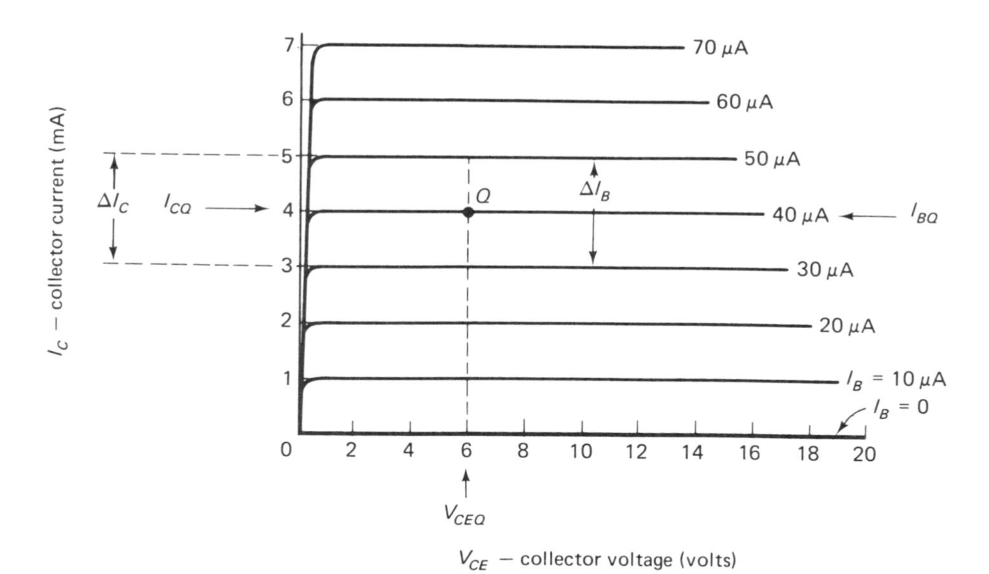

Figure 4-10. Graphical evaluation of h_{fe}.

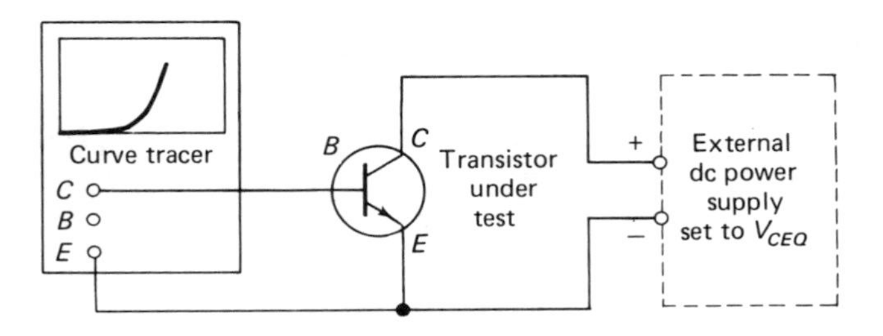

Figure 4-11. Connections for obtaining the *CE* input characteristics of a BJT.

Solution: We determine the first parameter, h_{ie}, from the inverse of the slope of the input characteristics at the Q-point, as shown in Fig. 4-9. Thus, with a change in V_{BE} from 0.55 to 0.65 V, we have a change in I_B from 0 to 80 μA. This change corresponds to:

$$h_{ie} = \frac{0.65-0.55}{80-0}\ \text{M}\Omega \cong 1.25\ \text{k}\Omega$$

We next evaluate h_{fe} from the output characteristics in Fig. 4-10. If the base current changes from 30 to 50 A (which is 10 A above and below I_{BQ}) for a constant V_{CE}, the corresponding change in I_C is from about 3 to 5 mA. Thus,

$$h_{fe} = \frac{5-3}{50-30} \times 10^3 \cong 100$$

We also evaluate h_{oe} from the output characteristics. It is the slope of the bias curve corresponding to $I_B = I_{BQ}$. To obtain somewhat better resolution for this calculation, the region of interest, shown in Fig. 4-12(a), is expanded. With the current scale expanded, as shown in Fig. 4-12(b), we see that for a variation in V_{CE} from 2 to 10 V (which is 4 V above and below the V_{CE} Q-point value), the corresponding I_C variation taken along the I_{BQ} characteristic curve is from about 3.95 to 4.05 mA. Thus,

$$h_{oe} = \frac{4.05-3.95}{10-2}\ \frac{\text{mA}}{\text{V}} = 12.5\mu\ \text{Mho}$$

This is an admittance corresponding to a resistance of 80 kΩ.

When we try to determine h_{re}, we run into problems. It should be measured by noting the difference in the base-emitter voltage (at $I_B = I_{BQ}$) on the input characteristics caused by a certain variation in V_{CE}. The setup shown in Fig. 4-11 is used for this measurement, with the variation in V_{CE} accomplished by the external power supply setting. However, the change in V_{BE} thus obtained is usually imperceptible. We conclude, therefore, that h_{re} is very small. Typically, it may be in the order of 10^{-4}. As we shall see, we do not need to make an actual determination for h_{re} in a majority of cases.

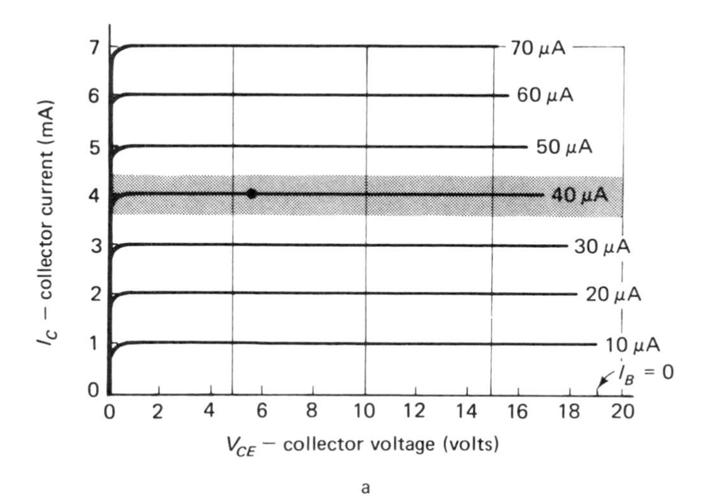

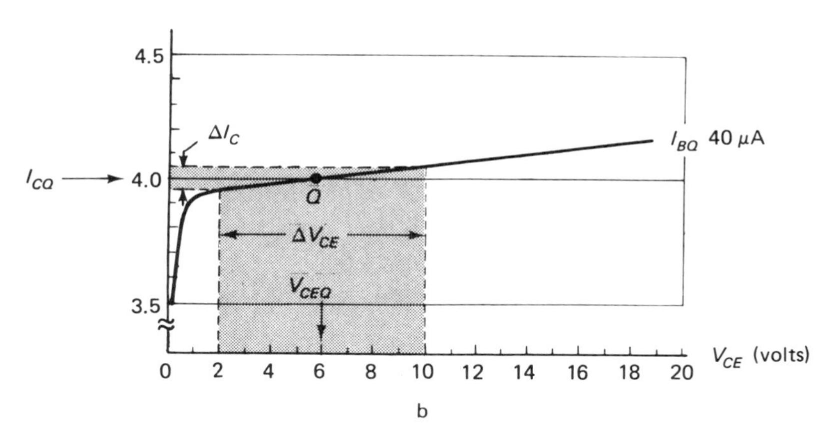

Figure 4-12. Graphical evaluation of h_{oe}: (a) output characteristics showing the region (shaded) of interest, (b) expanded current scale in the vicinity of Q-point.

If we use the typical values of the four h parameters, we need only make approximations for the transistor model. Note that the effect of h_{re} is usually extremely small. In most cases, we may consider the generator controlled by the output voltage negligible. In this case, the transistor may be approximated by the model shown in Fig. 4-13(b). A further approximation is possible in some cases. We neglect the effect of the output impedance (represented by h_{oe}), because in some cases it may be sufficiently larger than the load to make it unimportant. The resulting equivalent circuit for the transistor in the *CE* configuration, with both h_{re} and h_{oe} neglected (assumed zero), is shown in Fig. 4-13(c). Chapter 9 will discuss the specific cases when such approximations are valid.

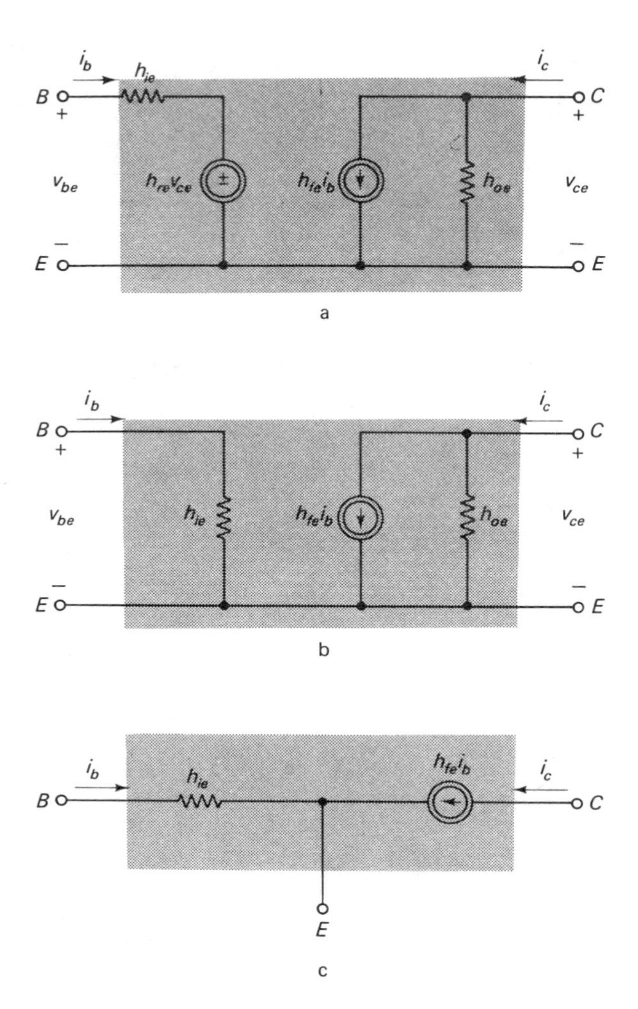

Figure 4-13. *CE* small-signal transistor model: (a) complete *h*-parameter model and successive approximate models with (b) h_{re} considered negligible and (c) both h_{re} and h_{oe} considered negligible.

We must point out that the *h* parameters are not constant for any particular BJT; that is, we cannot evaluate the *h* parameters for the BJT at one *Q*-point and hope to use these same values at another *Q*-point. The *CE h* parameters are very much a function of the specific operating point. For this reason, the manufacturers' data sheets may include a plot of the normalized *h*-parameter values as a function of the *Q*-point collector current. Such a plot is illustrated in Fig. 4-14(a). The *CE h* parameters are also sensitive to the junction operating temperature. The dependence on the junction temperature for a typical BJT is demonstrated in Fig. 4-14(b) and may also be given in the manufacturers' data sheets.

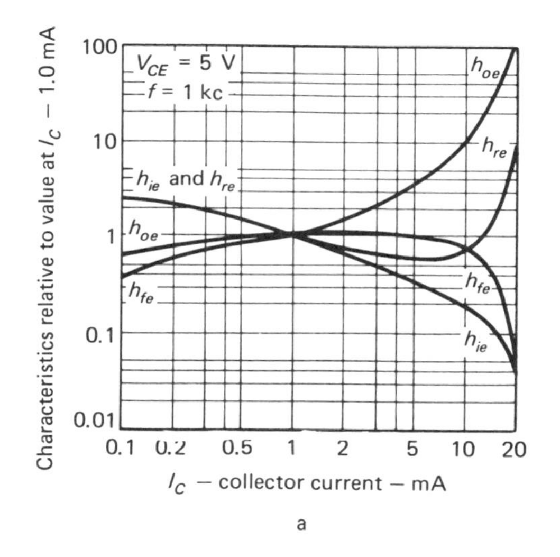

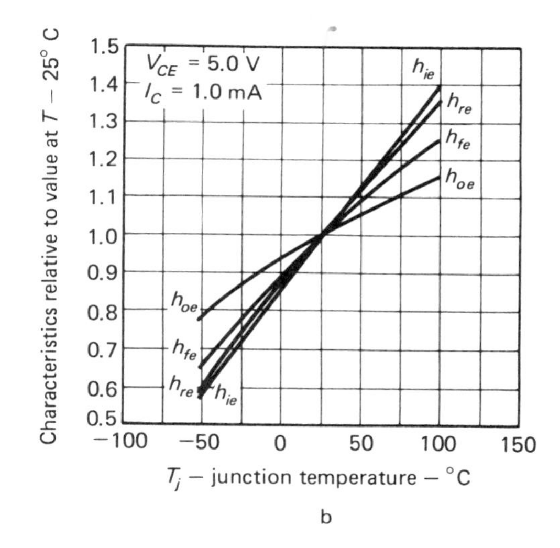

Figure 4-14. Typical variation of h parameter with (a) collector current, (b) temperature.

The use of these normalized curves can best be illustrated through the following example. Let us assume that a particular transistor is listed by the manufacturer to have $h_{ie} = 2.6\ \text{k}\Omega$, $h_{re} = 10^{-4}$, $h_{fe} = 60$, and $h_{oe} = 0.01$ mSiemens, with all values listed at 1 mA collector current and at 25°C. We can use Fig. 4-14(a) to predict the values of the parameters at a different operating point. For example, the same transistor operating at a collector current of 5 mA is seen from Fig. 4-14(a) to have multiplying factors of approximately 0.35, 0.6, 1.0, and 3.5 (listed in the same order as the parameters above). Therefore, at 5 mA, we would expect $h_{ie} = (0.35)(2.6\text{k}\Omega) = 910\Omega$, $h_{re} = (0.6)(10^{-4}) = 6 \times 10^{-5}$, $h_{fe} = 60$, and finally $h_{oe} = (3.5)(0.01) = 0.035$ mSiemens. Similarly, we can use Fig. 4-14(b) to predict the parameters at temperatures other than 25°C. If the same transistor were operated at 75°C, the parameters (listed in the same order as above) would change by 1.28, 1.24, 1.18, and 1.11.

The BJT almost always has a *CE* configuration. For completeness, however, the *CB* and *CC* configurations are mentioned here. The defining equations for the *CB* configuration may be seen from Fig. 4-15. They are:

$$v_{eb} = -h_{ib} i_e + h_{rb} v_{cb}$$
$$i_c = -h_{fb} i_e + h_{ob} v_{cb} \qquad (4\text{-}34)$$

Notice the minus signs in the above equations. They indicate that the definition of i_1 is in the opposite direction to the actual flow of emitter current in an *NPN* transistor. The *CB* equivalent circuit is shown in Fig. 4-15(b). However, in most cases, it is more convenient to use the approximate *CE* transistor model, as shown in Fig. 4-15(c).

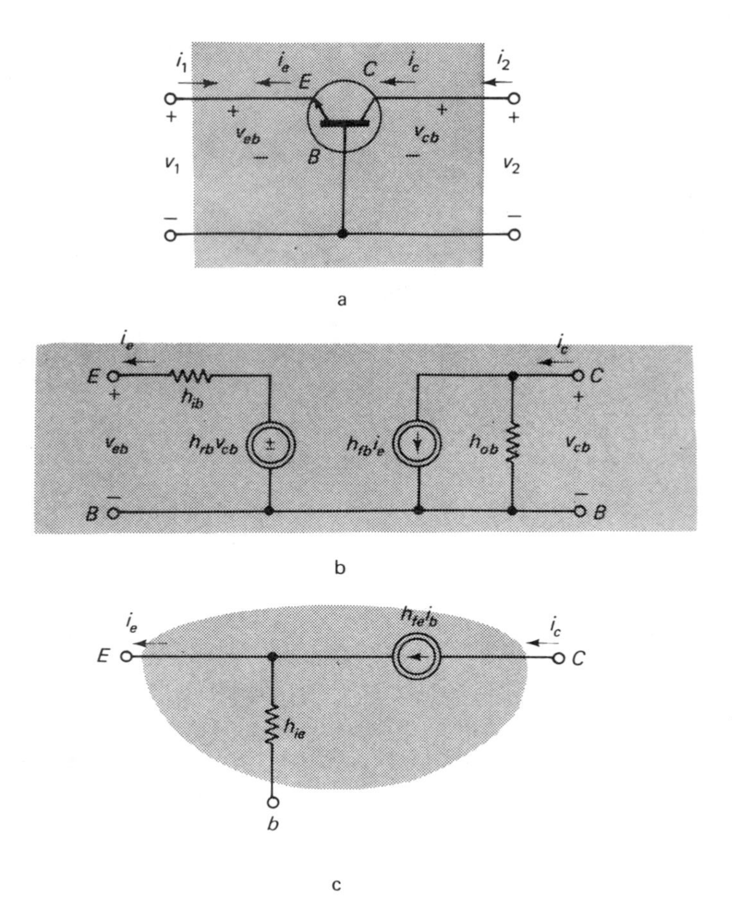

Figure 4-15. *CB* small-signal models: (a) *CB* configuration, (b) complete *CB* *h*-parameter model, (c) approximate *CB* model using *CE parameters.*

The *CC* configuration is illustrated in Fig. 4-16. The defining equations in this case are:

$$v_{bc} = h_{ic} i_b + h_{rc} v_{ec}$$

$$-i_e = h_{fc} i_b + h_{oc} v_{ec} \tag{4-35}$$

The BJT model in the *CC* configuration is shown in Fig. 4-16(b). As was the case in the *CB* configuration, however, it is more convenient to use the *CE* approximate model, as depicted in Fig. 4-15(c).

For the few cases where the actual *CB* and *CC* *h*-parameter equivalent circuits must be used, the needed parameter values can be determined by using the conversions listed in Tables 4-1, 4-2, and 4-3. First, calculate the *CE h* parameters as discussed in Example 4-6. Then use these values in conjunction with the conversion tables to determine the needed parameters.

Table 4-1. Conversion between *CB and CE h* parameters*

$h_{ib} = \dfrac{h_{ie}}{h_{fe}+1}$	$h_{rb} = \dfrac{h_{ie}h_{oe}}{h_{fe}+1} - h_{re}$
$h_{fb} = -\dfrac{h_{fe}}{h_{fe}+1}$	$h_{ob} = \dfrac{h_{oe}}{h_{fe}+1}$
$h_{ie} = \dfrac{h_{ib}}{h_{fb}+1}$	$h_{re} = \dfrac{h_{ib}h_{ob}}{h_{fb}+1}$
$h_{fe} = -\dfrac{h_{fb}}{h_{fb}+1}$	$h_{oe} = \dfrac{h_{ob}}{h_{fb}+1}$

*The conversions contain approximations.

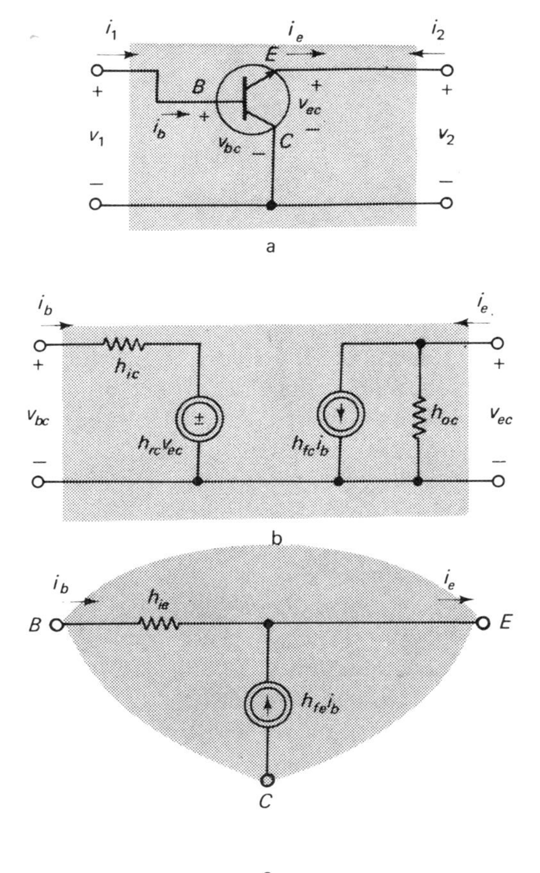

Figure 4-16. *CC* small-signal model: (a) *CC* configuration, (b) complete *CC h*-parameter model, (c) approximate *CC* model using *CE* parameters.

Table 4-2. Conversion between *CC* and *CE* *h* parameters*

$h_{ic} = h_{ie}$	$h_{rc} = 1 - h_{re}$
$h_{fc} = -(h_{fe} + 1)$	$h_{oc} = h_{oe}$
$h_{ie} = h_{ie}$	$h_{re} = 1 - h_{rc}$
$h_{fe} = -(h_{fc} + 1)$	$h_{oe} = h_{oc}$

Table 4-3. Conversion between *CB* and *CC* *h* parameters*

$h_{ic} = \dfrac{h_{ib}}{h_{fb} + 1}$	$h_{rc} = h_{rb} + 1 - \dfrac{h_{ib} h_{ob}}{h_{fb} + 1}$
$h_{fc} = -\dfrac{1}{h_{fb} + 1}$	$h_{oc} = \dfrac{h_{ob}}{h_{fb} + 1}$
$h_{ib} = -\dfrac{h_{ic}}{h_{fc}}$	$h_{rb} = h_{rc} - 1 - \dfrac{h_{ic} h_{oc}}{h_{fc}}$
$h_{fb} = -\dfrac{h_{fc} + 1}{h_{fc}}$	$h_{ob} = \dfrac{h_{oc}}{h_{fc}}$

*Certain conversions include approximations.

4.3 BJT HIGH-FREQUENCY MODEL

In the previous section, we discussed the BJT small-signal models. We can use these models in the analysis of transistor amplifiers at low frequencies and have good agreement between the predictions of performance thus obtained and those actually observed in the laboratory. However, at high frequencies, the behavior of transistors cannot be predicted from these equivalent circuits. The actual transistor behavior at high frequencies is quite different from that predicted by the use of the *h*-parameter models. At high frequencies, both the output current and voltage are actually lower than the models might lead us to expect.

The reason for the decrease in both current and voltage gain inside a transistor at high frequencies is relatively simple to understand. As we saw in our discussion of diodes, the effect of a forward bias on a *PN* junction is to inject large numbers of carriers from one region into the other. If this process is modulated by an ac signal, a slightly larger or smaller number of carriers is injected at any given time, depending on whether the ac voltage adds to the dc bias voltage or subtracts from it. In any case, although there is an increase in the frequency at which the ac signal is changing, the injection level at the forward-biased junction cannot change instantaneously. Instead, an averaging of the peak variation in the ac signal occurs at high frequencies. We call this effect *capacitive*. Such a capacitance is termed *diffusion capacitance* (see section 2.4). The base-emitter junction inside a transistor exhibits these effects.

In a reverse-biased junction (which is what the collector-base junction is) the capacitive effect is attributed to a transition region capacitance (again see section 2.4). The incorporation of these two capacitive effects

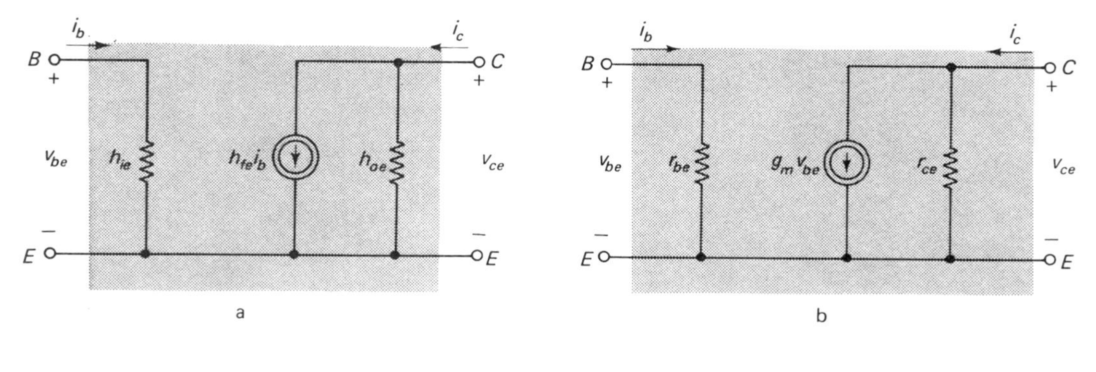

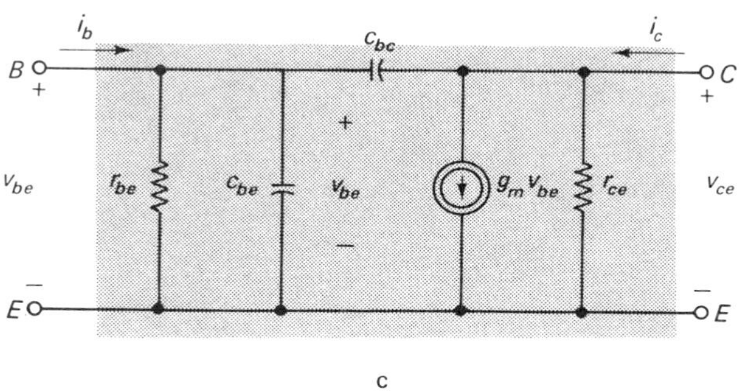

Figure 4-17. Development of the hybrid-π high-frequency model: (a) approximate low-frequency model, (b) approximate low-frequency model with relabeled parameters, (c) high-frequency hybrid-π model.

into the transistor h-parameter model is shown in Fig. 4-17. The components are renamed:

$$r_{be} = h_{ie} \tag{4-36}$$

$$r_{ce} = \frac{1}{h_{oe}} \tag{4-37}$$

Furthermore, because i_b can be seen to equal v_{be}/r_{be}, the output current generator can be replaced by

$$h_{fe} i_b = \frac{h_{fe} v_{be}}{r_{be}} = g_m v_{be}$$

where the *transconductance* g_m is defined as:

$$g_m = \frac{h_{fe}}{r_{be}} \tag{4-38}$$

The equivalent circuit is shown in Fig. 4-17(c). It is called the *hybrid-π circuit* and is the model of the BJT to be used for high-frequency calculations.

To verify that the hybrid-π circuit is valid at high frequencies, we calculate the short-circuit current gain. Under the condition that the

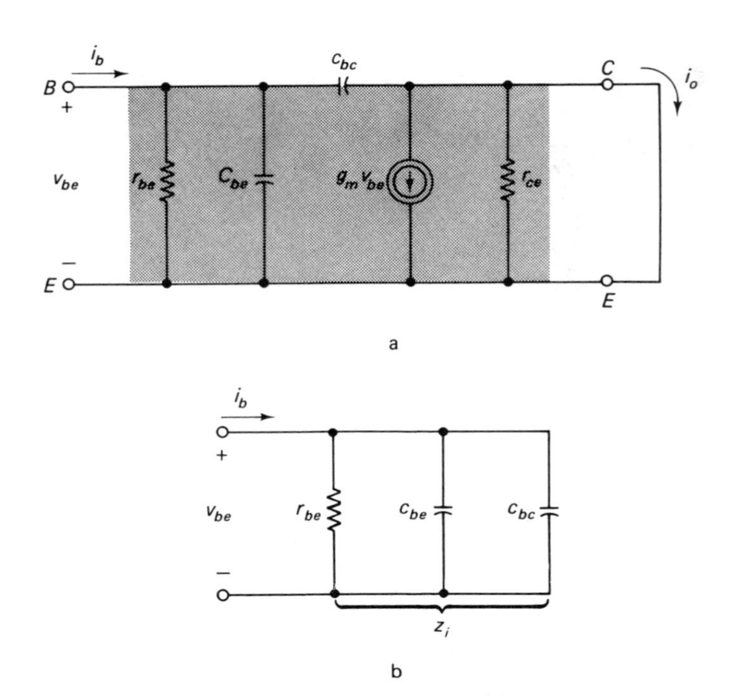

Figure 4-18. Determination of the ac short-circuit current gain.

output be shorted, the circuit is that shown in Fig. 4-18(a). Inspecting the output shows us that

$$i_o = -g_m v_{be}$$

If we label the input impedance seen between the base and emitter terminals by Z_i, we have:

$$v_{be} = Z_i i_b \tag{4-39}$$

You can see that the input impedance is the parallel combination of r_{be} and the two capacitors, C_{be} and C_{bc}:

$$Z_i = \frac{r_{be}}{1 + j\omega r_{be}(C_{be} + C_{bc})} \tag{4-40}$$

The short-circuit current gain, i_o/i_b, is then obtained by using Eqs. (4-38), (4-39), and (4-40):

$$A_{isc} = \frac{-g_m r_{be}}{1 + j\omega r_{be}(C_{be} + C_{bc})} \tag{4-41}$$

This result provides us with the same answer at low frequencies ($w = 0$) as that obtained from the h-parameter circuit, namely, $-h_{fe}$, which is the same as $-g_m r_{be}$ in the preceding equation. Further, as the frequency ω increases in this equation, the short-circuit current gain decreases, as is observed in the laboratory. Thus, the hybrid-equivalent circuit agrees well with observed results.

We may define a specific frequency, f_β:

$$f_\beta = \frac{1}{2\pi r_{be}(C_{be} + C_{bc})} \tag{4-42}$$

called the *β cutoff frequency*. At this frequency, the short-circuit current gain is down 3 dB. (See Chap. 9 for a discussion of decibels.) The manufacturer usually specifies the frequency f_T at which the magnitude of the short-circuit current gain is unity. Taking the magnitude of A_{isc} in Eq. (4-41) and setting it equal to 1, we obtain the relationship between f_T and f_β:

$$f_T \cong h_{fe} f_\beta \tag{4-43}$$

The quantity f_T is sometimes called the *gain-bandwidth product*, because it is the product of the midband current gain h_{fe} and the bandwidth f_β.

With the relationships developed here, the hybrid-parameter values may be calculated from the h parameters and from the data usually provided in the manufacturers' data sheets. Such a calculation is illustrated in the following example.

Example 4-7. A silicon *NPN* transistor is listed by the manufacturer as having: C_{bc}* = 12 pF and $f_T = 50$ MHz. It has the hybrid parameters as calculated in Example 4-6. We want to determine the hybrid-parameter values as well as f_β.

Solution: With $h_{ie} = 1.25$ kΩ, $h_{fe} = 100$, and $h_{oe} = 12.5$ μMho, we have:

$$r_{be} = 1.25 \text{ k}\Omega$$

$$r_{ce} = \frac{1}{12.5} \text{ M}\Omega \cong 80 \text{ k}\Omega$$

$$g_m = \frac{100}{1.25} \text{ milliMho} \cong 80 \text{ milliMho}$$

$$f_\beta = \frac{f_T}{h_{fe}} = \frac{50}{100} \text{ MHz} \cong 500 \text{ kHz}$$

$$C_{be} + C_{bc} = \frac{1}{2\pi f_\beta r_{be}} = \frac{1}{2\pi(5\times 10^5)(1.25\times 10^3)} \text{ F} \cong 250 \text{ pF}$$

$$C_{be} = 250 - C_{bc} = 250 - 12 \cong 238 \text{ pF}$$

To summarize, for a BJT to be operated as an amplifier, it must first be biased properly. Once a proper and stable operating point is achieved, the transistor amplifier may be analyzed by replacing the transistor symbol in the circuit by the appropriate small-signal model. At low frequencies (say, up to a few hundred kHz) the h-parameter model is applicable. At higher frequencies, we must use the hybrid-π model for the transistor.

---

* C_{bc} is sometimes listed as $C_{b'c}$.

Review Questions

1. What is meant by the phrase "biasing a transistor"?
2. How can neglecting one quantity with respect to another be justified?
3. How can we justify neglecting V_{BE} in the bias circuit shown in Fig. 4-1?
4. How can we justify *not* neglecting V_{BE} in the bias circuit of Fig. 4-2?
5. Why is the bias circuit shown in Fig. 4-1 called the "fixed-bias circuit"?
   What is "fixed" in this circuit?
6. How does the circuit shown in Fig. 4-1 bias the base-emitter junction? How does it bias the collector-base junction?
7. What is *thermal runaway*?
8. How is the thermal stability of a bias circuit measured?
9. How does the self-bias circuit achieve thermal stability? Explain.
10. What does the thermal stability of the fixed-bias circuit depend on?
11. What does the thermal stability of the self-bias circuit depend on?
12. What are the similarities and differences in designing and analyzing a bias circuit?
13. How is the dc load line plotted on the output characteristics?
14. What is meant by low-frequency model?
15. What are the factors that determine which model of the BJT (*CE*, *CC*, or *CB*) is appropriate in a specific application?
16. What factors influence the values for the *CE* h parameters of a BJT? List and explain why.
17. What are the ways in which the *CE* h parameters of a particular BJT can be determined?
18. What are the reasons why the h-parameter model of the BJT is not valid at high frequencies?
19. Under what conditions are the h-parameter and hybrid-π models used?
20. What is the gain-bandwidth product for the BJT? How is it related to the β cutoff frequency?

Problems

1. Repeat Example 4-1 for a silicon transistor with a minimum β of 20 and a maximum β of 100.
2. Design the bias circuit of Fig. 4-1 for an operating point of $V_{CEQ} = 7$ V and $I_{CQ} = 1$ mA if the transistor used has a β of 75 and the supply voltage is 15 V.
3. What is the highest value of R_C that can be used in Example 4-1 without causing the transistor to saturate? (Note: The minimum V_{CE} is approximately 0; V_{CE} cannot be negative for an *NPN* transistor.)

4. The circuit in Fig. 4-1 is used to bias a *PNP* transistor with a $\beta = 50$. The supply voltage is -20; $R_B = 200\ k\Omega$, $R_C = 1.8\ k\Omega$. Determine the operating point and compare your answers with those in Example 4-1.
5. Determine the operating point for the circuit in Fig. 4-19 if the transistor has a β of 40.
6. Repeat Problem 5 if the transistor β is 80.
7. Determine the value of R_E needed to set up the conditions shown in Fig. 4-20. Also determine the β of the transistor.
8. What is the thermal stability factor for the circuit in Problem 4?
9. What is the thermal stability factor for the circuit in Fig. 4-19? In Fig. 4-20?

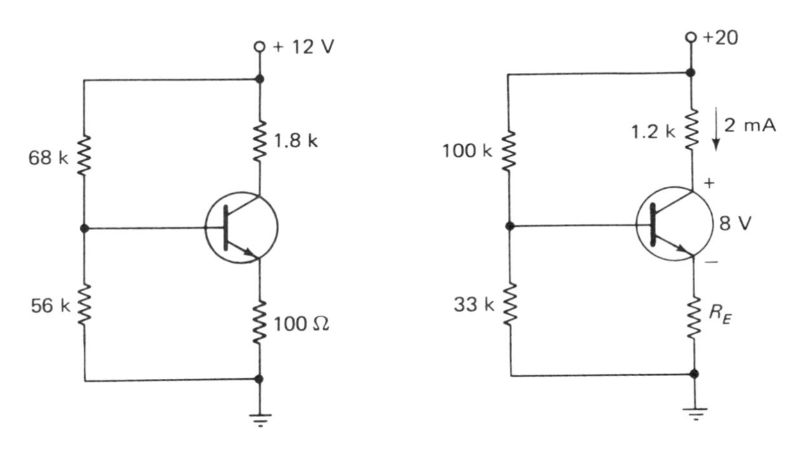

Figure 4-19. **Figure 4-20.**

10. Design the bias circuit of Fig. 4-2 for the same *Q*-point as in Example 4-4 but with $S = 5$.
11. Repeat Example 4-5 with a β spread from 20 to 130. Compare your answers with those in Example 4-5 and make a conclusion.
12. Draw the load line for the circuit values in Example 4-2. Repeat Example 4-2 and draw the load line if the transistor β is 75.
13. Determine h_{ie} for the transistor whose input characteristics are given in Fig. 4-9 for an operating-point base current of 10 μA.
14. Determine h_{fe} and h_{oe} for the transistor whose output characteristics are given in Fig. 4-10 if the *Q*-point is at $I_{CQ} = 1$ mA and $V_{CEQ} = 10$ V.
15. From the *CE h* parameters in Example 4-6, determine the *CB h* parameters.
16. From the *CE h* parameters in Example 4-6, determine the *CC h* parameters.

17. A transistor is listed as having $C_{bc} = 10$ pF and $f_T = 100$ MHz. Its h parameters are: $h_{ie} = 800\ \Omega$, $h_{fe} = 65$, $h_{oe} = 50\ \mu$A/V (h_{re} negligible). Determine the hybrid-parameters for this transistor.
18. For the transistor in Problem 17, what is the frequency at which the magnitude of the short-circuit current gain is down 3 dB from its low-frequency value?

5

Field-Effect Transistor Operation, Biasing, and Models

This chapter introduces the field-effect transistor, or FET. We shall examine the construction and the operation of a variety of FETs, leading to those characteristics observed at the terminals. Different biasing schemes as well as methods of overcoming operating-point instabilities will be covered. We shall also develop the small-signal model, using the FET terminal characteristics, and make appropriate modifications in the low-frequency model to extend its use to high frequencies, as we did for the BJT.

5.1 JUNCTION FIELD-EFFECT TRANSISTOR

The construction of the two possible types of junction field-effect transistors (J-FET) is depicted in Fig. 5-1. We shall confine our discussion to the *N*-channel J-FET. The operation of the *P*-channel J-FET is completely analogous to that of the *N*-channel.

An *N*-type semiconductor is formed, with leads attached to two of its ends. One end is called the *source*; the other, the *drain*. The semiconductor forms the *channel*, as shown in Fig. 5-1(a). A very narrow *P*-type region is diffused around the channel. This region is called the *gate*, and the third external connection is made to this region. The *N*-channel J-FET, therefore, consists of a single *PN* junction that was formed by the *N*-type channel and the *P*-type gate. In a *P*-channel J-FET, the channel is *P*-type and the gate is *N*-type.

Under normal operation, we make a current flow from the drain to the source by applying a potential between the two terminals, as shown in Fig. 5-2. To control this current, a reverse-biasing potential (V_{GG}) is applied between the gate and source. As a result of the reverse bias between gate and source, a depletion region extends into the channel. The shape of the depletion region is governed by the amount of reverse bias on the *PN* junction. If the drain is positive with respect to the source, the net reverse bias on the *PN* junction near the drain end of the channel is larger than it is near the source end, because the net voltage between drain and

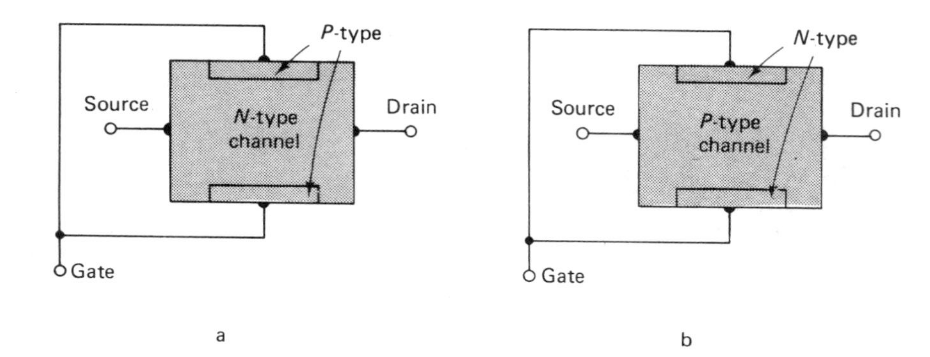

Figure 5-1. Schematic representation of J-FET types: (a) *N*-channel and (b) *P*-channel.

gate is $V_{DS} + V_{GS}$. Thus, the depletion region extends deeper into the channel toward the drain end of the channel.

The effective resistance between the drain and source is called the *channel resistance*. It depends on the resistivity of the channel and its volume. Note that the depletion region decreases the effective volume of the channel, thus increasing the effective resistance drain to the source. If we consider that the drain-to-source voltage is fixed, we can readily see that the drain current (I_D) is directly proportional to the channel resistance. In this manner, we may use the gate-source voltage to control the drain current. Increasing the reverse bias from gate-to-source causes the depletion region to occupy more of the channel, thus increasing the channel resistance, which in turn causes the drain current to decrease. Conversely, lowering the amount of reverse bias gate-to-source decreases the channel resistance and thereby increases the drain current. Observe

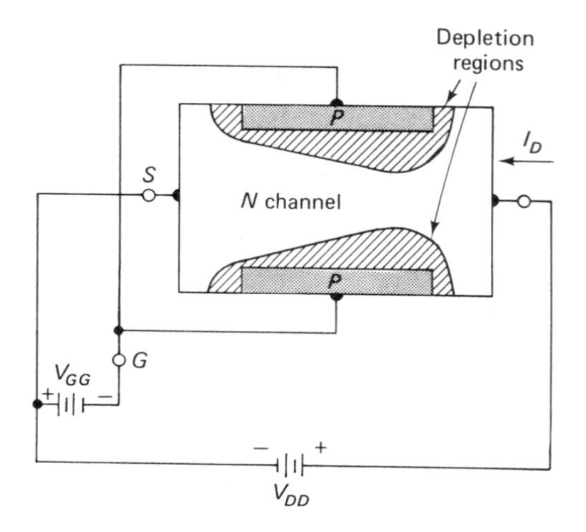

Figure 5-2. Normal operation of an *N*-channel J-FET.

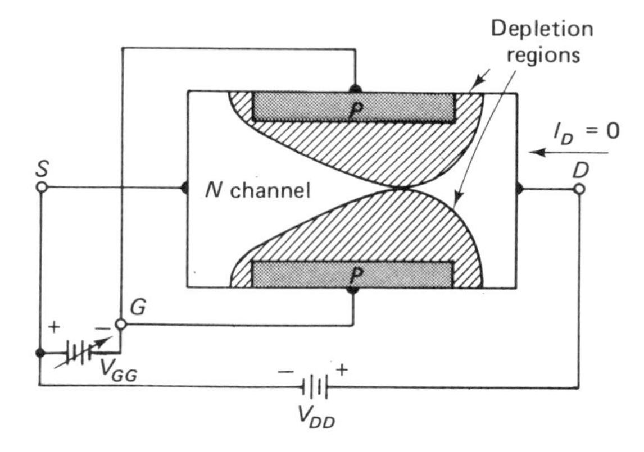

Figure 5-3. Cutoff in a J-FET.

that the gate-source junction is reverse biased; therefore, no gate current effectively flows.

If we apply a relatively low voltage between the drain and source, and if we make the gate increasingly negative, then we eventually cause the depletion region to *cut off* the channel, as shown in Fig. 5-3. Under these conditions, the channel resistance is extremely high, and, for all practical purposes, the drain current is zero. The gate voltage at which this condition occurs is called the *pinchoff voltage:* V_p.

If the gate-source voltage is kept constant at some value below the pinchoff voltage and the drain-to-source voltage is increased from zero, the resistance of the channel is essentially constant for V_{DS} less than a few tenths of a volt. Moreover, the drain current increases linearly with increasing V_{DS}. However, at a critical value of V_{DS}, given by $V_{DS} = V_{GS} - V_p$, the separation between the depletion regions reaches a minimum (labeled w), as shown in Fig. 5-4(a). Any further increase in the drain-to-source voltage causes more of the channel (toward the source end) to reach this minimum width w, as shown in Fig. 5-4(b). As more of the channel reaches the minimum width, the resistance of the channel is increasing at approximately the same rate that V_{DS} is increased. As a result, the drain current remains essentially unchanged even though V_{DS} is increased. Thus, this mode of operation provides us with essentially a constant current over a relatively large range of V_{DS}. The magnitude of the current is controlled only by the gate-source voltage.

If the gate-source voltage is zero, the channel is *pinched off* (that is, it has reached the minimum width w) when the drain-to-source voltage reaches V_p. The drain current under these conditions is called the *saturation drain current*, labeled I_{DSS}. An approximate relationship between this current and the drain current for any value of V_{GS} between zero and V_p is given by:

$$I_D = I_{DSS}\left(1 - \frac{V_{GS}}{V_p}\right)^2 \tag{5-1}$$

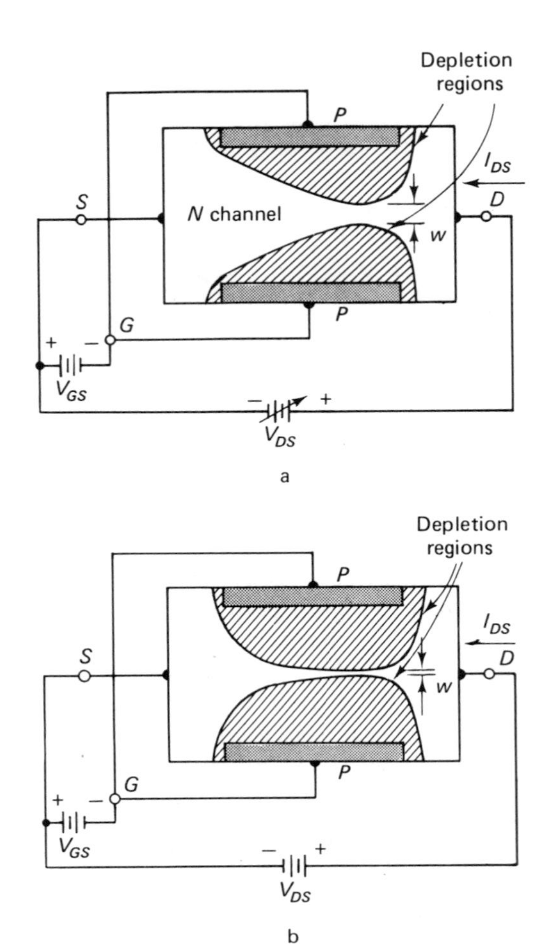

Figure 5-4. Current saturation in a J-FET (a) for $V_{DS} = V_{GS} - V_p$, (b) for $V_{DS} > V_{GS} - V_p$.

We can now summarize the operation of the J-FET. It has two distinct regions of operation. The first region, called the *constant resistance* region, is for very low drain-to-source voltages. In this mode of operation, the channel resistance is controlled by the amount of reverse bias at the gate. The second region, called the *constant current* region, has V_{DS} larger than a few tenths of a volt. It is characterized by drain currents that are only a function of the bias on the gate and essentially independent of V_{DS}. The voltage-controlled resistance of the J-FET has definite applications. But the more common use of the J-FET is in the constant current mode. We shall call this mode the normal operating region.

Figure 5-5 gives typical J-FET output characteristics and a plot of I_D as a function of V_{DS} for different values of gate voltage, which may be obtained on a curve tracer. The constant resistance region is to the left of

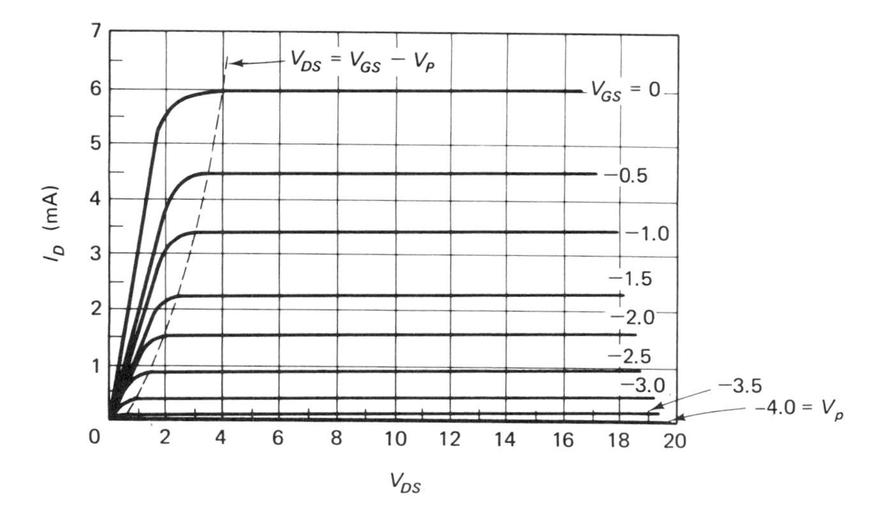

Figure 5-5. Typical J-FET output characteristics.

the dotted curve where $V_{DS} = V_{GS} - V_p$. The constant current region is between the origin and just to the right. Note the similarity between the *CE* output characteristics of the BJT and those of the J-FET. The most obvious difference is that the J-FET output current is controlled by the input *voltage* V_{GS}, whereas in the BJT the input *current* I_B controls the output current.

We can see another important characteristic of the J-FET if we plot Eq. (5-1). The transfer characteristics in the constant current region are shown in Fig. 5-6 for a J-FET with $I_{DSS} = 6$ mA and $V_p = -4$ V. Note that this transfer between the input voltage and output current is not linear. For example, if the input voltage changes from -3 to -2 V, the drain current

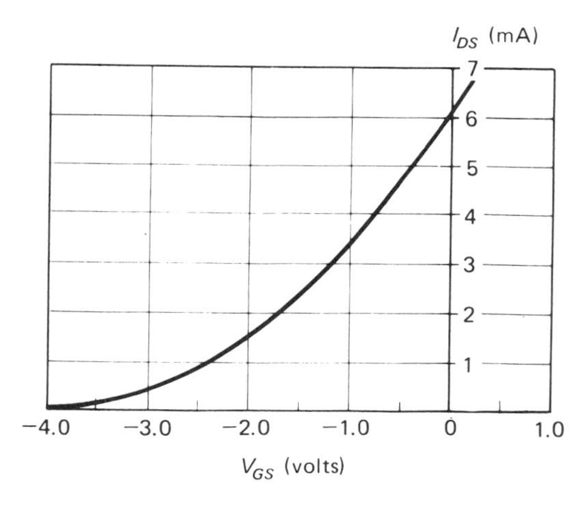

Figure 5-6. Typical J-FET transfer characteristics.

change is approximately 1 mA. However, if the input voltage changes from -2 to -1 V, the drain current change is almost 2 mA.

The J-FET may be operated with the gate-source junction slightly forward biased. But the amount of forward bias must be kept low enough to prevent the gate from drawing any appreciable current. Typically, the forward bias should be kept below 0.5 V.

The circuit symbols for both *N*-channel and *P*-channel J-FETs are given in Fig. 5-7. The arrow on the gate terminal is used to denote the type. If the arrow points in, an *N*-channel FET is indicated; if it points out, a *P*-channel FET is indicated.

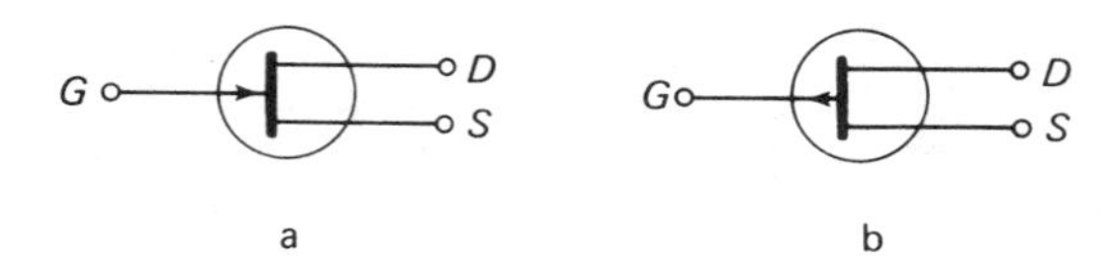

Figure 5-7. Circuit symbols for (a) *N*-channel J-FET and (b) *P*-channel J-FET.

5.2 BIASING THE J-FET

The analysis and design of J-FET bias circuits are very similar to their BJT counterparts. The bias circuit must establish the proper voltages for operating in the normal active region. A J-FET self-bias circuit is illustrated in Fig. 5-8. For this *N*-channel J-FET, the supply voltage V_{DD} is positive in order to set up a positive voltage between drain and source. Gate bias is accomplished by the voltage drop across R_S caused by the drain current. Equating the voltage supplied (V_{DD}) to the sum of the voltage drops in the output circuit, we obtain the load line equation:

$$V_{DD} = I_D(R_D + R_S) + V_{DS} \tag{5-2}$$

Because of the reverse bias between the *P*-type gate and the *N*-type channel, essentially no gate current flows. Thus, in the gate circuit, there is no dc voltage drop across R_G. Adding the voltages in the gate circuit gives us the *bias curve* equation for the circuit in Fig. 5-8:

$$V_{GS} = -I_D R_S \tag{5-3}$$

Equations (5-1), (5-2), and (5-3) may be used to determine the operating point for a J-FET in Fig. 5-8 if we assume that the J-FET characteristics as well as circuit component values are known. This procedure is illustrated in the following example.

Example 5-1. An *N*-channel J-FET with $I_{DSS} = 5$ mA and $V_p = -4$ V is used in the self-bias circuit of Fig. 5-8. The circuit values are: $V_{DD} = 12$ V, $R_D = 2.2$ kΩ, and $R_S = 470$ Ω. We want to determine the operating point V_{DS}, I_D, and V_{GS}.

Solution: Using the given values of I_{DSS} and V_p and Eq. (5-1), we make a plot of the transfer characteristics. We choose values of V_{GS} and calculate I_D in Eq. (5-1). The plot of the transfer characteristics is shown in Fig. 5-9.

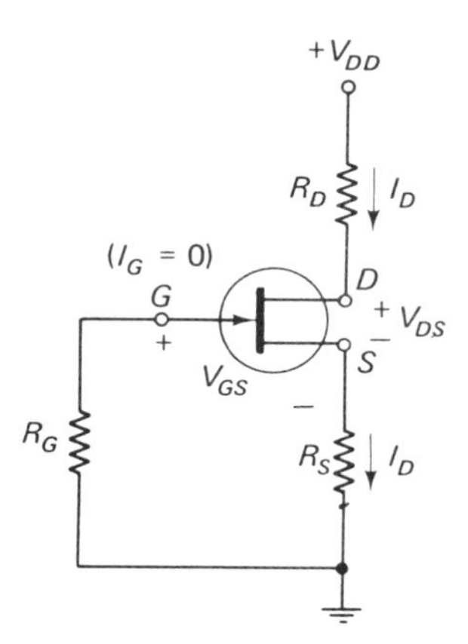

Figure 5-8. *N*-channel J-FET self-bias circuit.

We next plot the bias curve, Eq. (5-3), on the transfer characteristics by plotting two points. For $V_{GS}=0$, I_D is also zero. For $V_{GS}=-2$ V, $I_D \cong 2/0.47$ mA $\cong 4.25$ mA. The bias curve is now plotted as shown in Fig. 5-9. The Q-point is obtained at the intersection of the transfer curve and the bias curve. Thus, $I_{DQ} \cong 2.5$ mA and $V_{GSQ} \cong -1.2$ V.

The drain-source voltage at the Q-point is obtained from the load line equation:

$$V_{DSQ} = V_{DD} - I_{DQ}(R_D + R_S) \cong 12 - (2.5)(2.2 + 0.47) \cong 5.3 \text{ V}$$

The operating point, therefore, is given by:

$$V_{DSQ} \cong 5.3 \text{ V}, \qquad I_{DQ} \cong 2.5 \text{ mA}, \qquad V_{GSQ} \cong -1.2 \text{ V}$$

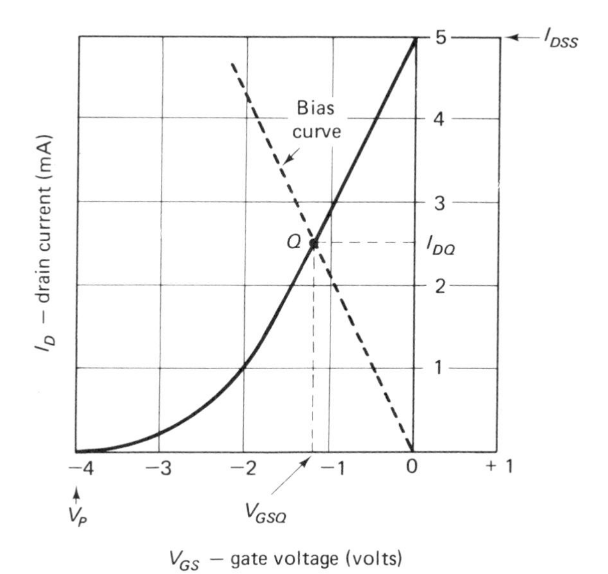

Figure 5-9. Graphical determination of the *Q*-point (see Example 5-1).

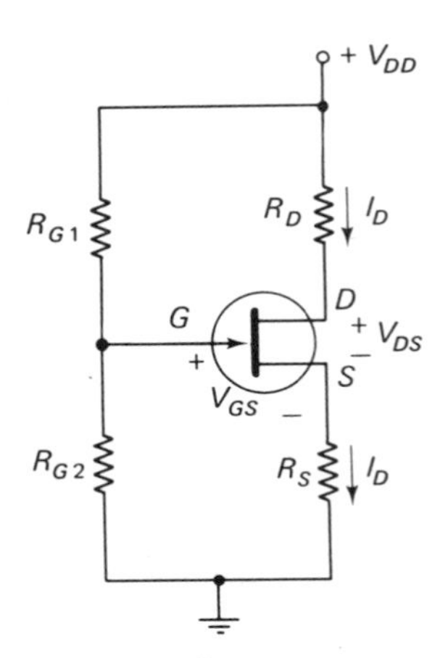

Figure 5-10. A superior bias circuit for the J-FET.

If we examine the problem of biasing a J-FET, we notice that, for a given J-FET, manufacturers usually specify a range of values for both I_{DSS} and V_p. The self-bias circuit provides little flexibility for accommodating such a spread in the characteristics. However, we may use the circuit shown in Fig. 5-10 to minimize the uncertainty in the Q-point because of the uncertainty in the FET parameters.

In the bias circuit of Fig. 5-10, the voltage divider set up by resistors R_{G1} and R_{G2} makes the voltage from gate to ground positive. In order to reverse bias the gate-source junction, the voltage drop across R_S must be larger than the open-circuit voltage across R_{G2}. This voltage, defined as V_{GG}, is given by:

$$V_{GG} = V_{DD}\frac{R_{G2}}{R_{G1}+R_{G2}} \tag{5-4}$$

For completeness, the equivalent resistance to the left of the gate, labeled R_G, is given:

$$R_G = \frac{R_{G1}R_{G2}}{R_{G1}+R_{G2}} \tag{5-5}$$

The resulting equivalent circuit with the voltage divider replaced by its Thevenin equivalent is illustrated in Fig. 5-11. From the output circuit, we obtain the load line equation:

$$V_{DD} = I_D(R_D + R_S) + V_{DS} \tag{5-6}$$

Note that it is the same as that for the circuit in Fig. 5-8. Note too the great similarities between the analysis of this J-FET bias circuit and that used for the BJT (Fig. 4-2).

Again we assume the gate-source junction to be reverse biased, so that no gate current flows. There is, therefore, no dc voltage drop across R_G in Fig. 5-11. The summation of voltages in the gate circuit of Fig. 5-11

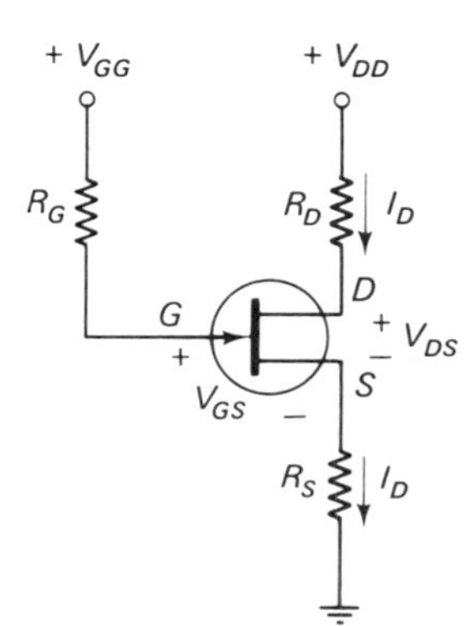

Figure 5-11. Equivalent circuit for the bias circuit shown in Fig. 5-10.

once again yields the bias curve equation:

$$V_{GG} = V_{GS} + I_D R_S \tag{5-7}$$

We can illustrate with the following example the additional flexibility in the stabilization of the Q-point offered by the bias circuit of Fig. 5-10.

Example 5-2. The bias circuit shown in Fig. 5-10 has the following circuit values: $V_{DD} = 25$ V, $R_{G1} = 470$ kΩ, $R_{G2} = 150$ kΩ, $R_D = 3.3$ kΩ, and $R_S = 3.9$ kΩ. The manufacturer lists the N-channel J-FET as having I_{DSS} between 2 and 5 mA and $V_{GS(\mathrm{OFF})}$ (another way of labeling V_p) between -4 and -2 V. We want to determine the worst-case limits on the operating-point voltages and current.

Solution: The first step is to plot the worst-case transfer curves that we might encounter. This is done by pairing the two maximum values of I_{DSS} and V_p to obtain the maximum transfer curve and by pairing the minimum values of I_{DSS} and V_p to obtain the minimum transfer curve. The two plots are accomplished in the same manner as in Example 5-1: They are depicted in Fig. 5-12. They are significant because they provide the limits for the transfer curve that we might have for any J-FET of the type specified. For example, a given J-FET may have an I_{DDS} of, say, 4.5 mA and a pinchoff voltage of -3 V. The transfer curve for this particular FET would lie between the two worst-case transfer curves shown in Fig. 5-12 and would therefore yield a Q-point within the limits to be determined below for the worst-case transfer curves.

The next step is to plot the bias curve of Eq. (5-7). We find one point by setting $V_{GS} = 0$ to obtain $I_D \cong V_{GG}/R_S$. From Eq. (5-4) we determine:

$$V_{GG} \cong 25 \frac{150}{470 + 150} \text{ V} \cong 6 \text{ V}$$

Thus, when $V_{GS} = 0$,

$$I_D \cong \frac{6}{3.9} \text{ mA} \cong 1.5 \text{ mA} \qquad \text{(point 1 in Fig. 5-12)}$$

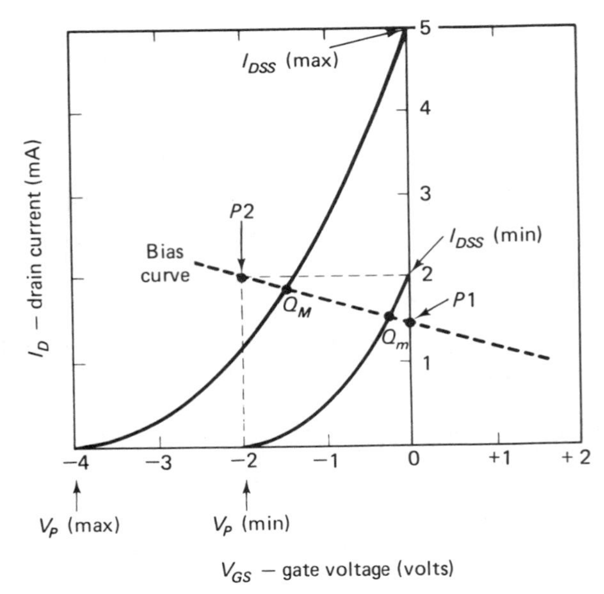

Figure 5-12. Graphical evaluation of worst-case *Q*-point for the J-FET bias scheme in Fig. 5-10.

For the second point needed to plot the bias curve, we choose a value of V_{GS}, for example, -2 V. The drain current is then:

$$I_D \cong \frac{6-(-2)}{3.9}\text{ mA} \cong 2\text{ mA} \qquad \text{(point 2 in Fig. 5-12)}$$

The bias curve is drawn between points 1 and 2 as shown in Fig. 5-12. The highest worst-case operating point Q_M is the intersection of the bias curve and the highest worst-case transfer curve, as shown. Similarly, the lowest worst-case operating point Q_m is obtained at the intersection of the bias curve and the minimum worst-case transfer curve, as shown.

Under these worst-case conditions, the operating point will be between the following limits:

$$\text{for } Q_M\text{:} \quad I_{DM} \cong 1.9\text{ mA and } V_{GSM} \cong -1.4\text{ V}$$
$$\text{for } Q_m\text{:} \quad I_{Dm} \cong 1.6\text{ mA and } V_{GSm} \cong -0.3\text{ V}$$

The corresponding limits on V_{DS} are determined from the load line equation. Note that when I_D is a minimum, V_{DS} will be a maximum and vice versa.

$$V_{DSM} \cong 25-(1.6)(3.3+3.9) \cong 13.5\text{ V}$$
$$V_{DSm} \cong 25-(1.9)(3.3+3.9) \cong 11.3\text{ V}$$

To summarize, any J-FET with I_{DSS} and V_p within the limits given, when used in the bias circuit of Fig. 5-10 (with the values listed) will have an operating-point drain current between 1.0 and 1.6 mA, V_{DS} between 11.3 and 13.5 V, and V_{GS} between −0.3 and −1.4 V.

We can design the bias circuit shown in Fig. 5-10 so that we can have specific variation in the Q-point drain current for a specified variation in the J-FET parameters. This design procedure is illustrated in the following example.

Example 5-3. The N-channel J-FET to be used in the bias circuit of Fig. 5-10 has the following parameters: I_{DSS} between 3 and 8 mA and V_p between −2 and −6 V. It is to be designed in such a way that if the I_{DSS} and V_p are within the limits given, the operating-point drain current will not be lower than 2.5 mA and not higher than 3.5 mA. The supply voltage is 20 V and $R_D = 1.8$ kΩ. We want to specify the remaining circuit components.

Solution: For the worst-case values of I_{DSS} and V_p given, the two transfer curves are plotted as shown in Fig. 5-13. Using the limits on the Q-point drain current of 3.5 and 2.5 mA, we project the transfer curves to obtain the worst-case operating points Q_M and Q_m, as shown.

We next have to choose a bias curve that will satisfy the requirements. We draw a straight line passing *below* Q_M and *above* Q_m, as indicated in Fig. 5-13. If the bias curve thus obtained is extended until it

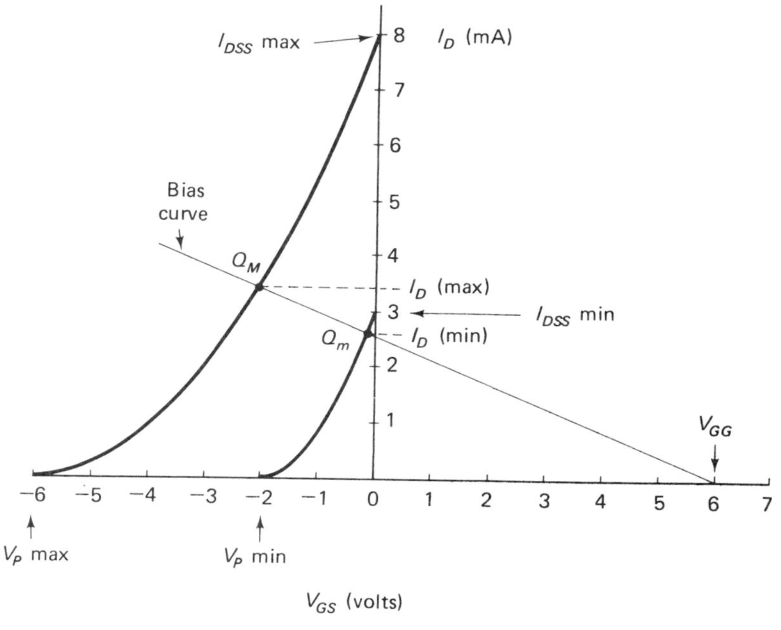

Figure 5-13. Designing for a specified maximum variation i the Q-point drain current.

intersects the V_{GS} axis, the intercept gives us the value of V_{GG} needed. In this case, it is 6 V. This calculation may be verified by noting that in the bias curve equation [Eq. (5-7)], when I_D is 0, $V_{GS} = V_{GG}$.

R_S is evaluated from the intersection of the bias curve and the I_D axis. When V_{GS} is 0, we see from the bias curve equation that:

$$R_S \cong \frac{V_{GG}}{I_D} \cong \frac{6}{2.1} \text{ k}\Omega \cong 2.8 \text{ k}\Omega$$

Next we choose R_{G1} and R_{G2} to give the desired value of $V_{GG} = 6$ V. One possible set of values that will provide the right V_{GG} is:

$$R_{G1} \cong 680 \text{ k}\Omega \text{ and } R_{G2} \cong 330 \text{ k}\Omega$$

We must point out that the answers obtained in this design example are not unique. There are many combinations of circuit values that would provide the desired performance. This is usually the case in most design problems.

5.3 INSULATED GATE FET

The J-FET is characterized by a high-input impedance (gate-to-source, typically 100 MΩ). Another form of the FET offers an even higher input impedance than that of a J-FET, typically 10^{15} Ω. It is called an *insulated gate FET*, abbreviated *IGFET*, or *metal-oxide-semiconductor* FET, abbreviated *MOSFET*. The latter name gives the construction of this type of FET. It is formed by a semiconductor channel, with a layer of (oxide) insulating material separating it from the metal gate, as depicted in Fig. 5-14.

The basic operation of the MOSFET is quite similar to that of a J-FET. The essential difference is the lack of any *PN* junctions in the MOSFET.

The *depletion mode* MOSFET, illustrated schematically in Fig. 5-14, has a shallow channel of lightly doped *N*-type semiconductor, with heavily doped wells under the drain and source terminals. This type of MOSFET may be operated with either positive or negative gate voltages. When we

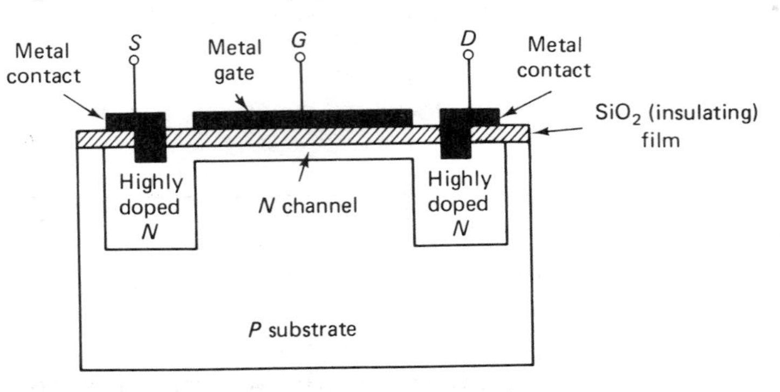

Figure 5-14. Depletion mode MOSFET.

apply a positive voltage to the gate, the N-type channel resistance decreases, because additional electrons from the drain and source regions are attracted into the channel. As a result, the drain current increases. When we apply a negative gate voltage, a portion of the channel is depleted because of the positive charge induced. Consequently, the drain current decreases.

Typical depletion mode MOSFET characteristics are shown in Fig. 5-15. Note that, like the J-FET, the depletion mode MOSFET has a nonlinear transfer characteristic between the input voltage and output current [Fig. 5-15(b)].

The *enhancement mode* MOSFET, shown schematically in Fig. 5-16(a), contains no channel between the source and drain. However, applying a positive voltage to the gate causes free electrons from the substrate and the source and drain regions to be attracted into the region just below the gate, thus forming an N-type channel. Such an operation is shown in

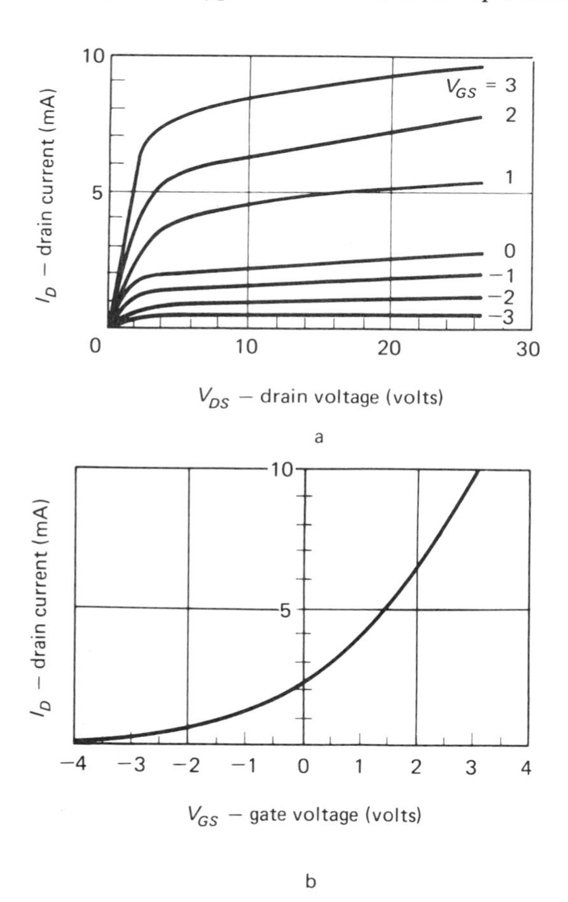

Figure 5-15. Typical depletion mode MOSFET characteristi--: (a) output characteristics and (b) transfer characteristics.

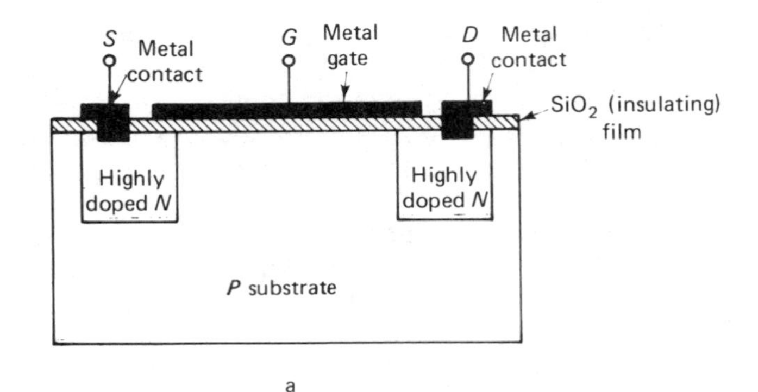

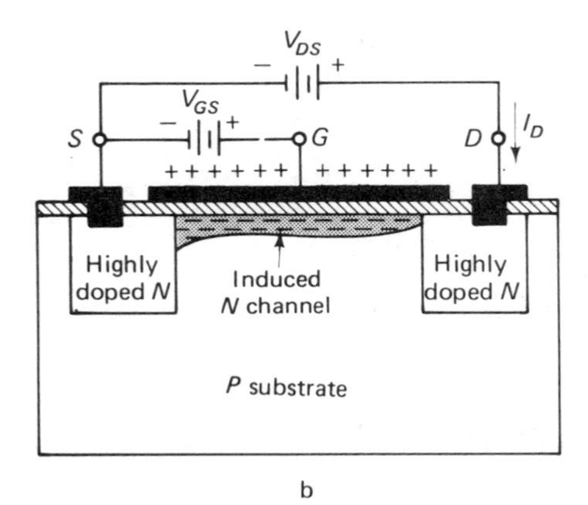

Figure 5-16. (a) Enhancement mode MOSFET and (b) induced channel as a result of bias on the gate.

Fig. 5-16(b). We call this type of operation *enhancement*, because the free electron number in the region below the gate is greatly enhanced (i.e., increased) by the positive bias on the gate. Applying a negative voltage to the gate causes no induced channel. Therefore, the enhancement mode MOSFET can only be operated with positive gate bias. Typical enhancement mode MOSFET characteristics are given in Fig. 5-17. Note that a minimum positive gate voltage, called the *threshold voltage* (V_T), is necessary in order to induce enough of a channel to form a conductive path between the drain and source. If the gate voltages are less positive than the gate threshold voltage, no drain current can flow.

Many varied circuit symbols are commonly used for MOSFETs. Some of these symbols are included in Fig. 5-18. As is the case with the J-FET, the type (*N*- or *P*-channel) is indicated by the arrow in the symbol. Circuit symbols for the *P*-channel MOSFETs are obtained from Fig. 5-18 by reversing all the arrows in the symbols.

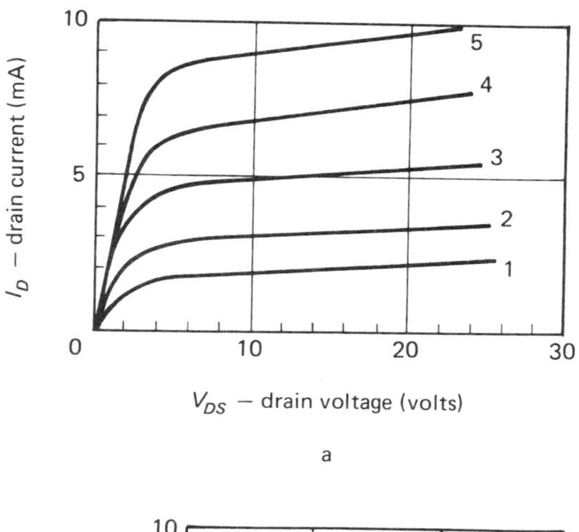

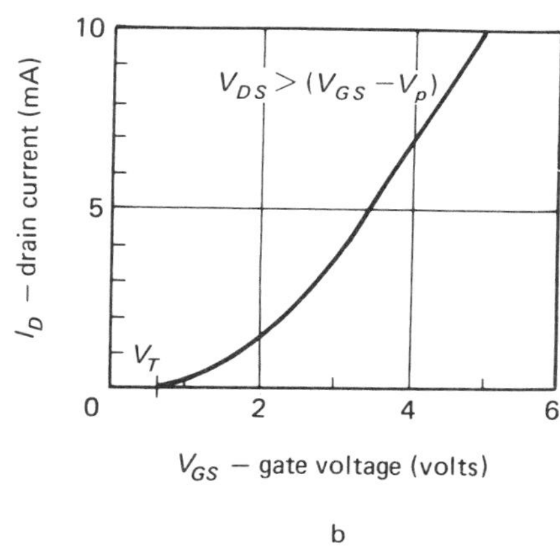

Figure 5-17. Typical enhancement mode MOSFET characteristics: (a) output characteristics and (b) transfer characteristics.

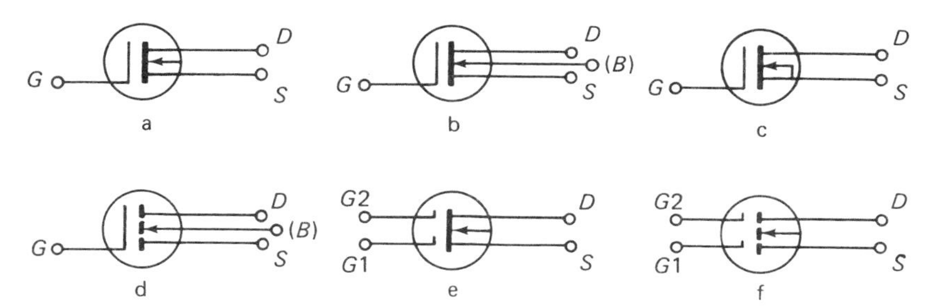

Figure 5-18. Various *N*-channel MOSFET symbols: (a) depletion type, (b) depletion type with external substrate (*B*) connection, (c) depletion type, substrate internally connected to source, (d) enhancement type with external substrate (*B*) connection, (e) dual-gate depletion type, (f) dual-gate enhancement type.

You must be careful in handling MOSFETs. Some MOSFETs contain protective diodes inside the device and may be handled in the same manner as any other device. Some, however, do not have internal protection and come supplied with a ring shorting all of its leads together. This ring should not be removed until the MOSFET is safely installed in a circuit. The purpose of the ring is to prevent static charges from building up on the gate. Because the insulating layer between the gate and the channel is extremely thin, even static charges are sufficient to destroy the insulating properties of the oxide layer.

The combination of metal, insulating layer, and semiconductor form an equivalent parallel plate capacitance. The equivalent capacitance between drain and gate has the largest effect on the MOSFET high-frequency performance. To minimize this capacitive effect, a second gate region is introduced between the control gate and the drain, as shown in Fig. 5-19. The resulting device is called a *dual-gate MOSFET*. Normally, the second gate is biased positively in order to reduce the resistance of the second channel and, more importantly, to break up the effective drain-to-gate capacitance. In all other respects, the operation of the dual-gate MOSFET is identical to that of the depletion mode MOSFET, with gate 1 used as the input. Figure 5-20 illustrates circuit symbols for N- and P-channel dual-gate MOSFETs.

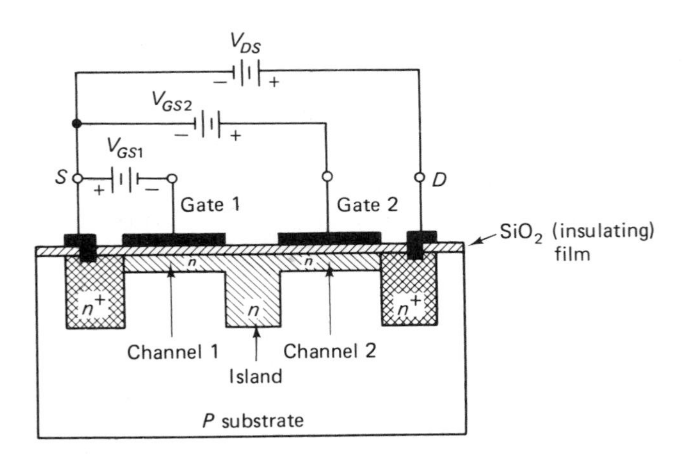

Figure 5-19. Dual-gate MOSFET.

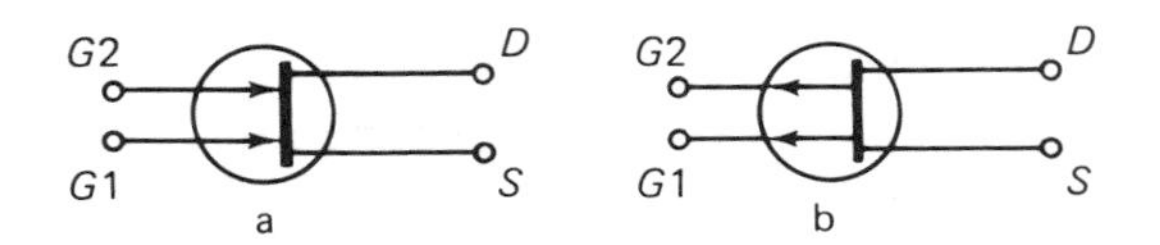

Figure 5-20. Alternate circuit symbols for a dual-gate MOSFET: (a) N-channel and (b) P-channel.

5.4 BIASING THE MOSFET

The J-FET bias circuit shown in Fig. 5-10 may also be used for biasing a MOSFET. As we discussed in the previous section, the enhancement mode MOSFET must have a positive gate-source voltage. Therefore, in the bias circuit of Fig. 5-10, the voltage developed across R_S must be somewhat smaller than that developed across R_{G2}. An alternate bias circuit for the enhancement mode MOSFET is indicated in Fig. 5-21. This circuit has R_S omitted; therefore, the gate-source voltage is equal to the open-circuit voltage across R_{G2}, which is always positive.

The depletion mode MOSFET may be operated without any gate bias or with either a negative or positive gate bias. Both of the J-FET bias circuits, Fig. 5-8 and Fig. 5-10, as well as the enhancement mode MOSFET bias circuit, Fig. 5-21, may be used for biasing a depletion mode MOSFET.

The analysis as well as design of MOSFET biasing circuits is almost identical to that for the J-FET. Consequently, we can use the same methods and procedures for the MOSFET circuits.

The dual-gate MOSFET may be biased as indicated in Fig. 5-22. The second gate can have a positive gate voltage with respect to the source if we choose the voltage divider resistors R_a and R_b so that the open-circuit voltage across R_b is larger than the voltage drop across R_S. The signal gate (gate 1) can be negatively biased with respect to the source by the voltage drop across R_S. The basic bias circuit of Fig. 5-8 is used for the bias of the signal gate.

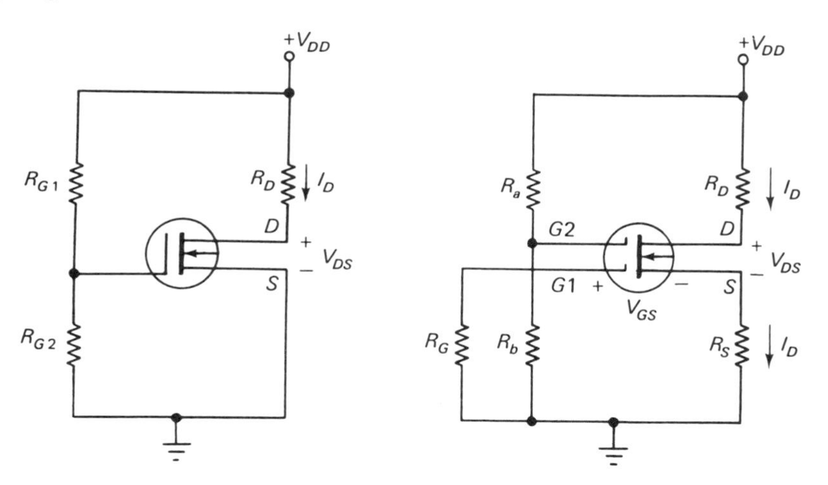

Figure 5-21. A possible MOSFET bias circuit.

Figure 5-22. Dual-gate MOSFET bias circuit.

5.5 SMALL-SIGNAL FET MODEL

The FET, be it a J-FET or a MOSFET, may be operated as an amplifier by applying a small time-varying signal to the gate and taking the amplified signal at the drain. The common terminal for both the input and output is therefore the source.

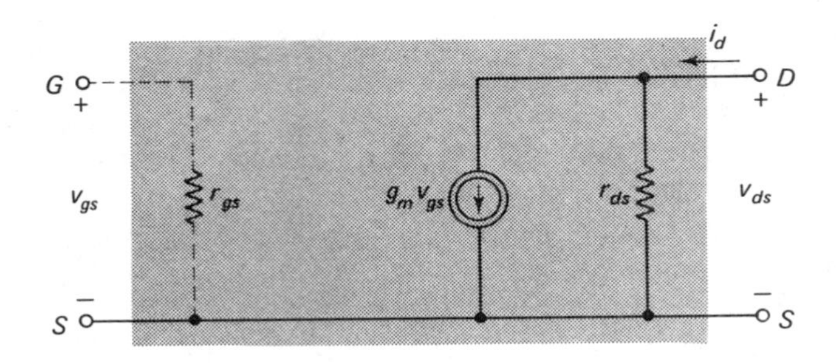

Figure 5-23. FET low-frequency small-signal model. (r_{gs} is usually considered infinite—see text.)

Note that the basic similarity in the operation of the different FETs discussed is indicated by the similarity in their terminal characteristics. As a result, we would expect that the ac small-signal models for the different FETs would be similar. In fact, the only difference between the ac performance of different FETs is in the slightly different magnitudes of some of the parameters.

The low-frequency small-signal model for the FET is shown in Fig. 5-23. At the drain terminal, the sum of the currents yields:

$$i_d = g_m v_{gs} + \frac{1}{r_{ds}} v_{ds} \tag{5-8}$$

where the FET small-signal parameters g_m and r_{ds} are defined and evaluated from:

$$g_m \equiv \text{transconductance} \equiv \frac{\text{change in } I_D}{\text{change in } V_{GS}} \text{ evaluated at } V_{DSQ} \tag{5-9}$$

$$r_{ds} \equiv \text{output resistance} \equiv \frac{\text{change in } V_{DS}}{\text{change in } I_D} \text{ evaluated at } V_{GSQ} \tag{5-10}$$

It is also useful to define the amplification of the FET:

$$\mu \equiv g_m r_{ds} \tag{5-11}$$

The quantity r_{gs} is the input resistance, which is very hard to measure, especially for a MOSFET where the value may be over 1 million megOhms. This quantity may be replaced by an open-circuit in most applications.

We determine the small-signal FET parameters from the FET terminal characteristics in a manner similar to that used for the BJT in Chap. 4. The graphical procedure is illustrated in the following example.

Example 5-4. An N-channel J-FET, whose characteristics are shown in Figs. 5-5 and 5-6, is operated at the Q-point of: $V_{GSQ} = -1$ V, $I_{DQ} = 3.5$ mA, and $V_{DSQ} = 10$ V. We want to determine the small-signal parameters for the FET model.

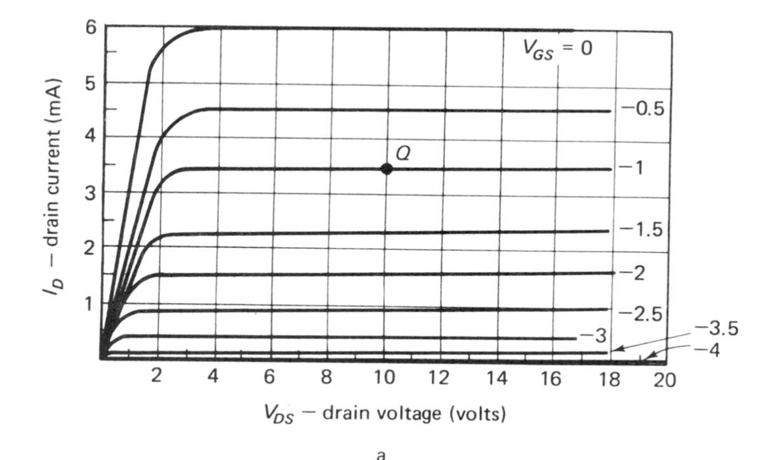

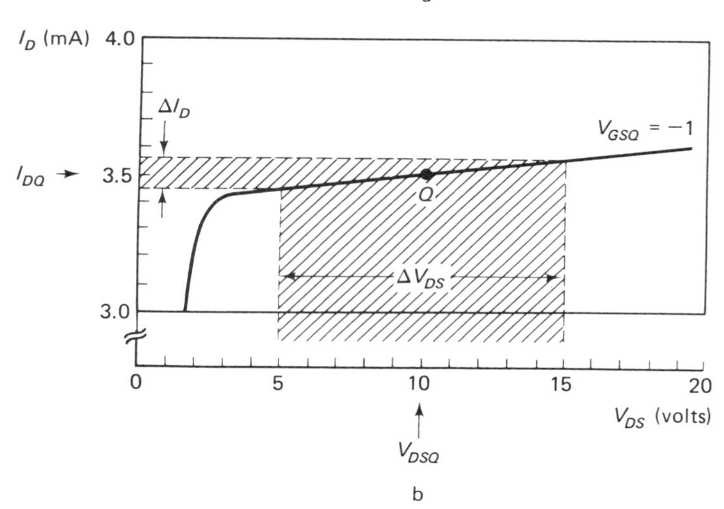

Figure 5-24. Graphical determination of r_{ds} for a J-FET from (a) the output characteristics and (b) the characteristics with expanded current scale in the vicinity of Q-point.

Solution: The output characteristics are redrawn with the operating point as specifically indicated on Fig. 5-24(a). In order to evaluate r_{ds}, we first expand the current scale in the vicinity of the operating point, as shown in Fig. 5-24(b). The inverse of the slope of the characteristic curve corresponding to V_{GSQ} is r_{ds}. We see that for a change in V_{DS} from 5 to 15 V, the drain current change is from about 3.45 to 3.55 mA.

$$r_{ds} \cong \frac{15-5}{3.55-3.45}\,\text{V/mA} \cong 100\ \text{k}\Omega$$

From the transfer characteristics, redrawn in Fig. 5-25, we obtain g_m as the

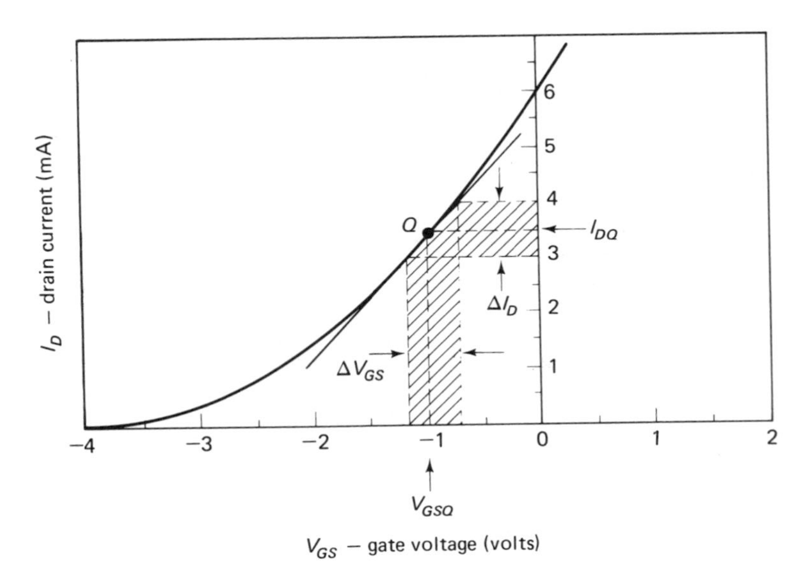

Figure 5-25. Graphical determination of g_m for the FET of Example 5-4.

slope of the curve at the Q-point. For a change in I_D from 3 to 4 mA, the corresponding change in V_{GS} is from -1.25 to -0.75 V.

$$g_m \cong \frac{4-3}{-0.75-(-1.25)} \cong 2 \text{ milliMho}$$

For completeness we can also determine $\mu \cong (100)(2) \cong 200$.

Typically, J-FET small-signal parameters values are: g_m between 1 and 10 milliMho, r_{ds} between 10 kΩ and 1 MΩ, and r_{gs} between 10 and 100 MΩ.

MOSFET small-signal parameters are determined in exactly the same manner as the J-FET parameters. Typically, MOSFETs have a somewhat larger range on the transconductance. The value for g_m may be between 1 and 30 milliMho; for r_{ds}, in the range of 1 to 100 kΩ. In a MOSFET, r_{gs} ranges from 10^{10} to 10^{14} Ω.

The FET small-signal model is adequate at low frequencies. As was the case with the BJT, however, so it is with the FET: At high frequencies, the FET and MOSFET both exhibit capacitive effects that must be accounted for in the model.

5.6 FET HIGH-FREQUENCY MODEL

The equivalent capacitors inside a J-FET appear as capacitors in the model. They are valid at high frequencies and are illustrated in Fig. 5-26. Note the similarity between the FET high-frequency model and the BJT hybrid-π model. The effective capacitance between drain and source has

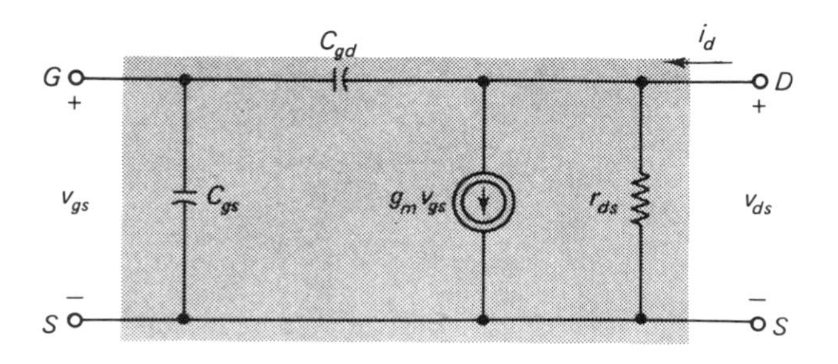

Figure 5-26. FET high-frequency small-signal model.

been left out, because its effects are only minor. Moreover, it would tend to complicate the use of the model.

The capacitive effects inside the J-FET are caused by the reverse-biased *PN* junction and the charge storage associated with it. In the MOSFET, the capacitive effects result from the two conductors (the metal gate and the channel) separated by an insulator (the oxide), which thus forms an effective parallel-plate capacitor.

Typical values of C_{gs} and C_{gd} for a J-FET are in the range from 1 to 20 pF. C_{gd} is usually the smaller of the two. In a MOSFET, C_{gs} may range from 2 to 20 pF, while C_{gd} ranges between 1 and 10 pF.

The FET high-frequency model may also be used for the dual-gate MOSFET. If the second gate potential is fixed, then C_{gs} in the high-frequency model is replaced by $C_i = C_{gs} + C_{gg2}$ (where C_{gg2} is the effective capacitance between the two gates). Typical values for a dual-gate MOSFET are C_{gs} somewhat larger than for the J-FET. There is also a greatly reduced drain-to-gate capacitance: in the range of from 0.005 to 0.02 pF.

In their data sheets some manufacturers use different symbols. We list here some of the more commonly used ones with the equivalents that we have used in our discussion. The transconductance may be labeled y_{fs}; the output impedance (r_{ds}) may be labeled y_{os}; the gate-to-drain capacitance (C_{gd}) as C_{rss}; and the effective input capacitance $C_{iss} = C_{gs} + C_{gd}$.

Review Questions

1. What are the three terminals in a J-FET?
2. What is the construction of a J-FET?
3. What is the difference between an *N*-channel and a *P*-channel J-FET?
4. How many *PN* junctions are there in a J-FET? What are they?
5. What is the mechanism for the constant resistance operation of a J-FET?
6. What is the mechanism for the constant current operation of a J-FET?
7. What are the conditions under which a J-FET may be used as a voltage controlled resistor? Explain.

8. What is the normal operating region for the J-FET? Explain in terms of the bias on it.
9. In what manner is the J-FET biased for operation as an amplifier?
10. What are the conditions in the J-FET described by pinchoff?
11. What is cutoff? How does it differ from pinchoff?
12. What are the conditions under which the drain saturation current I_{DSS} is measured?
13. Under what conditions is the pinchoff voltage V_p measured?
14. How is the negative gate-source voltage developed in the J-FET self-bias circuit?
15. What is an IGFET? A MOSFET?
16. Describe the construction that gives the MOSFET its name.
17. What is the operation of an enhancement mode MOSFET?
18. What is the operation of a depletion mode MOSFET?
19. Discuss the operation of the J-FET in terms of its capacitive behavior.
20. Discuss the operation of a MOSFET in terms of the capacitive effects inside it.
21. What is the difference between a regular MOSFET and a dual-gate MOSFET?
22. What is the purpose of the second gate in a dual-gate MOSFET?
23. How is the channel formed in an enhancement mode MOSFET?
24. How is the channel depleted in a depletion mode MOSFET?
25. What are the similarities and differences in the behavior of J-FETs, MOSFETs, and dual-gate MOSFETs?

Problems

1. An *N*-channel J-FET whose worst-case parameters are given in Example 5-2 is operated in the self-bias circuit of Fig. 5-8, with $V_{DD}=12$ V, $R_D=2.2$ kΩ, and $R_S=470$ Ω. Determine the worst-case *Q*-point.
2. Determine the bias circuit values for the self-bias circuit of Fig. 5-8 if the J-FET to be used has the following parameters: $I_{DSS}=10$ mA and $V_p=-3$ V. The desired *Q*-point is: $I_D=7$ mA and $V_{DS}=V_{DD}/3$. (Choose $V_{DD}=25$ V.)
3. Repeat Example 5-2, using an *N*-channel J-FET with I_{DDS} between 2 and 4 mA and $V_{GS(\mathrm{OFF})}$ between -1 and -6 V.
4. Repeat Example 5-3 for the *Q*-point between 3 and 3.5 mA (drain current).
5. For the J-FET whose transfer characteristics are shown in Fig. 5-6, determine the transconductance at the following operating points: $I_D=0$, 1, 2, 3, 4, 5, and 6 mA.
6. Make a plot of the variation of the FET g_m in Problem 5 as a function of the operating-point drain current. Make a conclusion about the desired location of the *Q*-point for maximum g_m.

7. Determine the FET transconductance for the transfer characteristics and Q-point shown in Fig. 5-9. Which way would the Q-point have to be moved in order to increase g_m? Why?
8. Determine the transconductance, minimum and maximum, for the J-FET whose transfer characteristics are shown in Fig. 5-12. (Do this procedure for the two operating points, Q_M and Q_m, shown.)
9. Determine the small-signal parameters for the depletion mode MOSFET whose characteristics are shown in Fig. 5-15 for an operating point: $I_D = 5$ mA and $V_{DS} = 15$ V.
10. Repeat Problem 9 for an operating point: $I_D = 2$ mA and $V_{DS} = 10$ V.
11. Design the MOSFET bias circuit of Fig. 5-21, using the MOSFET whose characteristics are shown in Fig. 5-15 to provide an operating point of $I_D = 4$ mA and $V_{DS} = 10$ V from a supply voltage of 20 V.
12. Repeat Problem 11, using the enhancement mode MOSFET whose characteristics are shown in Fig. 5-17.
13. Repeat Problem 9, using the enhancement mode MOSFET whose characteristics are given in Fig. 5-17.
14. Determine the minimum and maximum g_m for the enhancement mode MOSFET whose characteristics are shown in Fig. 5-17 for a gate-source voltage between the threshold value and 4 V.
15. What is the value of I_{DSS} for the J-FET whose characteristics are shown in Fig. 5-5?
16. What is the pinchoff voltage for the J-FET whose output characteristics are shown in Fig. 5-5?
17. Repeat Problems 15 and 16 for the depletion mode MOSFET whose characteristics are shown in Fig. 5-15.
18. What is the threshold voltage for the enhancement mode MOSFET whose characteristics are shown in Fig. 5-17? Explain.

6 Thyristors and Related Devices

This chapter describes the operation and terminal characteristics of an assortment of semiconductor devices. All of them have one or more *PN* junctions and are control devices called *thyristors*. In addition, we shall examine other devices whose properties are closely related to those of thyristors.

6.1 UJT TRANSISTOR

The construction of a *unijunction transistor* (UJT) consists of an *N*-type bar of silicon with ohmic (i.e., nonrectifying) contacts at either end. A single *PN* junction is formed by a small *P*-type insert near the middle of the bar, as shown in Fig. 6-1. The *N*-type region, called the *base*, is of high-resistivity silicon. It has two terminals connected to it: base 1 ($B1$) and base 2 ($B2$). There is some basic similarity between the UJT and the J-FET (discussed previously). However, the gate region of the J-FET surrounded most of the channel; whereas, in the UJT, the corresponding *P* region, called the *emitter*, is much smaller. The other characteristics of the two devices are decidedly dissimilar.

We can best explain the operation of the UJT by examining the circuit of Fig. 6-2. Base 2 is usually made positive with respect to base 1. The *interbase resistance* R_{BB} is the effective resistance between bases 1 and 2. It is usually high because of the light doping in the base region. In the equivalent circuit shown in Fig. 6-3, this interbase resistance is represented by series resistances R_{B1} and R_{B2}. The *PN* junction between the emitter and base regions is shown by a diode in the equivalent circuit. The interbase voltage V_{BB} divides between R_{B2} and R_{B1}. The *intrinsic standoff ratio*, η, is a measure of how much appears across R_{B1}. Thus,

$$\eta = \frac{R_{B1}}{R_{B1} + R_{B2}} \qquad (6\text{-}1)$$

Let us now examine the consequences of applying a positive voltage to the emitter. So long as the emitter-base 1 voltage is less than the voltage across R_{B1}, the emitter diode is reverse biased. So, essentially, no emitter

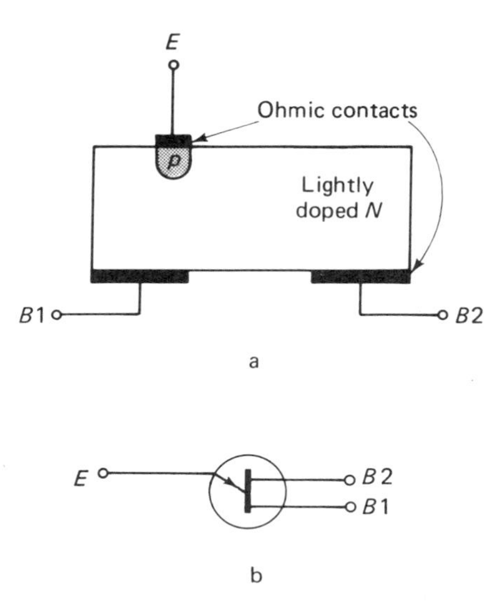

Figure 6-1. UJT: (a) construction and (b) circuit symbol.

current flows (except for a very low leakage current). Under the conditions of zero emitter current, the interbase resistance may be in the order of 5000 to 10,000 Ω. If V_{EB1} is made larger than ηV_{BB} (the voltage across R_{B1}), the diode becomes forward biased and conducts. Emitter current flows because of the motion of (1) holes from the emitter to base 1 and (2) electrons from the base region into the emitter. The increased carrier density in the region between base 1 and emitter decreases the effective value of R_{B1}. In turn, the emitter voltage necessary for maintaining the forward bias on the emitter-base junction is reduced.

Once the diode forward voltage reaches the cut-in voltage V_a, a large emitter current flows and the effective value of the interbase resistance is drastically decreased. At this point, the emitter voltage, called the *peak voltage* (V_p), is

$$V_p = \eta V_{BB} + V_a \tag{6-2}$$

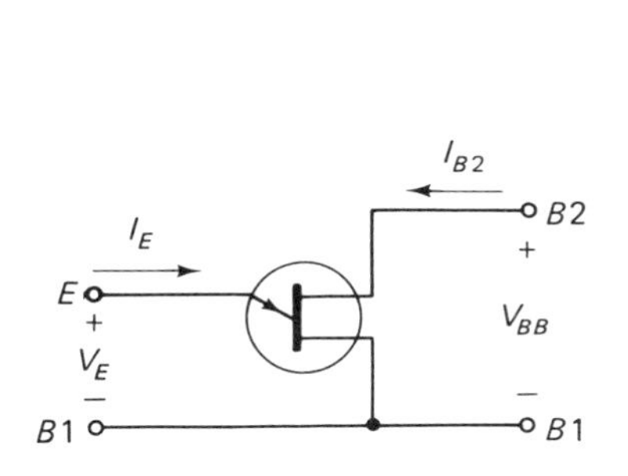

Figure 6-2. UJT voltages and currents.

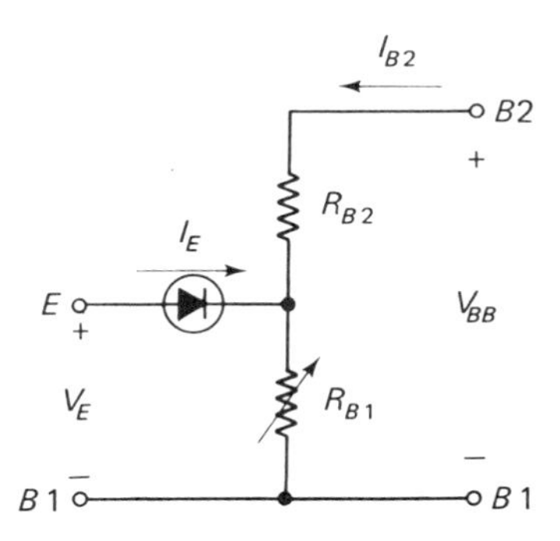

Figure 6-3. UJT equivalent circuit.

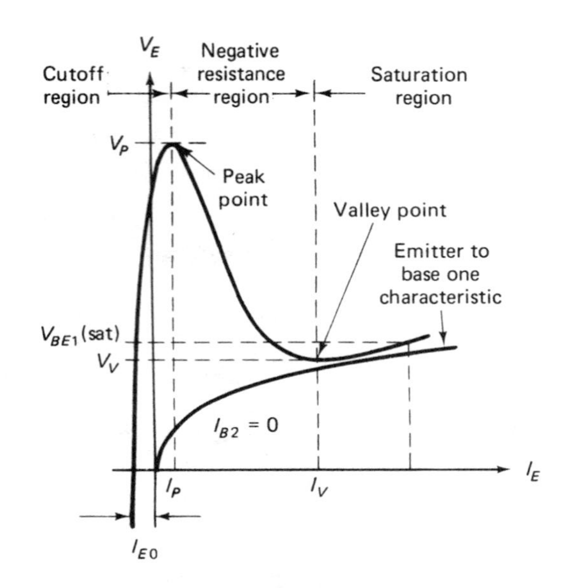

Figure 6-4. Typical UJT emitter characteristics.

Typical input characteristics are displayed in Fig. 6-4. The decrease in voltage resulting from an increase in current corresponds to a negative resistance. We can see that the UJT has a negative resistance region in its characteristics. It should be obvious from Eq. (6-2) that the peak point depends on the interbase bias voltage, V_{BB}. Typically, UJTs have V_a of between 0.5 and 0.7 V, η between 0.5 and 0.8, and a peak point current of between 5 and 25 μA.

The output or interbase characteristics of a typical UJT are shown in Fig. 6-5. The curve for zero emitter current corresponds to a fixed high interbase resistance. The successive curves for higher I_E show a region of

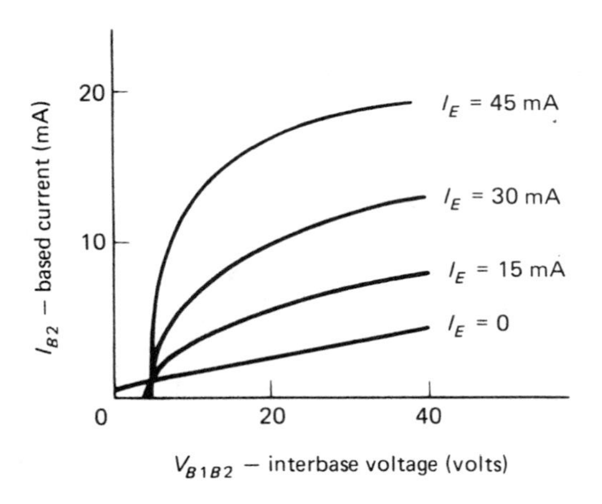

Figure 6-5. UJT output (interbase) characteristics.

low interbase resistance (where the curves are almost perpendicular to the voltage axis). As we pointed out, this low value of interbase resistance is caused by the large number of carriers injected into the base region. At higher interbase voltages, the characteristic curves level off and some transistor action is evident. However, the UJT α (defined as the ratio of the change in I_{B2} to the change in I_E at a fixed V_{BB}) is quite small, in the range of 0.1 to 0.5. Therefore, the UJT is not normally used as an amplifier. However, the two distinct regions of the emitter characteristic —one of extremely high resistance (V_E below V_p) and the other of very low resistance (V_E above V_p)–make the UJT very useful in a number of applications to be discussed later.

Complementary UJT (CUJT) devices, with *P*-type base and *N*-type emitter regions, are also available. Their circuit symbol is shown in Fig. 6-6.

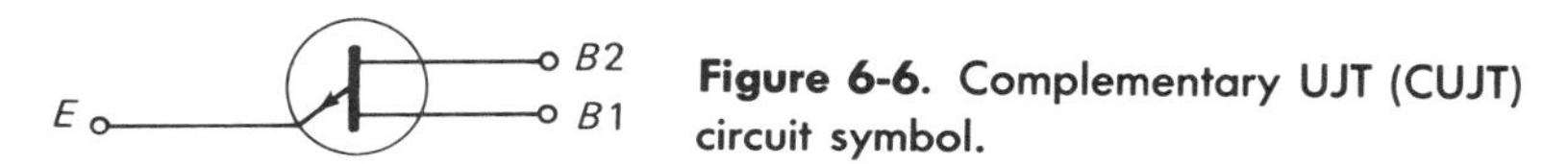

Figure 6-6. Complementary UJT (CUJT) circuit symbol.

6.2 MULTIPLE-LAYER DIODES

In Chap. 2 we discussed the simple *PN*-junction diode as well as some special types of diodes. All those diodes have the same basic construction: a single junction between their *P*- and *N*-type regions. In this section we shall discuss devices with two or more *PN* junctions that still qualify for the term "diodes" because they have only two external terminals.

6.2.1 AC Trigger Diode (Diac)

The ac trigger diode is a three-layer, two-junction device quite similar in construction to a BJT. But the region that would correspond to the base has no external connections. Furthermore, the ac trigger diode has identical *N* regions, thus distinguishing it from a BJT.

An *NPN* trigger diode, or *trigger diac*, is illustrated schematically in Fig. 6-7. It is a bilateral (or bidirectional) device; that is, it exhibits identical (typically within 10%) characteristics in both directions. Its two terminals are indistinguishable.

When a voltage of either polarity is applied to the trigger diac, one junction is forward biased, the other reverse biased. The current through the diode is limited to a small leakage current by the reverse-biased junction. However, when this junction is reverse biased to the extent that breakdown occurs, a large current through the diode results. Because the

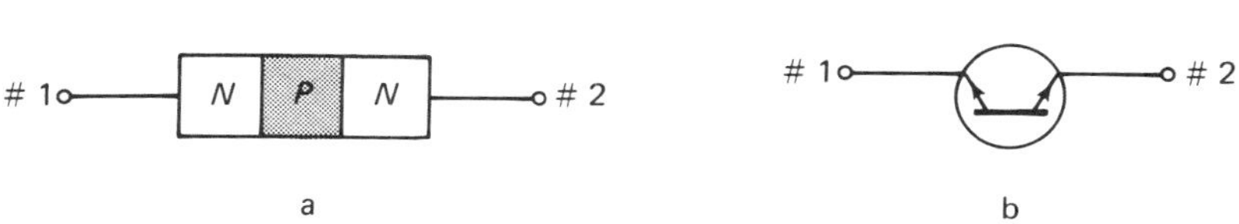

Figure 6-7. An ac trigger diode: (a) construction and (b) circuit symbol.

reverse-biased junction breaks down, a smaller total voltage across the diac is necessary to maintain the current. This corresponds to a negative resistance region.

Typical trigger diac characteristics are shown in Fig. 6-8. Notice the two distinct regions: one of low current (and high resistance), the other of high current (and low resistance). At the transition point between the two regions, the trigger diac is said to *fire* and begin to conduct. This action is that of a switch. The diac is an effective open circuit until the voltage across its terminals reaches the breakdown voltage, at which point the diac resistance becomes low and the switch is effectively closed. Because this action occurs in both directions, the device is sometimes called an *ac switch*.

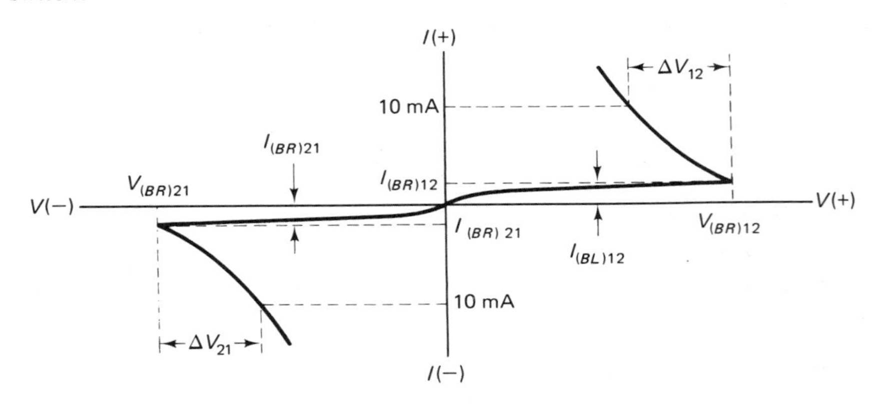

Figure 6-8. Typical characteristics for an ac trigger diode.

6.2.2 Four-Layer (Shockley) Diode

As the name indicates, this diode is a four-layer, three-junction device. It is illustrated in Fig. 6-9. Using conventional diode terminology, we call the connection to the P_1 layer the *anode* and the lead at the N_2 layer the *cathode*. But here the similarity between a conventional *PN* diode and the *PNPN* diode stops.

The *PNPN* structure of the four-layer diode can be considered the equivalent of two BJTs: one a *PNP*, the other an *NPN*, as shown in Fig. 6-10. With a forward bias voltage across the diode (anode positive with respect to the cathode), the $P_1N_1(J_1)$ and $P_2N_2(J_2)$ junctions are forward biased, while the $N_1P_2(J_2)$ junction is reverse biased. From the equivalent circuit in Fig. 6-10(b), we see that $I_A = I_{E1} = I_{E2}$ and also $I_{C1} = I_{B2}$. We can recall from Chap. 3 the relationship of transistor currents as:

$$I_{C1} = \alpha_1 I_{E1} + I_{CO1} \tag{6-3}$$

and for transistor 2:

$$I_{C2} = \alpha_2 I_{E2} + I_{CO2} \tag{6-4}$$

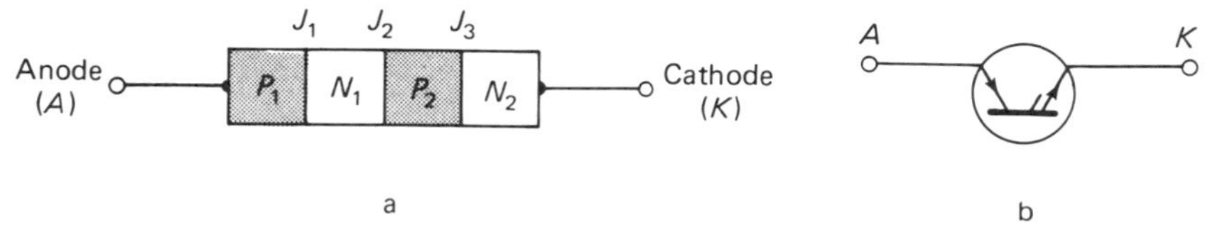

Figure 6-9. Four-layer (Shockley) diode: (a) construction and (b) circuit symbol.

Using the relationships from the equivalent circuit in Eq. (6-1), we obtain:

$$I_{B2} = \alpha_1 I_A + I_{CO1} \tag{6-5}$$

Remembering that $I_{B2} = I_{E2} - I_{C2}$ from Eq. (6-3), we get:

$$I_A - I_{C2} = \alpha_1 I_A + I_{CO1} \tag{6-6}$$

We now substitute Eq. (6-2) for I_{C2} in Eq. (6-4):

$$I_A - \alpha_2 I_A - I_{CO2} = \alpha_1 I_A + I_{CO1} \tag{6-7}$$

Rearranging and solving this equation for I_A yields:

$$I_A = \frac{I_{CO1} + I_{CO2}}{1 - (\alpha_1 + \alpha_2)} \tag{6-8}$$

Because of the reverse bias on J_2, the anode current is low. Remember that the transistor α is dependent on the current (emitter current

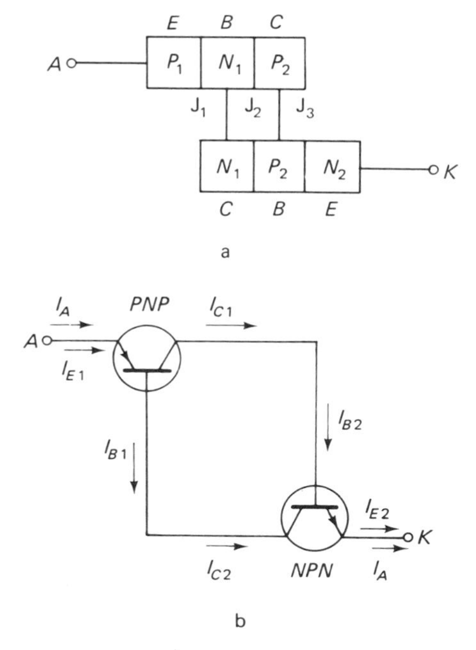

Figure 6-10. (a) Two-transistor representation for the four-layer diode and (b) the resulting circuit.

specifically). For low currents, α is very small, and so the anode current may be approximated by:

$$I_A \cong I_{CO1} + I_{CO2} \tag{6-9}$$

Thus, the anode current is quite small, being the sum of two reverse-leakage currents.

Let us now consider the results of further increasing the voltage in the forward direction. Almost all of the increase in voltage appears across the reverse-biased junction J_2. At a critical voltage, called the *breakover voltage*, this junction breaks down by the avalanche mechanism. The result is an abrupt increase in current and both α_1 and α_2 increase. Equation (6-8) tells us that obviously when $(\alpha_1 + \alpha_2)$ becomes equal to 1, the anode current can increase without limits. Consequently, we must use a series resistor to limit the anode current. Under the breakdown conditions described, the voltage required to maintain the increasing anode current rapidly falls off, and the diode resistance becomes drastically reduced.

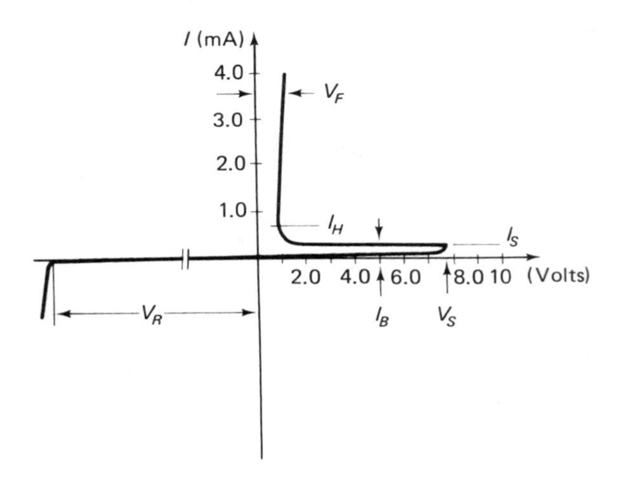

Figure 6-11. Typical characteristics of a *PNPN* (Shockley) diode.

Typical *PNPN* diode characteristics are depicted in Fig. 6-11. In the forward direction, the diode behaves like a solid-state switch with a high-resistance region (switch off) and a low-resistance region (switch on). In the reverse direction, if effectively blocks current for reverse voltages less than the reverse breakdown voltage, V_{RB}.

Once the *PNPN* diode is in its forward conducting state, we can turn it off either by interrupting or reducing the anode current (below the holding current I_H) or by removing or reducing the anode voltage (below the holding value V_H). Any one of these actions causes the *PNPN* diode to open up and return to its high-resistance state.

6.2.3 Bilateral Diode Switch (Diac)

The construction and circuit symbol for the bilateral or bidirectional diode are shown in Fig. 6-12. The term "diac" is applied to all two-terminal devices having identical (or nearly identical) characteristics in both directions.

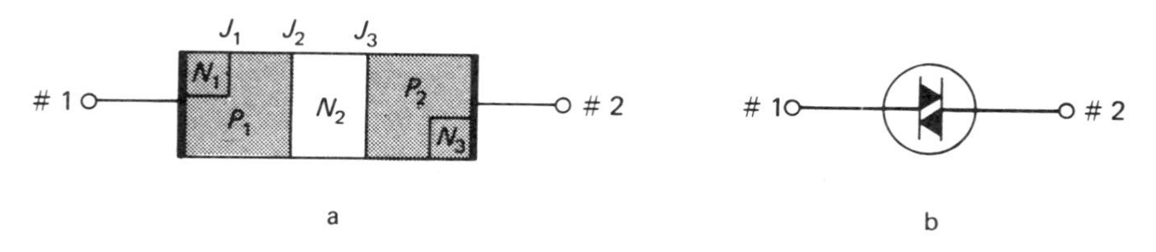

Figure 6-12. Bilateral diode (diac): (a) construction and (b) circuit symbol.

To understand the operation of the bilateral diode, consult the diagram in Fig. 6-13. We can represent the bilateral diode as two *PNPN* diodes connected in parallel. (The two are $P_2N_2P_1N_1$ and $P_1N_2P_2N_3$.) When terminal 1 is made positive with respect to terminal 2, junctions J_1 and J_3 are reverse biased, whereas junctions J_2 and J_4 are forward biased. The *PNPN* diode between K_1 and A_2 is in the off state, because K_1 is positive with respect to A_2. However, the other *PNPN* diode between A_1 and K_2 is forward biased and will conduct heavily when the voltage is high enough to cause avalanche breakdown at junction J_3. (The ohmic drop, caused by a current flow in the P_1 region, forces J_1 to be reverse biased.) Thus, with terminal 1 positive with respect to terminal 2, the characteristics are governed by the four-layer diode $P_1N_2P_2N_3$.

When terminal 2 is made positive with respect to terminal 1, the other four-layer diode (composed of $P_2N_2P_1N_1$) controls the characteristics; that is, it turns on when the voltage applied is high enough to cause avalanche breakdown at J_2. The bilateral diode thus exhibits turn-on characteristics in either direction, accompanied by a low resistance, in much the same way as the trigger diode. Figure 6-14 gives typical terminal characteristics for a bilateral diode.

Once the bilateral diode has turned on in either direction, we can turn it off in the same manner as for the simple *PNPN* diode: (1) by interrupting the current, (2) by reducing the current below I_H, or (3) by reducing the voltage below V_H. Commercially available bilateral diodes

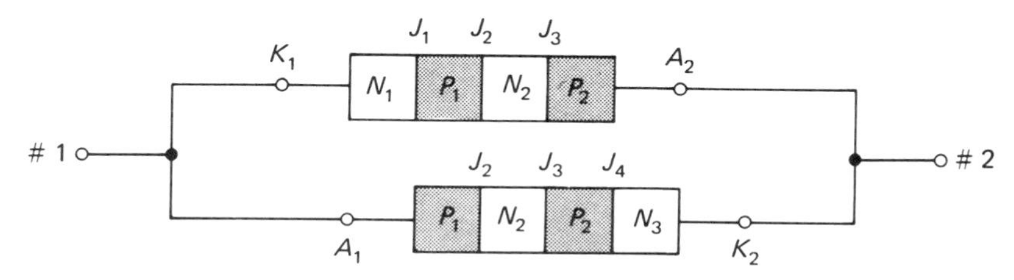

Figure 6-13. Two-*PNPN*-diode representation of a bilateral diode (diac).

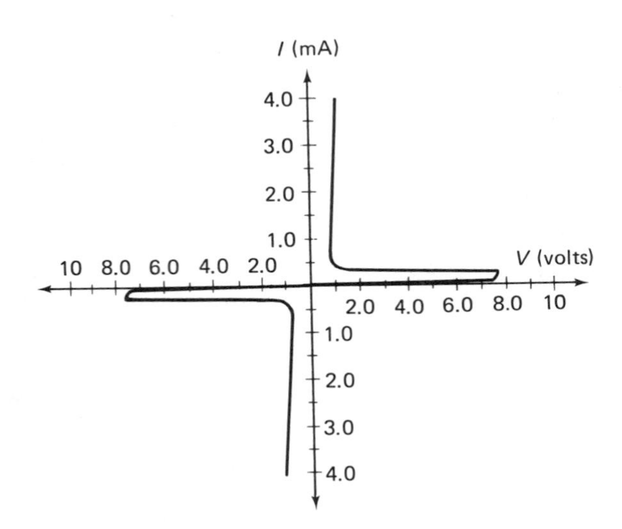

Figure 6-14. Typical characteristics for a bilateral diode (diac).

have forward and reverse characteristics that are closely matched. (See Appendix 3.)

6.3 THYRISTORS

The term "thyristor" is a name usually applied to a family of solid-state devices having turn-on characteristics that can be externally controlled by either current or voltage.

6.3.1 Silicon-Controlled Rectifier (SCR)

The silicon-controlled rectifier (SCR) is also called the *reverse blocking thyristor*. It is a four-layer device similar in construction to the Shockley diode. The difference is in the third terminal connected to the P_2 layer, as indicated in Fig. 6-15(a). Two forms of the circuit symbol are shown in Fig. 6-15(b).

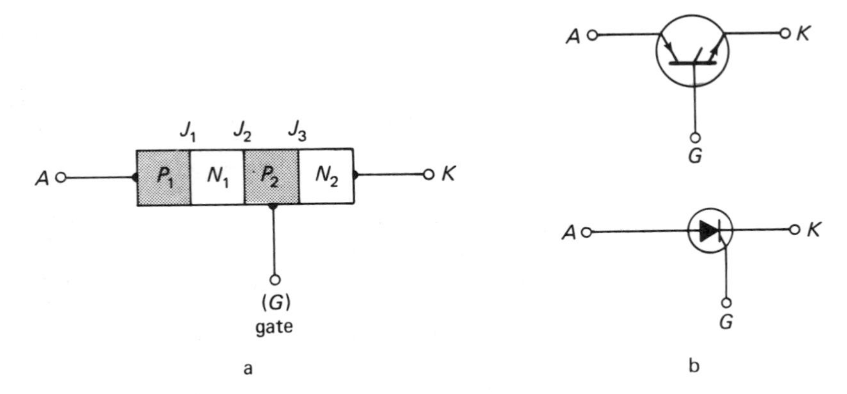

Figure 6-15. Silicon-controlled-rectifier (SCR): (a) construction and (b) alternate circuit symbols.

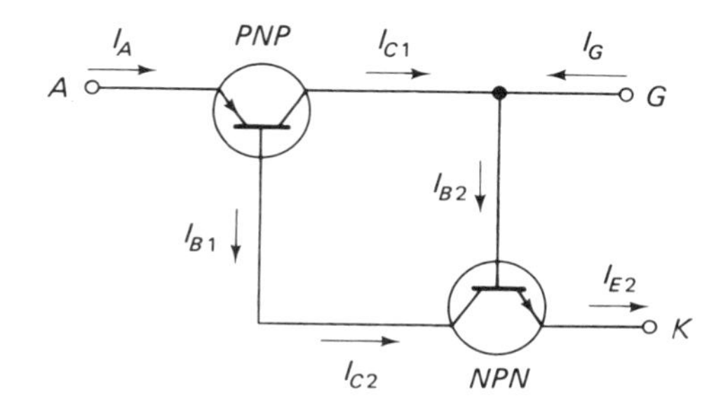

Figure 6-16. Two-transistor representation of an SCR.

The operation of the SCR is quite similar to that of the Shockley (four-layer) diode. With a forward voltage applied between anode and cathode, the SCR is in the normally off or high-resistance state. As the forward voltage is increased and reaches a critical value (the *breakover voltage*), J_2, which is reverse biased, avalanches. Then the anode current becomes high; the SCR has turned on. The third terminal is called the *gate*. It regulates the voltage at which breakover occurs–or the switching between the off and on states.

To understand how this control is achieved, let us examine the two-transistor analogy of the *PNPN* device. This analogy is shown in Fig. 6-16, where the gate lead is attached to the base of the *NPN* transistor. You might also refer to a similar discussion in section 6.2.3 concerning the Shockley diode. In this case, however,

$$I_{B2} = I_{C1} + I_G \tag{6-10}$$

It can be shown that the anode current for the SCR is given by

$$I_A = \frac{I_{CO1} + I_{CO2}}{1-(\alpha_1+\alpha_2)} + I_G \frac{\alpha_2}{1-(\alpha_1+\alpha_2)} \tag{6-11}$$

Obviously, where $I_G = 0$, the anode current for the SCR is the same as that for the Shockley diode [Eq. (6-8)]. In the normal forward-blocking (off) state, the anode current is low and the αs are small. The term containing I_G signifies that the anode current for a given forward voltage is higher than in the Shockley diode. Moreover, the quantity $(\alpha_1+\alpha_2)$ reaches unity for a lower value of forward voltage because of the more rapid increase in current. Thus, the gate current effectively controls the voltage at which the SCR switches from its nonconducting off state to its forward-conducting on state. We determine the breakover voltage by the magnitude of the gate current: The higher the gate current, the lower the breakover voltage. Figure 6-17 supplies these typical SCR characteristics.

The reverse characteristics of the SCR are essentially unaffected by the gate signal and resemble those of the Shockley diode.

Once the SCR is in its forward-conducting state, the gate loses control. In fact, to turn on an SCR, gate current need flow for only a few microseconds to a few milliseconds. So long as $(\alpha_1+\alpha_2)$ is unity (or J_2 is

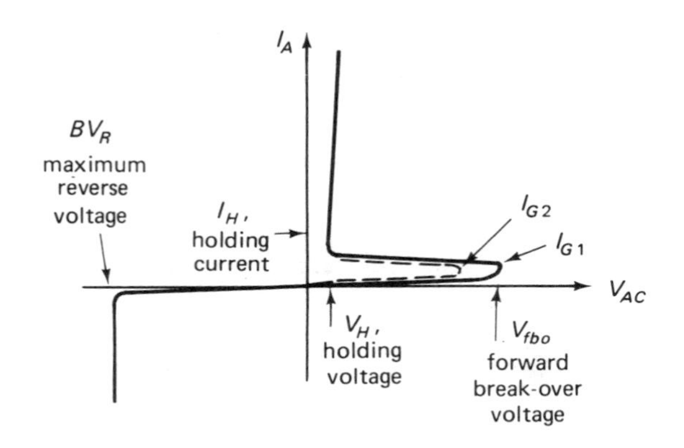

Figure 6-17. Typical anode characteristics of an SCR.

under avalanche breakdown), the anode current will be high and the SCR will remain on, regardless of the gate potential and current. Even a negative gate potential cannot turn off the SCR once it has fired. To turn off the SCR, the anode current must be interrupted or reduced below a certain minimum value. But this minimum value, called the *holding current* (I_H), is dependent on the gate bias, as shown in Fig. 6-17. With zero gate current, the minimum anode current to keep the SCR on is called the *latching current* (I_L).

In addition to the anode characteristics, the manufacturer may also specify the gate characteristics for an SCR. Figure 6-18 provides an example. Because of the variation between devices of even the same family, slightly different combinations of gate (to cathode) voltage and gate current are necessary to fire a given SCR. In normal applications, to insure firing of the SCR, the gate is usually overdriven; that is, we apply a gate current in excess of the minimum specified.

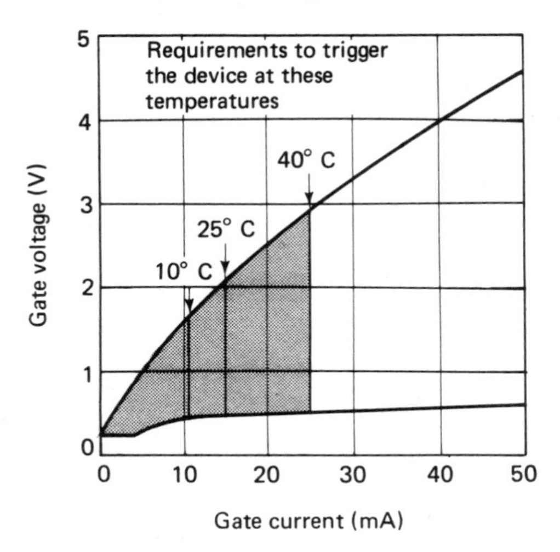

Figure 6-18. Typical gate characteristics for an SCR family.

Commercially available SCRs range from very low-power units to some capable of controlling anode currents in excess of 1000 Amperes.

6.3.2 Gate Turnoff Switch

As we stated earlier, the SCR cannot normally be turned off at the gate. A device very much similar to the SCR, called the *gate turnoff switch*, or *gate controlled switch* (GCS), can be turned off by a negative bias on the gate. In construction, the GCS is identical to the SCR. However, in the terminology of the two-transistor representation of an SCR in Fig. 6-16, the α of the *PNP* transistor is made small. The GCS is inherently a low-current device when compared to an SCR.

The turn-on characteristics of a GCS are identical to those of a comparable SCR. The difference between the two lies in the turnoff. To see how turnoff can be accomplished in a GCS, we use the two-transistor representation in Fig. 6-16. When the GCS is in its conducting state, anode current flows into the anode and essentially the same current flows out of the cathode. Turnoff can be accomplished by causing a negative gate current, equal to I_{C1}, to flow. So we apply a negative bias to the gate. (A negative gate current flows in the direction opposite to that shown in Fig. 6-16.) Equation (6-10) shows that this procedure in effect eliminates the base current, I_{B2}, from the *NPN* transistor. The *NPN* transistor emitter current is sufficiently decreased to cause turnoff. If the α of the *PNP* transistor is small, then I_{C1} will be relatively small and the negative gate current required for turnoff will also be small. The ratio of the anode current in the on state to the negative gate current required to turn off the GCS is called the *turnoff current gain*. Typical values for the turnoff current gain of commercially available GCSs are between 5 and 10. The GCS circuit symbol is shown in Fig. 6-19.

As a sidelight to the operation of a GCS, note that a high-current SCR cannot be turned off by a negative gate bias, but the application of a negative gate bias can significantly reduce the turnoff time of an SCR (that is being turned off at the anode).

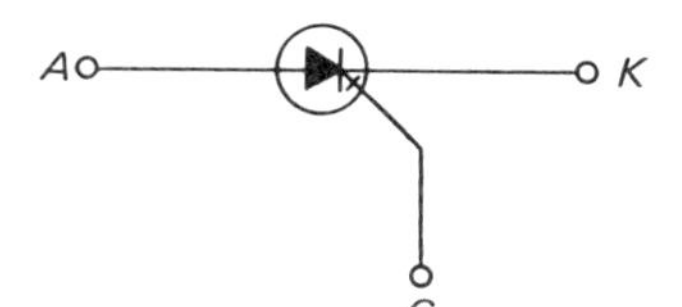

Figure 6-19. Gate turnoff or gate controlled switch (GCS) circuit symbol.

6.3.3 Silicon-Controlled Switch (SCS)

The silicon-controlled switch (SCS) is sometimes called the *tetrode thyristor*. It is a four-layer device and is very similar to the SCR. As the tetrode name would suggest, the SCS has four terminals, one on each of its four layers, as shown in Fig. 6-20(a). The SCS circuit symbol is shown in Fig. 6-20(b). In addition to the cathode gate, G_k, the SCS has an anode gate, G_a, connected to the N_1 region.

The SCS is a versatile device. It can be operated in numerous different ways. One obvious mode of operation is as an SCR with the

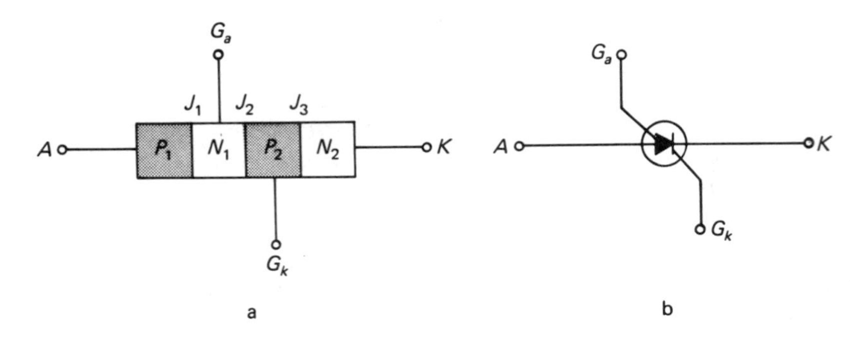

Figure 6-20. Silicon-controlled switch (SCS): (a) construction and (b) circuit symbol.

anode gate not connected. It may also be operated as a complementary SCR. In this case, a negative anode gate voltage causes the SCS to fire; the cathode gate is not connected, however. (Complementary SCRs, with only an anode gate, are manufactured, but the conventional SCR is by far the more common of the two.) Neither of the preceding applications warrants the existence of two gates. The full potential of the SCS is realized only when both gates are used. If we operate the SCS like an SCR, the anode gate can turn off the SCS once it has fired. We would apply a sufficiently positive potential to the anode gate junction J_1 to cause reverse biasing, thus turning off the SCS.

Perhaps the most important use of the anode gate is to make the SCS turn on quicker. In the off state, J_2 is reverse biased. As such, it has a capacitive effect that slows down the rate at which the SCS turns on when the anode becomes positive enough. If we apply a positive voltage to the anode gate, we charge this junction capacitance, so that when the anode signal exceeds V_{BO} and initiates the firing of the SCS, the turn-on rate is drastically reduced.

The SCS is a low-power device. As such, it may be turned off at the cathode gate by a negative voltage, in a way much like the GCS.

6.3.4 Bilateral Triode (Triac)

The name "triac" is applied to all bilateral (or bidirectional) triodes. To fully understand the operation of a triac, we first need to examine the principles involved in the operation of thyristors having a *remote gate* or a *junction gate*.

A remote gate thyristor is illustrated schematically in Fig. 6-21. It is a four-layer device comprised of $P_1N_1P_2N_2$, with the gate not at either N_1 or P_2 as in a conventional SCR, but at N_3, as indicated. In the forward-blocking state (where A is positive with respect to K) and a negative voltage applied to the gate, the N_3P_1 junction is forward biased. In addition, J_1 and J_3 are forward biased and J_2 is reverse biased. Because of the forward bias on J_4, electrons from N_3 are injected into P_1, diffuse through P_1, and are collected at J_1. The effect is to increase the forward bias on J_1, so additional holes are injected from P_1 into N_1. These holes diffuse across N_1

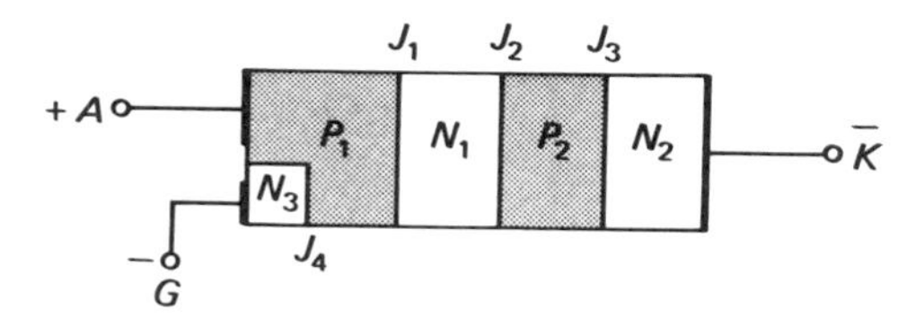

Figure 6-21. Schematic representation of a remote gate thyristor.

and are collected at P_2. As a result, the forward bias across J_3 is increased. Additional electrons are injected from N_2 into P_2. They are collected at J_2, thus further increasing the forward bias across J_1. We have then a cycle started by a negative bias on the gate, causing a larger forward bias across J_1, which, by the action described, is self-perpetuating; it increases the anode current until the thyristor fires in the same manner as a conventional SCR. The remote gate thyristor fires in the forward direction as a result of negative gate bias.

A schematic diagram of a junction gate thyristor is shown in Fig. 6-22. Let us assume that the device is off, with a forward bias on the anode and a negative bias on the gate. Because of the gate bias V_G, the gate is more negative with respect to the anode than the cathode. The four-layer device comprising $P_1N_1P_2N_3$ begins to turn on. As a result, gate current starts to flow in the direction shown. As the gate current increases, so does the voltage drop across the gate bias resistor R. When the drop across R exceeds V_G, the gate becomes positive and the large flow of holes from P_2 into N_3 is diverted to N_2. Thus, the forward bias of J_3 is significantly increased, and the four-layer $P_1N_1P_2N_2$ turns on. In effect, the negative gate bias initiates turn-on of $P_1N_1P_2N_3$. As voltage across R builds up, the large current from the anode is diverted to the cathode, accomplishing turn-on of $P_1N_1P_2N_2$.

We can now proceed with the discussion of the bilateral triode thyristor or triac, shown schematically in Fig. 6-23. The triac is truly a bidirectional device. It can fire with either positive or negative gate bias. Let us examine the four possible configurations:

Terminal 2 positive, gate positive (both with respect to $T1$): The active regions are $P_1N_1P_2N_2$, forming a conventional SCR with $T2$ as the anode, $T1$ as the cathode, and G connected to P_2 as the cathode gate.

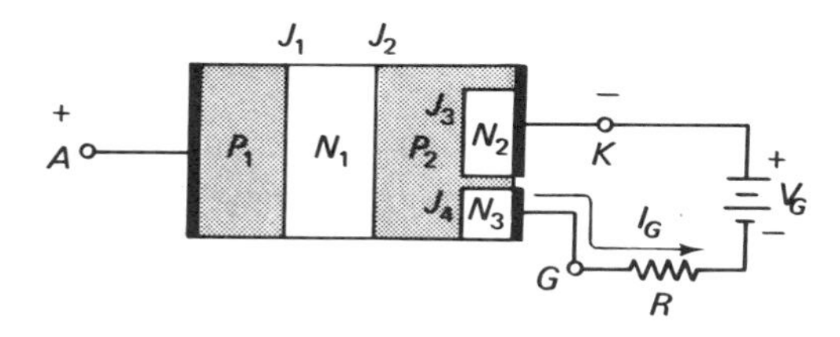

Figure 6-22. Schematic representation of a junction gate thyristor.

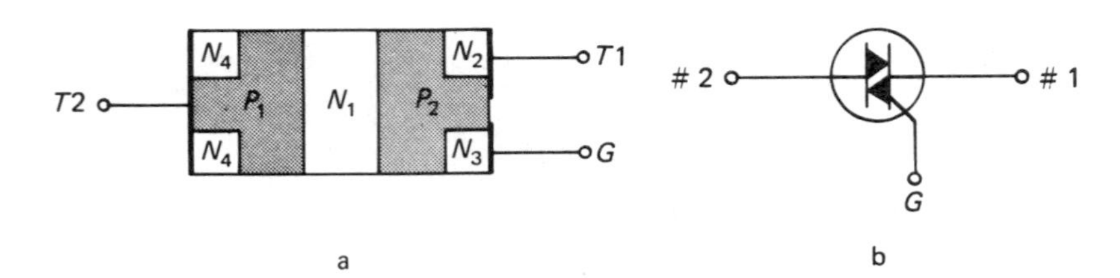

Figure 6-23. Bilateral triode (triac): (a) construction and (b) circuit symbol.

Terminal 2 negative, gate negative: The operation is that of a rėmote gate thyristor with active regions $N_4P_1N_1P_2$. $T2$ acts as the cathode, $T1$ as the anode, and the remote gate function is accomplished by the N_3P_2 junction.

Terminal 2 positive, gate negative: Operates as the junction gate thyristor, with active regions $P_1N_1P_2N_2$. $T2$ acts as the anode, $T1$ as the cathode, and the junction gate is formed by the junction of N_3 and P_2.

Terminal 2 negative, gate positive: The operation is that of a remote gate thyristor, shown in Fig. 6-24. The active regions are $P_2N_1P_1N_4$, with $T1$ acting as the anode, $T2$ as the cathode, and the remote gate function accomplished by the gate contact at P_2. The forward bias on the P_2N_2 junction initiates the firing.

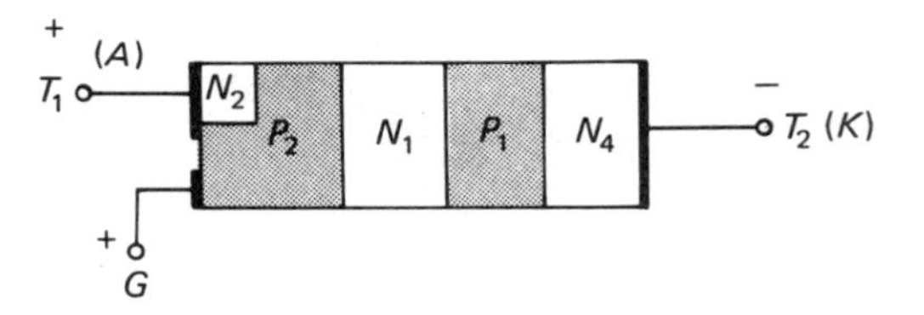

Figure 6-24. Remote gate operation of a triac with *T2* and *G* positive.

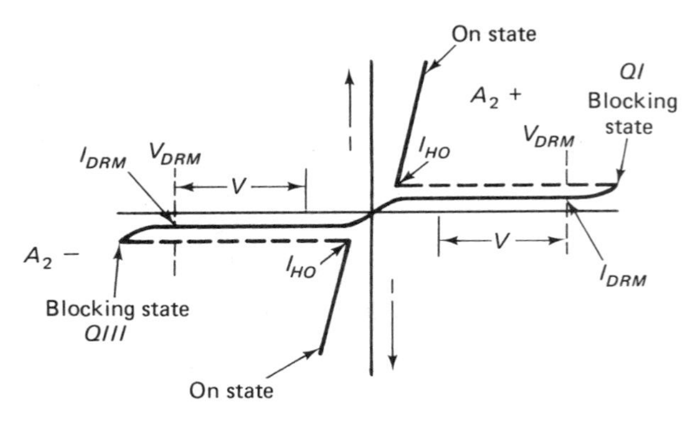

Figure 6-25. Typical triac characteristics.

The terminal characteristics of the triac are identical in both directions. They resemble the forward characteristics of an SCR. Typical characteristics are given in Fig. 6-25.

We can turn off a triac operating in any of the possible configurations by either interrupting or lowering the current below the holding value of the $T1T2$ current.

Review Questions

1. In terms of its construction, justify the naming of a UJT.
2. What are the three terminals of a UJT?
3. What is the intrinsic standoff ratio? How is it defined?
4. What is the relationship of the interbase resistance of a UJT to its intrinsic standoff ratio?
5. How are the peak, cut-in voltage and the intrinsic standoff ratio for a UJT related?
6. Briefly explain the "firing" of a UJT.
7. What is the structural difference between an ac trigger diode and a Shockley diode?
8. Briefly describe the firing of an ac trigger diode. Is it a unilateral or a bilateral device?
9. How does the operation of a Shockley diode differ from that of the ac trigger diode? How are they similar?
10. What is the construction of a bilateral diode? How do its characteristics differ from an ac trigger diode? How are they similar?
11. Once in conduction, how can we turn off the ac trigger diode, the Shockley diode, and the bilateral (diac) diode?
12. In what way does an SCR differ from the diodes (ac trigger, Shockley, and bilateral)? How is it similar?
13. Explain the conditions leading to triggering (firing) in an SCR?
14. Describe the two-transistor analogy for an SCR.
15. What determines the firing voltage (anode-to-cathode) in an SCR?
16. Once an SCR has been fired, what are the ways in which it can be turned off?
17. Is the SCR a unilateral or a bilateral device? Why?
18. How does a gate turnoff switch GTS differ from an SCR? In what ways are they similar?
19. What controls the firing characteristics for a GTS? How can it be turned off once it is in conduction?
20. What is the turnoff current gain in a GTS? What is it numerically?
21. How are an SCR and an SCS similar? In what ways are they different?

22. What are the ways in which an SCS can be turned on? What control is provided by the two gates?
23. What are the advantages of using an SCS with the additional anode gate over using an SCR?
24. What are the firing characteristics of a triac?
25. In what ways are the characteristics and operation of an SCR and a triac similar? In what ways are they different?
26. What are the ways in which a triac may be turned off once it is conducting?

7

Photoelectric Devices

In this chapter we shall briefly discuss semiconductor devices that are activated by light and that emit light. These devices are used for industrial control, information transmission, visual displays, and even power generation.

7.1 PHOTOCONDUCTIVE CELLS

Photoconductive cells are sometimes called *photoresistors*. They operate on the principle of *photoresistivity*. Certain semiconductors, when illuminated by light, show a decrease in resistance. Typical examples of photoresistive semiconductors are made of cadmium sulphide (CdS) and cadmium selenide (CdSe). The cells are fabricated by depositing a thin layer of semiconductor on a ceramic base, with two interleaved electrodes, as shown in Fig. 7-1(a). The enclosure contains a glass window or lens that allows light to reach the semiconductor.

As we noted in Chap. 1, the resistivity of a semiconductor is a function of the number of free charge carriers that are available for conduction. Before the semiconductor is illuminated, this number of free charge carriers is very small, so resistivity is high. When light, in the form of photons,* is allowed to strike the semiconductor, each photon delivers energy to the semiconductor. If the energy exceeds the energy gap of the semiconductor, free mobile charge carriers are liberated; and, as a result, the resistivity of the semiconductor is decreased.

The response of a given photoconductive cell is determined by its energy gap. For a specific energy gap, only photons with energies larger than E_G can liberate additional carriers, so the cell reacts only to those photons. The frequency (f) of light is related to the wavelength, λ, as

*For the purposes of our discussion, visualize the light striking the semiconductor as being made up of a stream of particles, called *photons*. Each photon carries with it an amount of energy hf, where h is Plank's constant (about 4×10^{-15} eV-sec) and f is the frequency of the light.

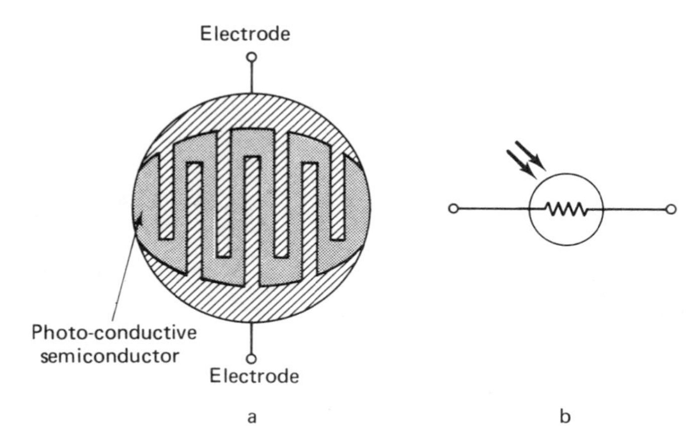

Figure 7-1. Photoconductive cell: (a) typical construction and (b) circuit symbol.

follows:

$$f = \frac{c}{\lambda} \tag{7-1}$$

where c is the speed of light (3×10^{10} cm/s). The energy of a photon can then be expressed in terms of the wavelength:

$$E = \frac{hc}{\lambda} \tag{7-2}$$

We calculate the minimum wavelength needed to decrease the resistivity as:

$$\lambda = \frac{hc}{E_G} = \frac{12}{E_G} \times 10^{-5} cm = \frac{12{,}000}{E_G} \mathring{A} \tag{7-3}$$

where E_G is in electronVolts and 1 cm is equal to 10^8 Angstron units (Å). Figure 7-2 shows the spectrum visible to the human eye. Typical photocon-

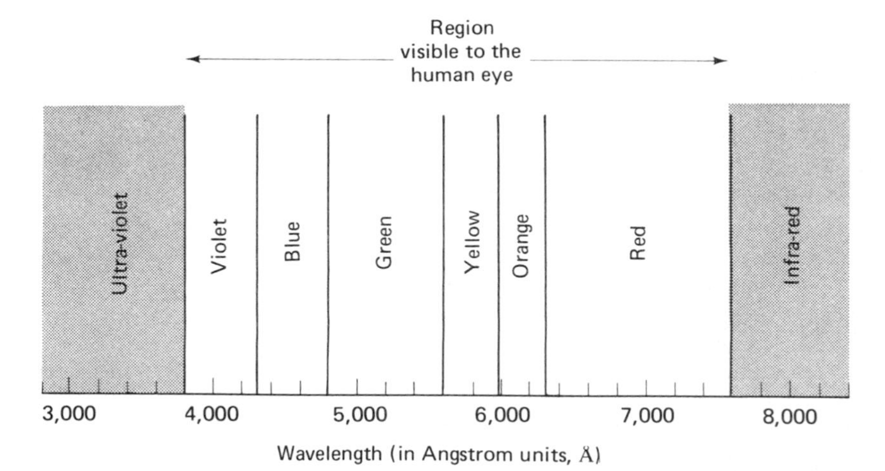

Figure 7-2. Visible frequency spectrum.

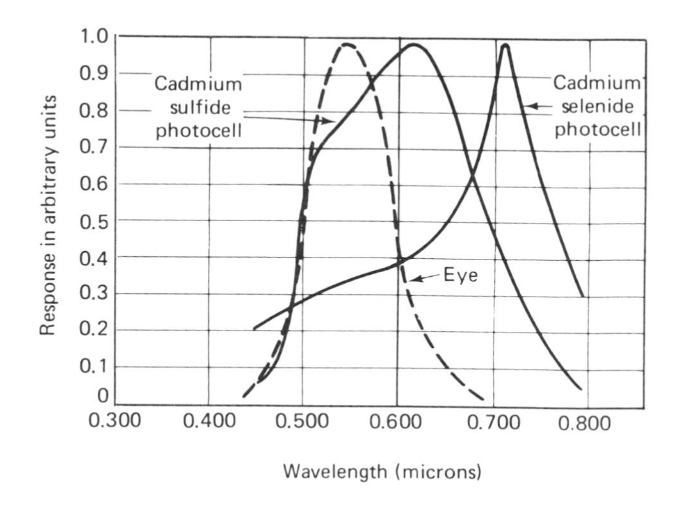

Figure 7-3. Photocell spectral response curves.

ductive semiconductors have energy gaps on the order of 2eV. The spectral response of CdS and CdSe is illustrated in Fig. 7-3, together with the response of the human eye. Cadmium sulphide has a response similar to the human eye, so CdS cells are often used to simulate the human eye. For example, they are utilized in light-metering circuits in photographic cameras.

The terminal characteristics of a photoconductive cell are a plot of cell resistance as a function of light intensity. Typically, cells have a *dark resistance* of 1 MΩ or higher. The dark resistance, as the name implies, is the cell resistance when there is no light striking it. Under illumination, the cell resistance drops to a value between 1 and 100 kΩ, depending on the light intensity. Units of light intensity are lumens per square foot, abbreviated lm/ft^2. Typical characteristics, showing the nonlinear decrease in resistance with increasing light intensity, are given in Fig. 7-4.

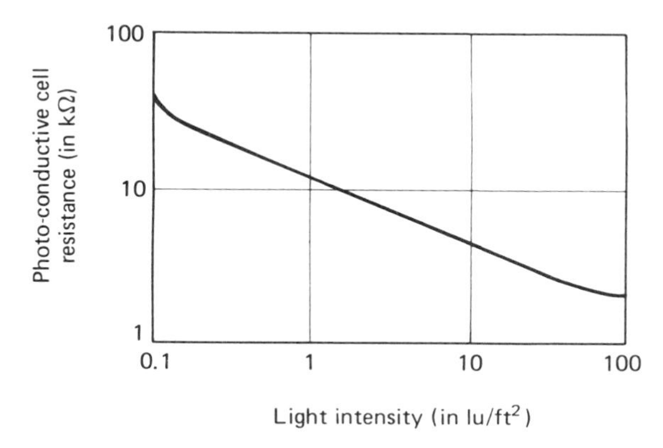

Figure 7-4. Typical characteristics for a photoconductive cell.

We must point out that the resistance value under a specific light intensity depends on the previous light intensity to which the cell was subjected. Moreover, there is a finite time required for the cell to react to a change in light intensity. The *rise time* is the time required for a cell starting in the dark condition to reach the lower resistance value (actually 62% of this value) at a higher illumination level. Typical cells have rise times of less than a millisecond.

7.2 PHOTOVOLTAIC CELLS

Photovoltaic, or *solar*, cells are devices characterized by the generation of a small voltage upon illumination. The solar cell contains a *PN* junction. When the junction is illuminated, holes and electrons are liberated in much the same way as in photoconductive cells. However, these newly liberated charge carriers are acted upon by the junction contact potential. Electrons are forced toward the *N* side, holes toward the *P* side. If external wiring is connected, we can observe that a current results. The displacement of holes toward the *P* side and electrons toward the *N* side lowers the junction contact potential. This effect is manifested externally by the appearance of a voltage across the terminals of the cell. The electrode connected to the *P* side becomes positive with respect to the electrode connected to the *N* side. Thus, a terminal voltage is generated as a result of light striking the *PN* junction of the cell. The voltage is directly proportional to the intensity of the light; it may be as high as $\frac{6}{10}$ or $\frac{7}{10}$ V, depending on the load placed across the cell.

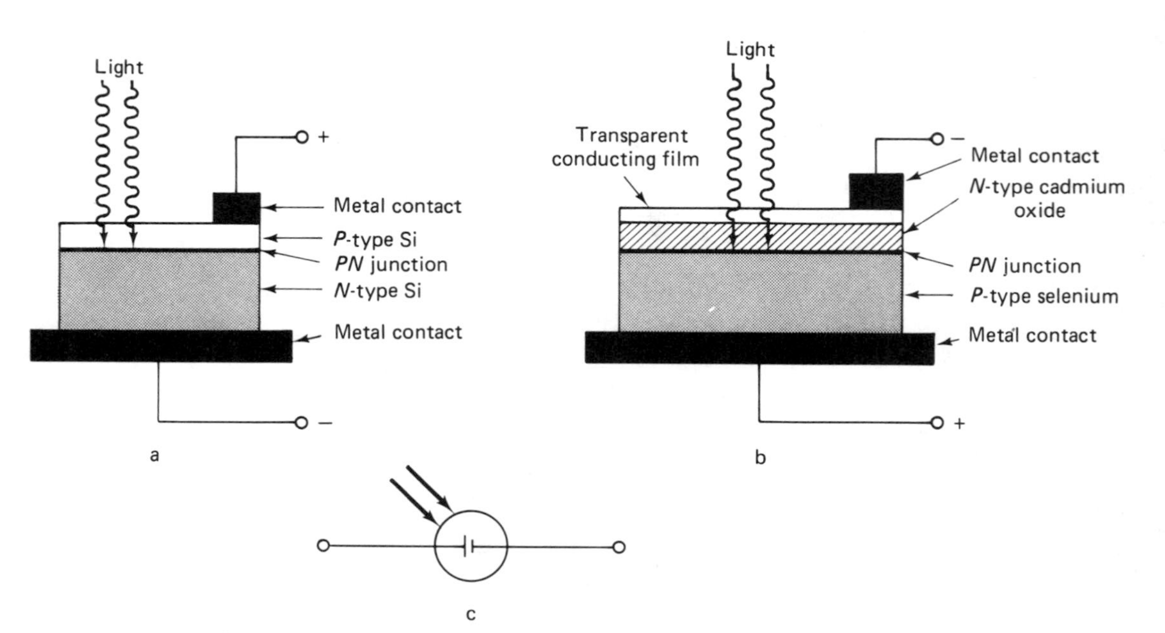

Figure 7-5. Construction of photovoltaic cells: (a) silicon, (b) selenium, (c) circuit symbol.

Silicon is a commonly used material in the fabrication of photovoltaic cells. Another fairly common cell construction features *P*-type selenium, with a layer of *N*-type cadmium oxide over it to form the *PN* junction. The construction of these two cells is depicted in Fig. 7-5. Typical characteristics of photovoltaic cells are shown in Fig. 7-6. Note that the rise in voltage is not linearly related to increases in illumination.

As the name "solar cell" might suggest, photovoltaic cells can be used to convert solar energy into electrical energy. In such applications, literally thousands of individual cells can be connected together to generate power levels in excess of a kilowatt.

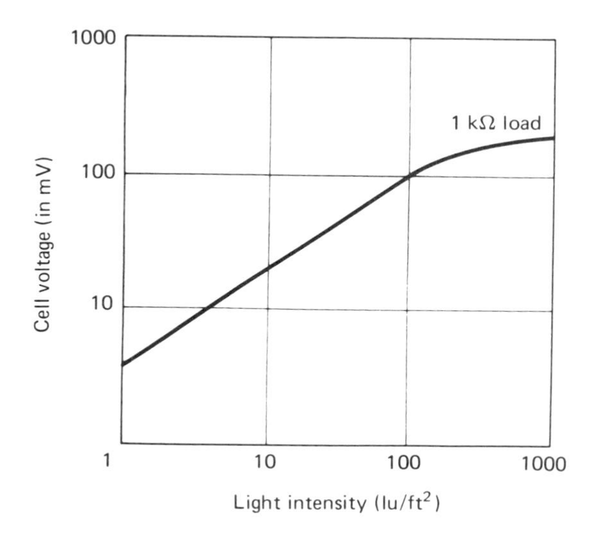

Figure 7-6. Typical selenium cell characteristics.

7.3 PHOTODIODES

Photodiodes are almost identical in construction to photovoltaic cells. However, photodiodes are operated in the reverse-biased condition. Under these conditions they behave similarly to photoconductive cells.

To understand the operation of photodiodes, recall that a conventional diode under reverse bias sustains a small (reverse-saturation) current that is the result of the flow of minority carriers. The same process occurs in a photodiode when it is not illuminated. So the dark current of a photodiode is low and essentially the same as the normal reverse-saturation current. When light is allowed to fall on the *PN* junction of a photodiode, those photons whose energy exceeds the energy gap of the semiconductor produce additional free charge carriers: holes and electrons. These newly created free carriers have no appreciable effect on the majority carrier concentrations in the *N* and *P* regions, but they greatly increase the minority carrier concentrations. The action of the reverse bias forces the newly created holes toward the cathode and the electrons toward the anode. The net effect is a large increase in the reverse current.

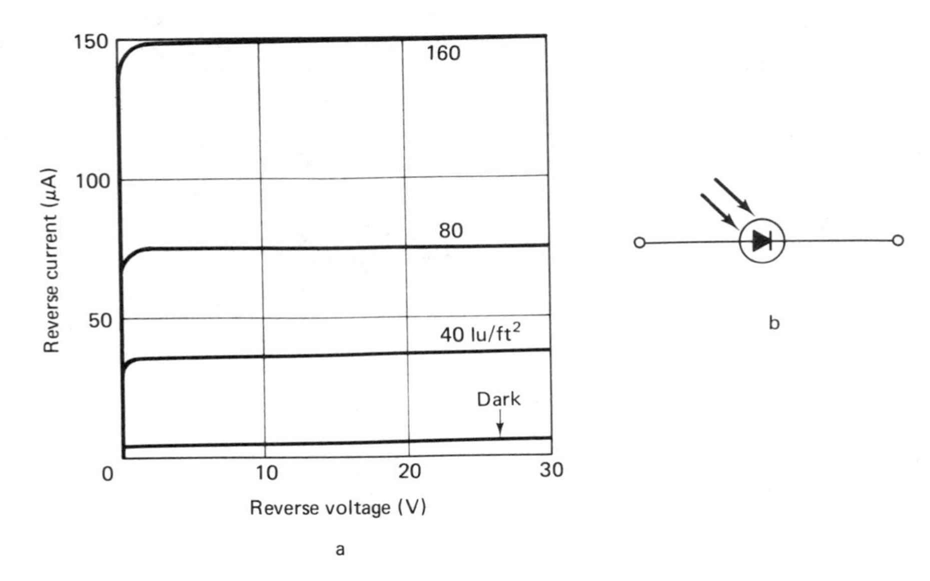

Figure 7-7. Photodiode: (a) typical characteristics and (b) circuit symbol.

In typical photodiodes, the increase in the reverse current is almost directly related to the light intensity. A typical set of characteristics, together with the circuit symbol for a photodiode, is given in Fig. 7-7.

7.4 PHOTO-TRANSISTORS

Phototransistors are made in much the same way as their conventional BJT counterparts, but they usually have no external connection to the base. Their operation is based on the photodiode that exists at the collector-base junction.

Silicon *NPN*s are the most common phototransistor configurations. Usually, the phototransistor is made by the diffusion process, with a transparent window or lens in the case. The light incident on the collector-base junction creates additional charge carriers. The phototransistor is operated with the *N*-type emitter negative with respect to the *N*-type collector. The base-emitter junction is then forward biased, whereas the collector-base junction is reverse biased. If the incident light falls on the collector-base junction and the photons have enough energy to create additional free charge carriers, the reverse bias at the junction causes electrons to go into the collector region. The remaining holes in the base region, being positive, raise the potential of the base with respect to the emitter. As a result, electrons from the emitter are injected into the base. Some of these electrons are lost through recombination in the base, but most of them do reach the collector. Electrons reaching the collector flow out through the collector lead and make up the output current.

The operation of a phototransistor is similar to that of a conventional BJT. However, in a phototransistor, light-liberating charge carriers constitute the input (base) current.

The dark current in a phototransistor is larger than in a photodiode. It is essentially equal to I_{CEO}, typically between a nanoAmpere and a microAmpere. However, the collector current in a phototransistor under illumination is essentially β times the photocurrent of a similar photodiode. Figure 7-8 displays typical output characteristics of a phototransistor, with illumination instead of base current as the parameter. The same figure also shows the circuit symbol.

We can utilize this photodiode action in a field-effect transistor. The result is called a *photoFET*. The main advantages of photoFETs over phototransistors are greater sensitivity and variable sensitivity. The photoFET may be biased as indicated in Fig. 7-8, with the gate-source junction reverse biased. When light is allowed to strike the junction, photons with enough energy create additional charge carriers. The reverse bias makes these carriers flow between source and gate, thus constituting a gate current. The flow of this photocurrent out of the gate causes a voltage drop across the bias resistor R_G, which opposes the bias voltage developed across R_S. As a result, the reverse bias of the gate-source junction is decreased, subsequently changing the drain (i.e., output) current. We can control the sensitivity to light by adjusting the size of R_G. When R_G is made large, even a small photocurrent at the gate causes an appreciable voltage change in the gate-source bias, thus resulting in a large variation of the drain current.

The construction of photoFETs is similar to that of conventional FETs. But a window, or lens, is incorporated in the case to allow light to strike the *PN* junction. The circuit symbol for a photoFET is shown in Fig. 7-9.

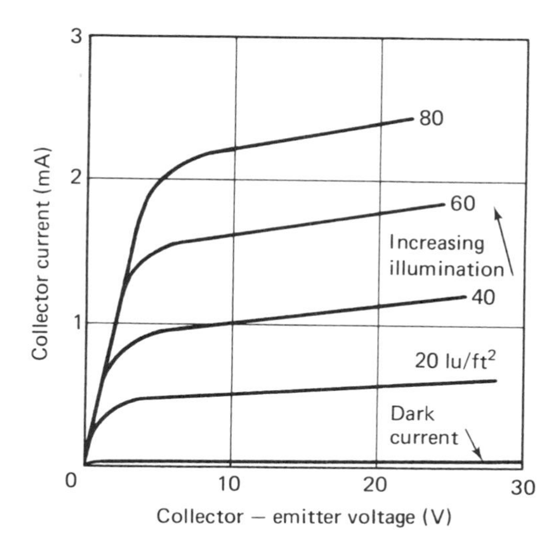

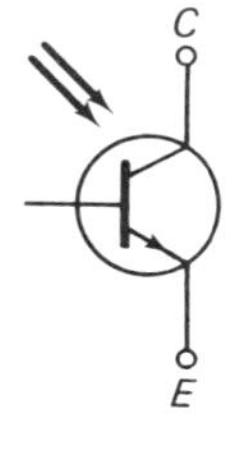

Figure 7-8. Phototransistor: (a) typical characteristics and (b) circuit symbol.

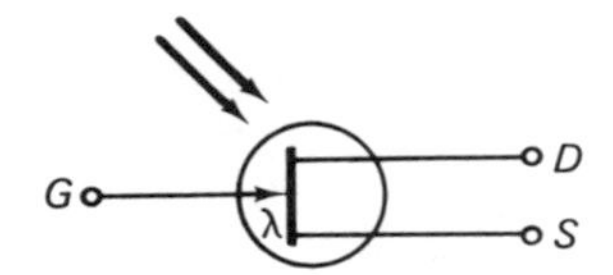

Figure 7-9. PhotoFET circuit symbol.

7.5 LIGHT-ACTUATED *PNPN* DEVICES

As we saw in Chap. 6, switching in an SCR or an SCS takes place when we increase the anode-to-cathode current. In a conventional SCR, this switching is controlled by current supplied to the gate. In a light-activated SCR (LASCR), the anode-to-cathode current is increased when light strikes near the middle junction. Photons with enough energy create additional free carriers near the reverse-biased middle junction (J_2 in Fig. 6-15). These additional carriers are acted upon by the reverse bias at J_2. Holes are forced toward the cathode; electrons, toward the anode. As a result, the SCR current increases, in turn increasing α_1 and α_2 in Eq. (6-11). Switching is initiated as if the increase in current were being provided by the gate.

The photocurrent at the middle junction is directly proportional to the illumination. Therefore, the rise in intensity of the illumination incident on a LASCR makes the switching occur at an ever-decreasing voltage. As is the case with a conventional SCR, once the LASCR is in its conducting state, it will continue to conduct, whether light is present or not. It can be turned off only by interrupting the anode current or by any of the other ways common to the SCR.

We have somewhat increased versatility of operation in a LASCS (light-activated silicon-controlled switch), which is also called a light-activated switch (LAS). The LASCS is a photoadaptation of the silicon-controlled switch. We turn it on in the same way as the LASCR. But the anode gate gives us additional flexibility in turning the switch off.

The circuit symbols for LASCR and LASCS are shown in Fig. 7-10.

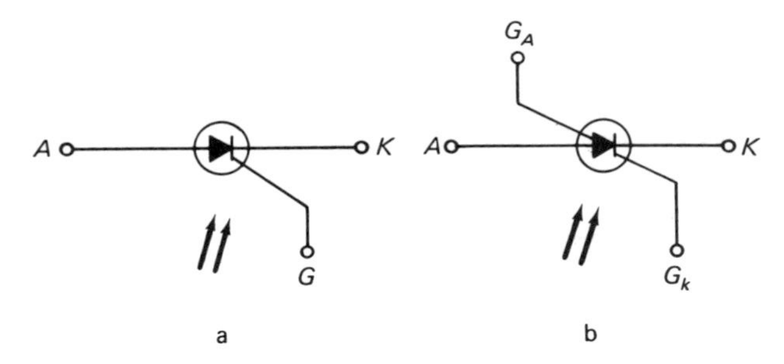

Figure 7-10. Circuit symbols for light-activated *PNPN* switches: (a) LASCR and (b) LASCS (LAS).

7.6 LIGHT-EMITTING DIODES

The *light-emitting diode* (LED) is quite different from the other devices discussed in this chapter. All the other devices use light as the *input* to control their electrical characteristics. The LED emits light or has light as its *output*.

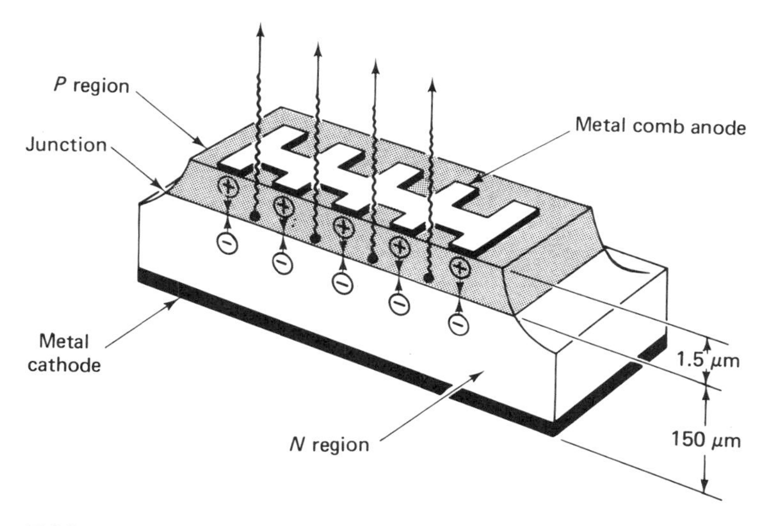

Figure 7-11. LED construction. The comb-type anode gives even current distribution while blocking less than 25 percent of the available light. (Courtesy of Hewlett-Packard)

One possible construction of an LED is depicted in Fig. 7-11. An alloy of gallium arsenide phosphide (*N*-type) is grown epitaxially on a gallium arsenide substrate. The *P* region is then diffused and topped off with a metal electrode. The electrode is shaped in such a way that it gives an even current distribution and allows most of the light to escape.

When the LED is forward biased, the junction potential is reduced and current flows. The current is made up of electrons flowing into the *P* region and holes into the *N* region. These carriers become the minority carriers in the respective regions. When electrons cross the junction, they may be lost through recombination. At the same time that they recombine they give off energy, which is the difference between their "free" or conduction-band state and their bound valence-band state. This energy is given off in the form of radiation. In some cases, the energy difference corresponds to radiation in the visible spectrum, i.e., to light. Depending on the LED material, the light may be in either the green or (more commonly) the red part of the spectrum.

The brightness of the light given off by the LED is contingent on the number of photons released by the recombination of carriers inside the LED: The higher the forward bias voltage, the larger the current and the larger the number of carriers that recombine. Therefore, we can increase the brightness by increasing the forward voltage. Typically, LEDs have a forward voltage rating of between 1 and 2 V and a current rating of about 50 mA. Figure 7-12 gives typical LED characteristics, showing the light intensity as a function of forward voltage, as well as the LED circuit symbol.

We can utilize light-emitting diodes (coupled to proper detectors) in the transmission of digital information. When used in arrays, they are

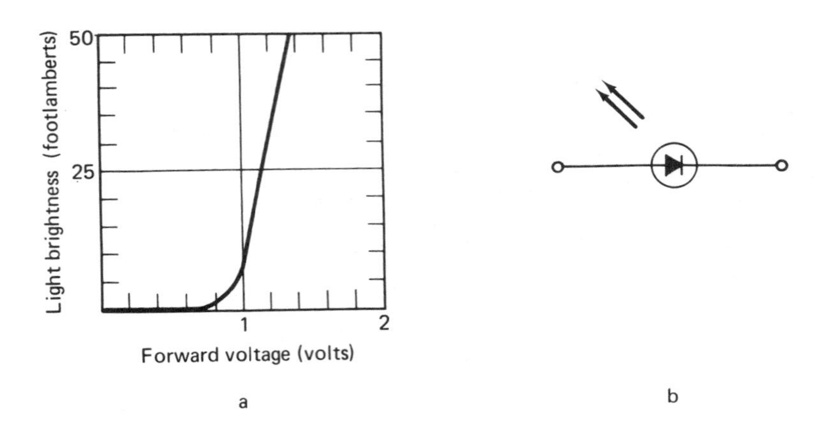

Figure 7-12. LED: (a) typical emitting characteristics and (b) circuit symbol.

suitable for visual alphanumeric (letter and number) readout displays in instrumentation and other applications. Perhaps the most important application of LEDs is still being developed: that of a solid-state color TV picture tube made of an array of diodes emitting green light, red light, and blue light.

Review Questions

1. What is the principle of photoresistivity? Name some substances exhibiting it.
2. How does a photoconductive cell behave when illuminated?
3. What is meant by the term "dark resistance" as applied to photoconductive cells?
4. What are the characteristics of photovoltaic cells?
5. In what ways are photoconductive cells different from photovoltaic cells?
6. What are some commonly used materials in the production of photovoltaic cells?
7. What is a solar cell? Name some applications for solar cells.
8. How does a photodiode differ from a photoconductive cell? In what ways are they similar?
9. What is the meaning of the term "dark current" as applied to a photodiode?
10. When a photodiode is illuminated, which of its properties changes and how?
11. What is a phototransistor and how is it different from a photodiode? In what ways are they similar?
12. What similarity and differences with a BJT does a phototransistor have?
13. What advantages does a phototransistor have over a photodiode?

14. Briefly describe the operation of a photoFET in terms of the operation of a regular type of J-FET.
15. What is the operation of a LASCR? How is it similar to the operation of an SCR? How is it dissimilar?
16. What advantages are offered by an LAS (LASCS) over the LASCR?
17. How is the LED different from all the other light-operating devices?
18. What is the mechanism that causes the LED to emit light?
19. What controls the brightness of the LED light output?

2

Circuits

This part of the book covers assorted electronic circuits. We shall discuss basic applications of the many devices introduced in Part 1, together with some of the most common integrated circuits.

In order to understand these electronic circuits, you will study such basic concepts as amplifier fundamentals, oscillators, differential and operational amplifiers, and digital circuits. Besides providing applications for the devices covered in Part 1, these circuits serve as the building blocks for the more complex systems treated in Part 3.

8

Rectifiers and Filters

We have previously seen that most devices need a supply of dc voltage to set up an appropriate operating point in order to operate on time-varying signals. In this chapter we shall examine some of the basic building blocks of a system that converts the readily available 60-cycle ac voltage into a dc voltage. Such a system is called a *power supply*, and it varies in complexity depending on the application. The complete power supply will be taken up in Chap. 16.

The first and most important component of a power supply is called the *recitifer*. It converts or rectifies the alternating current into a unidirectional current that has a time-varying component as well as a dc (or average) value. The time-varying component of the rectifier output, called the *ac ripple*, is undesirable; its effects are minimized by the use of appropriate *filters*. An ideal filter, in this context, is a two-port network that passes direct current but effectively blocks any alternating current. An ideal filter does not exist. We shall apply the term "filter" to any two-port (four-terminal) network that has a smaller ac ripple component at its output terminals than at its input terminals.

8.1 HALF-WAVE RECTIFIER

The half-wave rectifier circuit is shown in Fig. 8-1. The ac input is typically 115 V-rms at 60 cycles. It is usually coupled into the rectifier by a transformer, as shown. The transformer either steps up or brings down the ac voltage, depending on the magnitude of the dc voltage desired. The dot convention on the transformer indicates that if the current at the input is entering the side with the dot, the current at the output is also entering the side with the dot, and vice versa. Furthermore, if the input voltage is such that the terminal with the dot is positive, the voltage at the output terminals will be such that the dotted terminal is positive.

When a sinusoidal ac voltage, such as shown in Fig. 8-2(a), is impressed on the input, a similar ac voltage—either larger or smaller in magnitude—appears at the output terminals of the transformer. During the first half-cycle of the input voltage, a current will flow in the circuit,

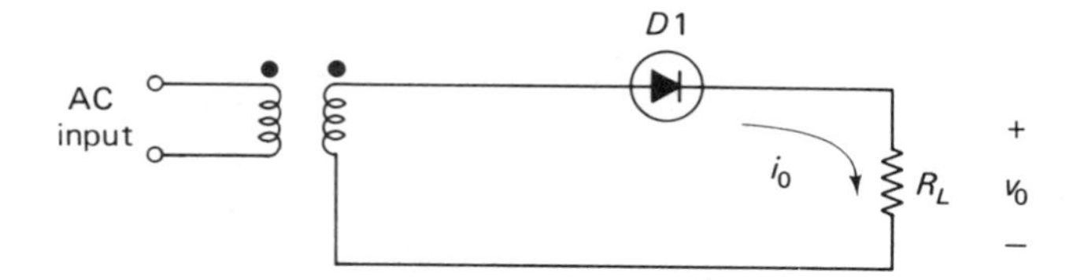

Figure 8-1. Half-wave rectifier.

because the anode of the diode is driven positive with respect to the cathode. There is a small voltage drop across the forward-biased diode, but it can, in most cases, be neglected. During the positive half-cycle, all of the applied voltage appears across the load resistor R_L and almost none across the forward-biased diode. The relationship between voltage and current in R_L is linear. During this time, therefore, the voltage and current waveforms are identical to the input waveform, as illustrated in Figs. 8-2(b) and 8-2(c).

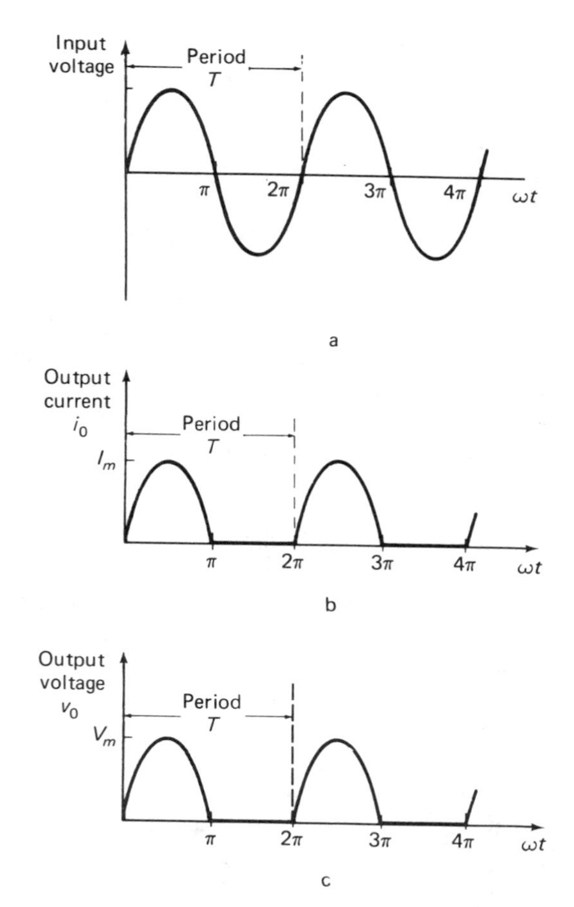

Figure 8-2. Waveforms in the half-wave rectifier circuit.

During the second half-cycle of the input voltage, the polarity is reversed; the anode of the diode is made negative with respect to the cathode. The diode is then reverse biased and acts as an effective open circuit. An extremely small reverse-saturation current does flow. But it is many orders of magnitude smaller than the forward current, so it can be neglected. All the voltage appears across the diode during the negative half-cycle of the input. There is no current through nor any voltage across the load resistor R_L.

As the cycle of the input voltage is repeated, the diode alternately conducts current (during the positive half-cycle) and blocks current (during the negative half-cycle). The current and voltage waveshapes at the load, as shown in Figs. 8-2(b) and 8-2(c), result. The output of a half-wave rectifier is a unidirectional time-varying current, also called *pulsating direct* current. It is also periodic or repetitive. We can express the output current over one period as

$$i_o = \begin{cases} I_m \sin(wt) & \text{for } 0 \leqslant wt \leqslant \pi \\ 0 & \text{for } \pi < wt < 2\pi \end{cases} \tag{8-1}$$

where I_m is the peak value of the current, $w = 2\pi f$, and usually $f = 60$ Hz. The period, T, of the waveshape appears at $wT = 2\pi$, so $T = 1/f$.

The dc or average value of the current is given by the area under the waveshape in one period divided by the period. We can calculate the area under the waveshape for one period by taking the integral of the current over one period, so

$$\begin{aligned} I_{dc} &= \frac{1}{T}\int_0^T i_o(wt)\,d(wt) \\ &= \frac{1}{2\pi}\int_0^{2\pi} i_o(wt)\,d(wt) \end{aligned} \tag{8-2}$$

The above integral can be broken into two parts. The first part is from 0 to π; the second, from π to 2π.

$$\begin{aligned} I_{dc} = &\frac{1}{2\pi}\int_0^{\pi} i_o(wt)\,d(wt) \\ &+ \frac{1}{2\pi}\int_{\pi}^{2\pi} i_o(wt)\,d(wt) \end{aligned} \tag{8-3}$$

However, the current is identically zero during the second half of the period, from π to 2π, so that the second integral in Eq. (8-3) is also zero. Using Eq. (8-1) for i_o, we get

$$\begin{aligned} I_{dc} &= \frac{1}{2\pi}\int_0^{\pi} I_m \sin(wt)\,d(wt) \\ &= \frac{1}{2\pi}\left[-I_m \cos(wt)\right]_0^{\pi} = -\frac{I_m}{2\pi}\left[(-1)+(-1)\right] \\ &= \frac{I_m}{\pi} \end{aligned} \tag{8-4}$$

A measure of the alternating or fluctuating component in the rectified waveshape, called the *ripple factor* (labeled r), is defined as

$$r=\frac{\text{effective (rms) value of alternating component of the wave}}{\text{average (dc) value of the wave}} \tag{8-5}$$

The effective value of a wave is the rms value. Rms stands for root-mean-square. It is determined by taking the square root of the average value of the square of the wave. It is not always easy to determine the rms value of the alternating component of the wave. We label the total instantaneous current by i_T and the ac component of the current by i_{ac}. Then we can write:

$$i_T=i_{ac}+I_{dc} \tag{8-6}$$

or

$$i_{ac}=i_T-I_{dc} \tag{8-7}$$

The effective or rms value of the alternating component of the current, labeled I_{ac}, is given by:

$$I_{ac}=\left[\frac{1}{T}\int_0^T (i_T-I_{dc})^2\,dx\right]^{1/2} \tag{8-8}$$

where T is the period of i_{ac}; and wt has been replaced by x for simplicity. Squaring the term inside the brackets, we have

$$I_{ac}=\left[\frac{1}{T}\int_0^T (i_T^2-2I_{dc}i_T+I_{dc}^2)\,dx\right]^{1/2} \tag{8-9}$$

The single integral in Eq. (8-9) can be broken up into three separate integrals. At the same time we can take I_{dc} outside the integral sine, because it is a constant. Thus,

$$I_{ac}=\left[\frac{1}{T}\int_0^T i_T^2\,dx-2I_{dc}\frac{1}{T}\int_0^T i_T\,dx+I_{dc}^2\frac{1}{T}\int_0^T dx\right]^{1/2} \tag{8-10}$$

The rms value of the total instantaneous current, labeled I_{rms}, is given by:

$$I_{rms}=\left[\frac{1}{T}\int_0^T i_T^2\,dx\right]^{1/2} \tag{8-11}$$

We can then recognize the first term in Eq. (8-10) as I_{rms}^2. Moreover, in the second term of Eq. (8-10)

$$\frac{1}{T}\int_0^T i_T\,dx$$

is the average value of i_T, previously defined as I_{dc}. Making these substitutions in Eq. (8-10), we get:

$$I_{ac}=[I_{rms}^2-2I_{dc}^2+I_{dc}^2]^{1/2}=[I_{rms}^2-I_{dc}^2]^{1/2} \tag{8-12}$$

The preceding expression for the rms value of the alternating component of the current is perfectly valid in general. We now rewrite Eq. (8-5) for the ripple factor, using symbols, and obtain:

$$r = \frac{\sqrt{I_{rms}^2 - I_{dc}^2}}{I_{dc}} = \sqrt{\left(\frac{I_{rms}}{I_{dc}}\right)^2 - 1} \tag{8-13}$$

For the half-wave rectifier circuit we can readily evaluate the rms value of the current from

$$\begin{aligned} I_{rms} &= \left[\frac{1}{2\pi}\int_0^{2\pi} i_0^2\,dx\right]^{1/2} \\ &= \left[\frac{1}{2\pi}\int_0^{\pi} I_m^2 \sin^2 x\,dx\right]^{1/2} \\ &= \left[\frac{1}{2\pi}(I_m^2)\left(\frac{\pi}{2}+0\right)\right]^{1/2} \\ &= \frac{I_m}{2} \end{aligned} \tag{8-14}$$

Using the rms value just calculated and the dc value given in Eq. (8-4), we can calculate the ripple factor for the half-wave rectifier as

$$r = \sqrt{\left(\frac{I_m/2}{I_m/\pi}\right)^2 - 1} = \sqrt{\left(\frac{\pi}{2}\right)^2 - 1} \cong 1.21 \tag{8-15}$$

Thus, the ripple factor for a half-wave rectifier is very high. Ideally, we want the alternating component to be as small as possible or the ripple factor to be as close to zero as possible.

8.2 FULL-WAVE RECTIFIER

In applications requiring low ripple and a high dc component, we find the half-wave rectifier unsuitable because of its rather high ripple factor and low value of the dc component. In such cases, we can obtain a somewhat improved performance from a full-wave rectifier, like the one shown in Fig. 8-3.

In the full-wave rectifier circuit, the ac input is applied to a transformer having a center-tapped secondary. The two-diode circuit functions like a tandem connection of two half-wave rectifier circuits; that is, during the positive half-cycle of the input, one half-wave rectifier is conducting; the other rectifier is off. The operation is reversed during the negative half-cycle of the input.

If we apply a sinusoidal input of the type illustrated in Fig. 8-4 during the positive half-cycle (between 0 and π), diode $D1$ is forward biased and diode $D2$ is reverse biased. Thus, $D1$ passes current, whereas $D2$ is effectively an open circuit. When the input voltage goes negative (between π and 2π), $D1$ is reverse biased and effectively an open, whereas

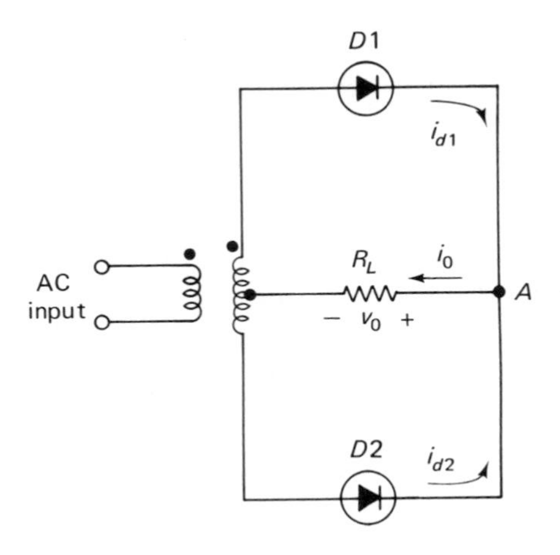

Figure 8-3. Full-wave rectifier.

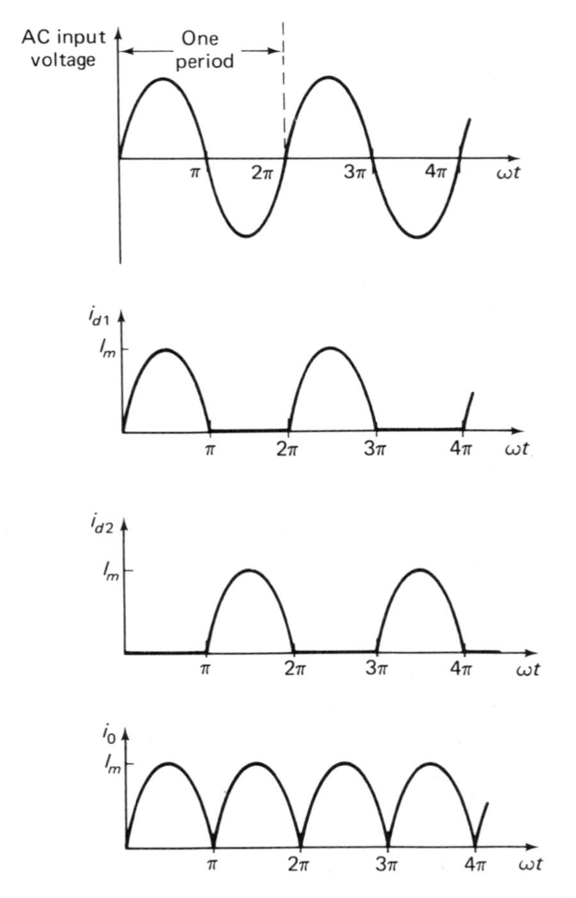

Figure 8-4. Waveforms in the full-wave rectifier circuit of Fig. 8-3.

$D2$ is forward biased and conducts. The output current, i_o, can be determined by adding the currents at node A in Fig. 8-3:

$$i_o = i_{d1} + i_{d2} \tag{8-16}$$

Because i_{d1} flows during the first half-cycle and i_{d2} flows during the second half-cycle, output current flows for the full cycle—thus, the name "full-wave rectifier." The pertinent waveshapes are also shown in Fig. 8-4.

Another form of the full-wave rectifier, called the *bridge rectifier*, is depicted in Fig. 8-5. The bridge rectifier offers some advantages over the two-diode full-wave rectifier. It has a lower reverse voltage per diode, and it eliminates a center-tapped transformer. But it does need four diodes as opposed to two for the regular full-wave rectifier.

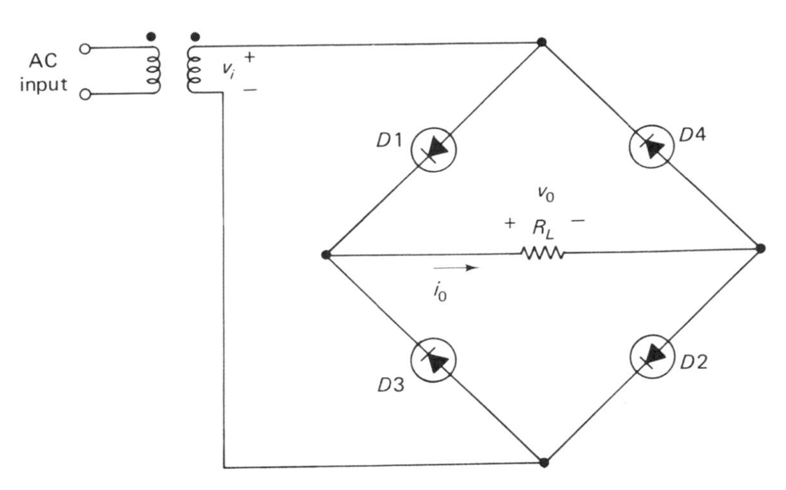

Figure 8-5. Bridge rectifier.

The operation of the bridge rectifier is similar to that of the conventional full-wave rectifier. Diodes operate in pairs. During the first half-cycle of the input while v_i is positive, $D1$ and $D2$ are forward biased, and $D3$ and $D4$ are reverse biased and effectively open. $D1$ and $D2$ provide a current path through the load resistor R_L, as shown in Fig. 8-6. When the input voltage reverses polarity during the second half-cycle, the situation is reversed. $D1$ and $D2$ are reverse biased and effectively open; $D3$ and $D4$ are forward biased and provide a current path through R_L, as shown in Fig. 8-7. The current through R_L, labeled i_o, is in the same direction (in this case, from left to right) during both the negative and positive half-cycles. The pertinent waveshapes for the bridge rectifier are shown in Fig. 8-8.

The load current, i_o, for either of the full-wave rectifier circuits is given by

$$i_o = \begin{cases} I_m \sin x & \text{for } 0 \leqslant x \leqslant \pi \\ -I_m \sin x & \text{for } \pi < x < 2\pi \end{cases} \tag{8-17}$$

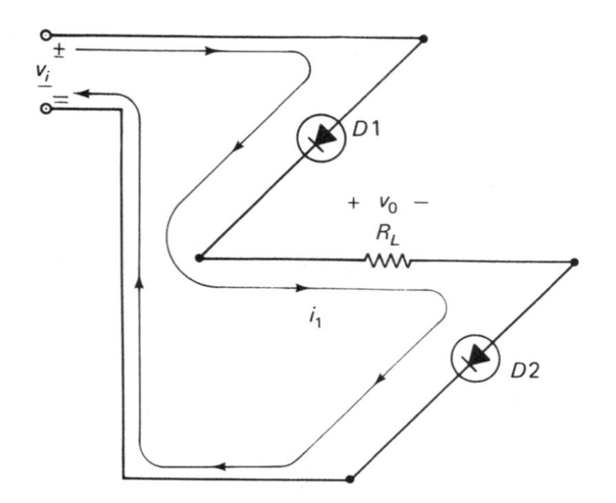

Figure 8-6. Bridge rectifier circuit during the positive half-cycle.

The dc and rms current can be determined by the methods outlined in the previous section. They are given by

$$I_{dc} = \frac{2I_m}{\pi} \tag{8-18}$$

$$I_{rms} = \frac{I_m}{\sqrt{2}} \tag{8-19}$$

Comparing Eqs. (8-4) and (8-18), we can readily see that for the same peak input current, the full-wave rectifier yields a somewhat larger dc output current. Another comparison between the two is afforded by the ripple factor. Using Eq. (8-13), we can obtain the ripple factor for the

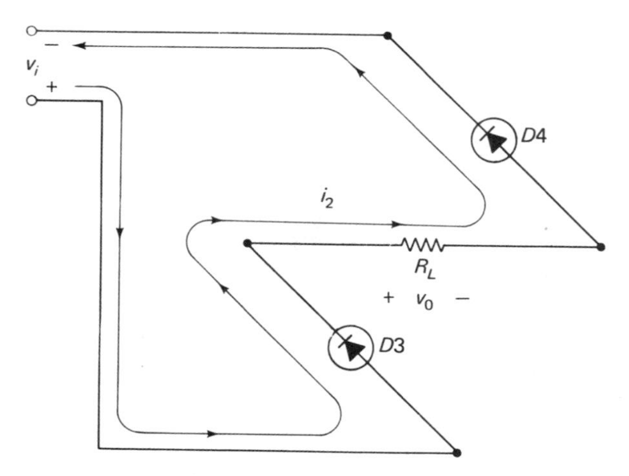

Figure 8-7. Bridge rectifier circuit during the negative half-cycle.

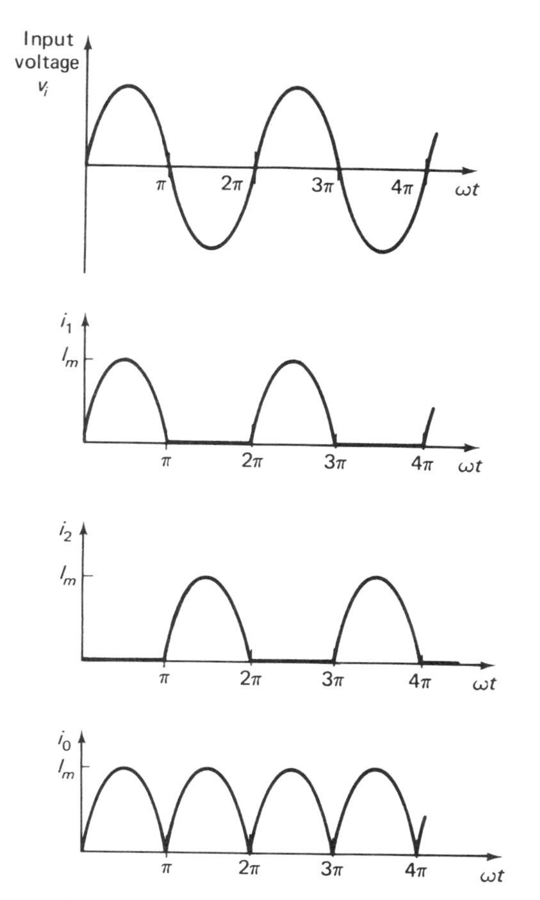

Figure 8-8. Waveshapes for the bridge rectifier circuit.

full-wave rectifier:

$$\begin{aligned} r &= \left[\left(\frac{I_m}{\sqrt{2}} \frac{\pi}{2I_m} \right)^2 - 1 \right]^{1/2} \\ &= \sqrt{1.2337 - 1} \\ &= 0.483 \end{aligned} \tag{8-20}$$

We can readily see that the ripple factor is lower for the full-wave rectifier, so the ac component in the rectified output will also be smaller for the full-wave rectifier than for the half-wave rectifier.

Even though the full-wave rectifier improves the ac component, the ac ripple is too high for most applications. With either a half-wave or a full-wave rectifier, we usually filter the rectified output to reduce the ac component.

8.3 RECTIFIER FILTERS

Many different two-port networks can be and are used to filter the output of a rectifier. No matter how complex or how simple these filters are, their operation is based on the action of certain reactive elements: capacitors, inductors, or both. Two simple single-element filters are shown in Fig. 8-9. In the inductor filter, an inductor is inserted in series with the load. The rectified voltage, containing both dc and ac components, is applied to the inductor-resistor combination. It divides between the two; the inductor having almost zero dc resistance has essentially negligible dc voltage across its terminals. Almost all of the dc voltage appears across the resistor. The impedance of the inductor is extremely high to the ac component of the current, however, so that most of the ac voltage appears across the inductor. We can also view the filtering action of the inductor in another way. Recall that the inductor tends to oppose changes in current, thereby smoothing out the current waveshape. In effect, this process reduces the ac component in the output current.

In the capacitor filter, a capacitor is inserted in parallel with the load resistor. The rectifier output current, containing both direct and alternating components, divides between the capacitor and resistor. The capacitor offers almost infinite impedance to the dc current. Consequently, almost all of the dc current flows through the resistor. The impedance of the capacitor is very low to the ac component of the current, so that a large portion of the alternating component of the current flows through the capacitor. Another way of viewing the filtering action of the capacitor is to remember that the capacitor acts to prevent the voltage across its terminals from changing. Because it is in parallel with the load resistor, the output voltage is smoothed by the action of the capacitor.

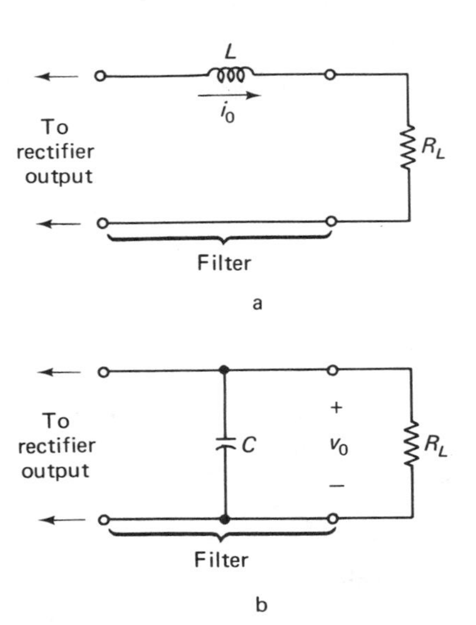

Figure 8-9. Examples of single-element filters: (a) inductor and (b) capacitor.

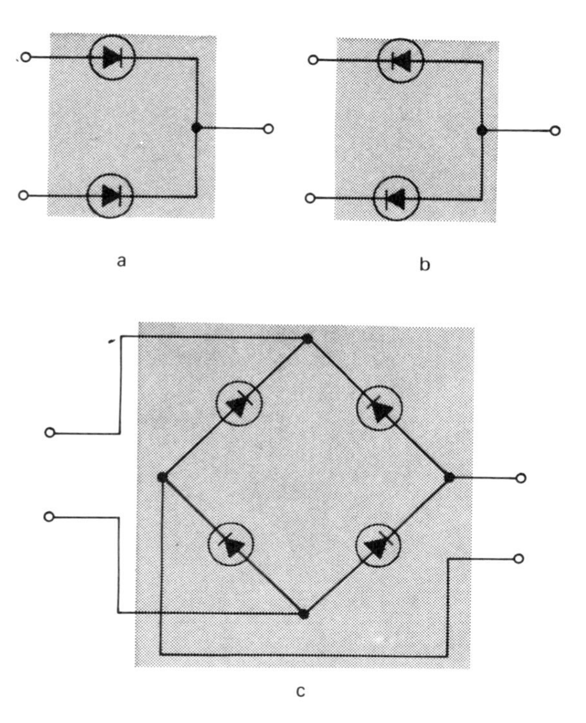

Figure 8-10. Commercially available full-wave rectifier assemblies: (a) for a positive dc voltage, (b) for a negative dc voltage, (c) bridge.

If we use either the inductor or capacitor filter with a full-wave rectifier, we reduce the ripple factor. The availability of bridge and conventional full-wave rectifier assemblies in a single package almost precludes the use of a half-wave rectifier for most applications. The circuits for some commercially available rectifier assemblies are shown in Fig. 8-10.

Combining the action of the choke (inductor) and capacitor in a single filter further improves the performance of the rectifier. Two filters in common use are the L-section and π-section filters, illustrated in Fig. 8-11. In both filters, the choke is employed in series with the load to smooth out the current. The capacitor (or capacitors) is placed in parallel to smooth out the voltage. Pertinent data for the four basic filters are summarized in Table 8-1.

Frequently, the choke that needs to be used in the π-section filter is quite high in value and, therefore, very bulky and expensive. In such cases, the choke may be replaced by a resistor whose resistance is equal to the reactance of the choke. However, this is possible only when the voltage drop across the series resistor (which replaces the choke) is not large and does not represent an unwarranted loss of power.

In some special applications, where the ripple must be reduced further, the filter is made up of multiple L-sections or π-sections.

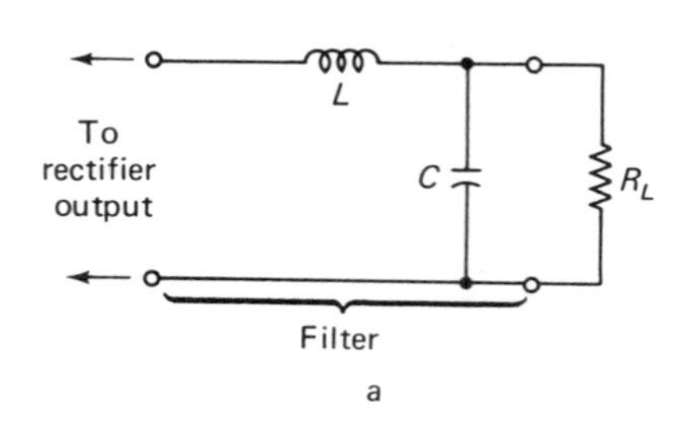

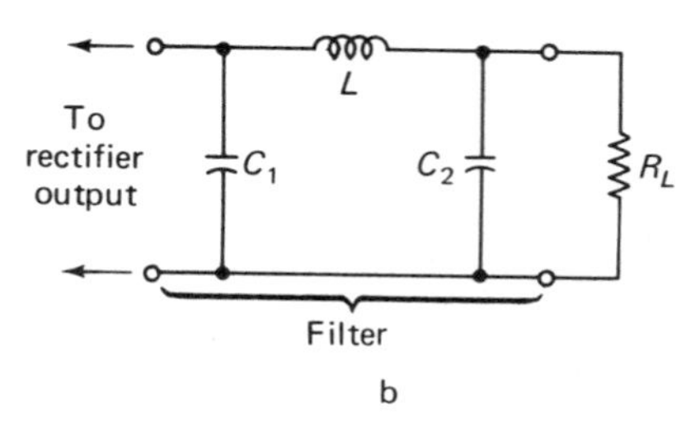

Figure 8-11. Multiple-element filters: (a) *L*-section and (b) π-section.

Table 8-1. Filter performance*

Filter type	*Ripple factor (r)*	*dc voltage (V_{dc})*
Choke (inductor)	$\frac{R_L}{1{,}600L}$	$0.636\ V_m$
Capacitor	$\frac{2{,}400}{R_L C}$	$V_m - \frac{4{,}200 I_{dc}}{C}$
L-section	$\frac{0.83}{LC}$	$0.636\ V_m$
π-section	$\frac{3{,}300}{C_1 R_L C_2 L}$	$V_m - \frac{4{,}200 I_{dc}}{C}$

*For a full-wave rectifier with resistance in Ohms, capacitance in μF, and inductance in Henries. The frequency is assumed to be 60 Hz.

Example 8-1. For the filtered rectifier circuit shown in Fig. 8-12, we want to determine: (1) the ripple factor and the ripple voltage, (2) the diode ratings, and (3) the transformer rating. The output dc voltage is 50 V at a current of 100 mA.

Solution: (1) From Table 8-1 we see that the ripple factor for the π-section filter is:

$$r = \frac{3300}{(20)(R_L)(20)(8)}$$

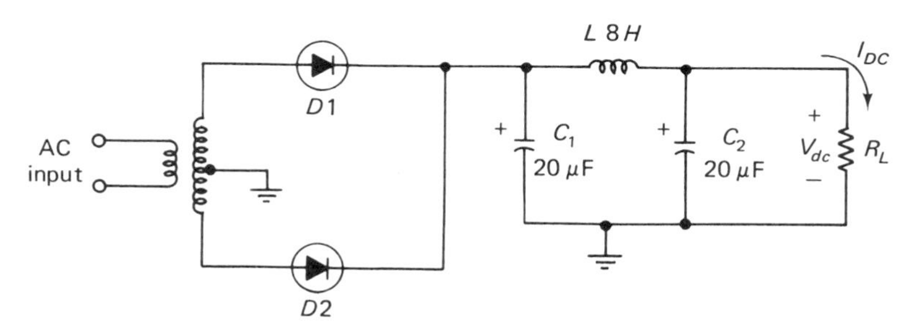

Figure 8-12. Full-wave rectifier with a π-section filter (see Example 8-1).

The load resistance R_L can be determined from the ratio of V_{dc} to I_{dc}: $R_L = 50/0.1 \Omega = 500\ \Omega$. Thus,

$$r = \frac{1.03}{500} \cong 0.002$$

From the definition of the ripple factor, we can determine the rms ripple voltage V_r as

$$V_r = rV_{dc} = (0.002)(50)\ \text{V} \cong 0.1\ \text{V}$$

(2) Each diode must be capable of sustaining a forward current of at least 100 mA and must have a peak-inverse voltage rating in excess of $2V_m$. From Table 8-1 we can determine

$$V_m = V_{dc} + \frac{4200 I_{dc}}{C} \cong 50 + 21 \cong 71\ \text{V}$$

The exact peak forward current that each diode experiences is $V_m IR_L$ or 142 mA. Diodes with a current rating of 150 mA and reverse voltage rating in excess of 150 V may be used. The 1N485 diode meets these requirements (see Appendix 3).

(3) The transformer rating is specified by the secondary voltage and current rating. (Assume that the primary voltage is 115 V-rms.) In this case, a 75 V-0-75 V transformer with a current rating of 500 mA might be used.

It is not possible to replace the choke by an appropriate resistor in the preceding example, because the resistance needed to keep the same ripple factor would be $2\pi fL \cong 3000\ \Omega$. The dc current would fall to about $V_{dc}/3\ \text{k}\Omega$, and the output dc voltage would be almost negligible.

Review Questions

1. What is the role of a rectifier in a dc power supply?
2. What devices are used in rectifiers?
3. What are the basic types of rectifiers?
4. What are the characteristics of a half-wave rectifier? How many diodes are necessary?

5. What is the waveshape output from a half-wave rectifier? Sketch it.
6. Sketch a half-wave rectifier circuit that provides a positive output voltage.
   Sketch a circuit for a negative output voltage.
7. How does a half-wave rectifier operate? Discuss in terms of the conduction and nonconduction of the diode for a sinusoidal input.
8. What is meant by *ripple*?
9. What provides a numerical measure of the ripple?
10. What are the different types of full-wave rectifiers?
11. In what ways are half-wave and full-wave rectifiers similar? How are they different?
12. What advantages does a full-wave rectifier offer over a half-wave rectifier?
13. At what cost does the full-wave rectifier offer advantages over a half-wave rectifier?
14. What is the difference between a full-wave rectifier and a bridge rectifier?
    What are the similarities?
15. What are the advantages of a bridge rectifier over a full-wave rectifier? What are the disadvantages?
16. Describe the operation of a full-wave rectifier (Fig. 8-3) in terms of the individual diodes.
17. What is the role of a filter in a dc power supply?
18. What basic properties of inductors and capacitor are utilized in filters?
19. What is the action of a capacitor filter?
20. What is the action of an inductor filter?
21. What are some other commonly used filters?
22. What are the advantages of multiple-element filters over the simple capacitor filter? What are the disadvantages?
23. What is the effect of any filter on the ripple?
24. What are the elements of an L-section filter?
25. What are the elements of a π-section filter?

Problems

1. The half-wave rectifier circuit shown in Fig. 8-1 has an input that is 115 V-rms. It is used to drive a 100 Ω load. Neglecting the voltage drop across the diode, determine the dc output current and the ripple.
2. The full-wave rectifier circuit shown in Fig. 8-3 has a center-tapped transformer delivering 50 V-rms from each end to the tap. The load resistance is 30 Ω. Determine the output dc voltage and the ripple factor.

3. Repeat Problem 2 with an input of 50 V-rms across the complete secondary of the transformer.
4. The transformer delivers 10 V-rms to the bridge rectifier circuit shown in Fig. 8-5. Use silicon diodes with a forward drop of 1 V and a 20 Ω load. Determine the dc output current and the ripple.
5. Repeat Problem 4, neglecting the diode forward drop. In terms of percentage, how much error is introduced?
6. The input to an *L*-section filter is a full-wave rectified voltage, 45 V peak. Determine the ripple factor if the inductor is 1 henry (H) and the capacitor is 100 microfarads (μF).
7. The output of the rectifier in Problem 2 is applied to a capacitor filter, with a dc load of 100 Ω. Determine the capacitor needed to produce a ripple factor of 0.01. For this value of capacitor, determine the output dc voltage.
8. The input to a π-section filter is a full-wave rectified voltage with a 50 V peak. The load is 1 kΩ. An 8 H inductor is available. Determine the capacitors needed to produce a ripple of 0.001. (Assume $C_1 = C_2$.)
9. A π-section filter driving a 100 Ω load produces a ripple factor of 0.05. Determine the ripple factor if the load is: (a) 50 Ω and (b) 200 Ω.
10. A 5000 μF capacitor is placed at the output of a full-wave rectifier with a voltage of 8 V peak. Determine the load resistance if the output dc voltage of 6 V is measured. Under these circumstances, what is the ripple factor?

9 Amplifier Fundamentals

One of the most basic and common blocks used in electronic systems is the amplifier. As the name implies, an amplifier is a circuit which raises the level of a signal. A small signal at the input of the amplifier is processed to provide a signal at the output which is identical to the input signal in all aspects except that it is larger in magnitude.

The complexity and type of circuit used in the amplifier is dictated by the type of signal to be amplified and by the amount of amplification (called *gain*) needed. Many different classifications are possible: current amplifiers, voltage amplifiers, power amplifiers (where both current and voltage are amplified), dc amplifiers, ac amplifiers (where time-varying signals are amplified), as well as numerous other designations. However, all amplifiers have certain properties in common. (1) They are to amplify; therefore, they all utilize at least one *active* device. (2) They all require a dc supply. (3) Their operation can be summarized by specifying the gain, input impedance, output impedance, and frequency response.

The gain is a measure of the amount of amplification. It is defined as the numerical ratio of the output signal to the input signal. We can refer to a current gain (the ratio of output current to input current), or a voltage gain (the ratio of output voltage to input voltage), or a power gain (the ratio of output power to input power).

The input impedance of an amplifier is the impedance seen by the source of the signal to be amplified. It is defined as the ratio of voltage to current at the input terminals of the amplifier. The output impedance of an amplifier is the impedance seen by the load, which is the recipient of the amplified signal. It is defined as the ratio of voltage to current at the output terminals of the amplifier. Both the input and output impedance may be *real*, i.e., purely resistive. On the other hand, they may be *complex*; that is, have capacitive and inductive components as well as resistive ones.

The frequency response of an amplifier is closely related to the gain. Usually the gain is not constant over all possible frequencies; that is, signals of different frequencies are not amplified by the same amount.

Therefore, when specifying the gain, we usually specify the range of frequencies (called the *bandwidth*) over which the gain is essentially constant. The bandwidth may be expressed in terms of frequency. For example, an audio amplifier may have a bandwidth of 20 kHz. The bandwidth may also be specified in terms of the lower and upper limits, called the lower and upper *cutoff frequencies*. In the case of the audio amplifier, these limits may be specified as a certain gain from 30 Hz to 20 kHz.

We can summarize our introduction of amplifiers by saying that our analysis will concern the gain (either current, voltage, or power), the input and output impedance, and the frequency response.

9.1 GAIN CALCULATIONS —SYSTEMATIC ANALYSIS

In the broadest sense, we can distinguish between two basic types of amplifiers according to the method of analysis that we find most suitable. In the previous discussion, we assumed that the gain of the amplifier did not depend on the amplitude of the input signal. If this is indeed the case, the amplifier properties can be determined by using the small-signal equivalent circuit or model of the active devices in the amplifier and then applying circuit analysis. This type of amplifier is said to be *linear* or to have a *linear gain*. For example, if the amplifier gain is 100 and the input voltage is 10 mV, the output voltage will be 100×10 mV, or 1 V. In the same amplifier, if the input is 20 mV, the output will still be 100 times larger, or 2 V. This type of amplifier is called a *small-signal amplifier*.

In the second type of amplifier, called *large-signal* or *power* amplifier, the input signal is so large that the device (or devices) inside the amplifier does not operate in a linear fashion. For example, the same amplifier that gives an output signal of 5 A for an input signal of 0.2 A may give an output of 6 A for an input of 0.5 A. (The current gains are 25 and 12, respectively.) The large-signal or power amplifier is also characterized by varying degrees of *distortion*. Distortion is an output-signal waveshape which is not a true reproduction of the input-signal waveshape. Graphical analysis is best suited to this kind of amplifier. We shall treat it in more detail in Chap. 12.

There is no clear-cut line separating small-signal and large-signal amplifiers. In general, amplifiers having an output power level up to a few hundred milliwatts may be considered small-signal amplifiers and analyzed as such. With output power levels around a watt or higher, graphical analysis should be used. However, consider these power levels as only the first steps in determining the type of analysis to be used. In the long run, the only valid analysis is the one that gives verifiable answers. Therefore, you must examine the characteristics of the devices used in the amplifier and ascertain whether the input signal is large enough to cause nonlinear operation. The test is not the characteristics of the devices alone nor the power or signal level alone; it is a combination of the two.

Let us assume for the moment that we are dealing with an amplifier operating under small-signal conditions. We then proceed with a systematic analysis as follows:

Step 1. Draw the schematic of the complete amplifier with the appropriate symbols for the active devices.

Step 2. Label all device terminals with a letter and number. For example, B for base, D for drain. If there are two or more of one type of device, use numbers as well. If there are two or more BJTs, use $B1$, $E1$, and $C1$ for the first, $B2$, $E2$, and $C2$ for the second, and so on.

Step 3. Redraw the schematic, keeping the relative positions of the components as close to the original as possible, with the following changes:
- (a) Replace all device symbols by the device model (or equivalent circuit).
- (b) Replace all dc supplies by their equivalent impedance. In most cases, this equivalent impedance can be considered zero. Therefore, replace all dc supplies by a short circuit.

Step 4. Determine the quantity of interest (i.e., gain) by the application of conventional circuit analysis. At this point, the circuit consists of ac voltage and current generators, capacitors, inductors, and resistors.

The procedure outlined here is quite valid; however, its implementation requires thought. The first stumbling block is determining which model or equivalent circuit for the device to use in Step 3(a). The answer is to use the model that is valid for the calculation to be made. For example, the model of the base-emitter part of the BJT used the calculation of the dc operating point in Fig. 4-4 was a dc battery. Obviously such a model is inappropriate for ac analysis. But which form of which ac model to use for the BJT in a particular circuit may not be so obvious. The first step in this case is to decide whether the BJT is connected in the *CE*, *CB*, or *CC* configuration. The model is then chosen from among those depicted in Figs. 4-13, 4-15, and 4-16. The specific form of the model to be used depends on the accuracy desired and the quantity desired. If, for example, we consider the *CE* configuration, we can choose among the three models in Fig. 4-13. The model in Fig. 4-13(a) is the most accurate, whereas the one in Fig. 4-13(c) is the least accurate and Fig. 4-13(b) somewhere in between.

In the final application of circuit analysis specified in Step 4, it is not always clear which way or method leads to the answer most directly. There is no best way. You may use many different methods to obtain the solution. The preferable method is one that is the shortest. However, you should begin by using the methods you are most familiar with and with which you feel the most comfortable. The following sections will be devoted to the application of the procedure outlined in the analysis of single-stage BJT and FET amplifiers.

9.2 SINGLE-STAGE BJT AMPLIFIER

A single-stage BJT (common-emitter) amplifier is shown in Fig. 9-1. Before we illustrate how this simple amplifier may be analyzed, let us discuss the reasons for the circuit components in the amplifier.

Resistors R_1, R_2, R_E, and R_C are bias resistors needed to set up an appropriate operating point as discussed in Chap. 4. Capacitors C_1 and C_2 are called *coupling capacitors*. Their role is to block direct current, so that the dc performance of the amplifier will not be disturbed by the source or the load. They also prevent direct current from entering the source and load. The capacitor C_B is called a *bypass* capacitor because it is usually chosen large enough to effectively short-out R_E at signal frequencies.

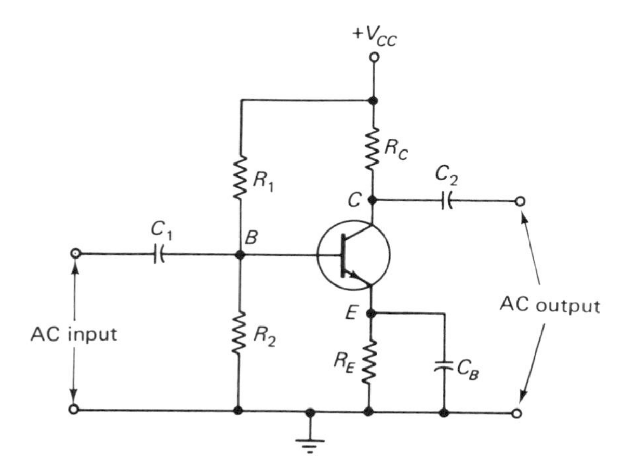

Figure 9-1. Single-stage CE BJT amplifier.

We can proceed with the steps as outlined, noting that the first two steps are already incorporated into Fig. 9-1. The results of Step 3 are shown in Fig. 9-2. If we replace the supply V_{CC} by a short, we place R_2 in parallel with R_1 and put R_C from collector to ground, as indicated in Fig. 9-2(a). The transistor symbol is replaced by the hybrid parameter equivalent circuit of Fig. 4-13(b). This model for the transistor is valid only at low and medium frequencies, not at high frequencies. Thus, Fig. 9-2(a) is an equivalent circuit for the amplifier shown in Fig. 9-1 at low frequencies (because the coupling and bypass capacitors are included). The coupling and bypass capacitors are large, usually in the microFarad range, so that at some frequencies, their impedance approaches a short. The equivalent circuit of the amplifier then becomes the one shown in Fig. 9-2(b), with C_1 and C_2, as well as C_B, replaced by short circuits. Note that this equivalent circuit contains no reactive elements, and it will yield a gain which is

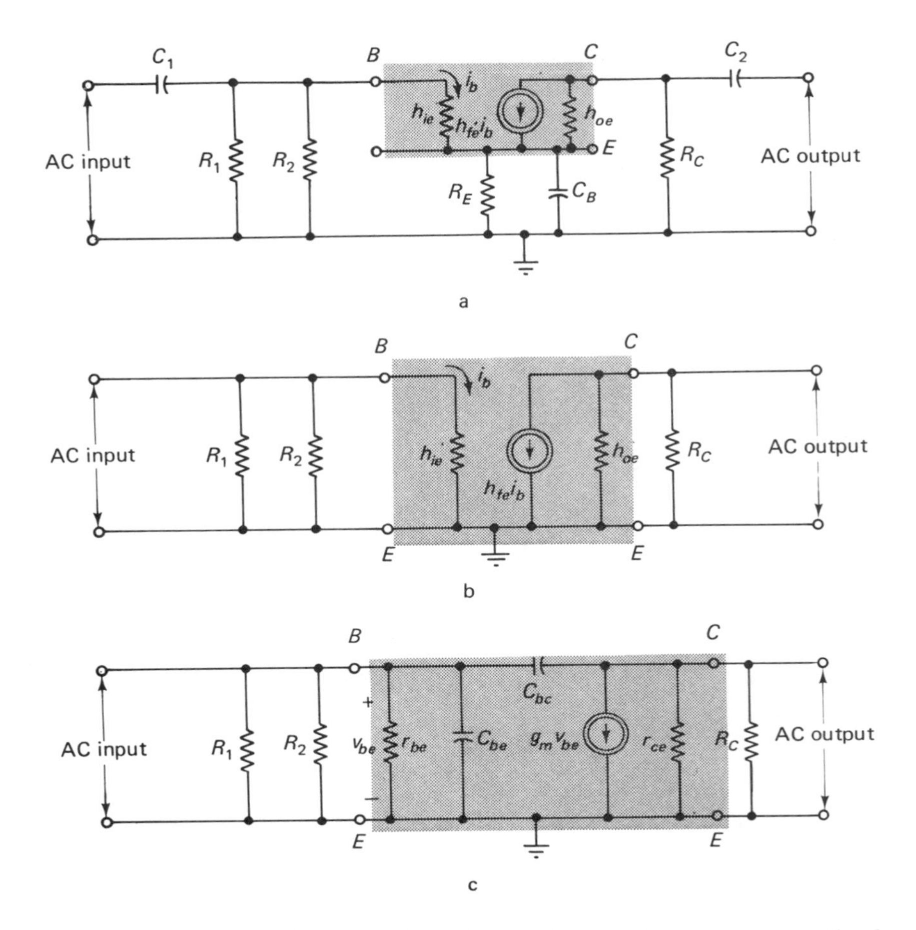

Figure 9-2. BJT amplifier equivalent circuits: (a) for low frequencies, (b) for midband, and (c) for high frequencies.

independent of frequency. The range of frequencies over which the equivalent circuit of Fig. 9-2(b) is valid is called the *midband region*, and the gain over these frequencies is called the *midband gain*. At some higher frequencies (perhaps in the megahertz range, depending on the transistor), the hybrid parameter model is no longer valid. Then the hybrid-π model must be used. This model yields the high-frequency equivalent circuit of the amplifier, shown in Fig. 9-2(c).

We now have three possible equivalent circuits for the amplifier illustrated in Fig. 9-1. Each equivalent circuit is valid and may be used to predict the properties of the amplifier. For example, if we want to determine the amplifier gain, we should use the midband equivalent circuit shown in Fig. 9-2(b). We can then use the equivalent circuit shown in Fig. 9-2(a) to determine the frequency at which the gain begins to fall off, i.e., the lower limit of the midband region. Similarly, we can use the equivalent circuit of Fig. 9-2(c) to predict the upper limit for the midband region.

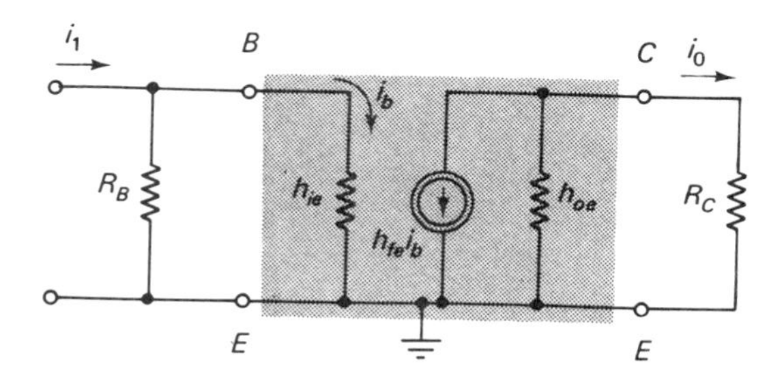

Figure 9-3. BJT amplifier simplified equivalent circuit.

Let us use the midband equivalent circuit of Fig. 9-2(b) to determine the midband gain. Combining R_1 and R_2 and labeling their parallel combination as R_B, we come up with the equivalent circuit shown in Fig. 9-3. We can now implement Step 4 and find the current gain of the amplifier. We first define the input current i_1 as shown and the output current i_o as shown. The current gain, labeled A_I, is then the ratio of i_o to i_1. The simplest way to evaluate the current gain is to express both i_1 and i_o in terms of i_b. At the input we have current division of i_1 between R_B and h_{ie}. Thus,

$$i_b = i_1\left(\frac{R_B}{R_B + h_{ie}}\right) \tag{9-1}$$

At the output we have current division between R_C and the resistance $1/h_{oe}$. Thus,

$$i_o = (-h_{fe}i_b)\left[\frac{\dfrac{1}{h_{oe}}}{\dfrac{1}{h_{oe}} + R_C}\right] \tag{9-2}$$

The minus sign is due to the direction of the $h_{fe}i_b$ current generator. Equation (9-2) can be rearranged by multiplying the numerator and denominator by h_{oe} to obtain:

$$i_o = i_b\frac{-h_{fe}}{1 + h_{oe}R_C} \tag{9-3}$$

We can now substitute Eq. (9-1) for i_b into Eq. (9-3):

$$i_o = i_1\left(\frac{R_B}{R_B + h_{ie}}\right)\left(\frac{-h_{fe}}{1 + h_{oe}R_C}\right) \tag{9-4}$$

Dividing both sides of the preceding equation by i_1 yields the desired result—the midband current gain:

$$A_I = \frac{i_o}{i_1} = \frac{-h_{fe}R_B}{(R_B + h_{ie})(1 + h_{oe}R_C)} \tag{9-5}$$

Equation (9-5) gives the midband current gain of tl amplifier in Fig. 9-1 in general form. For a specific case, to get a numerical answer for the midband gain, we simply need to insert the resistor values together with the transistor parameter values into the same equation.

Example 9-1. A BJT amplifier like the one in Fig. 9-1 is constructed with the following resistor values: $R_1 = 150\ \text{k}\Omega$, $R_2 = 100\ \text{k}\Omega$, $R_E = 1.5\ \text{k}\Omega$, and $R_C = 4.7\ \text{k}\Omega$. It has been determined that the transistor parameters at the resulting operating point are: $h_{ie} = 1\ \text{k}\Omega$; h_{re} is negligibly small; $h_{fe} = 80$; $h_{oe} = 20\ \mu\text{mho}$. We want to determine the midband current gain.

Solution: Before we can use Eq. (9-5), we need to determine R_B:

$$R_b = \frac{R_1 R_2}{R_1 + R_2} = \frac{(150)(100)}{150 + 100}\,\text{k}\Omega = 60\ \text{k}\Omega$$

Now, making use of Eq. (9-5), we find that

$$A_I = \frac{(-80)(60)}{(60+1)[1+(0.02)(4.7)]} \cong -72$$

You can gain additional insight into determining the gain if you examine how we obtained the gain expression in Eq. (9-5), considering the typical values in the preceding example. When obtaining Eq. (9-1), note that if R_B is very large compared to h_{ie} (which it usually is), we can to a good approximation neglect the effect of R_B. So Eq. (9-1) becomes

$$i_1 \cong i_b$$

The same kind of approximation can be made in Eq. (9-2); that is, the resistance corresponding to $1/h_{oe}$ may be much larger than R_C and, as such, may be neglected. We would then have

$$i_o \cong -h_{fe} i_b$$

The midband current gain can then be approximated by

$$A_I \cong -h_{fe} \tag{9-6}$$

In general, therefore, if one resistor is at least ten times larger than another which is in parallel with it, we can to a good approximation (no worse than 10% error) neglect the effect of the larger resistor.

9.3 SINGLE-STAGE FET AMPLIFIER

A single-stage FET amplifier is shown in Fig. 9-4. The basic similarity between the FET and BJT amplifiers is obvious. The procedure in determining the properties of the FET amplifier is essentially the same as for the BJT amplifier.

We can redraw the circuit, replacing the FET symbol by its low-frequency equivalent circuit, as shown in Fig. 9-5(a). This equivalent circuit, containing the coupling and bypass capacitors, is valid at low frequencies. As in the case of the BJT amplifier, we can replace all coupling and bypass capacitors by short circuits to obtain the midband equivalent circuit for the FET amplifier, as indicated in Fig. 9-5(b). For

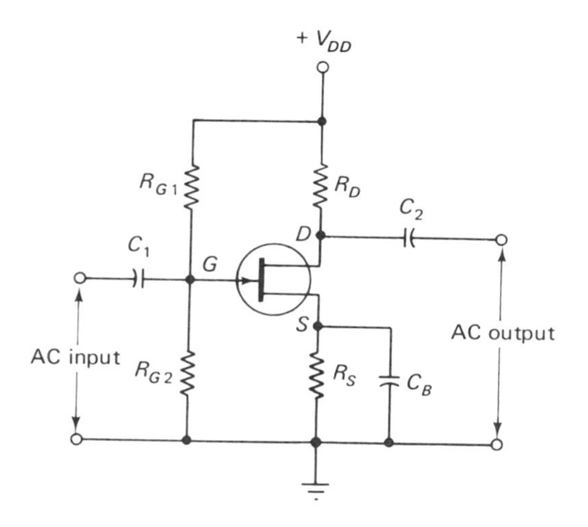

Figure 9-4. Single-stage FET amplifier.

completeness, the high-frequency equivalent circuit for the FET amplifier is illustrated in Fig. 9-5(c). It is distinguished by the use of the high-frequency model for the FET.

Let us now proceed with the analysis of the FET amplifier, i.e., the determination of the midband voltage gain. The midband equivalent circuit of Fig. 9-5(b) is redrawn, combining resistors R_{G1} and R_{G2} into a single resistor: R_G. In addition, the input and output voltage polarities, v_1 and v_o respectively, are defined. These steps are shown in Fig. 9-6. In this figure, we see that

$$v_1 = v_{gs} \tag{9-7}$$

At the output, the total resistance, R_t, is the parallel combination of r_d and R_D, so

$$R_t = \frac{r_d R_D}{r_d + R_D} \tag{9-8}$$

The output voltage v_o is then given by the product of the current and total resistance:

$$v_o = -g_m v_{gs} R_t \tag{9-9}$$

The minus sign is due to the direction of the current generator. The voltage gain A_v is defined as the ratio of the output to input voltage. Thus, by making use of Eq. (9-7) in Eq. (9-9), we can obtain:

$$A_V = -g_m R_t \tag{9-10}$$

where R_t is defined in Eq. (9-8). As was the case with the BJT amplifier, the gain expression obtained here is general. In a specific case, to obtain a numerical solution for the midband gain, we need only insert proper values into Eq. (9-10).

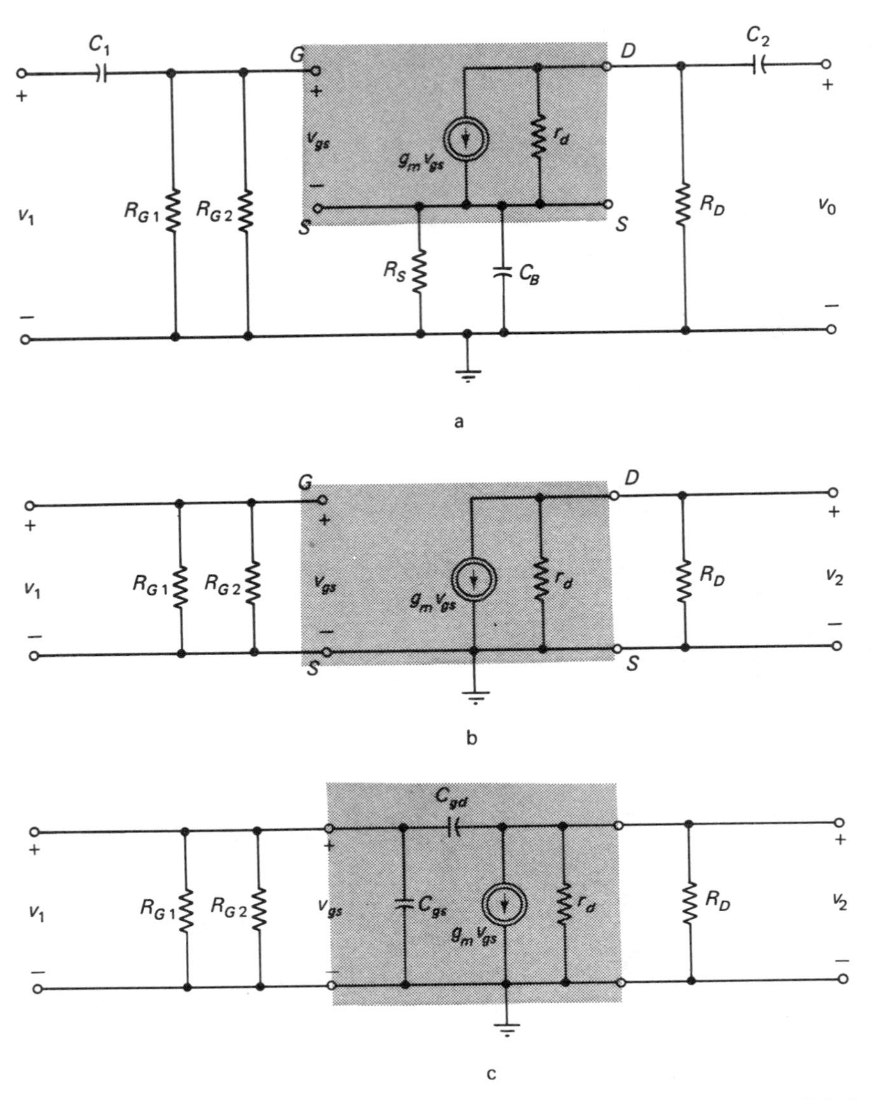

Figure 9-5. FET amplifier equivalent circuits: (a) for low frequencies, (b) for midband, and (c) for high frequencies.

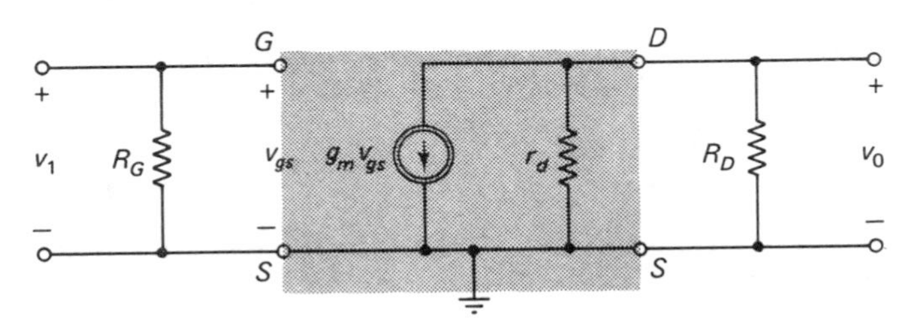

Figure 9-6. Simplified FET amplifier equivalent circuit.

Example 9-2. An FET amplifier, as illustrated in Fig. 9-4, is constructed with resistor values: $R_{G1}=R_{G2}=1$ MΩ, $R_S=5.6$ kΩ, and $R_D=6.8$ kΩ. At the resulting operating point, the FET parameters have been determined as $g_m=0.002$Mho and $r_d=160$ kΩ. We want to calculate the midband voltage gain of the amplifier.

Solution: We first find R_t. (Note that r_d is much larger than R_D, so we could neglect r_d and approximate the total resistance as essentially R_D.)

$$R_t=\frac{(160)(6.8)}{160+6.8}\text{ k}\Omega=6.52\text{ k}\Omega$$

Making use of Eq. (9-10), we obtain the midband voltage gain:

$$A_V=-(2)(6.52)=-13$$

Had we neglected r_d, we would have come out with a slightly greater gain, i.e., -13.6.

We can use the procedures outlined and illustrated here to evaluate the current or voltage gain of any small-signal amplifier stage, no matter how many stages there are or how complete the circuit is. The same general approach can also be followed to evaluate the current and voltage gain at either high or low frequencies. Just use the appropriate model of the device (BJT or FET) under consideration. The only difference is the complexity of the analysis.

You must be careful, however, not to try blindly applying set formulas that may appear elsewhere. There is nothing wrong with using available formulas, but you may not interpret the symbols correctly. Until you have a certain amount of experience, it is much safer to follow the procedures outlined in the previous sections.

9.4 FREQUENCY RESPONSE

Any amplifier provides gain over a certain range of frequencies. For a particular amplifier, the specific range of frequencies over which it can provide gain is a function of two quantities: (1) the response of the device (or devices) used in the amplifier and (2) the response forced by the choice of reactive elements used in conjunction with the amplifier.

The simple single-stage amplifier illustrated in Fig. 9-1 has three distinct frequency ranges: the midband (already discussed), the high, and the low. These ranges have definite boundaries: The gain begins to decrease from its midband value as the frequency is either increased (high-frequency end) or decreased (low-frequency end).

The high-frequency response is determined from the high-frequency equivalent circuit, which is obtained by following the rules just outlined and using the high-frequency model for the transistor (see Fig. 4-17). The high-frequency equivalent circuit is that shown in Fig. 9-2(c). Analysis of

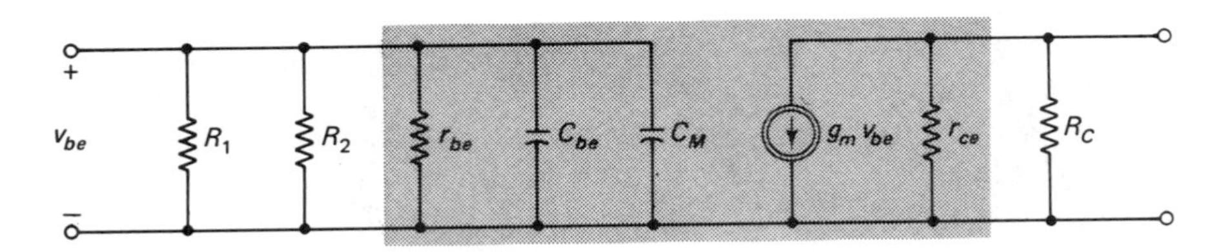

Figure 9-7. High-frequency equivalent circuit for BJT amplifier in Fig. 9-1.

this circuit gives us the following results:*

$$A_{i(hf)} \cong \frac{A_{i(\text{mid})}}{1 + j\dfrac{f}{f_2}} \tag{9-11}$$

where $A_{i(hf)}$ stands for the high-frequency current gain; $A_{i(\text{mid})}$ is the midband current gain: f is the frequency and can take on any value; f_2 is the upper cutoff frequency determined from the following relationship:

$$f_2 = \frac{1}{2\pi R_t C_t} \tag{9-12}$$

and R_t and C_t are the circuit equivalent resistance and capacitance calculated from the high-frequency equivalent circuit. In Fig. 9-2(c), the effect of C_{bc} may be included in the input circuit, as shown in Fig. 9-7, by modifying the capacitance to a value C_{bc} times the voltage gain from base to collector without C_{bc} in the circuit. Thus, this equivalent capacitance, called the *Miller capacitance*, is given by:

$$C_M \cong g_m R_L C_{bc} \tag{9-13}$$

where R_L is the total resistance at the output (r_{ce} in parallel with R_C). Observe, therefore, that the total equivalent capacitance at the input is the parallel combination of C_{be} and C_M:

$$C_t \cong C_{be} + g_m R_L C_{bc} \tag{9-14}$$

Similarly, the total equivalent resistance at the input is the parallel combination of R_1, R_2, and r_{be}:

$$\frac{1}{R_t} = \frac{1}{R_1} + \frac{1}{R_2} + \frac{1}{r_{be}} \tag{9-15}$$

The midband gain can be written in terms of the hybrid-π parameters:

$$A_{i(\text{mid})} = -g_m \frac{R_L R_t}{R_C} \tag{9-16}$$

This expression is exactly equivalent to the current gain expression obtained in Eq. (9-5). The equivalence may be shown by making the

*For a more detailed discussion, see: *Cirovic, Semiconductors*, pp. 220–229.

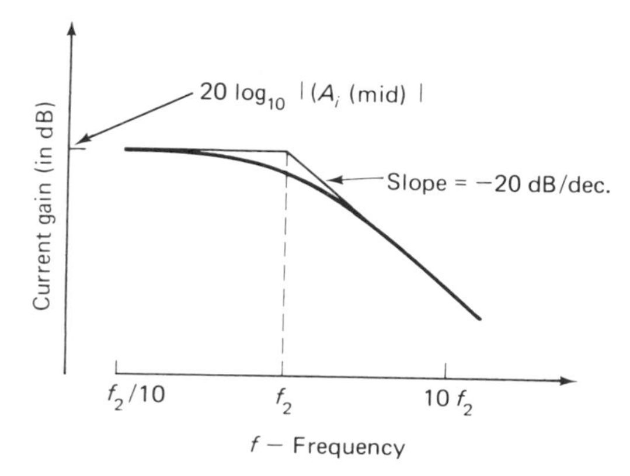

Figure 9-8. Magnitude response for the current gain.

appropriate substitutions from the hybrid-π parameters to h-parameters. In the case where r_{ce} is at least 10 times larger than R_C, r_{ce} may be neglected and $R_L = R_C$. We can then simplify the midband current gain expression to:

$$A_{i(\text{mid})} = -g_m R_t \tag{9-17}$$

A plot of the *hf* current gain given by Eq. (9-11) can be made if we allow the variable f to take on different values. This plot usually has two parts: (1) the magnitude of the gain and (2) the phase shift of the gain expression. Figure 9-8 shows the magnitude response as a function of frequency. The gain is seen to be essentially constant until the upper cutoff frequency f_2. Then it falls off at a rate of 20 dB per decade.

Let us explain these two units of measure for the magnitude of the response and its plots. The gain is usually measured in decibels (dB). Decibels measure the power gain, but the term may be applied to either a voltage or current gain. The number of decibels is determined from the following equation:

$$dB = 20 \log_{10} \frac{I_o}{I_{in}} \tag{9-18}$$

Instead of currents here, we could use voltages. The magnitude plot is made on semilogarithmic graph paper. This paper has one linear scale, on which the number of dB is plotted. The other scale is logarithmic; that is, the distance from 1 to 10 is the same as the distance from 10 to 100. We use this scale to plot the frequency. Such a graph is shown in Fig. 9-8. Because the frequency scale is logarithmic, the basic unit is a tenfold change in frequency, called a *decade*.

At the upper cutoff frequency ($f = f_2$), the magnitude of the *hf* current gain is equal to $0.707 A_{i(\text{mid})}$. In terms of dB (decibels), this figure corresponds to a value which is 3 dB below the midband dB gain. Thus, the upper cutoff frequency is also referred to as the upper 3 dB frequency.

Example 9-3. Assume that the amplifier shown in Fig. 9-1 has the following circuit values: $R_1 = R_2 = 47\ \text{k}\Omega$ and $R_C = 2.2\ \text{k}\Omega$. Transistor parameters are: $r_{be} = 1\ \text{k}\Omega$, $g_m = 50$ milliMhos, $r_{ce} = 60\ \text{k}\Omega$, $C_{be} = 200$ pF, and $C_{bc} = 10$ pF. We want to determine the upper cutoff frequency and sketch the current gain magnitude response.

Solution: First, we must decide whether or not any approximations can be made. Because R_1 and R_2 constitute a parallel resistance of over 20 kΩ, they may be neglected as compared to r_{be}. Similarly, because r_{ce} is more than 10 times larger than R_C, we may neglect r_{ce}. The midband current gain is now calculated:

$$A_{i(\text{mid})} = -(50)(1) = -50$$

where we have approximated R_t by r_{be}. The dB value of $A_{i(\text{mid})}$ is obtained next as:

$$20\log|(-50)| = 20\log(50) \cong 20(1.7) \cong 34\,\text{dB}$$

The total equivalent capacitance at the input is calculated from Eq. (9-13):

$$C_t = 200 + (50)(2.2)(10) \cong 1{,}300\ \text{pF}$$

The upper cutoff frequency is determined from Eq. (9-12):

$$f_2 = \frac{1}{2\pi(10^3)(1.3\times10^{-9})}\ \text{Hz} \cong 122\ \text{kHz}$$

The sketch of the magnitude response is shown in Fig. 9-9. Note that if the gain falls at 20 dB/decade after 122 kHz, it will be 14 dB at 1.22 MHz and -6 dB at 12.2 MHz. A negative dB value means that the gain is less than 1. Moreover, at 122 kHz, the gain is 3 dB down from its midband value of 34 dB, that is, the gain is at 31 dB.

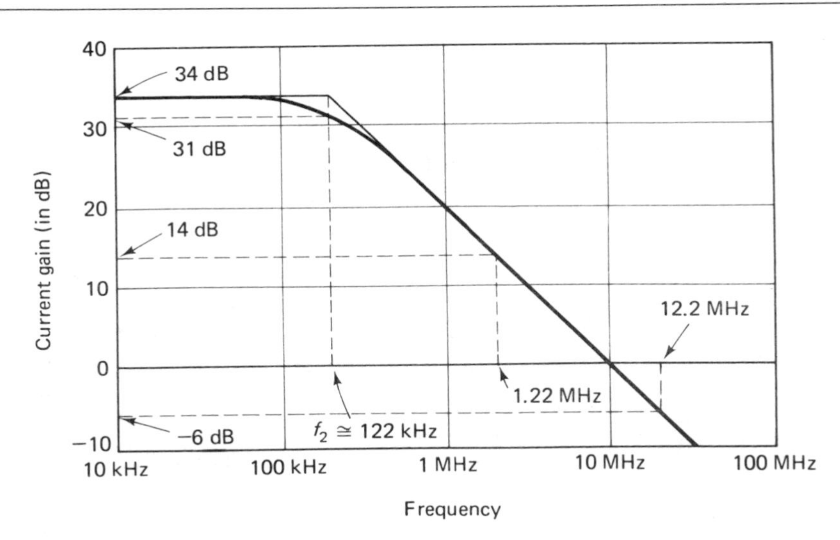

Figure 9-9. Magnitude response for the amplifier in Example 9-3.

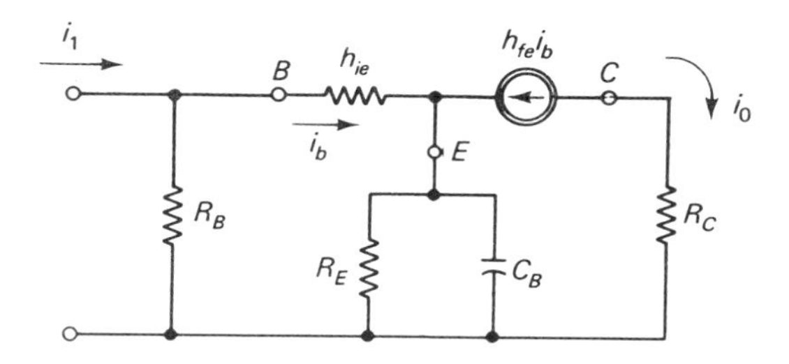

Figure 9-10. Approximate low-frequency equivalent circuit for the amplifier in Fig. 9-1.

We can find the high-frequency response for an FET amplifier in an analogous manner by using the FET high-frequency model.

As was stated earlier, the low-frequency response of the amplifier is governed by the coupling and bypass capacitors. For typical resistor and transistor parameter values, even for a relatively small coupling capacitor value (say, 1 μF), the lower cutoff frequency is determined by the bypass capacitor. If we assume that the coupling capacitors are shorted, the low-frequency equivalent circuit appears as shown in Fig. 9-10. The low-frequency current gain is then given by:

$$A_{i(lf)} \cong \frac{A_{i(\text{mid})}\dfrac{jf}{f_1}}{1+j\dfrac{f}{f_1}} \tag{9-19}$$

where the lower cutoff frequency f_1 is approximately given by:

$$f_1 = \frac{h_{fe}}{2\pi(R_B + h_{ie})C_B} \tag{9-20}$$

with the stipulation that $h_{fe}R_E$ be at least 10 times greater than R_B and $R_B > h_{ie}$ (which is usually the case). The midband current gain is calculated as before. The plot of the low-frequency response is illustrated in the following example.

Example 9-4. The amplifier shown in Fig. 9-1 has the following circuit and parameter values: $R_B = 10$ kΩ, $R_c = 3.3$ kΩ, $h_{ie} = 1$ kΩ, $h_{fe} = 100$, and $h_{oe} = 1/50$ kΩ. The bypass capacitor is 5 μF. We want to determine the lower cutoff frequency and sketch the current gain magnitude response at low frequencies.

Solution: The midband current gain is calculated from Eq. (9-5) by noting that the $h_{oe}R_c$ term is negligible with respect to 1. Thus,

$$A_{i(\text{mid})} \cong -\frac{100(10)}{10+1} \cong -91$$

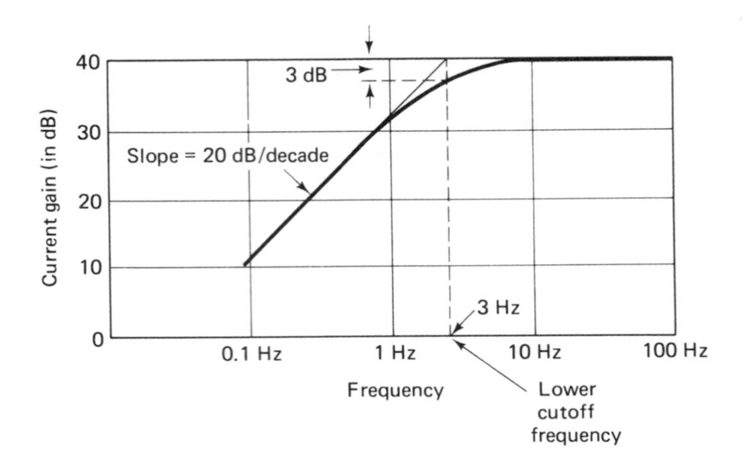

Figure 9-11. Low-frequency magnitude response for the amplifier in Example 9-4.

This value in dB is:

$$20\log(91) \cong 20(1.96) \cong 39.2 \text{ dB}$$

The lower cutoff frequency is now calculated:

$$f_1 = \frac{100}{2\pi(10+1)\times 10^3(5\times 10^{-6})} \text{ Hz} \cong 300 \text{ Hz}$$

The magnitude response for the low-frequency current gain is shown in Fig. 9-11.

The complete frequency response of an amplifier is obtained by combining the three individual responses: the midband response and the upper and the lower frequencies. This combination is shown in Fig. 9-12. The gain is constant over a range of frequencies called midband. It is 3 dB down from this value at both the upper and lower cutoff frequencies, and

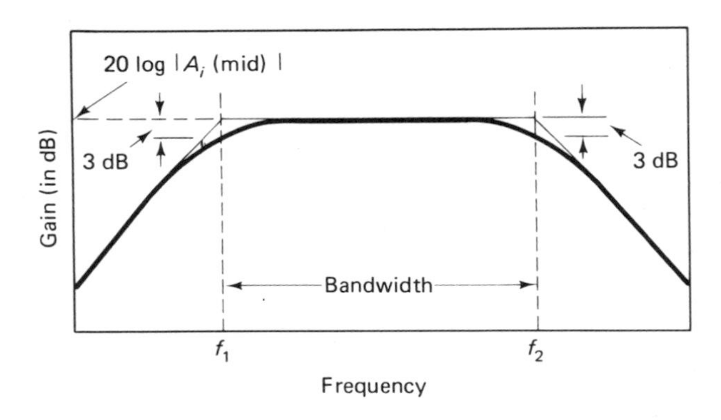

Figure 9-12. Complete frequency response.

it falls at an approximate rate of 20 dB/decade on either side of these frequencies. The bandwidth is the difference between the upper and lower 3-dB (cutoff) frequencies.

Review Questions

1. What is the role of an amplifier?
2. Name some of the important amplifier parameters.
3. What is a current amplifier? A voltage amplifier? A power amplifier?
4. What is the difference between a linear and a nonlinear amplifier?
5. How are small-signal amplifiers analyzed? How are nonlinear amplifiers analyzed?
6. What is the gain in a linear small-signal amplifier a function of?
7. What is the gain in a nonlinear amplifier a function of?
8. In what way can we distinguish between linear and nonlinear amplifiers? Give an example to illustrate.
9. What are the steps to be followed in the systematic analysis of small-signal amplifiers?
10. What are the steps to be followed in order to obtain the midband equivalent circuit for an amplifier? How is this different from the steps to obtain the low-frequency response? To obtain the high-frequency response?
11. In what ways is the systematic analysis of BJT amplifiers different from the analysis of FET amplifiers? In what ways are the two similar?
12. How is the midband region defined? Illustrate.
13. How is the high-frequency region defined?
14. How is the low-frequency region defined?
15. What transistor parameters determine the BJT amplifier midband response?
16. What FET parameters determine the midband response of a FET amplifier?
17. What circuit values determine the high-frequency response of a BJT amplifier?
18. What circuit values determine the low-frequency response of a BJT amplifier?
19. What is the amplifier bandwidth? What determines it?
20. What determines the complete response of an amplifier?

Problems

1. Identify the amplifier configuration in Fig. 9-13. Draw the midband equivalent circuit.

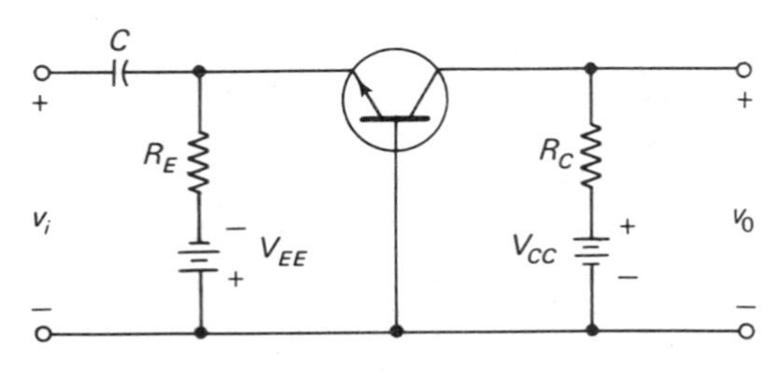

Figure 9-13.

2. Determine the midband voltage gain in Problem 1 if: $R_E = R_C = 10$ kΩ, $h_{ie} = 1$ kΩ, $h_{fe} = 80$, with h_{oe} and h_{re} neglected.
3. Draw the low-frequency equivalent circuit for the amplifier shown in Fig. 9-14.

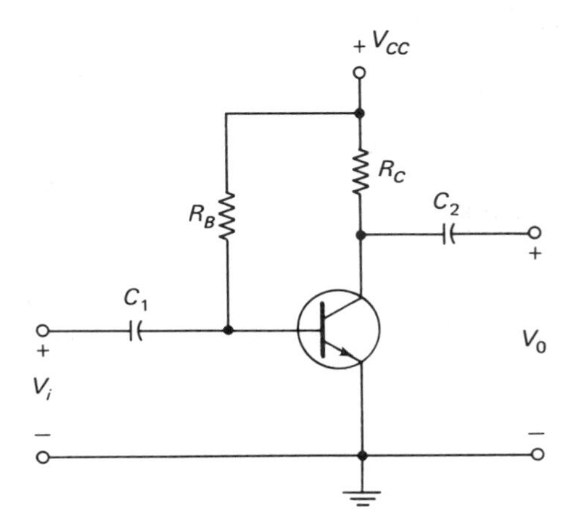

Figure 9-14.

4. Draw the high-frequency equivalent circuit for the amplifier in Fig. 9-14. Determine the upper cutoff frequency if the transistor parameters are the same as listed in Example 9-3 and $R_B = 220$ kΩ and $R_C = 1$ kΩ.
5. Repeat Problem 4 with $R_C = 10$ k.
6. Determine the midband current gain in Problem 4.
7. Identify the amplifier configuration in Fig. 9-15. Draw the midband equivalent circuit.
8. If the circuit parameters in Problem 7 are: $R_1 = R_2 = 33$ kΩ, $R_E = 1$ kΩ, and the transistor has the parameters listed in Example 9-4, determine the voltage gain v_o/v_i at midband.

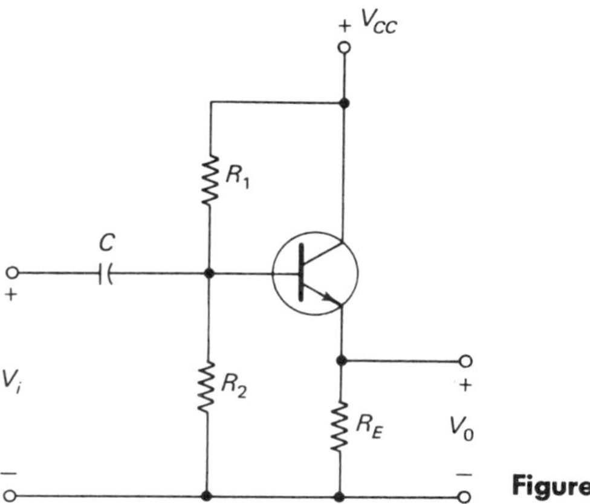

Figure 9-15.

9. Draw the low-frequency equivalent circuit for the amplifier shown in Fig. 9-15. Show that the low-frequency current gain expression has the same form as given in Eq. (9-19).
10. Determine the lower cutoff frequency in Problem 8 if $C = 0.1\ \mu\text{F}$.

Practical Amplifier Considerations

In this chapter we shall focus on some of the practical aspects of amplifiers. Specifically, these are the determination of input and output resistances, the real and apparent gain, the frequency response, and the effects of cascading amplifier stages.

10.1 INPUT AND OUTPUT IMPEDANCE

A practical transistor amplifier contains the active device, bias resistors, and sometimes coupling and bypass capacitors. The active device itself has an input and output impedance. These impedances are determined by the physical shape and makeup of the internal device and by the bias applied externally. However, as indicated in Fig. 9-3, external bias resistors appear in parallel with both the input and output terminals of the device. The actual input and output impedance (usually considered a resistance) of the amplifier contains both the effect of the device itself and the bias resistors. The midband equivalent circuit of the amplifier of Fig. 9-1 is repeated in Fig. 10-1, with the device outlined as before and the amplifier outlined with dotted lines.

We can determine the input impedance of the amplifier by finding the ratio of v_1 to i_1. In this case, the input impedance is just the parallel combination of h_{ie} and the bias resistor R_B. Therefore, looking into the input terminals of the amplifier, we see an effective impedance R_i, where R_i is defined as the input impedance. Here it is given by:

$$R_i = \frac{h_{ie} R_B}{h_{ie} + R_B} \tag{10-1}$$

At the output, the effective impedance is the parallel combination of the bias resistor R_C and the impedance corresponding to the transistor output admittance h_{oe}. This impedance is

$$R_o = \frac{R_C}{1 + h_{oe} R_C} \tag{10-2}$$

To complete the characterization of the amplifier in terms of its input and output impedances, we need to recalculate the gain.

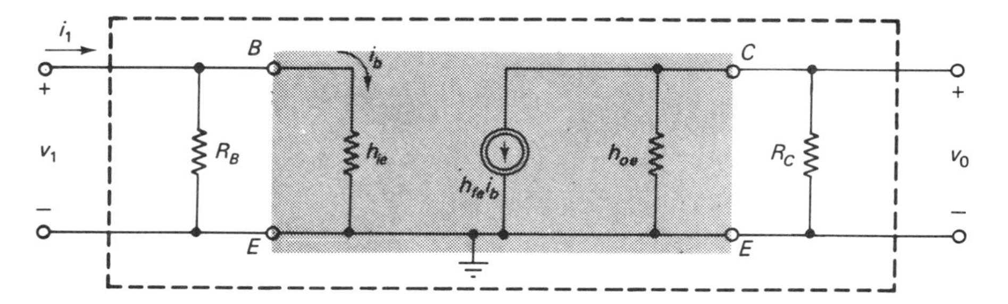

Figure 10-1. BJT amplifier equivalent circuit.

10.2 REAL AND APPARENT GAIN

Let us imagine that the amplifier of Fig. 10-1 has a short circuit across the output terminals, i.e., $v_o = 0$. Using the techniques discussed in Chap. 9, we can determine the current gain under these conditions. Note that the analysis of section 9.2 applies for the input but not the output, because no current will flow in either R_C or h_{oe}. Therefore, the current gain with a short circuit at the output is given by:

$$A_{ISC} = \frac{i_{osc}}{i_1} = -\frac{h_{fe} R_B}{h_{ie} + R_B} \tag{10-3}$$

We can now completely characterize the amplifier according to its input and output impedances and the short-circuit current gain, as indicated in Fig. 10-2. The current generator in this figure must be the short-circuit current gain multiplying the input current. In effect, we have found the Norton equivalent (see Introduction) of the output of the amplifier.

There are a number of advantages to representing an amplifier by the equivalent circuit shown in Fig. 10-2. First of all, *the short-circuit current gain is the largest possible current gain that any amplifier can provide*. Thus, it is important to know how large a number the short-circuit current gain for a particular amplifier actually is. From such a consideration, we can determine if a single amplifier is capable of providing sufficient gain or if additional stages of amplification will be necessary. It is also desirable to represent an amplifier as shown in Fig. 10-2 because it is a standard form into which any current amplifier can be placed.

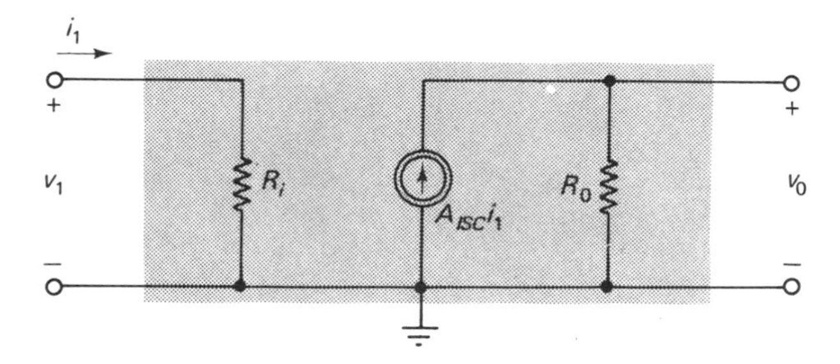

Figure 10-2 Current amplifier.

Consider next the case where we want to know the voltage gain of the original amplifier of Fig. 9-1. From Fig. 10-2 we can calculate the open-circuit voltage gain; that is, there is no resistance in parallel with the output. Note that the output voltage is given by:

$$v_o = A_{ISC} i_1 R_o \tag{10-4}$$

and also that the input voltage is given by:

$$v_1 = i_1 R_i$$

The open-circuit voltage gain (A_{VOC}) is then

$$A_{VOC} = \frac{v_0}{v_1} = A_{ISC} \frac{R_o}{R_i} \tag{10-5}$$

Using the open-circuit voltage gain, we can transform Fig. 10-2 into a voltage generator in series with a resistor at the output, as shown in Fig. 10-3. This procedure corresponds to finding the Thevenin equivalent circuit for the amplifier output terminals. The circuit of Fig. 10-3 is the voltage dual of the circuit in Fig. 10-2. Its importance is similar to that stated for Fig. 10-2. It has a general form and can represent any amplifier. Moreover, *the open-circuit voltage gain is the largest possible voltage gain for any amplifier*. Like the short-circuit current gain, the open-circuit voltage gain can be utilized in predicting the number of stages of amplification necessary in a particular application.

We can think of the short-circuit gain and open-circuit voltage gain as only apparent gains. In a real application, the actual gain will be lower; that is, the actual current gain will always be lower than A_{ISC}, and the actual voltage gain will always be lower than A_{VOC}.

Example 10-1. A BJT amplifier shown in Fig. 9-1 is to be used. The circuit values are listed in Example 9-1. We want to determine its input and output impedance, the short-circuit current gain, and the open-circuit voltage gain.

Solution: Use Eq. (10-1), so $R_i = (60)(1)/(60+1)\ \text{k}\Omega \cong 1\ \text{k}\Omega$. Equation (10-2) gives

$$R_o = \frac{4.7\ \text{k}\Omega}{1+(0.02)(4.7)} \cong 4.3\ \text{k}\Omega$$

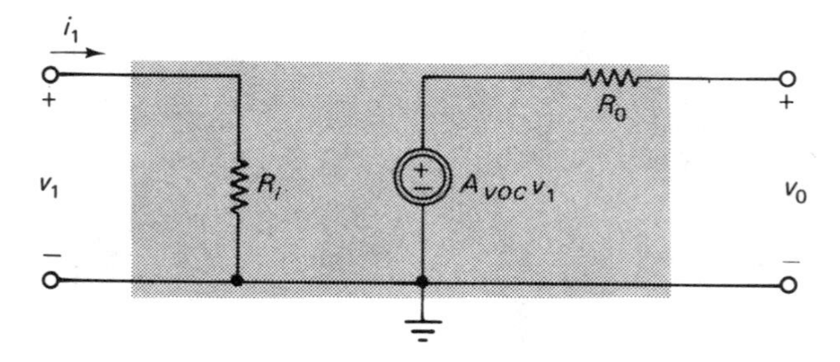

Figure 10-3. Voltage amplifier.

Calculating A_{ISC} from Eq. (10-3), we get

$$A_{ISC} = -\frac{(80)(60)}{1+60} \cong -79$$

Calculating A_{VOC} from Eq. (10-5), we get

$$A_{VOC} = (-79)\frac{(4.3)}{(1)} \cong -340$$

For comparison purposes, you may find it useful to note that the actual current gain (considering R_C as the load) found in Example 9-1 was -72, whereas the short-circuit current gain was -79.

The methods outlined in this section can be used to determine the input impedance, the output impedance, the short-circuit current gain, and the open-circuit voltage gain for any amplifier, be it a BJT in any of the three configurations (*CE*, *CB*, or *CC*), or a J-FET, or any MOSFET. Suffice it to say that the representations of an amplifier shown in Figs. 10-2 and 10-3 are universal and apply equally well to any amplifier.

10.3 AMPLIFIER LOADING

We next consider the effect of source impedance on the input current and the effect of load impedance on the output current. In any real application, the input signal is derived from a transducer or another amplifier. In any case, however, the source has a definite resistance (impedance) that must be accounted for in the analysis. This fact is illustrated in the following example.

Example 10-2. The amplifier of Example 10-1 is to be drawn from a source of 100 μA, which has an impedance of 20 kΩ. The amplified signal is to drive a load of 1 kΩ. We want to determine the output current.

Solution: The situation is shown in Fig. 10-4. At the input, the source current i_s divides between R_s and R_i. Only the component of i_s that flows into the amplifier (i_1) is amplified:

$$i_1 = i_s \frac{R_s}{R_i + R_s} \tag{10-6}$$

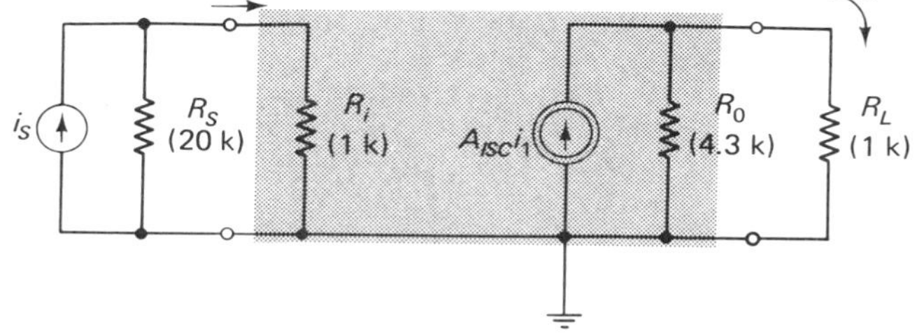

Figure 10-4. Loading in a current amplifier (see Example 10-2).

For the values listed, $i_1 = (100\ \mu A)\ 20/(20+1)$, or approximately 95 μA. Thus, of the 100 μA available from the source, the amplifier uses 95. With a short-circuit current gain of -79, the magnitude of the current generator is $(-79)(95\ \mu A) = -7.5$ mA. This available current divides at the output between R_o and R_L. The output current is given by

$$i_o = A_{ISC} i_1 \frac{R_o}{R_o + R_L} \tag{10-7}$$

For the values given $i_o = (-7.5\ \text{mA})\ 4.3/(4.3+1)$, or approximately -6 mA.

To review, with 100 μA available at the input, the output current is 6 mA. This figure represents a real current gain of 60.

If we examine the previous example, we can make some generalizations:

1. *The lower the input impedance of an amplifier, the higher the real current gain.*
2. *The higher the output impedance of an amplifier, the higher the real current gain.*

Thus, for given source and load impedances, we can improve the performance of a current amplifier (i.e., make the actual current gain larger) by either *decreasing* the input impedance or *increasing* the output impedance, or both.

We can also specify the points at which we call an amplifier a *current amplifier*: (1) when the input resistance of the amplifier is low in relation to the source impedance, (2) when the output impedance is high in relation to the load, and (3) when the amplifier itself has a significant current gain available (A_{ISC}).

The loading effect in a voltage amplifier is illustrated in the following example.

Example 10-3. An amplifier is characterized by an input impedance of 50 kΩ, an output impedance of 2 kΩ, and an open-circuit voltage gain of -100. It is driven from a source of 10 mV at an impedance of 5 kΩ. It is to drive a 10 kΩ load. We want to determine the output voltage and voltage gain.

Solution: The situation is depicted in Fig. 10-5. Only the part of the source voltage that appears across R_i is amplified. Thus,

$$v_1 = v_S \frac{R_i}{R_i + R_S} \tag{10-8}$$

With the values given, $v_1 = 9.1$ mV. The voltage available at the output is

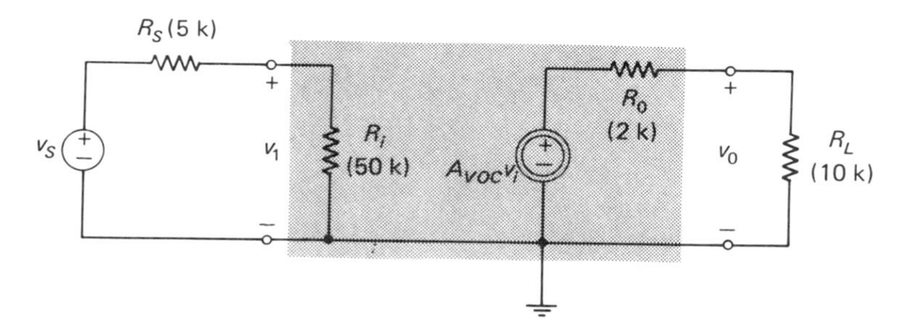

Figure 10-5. Loading in a voltage amplifier (see Example 10-3).

$A_{VOC}v_1$, which is -0.91 V. The output voltage is only the part of the available voltage that appears across R_L.

$$v_o = A_{VOC}v_1 \frac{R_L}{R_L + R_o} \tag{10-9}$$

In our example, we see that v_o will be approximately -0.76 V. The actual voltage gain is then v_o/v_S, which is about -76.

As was the case with the current amplifier, we can make some generalizations for the voltage amplifier:

1. *The higher the input impedance of an amplifier, the higher the actual voltage gain.*

2. *The lower the output impedance of an amplifier, the higher the actual voltage gain.*

Thus, for given source and load impedances, we can increase the voltage gain of an amplifier by either *increasing* the input impedance or *decreasing* the output impedance, or both.

We can also recognize the specific characteristics of what we term a *voltage amplifier*: It has an input impedance high in comparison to the source impedance and an output impedance low in comparison to the load impedance as well as an available voltage gain (A_{VOC}).

10.4 IMPEDANCE MATCHING

We saw in the previous section that for high current gain, R_i should be less than R_S and R_o greater than R_L. For high voltage gain, the reverse should be true, i.e., $R_i > R_S$ and $R_o < R_L$. In many applications, we are interested in both a high voltage as well as high current gain. Consequently, we would like to maximize the power gain.

Consider the circuit of Fig. 10-6. Let us assume for the time being that both v_S and R_S are fixed and specified. We would like to determine a value of R_i that will cause the largest power to be transferred to the input of our amplifier represented by R_i. The power supplied to the amplifier is the product of v_1 and i_1. It is this product that we need to maximize. Let us first examine the extreme values of R_i. If we first consider R_i to be zero, we

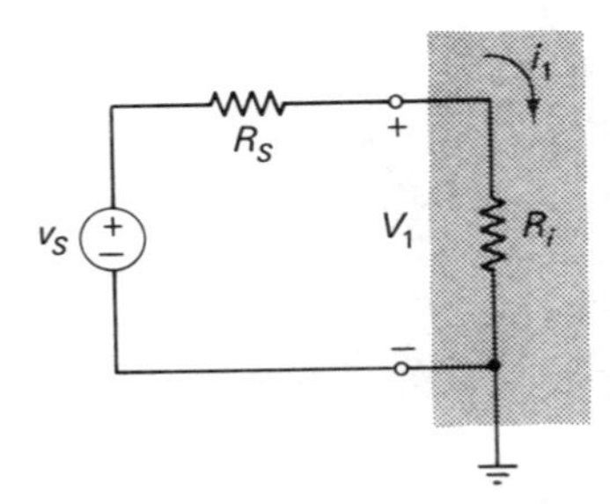

Figure 10-6. Maximum power transfer (voltage amplifier).

have obviously caused the largest current i_1 to flow [i_1 is given by $v_S/(R_S+R_i)$], while v_1 would be zero. Although the current is a maximum, the voltage is still zero, so no power is delivered to the amplifier.

We would have R_i at the other extreme when it is infinite. Under these conditions, i_1 would be zero, while v_1 would be a maximum and equal to v_S. Again, there is no power delivered to the amplifier, in this case, because the current is zero. Obviously, the maximum power to the input does not occur when either the voltage or the current is a maximum.

If we decide to make R_i equal to R_s or "match impedances," it turns out that the power delivered to the input is maximized. Under these conditions,

$$i_1=\frac{v_s}{2R_s} \quad \text{and} \quad v_1=\frac{v_s}{2} \tag{10-10}$$

The power delivered to the input P_1 is

$$P_1=i_1v_i=\frac{v_s^2}{4R_s} \tag{10-11}$$

The total power supplied by the v_s generator is twice the power delivered to the amplifier. (The proof of this statement is left as an exercise.) It may seem that one-half of the available power is being uselessly lost. This is not the case, because we cannot discount the power lost in R_s, which is an inseparable part of the source. The power in R_s is exactly the same as that in R_i, so that the power supplied by the source is exactly the same as that dissipated in R_i.

One of the exercises will show that maximum power is transferred to the amplifier in Fig. 10-7 when the impedances are matched, i.e., R_i made equal to R_s.

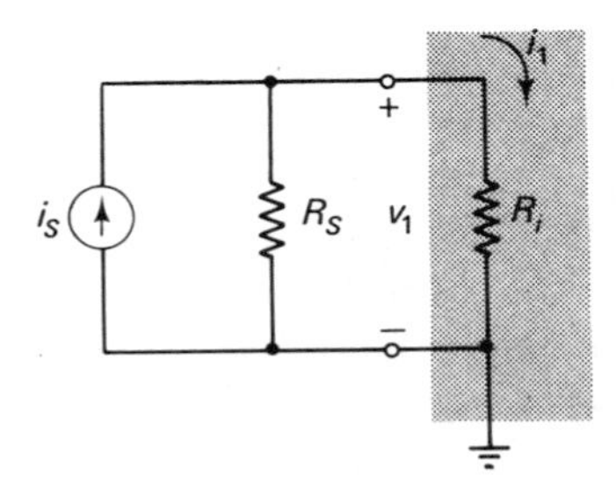

Figure 10-7. Maximum power transfer (current amplifier).

We can summarize by saying that if we want maximum power transfer, we must match impedances.

10.5 CASCADING OF AMPLIFIERS

In most applications, the gain provided by a single amplifier state is insufficient. Consequently, two or more amplifier stages are cascaded (or connected serially, from output 1 to input 2, etc.), as shown in Fig. 10-8. Although we shall confine our discussion to two stages, it should be apparent that the same procedure may be extended to handle any number of cascaded stages.

Example 10-4. Two amplifiers are cascaded as indicated in Fig. 10-8. The amplifier parameters are: $R_{i1}=1\ \text{k}\Omega$, $R_{o1}=5\ \text{k}\Omega$, $A_{ISC1}=-60$; $R_{i2}=1.4\ \text{k}\Omega$, $R_{o2}=6\ \text{k}\Omega$, $A_{ISC2}=-75$. $R_s=10\ \text{k}\Omega$, and $R_L=1\ \text{k}\Omega$. We want to determine the overall current gain (the ratio of i_o to i_s).

Solution: We start at the input and work our way toward the output. First,

$$i_1=i_s\frac{R_s}{R_s+R_{i1}}$$

next,

$$i_2=A_{ISC1}i_1\frac{R_{o1}}{R_{o1}+R_{i2}}$$

then,

$$i_o=A_{ISC2}i_2\frac{R_{o2}}{R_{o2}+R_L}$$

Now we are ready to put the separate parts together to determine the gain.

$$A_I=\frac{i_o}{i_s}=\frac{i_o}{i_2}\frac{i_2}{i_1}\frac{i_1}{i_s} \tag{10-12}$$

The three separate parts of A_I can be identified from the three preceding

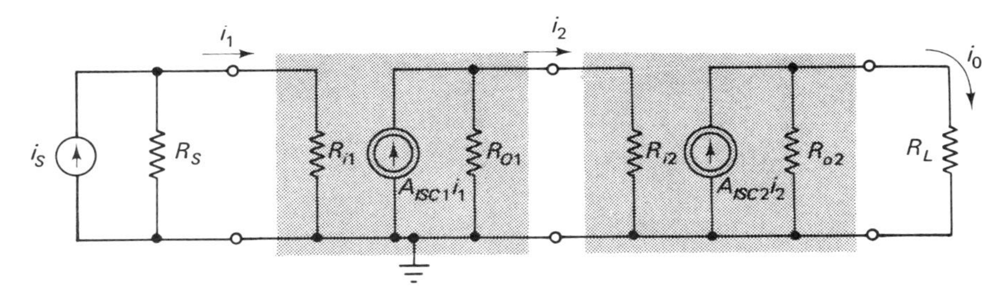

Figure 10-8. Cascading of two current amplifiers.

equations. Thus,

$$\frac{i_o}{i_2} = A_{ISC2}\frac{R_{o2}}{R_{o2}+R_L}$$

$$\frac{i_2}{i_1} = A_{ISC1}\frac{R_{o1}}{R_{o1}+R_{i2}}$$

$$\frac{i_1}{i_s} = \frac{R_s}{R_s+R_{i1}}$$

Using the above expressions in Eq. (10-12) yields the desired current gain:

$$A_I = (A_{ISC1})(A_{ISC2})\left(\frac{R_s}{R_s+R_{i1}}\right)\left(\frac{R_{o1}}{R_{o1}+R_{i2}}\right)\left(\frac{R_{o2}}{R_{o2}+R_L}\right) \qquad (10\text{-}13)$$

If we replace the symbols by the numbers for this particular example, we obtain:

$$A_I \cong (-60)(-75)(0.91)(0.78)(0.86) \cong 2747$$

Note one interesting aspect in this example. We could rewrite the gain expression of Eq. (10-13) in a slightly different form:

$$A_I = \left(\frac{R_s}{R_s+R_{i1}}\right)\left[(A_{ISC1})(A_{ISC2})\left(\frac{R_{o1}}{R_{o1}+R_{i2}}\right)\right]\left(\frac{R_{o2}}{R_{o2}+R_L}\right) \qquad (10\text{-}14)$$

If we now recognize that the quantity inside the square brackets is the effective short-circuit current gain of the cascaded pair of amplifiers, we can draw an equivalent circuit combining the two amplifiers into one, as shown in Fig. 10-9.

We could have arrived at this conclusion directly as well. We can now write:

$$A_{ISC} = A_{ISC1}A_{ISC2}\frac{R_{o1}}{R_{o1}+R_{i2}} \qquad (10\text{-}15)$$

where A_{ISC} denotes the short-circuit current gain for the cascade of two amplifiers.

The new single-stage equivalent amplifier shown in Fig. 10-9 has the input impedance of the first stage and the output impedance of the second stage.

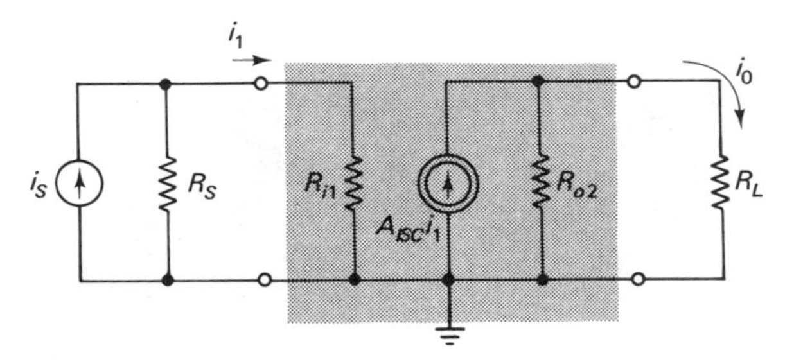

Figure 10-9. Single amplifier representation of the two-stage amplifier of Fig. 10-8.

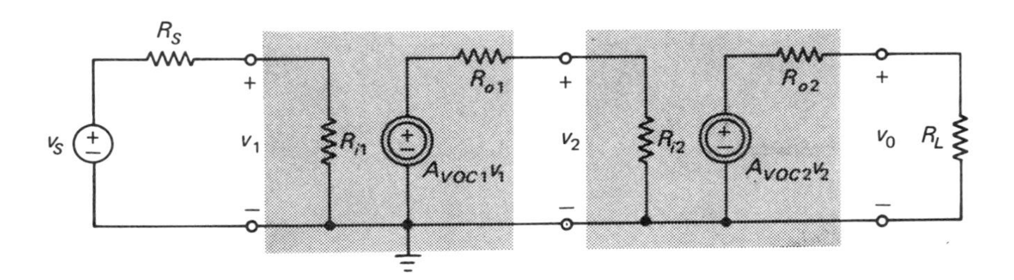

Figure 10-10. Cascading of two voltage amplifiers.

When two voltage amplifiers are cascaded as illustrated in Fig. 10-10, a similar procedure is used to determine the properties of the equivalent single-stage amplifier. We shall only list the results here, leaving the details as an exercise.

$$A_{VOC} = A_{VOC1} A_{VOC2} \frac{R_{i2}}{R_{i2} + R_{o1}} \tag{10-16}$$

where A_{VOC} stands for the open-circuit voltage gain of the equivalent single-stage amplifier. The input impedance is that of the first stage, and the output impedance is that of the second stage, as shown in Fig. 10-11.

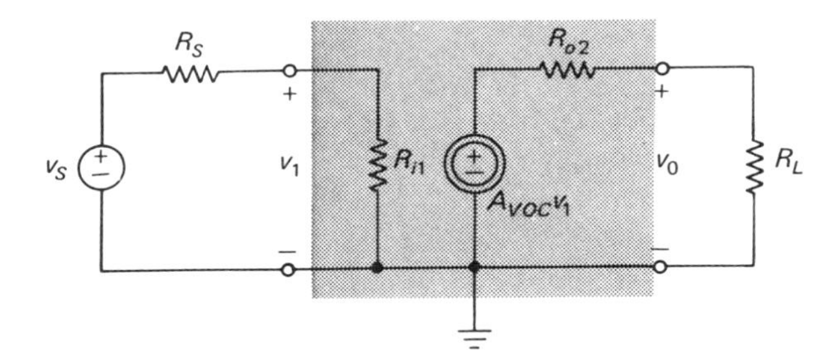

Figure 10-11. Single amplifier representation of the two-stage amplifier of Fig. 10-10.

The cascading of amplifiers obviously increases the gain. At the same time that it increases the gain, however, it also affects the bandwidth of an amplifier.

Suppose that a single amplifier stage has a gain of 20 dB (magnitude 10) and an upper cutoff frequency of 100kHz. If we start cascading additional amplifier stages to it, the combined amplifier will have a larger gain and an upper cutoff frequency which is lower. For example, if we cascade identical stages, each with a gain of 20 dB, the resulting amplifier will have the gain response curves shown in Fig. 10-12, where n is the number of similar stages cascaded.

Note two things. First, the 3 dB point becomes successively lower the more stages that we cascade. Secondly, the gain falls off more rapidly as the number of stages in cascade is increased. If a single amplifier gain falls

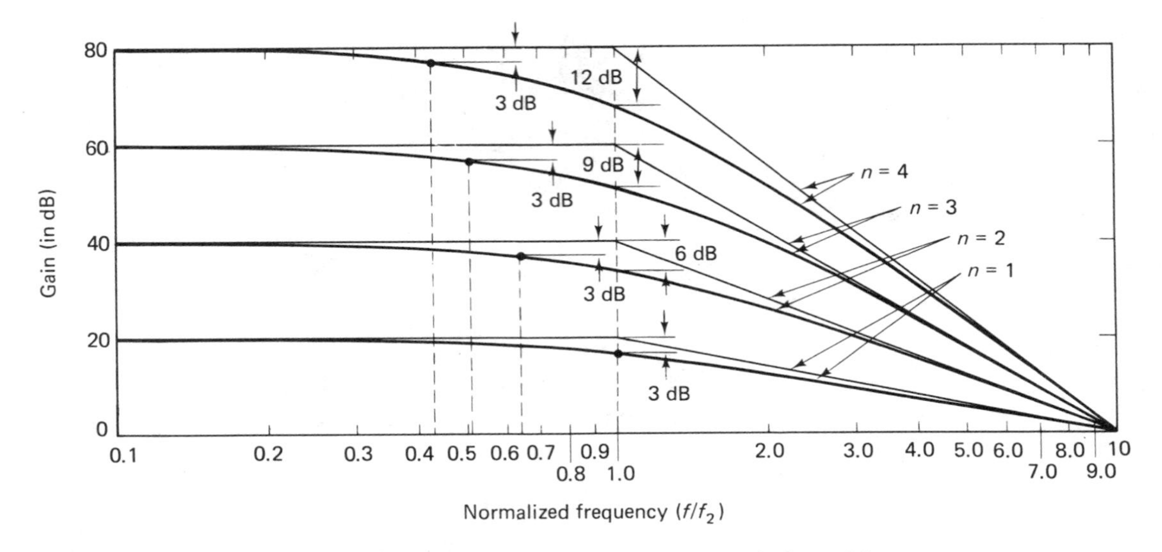

Figure 10-12. Bandwidth reduction in cascaded amplifiers.

off at a rate of 20 dB/decade (6 dB/octave), then the cascaded amplifier gain will fall off at a rate of $20n$ dB/decade ($6n$ dB/octave), where n is the number of stages in the cascade. The results of Fig. 10-12 are summarized in Table 10-1.

Table 10-1. Upper and Lower Cutoff Frequencies for a Cascaded Amplifier (only valid for identical stages)

Lower Cutoff Frequency (of cascaded amplifier)	*n (number of stages in cascade)*	*Upper Cutoff Frequency (of cascaded amplifier)*
$1.00 f_1$*	1	$1.00 f_2$†
$1.56 f_1$	2	$0.64 f_2$
$1.96 f_1$	3	$0.51 f_2$
$2.28 f_1$	4	$0.43 f_2$
$2.56 f_1$	5	$0.39 f_2$

*f_1 denotes the lower cutoff frequency of each stage.
†f_2 denotes the upper cutoff frequency of each stage.

In a similar manner, the lower cutoff frequency of a number of identical amplifier stages increases as the number of stages increases. This increase is also summarized in Table 10-1.

Example 10-5. Suppose that we construct an amplifier containing three identical stages, each with a gain of 20 dB, an upper cutoff frequency of 10 kHz, and a lower cutoff frequency of 50 Hz. We want to determine the gain and upper and lower cutoff frequencies of the cascaded amplifier.

Solution: We can determine the gain by adding the individual dB gains, or 60 dB for the cascaded amplifier.

We can calculate the upper cutoff frequency by looking up the shrinkage factor for $n=3$ in Table 10-1 to get 0.51. Therefore, the new upper cutoff frequency is (0.51)(10 kHz)$\cong$5.1 kHz. Similarly, we obtain the lower cutoff frequency as (1.96)(50 Hz)$\cong$98 Hz. Thus, the cascaded amplifier has a bandwidth that is roughly half of the single-stage amplifier.

Review Questions

1. How is the input impedance of an amplifier defined?
2. What is the practical significance of the input impedance?
3. How is the output impedance of an amplifier defined?
4. What is the practical significance of the output impedance of an amplifier?
5. What is meant by *real* gain? By *apparent* gain?
6. What are the conditions that yield the highest possible current gain? Explain.
7. What are the conditions that yield the highest possible voltage gain? Explain.
8. What are the universal equivalent circuits that apply equally well for any amplifier? What components do they contain?
9. What is meant by the term "loading" as applied to an amplifier?
10. What are the possible ways in which an amplifier may be loaded?
11. In order to maximize the current gain, what is the desired condition for the input impedance? For the output impedance?
12. In order to maximize the voltage gain, what is the desired condition for the input impedance? For the output impedance?
13. What is meant by the term "impedance matching" as applied to an amplifier?
14. Why is impedance matching important? When is it important?
15. What is meant by the term "cascading" as applied to an amplifier?
16. What is the effect of cascading a number of amplifiers on the gain? On the input impedance? On the output impedance?
17. What happens to the gain when two amplifiers are cascaded? What happens to the gain as expressed in decibels?
18. What is the effect of cascading a number of amplifiers on the upper cutoff frequency?
19. What is the effect of cascading a number of amplifiers on the lower cutoff frequency?
20. When a number of amplifiers are cascaded, what is the effect on the bandwidth of the overall amplifier?

Problems

1. The amplifier in Fig. 9-1 has the following circuit values: $R_1=56$ kΩ, $R_2=22$ kΩ, and $R_C=6.8$ kΩ. The transistor parameters are: $h_{ie}=2$ kΩ,

$h_{fe}=75$, and $1/h_{oe}=40$ kΩ. Determine the input impedance, the output impedance, the short-circuit current gain, and the open-circuit voltage gain.

2. Using the results of Problem 1, determine the actual circuit current and voltage gains.
3. The FET amplifier shown in Fig. 9-4 has the following circuit values: $R_{G1}=2.2$ MΩ, $R_{G2}=1$ MΩ, and $R_D=18$ kΩ. The FET parameters are: r_{gs} infinite, $g_m=2000$ micromhos, and $r_d=100$ kΩ. Determine the amplifier input and output impedances, the short-circuit current gain, and the open-circuit voltage gain.
4. Using the results of Problem 3, determine the actual circuit voltage gain.
5. Determine the output current and voltage in Problem 1, if the input is driven by a voltage of 1 mV with a source impedance of 10 kΩ.
6. Determine the output voltage in Problem 3, if the input is a 100 μV signal with a 100 kΩ source impedance.
7. Repeat Problem 6 if the source impedance is 1 MΩ.
8. If the amplifiers of Problems 1 and 3 are cascaded (the FET stage first), determine the input impedance, the output impedance, and the open-circuit voltage gain for the cascaded combination.
9. For the cascaded amplifier of Problem 8, what is the largest source impedance that may be applied without causing significant loading of the source? Explain.
10. A single-stage amplifier with a gain of 28 dB and upper and lower cutoff frequencies of 50 kHz and 10 Hz, respectively, is available. We wish to construct an amplifier containing two cascaded stages. It will have a gain of 60 dB; its upper and lower frequencies will be 0.64 and 1.56 times the respective cutoff frequencies of the available amplifier stage. Specify the parameters of the second stage that must be used.
11. Three identical amplifiers are cascaded. The parameters for each stage are: gain = 23, upper cutoff frequency = 1.1 MHz, and lower cutoff frequency = 40 Hz. Determine the gain (in decibels) and the upper and lower cutoff frequencies of the cascaded amplifier.
12. On semilogarithmic graph paper, sketch the results of Problem 11. Show also the frequency response of a single stage.

11

Tuned Amplifiers

In many communication circuits (such as radio and television), in instrumentation, and in other applications, we need amplifiers which amplify signals of only certain predetermined frequencies. Such amplifiers are called *tuned*, or *frequency-selective*, amplifiers. For example, in a radio receiver, we need to be able to distinguish between different stations, that is, tune to different stations. The amplifier, therefore, must pass signals of the desired frequency and reject all other frequencies. The rejection of undesired signals in many cases is as important as the amplification of desired signals.

Figure 11-1 shows the gain versus frequency characteristics of an ideal tuned amplifier. The gain is zero for all frequencies below f_1, becomes very high for frequencies in the desired pass band (between f_1 and f_2), and is again zero for all frequencies above f_2. The difference between the upper and lower cutoff frequencies is the *bandwidth (BW)*. The *center frequency* f_c is either arithmetically or geometrically the average of f_1 and f_2. Suffice it to say that an ideal tuned amplifier does not exist in the real world. Practical tuned amplifiers have characteristics which at best only approximate those of the ideal tuned amplifier shown in Fig. 11-1.

11.1 SINGLE-TUNED AMPLIFIERS

Single-tuned BJT and FET amplifiers are illustrated in Figs. 11-2 and 11-3, respectively. These circuits resemble the single-stage amplifiers discussed in Chap. 9. However, the resistive load of the single-stage amplifier is replaced by a *tuned*, or *tank*, circuit containing C and L. In both circuits, the three resistors establish the dc bias conditions. Capacitors C_1 and C_2 provide dc isolation between the source and load, respectively; C_B is the bypass capacitor, as discussed in section 9.2. All the capacitors (with the exception of C) are large enough to be effective short circuits at the frequencies we are interested in, so they will not enter into the analysis.

Using the generalized form of the voltage amplifier equivalent circuit developed in Chap. 10, we can draw a single equivalent circuit for both the BJT and the FET amplifiers, as indicated in Fig. 11-4. In the case of the

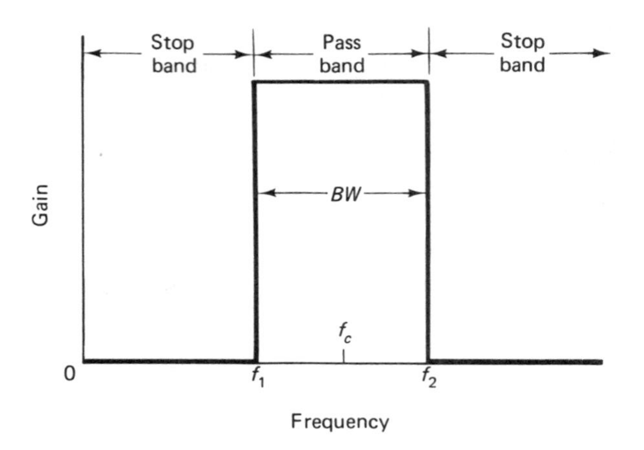

Figure 11-1. Characteristics of an ideal tuned amplifier.

BJT amplifier, R_i is the parallel combination of h_{ie}, R_1, and R_2, with A_{VOC} being given by:

$$A_{VOC} = -\frac{h_{fe}}{h_{ie}h_{oe}} \tag{11-1}$$

and R_o is $1/h_{oe}$.

In the case of the FET amplifier, R_i is the parallel combination of R_{G1} and R_{G2}, with A_{VOC} in this case given by:

$$A_{VOC} = -g_m r_d (= -\mu) \tag{11-2}$$

and R_o is r_d.

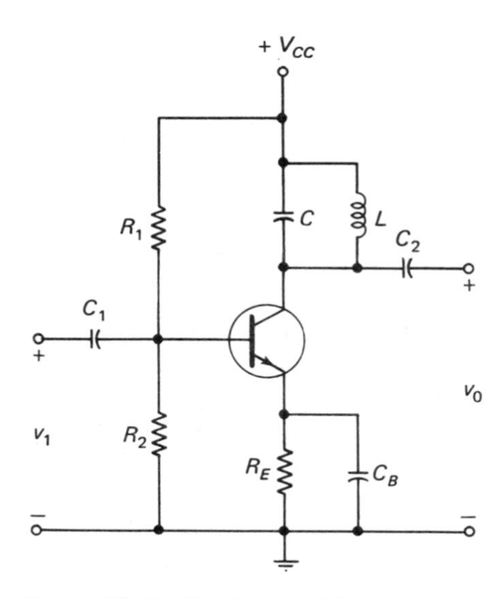

Figure 11-2. Single-tuned BJT amplifier.

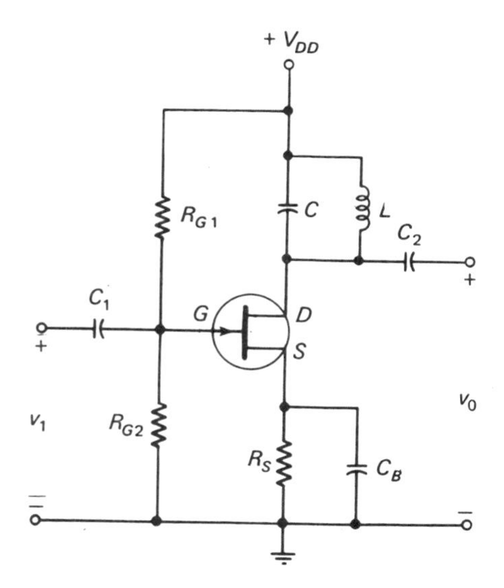

Figure 11-3. Single-tuned FET amplifier.

In the equivalent circuit of Fig. 11-4, the inductor is represented by an ideal inductor in series with a resistance R, which accounts for the nonzero dc resistance of the actual coil. Note that the impedance (Z_L) of the resonant circuit load in Fig. 11-4 is the parallel combination of a capacitor C and a series connection of an inductor L and resistor R. Thus,

$$Z_L = \frac{\dfrac{1}{j\omega C}(R + j\omega L)}{\dfrac{1}{j\omega C} + R + j\omega L} \qquad (11\text{-}3)$$

The *resonant frequency* (f_o) is given by:

$$f_o = \frac{\omega_o}{2\pi} = \frac{1}{2\pi\sqrt{LC}} \qquad (11\text{-}4)$$

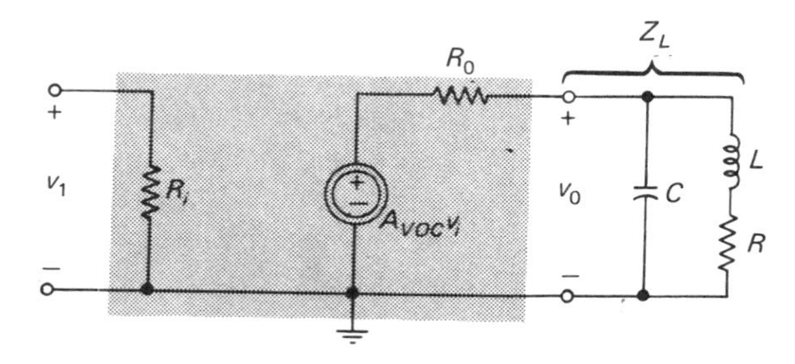

Figure 11-4. Equivalent circuit for single-tuned amplifiers of Figs. 11-2 and 11-3.

At this time, we must define another important parameter, called the *quality* factor Q:

$$Q_o = \frac{2\pi f_o L}{R} = \frac{1}{2\pi f_o RC} \tag{11-5}$$

The Q of the coil is a unitless measure of the quality of the coil. It is almost invariably designated at a specific frequency. So, if we know the inductance and the Q at a certain frequency, we can determine the effective series resistance of the coil.

Example 11-1. Assume that an IF radio coil has 0.1 mH inductance with a Q of 100 at 455 kHz. We want to determine the effective series resistance of the coil.

Solution: We can solve for R from Eq. (11-5):

$$R = \frac{2\pi f_o L}{Q_o} \cong \frac{(2\pi)(455\times 10^3)(0.1\times 10^{-3})}{10^2}\,\Omega \cong 2.86\Omega$$

For convenience and brevity, we use a new variable here: δ. It defines the normalized frequency deviation:

$$\delta = \frac{f - f_o}{f_o} \tag{11-6}$$

Using the definitions of Eqs. (11-4), (11-5), and (11-6), we rearrange the expression for the load impedance from Eq. (11-3) to read:*

$$Z_L = \frac{RQ_o^2}{1 + j2\delta Q_o} \tag{11-7}$$

Next we evaluate the load impedance Z_L at resonance, where $f = f_o$ and therefore $\delta = 0$. The equation for Z_L is then:

$$Z_{L(\text{resonance})} = RQ_o^2 \equiv R_{res} \tag{11-8}$$

Note that the load is purely resistive at resonance; this must be true because of the definition of resonance. Moreover, even though R itself is usually quite low, for typical values of Q, the load impedance at resonance offers a very high resistance, labeled R_{res} in Eq. (11-8).

At resonance the inductive and capacitive components of the load impedance exactly cancel. The resistive component is given by R_{res} in Eq. (11-8). Therefore, we can redraw the tuned circuit as shown in Fig. 11-5. Also at resonance this circuit is equivalent to the tuned circuit in Fig. 11-4

---

*There are certain approximations involved in the step between Eqs. (11-3) and (11-7). These approximations are called high-Q approximations and are valid near the resonant frequency f_o if $Q \gg 1$.

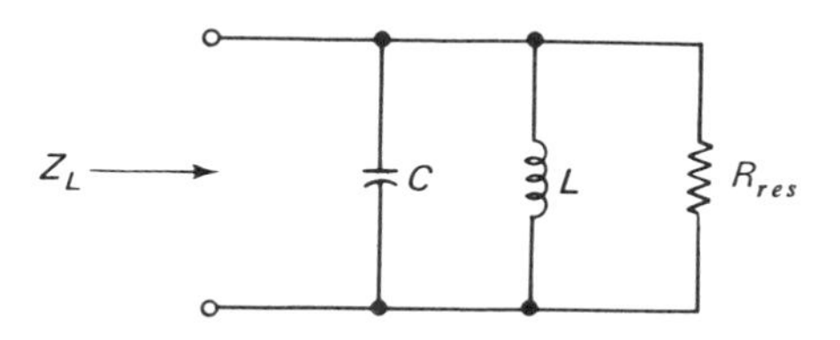

Figure 11-5. Approximate equivalent circuit of the load near resonance.

because of the load on the amplifier. Near resonance it is a good approximation; that is, δ is very small.

To see the effect of Q on the load impedance, we show a plot in Fig. 11-6 of the normalized load impedance as a function of δ for different values of Q. Observe that the higher the Q, the more selective (the narrower) the curve around $\delta=0$ (which corresponds to $f=f_o$). Note also that the curves are symmetrical about $\delta=0$ when δ is very small.

Returning to the amplifier of Fig. 11-4, we can write the voltage gain A_V (which is v_0/v_1):

$$A_V = A_{VOC}\frac{Z_L}{Z_L+R_o} \tag{11-9}$$

If we substitute Eq. (11-7) and making use of Eq. (11-8) in Eq. (11-9), after some manipulation, the gain can be rewritten:

$$A_V = A_{VOC}\left(\frac{R_P}{R_o}\right)\left(\frac{1}{1+j2\delta Q_e}\right) \tag{11-10}$$

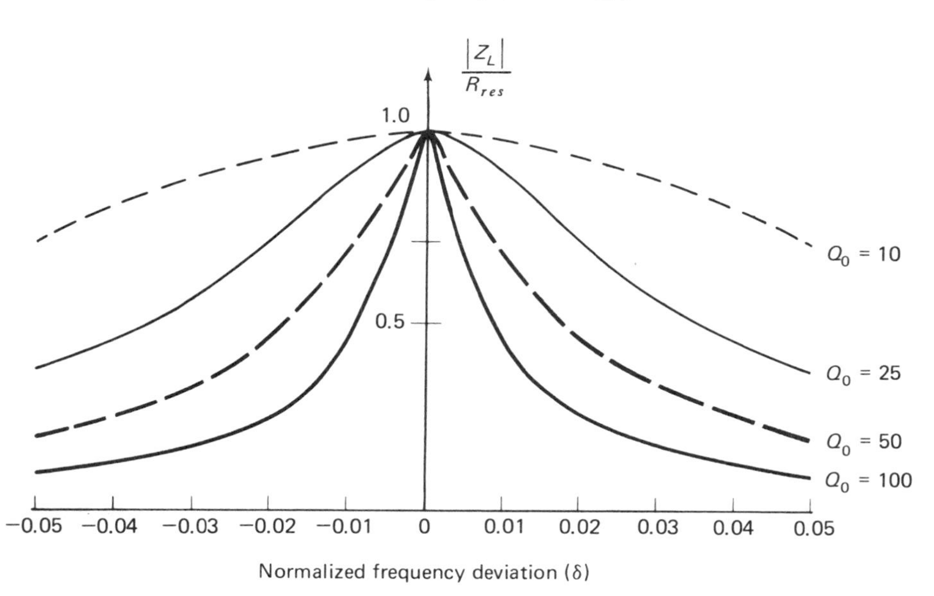

Figure 11-6. Variation of Z_L for different Q_o.

where the parallel resistance R_P and the effective Q have been defined as:

$$R_P = \frac{R_o R_{res}}{R_o + R_{res}} \tag{11-11}$$

$$Q_e = Q_o \frac{R_o}{R_o + R_{res}} = Q_o \frac{R_P}{R_{res}} \tag{11-12}$$

There are a few important points worth considering here. The parallel resistance R_P must always be smaller than either R_o or R_{res}, because it is the parallel combination of these two resistors. The ratio R_P/R_{res} in Eq. (11-12), therefore, is always less than 1. Furthermore, the *effective* Q (also called the *loaded* Q) must at all times be smaller than the unloaded or circuit Q (labeled Q_o). We say that the output of the amplifier loads down the tuned circuit and degrades the Q (i.e., lowers it).

Let us next examine the performance of the amplifier. At resonance, as we have already seen, δ is zero and the amplifier gain A_{Vres} is:

$$A_{Vres} = A_{VOC} \frac{R_P}{R_o} \tag{11-13}$$

This gain is the highest possible. Because R_P is smaller than R_o, the resonant amplifier gain is smaller than the open-circuit voltage gain. The gain response is the plot of the gain as a function of frequency. It has the same shape and characteristics as the impedance plot of Fig. 11-6.

We can determine both the cutoff frequencies (3 dB frequencies) and the bandwidth by determining the value of δ, for which the gain is 0.707 of the resonant gain. Equation (11-10) tells us that this occurs when $\delta = \pm 1/2Q_e$. From this fact, remembering the definition of δ, we can write:

$$f_1 = f_o\left(1 - \frac{1}{2Q_e}\right) \quad \text{and} \quad f_2 = f_o\left(1 + \frac{1}{2Q_e}\right) \tag{11-14}$$

Knowing the cutoff frequencies, we can determine the 3 dB bandwidth as follows:

$$BW_{3\text{ dB}} \cong f_2 - f_1 = \frac{f_o}{Q_e} \tag{11-15}$$

We specify here that it is the 3 dB bandwidth, because other bandwidths are also used to specify the performance of tuned amplifiers. For example, an ideal amplifier (as shown in Fig. 11-1) has a gain characteristic which falls and rises with infinite slope. It is customary to define a parameter, which specifies the steepness of the slopes in the characteristics of an amplifier. This parameter is sometimes called the *selectivity* of an amplifier. One possible definition for the selectivity is:

$$S \cong \frac{60\text{ dB bandwidth}}{3\text{ dB bandwidth}} \tag{11-16}$$

By the 60 dB bandwidth, we mean the bandwidth at the point where the gain is down 60 dB from its resonant value. It can be shown that for the single-tuned amplifier, the 60 dB bandwidth is given by $(1{,}000\, f_o/Q_e)$.

The selectivity of a single-tuned amplifier is then 1,000—a constant, irrespective of the resonant gain or the center frequency or either the loaded or unloaded Q. If we used the same definition for the selectivity of an ideal tuned amplifier, we would obviously get $S=1$, because the passband of an ideal amplifier is constant.

In practice, we would like as small a numerical value as possible for the selectivity. In other words, we would like the amplifier to be very or highly selective.

Example 11-2. A single-tuned BJT amplifier of the type shown in Fig. 11-2 is constructed to operate as an IF amplifier in an AM superheterodyne receiver. The center frequency is to be 455 kHz. The transistor parameters are: $h_{ie}=2$ kΩ, $h_{fe}=50$, and $h_{oe}=10$ μ℧ (micromho). The inductor of 1 mH with a Q of 100 at 455 kHz will be used in the tuned circuit. Determine the capacitor value, the resonant voltage gain, the cutoff frequencies, and the bandwidth.

Solution: First calculate C from Eq. (11-4):

$$C=\frac{1}{(2\pi f_o)^2 L}\text{ F}\cong\frac{1}{(6.28\times 4.55\times 10^5)^2(10^{-3})}\text{ F}\cong 122\text{ pF}$$

Just as in Example 11-1, we calculate the series inductor resistance to obtain $R=28.6$ Ω. From Eq. (11-8) we find:

$$R_{res}\cong 28.6\times 10^4\ \Omega\cong 286\text{ k}\Omega$$

We now turn our attention to the amplifier. With the parameters given:

$$R_o=\frac{1}{h_{oe}}=100\text{ k}\Omega$$

$$A_{VOC}=-\frac{50}{(2)(0.01)}\cong -2500$$

With R_o and R_{res} known, we calculate that R_P from Eq. (11-11) is approximately 74 kΩ. We can now calculate the resonant gain from Eq. (11-13):

$$A_{Vres}=(-2500)\frac{(74)}{(100)}\cong -1850$$

We can similarly calculate Q_e from Eq. (11-12):

$$Q_e=(100)\frac{(74)}{(286)}\cong 25.9$$

We next calculate the bandwidth from Eq. (11-16):

$$BW_{3\text{ dB}}\cong\frac{455}{25.9}\text{ kHz}\cong 17.6\text{ kHz}$$

To calculate the cutoff frequencies, it is best to rewrite Eq. (11-14) as:

$$f_1 = f_o - \frac{BW}{2} \cong 455 - 8.8 \cong 446.2 \text{ kHz}$$

$$f_2 = f_o + \frac{BW}{2} \cong 455 + 8.8 \cong 463.8 \text{ kHz}$$

This finishes the problem.

In practice, it is often necessary to adjust the tuned circuit for exactly the right center frequency. We can make this adjustment by tuning either the inductor or the capacitor, depending on which is variable. We can tune the inductor by turning the ferrite slug inside the coil. In the case of the capacitor, a small trimmer capacitor with a screw adjustment may be used. You must be careful to consider transistor internal capacitances. Our example and analysis have dealt with a "high-frequency transistor" where the internal capacitance is very low. In the previous example, however, if the amplifier stage were cascaded to an identical stage that typically might have an input shunting (internal) capacitance of a few hundred picofarads, the calculations would not hold.

Bandwidth adjustments in the single-tuned amplifier are also possible. In order to increase the bandwidth (it cannot be decreased), the effective Q needs to be decreased. So we add a parallel compensating resistor R_x, as shown in Fig. 11-7. As a result, the R_P is modified; it now is the parallel combination of R_{res}, R_o, and R_x. All other equations remain the same.

Example 11-3. Let us assume that we want to change the performance of the amplifier in Example 11-2 in such a way that it has a bandwidth of 25 kHz. The compensating resistor value has to be found.

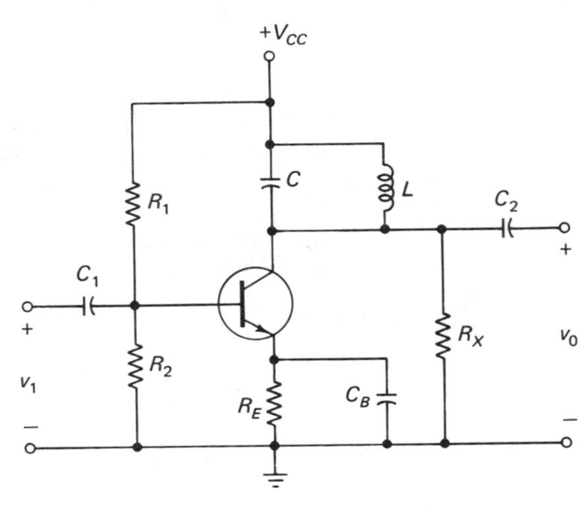

Figure 11-7. Modification of the circuit to increase the bandwidth (adding R_x).

Solution: From Eq. (11-15):

$$Q_e = \frac{f_o}{BW} = \frac{455}{25} \cong 18.2$$

From Eq. (11-12):

$$R_P = R_{res}\frac{Q_e}{Q_o} = (286)\frac{18.2}{100}\,\text{k}\Omega \cong 52\ \text{k}\Omega$$

$$\frac{1}{R_P} = \frac{1}{R_x} + \frac{1}{R_o} + \frac{1}{R_{res}}$$

Therefore, with R_P as calculated here (52 kΩ) and R_o and R_{res} the same as in the previous example, we can determine:

$$\frac{1}{R_x} = \frac{1}{R_P} - \frac{1}{R_o} - \frac{1}{R_{res}}$$

$$\therefore R_x = 165\ \text{k}\Omega$$

The only manner in which the bandwidth could be narrowed is by replacing the inductor with another one having a higher Q at the same frequency.

11.2 COUPLING OF TUNED AMPLIFIERS

When we cascade amplifiers, the input impedance of the second amplifier appears as part of the load to the first stage. In the case of tuned amplifiers, this procedure presents some special problems, especially when BJT amplifiers are involved. To see the full consequences, let us consider the amplifier in Example 11-2. Suppose there is another identical amplifier cascaded to it. In effect, we would place the input impedance (in this case about 2 kΩ) of the second stage in parallel with the output of the first stage, the equivalent of placing R_x as we did in Example 11-3. However, the input impedance of a BJT amplifier in the tuned circuit is extremely low, and the loading would destroy any selectivity of the amplifier. (It would in the case mentioned given an effective Q of less than 1.) We could not even truthfully call such a cascade a tuned amplifier.

To prevent this loading effect in BJT amplifiers, we have three basic methods available to us. In the first method, we can use an impedance matching transformer, as illustrated in Fig. 11-8. The transformer raises the effective impedance seen by the tuned circuit and thus prevents undesirable loading and bandwidth broadening that would otherwise result. Although a successful method, the interstage matching transformer is not used frequently, because other more efficient methods offer the same benefits at lower cost.

A more popular scheme for minimizing, or in some cases eliminating, the loading effect is shown in Fig. 11-9. It utilizes tapped inductors. The total inductance in the output of the first stage is $L_1 + L_2 + 2M$ (where M is the mutual inductance between the two windings). If the number of turns

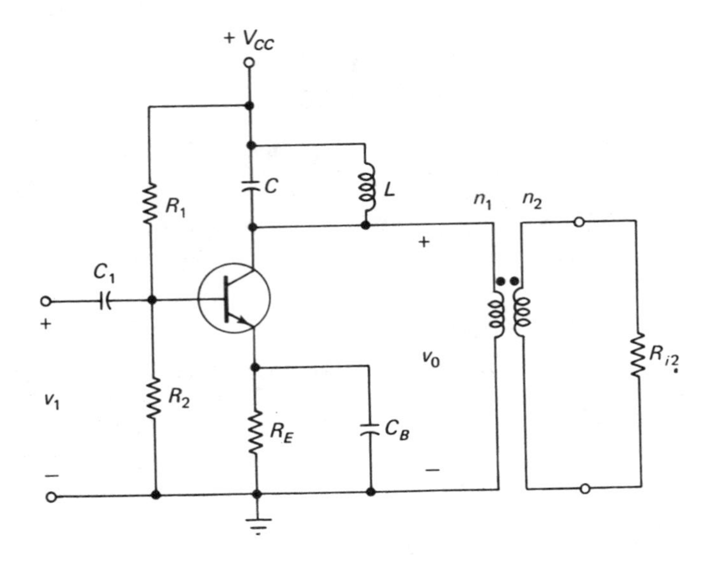

Figure 11-8. Impedance matching with an interstage transformer ($n_1 \gg n_2$).

in L_2 is sufficiently smaller than the number of turns in L_1, the low input impedance of the second stage is stepped up, and loading is minimized.

The third method is similar, but it uses tapped capacitors, as indicated in Fig. 11-10.

In the case of FET tuned amplifiers, stages can be cascaded directly without loading problems because of the extremely high input impedance of FETs, either J-FETs or MOSFETs.

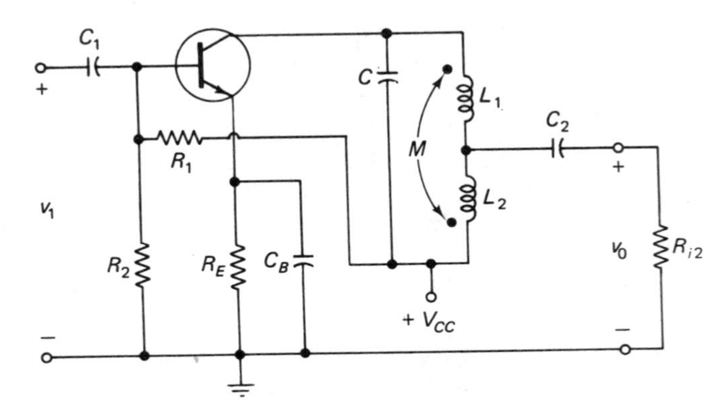

Figure 11-9. Impedance matching with tapped inductors.

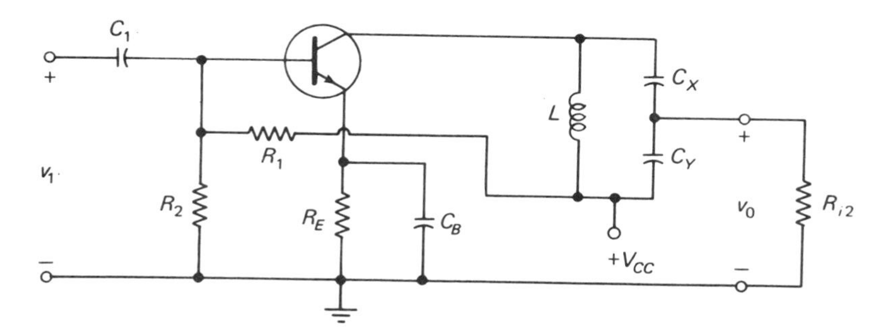

Figure 11-10. Impedance matching with tapped capacitors.

11.3 DOUBLE-TUNED AMPLIFIERS

In many applications the selectivity of a single-tuned amplifier stage is insufficient. In these cases, two or more tuned amplifier stages may be cascaded, as indicated in Fig. 11-11, or one amplifier stage may contain two tuned circuits, as shown in Fig. 11-12. In either case, two possibilities exist. The tuned circuits may be adjusted to the same center frequency (called *synchronous tuning*), or they may be adjusted to either side of the desired center frequency (called *stagger tuning*).

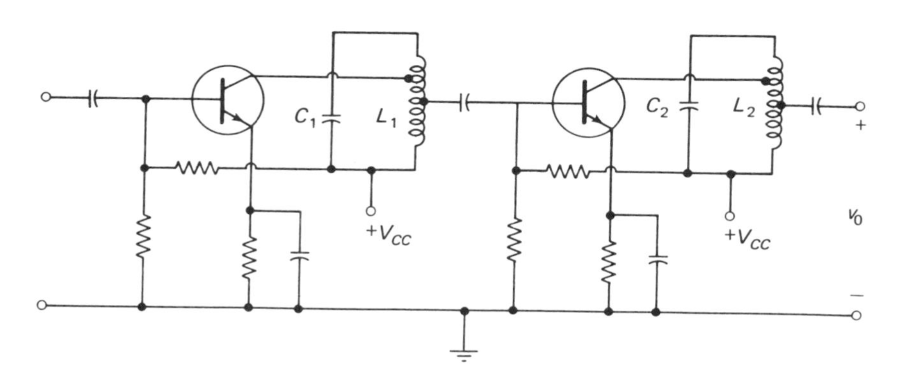

Figure 11-11. Double-tuned two-stage amplifier.

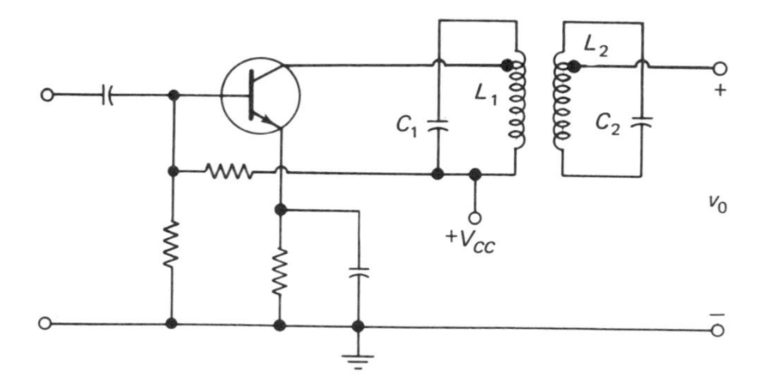

Figure 11-12. Double-tuned one-stage amplifier.

11.3.1 Synchronously Tuned Amplifiers

Let us consider the circuit of Fig. 11-11. Assume that the amplifiers are identical; that is, they have the same parameters and are tuned to the same frequency. The results are tabulated (for up to five cascaded stages) in Table 11-1 and shown graphically (for up to three cascaded stages) in Fig. 11-13. Note that as the number of stages is increased, the bandwidth of the overall amplifier is reduced and the selectivity is improved. Improvement in the selectivity is especially marked (by a factor of over 30) when we go from a single stage to two stages, so two stages are sufficient in most applications.

Table 11-1. Bandwidth Reduction in Cascaded Tuned Amplifiers

n (*Number of identical amplifier stages in cascade*)	BW_n (*3 dB bandwidth of overall n-stage amplifier*)	S (*Selectivity, see Sect. 11.2*)
1	BW*	1000.
2	0.64 BW	31.6
3	0.51 BW	10.
4	0.43 BW	5.5
5	0.39 BW	3.8

*BW denotes the 3 dB bandwidth (f_o/Q_e) of each individual stage, with all stages assumed identical.

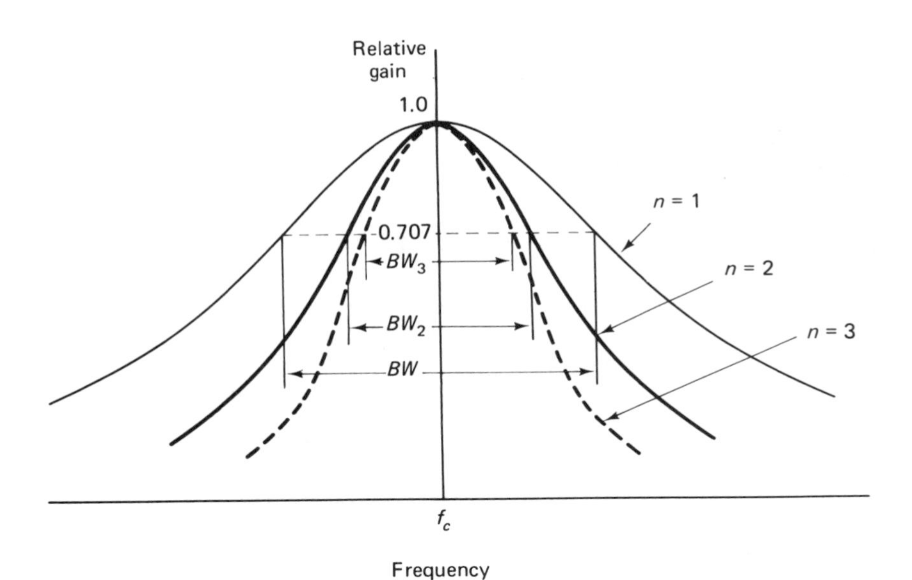

Figure 11-13. Response of multiple, synchronously tuned, cascaded (single-tuned) amplifiers showing the increased selectivity with an increase in the number of stages.

11.3.2 Stagger-Tuned Amplifiers

In order to achieve a stagger-tuned arrangement, we need to tune each of the tuned circuits depicted in either Fig. 11-11 or Fig. 11-12 slightly to one side of the desired center frequency. For example, we could tune L_1C_1 to slightly below the desired center frequency, with L_2C_2 slightly above the desired frequency. In order to achieve the optimum response (see Fig. 11-14), the tuning is critical. As shown in Fig. 11-14, if the two tuned circuits are tuned too far away from the desired center frequency, the response has a hump or dip in it. On the other hand, if the two tuned circuits are tuned too close to the center frequency, there is a loss in the selectivity possible with critical tuning. When the two tuned circuits are tuned too far away, we say that they are *loosely coupled*, or *undercoupled*. When the tuned circuits are tuned too closely, we say that they are *overcoupled*.

For critical coupling, assume that the overall bandwidth is known (labeled BW_t for total bandwidth). Then the two tuned circuits should be adjusted as follows:

$$f_{o1} = f_c - 0.35\ BW_t \tag{11-17}$$

$$f_{o2} = f_c + 0.35\ BW_t \tag{11-18}$$

$$BW_1 = BW_2 = 0.7\ BW_t \tag{11-19}$$

where f_{o1} and f_{o2} are the resonant frequencies of the two tuned circuits, BW_1 and BW_2 are the 3 dB bandwidths of the two tuned circuits respectively, and f_c and BW_t are the desired center frequency and bandwidth respectively.

Note that one resonant circuit is tuned exactly the same amount below the center frequency as the other resonant circuit is above the center frequency. Obviously, the center frequency will be exactly between the two resonant frequencies of the two tuned circuits.

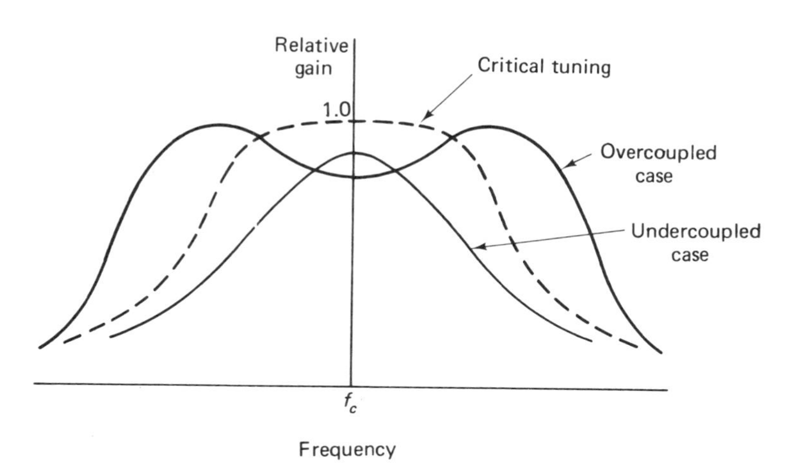

Figure 11-14. Possible cases for the response of a two-stage stagger-tuned amplifier.

Practical adjustments of stagger-tuned amplifiers are quite tricky and involve iterative trial-and-error procedures. Tuning one of the two tuned circuits changes the overall bandwidth and the center frequency. We shall illustrate the design procedure in the following example.

Example 11-4. We wish to design a two-stage stagger-tuned IF amplifier with a bandwidth of 10 kHz and a center frequency of 455 kHz. Specify the amplifier parameters.

Solution: For critical tuning, we use Eqs. (11-17) through (11-19):

$$f_{o1} = 455 - (0.35)(10) \cong 451.5 \text{ kHz}$$
$$f_{o2} = 455 + (0.35)(10) \cong 458.5 \text{ kHz}$$
$$BW_1 = BW_2 = (0.7)(10) \cong 7.0 \text{ kHz}$$

We can also proceed to calculate the loaded Qs for each stage:

$$Q_{e1} = \frac{f_{o1}}{BW_1} \cong \frac{451.5}{7.0} \cong 64.5$$
$$Q_{e2} = \frac{f_{o2}}{BW_2} \cong \frac{458.5}{7.0} \cong 65.5$$

Note that the two effective (loaded) Qs are not exactly the same.

For the preceding numbers, the individual stage responses are shown in Fig. 11-15. The response of the overall amplifier is shown in Fig. 11-16. Note that the selectivity for the stagger-tuned amplifier is better (that is, the skirt of the response is steeper) than for the single-tuned amplifier.

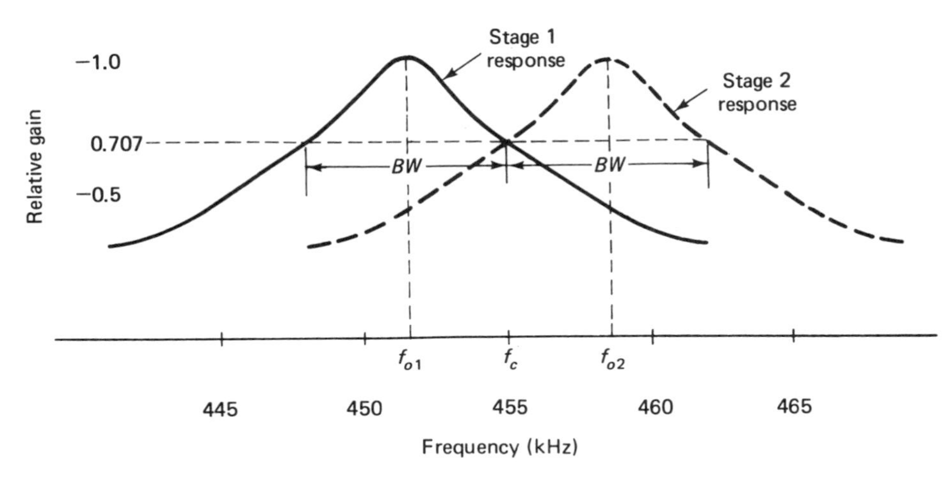

Figure 11-15. Individual response curves for stagger-tuned stages.

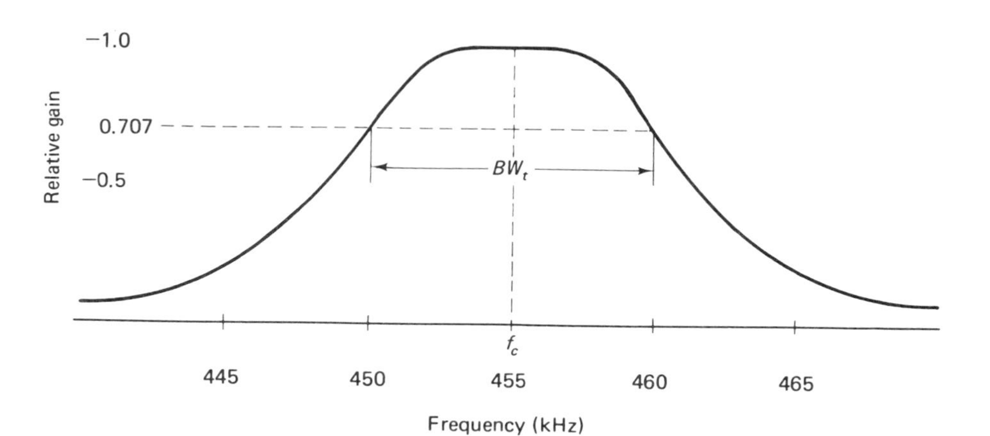

Figure 11-16. Response of overall IF stagger-tuned amplifier.

Review Questions

1. What is a tuned amplifier? Describe it in comparison to other types of amplifiers.
2. What are the characteristics of an ideal tuned amplifier?
3. What is meant by center frequency? By bandwidth?
4. What is the difference between single-tuned and double-tuned amplifiers?
5. What are the circuit components that distinguish a tuned amplifier from any other amplifier?
6. What is the significance of a tank circuit in a tuned amplifier? (Discuss in terms of its function.)
7. What is the quality factor Q? What is it a function of and what does it indicate?
8. What is the relative magnitude of the impedance of a tuned circuit at resonance with respect to its magnitude away from resonance?
9. What are the conditions for maximum gain in a tuned amplifier? Why?
10. What is the sensitivity of a tuned amplifier? How is it defined?
11. What would be the sensitivity of an ideal tuned amplifier if it existed?
12. What are the different methods of double-tuning an amplifier?
13. Describe what is meant by synchronous tuning.
14. What is meant by stagger tuning?
15. What is the effect of resistive loading on a tuned circuit? (Discuss in terms of the response.)
16. What is meant by critical coupling as applied to a double-tuned amplifier?

Problems

1. A coil of 1 mH has an equivalent series resistance of 1 Ω. Determine its Q at the following frequencies: 10, 100, 455, and 560 kHz. At what frequency is the Q equal to 100?
2. The coil specified in Example 11-1 is to be used in a tuned amplifier with a center frequency of 910 kHz. Determine the circuit Q.
3. Repeat Problem 2 if the tank circuit is loaded by an equivalent parallel resistance of 60 kΩ.
4. The single-tuned BJT amplifier of Example 11-2 is to be modified to provide a 3 dB bandwidth of 10 kHz by using a different inductor. With all other circuit values the same, can this modification be accomplished? If so, what are the parameters of the coil needed?
5. A tuned circuit uses a 200 pF capacitor and a 2.5 mH inductor. At its resonant frequency, it offers an equivalent shunt resistance of 100 kΩ. Determine the Q_o, f_o, and BW.
6. If the tuned circuit in Problem 5 is used at the output of a FET amplifier (illustrated in Fig. 11-3), which has $r_{ds} = 100$ kΩ and $g_m = 500$ μMho, determine the resonant gain, the effective Q, and the bandwidth.
7. What are the ways in which the resonant gain of the amplifier in Problem 6 could be increased? Explain.
8. The bandwidth of a tuned amplifier depends on the center frequency. If the capacitor in Problem 6 is varied from 200 to 400 pF (in steps of 50 pF), with all other circuit values constant, make a plot of the bandwidth as a function of the resonant frequency.
9. A double-tuned amplifier is formed by cascading two single-tuned amplifiers. Each has a resonant gain of 26 dB, a center frequency of 1 MHz, and an effective Q of 50. What is the response of this synchronously tuned amplifier? Sketch it.
10. A two-stage FET amplifier is shown in Fig. 11-17, with its equivalent circuit given in Fig. 11-18. We want to make it into an IF amplifier ($f_o = 455$ kHz). The two tank circuits are to be synchronously tuned. The circuit values are: $R_G = 1$ MΩ, $L_1 = L_2 = 10$ μH with $Q = 150$ at 455

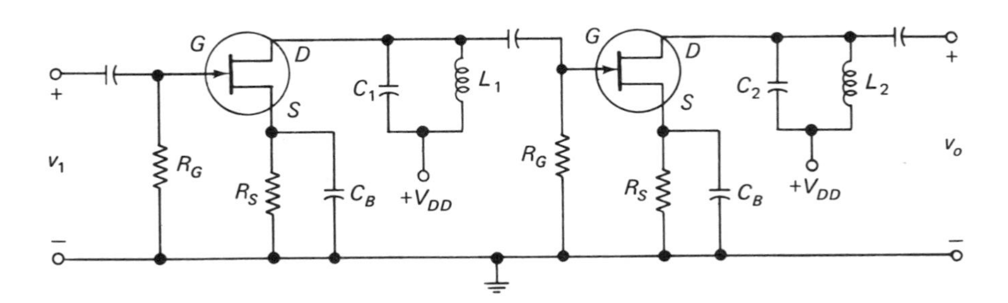

Figure 11-17.

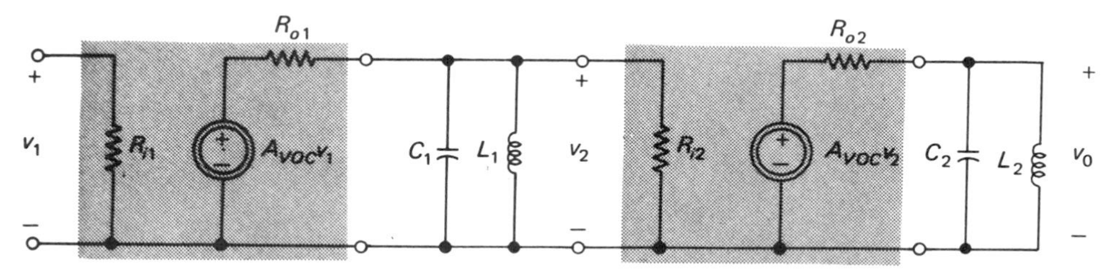

Figure 11-18.

kHz. Identical FETs are used; r_{gs} is infinite; $r_{ds} = 100\ \text{k}\Omega$; and $g_m = 500$ micromho. Determine the resonant gain, the capacitance values needed ($C_1 = C_2$), and the bandwidth.

11. A stagger-tuned amplifier is constructed with the desired bandwidth of 6 MHz and a center frequency of 60 MHz. Determine the parameters (f_o, BW, and Q_e) for each stage.
12. Repeat Problem 10 if $C_1 \neq C_2$; that is, the amplifier is stagger-tuned with critical coupling. $L_1 = L_2 = 0.1$ mH with $Q = 250$ at 455 kHz if total bandwidth is 3.03 kHz.

Power Amplifiers

In this chapter we study power amplifiers, which are characterized by two distinctive properties: (1) large signals and (2) nonlinear operation. As we discussed in Chap. 10, when the signal swing around the operating point is large, small-signal analysis is no longer valid. We must use other means. We shall introduce here graphical techniques that can be utilized in the analysis of power amplifiers.

12.1 CLASSES OF POWER AMPLIFIERS

Let us begin by examining the different classes of operation. The first class has already been discussed in terms of all small-signal amplifiers: namely, class A. As shown in Fig. 12-1(a), a class-A amplifier has an output signal that is present during the complete input-signal cycle. All linear, and some power, amplifiers are operated in class A.

In a class-B amplifier, the input signal is actually half-wave rectified, as indicated in Fig. 12-1(c). A class-B amplifier usually consists of two transistors operated in tandem, as we shall see in a later section.

A class-AB amplifier, as the name suggests, is a compromise between a class-A and class-B amplifier. The output waveshape resembles the input waveshape for almost the complete cycle, as shown in Fig. 12-1(b).

The class-C amplifier waveshapes are illustrated in Fig. 12-1(d). A class-C amplifier usually has a tuned circuit as a load and is characterized by very high power efficiencies. Its applications include radio and TV transmitters.

12.2 SERIES-FED CLASS-A AMPLIFIERS

Figure 12-2 demonstrates how the simple fixed-bias circuit discussed in Chap. 4 can be used for a power amplifier. We determine the dc Q-point by drawing the load line and then calculating the base current from the circuit values. This procedure is illustrated in the following example.

Example 12-1. A silicon *NPN* transistor TIP29 is used in the amplifier circuit of Fig. 12-2 with $V_{CC}=10$ V, $R_C=4\ \Omega$, and $R_B=470\ \Omega$. The output

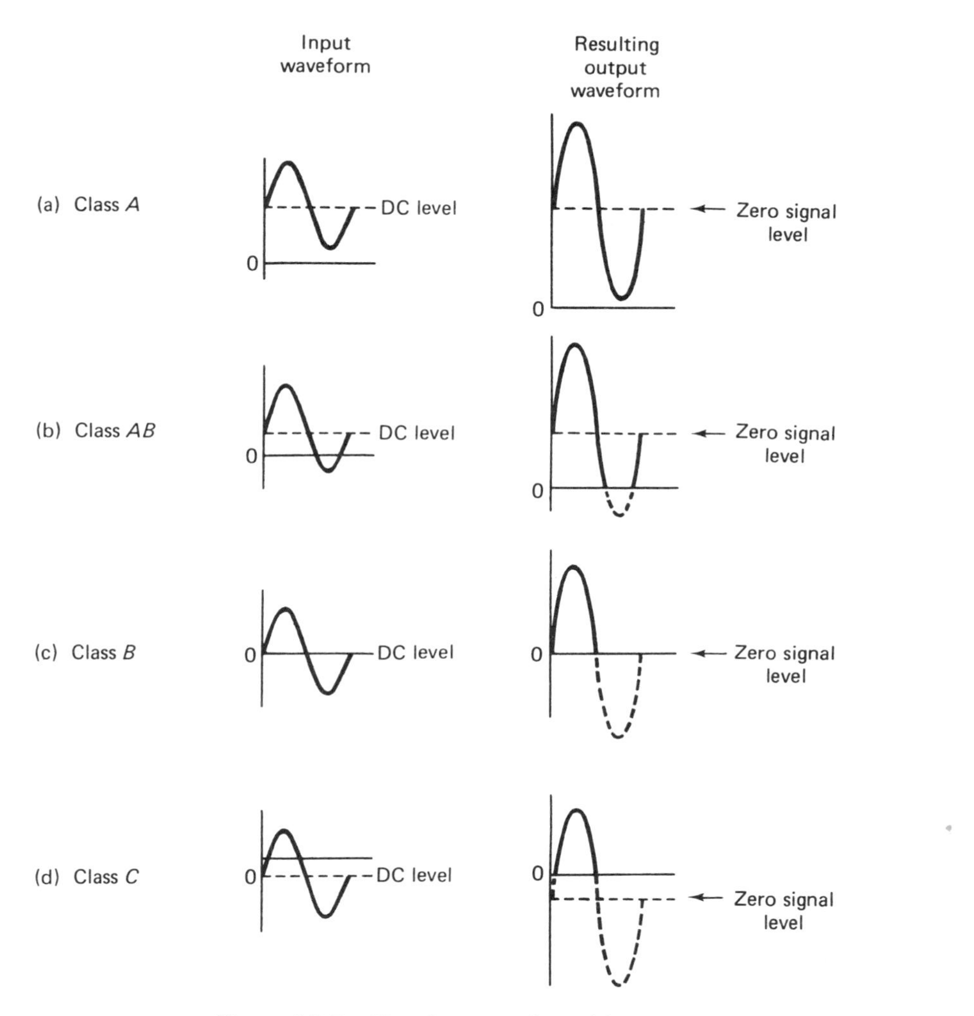

Figure 12-1. Classification of amplifiers.

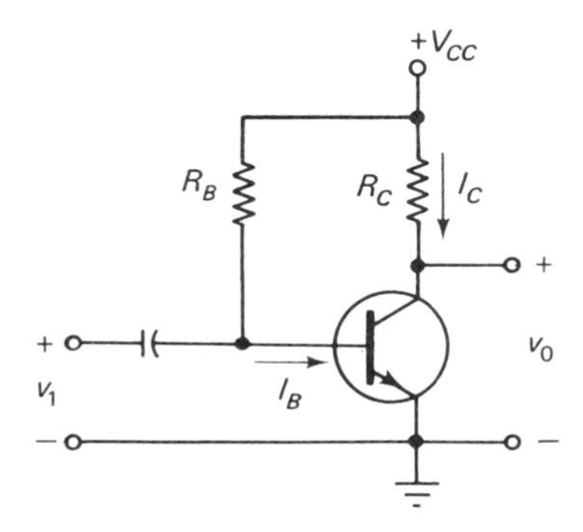

Figure 12-2. Series-fed class-A amplifier.

characteristics for the transistor may be displayed on a Transistor Curve Tracer. A typical set of these characteristics is shown in Fig. 12-3. We want to determine the dc Q-point.

Solution: To plot the load line, we observe that when no collector current flows, the collector voltage is equal to the supply voltage. Thus, we have point $P1$ as shown. We get point $P2$ by determining that when the collector current is 1 A, the voltage drop across R_C is 4 V. The collector voltage must be $10-4=6$ V. The load line is then drawn by connecting points $P1$ and $P2$, as indicated on Fig. 12-3. We calculate the base current by assuming a voltage drop of about 0.6 V between base and emitter. (It is a forward-biased junction of a silicon transistor.) Thus,

$$I_B = \frac{V_{CC} - V_{BE}}{R_B} \cong \frac{10-0.6}{0.47} \text{ mA} \cong 20 \text{ mA}$$

The Q-point is now established where the load line intersects the $I_B=20$ mA characteristic curve. The Q-point is at $I_C \cong 0.65$ A and $V_{CE} \cong 7.4$ V.

From the previous example it should be evident that we determine the dc Q-point for a power amplifier the same way as we did for small-signal amplifiers in Chap. 4.

Example 12-2. A sinusoidal ac signal is applied to the amplifier described in Example 12-1. It causes a 20 mA peak (40 mA peak-to-peak) base current. We want to determine the collector current waveshape.

Solution: The first step is to determine graphically the relationship between the base and collector currents. We make a table of two columns: one for I_B, one for I_C. We fill in the I_B column to correspond to values for which

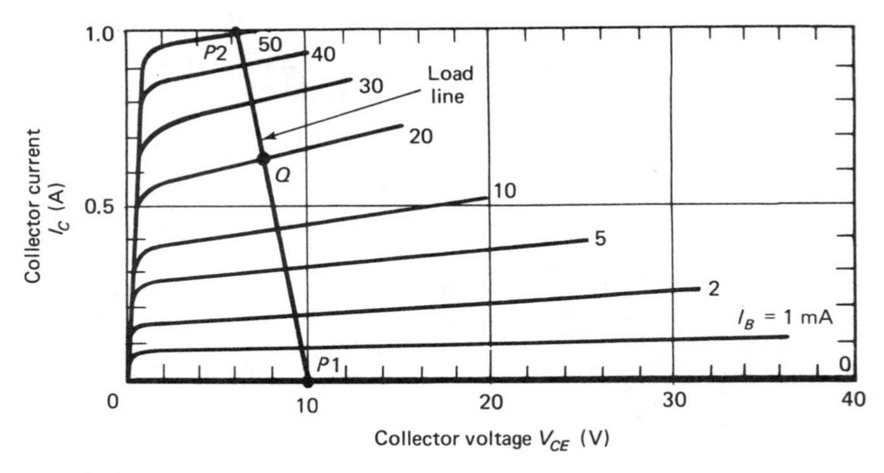

Figure 12-3. Typical power transistor output characteristics drawing the load line for Example 12-1.

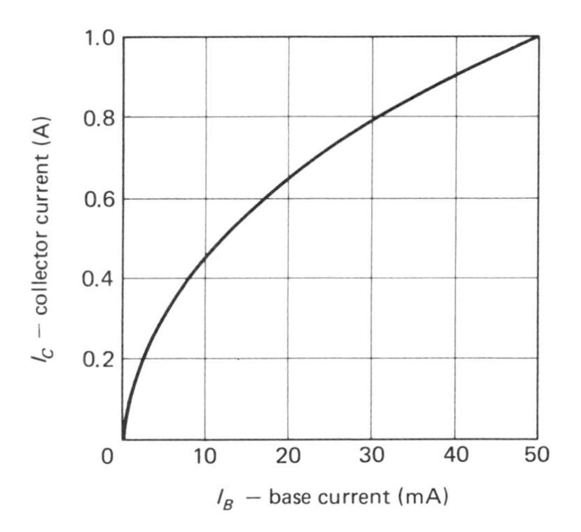

Figure 12-4. Current transfer curve for Example 12-2.

we have characteristic curves. In our example, we would choose 0, 1, 2, 5, 10, ... mA for the I_B column. For these specific values of I_B, we look up the intersections of the characteristic curves (on the output characteristics) with the load line; then we read off the corresponding values of I_C. As an example, the characteristic curve for I_B of 10 mA intersects the load line at approximately 0.45 A for I_C. The rest of the table is filled in similarly. The points from the table can now be plotted and connected by a smooth curve, as shown in Fig. 12-4. This curve is the *current transfer* curve.

The next step is to make a plot of a sinusoidal waveshape for I_B below the transfer curve, as indicated in Fig. 12-5. (When you plot a sinusoidal waveshape, it should be sufficient to plot the nine points A through I, as shown.) The important thing to remember is to line up the zero for the waveshape with the dc value of I_B (in this case, 20 mA). Because, in a manner of speaking, the ac part of the waveshape "rides" on top of the dc level, this procedure must be followed. Once the complete base current signal is plotted below the transfer curve as shown in Fig. 12-5, a second time scale is drawn to the right of the transfer curve for the collector current.

To obtain the collector current waveshape, we use the following construction: With a straight edge, project the point A on the base current waveshape directly upwards until it intersects the transfer curve (also point A). Now project this point directly to the right until it intersects the $\omega t = 0$ point on the collector current scale. This point is also labeled A. The time reference (value of ωt) on both the base current and collector current scales must be observed. In a similar manner, project point B from the base current graph upwards to the transfer curve, then to the right to the point $\omega t = 30$ on the collector current graph, also point B. To complete the

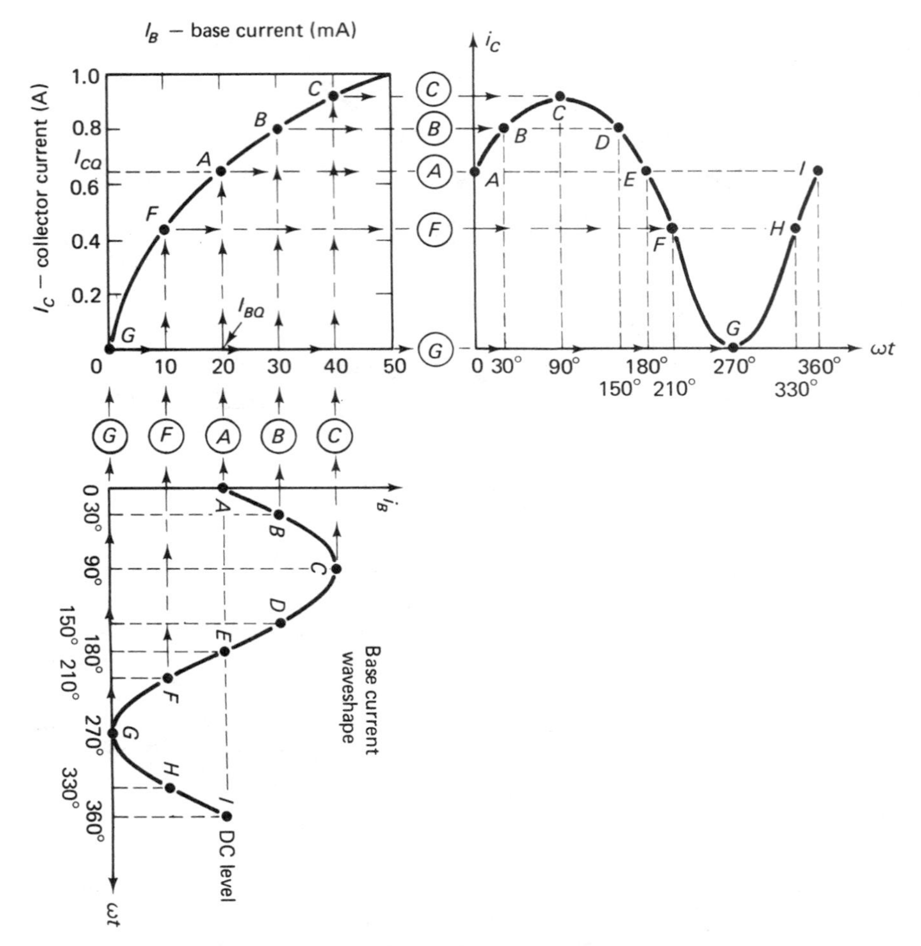

Figure 12-5. Constructing the collector current waveshape using the transfer characteristic (see Example 12-2).

construction, proceed with the projections of points *C, D, E, F, G, H,* and *I*. Note that, in reality, points *D, E, H,* and *I* do not need to be projected. They give the same intercepts on the transfer curve as points *B, A, F,* and *A*, respectively. Therefore, they may be plotted on the I_c graph without projection, still observing the time reference.

Once the points *A* through *I* are obtained on the graph to the right, the collector current waveshape may be sketched in by connecting the points with a smooth curve. We note that the collector current is 0.9 A peak-to-peak and that the waveshape is not an exact reproduction of the input waveshape. In other words, the output waveshape is distorted.

It is imperative to realize that the points *A* through *I* in the previous example were not chosen at random. Points *A, E,* and *I* were chosen where the sinusoid is zero—at 0°, 180°, and 360°, respectively. Points B and D are located where the sinusoid is one-half of its positive peak value—at 30° and 150°, respectively. Points F and H indicate where the sinusoid is one-half of the negative peak value—at 210° and 330°, respectively. Lastly, points C and G were chosen where the sinusoid has its positive and negative peaks—at 90° and 270°, respectively. It is most important to observe these choices, especially in the calculation of harmonic distortion.

12.3 POWER EFFICIENCY AND DISSIPATION

In a power amplifier, unlike a small-signal amplifier, we are interested in the average ac power delivered to the load rather than in voltage or current gain. Moreover, we want to see how efficiently we are delivering power to the load. Lastly, we must concern ourselves with the power dissipated in the transistor to insure that it is not excessive. Excessive power would not only cause permanent damage to the transistor but also make it useless.

12.3.1 Power Efficiency

The power efficiency of an amplifier is given by the ratio between the ac power delivered to the load and the dc power that the power supply delivers:

$$\%\ \text{efficiency} = \frac{P_{o(ac)}}{P_{i(dc)}} \times 100 \tag{12-1}$$

The dc input power is calculated as the product of the power supply voltage (V_{CC}) and the average current delivered (approximately given by I_{CQ}). Thus,

$$P_{dc} = V_{CC} I_{CQ} \tag{12-2}$$

The ac power delivered to the load is the product of the rms values of the output voltage and currents. The rms value for any sinusoidal signal may be found by dividing the peak-to-peak value by $2\sqrt{2}$. Thus,

$$P_{ac} = \frac{(V_{p-p})(I_{p-p})}{(2\sqrt{2})(2\sqrt{2})} = \frac{(V_{p-p})(I_{p-p})}{8} \tag{12-3}$$

where V_{p-p} and I_{p-p} are the output peak-to-peak voltage and current, respectively. In the case where the load is resistive, we can express the voltage in terms of the current:

$$V_{p-p} = I_{p-p} R_C \tag{12-4}$$

and rewrite the ac power equation accordingly:

$$P_{ac} = \frac{(I_{p-p})^2 R_C}{8} \tag{12-5}$$

With the preceding equations, it is possible to determine the ac output power, the dc input power, and the power efficiency for any given power amplifier.

Efficiency in power amplifiers is important because power levels are in the watt range; whereas in small-signal amplifiers the power involved does not exceed a few hundred milliwatts at most. Calculation of power efficiency is illustrated in the following example.

Example 12-3. Calculate the ac and dc power as well as the power efficiency for the series-fed class-A amplifier discussed in Examples 12-1 and 12-2.

Solution: The Q-point collector current was determined in Example 12-1 to be 0.65 A. Using Eq. (12-2) we get:

$$P_{dc} = (0.65)(10)\ \mathrm{W} \cong 6.5\ \mathrm{W}$$

From Fig. 12-5 and Example 12-2 we see that the peak-to-peak collector current is 0.9 A. This fact, together with an R_C of 4 in Eq. (12-5), gives us:

$$P_{ac} = \frac{(0.9)^2(4)}{8}\ \mathrm{W} \cong 0.405\ \mathrm{W}$$

The power efficiency is now calculated from Eq. (12-1):

$$\%\ \text{efficiency} = \frac{(0.405)}{(6.5)} \times 100 \cong 6.23\%$$

As we can readily see, the power efficiency of a series-fed class-A amplifier is typically very low, making this circuit rather impractical for high-power applications.

As a sidelight to the preceding example note that the power efficiency of a series-fed class-A amplifier has 25% as the absolute maximum, again making this type of amplifier a poor choice where power efficiency is a serious consideration.

12.3.2 Transistor Maximum Power Dissipation

How do we choose a transistor to use in a given situation? First, we are usually limited to those transistors that are already in stock. Then we must decide on the specific transistor that would be suitable for a certain application. The answer is usually contained in the manufacturers' data sheets. Without implying that it is the best, we have chosen a Texas Instruments TIP29 silicon *NPN* transistor as being representative for the examples here. Full data sheets are given in Appendix 3.

The first step in determining the suitability of a transistor for a certain application is to establish its permissible region of operation. The permissible region of operation for a TIP29 transistor is shown as the shaded area in Fig. 12-6. It is determined by the saturation and cutoff regions, as shown, and by three quantities specified by the manufacturer. These quantities are (1) the absolute maximum collector current, (2) the absolute maximum collector voltage, and (3) the continuous device power dissipation. For a TIP29, the manufacturer lists these as 1 A, 40 V, and 2 W, respectively. The manufacturer also supplies two power ratings. One

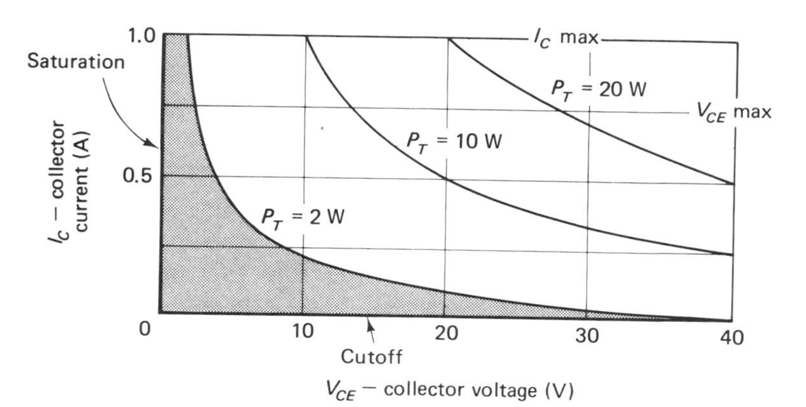

Figure 12-6. **Permissible region of operation (shaded area) for a typical power transistor (Texas Instruments TIP29) without a heat sink. The maximum power capability increases with the use of a heat sink.**

that is usually specified as "continuous device dissipation at (or below) 25°C free-air temperature" is the maximum power rating *without a heat sink*. The second rating, "continuous device dissipation at (or below) 25°C case temperature," is the maximum power rating with additional cooling provided by a fan or a heat sink, or both.

Under normal operation, the emitter junction is forward biased and thus has a voltage of less than 1 V across it. The collector junction, on the other hand, usually has a much higher voltage across it. With the current through the two junctions approximately the same, the collector junction must be dissipating essentially all of the power given out by the transistor. If we make that approximation, then the total power dissipated in the transistor is given by:

$$P_T = I_C V_{CE} \tag{12-6}$$

We can plot this equation for $P_T = 2$ W in Fig. 12-6 by choosing values of either I_C or V_{CE} and calculating the other.

The manufacturer also specifies the "operating collector junction temperature range." For the TIP29 this range is listed as −65°C to 150°C. The more important quantity here is the upper limit. Typically, the maximum collector junction (or simply junction) temperature is 150°C for silicon transistors and 75°C for germanium transistors. This maximum junction temperature should not be exceeded; it determines the amount of power that can safely be dissipated with a given heat sink mounted on the transistor.

Manufacturers usually provide power dissipation derating information or curves of the type depicted in Fig. 12-7. With ambient or room temperature assumed at 25°C, the relationship between the junction temperature and the total power dissipated is:

$$T_J - 25 = \theta_{JA} P_T \tag{12-7}$$

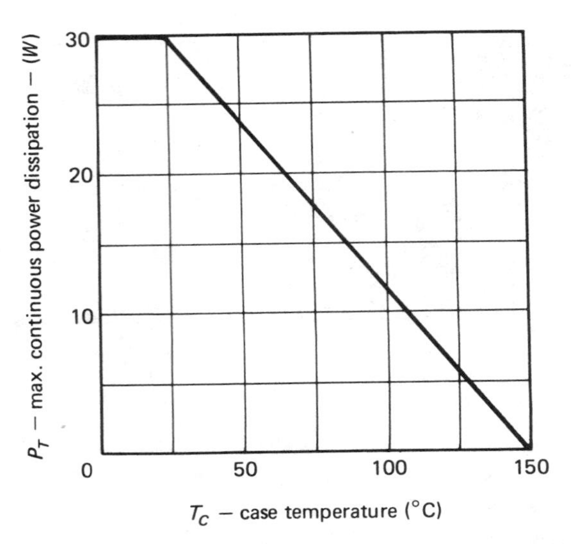

Figure 12-7. Power dissipation derating curve for Texas Instruments TIP29 silicon *NPN* transistor.

where T_J is the collector junction temperature in degrees centigrade and θ_{JA} is the junction-to-ambient thermal resistance in degrees centigrade per watt. Without a heat sink, θ_{JA} is a property of the transistor and indicates the increase in junction temperature (above room temperature) for each additional watt of power dissipated in the transistor. For example, suppose that we dissipate 1 W in a TIP29 transistor, which has θ_{JA} listed as 62.5°C/W. Then the junction temperature will rise 62.5°C above room temperature, or will be 87.5°C. This temperature is well below the 150°C maximum, so safe operation would result. If, however, we tried to dissipate 3 W (without a heat sink), the transistor would probably burn out, because the junction temperature would reach upwards of 200°C, which is well in excess of the 150°C maximum for this transistor.

By using a thermally conductive, electrically insulating washer (made of mica or other suitable material) and a heat sink, we can increase the power dissipation in the transistor without exceeding the maximum temperature. In this case, we can rewrite Eq. (12-7):

$$T_J - 25 = P_T(\theta_{JC} + \theta_{CS} + \theta_{SA}) \tag{12-8}$$

where the single junction-to-ambient thermal resistance has been replaced by three separate thermal resistances: θ_{JC} is the thermal resistance between the junction and the case of the transistor; θ_{CS} is the thermal resistance between the case and heat sink (i.e., the thermal resistance of the washer); and θ_{SA} is the thermal resistance between the heat sink and the air around it (i.e., the thermal resistance of the heat sink). If no washer is used, θ_{CS} and θ_{SA} are replaced by θ_{CA}, which is the thermal resistance of the heat sink by itself.

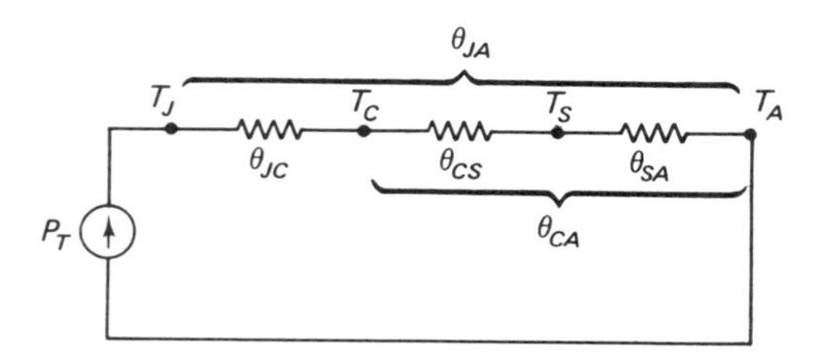

Figure 12-8. Electrical analog for determining power dissipation and temperature for power transistors.

To help you visualize the operation, we show an electrical analog for Eq. (12-8) in Fig. 12-8. The total power is analogous to current; the different temperatures are analogous to voltages; and thermal resistances are analogous to electrical resistances.

Example 12-4. Without a heat sink, the TIP29 transistor can safely dissipate 2 W. Its thermal resistances are listed as: $\theta_{JC} = 4.17°\text{C/W}$ and $\theta_{JA} = 62.5°\text{C/W}$. Let us determine the thermal resistance of the heat sink needed to safely dissipate (1) 10 W and (2) 20 W, assuming the thermal resistance of the mica washer to be 1°C/W. (This is a typical value.)

Solution: For 10 W and with T_J a maximum of 150°C, we calculate that the θ_{JA} needed is

$$\theta_{JA} = \frac{150-25}{10} = 12.5°\text{C/W}$$

The heat sink thermal resistance θ_{SA} is found next:

$$\theta_{SA} = \theta_{JA} - \theta_{JC} - \theta_{CS} = 12.5 - 4.17 - 1 = 7.3°\text{C/W}$$

As a matter of interest, we can determine the temperatures T_C and T_S in Fig. 12-8 to be 108°C and 98°C, respectively. There is then a 42° temperature differential between the junction and the transistor case, a 10° differential across the washer, and a 73° differential across the heat sink. Thus, the effect of the heat sink and washer is to reduce the total temperature differential between the transistor collector junction and its case, or to help remove the heat from the junction.

We proceed in a similar manner to calculate what thermal resistance is needed in the heat sink to dissipate 20 W. We obtain:

$$\theta_{JA} = \frac{150-25}{20} = 6.25°\text{C/W}$$

Thus, the heat sink thermal resistance must then be:

$$\theta_{SA} = 6.25 - 4.17 - 1 = 1.08°\text{C/W}$$

Heat sink information is available from manufacturers of heat sinks and in the *Motorola Semiconductor Data Book*, 4th edition. As a general guide, the larger the lateral surface *area*, the lower the thermal resistance of the sink and the better it dissipates heat.

Figure 12-6 plots the power limits of 10 and 20 W with the heat sinks as determined in the preceding example. The permissible region of operation, therefore, can be increased with the use of proper heat sinks.

We must mention one important fact in connection with maximum power considerations. You have to take care when mounting and installing power transistors in a case so that there is sufficient ventilation and air circulation around the transistor and its heat sink (if used). This is to insure that the air in contact with the case or heat sink will be at about 25°C, and not some higher value, which would invalidate all the calculations carried out to determine the proper heat sink.

One additional specification in the manufacturer's data sheet needs further clarification: the "peak collector current," which usually has an accompanying note. This quantity is not the same as the maximum collector current discussed earlier. For example, in the TIP29 data sheet, the manufacturer specifies "peak collector current...3 A" with a note that "this value applies for $t_w \leqslant 0.3$ ms, duty cycle $\leqslant 10\%$." This specification means that the collector current may go as high as 3 A for a maximum time of 0.3 millisecond (ms) or for less than 10% of one cycle, whichever is less. Thus, if the signal frequency is 1 kHz, the collector current could go as high as 3 A for a time not to exceed 0.1 ms during each cycle.

12.4 HARMONIC DISTORTION

As we noted earlier (see Fig. 12-5), nonlinear operation results in distortion of the waveshape at the output. From the graphic analysis discussed in section 12.2, we can determine the harmonic content and distortion in the output waveshape. We can express the output waveshape in the form of a truncated series:

$$I_C = M_0 + M_1\cos(\omega t) + M_2\cos(2\omega t) + M_3\cos(3\omega t) + M_4\cos(4\omega t) \tag{12-9}$$

where ω is the frequency of the input signal (called the *fundamental frequency*); M_1 is the amplitude of the fundamental component of the output wave; M_2, M_3, and M_4 are the amplitudes of the second, third, and fourth harmonics, respectively. These M coefficients can be determined from the graphic analysis:

$$\begin{aligned}
M_0 &= \frac{1}{6}(I_M + I_m) + \frac{1}{3}(I_1 + I_2) - I_Q \\
M_1 &= \frac{1}{3}(I_M - I_m) + \frac{1}{3}(I_1 - I_2) \\
M_2 &= \frac{1}{4}(I_M + I_m) - \frac{1}{2}I_Q \\
M_3 &= \frac{1}{6}(I_M - I_m) - \frac{1}{3}(I_1 - I_2) \\
M_4 &= \frac{1}{12}(I_M + I_m) - \frac{1}{3}(I_1 + I_2) + \frac{1}{2}I_Q
\end{aligned} \tag{12-10}$$

The symbols have the following definitions:

I_Q—no-signal collector current	(point A, Fig. 12-5)
I_M—peak collector current	(point C, Fig. 12-5)
I_m—minimum collector current	(point G, Fig. 12-5)
I_1—half-peak collector current	(point B, Fig. 12-5)
I_2—half-minimum collector current	(point F, Fig. 12-5)

Once the M coefficients are determined, the harmonic content of the output wave is known and the amount of distortion may be calculated as follows:

$$D_2 \cong \left| \frac{M_2}{M_1} \right| \times 100\%$$

$$D_3 \cong \left| \frac{M_3}{M_1} \right| \times 100\% \qquad (12\text{-}11)$$

$$D_4 \cong \left| \frac{M_4}{M_1} \right| \times 100\%$$

where D_2, D_3, and D_4 are the second, third, and fourth harmonic distortions in percentage, respectively. We find the total harmonic distortion, D_T, from the following equation:

$$D_T = \sqrt{D_2^2 + D_3^2 + D_4^2} \qquad (12\text{-}12)$$

These calculations are best illustrated in the following example.

Example 12-5. For the series-fed class-A amplifier described in Example 12-2, we want to determine the harmonic content and distortion in the output waveshape.

Solution: The numbers to be used in Eq. (12-10) are determined by inspecting the output waveshape shown in Fig. 12-5. They are:

$I_Q \cong 0.65$ A	(from point A on Fig. 12-5)
$I_M \cong 0.9$ A	(from point C on Fig. 12-5)
$I_m \cong 0.0$ A	(from point G on Fig. 12-5)
$I_1 \cong 0.8$ A	(from point B on Fig. 12-5)
$I_2 \cong 0.45$ A	(from point F on Fig. 12-5)

We then insert these values into Eq. (12-10) to find the M coefficients:

$$M_0 = \frac{1}{6}(0.9) + \frac{1}{3}(0.8 + 0.45) - 0.65$$
$$\cong -0.083 \text{ A} \cong -83 \text{ mA}$$
$$M_1 = \frac{1}{3}(0.9) + \frac{1}{3}(0.8 - 0.45) \cong 0.42 \text{ A}$$
$$M_2 = \frac{1}{4}(0.9) - \frac{1}{2}(0.65) \cong -0.1 \text{ A}$$
$$M_3 = \frac{1}{6}(0.9) - \frac{1}{3}(0.8 - 0.45)$$
$$\cong 0.03 \text{ A} \cong 30 \text{ mA}$$
$$M_4 = \frac{1}{12}(0.9) - \frac{1}{3}(0.8 + 0.45) + \frac{1}{2}(0.65)$$
$$\cong -0.017 \text{ A} \cong -17 \text{ mA}$$

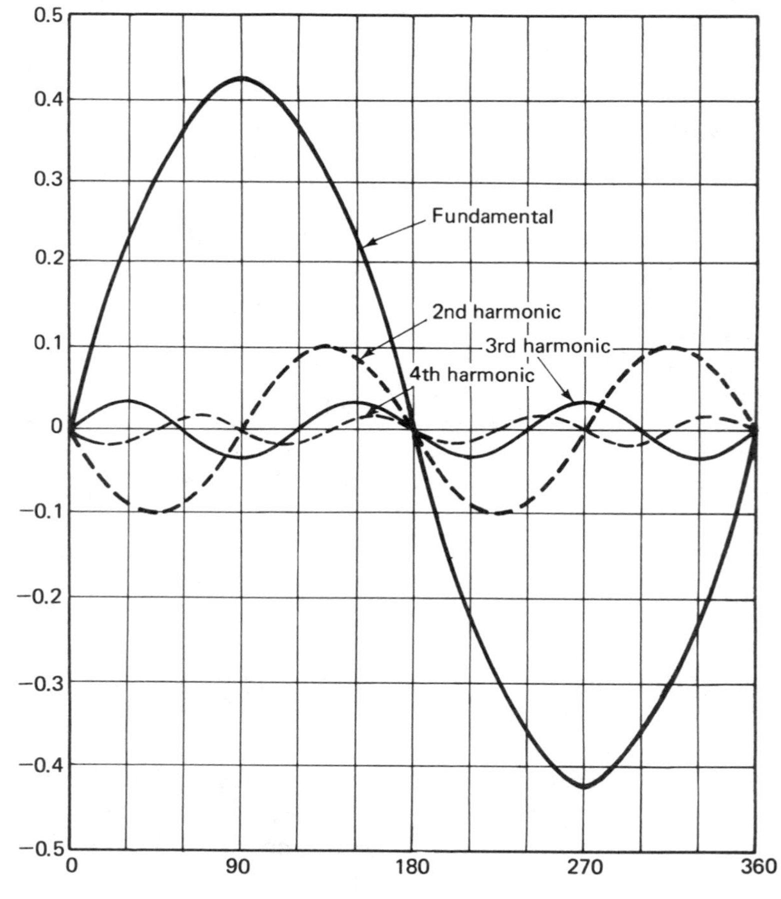

Figure 12-9. Approximate harmonic content of the collector waveshape of Fig. 12-5. (The phase relationship among harmonics is not shown.)

The separate and total harmonic distortion can now be calculated:

$$D_2 = \frac{0.1}{0.42} \times 100\% \cong 24\%$$

$$D_3 = \frac{0.03}{0.42} \times 100\% \cong 7\%$$

$$D_4 = \frac{0.017}{0.42} \times 100\% \cong 4\%$$

$$D_T = \sqrt{(24)^2 + (7)^2 + (4)^2} \cong 25\%$$

Obviously, harmonic distortion is high (as we might have expected from the actual waveshape). In essence, the total harmonic distortion is mainly due to the second harmonic.

To see the meaning of these harmonic calculations, you must plot the individual waveshapes, which add up to produce the collector current of Fig. 12-5, for the numbers obtained in Example 12-5 (see Fig. 12-9).

In practice, harmonic distortion analysis may be carried out either (1) by observing the actual waveshape on an oscilloscope and proceeding as just outlined or (2) by using a special instrument called a *wave,* or *distortion, analyzer* that gives the harmonic amplitudes directly. Another method involves using a spectrum analyzer to make the measurements.

12.5 SINGLE-ENDED CLASS-A AMPLIFIERS

As we have seen in the previous sections, the series-fed class-A amplifier performs poorly, both in terms of power efficiency and distortion. Consequently, it is not commonly used, but it does serve to illustrate some of the basic graphic techniques. Let us next examine the transformer-coupled single-ended class-A amplifier, whose typical circuit is illustrated in Fig. 12-10.

The analysis of the circuit shown in Fig. 12-10 is quite similar to that of the series-fed amplifier, with two important differences. First, the dc load line of the transformer-coupled amplifier has a slope proportional to the negative reciprocal of $R_E + R$, where R is the effective series resistance of the primary (input winding) of the transformer. The other difference is that the ac load line of the transformer-coupled amplifier is not the same as the dc load line; it is governed by the turns ratio n and the load resistance R_L. The effective or reflected ac load on the amplifier is given by:

$$R_L' = \frac{1}{n^2} R_L \tag{12-13}$$

where the transformer turns ratio is defined in terms of the number of turns in the primary (N_1) and secondary (N_2) winding:

$$n = \frac{N_2}{N_1} \tag{12-14}$$

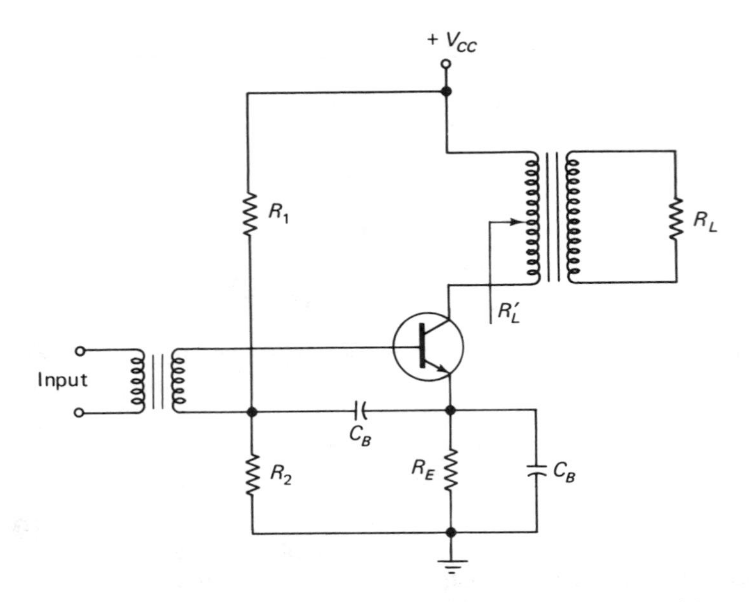

Figure 12-10. Transformer-coupled single-ended class-A amplifier.

The transformer is used as an impedance-matching device to match the actual load, R_L, to the output of the transistor. The actual load might typically be 4 or 8 Ω (for a speaker). By using the transformer, we can increase the load of the transistor to a few hundred or even a few thousand ohms.

In analyzing the transformer-coupled circuit, we may proceed to draw the dc load line and establish the Q-point as we did for the series-fed amplifier. The ac load line is then plotted to pass through the Q-point and to have the slope of $-1/R'_L$. This plot is shown on a typical set of output characteristics in Fig. 12-11.

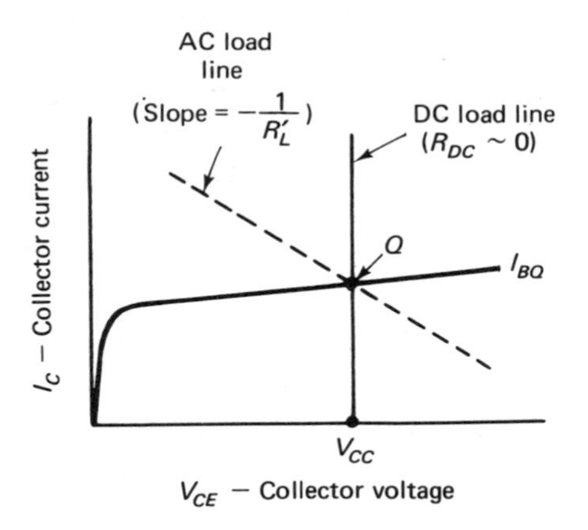

Figure 12-11. Ac and dc load lines for transformer-coupled class-A amplifier.

From that point on, the graphical analysis parallels the one used in the previous sections. However, we use the ac load line, which is different from the dc load line in this case.

Typical performance comparisons of the series-fed and transformer-coupled amplifiers, we shall see, yield somewhat lower distortion and somewhat higher power efficiency for the transformer-coupled amplifier. The absolute maximum power efficiency for the transformer-coupled amplifier is 50%, as compared to 25% for the series-fed amplifier.

12.6 TRANSFORMER-COUPLED PUSH-PULL AMPLIFIERS

A typical transformer-coupled push-pull amplifier circuit is depicted in Fig. 12-12. We can explain its operation by drawing waveshapes, as shown. If the input signal is sinusoidal, the input transformer applies it to the two transistors. But the two base signals are 180° out of phase. Let us consider that the two transistors are biased so that each will operate in class B (with R_1 removed and R_2 shorted). When the base of $Q1$ goes positive, it conducts and amplifies the input signal. At the same time, the base of $Q2$ is being driven negative, and $Q2$ is cut off. Thus, the output is being supplied by $Q1$. During the other half of the cycle, the base of $Q1$ is driven negative, while the base of $Q2$ is driven positive. The situation is exactly reversed, with $Q1$ being cut off and $Q2$ conducting. The collector signals are coupled through the output transformer to the load.

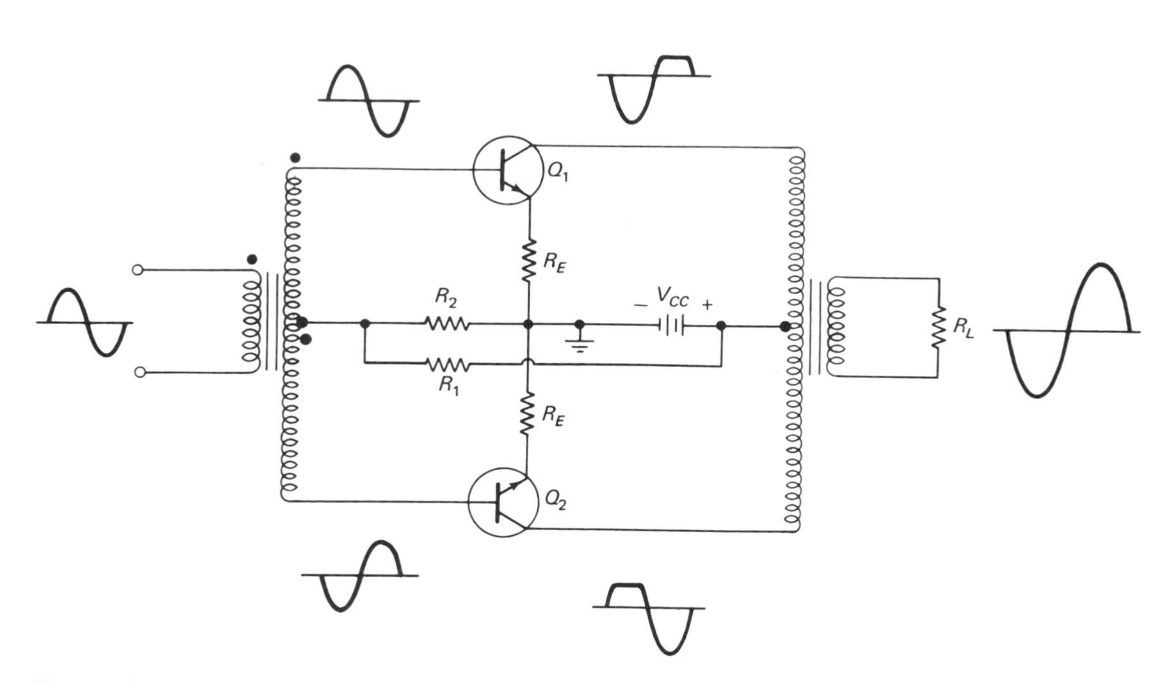

Figure 12-12. Transformer-coupled push-pull amplifier (shown with waveshapes for class-AB operation).

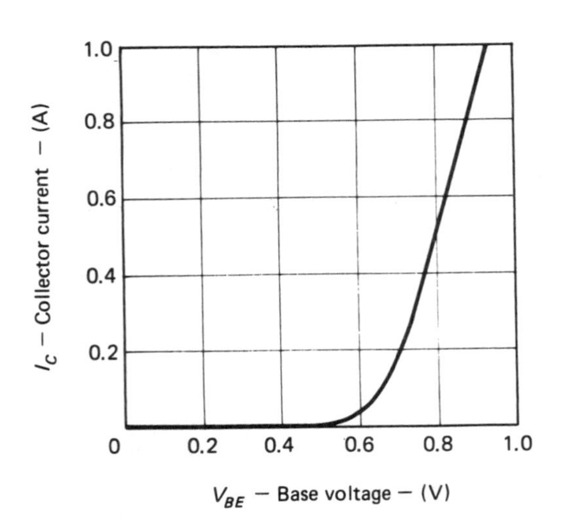

Figure 12-13. Typical voltage transfer curve for a power transistor.

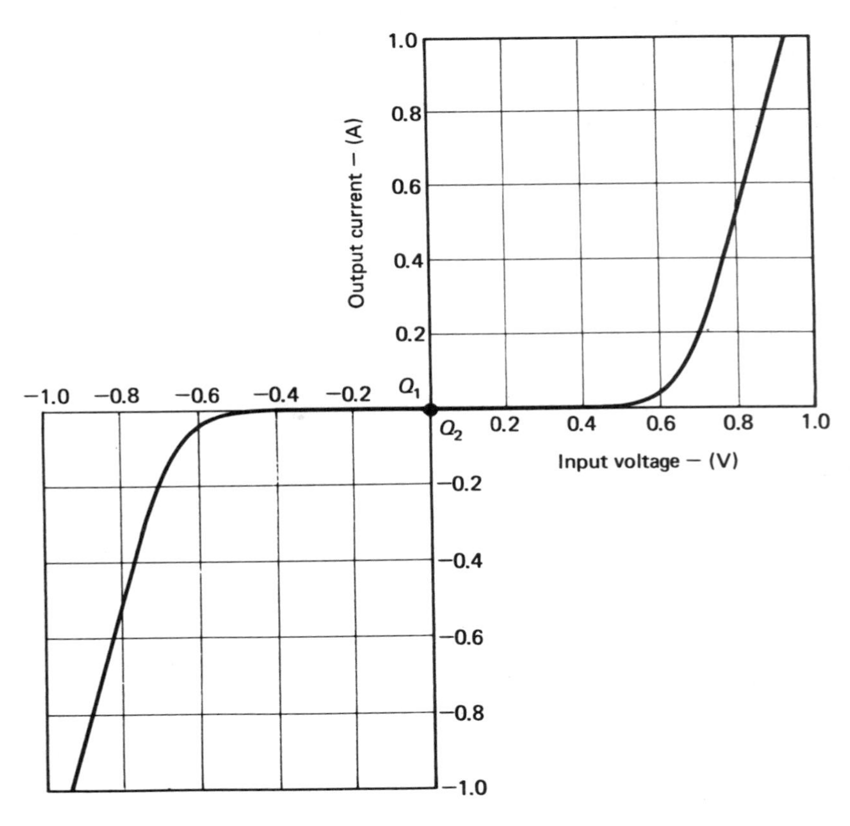

Figure 12-14. Voltage-to-current transfer curve for a class-B push-pull (or complementary-symmetry) amplifier.

The name for the amplifier is indicative of circuit operation. One transistor is conducting while the other one is off, and vice versa.

The waveshapes shown along with the push-pull circuit in Fig. 12-12 are for class-AB operation. Before we go on to class-AB operation, however, let us examine the characteristics of class-B amplifiers. Then you will understand the need for class-AB.

Let us assume that the input signal is from a very low impedance source, so that the input to the transistors is a voltage. A set of voltage-to-current transfer characteristics for a typical transistor is shown in Fig. 12-13. We can also assume that the transistors used in the push-pull amplifier are matched and have exactly the same properties. The composite transfer curve for the push-pull amplifier as a whole is obtained (in class-B operation) by placing the individual transistor transfer curves as indicated in Fig. 12-14. The horizontal axis (which was V_{BE} for the individual transistors) now becomes the input voltage axis for the composite device: the push-pull amplifier. Above the horizontal axis is the collector current for $Q1$, below is the collector current for $Q2$. The output current is the algebraic difference of the two collector currents, thus accounting for the inversion of the $Q2$ characteristic. The vertical axis is then the output current for the push-pull amplifier.

One of the advantages of the push-pull amplifier is that the second and fourth harmonics of the two transistors are exactly in phase and cancel in the output transformer. The output current contains only the third harmonic. We, therefore, would expect very low distortion from a push-pull amplifier. However, this is not the case. Consider the following example.

Example 12-6. The transformer-coupled amplifier in Fig. 12-12 is operated in class B. The voltage-current transfer curve for the composite device is illustrated in Fig. 12-14. The input voltage is a 0.9 V peak sinusoid with zero dc component, as shown in Fig. 12-15(a). Determine the output waveform.

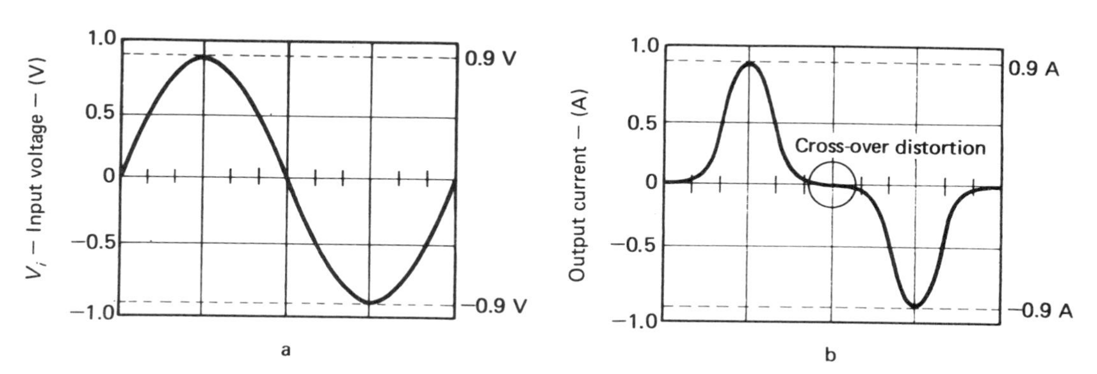

Figure 12-15. Class-B push-pull amplifier: (a) input waveshape and (b) output waveshape.

Solution: We use the same graphic techniques as in Example 12-2; that is, we sketch the input wave below the transfer characteristics and project it up to the characteristics and across to the right to get the output waveshape. Although this construction is not shown, the result (i.e., the output waveshape) appears in Fig. 12-15(b). Note the extreme flattening near the origin; this characteristic is called *crossover distortion*.

Although the second and fourth harmonics are absent for a class-B push-pull amplifier, distortion is not low. There is a very large third harmonic component.

To eliminate some of this distortion, we operate the transistors in class AB by biasing slightly. The dc voltage across R_2 is allowed to exceed the dc voltage across R_E. For example, let us assume that the base-emitter voltage of each transistor is raised to, say, 0.65 V at the operating point. The composite characteristics for this case (class AB) are found by matching the individual characteristics at the V_{BE} operating-point value; we obtain 0.65 V, as shown in Fig. 12-16. To determine the composite curve,

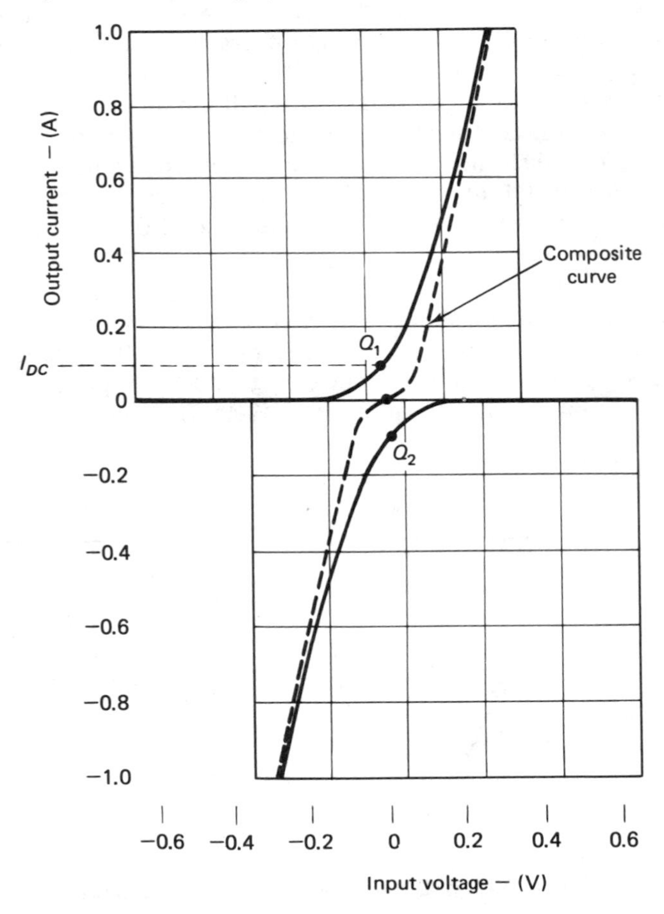

Figure 12-16. Transfer characteristics for class-AB operation.

we calculate the algebraic sum of the individual curves at a particular voltage. The result is indicated by the dashed curve in Fig. 12-16. Note that the class-AB composite curve is nearly a straight line; in any case, it is much closer to a straight line than the transfer curve for the class-B amplifier in Fig. 12-14.

By biasing the bases even more positive, we can make the composite transfer curve almost perfectly linear, as shown in Fig. 12-17. This special case of class AB is sometimes called class ABB. The linearity of the transfer curve indicates that distortion will be extremely small. However, this improvement is not completely free. As we can see in Fig. 12-17, the Q-point is fairly high; that is, there is a fairly large dc (idling) current which decreases the power efficiency. In general, class-B amplifiers have the best power efficiency, but large-distortion class-AB amplifiers have fairly good (i.e., high) power efficiency and very low distortion. The choice is a trade-off between power efficiency and harmonic distortion.

We must mention another difference between class-B and class-AB amplifiers. From Fig. 12-17, it becomes evident that only a 0.2 V peak

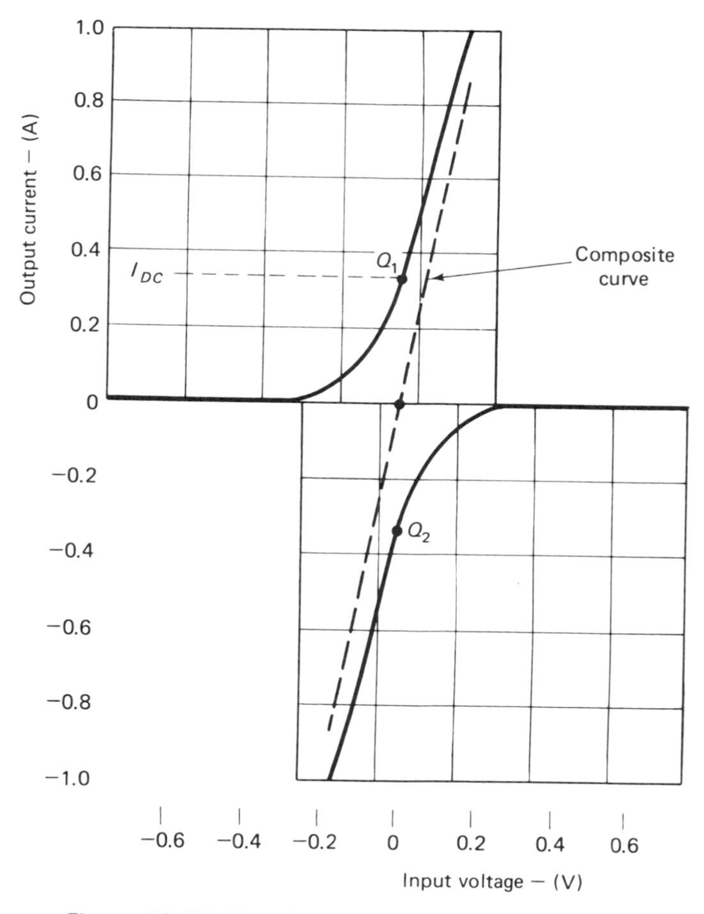

Figure 12-17. Transfer curve for class-ABB operation.

input voltage is necessary to cause approximately the same output current as the 0.8 V peak signal in the class-B amplifier. Thus, the gain is increased when the transistors are biased upwards. In fact, even higher gain and lower harmonic distortion are possible when the amplifier is operated in class A. The transfer curve for class-A operation is shown in Fig. 12-18. Class-A operation is very low in power efficiency—so low, in fact, that it is uneconomical.

Among the four operating points (Figs. 12-14, 12-15, 12-17, and 12-18), perhaps the best compromise is shown in Fig. 12-16. The power efficiency is quite good, and the distortion is low enough that it can be brought within acceptable levels for use in stereo amplifiers through negative feedback. We shall discuss feedback in detail in Chap. 13.

Transformer-coupled push-pull amplifiers can offer very high-quality performance as the final stage in audio and servo amplifiers. However, they have one major drawback. They need large, heavy, and expensive transformers, both at the input and output. A circuit that replaces the input transformer, together with the ensuing waveforms, is depicted in Fig.

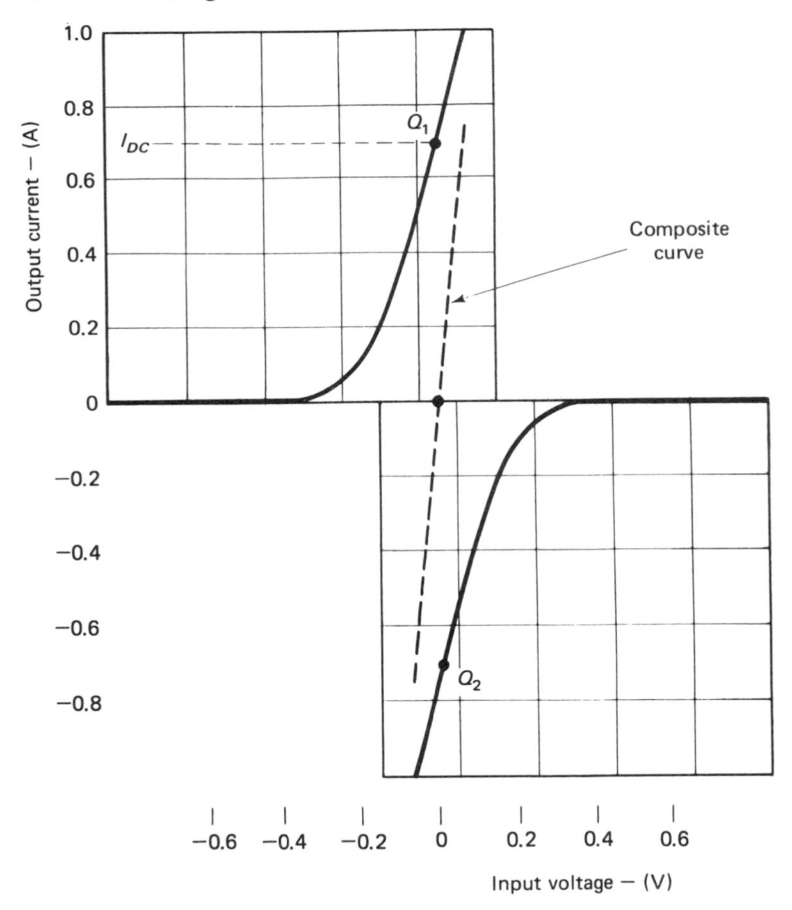

Figure 12-18. Transfer curve for class-A operation.

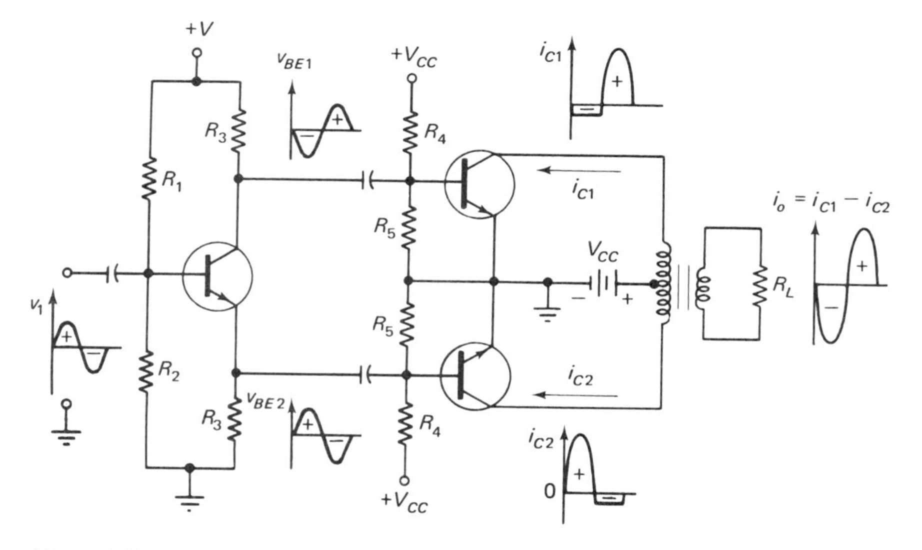

Figure 12-19. Split-load phase-inverter input circuit for a push-pull amplifier.

12-19. The operation of this circuit is identical to that already discussed, with the exception that the driver stage is a class-A operated transistor.

12.7 OTHER PUSH-PULL AMPLIFIERS

There are many schemes for eliminating the need for the transformers at both the input and output. Basically, such schemes fall into two categories. One encompasses push-pull amplifiers, which utilize the same type of transistors for the output stage, i.e., either both *NPN* or both *PNP*. The second system includes complementary-symmetry amplifiers where two different types of transistors are used in the output stage.

Figure 12-20 gives an example of these two different types of transformerless push-pull amplifier output stages. Their operation is almost

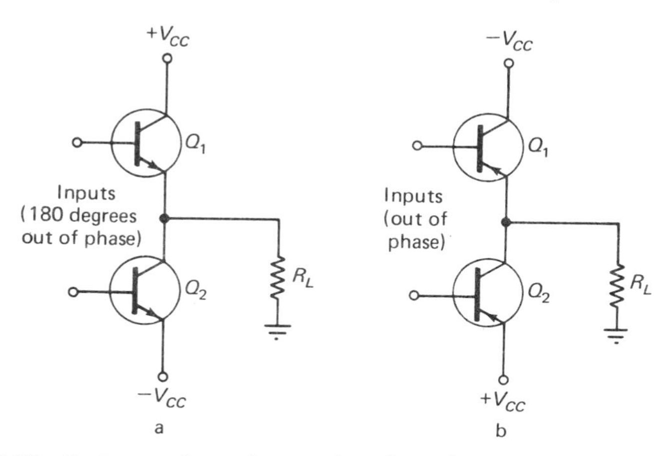

Figure 12-20. Basic transformerless push-pull configuration with (a) *NPN* and (b) *PNP* transistors.

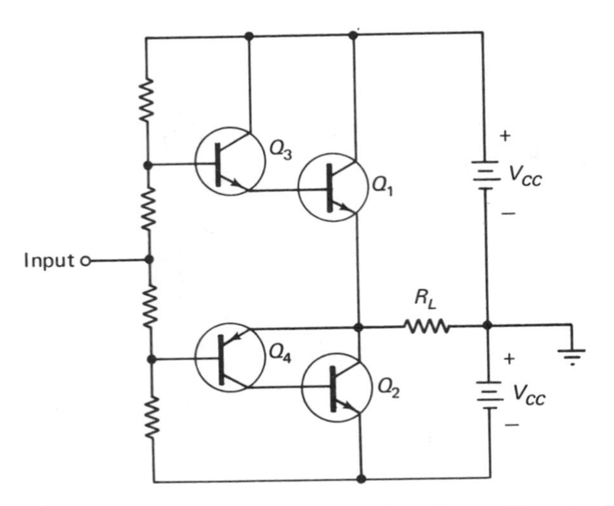

Figure 12-21. Transformerless push-pull amplifier circuit.

identical to that of the transformer-coupled push-pull amplifier, with a few exceptions. First, the amplifiers in Fig. 12-20 are dc-coupled to the load; that is, there may be a direct current through the load if any imbalance exists between the two transistors. Secondly, these amplifiers require a dual (+ and −) supply, whereas the transformer-coupled amplifiers only require a single supply.

The inputs to the output stage still need signals which are 180° out of phase, just like the transformer-coupled amplifier. We can furnish such signals by an input (center-tapped) driver transformer or by the driver circuit shown in Fig. 12-21. Driver transistors $Q3$ and $Q4$ are of comple-

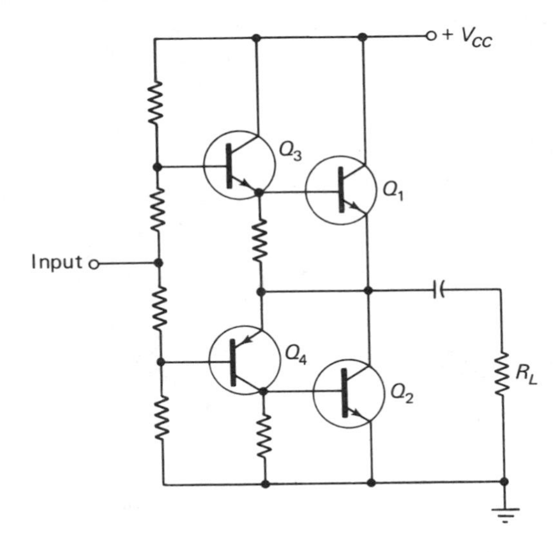

Figure 12-22. Single-supply transformerless push-pull amplifier circuit.

mentary types, one a *PNP* and the other an *NPN*, thus providing outputs which are out of phase while having a common input.

The transformerless push-pull amplifier circuit illustrated in Fig. 12-22 needs only one power supply to operate. It may be a more desirable version for certain applications. Note that in this circuit the load is capacitively coupled to the amplifier.

12.8 COMPLEMENTARY-SYMMETRY AMPLIFIERS

Push-pull operation may be accomplished by using complementary transistors (one *NPN* and one *PNP*), as depicted in Fig. 12-23. The use of complementary matched transistors in the output stage eliminates the need for two out-of-phase input signals.

The circuit operates as follows: When the input signal is positive, $Q1$ is biased on and amplifies it, while $Q2$ is essentially cut off. This action gives the positive output signal. When the input is negative, $Q1$ is cut off and $Q2$ is biased on, so that the negative part of the output signal is provided by $Q2$.

Class-AB complementary-symmetry amplifiers are shown in Figs. 12-24 and 12-25. Resistors R_B together with power supplies V provide the AB base bias in Fig. 12-24. Resistors R_E provide current limiting at the output (when the transistors saturate) as well as bias stability. The waveshapes, assuming a sinusoidal signal, are shown.

The class-AB configuration shown in Fig. 12-25 is especially useful as the power stage in a dc-coupled high-voltage-gain amplifier, that is as the output power stage to be used with a monolithic operational amplifier. Here, the bias for the base-emitter junctions is provided by the forward drop across the diodes. Typically only $D1$ and $D2$ would be used where the dc drop across the limiting resistors R_2 is quite small. Under these

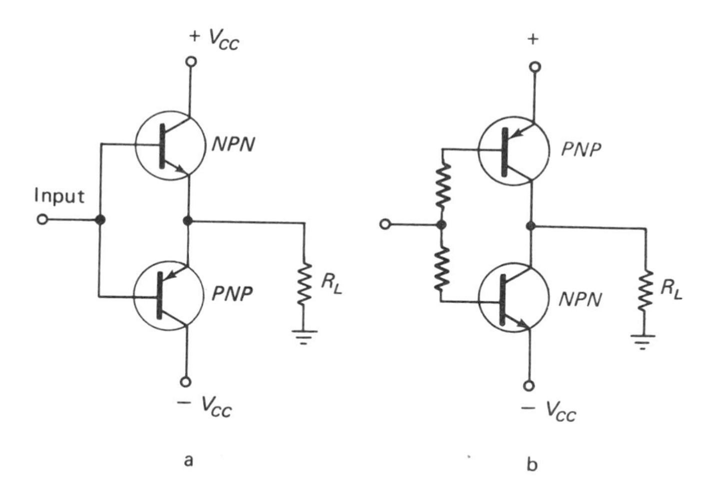

Figure 12-23. Basic complementary-symmetry configurations.

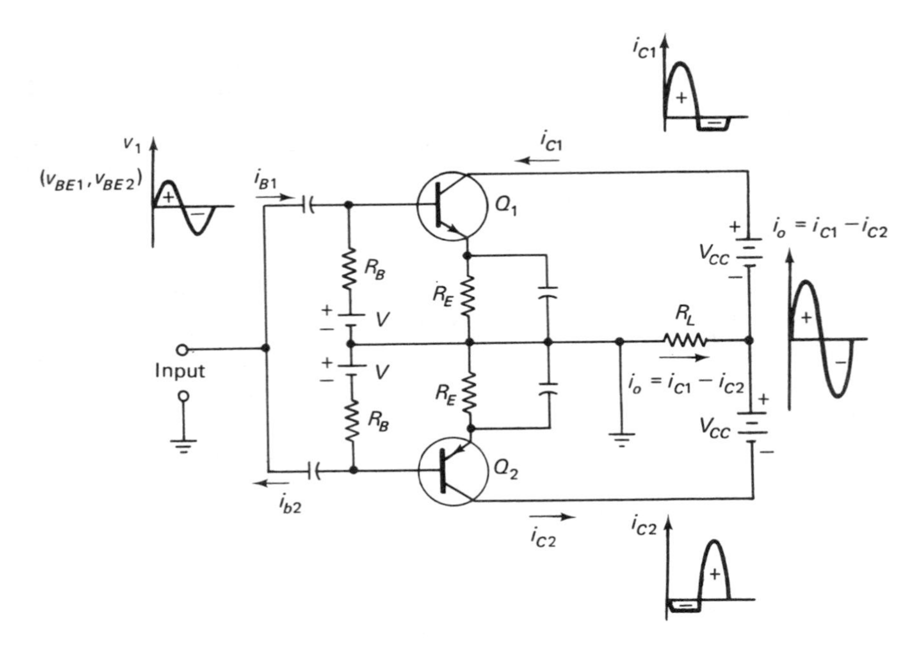

Figure 12-24. Class-AB complementary-symmetry amplifier.

conditions, the input can be dc-coupled. However, if the quiescent collector current needs to be significant, $D3$ or additional diodes are included to provide sufficient base drive for the power transistors.

Graphical analysis techniques apply equally well to all forms of amplifiers: transformer-coupled and transformerless push-pull, as well as the complementary-symmetry amplifiers.

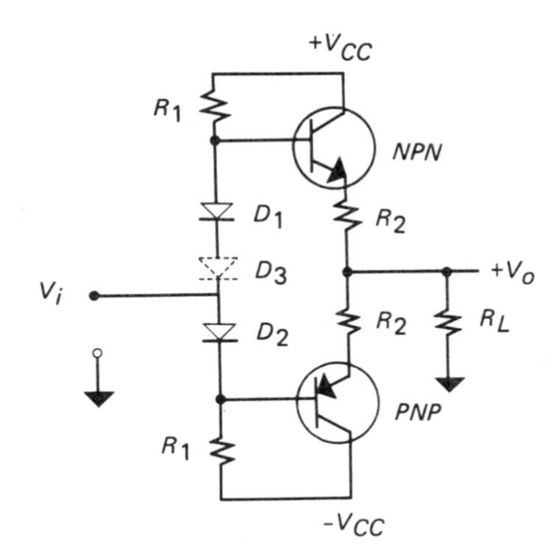

Figure 12-25. Class-AB complementary-symmetry amplifier.

There is no "best" power amplifier circuit, because most of the circuits have some good and some bad points. The specific application should be the deciding factor, as well as the availability or nonavailability of a dual power supply and of matched complementary transistors, etc.

12.9 SUMMARY

The design of power amplifiers is by no means an easy task. There are trade-offs between output power, power efficiency, and harmonic distortion.

We can analyze power amplifiers most easily by using graphical techniques. Remember, however, that these techniques are at best only good approximations. Moreover, an analysis carried out for one transistor, or a pair of transistors, may not hold for another transistor of the same type or another pair of similar transistors. The circuits included in this chapter should give you a good start in the design of many power amplifiers. We shall examine a complete power amplifier circuit in Chapter 14.

Review Questions

1. In what way are power amplifiers different from small-signal amplifiers?
2. What is the main method of analyzing power amplifiers?
3. What is a class-A power amplifier? Discuss in terms of the input and output signals.
4. What is a class-B amplifier?
5. What is a class-AB amplifier?
6. What gives rise to harmonic distortion in a power amplifier?
7. What is meant by the efficiency in a power amplifier?
8. Why is the efficiency of a power amplifier important, whereas that of a small-signal amplifier is not?
9. What is the role of a transformer used in the output of a power amplifier?
10. In a series-fed class-A power amplifier, what is the relationship between the output power and harmonic distortion?
11. What is the maximum power efficiency possible from a class-A amplifier?
12. What are the advantages and disadvantages of a push-pull amplifier over the single-ended amplifier?
13. What classes of amplifiers are operated in push-pull? Why these classes and not others?
14. What is the action of the two transistors used in a push-pull amplifier?
15. The input stage of a push-pull amplifier may have a center-tapped transformer. Why is this transformer necessary?

16. What other circuits may be used to feed a push-pull amplifier besides transformer-coupled? Explain.
17. What are the differences and similarities in the performance of class-B and class-AB push-pull amplifiers?
18. What are the advantages and disadvantages of a complementary-symmetry amplifier as compared to a push-pull amplifier?
19. What types of transistors must be used in a complementary-symmetry amplifier? Why?
20. What are the advantages and disadvantages of a quasi-complementary-symmetry amplifier over a regular complementary-symmetry amplifier?

Problems

1. A TIP29B silicon *NPN* transistor (see Appendix 3) is to be used in a power amplifier. Determine and sketch its permissible region of operation.
2. Repeat Problem 1 for the TIP29C transistor.
3. The TIP29B transistor is used in conjunction with a mica washer and heat sink, which have a combined thermal resistance of 5°C/W. Determine the maximum power that can be dissipated without exceeding the maximum junction temperature.
4. What thermal resistance of the heat sink will be needed if a TIP29A transistor is to dissipate 15 W? (Assume that no washer is used.)
5. Repeat Problem 4 for a TIP29C transistor if a mica washer with a thermal resistance of 1.2°C/W is used.
6. What is the approximate operating junction temperature of the transistors in Problems 4 and 5?

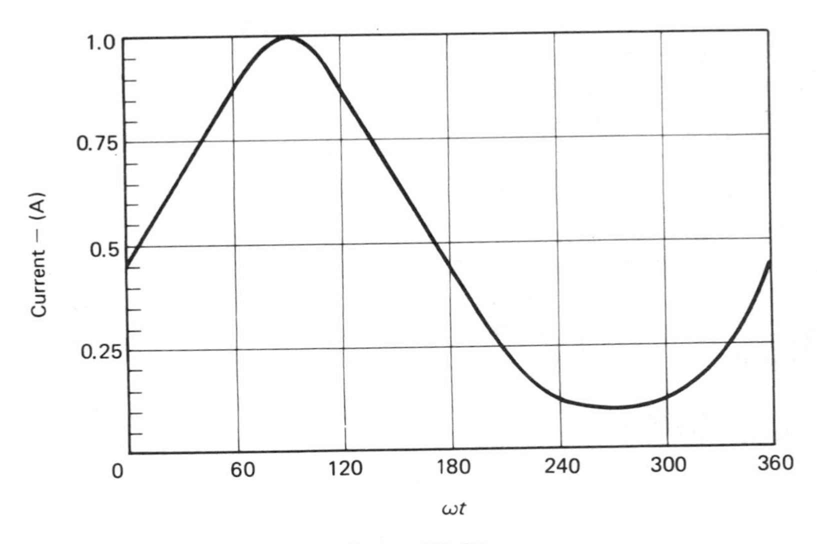

Figure 12-26.

7. A class-A amplifier driving a 4 Ω load produces the current waveshape shown in Fig. 12-26. If the dc power supplied is 4 W, determine (a) output power, (b) total harmonic distortion, and (c) power efficiency.
8. For the load line and *Q*-point as shown in Fig. 12-3, the input signal is 20 mA peak-to-peak. Determine (a) the output waveshape, (b) output power, (c) total harmonic distortion, and (d) power efficiency.
9. Repeat Problem 8 if the load resistance is 8 Ω. (Note that a new load line must be drawn.)
10. Repeat Example 12-6 if the input voltage is 1.6 V peak-to-peak.
11. If the load in Problem 10 is 16 Ω, determine the output power.
12. A complementary-symmetry amplifier has the composite transfer characteristics shown in Fig. 12-17. The input voltage is 0.3 V peak-to-peak. Sketch the output (current) waveshape. What is the power delivered into an 8 Ω load?

Negative Feedback Amplifiers

In this chapter we introduce negative feedback amplifiers and discuss their basic properties. We shall also be concerned with the advantages and disadvantages of negative feedback amplifiers as compared to the basic non-feedback type of amplifier.

Two types of feedback are possible. The first is called *negative* feedback: A part or all of the output signal (voltage or current) is diverted and applied at the input so as to subtract from it. In this manner, the apparent input signal to the original amplifier is reduced, and thus the output signal is reduced accordingly. Negative feedback amplifiers are typically characterized as having lower gain than the same amplifier without feedback.

The second type of feedback is called *positive* feedback. It applies a part or all of the output signal to the input in such a fashion as to add to it. In terms of amplifiers, positive feedback is undesirable because it usually causes the amplifier to be unstable and oscillate. This property, however, is utilized fully in oscillator circuits. In this chapter we shall confine ourselves to negative feedback amplifiers. Chapter 15 will treat the positive feedback amplifier, or oscillator.

13.1 GENERAL FEEDBACK CONCEPTS

We first saw the advantage of using feedback constructively in the discussion of Q-point stability in Chap. 4. Feedback is an extremely useful tool in numerous applications. We can see a typical use of negative feedback in control systems. These systems encompass all circuits where the output is sensed and used to control or correct the input to provide the desired output. Other uses of feedback include the sensing of the output, then comparing it to some reference signal, and finally controlling the input (and thus the output) in accordance with the difference between the output and reference signal. Specifically, negative feedback in an amplifier can be used to:

1. Stabilize the gain (either voltage or current)
2. Achieve more linear operation

3. Broaden the bandwidth
4. Lower or raise input impedance
5. Lower or raise output impedance
6. Reduce the noise in the amplifier
7. Reduce thermal effects

By stabilization of the gain, we mean making the gain less dependent on the specific device parameters. Linearity of operation is important for all amplifiers, but an improvement in the linearity (i.e., lower distortion) is especially important in power amplifiers, discussed in the last chapter. Items 3, 4, and 5 are self-explanatory. The noise (spurious electrical signals generated within the amplifier) is especially bothersome in amplifiers with extremely small signal levels. In these cases, negative feedback can be used to decrease the amount of noise in the amplifier. We have already discussed in Chap. 4 the reduction in thermal effects.

We shall classify types of feedback according to the action of the feedback on the gain. Two types are *current feedback* and *voltage feedback* circuits; they are characterized by a decrease in gain. Two other feedback types, termed *shunt* and *series feedback* circuits, will also be treated.

The basic feedback amplifier block diagram is shown in Fig. 13-1, with signal paths as shown. The signal at any point in Fig. 13-1 could be either a voltage or a current, depending on the type of performance desired.

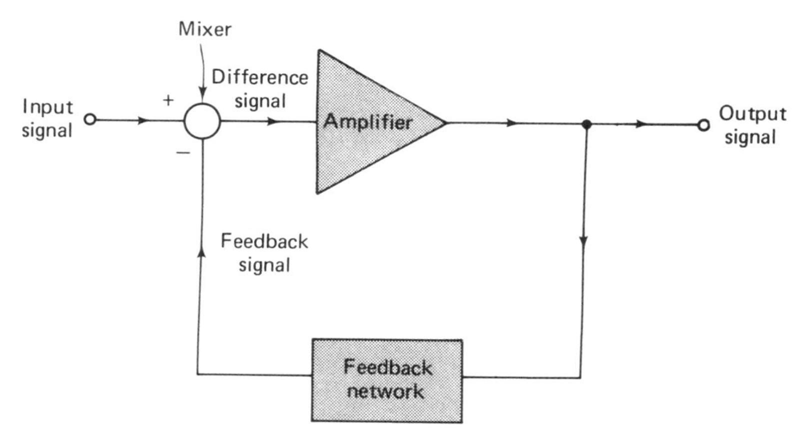

Figure 13-1. General block diagram of a feedback amplifier.

13.2 VOLTAGE-FEEDBACK AMPLIFIERS

Referring to Fig. 13-1, we see that when all the signals are voltages, the circuit is a voltage-feedback amplifier. The general form of a voltage-feedback amplifier is shown in Fig. 13-2. Negative feedback is accomplished by causing a part of the output voltage to subtract from the input voltage.

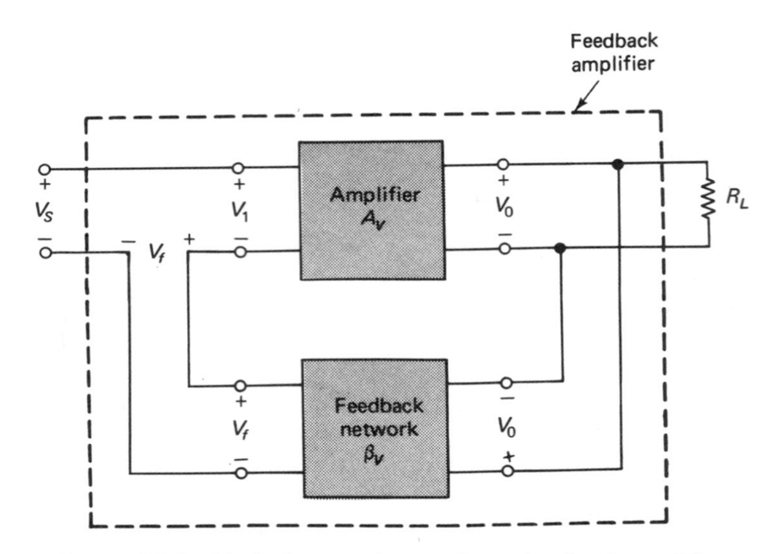

Figure 13-2. Block diagram for a voltage-feedback amplifier.

Voltage Gain. In Fig. 13-2, the output voltage V_o appears across both the external load R_L and the feedback network. The reverse voltage gain of the feedback network β_v is defined as

$$\beta_v = \frac{V_f}{V_o} \tag{13-1}$$

(Note: The β as defined here is *not* the same as the transistor dc β defined previously and should not be confused with it.)

The open-circuit voltage gain of the amplifier A_v is defined by

$$A_v = \frac{V_o}{V_1} \tag{13-2}$$

Setting to zero the sum of the voltages at the input indicated in Fig. 13-2, we obtain

$$V_s = V_1 + V_f \tag{13-3}$$

The open-circuit gain of the feedback amplifier A_{vf} is given by:

$$A_{vf} = \frac{V_o}{V_s} = \frac{V_o}{V_1 + V_f} = \frac{\dfrac{V_o}{V_1}}{1 + \dfrac{V_f}{V_1}} \tag{13-4}$$

From Eq. (13-1) we see that $V_f = \beta_v V_o$. Also noting that $A_v = V_o / V_1$, we obtain:

$$A_{vf} = \frac{A_v}{1 + \beta_v A_v} \tag{13-5}$$

We define negative feedback for $(1+\beta_v A_v)$ greater than 1 and positive feedback for $(1+\beta_v A_v)$ less than 1.

Usually, $|A_v|$ is much larger than 1, so that we can approximate:

$$A_{vf} \cong \frac{1}{\beta_v} \tag{13-6}$$

Input Resistance. The input resistance for the feedback amplifier is defined as the ratio of V_s to I_1. Substituting Eq. (13-1) for V_f into Eq. (13-3), we obtain:

$$V_s = V_1 + \beta_v V_o = V_1(1+\beta_v A_v) \tag{13-7}$$

We can substitute $V_1 = R_i I_1$. Thus,

$$R_{if} = \frac{V_s}{I_1} = R_i(1+\beta_v A_v) \tag{13-8}$$

If negative voltage feedback is used, therefore, the input resistance is increased.

Output Resistance. Assuming that the current drawn off by the feedback network in Fig. 13-2 is negligibly small, we can write:

$$V_o = A_v V_1 - I_o R_o \tag{13-9}$$

Substituting for V_1 from Eq. (13-3), we have:

$$V_o = A_v V_s - A_v V_f - I_o R_o \tag{13-10}$$

We then rearrange the equation:

$$V_o(1+\beta_v A_v) = A_v V_s - I_o R_o \tag{13-11}$$

Dividing both sides by $(1+\beta_v A_v)$, we obtain:

$$V_o = A_{vf} V_s - I_o \frac{R_o}{1+\beta_v A_v} \tag{13-12}$$

We solve for the feedback amplifier output resistance by setting $V_s = 0$. Thus,

$$R_{of} = \frac{V_o}{-I_o} = \frac{R_o}{1+\beta_v A_v} \tag{13-13}$$

If the feedback is negative, the output resistance with feedback is lower than the output resistance without feedback.

Equivalent Circuit. An examination of Eq. (13-12) suggests an equivalent circuit for the output of the feedback amplifier. The complete equivalent circuit for the feedback amplifier is given in Fig. 13-3.

Example 13-1. The amplifier shown in Fig. 13-4 is a voltage-feedback amplifier. The feedback network consists of voltage-divider resistors R_9 and R_{10}. The amplifier without feedback has the following parameters:

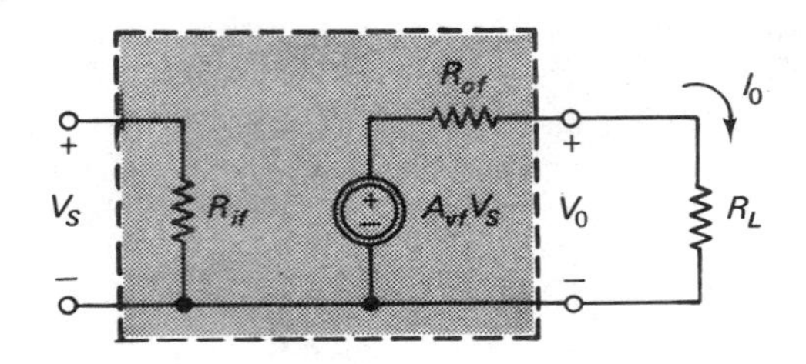

Figure 13-3. Equivalent circuit for the voltage-feedback amplifier.

$A_v = 100$, $R_i = 2$ kΩ, and $R_o = 5$ kΩ. We want to determine the parameters of the amplifier with feedback.

Solution: The feedback factor β_v is calculated from the resistor ratio:

$$\beta_v = \frac{R_{10}}{R_{10} + R_9} = \frac{0.1}{0.1 + 2.2} = \frac{1}{23}$$

We next find the amount of feedback:

$$1 + \beta_v A_v = 1 + \frac{100}{23} = 5.35$$

The voltage-feedback amplifier parameters can now be calculated as follows:

$$R_{if} \cong (5.35)(2)\ k\Omega \cong 10.7\ k\Omega$$

$$R_{of} \cong \frac{5}{5.35}\ k\Omega \cong 0.93\ k\Omega \cong 930\Omega$$

$$A_{Vf} \cong \frac{100}{5.35} \cong 18.7$$

Note that by using the approximation in Eq. (13-6) we obtain $A_{Vf} \cong 23$, which is a crude approximation in this case. The approximation becomes usable when $\beta_v A_v$ is greater than 10, but it is not so in this case.

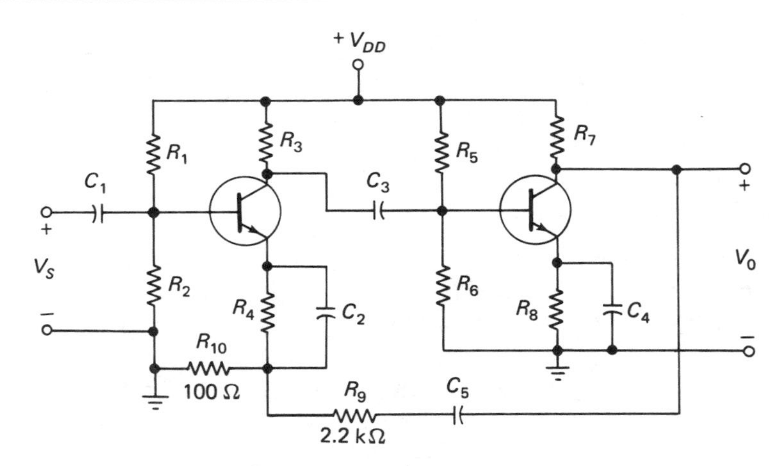

Figure 13-4. Example of a voltage-feedback amplifier.

The calculation of the gain, input impedance, and output impedance for the amplifier without feedback must be carried out with care, because the feedback network cannot be ignored completely. For input calculations, the amplifier in Fig. 13-4 without feedback must be considered as having $V_o = 0$ (output short-circuited). For output calculations, the amplifier must be considered as having $V_f = 0$(in this case, R_{10} shorted). You must realize, however, that the input impedance in this example is considered without the parallel combination of R_1 and R_2. The total input impedance contains these two resistors.

13.3 CURRENT-FEEDBACK AMPLIFIERS

When all the signals in Fig. 13-1 are currents, the circuit is a current-feedback amplifier. Such a block diagram is depicted in Fig. 13-5. Negative feedback is accomplished by causing the output current to subtract from the input current, as shown.

Current Gain. In Fig. 13-5, the output current I_o is supplied to the load R_L and to the feedback network. The reverse current gain of the feedback network, β_I, is defined as

$$\beta_I = \frac{I_f}{I_o} \tag{13-14}$$

Therefore, when the output current I_o flows into the feedback network, the component that reaches the input of the amplifier is:

$$I_f = \beta_I I_o \tag{13-15}$$

The amplifier input current I_1 is given by:

$$I_1 = I_s - I_f = I_s - \beta_I I_o \tag{13-16}$$

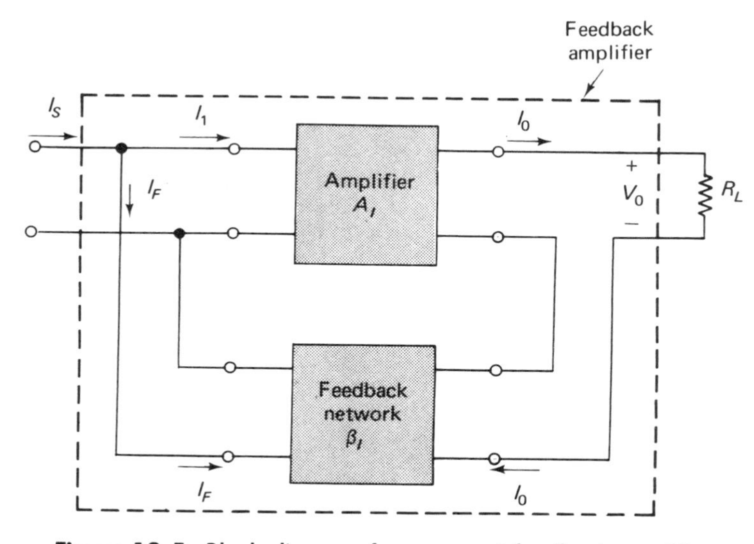

Figure 13-5. Block diagram for a current-feedback amplifier.

The feedback amplifier input current is I_s and can be obtained from Eq. (8-3):

$$I_s = I_1 + \beta_I I_o \tag{13-17}$$

If the forward short-circuit current gain of the amplifier is

$$A_I = \frac{I_o}{I_1} \tag{13-18}$$

we can write:

$$I_o = A_I I_1 \tag{13-19}$$

The short-circuit current gain of the feedback amplifier A_{If} is given by the ratio of I_o to I_s. Using Eqs. (13-17) and (13-19) we obtain:

$$A_{If} = \frac{I_o}{I_s} = \frac{A_I I_1}{I_1 + \beta_I I_o} \tag{13-20}$$

Dividing the numerator and denominator of Eq. (13-20) by I_1 gives us:

$$A_{If} = \frac{A_I}{1 + \beta_I A_I} \tag{13-21}$$

This equation relates the short-circuit current gain of the feedback amplifier, A_{If}, to the short-circuit current gain of the amplifier without feedback, A_I.

We can approximate the current gain of the feedback amplifier if we note that usually $|A_I|$ is much larger than 1. Thus, if we divide both the numerator and the denominator of Eq. (13-21) by A_I, we have:

$$A_{If} \cong \frac{1}{\beta_I} \tag{13-22}$$

Consequently, the short-circuit current gain of the feedback amplifier can be made independent of the device parameters and dependent only on the feedback network component values.

Input Resistance. The input resistance of the feedback amplifier R_{if} is defined as the ratio of V_s to I_s, where V_s is the voltage across the input terminals in Fig. 13-5.

$$R_{if} = \frac{V_s}{I_s} = \frac{I_1 R_i}{I_f + I_1} = \frac{R_i}{1 + \frac{I_f}{I_1}} \tag{13-23}$$

Noting that $I_f = \beta_I I_o$, we have:

$$R_{if} = \frac{R_i}{1 + \beta_I A_I} \tag{13-24}$$

If the feedback is negative, $(1 + \beta_I A_I)$ is greater than 1 and the input resistance is lowered as a result of the feedback.

Output Resistance. According to Fig. 13-5, the output resistance of the feedback amplifier is defined as the ratio of V_o to $-I_o$ for the condition $I_s = 0$. If we assume that the voltage developed across the feedback network in the output loop is negligibly small when compared to either V_o or the voltage across R_o, then we can say that the voltage across R_o is approximately equal to V_o.

$$V_o = (A_I I_1 - I_o) R_o \tag{13-25}$$

Substituting Eq. (13-16) for I_1, we have:

$$V_o = (A_I I_s - \beta_I A_I I_o - I_o) R_o \tag{13-26}$$

We can now factor out $(1 + \beta_I A_I)$:

$$V_o = \left[\left(\frac{A_I}{1 + \beta_I A_I} \right) I_s - I_o \right] R_o (1 + \beta_I A_I) \tag{13-27}$$

When $I_s = 0$, we can solve for the output resistance of the feedback amplifier

$$R_{of} = \frac{V_o}{-I_o} = R_o (1 + \beta_I A_I) \tag{13-28}$$

Thus, we see that the effect of negative current feedback is to increase the output resistance.

Equivalent Circuit. We have noted the effect of negative current feedback on the current gain, input resistance, and output resistance. We can summarize these factors in an equivalent circuit for the feedback amplifier. In Eq. (13-27), we recognize the coefficient of I_s as A_{If}. If we use the definition of R_{of}, we can write:

$$V_o = (A_{If} I_s - I_o) R_{of} \tag{13-29}$$

This equation suggests an equivalent circuit of the output with a current generator $A_{If} I_s$ and output resistance R_{of}. The input current is I_s and the input resistance is R_{if}. This feedback amplifier equivalent circuit is shown in Fig. 13-6.

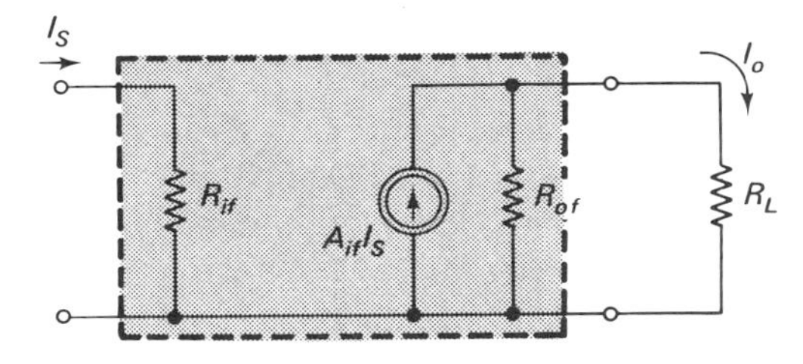

Figure 13-6. Equivalent circuit for the current-feedback amplifier.

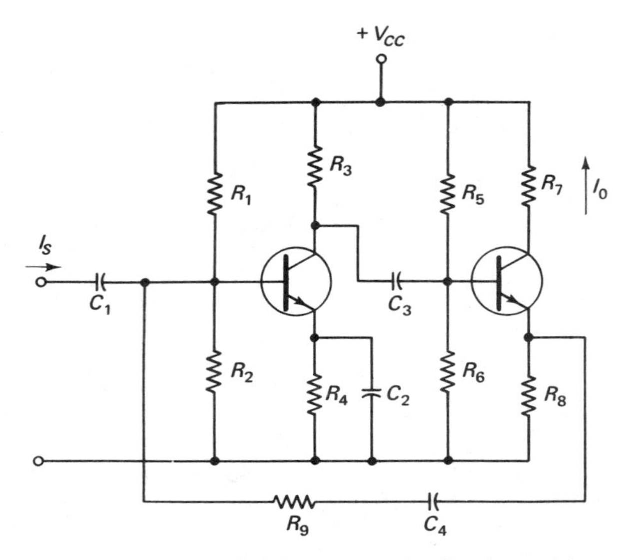

Figure 13-7. Example of a current-feedback amplifier.

Example 13-2. The circuit shown in Fig. 13-7 is a current-feedback amplifier. Without feedback, the amplifier parameters are: $A_I = 800$, $R_i = 1\ \text{k}\Omega$, and $R_o = 10\ \text{k}\Omega$. Feedback is applied through the feedback network consisting of R_8 and R_9 (220Ω and 4.7 kΩ, respectively). We want to determine the amplifier parameters with feedback.

Solution: The feedback factor β_I is found from the resistor ratio:

$$\beta_I \cong \frac{R_8}{R_8 + R_9} \cong \frac{0.22}{0.22 + 4.7} \cong \frac{1}{22.4}$$

We next calculate the amount of feedback:

$$1 + \beta_I A_I \cong 1 + \frac{800}{22.4} \cong 36.7$$

The current-feedback amplifier parameters can be calculated as follows:

$$R_{if} \cong \frac{1000}{36.7}\ \Omega \cong 27\ \Omega$$

$$R_{of} \cong (10)(36.7)\ \text{k}\Omega \cong 367\ \text{k}\Omega$$

$$A_{If} \cong \frac{800}{36.7} \cong 21.8$$

The approximation for the current gain given in Eq. (13-22) is quite valid in this case because $\beta_I A_I$ is greater than 10. Using the approximation, we get $A_{If} = 22.4$, which compares quite favorably with the value just calculated. This equation means that in the current-feedback amplifier in this example the current gain is insensitive to the transistor parameters and depends on the values of the feedback resistors R_8 and R_9.

The determination of the amplifier parameters without feedback should be carried out with care. If we want to determine the input parameters, the output current should be zero (output open-circuited at the second emitter in Fig. 13-7). When we calculate the output parameters, the input current should be zero (input open-circuited at the first base). In this way, feedback is eliminated, although the loading of the feedback circuit on the amplifier without feedback is taken into account.

13.4 EFFECT OF FEEDBACK ON FREQUENCY RESPONSE

As we have seen in the previous two sections, feedback changes the gain and input and output impedances of an amplifier. As might be expected, it also modifies the frequency response of an amplifier. The following discussion is applicable for the current-gain response of the current-feedback amplifier and the voltage-gain response of the voltage-feedback amplifier. We shall make no distinction between the two responses in this section.

An amplifier without feedback has lower and upper 3 dB frequencies labeled f_1 and f_2, respectively. The same amplifier with voltage feedback will have lower and upper 3 dB frequencies (labeled f_{1f} and f_{2f}, respectively) given by:

$$f_{1f}=\frac{f_1}{1+\beta A} \tag{13-30}$$

$$f_{2f}=f_2(1+\beta A) \tag{13-31}$$

where β and A would have the appropriate subscripts (I or V) depending on whether it was a current-feedback or voltage-feedback amplifier. The effect of the feedback is to *decrease* the lower 3 dB frequency and to *raise* the upper 3 dB frequency.

The bandwidth of the amplifier with voltage feedback is modified accordingly. If we assume that the lower 3 dB frequency is quite small compared to the upper 3 dB frequency, the bandwidth with feedback is given by:

$$BW_f \cong BW(1+\beta A) \tag{13-32}$$

The effect of feedback on the amplifier frequency response is illustrated in the following example.

Example 13-3. An amplifier (without feedback) has a voltage gain of 1,000 and lower and upper 3 dB frequencies of 100 Hz and 100 kHz, respectively. It is made into a feedback amplifier having 20 dB of feedback. We want to determine the frequency response of the feedback amplifier.

Solution: The frequency response of the amplifier is shown in Fig. 13-8. The amount of feedback is:

$$\text{dB of feedback} = 20\log|1+\beta A| = 20\,\text{dB}$$

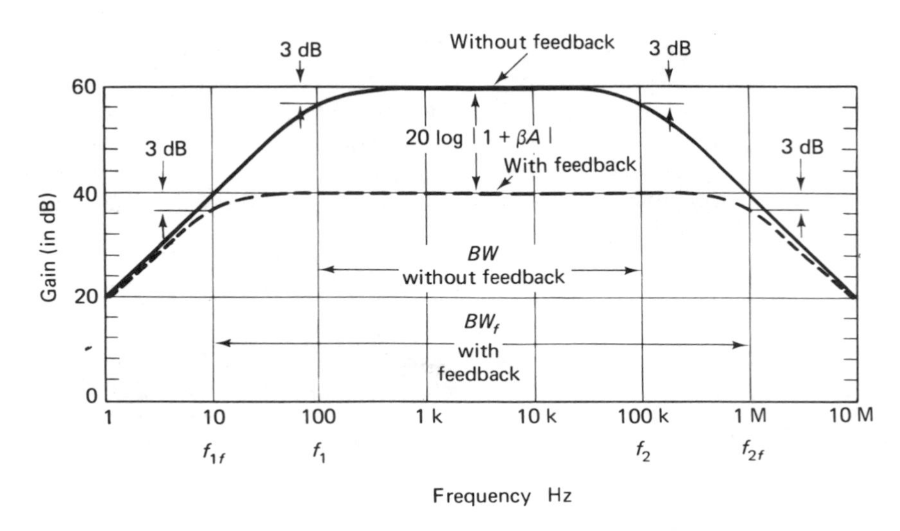

Figure 13-8. Effect of feedback on the frequency response of an amplifier.

Therefore:

$$1+\beta A=10$$

With feedback, the gain of the amplifier is:

$$A_{Vf}=\frac{1000}{10}=100 \text{ or } 40 \text{ dB}$$

The lower and upper 3 dB frequencies are:

$$f_{1f}=\frac{100}{10}\text{ Hz}=10 \text{ Hz}$$
$$f_{2f}=(100)(10)\text{ kHz}=1 \text{ MHz}$$

These results are plotted in Fig. 13-8. Note that the increase in bandwidth is always the same as the decrease in gain. In this case, the bandwidth is increased tenfold, whereas the gain is decreased tenfold.

13.5 SERIES- AND SHUNT-FEEDBACK AMPLIFIERS

Block diagrams for series-feedback and shunt-feedback amplifiers are shown in Figs. 13-9 and 13-10, respectively. In a series-feedback circuit, the output current is sampled and fed back to the input as a voltage. This connection is sometimes called a *voltage-series feedback* amplifier.

13.5.1 Series-Feedback Amplifiers

The series-feedback amplifier modifies the effective transconductance of the basic amplifier without feedback. Any amplifier can be represented by the equivalent circuit shown in Fig. 13-11. A typical example might be a single-stage FET amplifier.

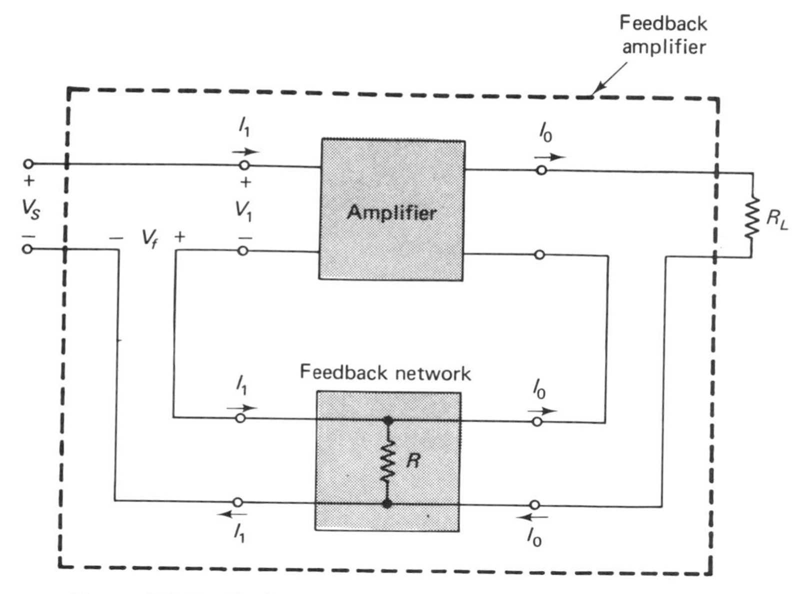

Figure 13-9. Block diagram for a series-feedback amplifier.

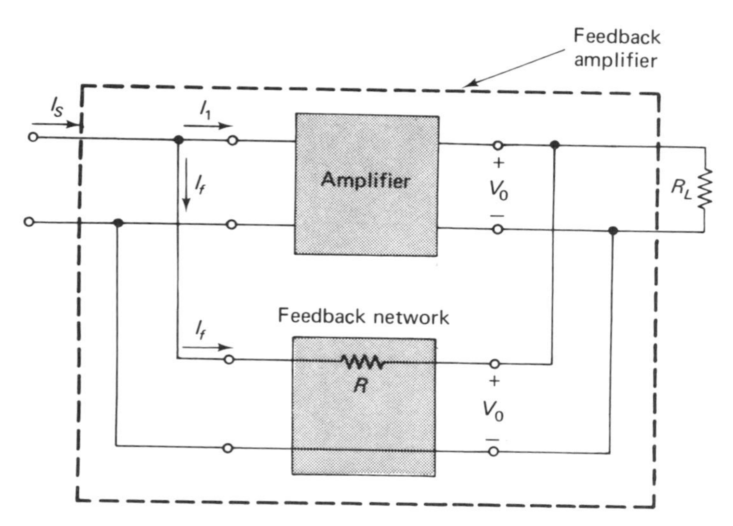

Figure 13-10. Block diagram for a shunt-feedback amplifier.

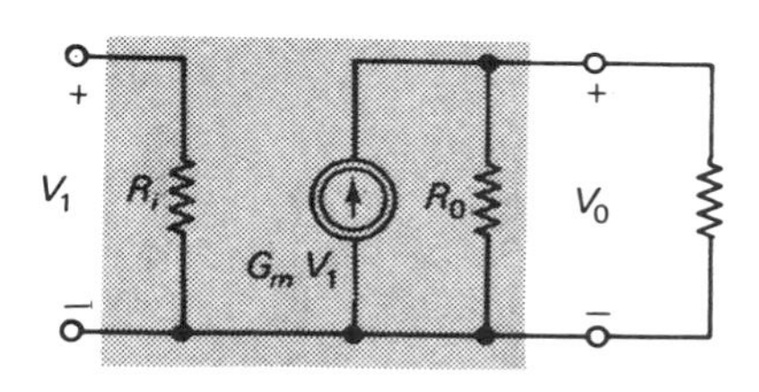

Figure 13-11. A transconductance amplifier.

The transconductance of the amplifier without feedback is modified, once series-feedback is applied, according to the following equation:

$$G_{mf} = \frac{G_m}{1+\beta_m G_m} \tag{13-33}$$

where G_{mf} and G_m are the respective transconductances, with and without feedback, and β_m is defined as:

$$\beta_m = \frac{V_f}{I_o} \tag{13-34}$$

and has units of conductance.

Table 13-1 summarizes the effect of the series-feedback circuit on the voltage and current gains and gives the input and output impedances.

Table 13-1. Summary of Equations for Different Feedback Circuits

Type of Feedback	A_{Vf} *(open circuit)*	A_{If} *(short circuit)*	R_{if}	R_{of}
Voltage (Fig. 13-2)	$\frac{A_V}{1+\beta_V A_V}$	A_I	$R_i(1+\beta_V A_V)$	$\frac{R_o}{1+\beta_V A_V}$
Current (Fig. 13-5)	A_V	$\frac{A_I}{1+\beta_I A_I}$	$\frac{R_i}{1+\beta_I A_I}$	$R_o(1+\beta_I A_I)$
Series (Fig. 13-9)	$\sim -\frac{R_i}{R_m}$*	A_I	$R_i + R_m(1-A_I)$	$R_o + R_m(1-A_V)$
Shunt (Fig. 13-10)	A_V	$\sim -\frac{R_i}{R_m}$†	$\left(\frac{R_m}{1-A_V}\right)\lVert(R_i)$	$(R_o)\lVert\left(\frac{A_V R_m}{A_V-1}\right)$

*For $A_V R_m \gg (R_o + R_i)$.

†For $A_I \gg \left(1-\frac{R_i}{R_m}\right)$.

Example 13-4. The circuit in Fig. 13-12 is an example of a series-feedback amplifier. The circuit values are: $R_1 = R_2 = 100$ kΩ, $R_C = 2.2$ kΩ, and $R_E = 1$ kΩ. The transistor parameters are: $h_{ie} = 1$ kΩ, $h_{fe} = 100$, with h_{re} and h_{oe} negligible. We want to find the parameters for the amplifier as shown.

Solution: For the amplifier without feedback:

$$R_i = h_{ie} \cong 1 \text{ k}\Omega$$

$$A_I = -h_{fe} \cong -100$$

$$A_V \cong -h_{fe}\frac{R_C}{h_{ie}} \cong -100\frac{2.2}{1} \cong -220$$

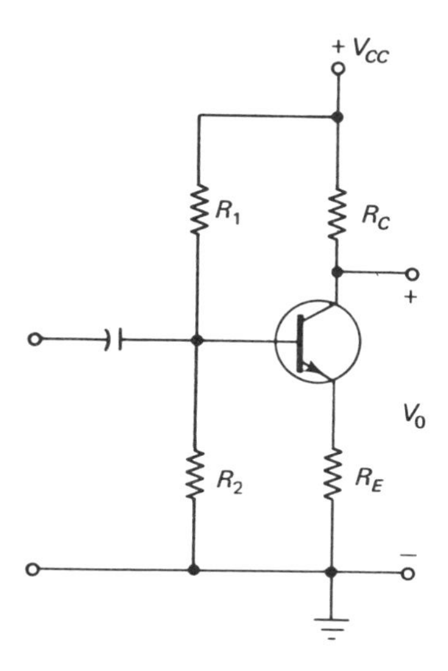

Figure 13-12. Example of a series-feedback amplifier.

We can assume that the output impedance is infinite. From Table 13-1, for the amplifier with feedback, we see that

$$R_{if} \cong 1 + 1(1+100)\ \text{k}\Omega = 102\ \text{k}\Omega$$

$$R_{of} \cong \infty$$

$$A_{Vf} \cong -\frac{2.2}{1} = -2.2$$

$$A_{If} \cong -100$$

The preceding results can also be obtained through simple circuit analysis by drawing the equivalent circuit of the amplifier shown.

13.5.2 Shunt-Feedback Amplifiers

The block diagram in Fig. 13-10 shows the connection for a shunt-feedback amplifier. The output voltage is sampled by the feedback network and fed back to the input in the form of a current.

We can choose to represent an amplifier (with or without feedback) in the form of a transresistance amplifier, as depicted in Fig. 13-13. Once shunt feedback is applied, the transresistance of the amplifier is modified according to:

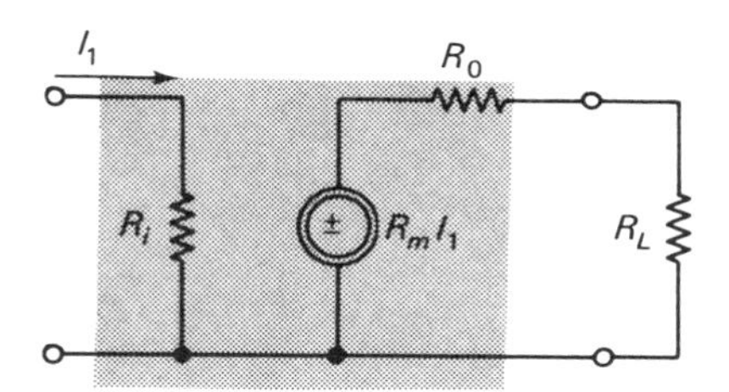

Figure 13-13. A transresistance amplifier.

$$R_{mf} = \frac{R_m}{1+\beta_r R_m} \tag{13-35}$$

where R_{mf} and R_m are the transresistances of the amplifier, with and without feedback, respectively. The feedback factor is defined as:

$$\beta_r = \frac{I_f}{V_o} \tag{13-36}$$

Table 13-1 lists the voltage and current gains as well as the input and output impedance transformations for the shunt-feedback amplifier illustrated in Fig. 13-10.

Example 13-5. The circuit in Fig. 13-14 is a simple example of a shunt-feedback amplifier. The circuit values are: $R_C = 3.3\ \text{k}\Omega$ and $R_B = 56\ \text{k}\Omega$. The transistor parameters are: $h_{ie} = 1.5\ \text{k}\Omega$, $h_{fe} = 75$; h_{re} and h_{oe} are negligible. We want to determine the parameters for the amplifier shown in Fig. 13-14.

Solution: For the amplifier without feedback:

$$R_i \cong h_{ie} \cong 1.5\ \text{k}\Omega$$

$$A_V \cong -h_{fe}\frac{R_C}{h_{ie}} \cong -165$$

$$A_I \cong -h_{fe} \cong -75$$

The output impedance is infinite.

From Table 13-1 we can determine the parameters of the amplifier with feedback:

$$R_{if} \cong \frac{56}{1+165} \| 1.5\ \text{k}\Omega \cong 275\Omega$$

$$R_{of} \cong R_B \cong 56\ \text{k}\Omega$$

$$A_{Vf} \cong -165$$

$$A_{If} \cong -\frac{3.3}{56} \cong -0.06$$

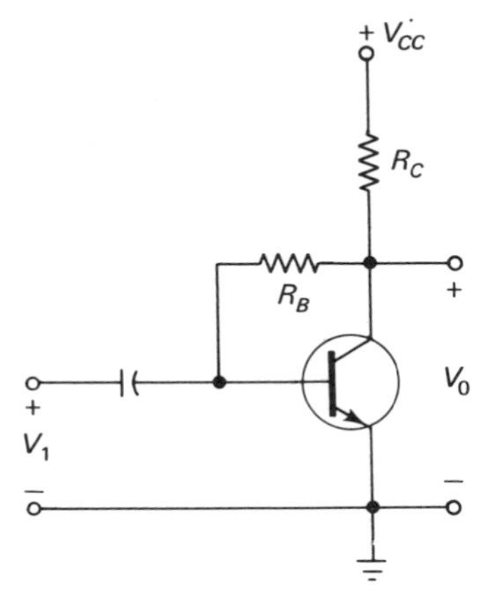

Figure 13-14. Example of a shunt-feedback amplifier.

The preceding values may also be obtained by more conventional circuit analysis after the small-signal equivalent circuit for the amplifier in Fig. 13-14 is drawn.

The effect of different types of feedback on circuit performance is summarized in Table 13-2.

Table 13-2. Summary of the Effects of Different Feedback Types on Circuit Performance

Type of Feedback	Effect on: Gain		Effect on: Impedance	
	Voltage†	*Current*‡	*Input*	*Output*
Voltage (Fig. 13-2)	Reduced	Unchanged	Increased	Decreased
Current (Fig. 13-5)	Unchanged	Reduced	Decreased	Increased
Series (Fig. 13-9)	*	Unchanged	Increased	Increased
Shunt (Fig. 13-10)	Unchanged	*	(a) Decreased for R_i large (b) Unchanged for R_i small	(a) Decreased for R_o large (b) Unchanged for R_o small

*See Table 13-1.
†Open-circuit voltage gain.
‡Short-circuit current gain.

13.6 EFFECT OF FEEDBACK ON NONLINEAR DISTORTION AND NOISE

As one result of negative feedback, nonlinear distortion in amplifiers is decreased. In particular, the amplitude of the distorted signal is reduced (for a discussion of harmonics, see Chap. 12):

$$D_f = \frac{D}{1+\beta A} \tag{13-37}$$

where β and A refer to voltage feedback and D_f and D are the amplitudes of the distortion signals with and without feedback.

Because all signals in a negative feedback amplifier are fed back to the input, their own amplitude at the output is reduced. Furthermore, distortion signals as well as internally generated noise signals are also reduced in amplitude.

Review Questions

1. What is *negative* feedback?
2. What is *positive* feedback?
3. What type of feedback (positive or negative) is used in a feedback amplifier?

   Why?
4. What is the effect of negative feedback on the amplifier gain stability?
5. What is the effect of negative feedback on the amplifier bandwidth?
6. What is the effect of negative feedback on the nonlinearity and noise in an amplifier?
7. In a voltage feedback circuit, how is the voltage gain stabilized against transistor parameter variations? Explain.
8. What is the effect of negative voltage feedback on the amplifier input resistance? On the output resistance?
9. In a current feedback circuit, how is the current gain stabilized against transistor parameter variations? Explain.
10. What is the effect of negative current feedback on the amplifier input resistance? On the output resistance?
11. What is the effect of negative feedback on the frequency response of an amplifier?
12. What does shunt feedback modify?
13. What does series feedback modify?
14. What are the basic advantages of a negative feedback amplifier over one using no feedback?

Problems

1. Suppose that we construct a voltage feedback amplifier, using an amplifier with: $A_V = 200$, $R_i = 100\ \text{k}\Omega$, and $R_o = 10\ \text{k}\Omega$. Determine the feedback factor β to give a gain (with feedback) of 26 dB. Also calculate the input and output resistances of the feedback amplifier.
2. An amplifier is characterized by $A_{Isc} = 80$, $R_i = 1.8\ \text{k}\Omega$, and $R_o = 40\ \text{k}\Omega$. What type of feedback must be used to achieve an input resistance of 500Ω? Determine the current gain and output resistance of the amplifier with feedback driving a 1 kΩ load.
3. The feedback amplifier shown in Fig. 13-12 is constructed with $R_1 = R_2 = 100\ \text{k}\Omega$, $R_C = 1.2\ \text{k}\Omega$, and $R_E = 680\ \Omega$. The transistor parameters are: $h_{ie} = 1\ \text{k}\Omega$, $h_{fe} = 85$, and $1/h_{oe} = 55\ \text{k}\Omega$. Determine the input and output resistances as well as the voltage gain.
4. The feedback amplifier shown in Fig. 13-14 is constructed with $R_B = 220\ \text{k}\Omega$ and $R_C = 5.6\ \text{k}\Omega$. Transistor parameters are the same as in Problem 3. Determine the input and output resistances and the current gain if a 1 kΩ load is placed from collector to ground.

5. An amplifier with a gain of 66 dB and upper and lower cutoff frequencies of 100 kHz and 100 Hz, respectively, is made into a feedback amplifier. We want bandwidth of 2 MHz. Specify the amount of feedback needed and the new cutoff frequencies.
6. Repeat Problem 5 if a gain of 20 (magnitude) is desired.
7. Repeat Problem 3 if the output voltage is taken across R_E.
8. Repeat Example 13-2 if the gain without feedback is changed to 1,000. Make a comparison of your answers with those in Example 13-2 and make a conclusion.
9. If the open-loop (no feedback) gain of the amplifier in Example 13-1 changes by 5%, how much does the gain with feedback change? How could this change be minimized still further?
10. If the resistors used in the feedback network in Example 13-2 have a tolerance (i.e., an uncertainty in the value) of $\pm 10\%$, what is the uncertainty of the gain with feedback?
11. Determine the values of resistors R_8 and R_9 in Example 13-2 to provide a current gain of 40. What are the new input and output resistance values?
12. If the input voltage in Example 13-4 is 100 mV, and $R_C = R_E = 1\ \text{k}\Omega$, determine the output voltage across R_C. Also determine the voltage across R_E. How could we use this circuit?

14

Differential and Operational Amplifiers

We introduce a very important group of amplifier circuits in this chapter. Differential and operational amplifiers are dc-coupled amplifiers that are used as building blocks in a multitude of applications. They are becoming more important as the cost of integrated circuits (IC) utilizing these devices decreases. We shall especially emphasize, therefore, IC differential and operational amplifiers.

14.1 EMITTER-FOLLOWER

The emitter-follower circuit was discussed previously. Here we want to reemphasize its usefulness and apply it in more complicated circuits. Figure 14-1 shows an emitter-follower circuit with two power supplies. This connection allows the input voltage to be at or near ground; or, if so desired, the output voltage could be maintained at or near ground potential. These conditions may be necessary in a dc-coupled amplifier consisting of many stages of amplification.

In some applications, the input impedance of the simple emitter-follower circuit depicted in Fig. 14-1 may be insufficient even though it is high (up to 50 kΩ). For these applications the circuit shown in Fig. 14-2 is used. Two transistors are interconnected as indicated in Fig. 14-3. The two collectors are common, and the emitter of one is connected to the base of the other; this interconnection constitutes what is called a *Darlington configuration*.

The Darlington-transistor pair may also be connected in the emitter-follower configuration, as shown in Fig. 14-2. To analyze the circuit, we can combine the Darlington pair into one composite device. In fact, Darlington-connected transistor pairs are available in a single, three-lead package. The ac small-signal equivalent circuit for the Darlington configuration is illustrated in Fig. 14-4. If the Darlington configuration is to be used in an emitter-follower configuration, h_{oe1} usually cannot be neglected because of the high effective input resistance at $B2$.

The small-signal equivalent circuit for the Darlington emitter-follower is shown in Fig. 14-5. Note that at $B2$ (which is the same point as

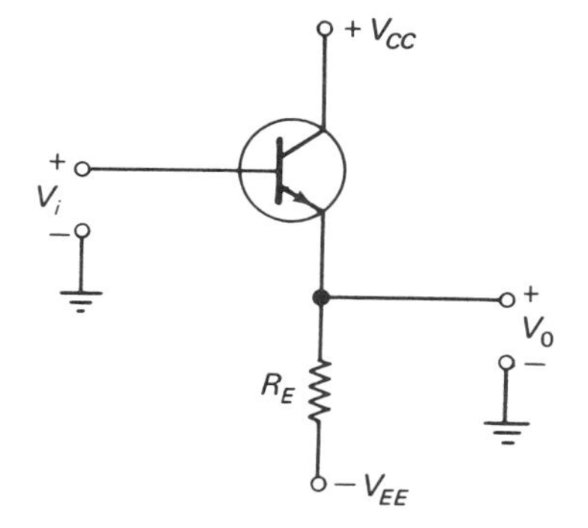

Figure 14-1. Emitter-follower with two power supplies.

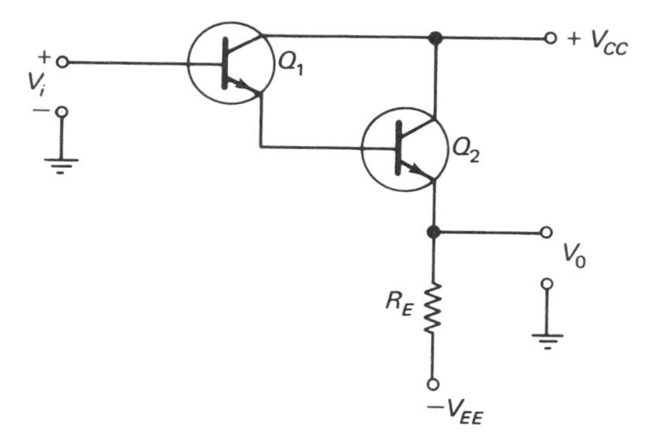

Figure 14-2. Darlington emitter-follower.

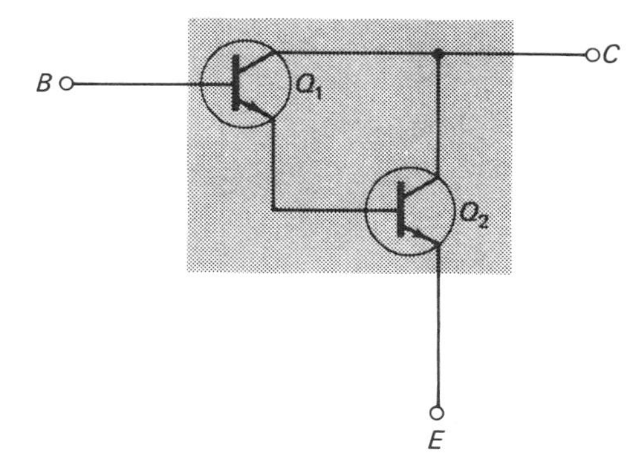

Figure 14-3. Darlington transistor configuration.

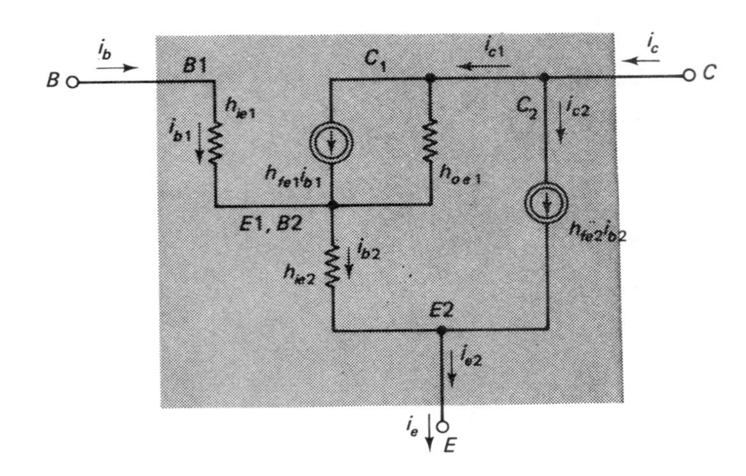

Figure 14-4. Small-signal equivalent circuit for the Darlington circuit.

$E\,1$), the impedance to ground is

$$R_{i2}=h_{ie}+(h_{fe}+1)R_E \cong h_{fe}R_E \tag{14-1}$$

There is current division between R_{i2} and h_{oe}, so that i_{b2} is given by:

$$i_{b2}=(h_{fe}+1)i_{b1}\frac{1}{1+h_{oe}R_{i2}} \tag{14-2}$$

The output current (through R_E) is i_e, that is,

$$i_e=(h_{fe}+1)i_{b2} \tag{14-3}$$

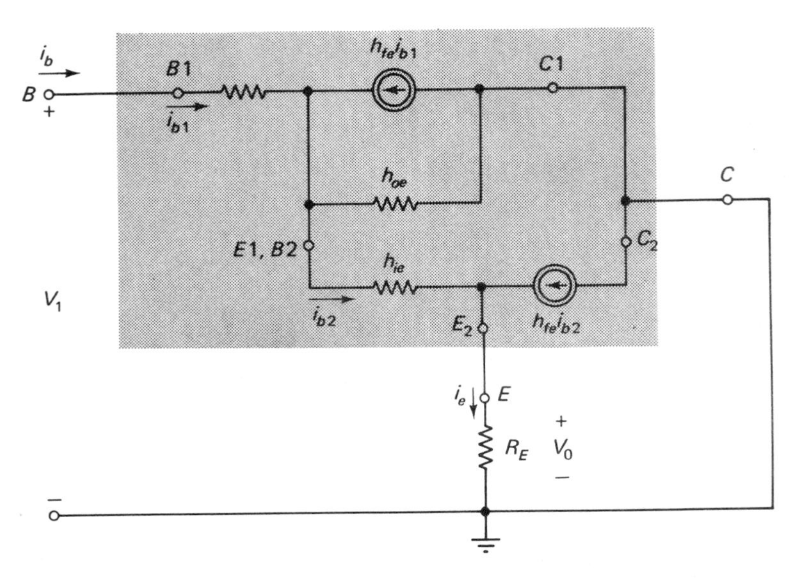

Figure 14-5. Equivalent circuit for the Darlington emitter-follower.

We can calculate the current gain (i_e/i_{b1}) from Eqs. (14-2) and (14-3):

$$A_I = \frac{(h_{fe}+1)^2}{1+h_{fe}h_{oe}R_E} \tag{14-4}$$

The input impedance may also be calculated. Note that R_{i2} is the effective load on transistor 1:

$$R_i = \frac{h_{fe}^2 R_E}{1+h_{fe}h_{oe}R_E} \tag{14-5}$$

Although the voltage gain of the Darlington emitter-follower circuit is just slightly less than 1, it may be assumed to be 1 for any practical considerations.

Example 14-1. The Darlington emitter-follower circuit shown in Fig. 14-2 is constructed with transistors having the following parameters: $h_{ie}=1$ kΩ, $h_{fe}=60$, and $h_{oe}=1/40$ kΩ. We want to determine the amplifier parameters for R_E of 100 Ω and 1 kΩ.

Solution: We calculate the input impedance and the current gain for the two values of R_E as follows:

$$R_i = \frac{(60)^2 R_E}{1+\dfrac{60R_E}{40\text{ k}}}$$

For $R_E=100$ Ω, this calculation is 310 kΩ. When $R_E=1$ kΩ, it is 1.45 MΩ.

$$A_I = \frac{(60+1)^2}{1+\dfrac{60R_E}{40\text{ k}}}$$

For $R_E=100$ Ω, this calculation is 3236. When $R_E=1$ kΩ, it becomes 1488.

From this example we can see the effect of R_E on both the input impedance and current gain and the increase in input impedance over a comparable single-transistor emitter-follower circuit.

Both the emitter-follower and Darlington emitter-follower circuits are used in applications where a high-input impedance, low-output impedance, and high-current gain are needed.

14.2 DIFFERENTIAL AMPLIFIERS

Differential amplifiers are characterized by the use of two matched transistors, having two possible inputs and two possible outputs. They are almost invariably dc-coupled and require two power supply voltages, one positive and one negative.

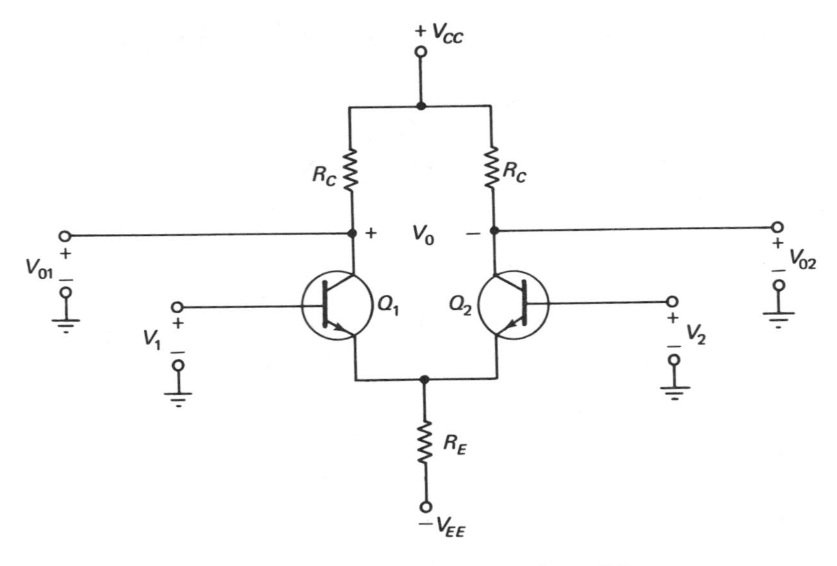

Figure 14-6. Basic differential amplifier.

14.2.1 Basic Differential Amplifiers

The basic differential amplifier circuit is illustrated in Fig. 14-6. It features two matched transistors with the emitters tied together. The input signals V_1 and V_2 are applied to the two bases, with two possible outputs taken at the two collectors.

Ideally, R_E should be infinite in order to provide infinite input impedance and to make the circuit operate as an ideal differential amplifier (DIFF AMP). The characteristics of an ideal DIFF AMP are as follows: (1) for identical inputs, the output is zero; and (2) for different inputs, the output is proportional to the difference between the inputs. You can see the need for infinite R_E by considering the equivalent circuit in Fig. 14-7. If R_E were indeed infinite, it would not draw current and i_{e1} would equal $-i_{e2}$. Under these conditions i_{c1} would equal $-i_{c2}$ and V_{o1} would equal $-V_{o2}$. Thus, the differential output V_o would be zero.

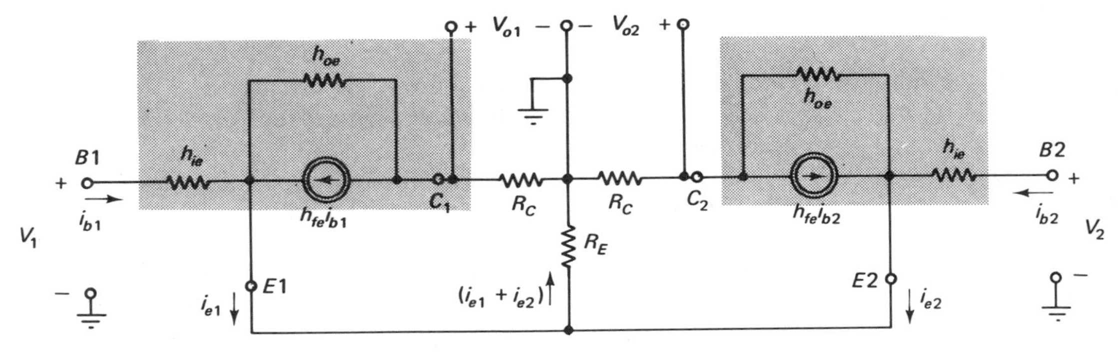

Figure 14-7. DIFF AMP equivalent circuit.

Because $Q1$ and $Q2$ must have dc bias, the size of R_E is limited. If R_E is increased, the negative supply voltage must also be increased in order to maintain the same dc bias current for the two transistors.

14.2.2 Improved DIFF AMP Circuit

Let us study the circuit shown in Fig. 14-8, where the resistor has been replaced with an ideal constant current source. This arrangement offers the dc bias current needed but still maintains an infinite resistance between the two emitters and ground.

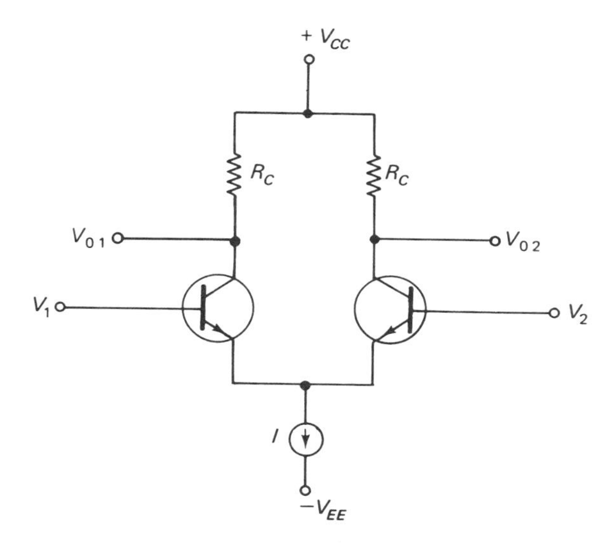

Figure 14-8. DIFF AMP with a constant current source.

In practice, a current generator can be approximated by a transistor, as shown in Fig. 14-9. The current I is controlled by bias resistors R_1, R_2, and R_3. Diodes $D1$ and $D2$ act to regulate the temperature of transistor $Q3$. If we assume forward diode voltages of 0.6 V and neglect the base current of $Q3$, the base voltage of $Q3$ with respect to ground is:

$$V_{B3} = -\frac{(V_{EE}-1.2)R_1}{R_1+R_2} \tag{14-6}$$

The bias current I may be calculated from the voltage across R_3:

$$V_{R3} = V_{B3} - V_{BE3} - V_{EE} = IR_3 \tag{14-7}$$

If R_1 and R_2 are equal, then:

$$I = \frac{V_{EE}}{2R_3} \tag{14-8}$$

The bias current I is, therefore, independent of the base-emitter voltage, which is quite sensistive to temperature; instead, it is a constant determined only by the supply voltage and R_3.

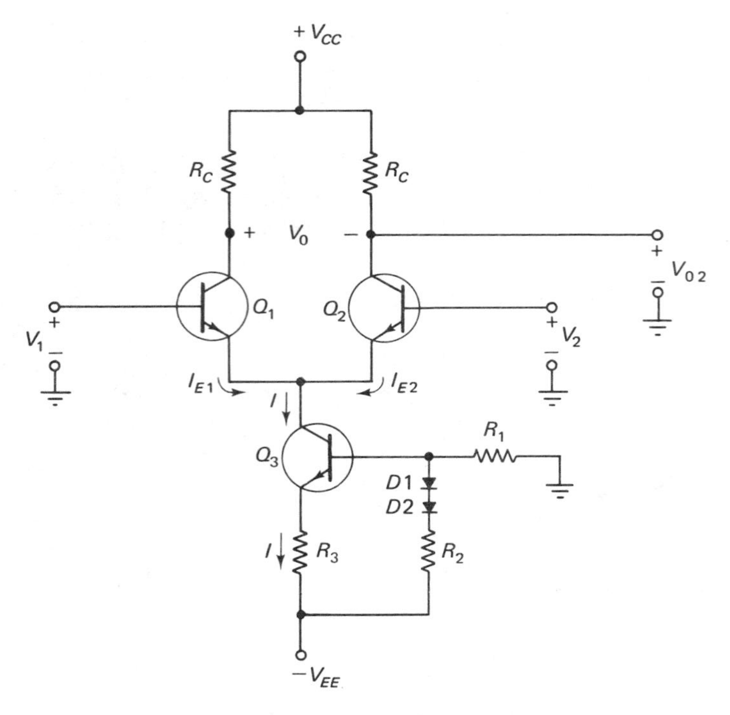

Figure 14-9. A practical DIFF AMP circuit.

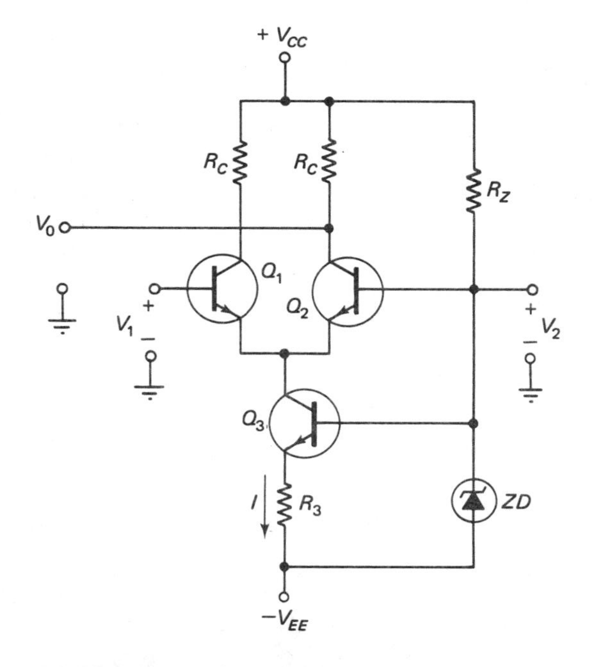

Figure 14-10. Zener-referenced current source in a DIFF AMP.

An alternate method of obtaining a temperature-stable bias current is shown in Fig. 14-10. A Zener diode is used to provide the bias to the current source transistor $Q3$. The series resistor R_Z is a current-limiting resistor used to keep the current through the Zener diode to a safe limit. If the Zener diode temperature characteristics match those of the base-emitter diode of $Q3$, then the bias current I will be insensitive to temperature and given by:

$$I = \frac{V_Z - V_{BE3}}{R_3} \tag{14-9}$$

where V_Z is the Zener diode (reference) voltage.

The ac performance of the two improved DIFF AMP circuits illustrated in Figs. 14-9 and 14-10 is quite similar. The small-signal equivalent circuit of Fig. 14-7 may be used with R_E replaced by the output impedance of $Q3$, as shown in Fig. 14-11. Note that the base of $Q3$ has no ac drive; therefore, all of its terminal currents have zero alternating current, and its ac equivalent circuit contains only the output impedance in the collector circuit, h_{oe}.

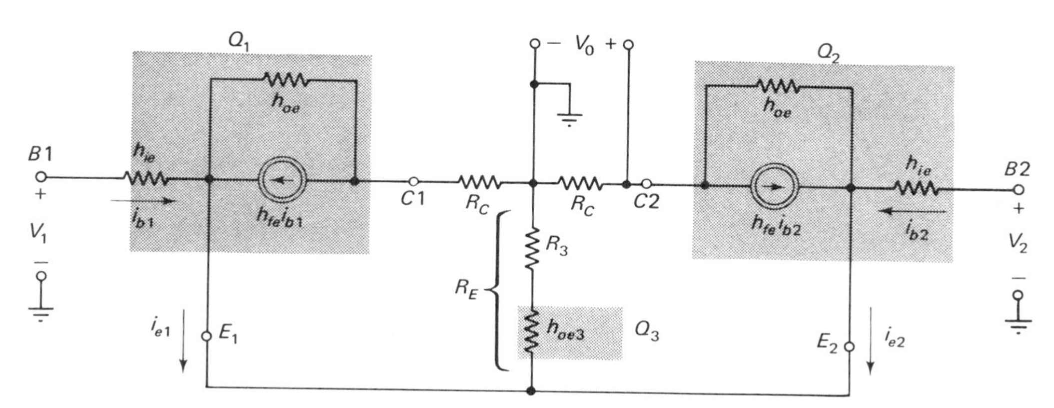

Figure 14-11. An ac small-signal equivalent circuit for improved DIFF AMP (Figs. 14-9 and 14-10).

Difference Gain. Let us consider the DIFF AMP circuit of either Fig. 14-9 or 14-10. If we apply input signals V_1 and V_2 (shown in Fig. 14-12) in such a way that:

$$V_1 = -V_2 = \frac{V_i}{2} \tag{14-10}$$

the resulting output voltage divided by V_i is called the *difference gain*, A_d. Thus,

$$A_d \cong \frac{V_o}{V_i} = \frac{V_o}{V_1 - V_2} \tag{14-11}$$

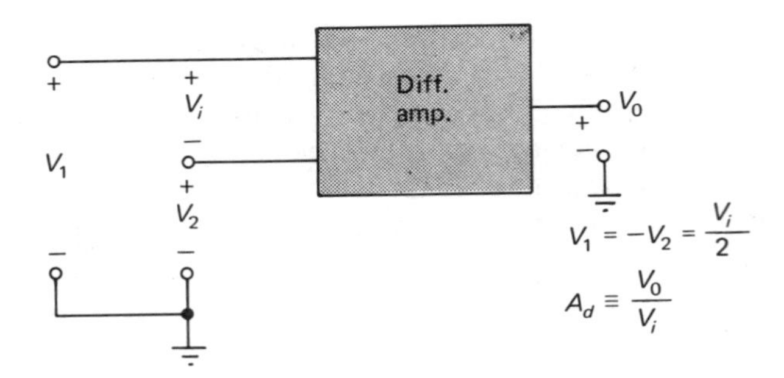

Figure 14-12. Configuration for determining DIFF AMP difference gain, A_d.

From the DIFF AMP equivalent circuit in Fig. 14-11, we can find that the single-ended difference gain is given by:

$$A_d \cong -\frac{h_{fe}R_C}{2h_{ie}} \tag{14-12}$$

Obviously the difference gain for this amplifier is not very high. The differential output gain is $-2A_d$ of Eq. (14-12).

Common-Mode Gain. When the input signals applied to the DIFF AMP are the same, $V_1 = V_2 = V_i$, as shown in Fig. 14-13, the output ideally should be zero, since there is no difference in the input signals. In an actual amplifier, however, this is not the case. Consequently, we define the common-mode gain as the ratio of the output to the input when the two inputs are equal. Thus,

$$A_c \cong \frac{V_o}{V_i} = \frac{V_o}{\frac{1}{2}(V_1 + V_2)} \tag{14-13}$$

It can be shown from the equivalent circuit that the common-mode gain is given by:

$$A_c \cong -\frac{R_C}{2R_E} \tag{14-14}$$

The common-mode gain appears to be quite small if R_E is made very large.

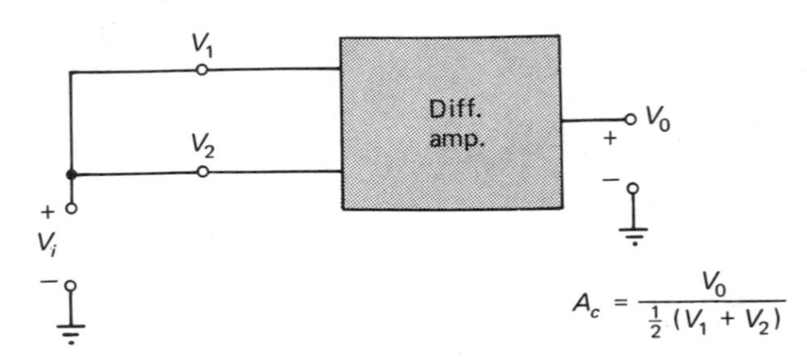

Figure 14-13. Configuration for determining DIFF AMP common-mode gain, A_c.

Common-Mode Rejection Ratio. The common-mode rejection ratio ($CMRR$) is a figure of merit for DIFF AMPs—a numerical indication of quality—(and for OP AMPs, as we shall subsequently see). It is defined as the ratio of the difference gain to the common-mode gain:

$$CMRR \cong \left| \frac{A_d}{A_c} \right| \tag{14-15}$$

The $CMRR$ measures the amplifier's ability to reject signals that are unwanted but are common to both inputs. Such signals might be noise or other undesirable signals. The larger the $CMRR$, the closer the proportion of the DIFF AMP output will be to the difference between the two input signals. The output voltage can, in general, be expressed in terms of the $CMRR$:

$$V_o = A_d V_d \left(1 + \frac{1}{CMRR} \frac{V_c}{V_d} \right) \tag{14-16}$$

where V_d is the difference input signal $(V_1 - V_2)$, and V_c is the common-mode input signal $\frac{1}{2}(V_1 + V_2)$.

For the two DIFF AMP circuits shown in Figs. 14-9 and 14-10, the $CMRR$ is obtained by dividing Eq. (14-12) by Eq. (14-14):

$$CMRR \cong \frac{h_{fe} R_E}{h_{ie}} \tag{14-17}$$

This analysis is illustrated in the following example.

Example 14-2. The DIFF AMP circuit in Fig. 14-10 is constructed with: $R_C = 100$ kΩ, $R_3 = 50$ kΩ, $R_Z = 1.8$ kΩ, $V_{cc} = V_{EE} = 12$ V. All the transistors may be assumed to be identical with $\beta = 100$, $h_{ie} = 5$ kΩ, $h_{fe} = 50$, and $h_{oe} = 1/200$ kΩ. The Zener diode is a 1N754 (6.8 V) type. We want to determine (1) the dc voltages and currents in the DIFF AMP, (2) the difference and common-mode gains in the circuit, and (3) the $CMRR$.

Solution: Because $\beta = 100$, we may safely neglect all base currents in comparison to either emitter or collector currents. From Eq. (14-9):

$$I \cong \frac{6.8 - 0.6}{50} \text{ mA} \cong 124\ \mu\text{A}$$

If transistors $Q1$ and $Q2$ are identical, then I divides equally between the two emitters:

$$I_{E1} = I_{E2} = \frac{I}{2} \cong 62\ \mu\text{A}$$

If we assume that no dc input is being applied to either input terminal, then $V_{B1} = V_{B2} = 0$ V. The emitters of both $Q1$ and $Q2$ must then be at -0.6 V.

Because β is large, we can say that the collector and emitter currents in $Q1$, $Q2$, and $Q3$ are equal. The voltage drops across the collector resistors are:

$$V_{Rc} \cong (62\ \mu A)(100\ k\Omega) \cong 6.2\ V$$

The collectors of $Q1$ and $Q2$ must then be 6.2 V below V_{CC}, or at $12-6.2=5.8$ V with respect to ground. The collector-to-emitter voltages for the three transistors are:

$$V_{CE1} = V_{CE2} = 5.8 - (-0.6) = 5.2\ V$$
$$V_{CE3} = -0.6 + 12 - (124\ \mu A)(50\ k\Omega) \cong 5.2\ V$$

The difference and common-mode gains may be calculated from Eqs. (14-12) and (14-14), respectively, noting that R_E is $1/h_{oe}$ in series with R_3:

$$A_d \cong -\frac{50(100)}{2(5)} \cong -500$$
$$A_c = -\frac{100}{2(200)} \cong -0.25$$

The $CMRR$ can now be determined as

$$CMRR = \frac{500}{0.25} \cong 2000 \text{ or } 66 \text{ dB}$$

In the previous example, if we had applied two input signals, one of 2 mV and the other of 5 mV, we would have expected an output voltage given by Eq. (14-16) of:

$$V_o = (-500)(3)\left[1 + \frac{7/2}{(2000)(3)}\right] - 1500(1.0005)\text{mV}$$

This number is almost exactly -500 times the difference input signal of 3 mV. If transistor $Q3$ had had a larger output impedance (i.e., a lower h_{oe}), the $CMRR$ would have been even higher and the error in the output would have been even lower than the 0.05% of Example 14-2.

14.3 INTEGRATED CIRCUIT DIFF AMPS

Integrated circuit (IC) DIFF AMPs can be manufactured with matched transistors on the same substrate at low cost. Consequently, they are widely used in many circuit applications. An example of a commercially available IC DIFF AMP is the RCA CA3000, whose circuit diagram is shown in Fig. 14-14. Some of the applications listed by the manufacturer are: RC-coupled feedback amplifier, crystal oscillator, modulator, Schmitt trigger, comparator, and sense amplifier.

The circuit diagram is a little more complicated than the configurations examined previously. However, we can recognize transistors $Q1$ and $Q2$ connected in the differential mode; transistor $Q3$ together with its resistors and diodes $D1$ and $D2$ acts as the constant current source. Transistors $Q4$ and $Q5$ are connected in the emitter-follower configuration and are the input transistors.

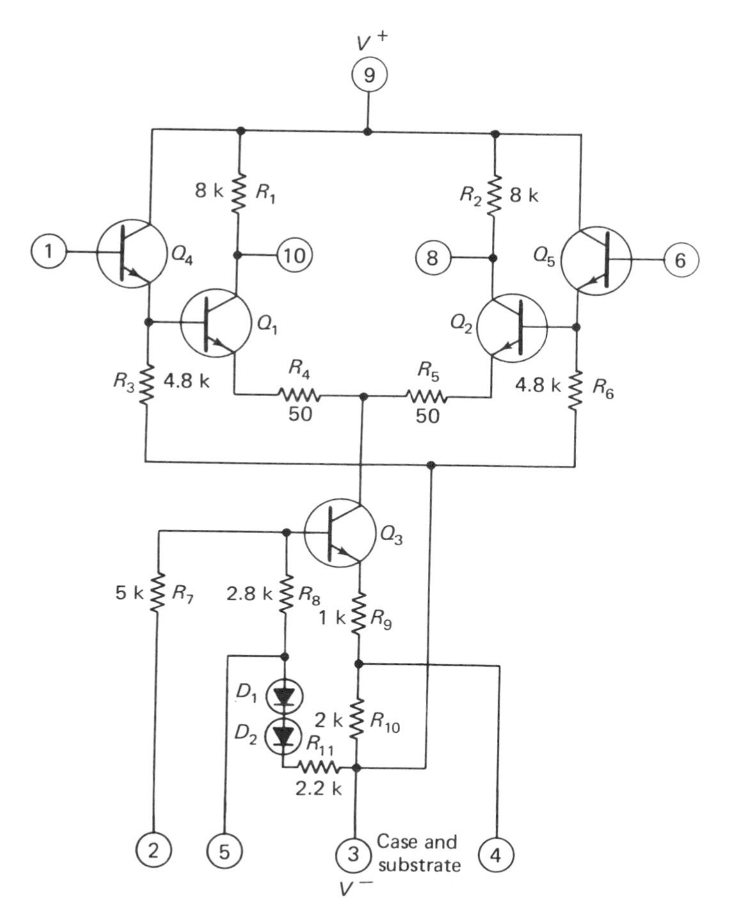

Figure 14-14. Example of an integrated circuit differential amplifier (RCA CA3000).

The manufacturer lists the following parameters for the CA3000:

$$A_d = 32 \text{ dB} \qquad \text{(magnitude 40)}$$
$$CMRR = 98 \text{ dB} \qquad (\text{magnitude} \sim 10^5)$$
$$R_i = 195 \text{ k}\Omega$$

From these values we can get some idea as to how this IC DIFF AMP compares with the discrete component DIFF AMP of Example 14-2. For example, we can estimate the common-mode gain for the CA3000 as 4×10^{-4}.

We shall postpone the discussion of other DIFF AMP parameters until we treat OP AMPs, where the discussion is just as applicable to DIFF AMPs.

14.4 OPERATIONAL AMPLIFIERS (OP AMPS)

Modern fabrication techniques and volume production have lowered the cost of high-quality monolithic (IC) OP AMPs and increased the importance of OP AMPs in many circuit applications. An operational amplifier is a high-gain differential amplifier which typically has very high input impedance, very low output impedance, and an open-loop gain in excess of 1000. In fact, open-loop gains of one million or higher are possible.

14.4.1 Properties of an Ideal OP AMP

We begin our study of OP AMPs by considering the properties of an ideal OP AMP. These are listed below:

1. Infinite gain
2. Infinite input resistance
3. Zero output resistance
4. Infinite bandwidth
5. Infinite *CMRR*

The OP AMP symbol is illustrated in Fig. 14-15. The two inputs, inverting and noninverting, are designated by − and +, respectively. The equivalent circuit for the OP AMP is shown in Fig. 14-16. Note that the input resistance is defined between the two inputs and not from one of the inputs to ground. The output is referenced to ground.

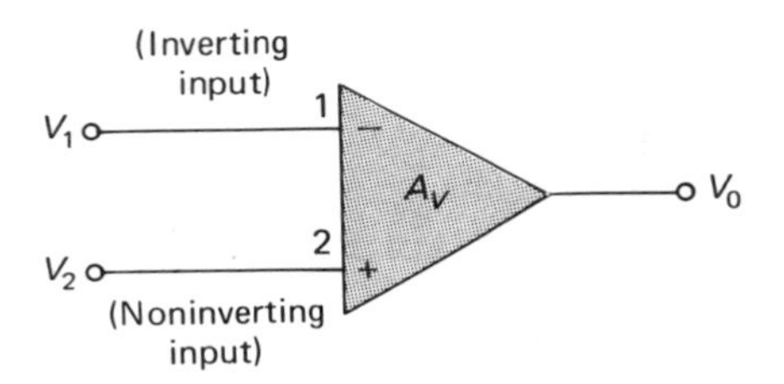

Figure 14-15. OP AMP symbol.

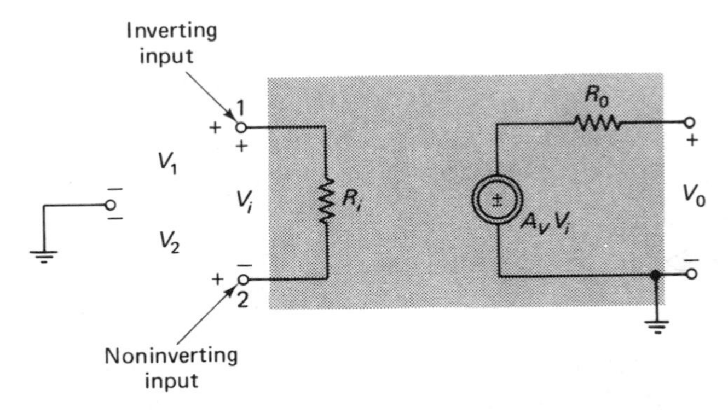

Figure 14-16. OP AMP equivalent circuit.

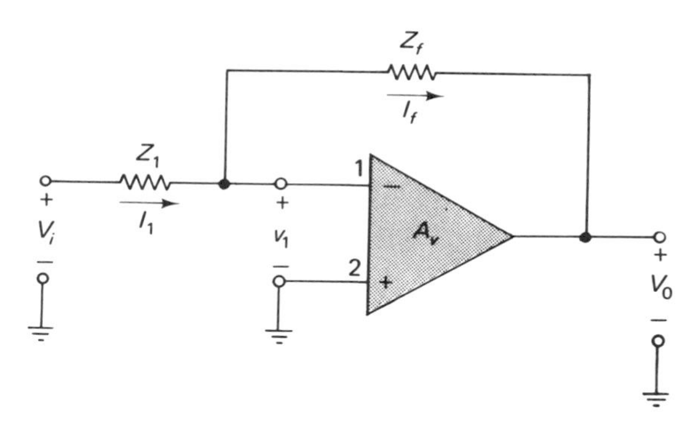

Figure 14-17. OP AMP with shunt feedback.

Consider the shunt-feedback configuration illustrated in Fig. 14-17. (Observe that this circuit is the same as that in Fig. 13-14, with the single transistor replaced by an OP AMP.) This diagram is the basic inverting-amplifier configuration of the OP AMP, with the noninverting input grounded. With impedances Z_f and Z_1 finite and the OP AMP input impedance infinite, obviously the current that flows in Z_1 (I_1) must be equal to the current through Z_f (I_f), because R_i is infinite. Moreover, if the output voltage is finite and the gain is infinite, the input voltage to the OP AMP (V_1) must be zero. Such a figure agrees with the fact that the OP AMP draws zero current.

This situation is depicted in Fig. 14-18. The input of the OP AMP draws zero current and has no voltage across it. It is called a *virtual ground*, having the symbol shown in Fig. 14-18. Were it a conventional ground, it would act like a short and draw the maximum current. The input of an ideal OP AMP acts like a dead short in terms of voltage and an open in terms of current. The result is, therefore, a virtual ground.

Making use of the fact that $I_1 = I_f$ and that $V_1 = 0$, we have:

$$I_1 = \frac{V_i}{Z_1} = I_f = -\frac{V_o}{Z_f} \tag{14-18}$$

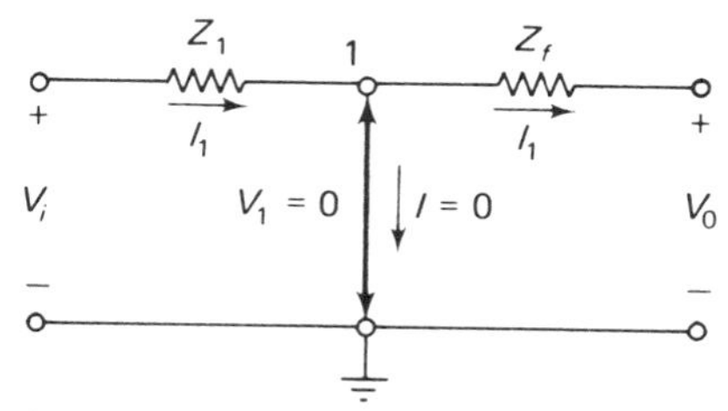

Figure 14-18. Equivalent circuit for Fig. 14-17 showing the virtual short at the input.

The voltage gain in this configuration is given by:

$$A_V = \frac{V_o}{V_i} = -\frac{Z_f}{Z_1} \tag{14-19}$$

This result is very important. It tells us that the gain is dependent only on the impedances used in the circuit and not on any amplifier parameters as long as (1) the open-loop gain of the amplifier is large, (2) its input impedance is very large, and (3) its output impedance is very small.

14.4.2 Practical IC OP AMPs

We now consider practical IC OP AMPs, which come quite close to matching many of the properties of the ideal OP AMP.

Typically, an IC OP AMP may contain one or two stages of differential amplifiers, an impedance matching emitter-follower stage, a dc level-translator stage, and an output driver stage, as shown in the block diagram in Fig. 14-19. Thus, we see that the differential amplifier, which has many applications of its own, is also used as a building block in an OP AMP. We have already examined the operation of the DIFF AMP and emitter-follower stages. Let us now examine the operation of the dc level-translator stage.

A typical circuit diagram for a dc level-translator is shown in Fig. 14-20. This type of circuit is important in dc-coupled amplifiers because it accomplishes much the same function as a coupling capacitor; that is, it removes the dc component of the signal. To see how this operation is accomplished, assume that the input V_i to the level-translator circuit has an ac component that is riding on top of some dc level. The desired output is an ac signal centered around ground. (There is no dc level.) The circuit components in Fig. 14-20 are adjusted so that the base of $Q3$ is approximately 0.6 V above ground, thus causing the average or dc level at V_o to be at ground (i.e., there is no direct current). The operation of this circuit is illustrated in the following example.

Example 14-3. The level-translator (shifter) circuit of Fig. 14-21 uses silicon transistors that are all on the same substrate and identical with $\beta = 100$. The dc component at the input is +6 V. We want to show that the dc component at the output is zero.

Solution: With silicon transistors, we can assume all base-emitter voltages

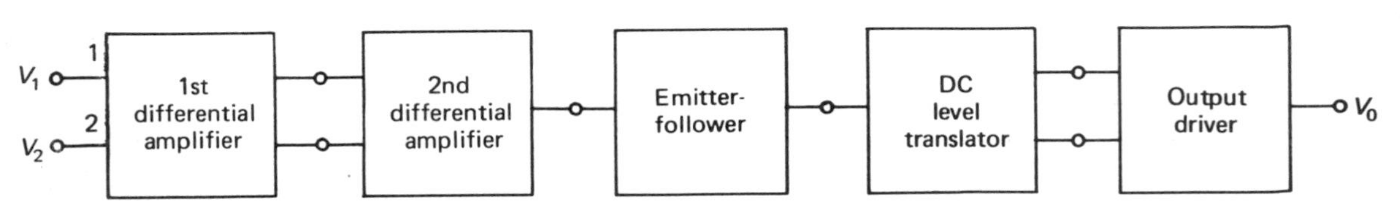

Figure 14-19. Block diagram of a typical IC OP AMP.

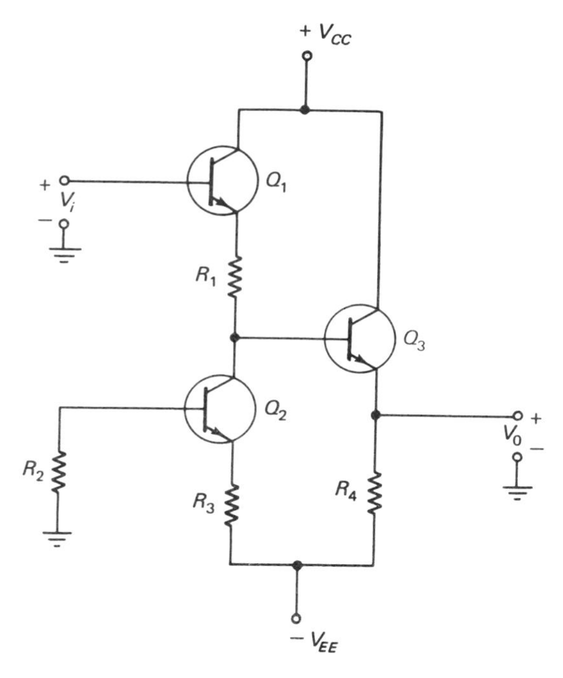

Figure 14-20. A dc level-translator circuit.

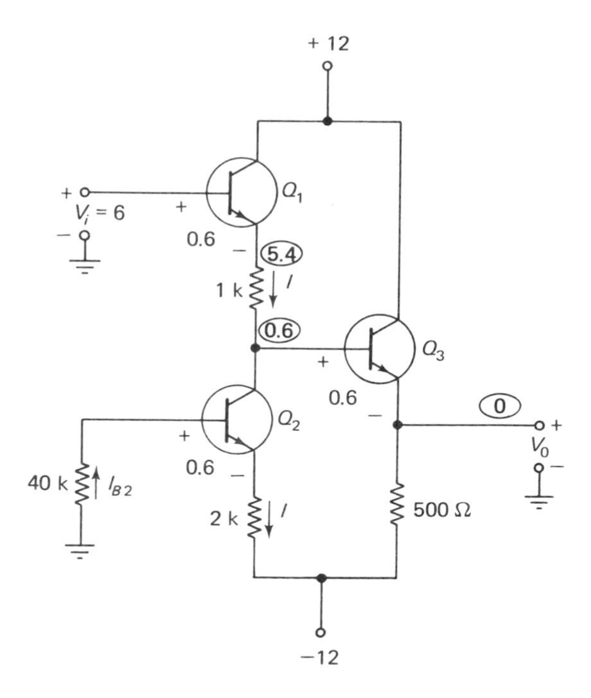

Figure 14-21. Example of a dc level-translator circuit.

to be 0.6 V. The base current of $Q2$ is

$$I_{B2} \cong \frac{12-0.6}{40+(\beta+1)2} \text{ mA} \cong 0.048 \text{ mA}$$

We next obtain $I_{C2}=\beta\ I_{B2} \cong 4.8$ mA. This current is approximately the same as in the emitter of $Q2$ and in both the collector and emitter of $Q1$ (where we have neglected the base currents as compared to either the collector or emitter current of any of the transistors). The voltage across R_1 (1 kΩ) is $V_{R1} \cong 4.8$ V. Because the base of $Q1$ is at 6 V, its emitter must be at 5.4 V. The base of $Q3$ (also the collector of $Q2$) is then at

$$V_{B3} = V_{E1} - V_{R1} = 5.4 - 4.8 = 0.6 \text{ V}$$

With 0.6 V base-emitter, the emitter of $Q3$, which is the output, must be at a zero dc level. The complete operation can be verified by ascertaining that none of the transistors is saturated. However, we leave verification to an exercise.

The ac performance of the level-translator circuit may be examined to verify that the ac signal attenuation is minimal. This type of circuit is practical in IC amplifiers because the additional cost of the transistors required is very low compared to the overall cost of the amplifier. In fact, a transistor can be manufactured on an IC at roughly the same cost as a resistor or a diode. In discrete circuits, this dc blocking function would be provided by a coupling capacitor. The very size and cost of such a capacitor rule out its use in ICs.

An example of a high-quality, internally compensated IC OP AMP is shown in Fig. 14-22. The circuit utilizes 20 transistors, three of which ($Q11$, $Q12$, $Q8$) are used as diodes, 11 resistors, and one compensating capacitor. It is available in an overall package just slightly larger than that of a single transistor.

Without attempting to go into the details of circuit operation, we can note the input differential amplifier stage consisting of $Q1$ and $Q2$, the Darlington configuration of $Q16$ and $Q17$, and the use of complementary transistors throughout. The level translation of direct current is accomplished through the series combination of $Q13$, $Q18$, and the Darlington-pair feeding the complementary output driver transistors $Q14$ and $Q20$ so that the output (pin 6) is at a ground dc level.

The practical IC OP AMP differs from the ideal OP AMP in many ways. Before we examine the properties of practical OP AMPs in comparison to ideal ones, we need to examine the meaning of the terms used to specify these properties.

Large-Signal Voltage Gain: The ratio of the output voltage swing to the change in the input voltage required to cause the output voltage swing.

Output Voltage Swing: The peak output voltage swing, referred to ground, that can be obtained without clipping.

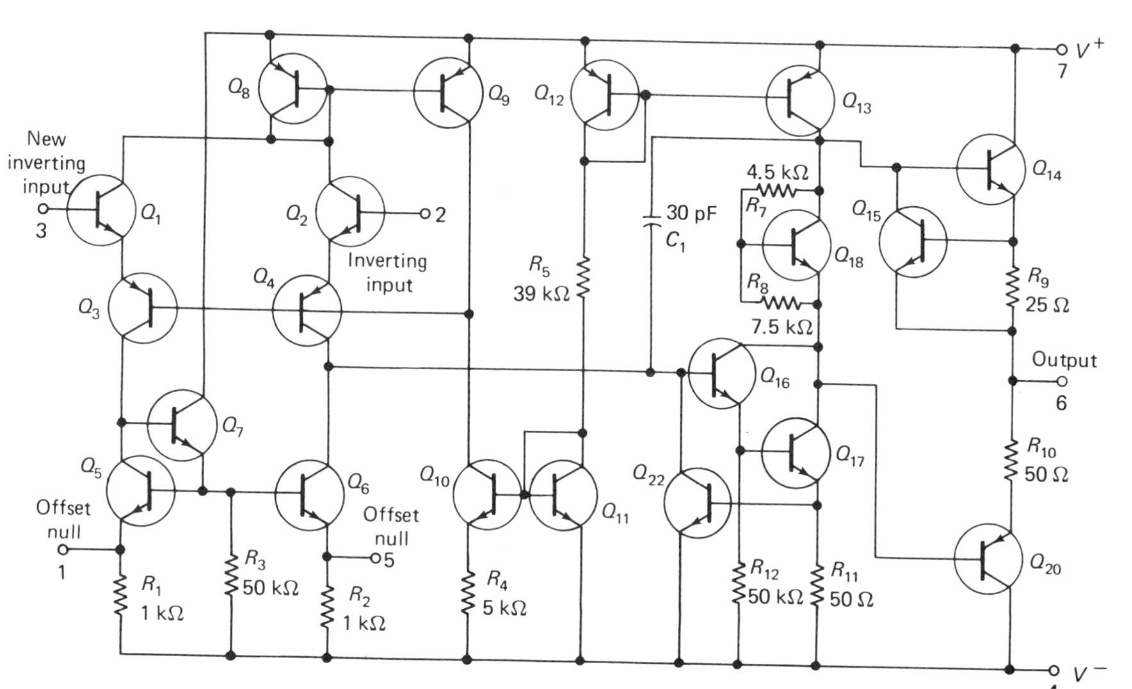

Figure 14-22. Example of an integrated circuit OP AMP with internal compensation (Fairchild μA741).

Input Offset Voltage: The voltage that must be applied between the two input terminals (through two equal resistors) in order to obtain zero output voltage.

Input Offset Current: The difference between the two input currents (flowing into the input terminals) when the output voltage is at zero.

Input Bias Current: The average value of the two input currents.

Input Voltage Range: The maximum range of input voltages that may be applied to the input terminals and still have the amplifier operate within its specifications.

Input Resistance: The ratio of the change in input voltage to the change in input current at one of the input terminals when the other input terminal is grounded.

Supply Current: The total current drain from the power supply needed to operate the OP AMP with no load and zero volts at the output.

Common-Mode Rejection Ratio: The ratio of the input voltage range to the peak-to-peak change in input offset voltage over this range of input voltages.

Power Supply Rejection Ratio: The ratio of the change in input offset voltage to the change in power supply voltage producing the change in input offset voltage.

Slew Rate: The internally set limit in the rate of change in the output voltage with a large-amplitude step function applied to the input.

Review Questions

1. What is a differential amplifier?
2. If the two input signals to a differential amplifier are equal, what should the output ideally be? Explain why.
3. What should be the input impedance of an ideal differential amplifier? Why?
4. What is the improvement in performance caused by the current generated ($Q3$) in Fig. 14-9?
5. What is the difference gain in a DIFF AMP?
6. What is the common-mode gain in a DIFF AMP?
7. What is the common-mode rejection ratio a measure of in a DIFF AMP?
8. What should the *CMRR* be for an ideal DIFF AMP? What is it for a practical IC DIFF AMP?
9. What is the advantage of IC DIFF AMPs as compared to discrete component ones?
10. What are the differences and similarities between DIFF AMPs and OP AMPs?

11. What are the properties of an ideal OP AMP? Compare these properties with those of a practical IC OP AMP.
12. What is a virtual ground? How is it different from a conventional ground?

Problems

1. The transistor used in Example 14-1 is replaced by one having: $h_{ie}=500\ \Omega$, $h_{fe}=50$, and $h_{oe}=12.5\ \mu$Mho. Determine the current gain and the input impedance.
2. The DIFF AMP in Fig. 14-10 has the following circuit values: $R_C=560\ \Omega$, $R_3=400\ \Omega$, $R_Z=1.5\ \text{k}\Omega$, with $+6$ and -6 V supplies. All transistors are identical silicon *NPN*s with: $\beta=h_{fe}=50$, $h_{ie}=800\ \Omega$, and negligible h_{oe}. Determine the dc voltages and currents in the circuit, for a 4 V Zener diode.
3. In Problem 2, determine the difference and common-mode gains, as well as the *CMRR*.
4. Repeat Problem 2 if the Zener diode is replaced by a 5.6 V Zener diode.
5. The circuit in Fig. 14-20 has the following values: $R_2=25\ \text{k}\Omega$, $R_3=1\ \text{k}\Omega$, $R_4=100\ \Omega$, with $V_{CC}=V_{EE}=6$ V. The output voltage is to be 0 V using silicon transistors. (Assume that all transistors are identical, $\beta=50$ and $V_{BE}=0.5$ V.) Determine the value of R_1 needed for an input voltage of 3 V.
6. Repeat Problem 5 above if the input voltage is 4 V and the transistor βs are 50.

15

Sinusoidal Oscillators

Oscillators are basic electronic circuits. They have no ac input, but they provide an ac output at a specific frequency. The only input needed for an oscillator is the power supply voltage to bias the active device or devices used in an oscillator. Oscillators in general are feedback amplifiers with positive or regenerative feedback.

15.1 CRITERIA FOR OSCILLATION

Consider the general oscillator circuit shown in Fig. 15-1. The amplifier (not necessarily an OP AMP) is characterized by a negative voltage gain A_V, output impedance R_o, and an extremely large input impedance R_i. In Fig. 15-2, we have redrawn the circuit to show more clearly the feedback network comprised of Z_1 and Z_2. The circuit is a form of voltage feedback (see section 13.2). We can rewrite the gain of the circuit as:

$$G = \frac{A}{1-\beta A} \qquad (15\text{-}1)$$

where β is the feedback factor as defined in section 13.2. However, if the circuit is to oscillate, the gain must be infinite. For the gain to be infinite, the denominator in Eq. (15-1) must go to zero. Thus,

$$|1-\beta A| = 0$$

or

$$|\beta A| = 1 \quad \text{and} \quad \text{Phase angle of } (\beta A) = 0 \qquad (15\text{-}2)$$

where βA is called the *loop gain* and both (or either) β and A are functions of frequency and so are complex numbers.

The condition in Eq. (15-2) is called the *Barkhausen criterion*; it specifies the conditions that must be met in order for oscillations to be sustained. According to the Barkhausen criterion, the frequency of oscillation is the frequency at which the signal travels around the loop. As indicated in Fig. 15-2, it starts at the input terminal, must remain in phase (insuring positive feedback), and the amplitude must not diminish in its trip around the loop. The frequency of oscillation is determined by the proper phase shift of the feedback loop; oscillation itself is insured by

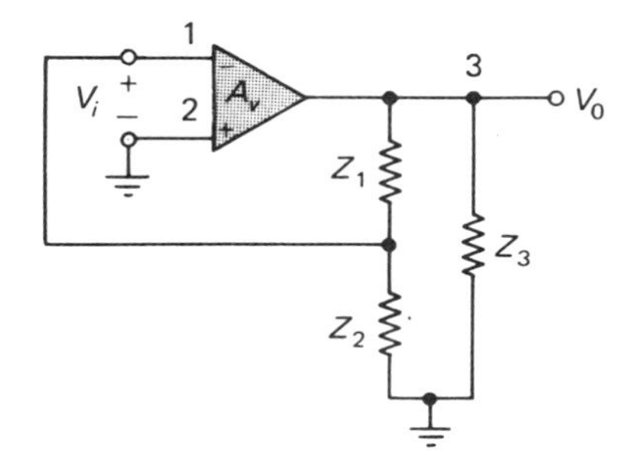

Figure 15-1. General oscillator circuit.

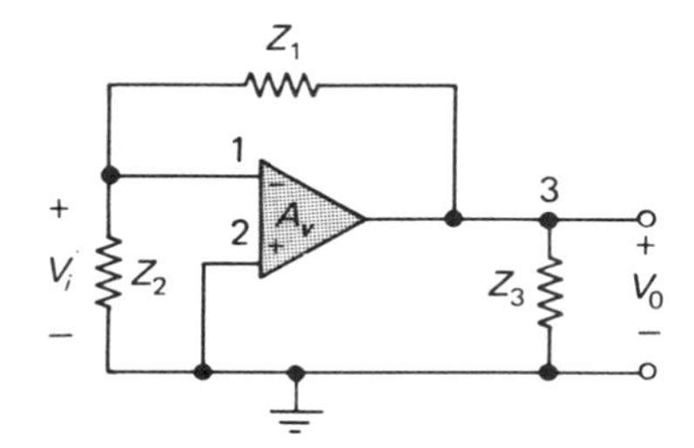

Figure 15-2. General oscillator circuit rearranged.

sufficient loop gain. Note that a loop gain much larger than 1 will cause distortion in the signal and the output will not be sinusoidal.

Replacing the amplifier by its equivalent circuit yields the circuit shown in Fig. 15-3. When the circuit of Fig. 15-3 is redrawn, as shown in Fig. 15-4, we can obtain the gain without feedback A:

$$A = A_v \frac{Z_L}{Z_L + R_o} \tag{15-3}$$

where we have defined Z_L as the effective load without feedback:

$$Z_L = \frac{(Z_1 + Z_2)Z_3}{Z_1 + Z_2 + Z_3} \tag{15-4}$$

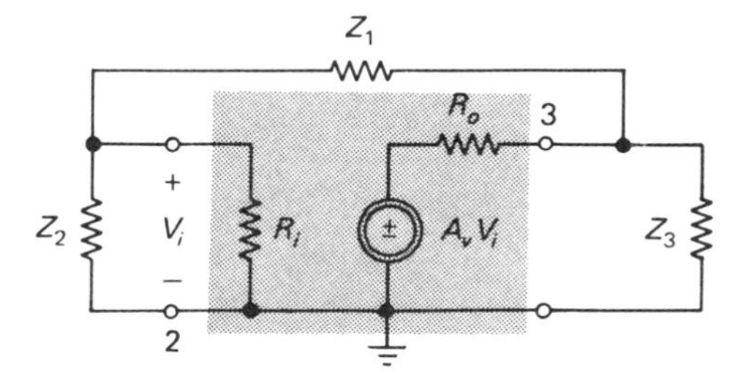

Figure 15-3. Equivalent circuit for Fig. 15-2.

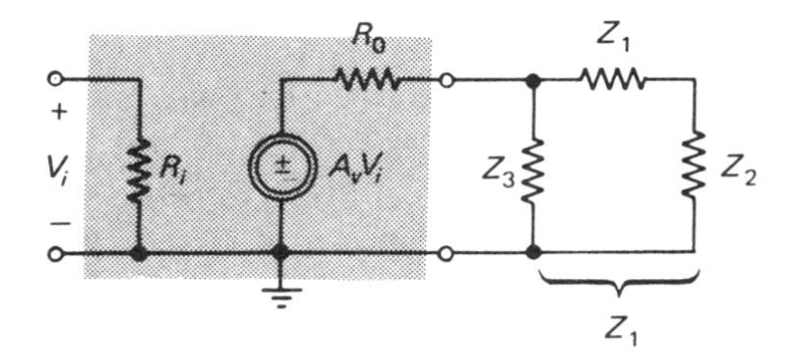

Figure 15-4. Determining the gain without feedback.

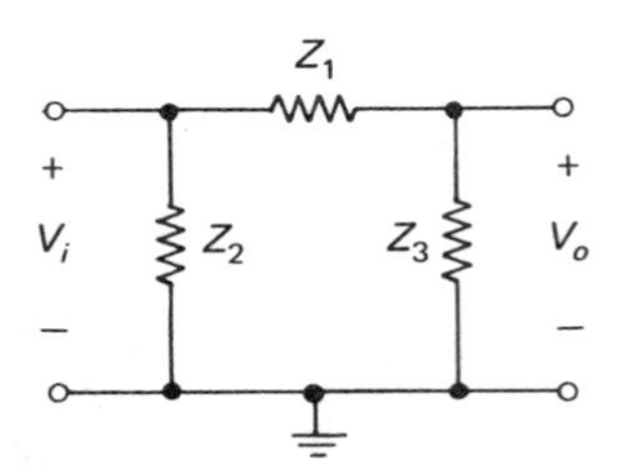

Figure 15-5. Determining the feedback factor.

Similarly, we can determine the feedback factor β from Fig. 15-5:

$$\beta = \frac{Z_2}{Z_1 + Z_2} \tag{15-5}$$

Substituting Eqs. (15-3), (15-4), and (15-5) into the Barkhausen criterion, we find that Eq. (15-2) gives us the frequency of oscillation and the amplifier gain needed. We shall consider the special case where all three impedances are purely reactive:

$$Z_1 = jX_1 \qquad Z_2 = jX_2 \qquad Z_3 = jX_3 \tag{15-6}$$

Using these relationships in the loop-gain equation, we obtain:

$$\beta A = A_V \frac{-X_2 X_3}{-X_3 X_1 - X_2 X_3 + jR_o(X_1 + X_2 + X_3)} \tag{15-7}$$

For the phase angle of βA to be zero requires that the imaginary part in the denominator of Eq. (15-7) be zero. Thus,

$$X_1 + X_2 + X_3 = 0 \tag{15-8}$$

In a given a circuit, Eq. (15-8) yields the frequency of oscillation. If we set the magnitude of the loop gain to 1, we have:

$$|A_V| = \frac{X_3}{X_2} \tag{15-9}$$

Equation (15-9) gives the critical value of the magnitude of the amplifier gain required for oscillation. Note that in reality the amplifier gain is negative.

15.2 HARTLEY OSCILLATORS

A Hartley oscillator using an OP AMP is illustrated in Fig. 15-6. We can verify that the circuit is of the basic form shown in Fig. 15-1 if we consider the amplifier to be the OP AMP together with its gain-setting resistors R_1 and R_f. The voltage gain from V_i to V_o is given by:

$$A_V = -\frac{R_f}{R_1} \tag{15-10}$$

We can match the inductors and capacitor of Fig. 15-6 with the impedances in Fig. 15-1 to get:

$$X_1 = \frac{-1}{\omega C_1} \qquad X_2 = \omega L_2 \qquad X_3 = \omega L_3 \tag{15-11}$$

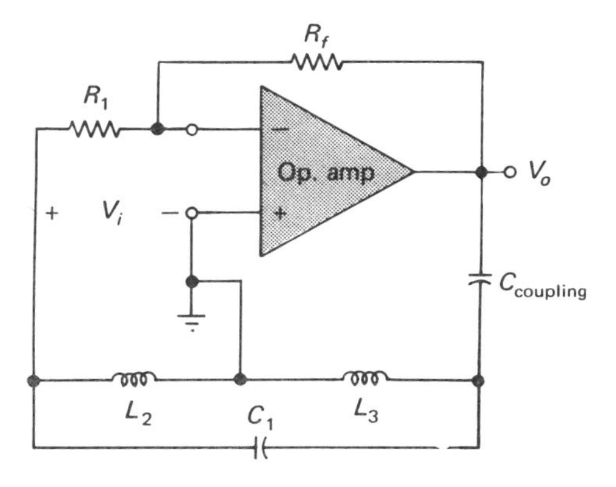

Figure 15-6. OP AMP Hartley oscillator.

The frequency of oscillation can be determined if we substitute Eq. (15-11) into Eq. (15-8):

$$\frac{-1}{\omega C_1} = \omega(L_2 + L_3) \tag{15-12}$$

which yields the frequency of oscillation f_o:

$$f_o = \frac{1}{2\pi\sqrt{(L_2 + L_3)C_1}} \tag{15-13}$$

The minimum gain needed is evaluated from Eq. (15-9):

$$|A_V| = \frac{L_3}{L_2} \tag{15-14}$$

An FET may be used for the amplifier to form the Hartley oscillator circuit shown in Fig. 15-7. Operation is the same as for the OP AMP Hartley oscillator with one exception: The gain in the FET circuit is given by $g_m r_d$. In Fig. 15-7, C_c is a coupling capacitor which is a short circuit at the desired frequency of oscillation. It blocks the dc drain voltage from being fed to the gate through the low dc resistance of L_2 and L_3. The radio-frequency choke (RFC) chosen must have a very high impedance at the frequency of oscillation, because it provides a dc path between the battery and the drain. FET bias is accomplished through R_s, which is

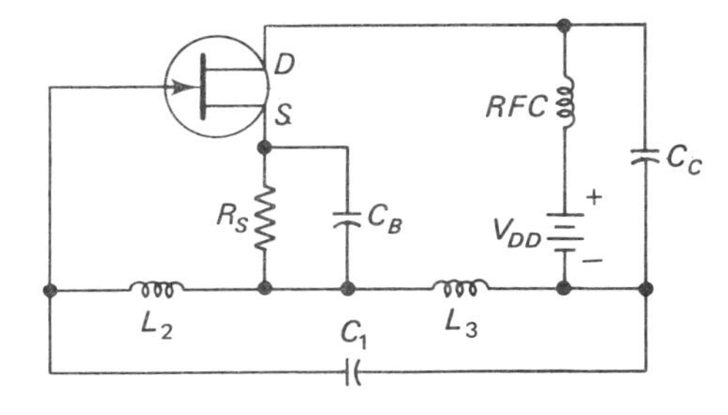

Figure 15-7. FET Hartley oscillator.

bypassed by C_B at the frequency of oscillation. Equations (15-13)—for the frequency of oscillation—and (15-14)—for the gain—are still valid.

A Hartley oscillator may be constructed using a BJT. Because of the very low input impedance, however, the analysis is complicated and will not be attempted here.

Example 15-1. We wish to construct a Hartley oscillator, as shown in Fig. 15-6, with $L_3 = 0.4$ mH, $L_2 = 0.1$ mH, and $C_1 = 0.002$ μF. We want to determine the frequency of oscillation and the values of R_1 and R_f to insure oscillation.

Solution: The frequency of oscillation is given in Eq. (15-13):

$$f_o = \frac{1}{2\pi\left[(0.4+0.1)(2\times10^{-12})\right]^{1/2}}\ \text{Hz} \cong 160\ \text{kHz}$$

The minimum gain needed is given by Eq. (15-14):

$$|A_V| = \frac{0.4}{0.1} = 4$$

Therefore, if we choose R_1 to be, say, 100 kΩ, we can use R_f of 430 kΩ to give a voltage gain of 4.3, which should insure oscillation.

15.3 COLPITTS OSCILLATORS

If the basic oscillator circuit is connected with capacitors and inductors reversed from the Hartley configuration, a Colpitts oscillator (shown in Fig. 15-8) results. It should be obvious that the Hartley and Colpitts oscillators are duals. We can analyze the Colpitts oscillator by using the results of the general discussion in section 15-1. Note that

$$X_1 = \omega L_1 \quad X_2 = \frac{-1}{\omega C_2} \quad X_3 = \frac{-1}{\omega C_3} \tag{15-15}$$

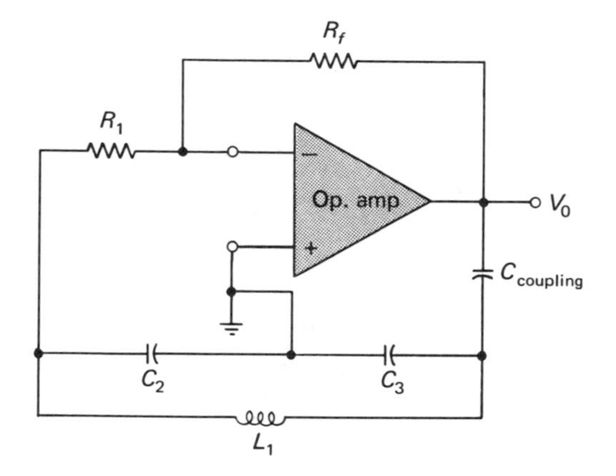

Figure 15-8. OP AMP Colpitts oscillator.

Setting the sum of these reactances equal to zero, we can determine that the frequency of oscillation is:

$$f_o = \frac{1}{2\pi\sqrt{L_1 C_S}} \tag{15-16}$$

where C_S is the effective series combination of C_2 and C_3; then

$$C_S \equiv \frac{C_2 C_3}{C_2 + C_3}$$

The minimum gain is determined from Eq. (15-9):

$$A_V = \frac{C_2}{C_3} \tag{15-17}$$

Because the gain is larger than 1, obviously C_2 must be larger than C_3.

Figure 15-9 gives an example of an FET Colpitts oscillator. You can easily see the basic similarity between this circuit and that of Fig. 15-7. The equations determined for the OP AMP Colpitts oscillator hold for the FET version, with the exception that the FET gain is given by $g_m r_d$.

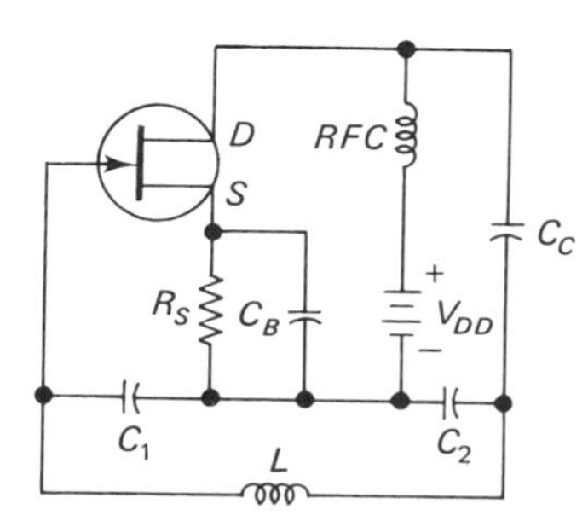

Figure 15-9. FET Colpitts oscillator.

Example 15-2. The OP AMP Colpitts oscillator of Fig. 15-8 is constructed with $L_1 = 0.1$ mH, $C_2 = 800$ pF, and $C_3 = 400$ pF. We want to determine the frequency of oscillation and the minimum gain needed.

Solution: We first calculate the equivalent capacitance:

$$C_S = \frac{(800)(400)}{800+400}\ \text{pF} \cong 270\ \text{pF}$$

The frequency of oscillation can now be found:

$$f_o = \frac{1}{2\pi\sqrt{(26 \times 10^{-15})}}\ \text{Hz} \cong 0.97\ \text{MHz}$$

The minimum gain needed is:

$$A_V = \frac{800}{400} = 2$$

We would, therefore, have $R_1 = 100$ kΩ and perhaps make R_f slightly higher than the required 200 kΩ—say, 220 kΩ—to insure oscillation.

15.4 RC PHASE-SHIFT OSCILLATORS

Basically, both the Hartley and Colpitts oscillators operate in such a way that at the frequency of oscillation, the feedback network provides a 180° phase shift. The amplifier provides the other 180° to insure that the net phase shift around the loop is zero (or any multiple of 360°). We can obtain similar operation with the RC phase-shift circuit, illustrated in Fig. 15-10. But a minimum of three RC circuits is necessary to provide the desired 180° phase shift. (At best, a single capacitor can provide only 90° —hence the need for at least three RC circuits, although the circuit may be built with four or even more.)

With three identical RC networks, as shown in Fig. 15-10, we can see that the feedback network provides 180° of phase shift at a frequency given by:

$$f_o = \frac{1}{2\pi(2.45)RC} \tag{15-18}$$

At this frequency the feedback network gain is $\frac{1}{29}$. The amplifier must then have a gain of at least 29 to insure oscillation. Thus, the equation

$$|A_V| = 29 = \frac{R_f}{R_1} \tag{15-19}$$

allows us to determine the resistors needed. We may choose $R_1 = 100\ \text{k}\Omega$ and $R_f = 3.3\ \text{M}\Omega$.

Example 15-3. We want to design the circuit of Fig. 15-10 for a frequency of oscillation of 10 kHz.

Solution: We start by choosing the capacitor value $C = 0.001\ \mu\text{F}$. We next

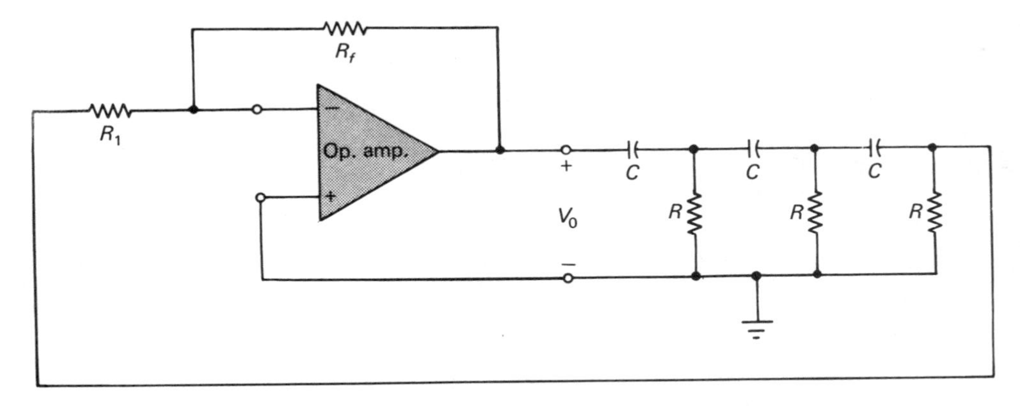

Figure 15-10. OP AMP RC phase-shift oscillator.

calculate the value of R from Eq. (15-18):

$$R = \frac{1}{2\pi(2.45)(0.01\times 10^{-3})} \cong 6.54\ \text{k}\Omega$$

To provide the proper gain and prevent loading down the feedback network, we can make $R_1 = 100\ \text{k}\Omega$ and $R_f = 3.3\ \text{M}\Omega$.

An FET RC phase-shift oscillator is depicted in Fig. 15-11. The operation is the same as the OP AMP version, with the exception of the FET gain. This gain is given by $g_m R_L$, where R_L is the total effective load resistance seen by the output of the FET; it includes r_d, R_D, and the loading effect of the feedback network. The equations developed for the frequency of oscillation and minimum gain are valid.

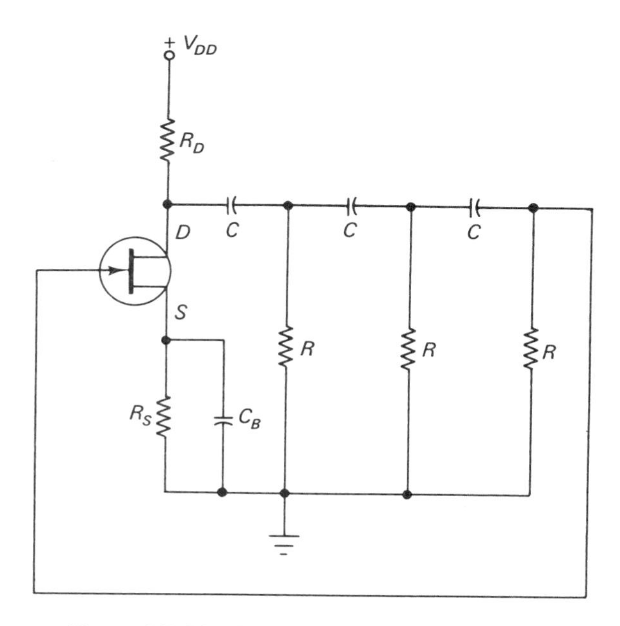

Figure 15-11. FET RC phase-shift oscillator.

Figure 15-12 shows BJT RC phase-shift oscillator circuit. To obtain three matched RC sections, we have to break up the last R because of the heavy loading of the feedback network by the input impedance (h_{ie}) of the BJT. To preserve the frequency of oscillation, the parallel combination of R_1, R_2, and h_{ie}, all in series with R', is made equal to R. Otherwise, the analysis is similar; that is, the gain of the BJT is determined in the same manner as for the FET. But we have to take into account the loading of the feedback circuit on the output of the BJT.

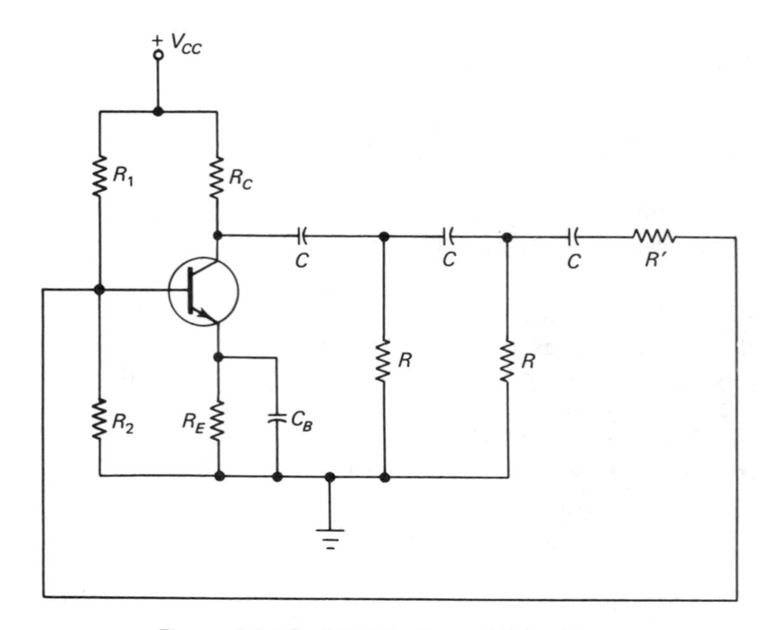

Figure 15-12. BJT RC phase-shift oscillator.

15.5 TUNED OUTPUT OSCILLATORS

We can also achieve positive feedback by using a tuned transformer, illustrated in Figs. 15-13 and 15-14. Capacitors C_B are bypass capacitors and are large enough to be short circuits at the frequency of oscillation. Bias is provided by resistors R_1, R_2, and either R_E or R_S (depending on which circuit is used). Note that the transformer coupling is designed in such a way that it provides a 180° phase shift. The frequency is selected by

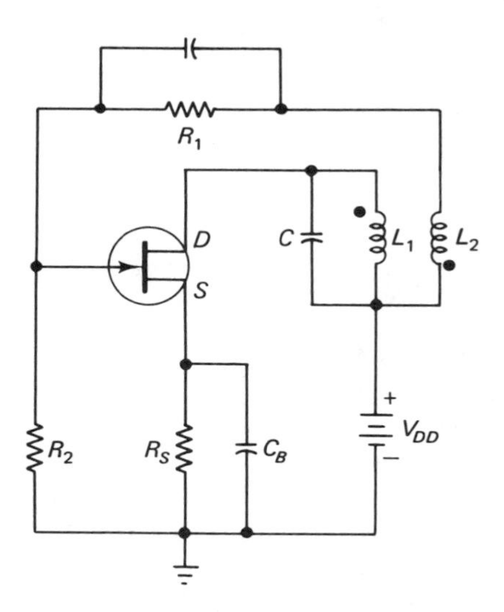

Figure 15-13. FET tuned-output oscillator.

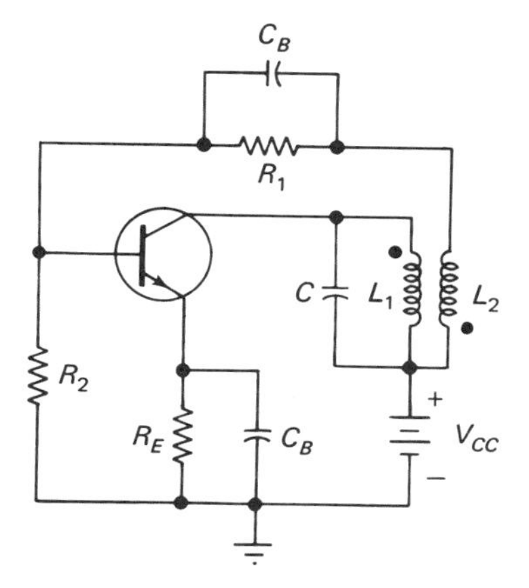

Figure 15-14. BJT tuned-output oscillator.

the tuning of the L_1C tank circuit. In effect, we have a tuned amplifier (with a maximum gain at resonance) and a transformer feedback network that provides the necessary phase shift. The frequency of oscillation is given by:

$$f_o = \frac{1}{2\pi\sqrt{L_1C}} \tag{15-20}$$

The gain may be calculated by the methods discussed in connection with tuned amplifiers in Chap. 11.

15.6 TWIN-T OSCILLATORS

An OP AMP twin-T oscillator circuit is shown in Fig. 15-15. The feedback network in this case is comprised of a dual (twin) T-section filter, which is sometimes referred to as a *notch* filter. This name describes its

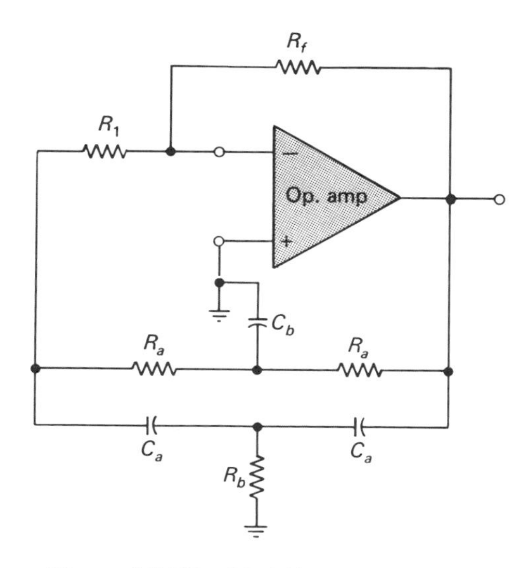

Figure 15-15. OP AMP twin-T oscillator.

transfer impedance frequency characteristic. The impedance is very low except in the vicinity of the resonant frequency, where it is very high. At the null or resonant frequency (the twin-T circuit response resembles that of a tank circuit), the phase shift through the feedback network is 180°, so that if the amplifier gain is adjusted to compensate for the signal loss in the feedback network, oscillations will occur at:

$$f_o \cong \frac{1}{2\pi RC\sqrt{8}} \cong \frac{1}{17.8RC} \tag{15-21}$$

where $R_a = 4R_b = R$ and $C_a = \frac{1}{2}C_b = C$. The gain needed for oscillation may be determined experimentally by adjusting the ratio of R_f to R_1. Typically, the minimum gain may be approximately 25, i.e., R_f should be at least 25 times larger than R_1.

15.7 WIEN-BRIDGE OSCILLATORS

An OP AMP Wien-bridge oscillator circuit is illustrated in Fig. 15-16. The positive feedback network consists of resistor-capacitor combinations: R_1 in parallel with C_1 and R_2 in series with C_2. Resistors R_3 and R_4 set the gain of the OP AMP. We have redrawn the circuit in Fig. 15-17 to make the analysis somewhat simpler; it is easier to recognize different parts of the circuit from Fig. 15-17 than from Fig. 15-16. The two circuits, nevertheless, are identical.

Oscillations occur when the phase shift through the feedback network is zero and the gain provided by the $R_3 - R_4$ combination is large enough to overcome the loss in signal in the feedback network.

The frequency of oscillation is determined from the condition that the impedance of the $R_1 - C_1$ branch be the same as the impedance of the $R_2 - C_2$ branch. Thus,

$$f_o = \frac{1}{2\pi\sqrt{R_1R_2C_1C_2}} \tag{15-22}$$

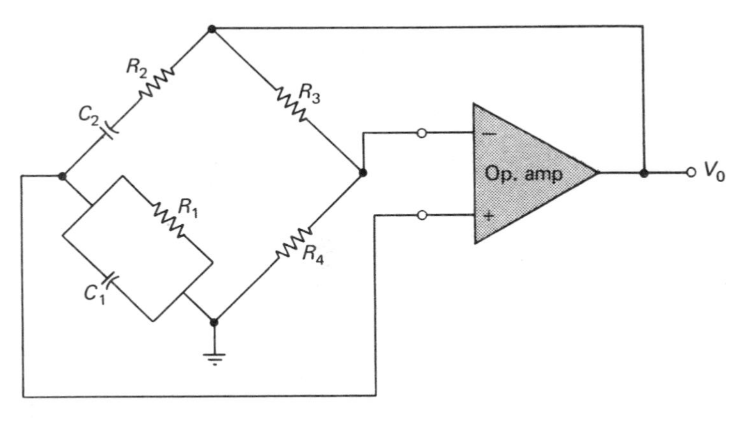

Figure 15-16. OP AMP Wien-bridge oscillator.

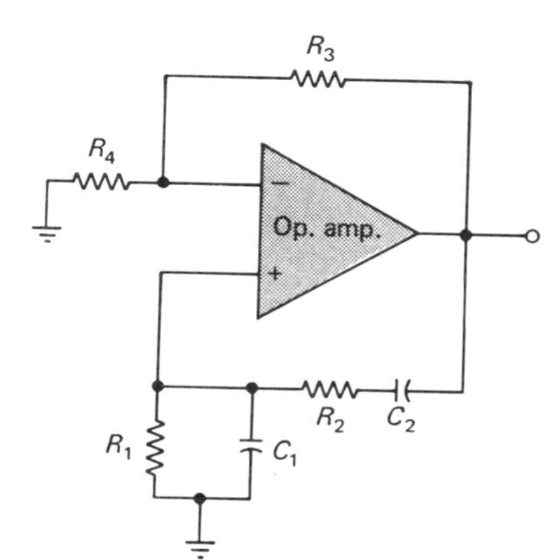

Figure 15-17. Redrawn OP AMP Wien-bridge oscillator.

At this frequency, in order to provide a loop gain of 1, the amplifier gain has to be:

$$K \geqslant \frac{R_1}{R_2} + \frac{C_2}{C_1} + 1 \tag{15-23}$$

Resistors R_1 and R_2 and capacitors C_1 and C_2 are chosen, therefore, to provide the desired frequency in accordance with Eq. (15-22). Resistors R_3 and R_4 are chosen to provide the gain in Eq. (15-23). The gain in terms of R_3 and R_4 is:

$$K = \frac{R_3 + R_4}{R_4} = 1 + \frac{R_3}{R_4} \tag{15-24}$$

Thus, we obtain the required values of R_3 and R_4 by combining Eqs. (15-23) and (15-24):

$$\frac{R_3}{R_4} \geqslant \frac{R_1}{R_2} + \frac{C_2}{C_1} \tag{15-25}$$

In the special case where $R_1 = R_2 = R$ and $C_1 = C_2 = C$, the frequency of oscillation is given by:

$$f_o = \frac{1}{2\pi RC} \tag{15-26}$$

and the minimum gain for oscillation is 3. Consequently, $R_3 \geqslant 2R_4$ will insure that the circuit will oscillate.

15.8 AMPLITUDE-STABILIZED OSCILLATORS

In many applications the oscillator amplitude is critical; therefore, it must be precisely controlled. The basic principle in stabilizing the amplitude of oscillation involves the use of an additional feedback loop employing gain control for the device in the oscillator. We may follow this procedure with any of the oscillators discussed. For example, we can stabilize the FET phase-shift oscillator amplitude by rectifying the output signal and feeding the equivalent dc voltage to bias the FET. In this way, should the amplitude at the output increase, so does the dc level feedback, which biases the FET for a slightly lower gain. Consequently, an essentially constant output voltage is maintained.

A somewhat more involved example, which utilizes the same principle, is given in Fig. 15-18. This is the Wien-bridge oscillator, with additional circuit components for amplitude stabilization. The output signal in excess of the Zener voltage of diode $D2$ is rectified by diode $D1$ and applies a negative dc voltage to the gate of FET $Q1$. This dc voltage is developed across the $R_5 - C_4$ combination. If the output voltage should increase, so does the negative bias on the gate of $Q1$. As a result, the drain-to-source resistance is increased, thus effectively reducing the gain of the OP AMP. Should the output voltage decrease, the negative bias on the gate of Q also decreases, in turn decreasing the drain-to-source resistance. As a result, the gain of the OP AMP is increased. Note that the gain of the OP AMP is set by R_3 and the parallel combination of R_4 and the drain-to-source resistance of the FET.

The frequency of oscillation is still determined by the positive feedback path provided by the $R_2 - C_2$ and $R_1 - C_1$ branches.

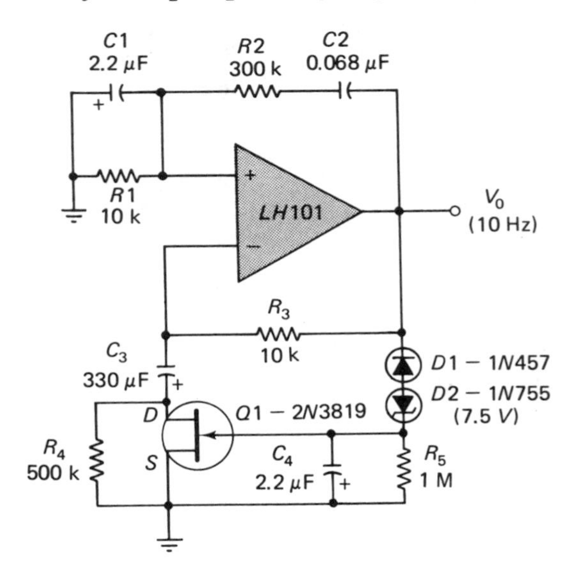

Figure 15-18. Practical amplitude-stabilized Wien-bridge oscillator. (Courtesy of National Semiconductor Corp.)

15.9 CRYSTAL OSCILLATORS

There are many different versions of crystal oscillators. A crystal provides a very temperature-stable tuned circuit, around which an oscillator may be built.

The frequency of oscillation is controlled by the crystal. It is stable over a long period of time and over a relatively wide temperature range.

An example of a crystal oscillator using an IC DIFF AMP is shown in Fig. 15-19. The crystal is connected between the inverting input and the

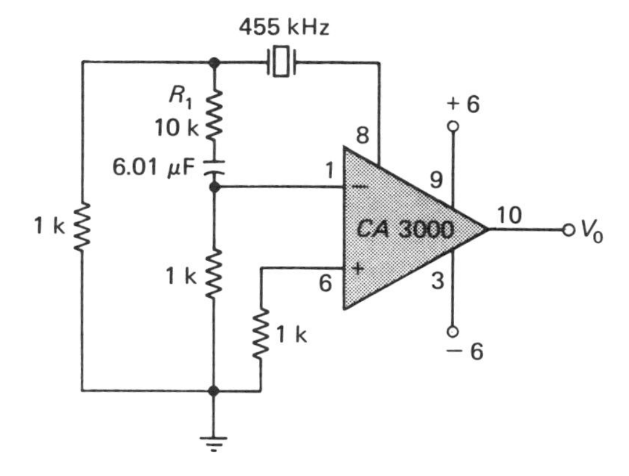

Figure 15-19. Practical DIFF AMP crystal oscillator(f_o=455 kHz).

noninverted output, supplying positive feedback so that oscillation can take place. The amount of feedback is controlled by the variable resistor R_1. It can also be adjusted for a sinusoidal output. The circuit shown is adjusted for oscillation at 455 kHz; however, the CA3000 DIFF AMP may be used in a crystal oscillator providing frequencies up to 1 MHz.

Review Questions

1. What is an oscillator? What is the input to an oscillator? What is its output?
2. What are the criteria for oscillation to be sustained?
3. What determines the frequency of oscillation?
4. What comprises the feedback network in a Hartley oscillator?
5. What circuit values determine the frequency of oscillation in a Hartley oscillator?
6. What is the feedback network in a Colpitts oscillator?
7. What circuit values determine the frequency of oscillation in a Colpitts oscillator?
8. What is the feedback network in an RC phase-shift oscillator?
9. What circuit values determine the frequency of oscillation in a phase-shift oscillator?
10. What is the feedback network in a tuned output oscillator?
11. What circuit values determine the frequency of oscillation in a tuned output oscillator?
12. What is the feedback network in a twin-T oscillator?
13. What circuit values determine the frequency of oscillation in a twin-T oscillator?
14. What is the feedback network in a Wien-bridge oscillator?
15. What circuit values determine the frequency of oscillation in a Wien-bridge oscillator?

16. In what manner can the amplitude of oscillation be stabilized in an oscillator using an FET as the amplifying device?
17. In what way may an FET be used to stabilize the amplitude of oscillation in an OP AMP oscillator?
18. What properties of a crystal make it suitable for use in very stable oscillator circuits?
19. What are the basic similarities and differences between the oscillator circuits in this chapter?
20. What conditions must be satisfied for the oscillator output to be sinusoidal with a minimum of harmonic distortion?

Problems

1. The FET Hartley oscillator of Fig. 15-7 is to be designed for oscillation at 455 kHz. The FET parameters are: $g_m = 6000$ μmho and $r_{ds} = 100$ kΩ. Determine the circuit values.
2. Repeat Problem 1, using the OP AMP version of the Hartley oscillator shown in Fig. 15-6.
3. The OP AMP Colpitts oscillator of Fig. 15-8 is constructed using $C_2 = 5000$ pF, $C_3 = 500$ pF, and $L = 1$ mH. Determine the frequency of oscillation and the resistor values needed to insure oscillation.
4. Repeat Problem 3 for the FET version of the Colpitts oscillator if the FET has $g_m = 1$ millimho and $r_{ds} = 50$ kΩ.
5. We wish to design the OP AMP phase-shift oscillator for a frequency of 18 kHz. Specify the circuit values. (Use all capacitors equal to 0.1 μF.)
6. Repeat Problem 5 for the FET phase-shift oscillator shown in Fig. 15-11 if the FET parameters are: $g_m = 5$ millimho and $r_{ds} = 80$ kΩ. Specify the value of R_D.
7. Repeat Problem 5 for the BJT phase-shift oscillator shown in Fig. 15-12 if the BJT parameters are: $h_{ie} = 2$ kΩ, $h_{fe} = 100$ with h_{re} and h_{oe} negligible. Specify the value of R' needed.
8. The twin-T oscillator of Fig. 15-15 is connected with $C_a = 0.01$ μF, $C_b = 0.02$ μF, $R_a = 10$ kΩ, and $R_b = 2.5$ kΩ. Determine the frequency of oscillation and the approximate values of R_3 and R_4 needed for oscillation.
9. Design the twin-T oscillator of Fig. 15-15 to provide an output at 50 kHz. Specify the approximate resistor and capacitor values needed.
10. The Wien-bridge oscillator of Fig. 15-16 is constructed with $R_1 = R_2 = 1$ kΩ, $C_1 = C_2 = 0.02$ μF. What is the frequency of oscillation and the values of the gain-setting resistors to insure oscillation?
11. Assuming that the gain of the amplitude-stabilized Wien-bridge oscillator in Fig. 15-18 is sufficient to provide oscillation, determine the values of R_1, R_2, C_1, and C_2 to give an output at a frequency of 100 Hz.

16

Digital Logic Circuits

Digital circuits process and store *binary* electrical signals. The importance of digital circuits becomes apparent even from a shortened listing of their uses: computers, data processing, digital communications, and control systems.

Although digital systems tend to be extremely large when compared to, say, a radio or even a TV, they use a fairly limited number of basic building blocks many times over. These basic devices can be classified into two groups. One group is made up of *gates*; they process digital signals. The second group consists of *multivibrators* (or multis for short); they can generate or store (i.e., remember) digital signals. We refer to gates as *combinational* logic circuits; we call multivibrators *sequential* logic circuits. This chapter deals only with combinational logic building blocks. We shall take up the basics of sequential circuits in Chap. 18.

16.1 THE BINARY SYSTEM

Most digital circuits use what is called the *binary*, or *base*-2, system. To understand the binary system, we shall reexamine the more familiar decimal system. In this system, also called a base-10 system, the numerals 0,1,2,3,4,5,6,7,8, and 9 are required to express a number. If the number is larger than 9, the same numerals are used, with the additional information contained in the *position* of the additional numerals. For example, the number thirteen hundred and fifty seven is written in the base-10 system as 1357. The true meaning of this array of numerals is:

$$1357 = (1 \times 10^3) + (3 \times 10^2) + (5 \times 10^1) + (7 \times 10^0)$$

where the position of a numeral determines the power of the base (in this case, 10) that multiplies it. In the number 1357, the 7 is in the position of zero, so it is multiplied by 10^0 (which is defined to be 1); the 5 is in the 1's position, so it is multiplied by 10^1; and so on.

In the binary system, the base is 2; and only two numerals, usually 0 and 1, are necessary to represent a number. The numerals are used in much the same manner as in the decimal system, but the position of the individual digits in the binary system indicates the power of 2 multiplying the numerals 0 or 1. The decimal number 1357 becomes 10101001101 in the binary system. To obtain this binary form of 1357, we successively divide it by 2 and set down the remainder, which will always be either 0 or 1. We keep dividing the answer by 2 until the answer becomes zero:

$1357 \div 2 = 678$	Remainder 1
$678 \div 2 = 339$	" 0
$339 \div 2 = 169$	" 1
$169 \div 2 = 84$	" 1
$84 \div 2 = 42$	" 0
$42 \div 2 = 21$	" 0
$21 \div 2 = 10$	" 1
$10 \div 2 = 5$	" 0
$5 \div 2 = 2$	" 1
$2 \div 2 = 1$	" 0
$1 \div 2 = 0$	" 1

The binary form of the number 1357 is the remainder column, the least significant digit on top. In the same manner, the binary equivalent for any decimal number may be obtained. The meaning of the series of 1s and 0s in a binary number can be seen with the aid of Table 16-1. The ones place corresponds to 2^0, the tens place to 2^1, and so forth. If a 1 is present, you count its decimal equivalent; if a 0 is present, you do *not* count its decimal equivalent. The binary form of the number 1357 is indicated in Table 16-1 and is read as:

$$N = 1024 + 0 + 256 + 0 + 64 + 0 + 0 + 8 + 4 + 0 + 1 = 1357$$
$$1 \quad 0 \quad 1 \quad 0 \quad 1 \quad 0 \quad 0 \quad 1 \quad 1 \quad 0 \quad 1$$

Table 16-1. The Decimal Number 1357 in Binary Form

2^{10}	2^9	2^8	2^7	2^6	2^5	2^4	2^3	2^2	2^1	2^0
1024	512	256	128	64	32	16	8	4	2	1
1	0	1	0	1	0	0	1	1	0	1

Binary equivalents of the numbers 0 through 10 are indicated in Table 16-2. It becomes a relatively simple matter to represent any decimal-based number as a binary number or any binary number as a decimal number. (In both the decimal system and the binary system, leading 0s carry no significance. The number 0005 is the same as the number 5 in the decimal system, and the number 000101 is the same as 101 in the binary system.)

The two-state system of either 0 or 1 also can be used to indicate yes or no, on or off, closed or open, true or false, high or low, and many other conditions. In terms of an electronic circuit, the two states are usually represented by voltages. In a real system there is uncertainty in the exact value of a voltage; therefore, a range of voltages is allowed to represent the 0 and a different range to represent the 1.

Table 16-2. Binary Equivalents of Decimal Numbers

Decimal	*Binary*
0	0000
1	0001
2	0010
3	0011
4	0100
5	0101
6	0110
7	0111
8	1000
9	1001
10	1010

16.2 TRANSISTOR SWITCH OR INVERTER

Consider the circuit shown in Fig. 16-1. Let us assume that terminal A is grounded for the time being. The negative base supply voltage V_{BB} reverse biases the base-emitter junction. The transistor is not conducting; it is *off*. Because no collector current flows, the output voltage at point Y is at the supply value, V_{CC}. If we choose to define voltages near zero as "logical 0" and voltages near the supply voltage as "logical 1," the circuit in Fig. 16-1 gives a logical 1 at the output (terminal Y) when a logical 0 input is applied (that is, terminal A is grounded). Suppose that, instead of grounding the input, we connected it to a positive voltage, say V_{CC}. If the conditions are right, the transistor will saturate and the output voltage will be near 0 V. In this case, a logical 1 input gives a logical 0 output. We can then say that the circuit *inverts* the input signal; that is, it causes a 0 output for a 1 input and a 1 output for a 0 input. We are obviously allowing only two kinds of signals as inputs: a signal that insures the transistor to be saturated (logical 1 input) and a signal that causes the transistor to be cut off (logical 0 input).

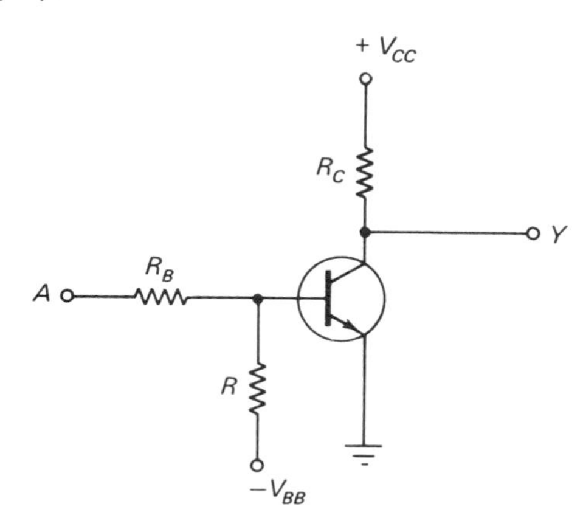

Figure 16-1. Transistor inverter.

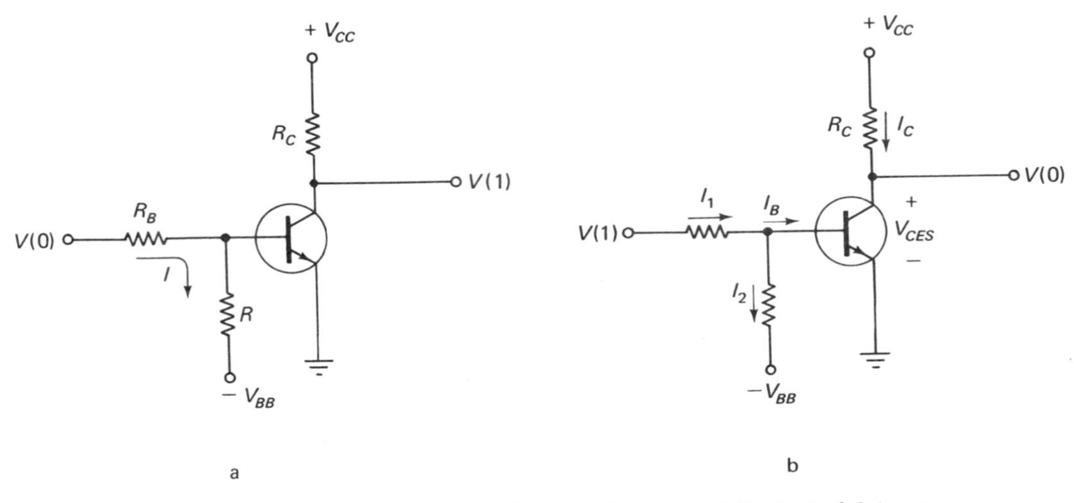

Figure 16-2. Transistor inverter: (a) logical 0 input and (b) logical 1 input.

Let us now examine the circuit requirements for such operation. We label the voltage corresponding to a logical 0 as $V(0)$ and the voltage corresponding to a logical 1 as $V(1)$. With $V(0)$ as an input [shown in Fig. 16-2(a)] we assume the transistor to be cut off and have:

$$I = \frac{V(0) - (-V_{BB})}{R_B + R} \tag{16-1}$$

The base-emitter voltage then is given by:

$$V_{BE} = -V_{BB} + IR \tag{16-2}$$

For the transistor to be cut off, V_{BE} must be negative. This is the case if the voltage drop across R (IR) is less than V_{BB}. Substituting Eq. (16-1) into Eq. (16-2), we have the following requirement:

$$V(0) \leqslant V_{BB}\frac{R_B}{R} \text{ (for cutoff)} \tag{16-3}$$

Thus, for a given V_{BB}, R_B, and R, Eq. (16-3) gives the maximum voltage that can correspond to a logical 0.

When a logical 1 voltage $V(1)$ is applied [as indicated in Fig. 16-2(b)] the transistor should saturate. The condition that characterizes saturation for a given transistor is:

$$\beta I_B \geqslant I_{CS} \tag{16-4}$$

where I_{CS} is the maximum collector current that can flow in the circuit and is determined by V_{CC} and R_C. When a transistor is saturated, its collector-emitter voltage is small, typically 0.2 to 0.3 V. We determine I_{CS} by adding voltages in the collector circuit:

$$I_{CS} = \frac{V_{CC} - V_{CES}}{R_C} \tag{16-5}$$

where V_{CES} is the collector-emitter saturation voltage.

If the transistor in Fig. 16-2(b) is saturated, its base-emitter voltage, labeled V_{BES}, is typically 0.7 V for silicon transistors and 0.3 V for germanium transistors. Currents I_1 and I_2 can be determined as follows:

$$I_1 = \frac{V(1) - V_{BES}}{R_B} \qquad I_2 = \frac{V_{BES} - (-V_{BB})}{R} \tag{16-6}$$

The base current is the difference between I_1 and I_2, so that the condition to insure saturation can be determined by combining Eqs. (16-4), (16-5), and (16-6):

$$V(1) \geqslant \frac{R_B}{\beta R_C}(V_{CC} - V_{CES}) + \frac{V_{BES} R}{R + R_B} + V_{BB}\frac{R_B}{R} \tag{16-7}$$

The output voltage is then given by V_{CES}. As a result, the inverter circuit has two possible states (depending on the input): V_{CC} and V_{CES}. As we noted earlier, these correspond to logical 1 and 0 levels respectively. Thus,

$$\begin{aligned} V(1) &\cong V_{CC} \\ V(0) &\cong V_{CES} \end{aligned} \tag{16-8}$$

With these definitions, the required circuit components can be calculated from Eqs. (16-3) and (16-7). This calculation is illustrated in the following example.

Example 16-1. Suppose that we are to design an inverter using a silicon transistor with: $V_{CES} = 0.3$ V, $V_{BES} = 0.7$ V, and β no less than 25. The supply voltages are: $V_{CC} = V_{BB} = 10$ V. Calculate the resistors needed to insure proper operation.

Solution: From Eq. (16-3) we see that we must choose

$$\frac{R_B}{R} \leqslant \frac{V(0)}{V_{BB}} \cong \frac{0.3}{10} = 0.03$$

We are free to choose $R = 100$ kΩ; then

$$R_B \leqslant 100(0.03) \cong 3\ k\Omega$$

We might choose $R_B = 2.2$ kΩ. From Eq. (16-7) we can determine the minimum value of R_C needed by noting that usually $R \gg R_B$ and

$$\frac{R}{R + R_B} \cong 1 \qquad \text{and} \qquad \frac{R_B}{R} \cong 0$$

When these approximations are made in Eq. (16-7), we solve for R_C.

$$R_C \geqslant \frac{1 R_B(V_{CC} - V_{CES})}{\beta V(1)} \cong \frac{2.2(10 - 0.3)}{25(10)}\ k\Omega \cong 850\Omega$$

We might use $R_C = 1$ kΩ.

Note that when $V_{BB}=V_{CC}$, proper circuit operation will result if $R>R_B$ and $R_C>R_B/\beta_{\min}$ (where $\beta_{\min}$ is the minimum guaranteed β specified by the manufacturer for the transistor to be used).

The transistor inverter circuit may be used as an electronic switch. Suppose that we replace the collector resistor by a pilot light having the same resistance as R_C. When the input is low (logical 0), no collector current flows and the light is off. When the input voltage is high (logical 1), the transistor saturates, collector current flows, and the light turns on. Thus, we can turn the light on and off by electrical signals applied to the input of the electronic switch in the same manner as flipping a mechanical switch.

16.3 DIODE GATES

Let us consider next the diode circuit shown in Fig. 16-3. If we maintain our convention of logic voltages as specified in Eq. (16-8), we can determine the logical behavior of the two-input circuit shown. Because there are two inputs, A and B, there are four possible combinations in which the inputs can be applied. These combinations are shown in Table 16-3, which is called a *truth table* for this particular circuit. If both inputs A and B are grounded, both diodes are forward biased by the supply voltage V and conduct. The output voltage (taken with respect to ground) is low; it is equal to the diode forward drop, usually about $\frac{1}{2}$ V. This voltage may be neglected, so we say that the output is essentially at ground. This position

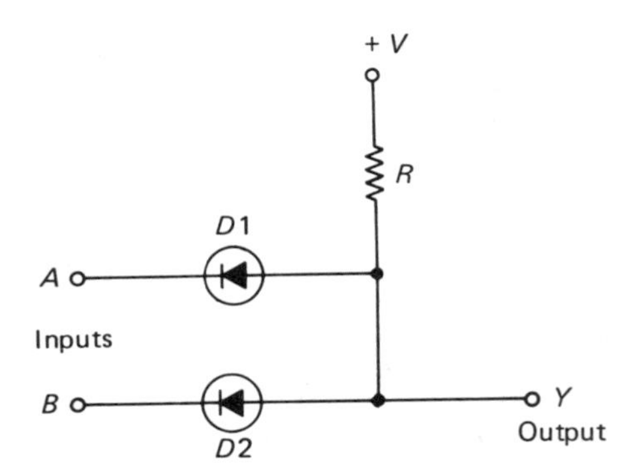

Figure 16-3. Diode AND gate.

Table 16-3. Voltage and Logic Truth Table for Two-Input AND Gate

Inputs				*Output*	
A		*B*		*Y*	
Voltage	*Logic*	*Voltage*	*Logic*	*Voltage*	*Logic*
0(gnd)	0	0	0	0	0
0	0	+V	1	0	0
+V	1	0	0	0	0
+V	1	+V	1	+V	1

is shown in the first line of Table 16-3. When either of the inputs is at ground with the other one high (at $+V$), the output still remains low because one of the two diodes (the one that is connected to ground) is forward biased and conducts, thus pulling the output down to essentially ground. These conditions are summarized in the two middle lines in Table 16-3. Lastly, when both inputs are high (at $+V$), both diodes are reverse biased and neither conducts. The current through R is zero, and the output voltage is at the supply voltage $+V$. This position is the last line in Table 16-3. The circuit in Fig. 16-3 is called an *AND gate* since both the *A AND B* inputs must be high in order for the output to be high. The logic symbol for an *AND* gate is shown in Fig. 16-4.

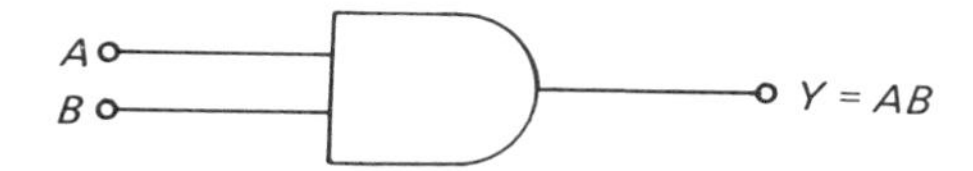

Figure 16-4. Logic symbol for an AND gate.

For illustrative purposes consider the following situation. There are two doors in a room with sensors attached to them. When the doors are closed, no voltage is available at the sensor; when a door is opened, its sensor puts out a voltage $+V$. If the sensors from the two doors are connected, one to each input of the AND gate, a high voltage at the output would indicate that both doors are open, whereas a low output could indicate that either door is open or neither door is open.

Suppose that we wanted an indication (i.e., high output) when either door was opened, as might be the case in a burglar-alarm system. The circuit of Fig. 16-5, called an *OR gate*, would be used. Its operation will be explained with the aid of the truth table given in Table 16-4. With both A and B inputs low (connected to ground), the output is at ground because there is no voltage source in the circuit (line 1 in Table 16-4). When either input is high (middle two lines in Table 16-4), connected to $+V$, its diode is forward biased and conducts. The output voltage then differs from $+V$ by the small forward diode drop and is high. This is also the case when

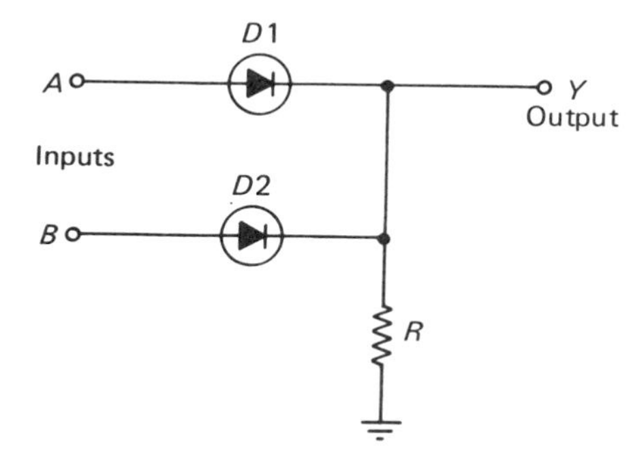

Figure 16-5. Diode OR gate.

Table 16-4. Voltage and Logic Truth Table for Two-Input OR Gate

Inputs				*Output*	
A		*B*		*Y*	
Voltage	*Logic*	*Voltage*	*Logic*	*Voltage*	*Logic*
0	0	0	0	0	0
0	0	+V	1	+V	1
+V	1	0	0	+V	1
+V	1	+V	1	+V	1

both inputs are connected to $+V$ (last line in Table 16-4); that is, the output is high (at $+V$). This circuit is called an OR gate because the output is high (logical 1) when input *A OR* input *B* is high. The logic symbol for an OR gate is shown in Fig. 16-6.

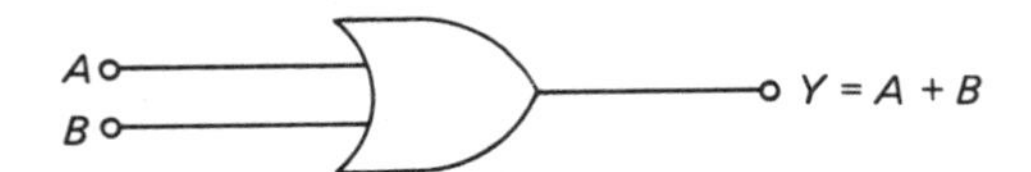

Figure 16-6. Logic symbol for an OR gate.

These diode gates are not used any more but serve well to introduce the basic concepts of AND and OR gates.

16.4 RESISTOR-TRANSISTOR LOGIC GATES

We next consider a *resistor-transistor logic (RTL)* gate, illustrated in Fig. 16-7. As may be obvious, the *RTL* gate shown is nothing more than the parallel connection of two inverter circuits.

If the circuit values are chosen appropriately, the transistors will saturate when their inputs are high and will cut off when their inputs are low, just like an inverter.

Consider the truth table given in Table 16-5 in the operation of the circuit in Fig. 16-7. When both the *A* and *B* inputs are low, both $Q1$ and

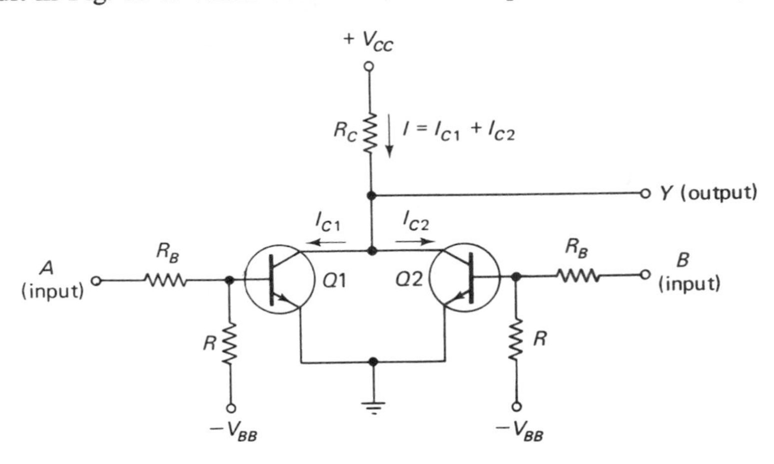

Figure 16-7. RTL NOR gate.

Table 16-5. Truth Table for a NOR Gate

Inputs		*Output*
A	*B*	*Y*
0	0	1
0	1	0
1	0	0
1	1	0

$Q2$ are cut off, and neither conducts. The current I is zero, and the output is at $+V_{CC}$ (logical 1). When either input is high ($+V_{CC}$), its transistor saturates, current I flows, and the output (Y) drops to near zero (logical 0). This is also the case when both inputs are high and both transistors are saturated. Note that the current I is essentially V_{CC}/R_C no matter if one or both of the transistors are saturated. Thus, the transistor collector current is half as large if the other transistor is saturated than if the other transistor is cut off.

We can say that the output of the *RTL* gate in Fig. 16-7 is not high when input *A OR B* is high. In fact, the only time the output is high is when inputs *A* and *B* are *NOT* high. Such a gate is termed a *NOT-OR* or simply *NOR gate*. The truth table for the OR and NOR gates indicates that for the same input conditions, the output of the NOR gate is the inverse of the OR gate; that is, a logical 1 becomes a logical 0 and vice versa.

Consider the case of connecting two inverter circuits in series, as shown in Fig. 16-8. Such a configuration is called an *RTL NAND* gate. Its operation is explained with the aid of the truth table in Table 16-6.

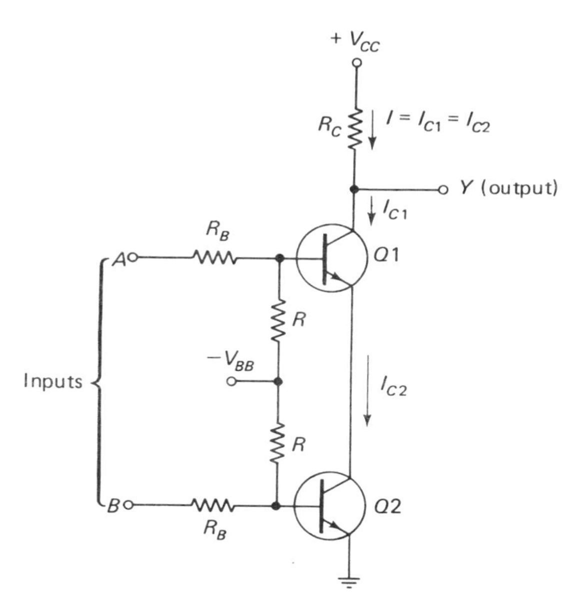

Figure 16-8. RTL NAND gate.

Table 16-6. Truth Table for a NAND Gate

Inputs		*Output*
A	*B*	*Y*
0	0	1
0	1	1
1	0	1
1	1	0

If both inputs to the circuit in Fig. 16-8 are low, both transistors are cut off, no collector current flows, and the output is high (logical 1). If one of the inputs is high with the other low, one transistor tries to saturate, while the other is cut off. The transistor in cutoff does not allow any collector current to flow. Because it is in series with the other transistor, it prevents the other transistor from passing any current. Thus, one I_c is zero; therefore, the current I is zero, and the output is at $+V_{CC}$ (logical 1). Only when both inputs are high and both transistors are saturated does current flow in the output and the output voltage is low. Note that the output voltage will be $2V_{CES}$ (still logical 0) when both transistors are saturated.

This circuit is called a *NOT-AND*, or *NAND, gate* because the output is *NOT* high only when both the *A AND B* inputs are high. A comparison of the respective truth tables for the AND and NAND gates reveals further that for the same input conditions the two outputs are the inverse of each other.

Logic symbols for the NOR and NAND gates are given in Figs. 16-9 and 16-10, respectively. The inversion or negation of the OR and AND gates is indicated by the little circle at the output of the symbol. It is also symbolized by the bar over the letters for the input.

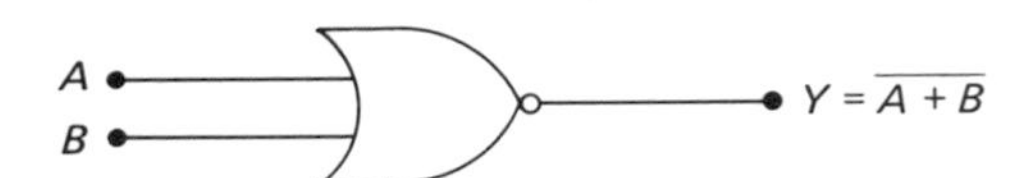

Figure 16-9. NOR gate logic symbol.

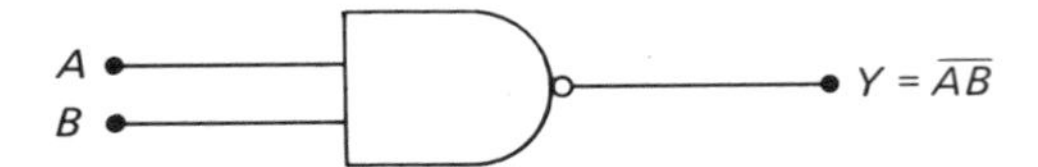

Figure 16-10. NAND gate logic symbol.

16.5 DIODE-TRANSISTOR LOGIC GATES

The NOR and NAND gates discussed in the previous section can also be constructed by cascading the diode OR and AND gates with the inverter (or NOT) circuit. In this case, we have the *diode-transistor logic (DTL)* gates shown in Figs. 16-11 and 16-12.

Let us examine the DTL NOR gate first. We can verify the NOR truth table for Fig. 16-11 by considering both inputs at ground. The

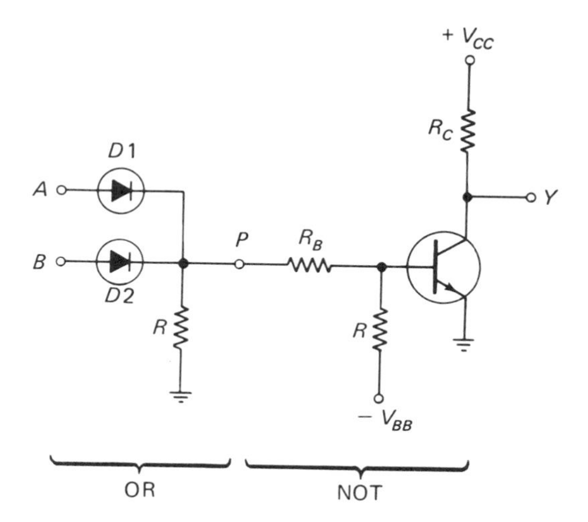

Figure 16-11. DTL NOR gate.

transistor must be cut off because there is no positive supply in the base circuit. The output is high because no collector current flows. When either input or both inputs are high ($+V_{CC}$), one or both diodes conduct and point P is at almost $+V_{CC}$. This result insures that the transistor will saturate and the output will be low. Thus, the circuit of Fig. 16-11 satisfies NOR logic (given in Table 16-5).

In the circuit of Fig. 16-12, with one or both inputs low, one or both diodes are forward biased and conduct, thus causing point P to be low (at about 0.7 V). This voltage is insufficient to turn on the transistor because of the negative supply V_{BB}, and so the transistor is cut off. The output is high. When both inputs are high, both diodes are reverse biased and draw

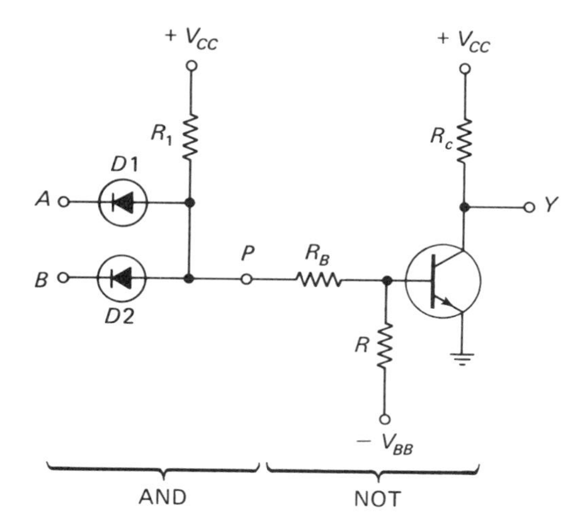

Figure 16-12. DTL NAND gate.

no current. The base of the transistor is then connected to $+V_{CC}$ through resistors R_B and R_1. If R is greater than (R_B+R_1), the transistor will saturate and the output will be low. Thus, the circuit satisfies NAND logic given in Table 16-6. Note that point P is at some voltage below V_{CC} when both inputs are high. This results from the drop across R_1 caused by I_B of the conducting transistor. Because the two inputs are at V_{CC}, we are sure that the two diodes are reverse biased.

Figure 16-13 depicts a modified DTL NAND gate that is used in ICs. The resistor R_B is replaced by diodes $D3$ and $D4$, and the negative voltage supply is eliminated. We include the additional diodes in order to prevent the transistor from turning on when it is not supposed to. For example, when both or either of the inputs is low, point P may be at about 0.9 V (assuming that the input is derived from a transistor in saturation with 0.3 V and the forward drop across the diode of 0.6 V). Suppose that the circuit is to perform a NAND operation. Then with both inputs or either input low, the output should be high, meaning that the transistor should be off. But with 0.9 V at point P, without diodes $D3$ and $D4$, the transistor would surely turn on. With $D3$ and $D4$ in place as shown, the transistor does not turn on until point P rises to roughly three times the forward drop of a diode or 2 V. Thus, the diodes act to keep the transistor off when it should be off.

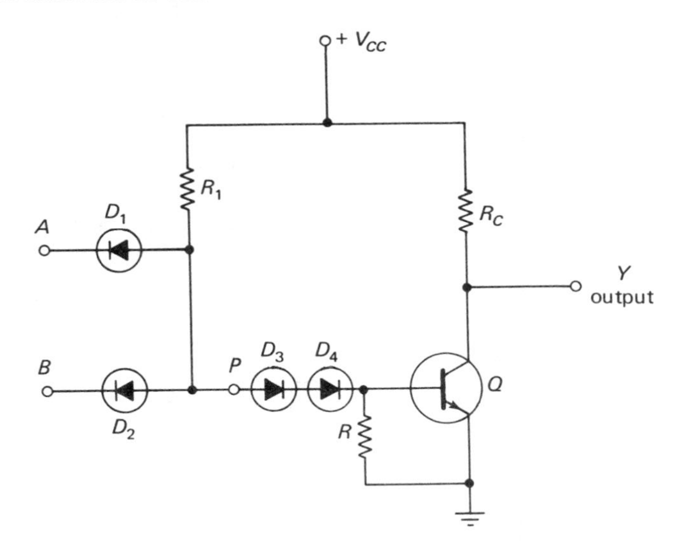

Figure 16-13. IC DTL NAND gate.

16.6 TRANSISTOR-TRANSISTOR LOGIC GATES

A family of gates that offers excellent performance is called *transistor-transistor logic (TTL or T^2L)*. These types of gates were not possible without an integrated circuit technology because they utilize a multiple-emitter transistor. A typical basic TTL NAND gate is depicted in Fig. 16-14. Transistor $Q1$ is a multiple-emitter transistor and may have as

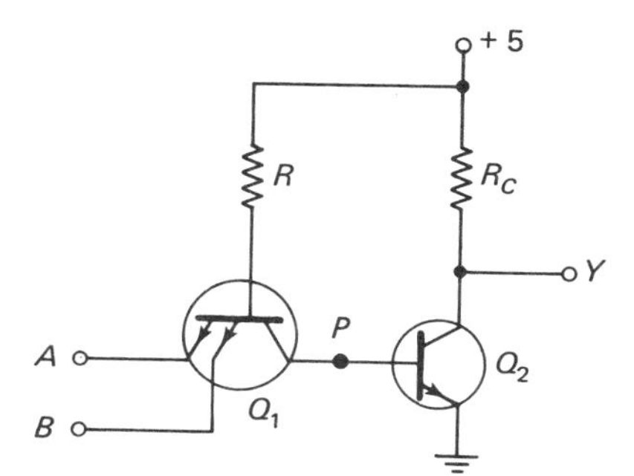

Figure 16-14. TTL NAND gate.

many emitters as inputs are needed. Fig. 16-15 shows how the multiple-emitter transistor is fabricated. In effect, we take a number of discrete transistors and connect all their bases and all their collectors together.

The TTL gate shown in Fig. 16-14 operates much the same as the DTL gate described in Fig. 16-13. The input diodes are replaced by the emitter-base junctions of the multiple transistor, and the collector-base junction of $Q1$ acts much like diodes $D3$ and $D4$ in Fig. 16-13. One important difference exists. The transistor action caused by $Q1$ in the TTL gate is not available in the DTL gate. By transistor action we mean the fact that, in a transistor, the collector and emitter currents are β times larger than the base current. When a junction is biased (either forward or reverse), charges are stored at and around the junction. When the biasing voltage is removed, there is a nonzero time required for these charges to be removed.

In the DTL gate with both inputs high, Q is saturated. When one of the inputs goes low, the charges stored at the base-emitter junction of Q must be removed before Q turns off. (This effect is measured by storage time of the transistor.) The turn-off current flows through the input diode that is conducting. In the TTL gate under similar conditions during the

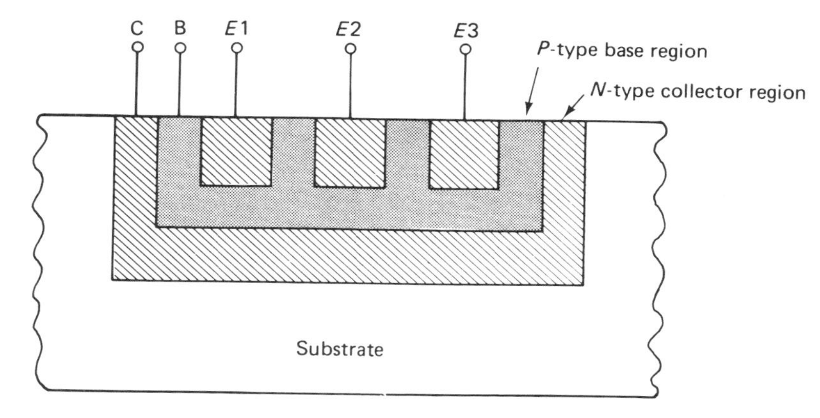

Figure 16-15. IC fabrication of a multiple-emitter transistor.

turn-off time of $Q2$, transistor $Q1$ is momentarily in its active region, and its collector current quickly removes the charge from the base of $Q2$.

A practical TTL NAND gate is illustrated in Fig. 16-16. Transistors $Q3$ and $Q4$ are added for increasing the drive capability at the output. If either input or both of the inputs in Fig. 16-16 are low, the base-emitter diodes of $Q1$ conduct and point M is at approximately 1 V. This amount is insufficient to turn on the collector-base diode of $Q1$ and $Q2$ at the same time. Therefore, $Q2$ is off, point T is at essentially 0 V, and point S is close to 5 V. Although $Q3$ is trying to turn on, because the base of $Q4$ is at ground, $Q4$ is off, and the output is high.

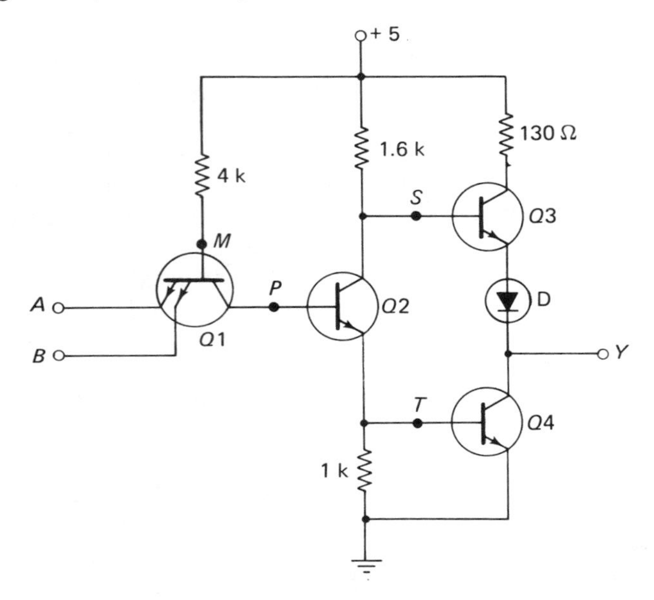

Figure 16-16. Practical IC TTL NAND gate.

When both inputs are high, point P is high and $Q2$ saturates. Consequently, a relatively large current flows through the 1-kΩ resistor, thus raising point T until $Q4$ saturates. The voltages are then: 0.7 V at T, 1.4 V at P, and 2.1 V at M. Point S is at 0.3 V (the saturation voltage of $Q2$) above point T, or at 1.0 V. With $Q4$ saturated, the output is at approximately 0.3 V; that is, there is about 1.0 to 0.3 V between the collector of $Q2$ and the output. If it were not for the diode D, this voltage would cause $Q3$ to saturate, and the power supply drain would be needlessly high. However, with a 0.7 V total across the base-emitter junction of $Q3$ and diode D, $Q3$ does not saturate and is in its active region allowing only sufficient current to insure the saturation of $Q4$.

Figure 16-17 shows a practical IC TTL NOR gate, with a similar output pair of driver transistors. With both inputs low, the base-emitter diodes of input transistors $Q1$ and $Q2$ are forward biased and pull the respective bases to roughly 1 V. In this case, the 1 V is not enough to turn

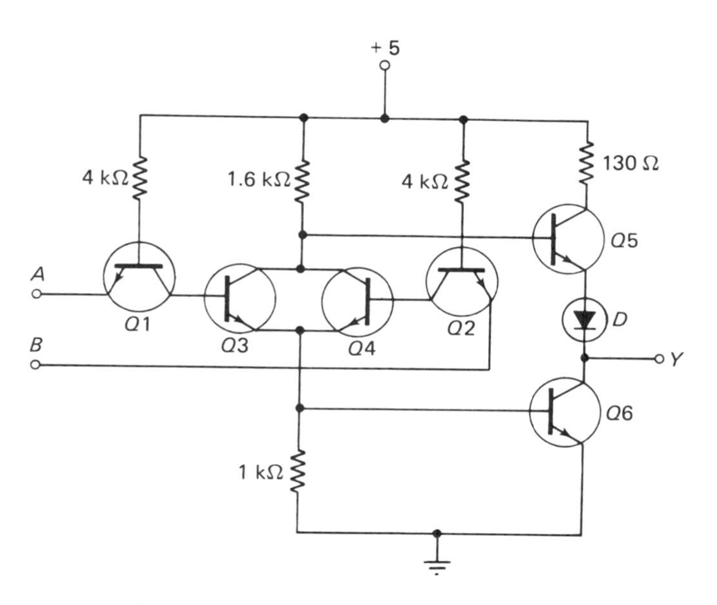

Figure 16-17. Practical IC TTL NOR gate.

on either $Q3$ or $Q4$, and so no current flows through the 1-kΩ resistor. The base of $Q6$ is essentially at ground; therefore, $Q6$ is off and the output is high.

With a high input to either or both of the input transistors, either or both $Q1$ and $Q2$ are off. Their respective bases are then high, causing transistors $Q3$ and $Q4$ to turn on, depending on which input was high.

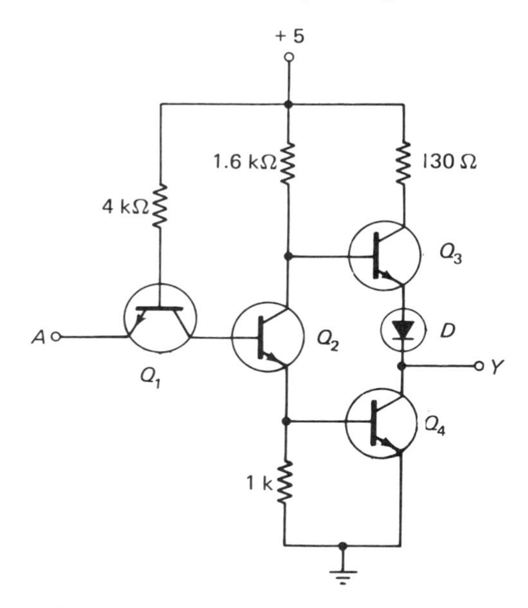

Figure 16-18. Practical TTL inverter gate.

With either or both $Q3$ and $Q4$ saturated, the emitter current causes a sufficiently high voltage across the 1-kΩ resistor to turn on $Q6$, thus making the output low. The operation of the $Q5$–D circuit is the same as that described for the NAND gate.

The basic TTL family of gates is completed with the inverter circuit shown in Fig. 16-18. The circuit is almost identical to that of the NAND gate. However, only one input A is available. Using the same analysis as for the TTL NAND gate, we can see that a high input causes a low output and a low input causes a high output. This is the case for all inverter circuits.

16.7 IC TTL GATES

A representative data sheet for a TTL NAND gate is shown in Table 16-7. The Texas Instruments SN7400 integrated circuit contains four individual two-input NAND gates and is called a quadruple, or simply quad, two-input NAND gate. As we can see from the data sheet, the circuit is designed to operate with a supply voltage between 4.75 and 5.25 V.

The first four quantities in the table refer to the *noise margin* for this circuit. The noise margin is the maximum error in the logical 0 and 1 voltages that can be tolerated. The SN7400 requires a minimum voltage of 2 V to correspond to a logical 1. It can tolerate a voltage as high as 0.8 V for a logical 0. The noise margins are shown in Fig. 16-19. The noise immunity of the circuit is the difference between the minimum logical 1 voltage and the maximum logical 0 voltage.

The logical 0 level input current for each input is the amount of current that *leaves* each emitter terminal (of the input transistor). The logical 1 level input current for each input is the amount of leakage current that *enters* the emitter of each input transistor.

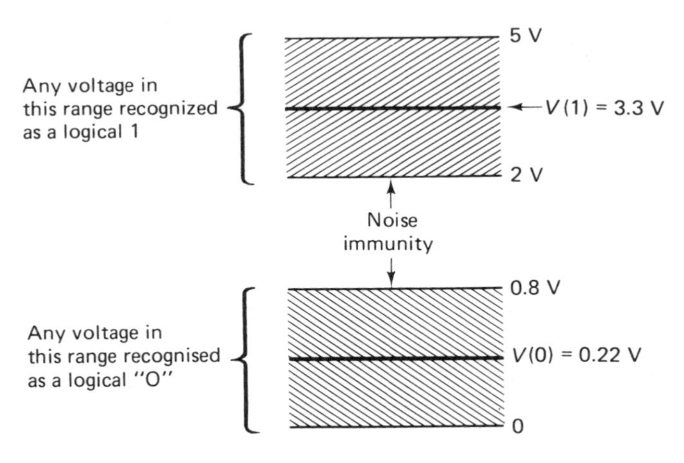

Figure 16-19. Noise margins for a TTL gate.

Table 16-7. Typical TTL Gate Characteristics (Texas Instruments SN7400)

CIRCUIT TYPES SN5400, SN7400
QUADRUPLE 2-INPUT POSITIVE NAND GATES

schematic (each gate)

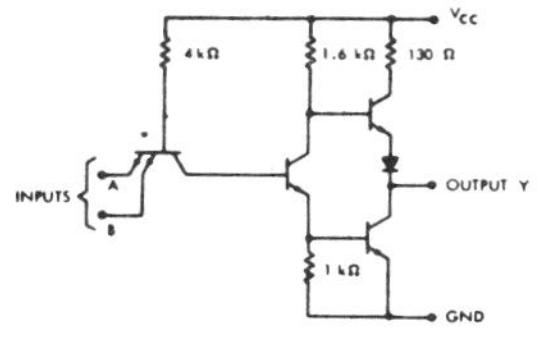

S FLAT PACKAGE (TOP VIEW)

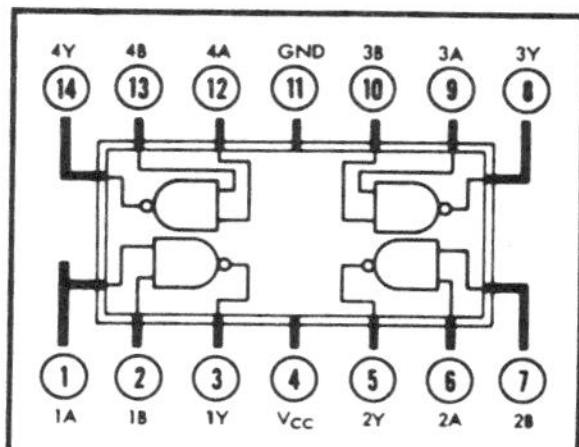

J OR N DUAL-IN-LINE PACKAGE (TOP VIEW)

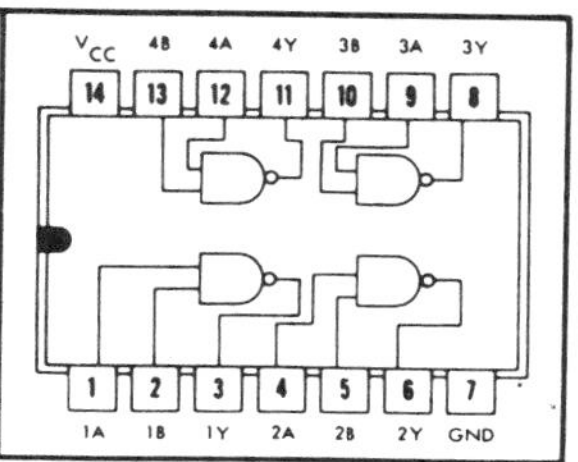

NOTE: Component values shown are nominal.

positive logic: $Y = \overline{AB}$

recommended operating conditions

		MIN	NOM	MAX	UNIT
Supply Voltage V_{CC}:	SN5400 Circuits	4.5	5	5.5	V
	SN7400 Circuits	4.75	5	5.25	V
Normalized Fan-Out From Each Output, N				10	
Operating Free-Air Temperature Range, T_A:	SN5400 Circuits	−55	25	125	°C
	SN7400 Circuits	0	25	70	°C

electrical characteristics over recommended operating free-air temperature (unless otherwise noted)

PARAMETER		TEST FIGURE	TEST CONDITIONS†			MIN	TYP‡	MAX	UNIT
$V_{in(1)}$	Logical 1 input voltage required at both input terminals to ensure logical 0 level at output	1	V_{CC} = MIN			2			V
$V_{in(0)}$	Logical 0 input voltage required at either input terminal to ensure logical 1 level at output	2	V_{CC} = MIN					0.8	V
$V_{out(1)}$	Logical 1 output voltage	2	V_{CC} = MIN, I_{load} = −400 μA	V_{in} = 0.8 V,		2.4	3.3		V
$V_{out(0)}$	Logical 0 output voltage	1	V_{CC} = MIN, I_{sink} = 16 mA	V_{in} = 2 V,			0.22	0.4	V
$I_{in(0)}$	Logical 0 level input current (each input)	3	V_{CC} = MAX,	V_{in} = 0.4 V				−1.6	mA
$I_{in(1)}$	Logical 1 level input current (each input)	4	V_{CC} = MAX,	V_{in} = 2.4 V				40	μA
			V_{CC} = MAX,	V_{in} = 5.5 V				1	mA
I_{OS}	Short-circuit output current§	5	V_{CC} = MAX		SN5400	−20		−55	mA
					SN7400	−18		−55	
$I_{CC(0)}$	Logical 0 level supply current	6	V_{CC} = MAX,	V_{in} = 5 V			12	22	mA
$I_{CC(1)}$	Logical 1 level supply current	6	V_{CC} = MAX,	V_{in} = 0			4	8	mA

switching characteristics, V_{CC} = 5 V, T_A = 25°C, N = 10

PARAMETER		TEST FIGURE	TEST CONDITIONS		MIN	TYP	MAX	UNIT
t_{pd0}	Propagation delay time to logical 0 level	65	C_L = 15 pF,	R_L = 400 Ω		7	15	ns
t_{pd1}	Propagation delay time to logical 1 level	65	C_L = 15 pF,	R_L = 400 Ω		11	22	ns

† For conditions shown as MIN or MAX, use the appropriate value specified under recommended operating conditions for the applicable device type.

‡ All typical values are at V_{CC} = 5 V, T_A = 25°C.

§ Not more than one output should be shorted at a time.

Fan-Out. The fan-out of a digital gate is the maximum number of similar gates that may be connected to the output with proper operation preserved. For example, from Table 16-7 we see that the 7400 NAND gate supplies 1.6 mA (max) from each of its inputs when the input is a logical 0. If the preceding stage is an identical gate, it can safely sink 16 mA at its output when it is in a logical 0 stage. Consequently, we could safely connect up to 10 (10×1.6 mA = 16 mA) TTL inputs to the output. The fan-out then is 10.

Propagation Delay. The manufacturer usually specifies two propagation times, as seen in Table 16-7. These times are the propagation delay for the output to reach a logical 0 state, labeled t_{pd0}, and the propagation delay for the output to reach the logical 1 level, labeled t_{pd1}. These values are specified in terms of an input pulse going from a logical 0 level to the logical 1 level and back to the logical 0 level, as indicated in Fig. 16-20. The propagation delays are defined in terms of the response to this input. Typical total propagation delay (defined as the average of $t_{pd}0$ and t_{pd1}) for TTL gates is in the order of 10 nanoseconds, which is significantly lower than for any of the other logic families discussed here. This net delay in the propagation of a pulse through any sort of gate is caused by capacitive effects that are the storage of charge at or near a *PN* junction under both forward-biased and reverse-biased conditions.

Lastly, we consider the question of *positive* and *negative* logic. The concept is quite simple but can be confusing, so it has been postponed to the last. Basically, positive logic is what we have been using throughout this chapter: A logical 1 corresponds to the *higher* circuit voltage, and logical 0 corresponds to the *lower* circuit voltage. In negative logic, the logical 1 is defined as the lower of the two circuit voltages and the logical 0 is defined as the higher of the two circuit voltages. So,

$$\text{negative logic: } V(0) > V(1)$$
$$\text{positive logic: } V(1) > V(0)$$

This definition is quite straightforward. What may at times be confusing is that the same circuit can be one type of gate for positive logic and another type of gate for negative logic. For example, any positive logic NAND gate is also a negative logic NOR gate. Similarly, any positive logic NOR gate is also a negative logic NAND gate. The inverter stays an inverter in both logic schemes.

The verification of this scheme is simple if you remember that any circuit functions in the same way irrespective of our definitions; that is, the circuit does not "know" about positive and negative logic. To show that these statements are true, you need only to write the *voltage* truth table for any circuit and then apply the definitions of $V(0)$ and $V(1)$.

Another easy way to verify the statements concerning positive and negative voltage is to redraw all the circuits using *PNP* transistors instead of *NPN*. You would reverse all diodes and reverse the polarities of all supply voltages. The voltage truth tables will be the same except that when

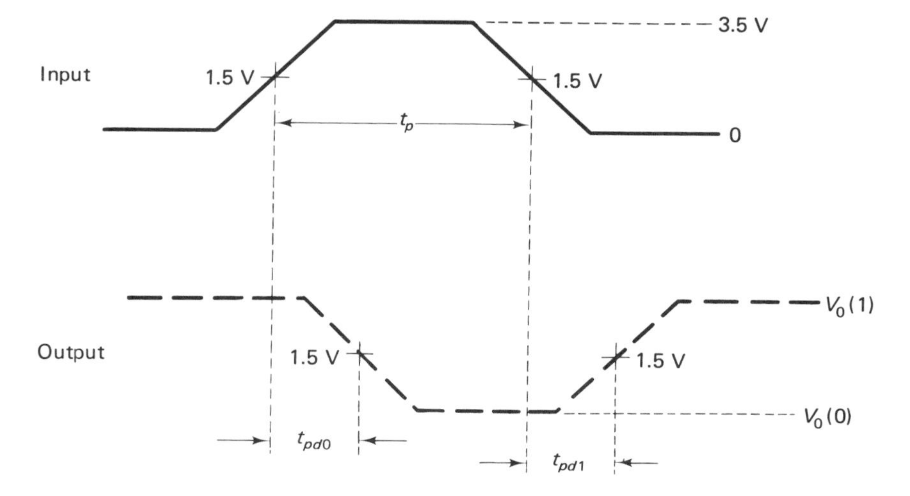

Figure 16-20. Propagation delays for a TTL gate.

a transistor is off, the voltage will be $-V_{CC}$ (which is lower than 0). In addition, you will have a 1 instead of a 0, and vice versa, in the logic truth tables for each circuit.

16.8 CMOS GATES

The CMOS logic family offers an alternative to bipolar (TTL) implementation of logic functions. The basic CMOS logic families are the 4000 line and the later 54C/74C line, which is a pin-for-pin direct replacement for the bipolar TTL line in most respects except for its lower speed.

Consider the inverter formed by interconnecting an N- and a P-channel MOSFET, as shown in Fig. 16-21. If V_{SS} is connected to ground and $V_{DD} = V_{CC}$, a positive supply voltage, we have the following operation: with V_i at or near ground, the gate of $Q1$ is negative with respect to the source, and assuming V_{CC} greater than V_T, $Q1$ is on. At the same time, the

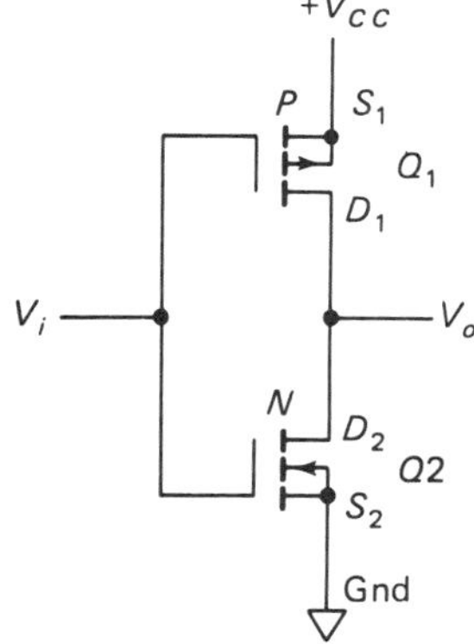

Figure 16-21. CMOS Inverter

gate-source voltage of $Q2$ is below V_T, thus causing $Q2$ to be off. With the high resistance of $Q2$, the P-channel MOSFET saturates ($V_{DS} \cong 0$), and V_0 is approximately equal to V_{CC}. Thus a low input causes a high output. Now suppose that the input is high, near V_{CC}: the gate of $Q2$ is now positive with respect to its source, and again assuming the input to be in excess of V_T, $Q2$ is on. At the same time, the gate-source voltage of $Q1$ is almost zero, thus causing $Q1$ to be off. With the high resistance of $Q1$ in series with it, $Q2$ saturates, and the output is essentially 0 V. Thus a high input causes a low output—the inverter function has been established.

CMOS devices typically have threshold voltages of 2 V. The resulting transfer characteristics for three different supply voltages are shown in Fig. 16-22.

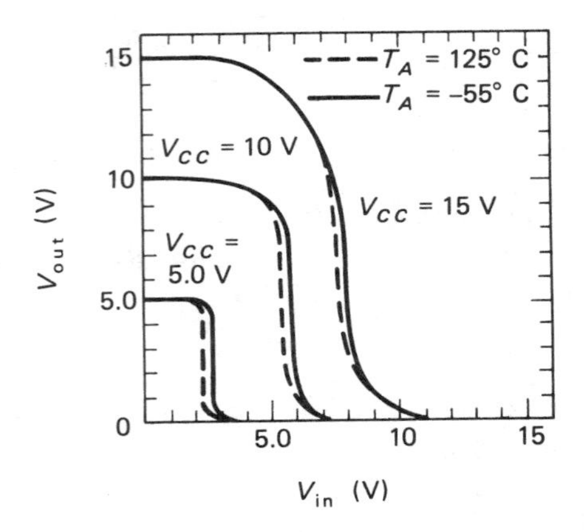

Figure 16-22. Typical inverter transfer characteristics

When the input is between 0 and approximately 2 V, $Q1$ is on and $Q2$ is off. Also, with the input between V_{CC} and V_{CC}−2 V, $Q2$ is on with $Q1$ off. However, with an input between 2 V and V_{CC}−2 V, both $Q1$ and $Q2$ are on. For digital applications, it is essential that the low-to-high or high-to-low input transition occurs as quickly as possible to minimize the time during which both the N- and P-channel devices are in conduction, thus minimizing power dissipation.

Note that because applying either a high or a low gives us one device on while the other is off, the net current drain from the power supply under static operation is almost insignificant. It is only during switching that significant amounts of current are drawn from the supply. With V_{CC} greater than $2V_T$, the supply current waveshape due to overlap in conduction is shown in Fig. 16-23. The resulting supply power drain is obviously directly proportional to the slope of the input waveshape, the power supply voltage, and the frequency.

The other part of the unloaded power dissipation of a CMOS inverter (or any other gate, for that matter) is the power required to charge the effective capacitance at the input. Note that the gate structure forms a

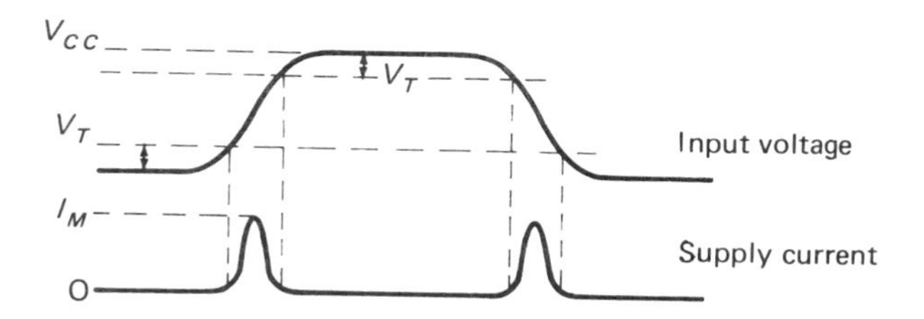

Figure 16-23. Inverter supply current waveform

capacitor, and in order to charge and discharge this capacitance, current must be supplied. The net effect of the gate capacitance as well as the current drawn due to the overlap in conduction can be modelled by a net effective capacitance C_{PD}. In one cycle of the input, this capacitance must be charged and discharged: The energy stored in the capacitor charged to V_{CC} is $\frac{1}{2}C_{PD}V_{CC}^2$. Thus, since the capacitor is both discharged and charged for each cycle, the net energy supplied (per cycle) must be twice the amount stored, or $C_{PD}V_{CC}^2$. The power is ascertained by determining the energy per unit time or by multiplying the energy per cycle by the time of one cycle: $P_D = C_{PD}V_{CC}^2 f$ (where f is the frequency of the input).

To determine the total dissipation per inverter, we need only include the effect of driving the load capacitance C_L. Thus the total power dissipation is $(C_{PD} + C_L)V_{CC}^2 f$. The effective power dissipation capacitance C_{PD} is listed by the manufacturer, and depends on the particular device. The load capacitance may be determined by using the value of input capacitance (also listed by the manufacturer) and multiplying it by the number of such gates connected to the output. For example, the manufacturer lists $C_{PD} = 12$pF and $C_{IN} = 6$ pF for each inverter in the 74C04 hex inverter. Thus we can estimate the power dissipation per inverter, if loaded by two other inverters operating from a 5 V supply with an input frequency of 10 kHz, to be approximately 6μW. If the input frequency is 1 MHz, the power dissipation is 0.6 mW. If the power supply is raised to 15 V, the power dissipation is 54μW at 10kHz and 5.4 mW at 1 MHz.

From the above discussion we can conclude that to minimize power dissipation, a low power-supply voltage should be used. However, the choice of power-supply voltage is not quite that simple. If we consider the switching characteristics of CMOS, we note that the propagation delay (time to cause the output of the gate to switch from a high to a low or vice versa) is a function of the supply voltage. That is, the charging of the effective internal capacitance is accomplished by passing a constant current through the conducting MOSFET—this current is a function of the geometry of the MOSFET and the supply voltage: The higher supply voltage, the higher the charging current and the lower the propagation delay. A typical set of characteristics is shown in Fig. 16-24. Note that the decrease in propagation delay is proportional to the increase in the supply voltage. Also, for a given supply voltage, the larger the load capacitance (the sum of the input capacitances of gates connected to the output), the longer the propagation delay.

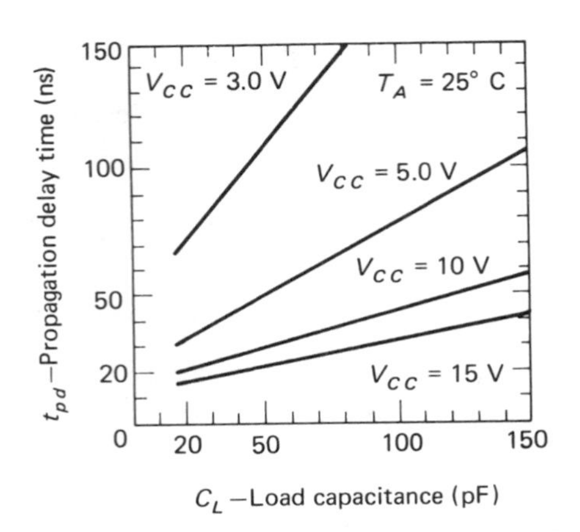

Figure 16-24. Typical CMOS propagation delay characteristics

The choice of power supply voltage for a particular application is dictated by the compromise between propagation delay (which determines the maximum signal frequency) and power dissipation. It is not possible to optimize both—that is, to minimize both the power dissipation and the propagation delay. A good approach is to make the propagation delay approximately one-tenth (or smaller) of the shortest signal period (highest signal frequency) and determine the power supply voltage accordingly, accepting the resulting power dissipation.

The structure of the CMOS inverter is shown in Fig. 16-25; this corresponds to the schematic of Fig. 16-21. Although not indicated in the figure, the chip substrate, which is the channel of the *P* MOSFET, is connected to its drain which is at $+ V_{CC}$. Similarly, the *P*–substrate, which

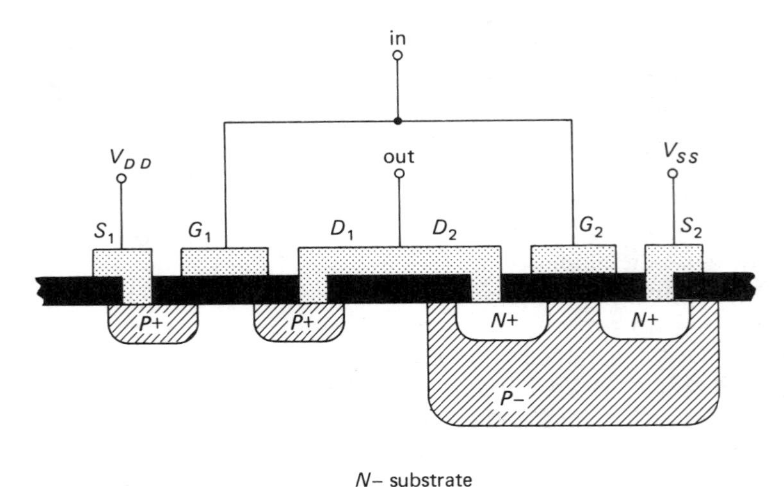

Figure 16-25. CMOS inverter structure

is the channel of the N MOSFET, is connected to its source, which is at ground. The resulting structure has a parasitic SCR (silicon-controlled rectifier) between the supply and ground. To see this, consider the two-transistor representation of the SCR as shown in Fig. 16-26. The NPN transistor is formed by the source of $Q2$, the P–substrate, and the N–substrate; the PNP transistor is formed by the source of $Q1$, the N–substrate, and the P–substrate. The relatively low bulk resistances of the two substrates, R_N and R_P, are in parallel with the base-emitter junctions. This has the effect of degrading the transistor β, which in turn degrades the parasitic SCR—in other words, we have a low-quality SCR. This means that the gate current required to turn on the SCR is quite large. Under normal operation of the inverter, such a large current is highly unlikely, so the parasitic SCR never turns on. However, it is not impossible for the SCR to be activated. Should the SCR fire, it places an extremely low effective resistance across the power supply and draws significant current. Under these conditions, the output is no longer controlled by the input (the gate is malfunctioning); the resulting power dissipation may permanently damage the IC. Assuming that the gate has not been damaged, the only way to turn off the parasitic SCR is to disconnect the power supply. In order to prevent this undesired circuit action, it is essential that the power supply voltage be on prior to any input signal. This is especially important to remember in bench testing where the input signal may be applied from a pulse generator: the power supply must be turned on before the pulse generator is connected.

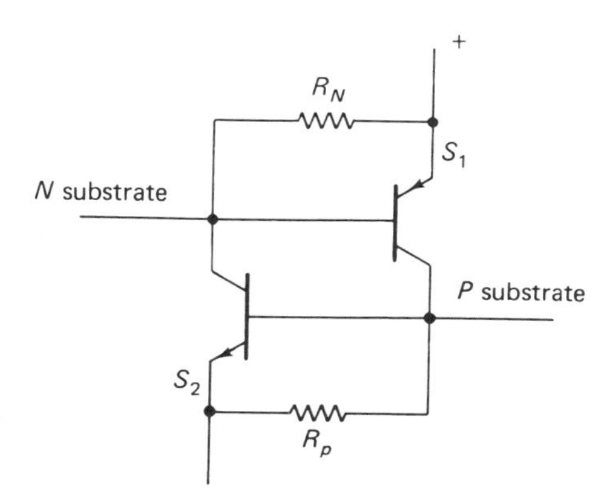

Figure 16-26. Equivalent parasitic SCR formed in CMOS inverter

The fan-out of CMOS driving CMOS is essentially infinite under static conditions since the gate draws negligible current. It is only during switching that the output of a gate has to supply (or sink) current to charge the effective load capacitance. However, the input capacitance of CMOS gates is not standardized. It depends on the physical dimension of the gate as well as the circuit configuration (type of gate). The fan-out then is limited by the signal frequency (the larger the load capacitance, the longer the propagation delay) and the maximum power dissipation of the particular device and/or IC package.

16.8.1 Special Considerations

All CMOS devices incorporate some sort of protection. This is essential since the thickness of the oxide layer is so small that even static discharge could break down the oxide and destroy the device. A typical protection scheme is shown in Fig. 16-27. Diode $D3$ is formed by the two substrates and only becomes activated should the power supply be reversed. Input protection is provided by $D1$ and $D2$: should the input become more positive than V_{CC}, $D1$ conducts and clamps the input at $V_{CC}+0.6$. The gates are further protected by the series resistance R (typically 500Ω), which limits any input current. Should the input become negative, $D2$ conducts and clamps the gates to -0.6V. Alternately, $D2$ can be connected between the input and ground, with basically the same operation.

Although all gates have this internal protection, it is still a good idea to exercise caution in order to prevent static discharge through the gates of the devices. A few simple rules should be observed:

1. Always store CMOS devices in a low resistance manner.
2. Never insert or remove a CMOS device while power is on.
3. Always reference unused inputs to either $+V_{CC}$ or ground (depending on the desired logic).

The first of these rules means that the leads of any CMOS devices should be electrically interconnected with minimum resistance. Typically, this is accomplished by inserting the IC pins into a special conductive foam or styrofoam covered by aluminum foil (the manufacturers usually supply the devices in an antistatic plastic tube or conductive foam). It is a

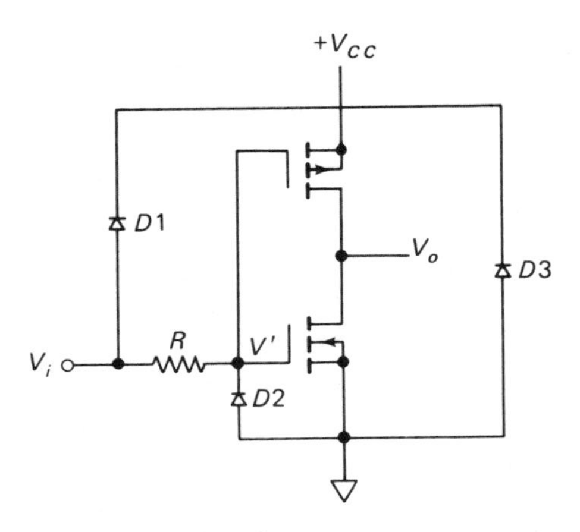

Figure 16-27. CMOS input protection

good idea to use this method whenever the IC is not in the circuit. When transporting ICs across a room, do not carry them by hand or in a plastic container, especially if the room is carpeted (carpeting tends to build up tremendous static). The ICs should be either in the conductive foam or in aluminum foil or should be carried in a tin tray with the leads in contact with the tin. Similarly, do not store unprotected devices in plastic drawers—either line the plastic drawers with aluminum foil or have the leads inserted into the conductive foam.

While soldering the leads of any CMOS device, or while a CMOS device is in the circuit and other components are being soldered, make certain that the tip of the soldering iron (30 W max.) is properly grounded.

While the above precautions sometimes may be disregarded with impunity, nevertheless it is worthwhile to observe these few simple rules to insure the integrity of CMOS/ICs.

16.8.2 CMOS Gates

The CMOS family of ICs is complete to the extent that inverters and NAND and NOR gates as well as flip flops, counters, etc., are available. In fact, with the introduction of the 54C/74C line, most of the functions available in TTL are duplicated in CMOS.

Consider the circuit diagram for the two-input NAND gate, shown in Fig. 16-28. Note that the basic complementary arrangement of the inverter is maintained. A high output, essentially V_{CC}, results if either input or both inputs are low, since $Q1$ or $Q2$ or both are on, with both $Q3$ and $Q4$ off. If both inputs are high, $Q1$ and $Q2$ are both off, while $Q3$ and $Q4$ are both on, causing a low output (essentially ground). The function implemented is a positive logic NAND. This scheme obviously can be extended to any number of inputs simply by adding one N- and one P-channel MOSFET for each additional input–the P-channel in parallel with $Q1$ and $Q2$ and the N-channel in series with $Q3$ and $Q4$. Thus a NAND gate with any number of inputs can be constructed.

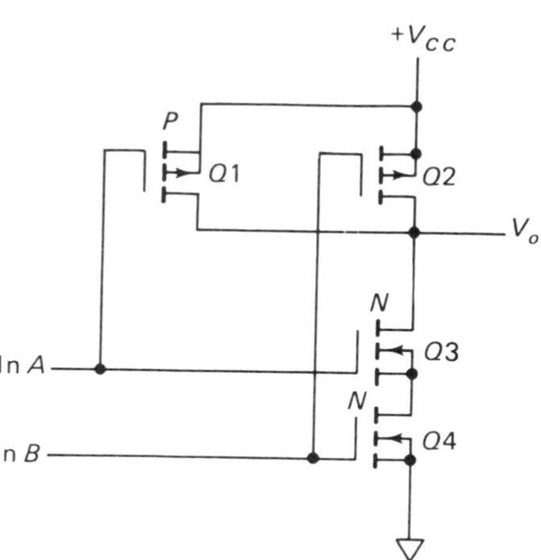

Figure 16-28. CMOS two-input NAND gate

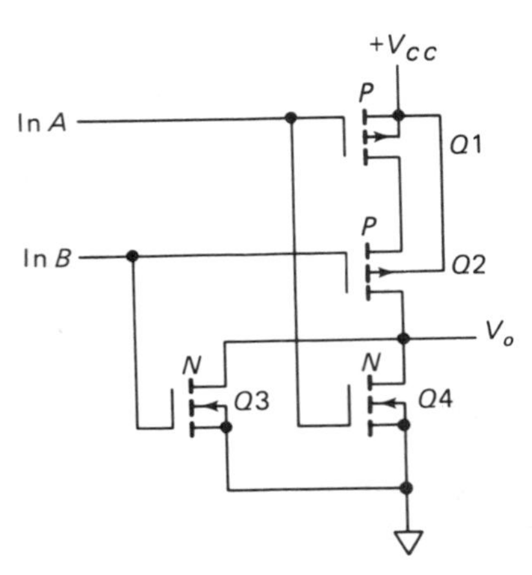

Figure 16-29. CMOS two-input NOR gate

A two-input positive logic NOR gate is shown in Fig. 16-29. Here the *P*-channel MOSFETs are connected in series, while the *N*-channel devices are connected in parallel. If either or both inputs are high, $Q3$ or $Q4$ or both conduct, and the output is low. If both inputs are low, $Q3$ and $Q4$ are both off, while $Q1$ and $Q2$ are both on, thus giving a high output. This is then the NOR operation. Again the scheme can be extended to provide a NOR gate with any number of desired inputs simply by adding a complementary pair of MOSFETs for each additional input: The additional *P*-channel MOSFETs would be placed in series with $Q1$ and $Q2$, while the *N*-channel devices would go in parallel with $Q3$ and $Q4$.

Although no protection circuitry is shown in Figs. 16-28 and 16-29, all CMOS gates have protective diodes and the series resistor incorporated on the chip.

16.8.3 Transmission Gate

The transmission gate has no counterpart in bipolar (TTL) circuitry. Besides allowing for different implementation of flip flops and other digital functions, it is very useful in analog applications.

The basic transmission gate is shown in Fig. 16-30. It consists of an inverter formed by $Q1$ and $Q2$, *P*- and *N*-channel MOSFET as before. In addition, two complementary MOSFETs are included with uncommitted source and drain terminals. Note that the control input signal is applied directly to the gate of $Q3$, while the gate of $Q4$ is driven by the control input after it has been inverted. Thus $Q3$ and $Q4$ are on together while the control input is high, and are off together while the control input is low. Consider the effect of this action on the input connected to $Q3$ and $Q4$: If the control input is high, $Q3$ and $Q4$ are on (the parallel connection reduces the on-resistance), and the input is transmitted to the output. If, on the other hand, the control input is low, both $Q3$ and $Q4$ are off (in a high resistance state), and the input is effectively isolated from the output. Thus a high on the control line transmits the input to the output, while a low isolates the output from the input.

For digital applications, $V_{DD} = V_{CC}$ (3 to 15 V), and V_{SS} is connected to ground. By applying a control signal from another CMOS gate con-

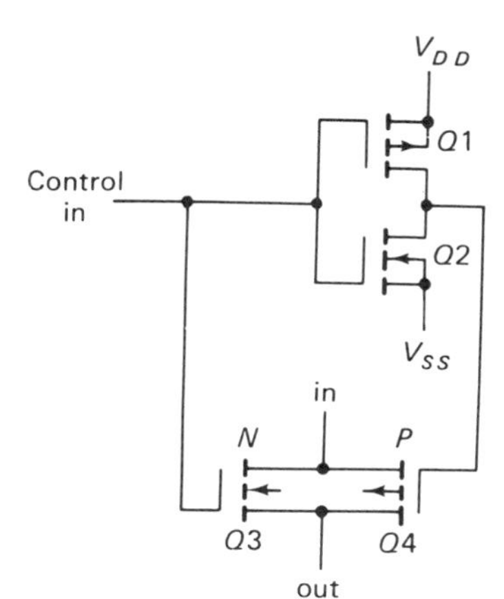

Figure 16-30. CMOS transmission gate

nected to the same supply, another digital signal can be switched on and off through $Q3$ and $Q4$. Since these MOSFETs are symmetrical, the signal input and output are interchangeable, that is, the transmission gate passes signals in either direction. We shall examine a typical digital application in the next section.

Another important application for the transmission gate is the digitally controlled analog switch: a switch which can pass analog signals. For these applications it is essential that the digital control signal, the power supply voltages, and the range of analog signals be properly coordinated. By using the same supply for the driving CMOS gate as for the transmission gate, the proper relationship between the control input and the power supply is automatically established. For positive turn-on and turn-off of the transmission gate, the control input in the high state should be equal to V_{DD}, and in the low state, equal to V_{SS}. This insures the proper operation of the $Q1$-$Q2$ inverter. For proper operation of the switch ($Q3$ and $Q4$), the maximum analog voltages at the input or output must be more positive than V_{SS}, but less than V_{DD}. For example, if the analog signal is between 0 and +5 V, the transmission gate can be operated from any supply voltage (V_{DD}) between +5 and +15 V with V_{SS} at ground. If, on the other hand, the analog input is between ±7.5 V, V_{DD} should be +7.5 V with V_{SS} equal to −7.5 V.

The CD4016 chip contains four such transmission gates or analog switches, and provides typically 200 Ω on-resistance and leakage of 100 pA in the off state. For applications requiring lower on-resistance and one more constant over the analog input range, the CD4066, which is a pin-for-pin replacement for the CD4016, should be used. In either case, using the CD4016 or the CD4066, the on-resistance can be lowered simply by paralleling two or more transmission gates. This is accomplished simply by connecting all the control inputs, all the analog inputs, and all the analog outputs together. (Remember that this will increase the capacitive loading on the digital drive to the control input, as well as increase the off-state leakage.)

Review Questions

1. What is meant by *combinational* logic circuits?
2. What is meant by *sequential* logic circuits?
3. What is the binary system?
4. What is the base of the binary system?
5. What are some of the conditions that may be indicated by the two-digit binary notation?
6. What is an inverter?
7. What condition inside a transistor indicates a logical 1? A logical 0?
8. What must be the input to an inverter if the output is a logical 1? A logical 0?
9. Why is the transistor inverter circuit often referred to as a transistor switch? Explain.
10. What is a truth table? What does it indicate?
11. What is an OR gate? What is its truth table?
12. What is an AND gate? What is its truth table?
13. How does the diode OR gate satisfy the OR gate truth table?
14. How does the diode AND gate satisfy the AND gate truth table?
15. What is meant by RTL?
16. What is a NOR gate? Give its truth table.
17. What is a NAND gate? Give its truth table.
18. Explain the operation of the RTL NOR gate.
19. Explain the operation of the RTL NAND gate.
20. What is meant by DTL?
21. Explain the operation of the DTL NAND gate.
22. What is meant by TTL?
23. Why is the implementation of TTL only possible using ICs and not possible in discrete component form?
24. Describe the operation of the TTL NAND gate shown in Fig. 16-14.
25. Describe the operation of the TTL inverter in Fig. 16-18.
26. Describe the operation of the TTL NOR gate shown in Fig. 16-17.
27. Describe the operation of a CMOS inverter.
28. Assuming that both the CMOS and TTL two-input NAND gates are operated from a 5V supply, discuss the differences.

Problems

1. The transistor inverter circuit of Fig. 16-1 is constructed with $V_{CC} = 15$ V, $R_B = 10$ kΩ, and $R = 100$ kΩ. The transistor used has the following properties: $\beta = 50$, $V_{BES} = 0.7$ V, $V_{CES} = 0.2$ V. Determine the value of R_C for the circuit to function as an inverter. Also determine the collector voltages corresponding to the two logic levels.

2. Determine the voltage truth table for the diode AND gate in Fig. 16-3 if $V=10$ V and $R=1$ kΩ.
3. Determine the voltage truth table for the diode OR gate in Fig. 16-5 if $V(1)=6$ V and $R=1.5$ kΩ.
4. Determine the voltages and currents in the RTL NOR gate in Fig. 16-7 for: $R_B=12$ kΩ, $R=68$ kΩ, $R_C=1$ kΩ, and $V_{BB}=V_{CC}=9$ V. The transistors are identical with $\beta>50$; V_{CES} and V_{BES} may both be neglected. The input conditions are: $V_A=V(1)$ and $V_B=V(0)$.
5. Repeat Problem 4 if the input conditions are: $V_A=V_B=V(0)$.
6. Repeat Problem 4 for the RTL NAND gate in Fig. 16-8.
7. The circuit values for the DTL NOR gate in Fig. 16-11 are: $R=4.7$ kΩ, $R_B=10$ kΩ, $R_C=560$ Ω, $V_{BB}=2$ V, and $V_{CC}=10$ V. The transistor has the following properties: $V_{BES}=0.6$ V and $V_{CES}=0.2$ V. The diode on-voltage is 0.6 V. Determine the minimum β needed for the circuit to obey NOR logic.
8. The inputs to the NOR gate in Problem 7 are: $V_A=V(1)$ and $V_B=V(0)$. Determine the circuit conditions (currents and voltages) if the transistor β is larger than the minimum value calculated in Problem 7.
9. Using the same circuit values as in Problem 7 in the DTL NAND gate of Fig. 16-12 (with $R=R_1$), determine the minimum β needed for the circuit to obey NAND logic.
10. When both inputs are low in the DTL NAND gate in Fig. 16-13, determine (a) the current through R_1 and (b) the output voltage if: $R_1=10$ kΩ, $R=2$ kΩ, $R_C=1$ kΩ, $V_{CC}=12$ V, $\beta=100$, and the base-emitter junction is represented by a 10 kΩ resistance when forward biased less than 0.5 V.
11. For the values shown in the TTL NAND gate in Fig. 16-16, give an approximate voltage truth table indicating the voltages at points P, S, T, and Y. Assume that all forward-biased junctions have 0.7 V across them, that collector saturation voltages equal 0.2 V, and that all transistor βs equal 50.
12. Obtain a manufacturer's data sheet for a 74CO4 hex CMOS inverter. Estimate the maximum input low voltage, the minimum input high voltage, the output low and high voltages if the supply is: (a) 5V, (b) 10V, (c) 15V.

17 Clipping, Clamping, and Wave-Shaping Circuits

This chapter introduces a variety of relatively simple circuits used for signal processing. In many applications it is necessary to pass one portion of the signal and reject another portion. Clipping circuits perform this function. Sometimes we must clamp or latch the output at a certain level—a function accomplished by clamping circuits. The general wave-shaping circuits are used to modify the input waveshape into another waveshape or to prevent the input waveshape from being modified.

17.1 SINGLE-LEVEL CLIPPING CIRCUITS

In general, we use clipping circuits to select either the part of the input waveshape which lies below or above a certain reference level (single-level clippers) or the part of the input that lies between two selected levels (two-level clippers).

17.1.1 Positive Clipping Circuits

The circuit shown in Fig. 17-1 clips the input waveshape at a positive voltage determined by the reference voltage V. In order to understand the operation of this circuit, suppose that we have a sinusoidal input waveshape whose peak amplitude is larger than the reference voltage V. When the input is positive and less than V, the diode is reverse biased and therefore does not conduct. We may assume that a reverse-biased diode offers an open circuit. The output waveshape then follows the input because no current flows through R. When the input voltage exceeds the reference voltage V (in the positive direction), the diode is forward biased and conducts. In the forward direction, we can assume that the diode acts as a short circuit. During this time, the difference between the input voltage and the reference voltage appears across R, and the output is fixed at V. (Note that for low-amplitude signals, the forward drop across the diode should not be neglected.)

Thus, the operation of the positive clipping circuit of Fig. 17-1 may be summarized as follows. The output voltage follows the input voltage as long as the reference voltage is greater than the input (diode, reverse-

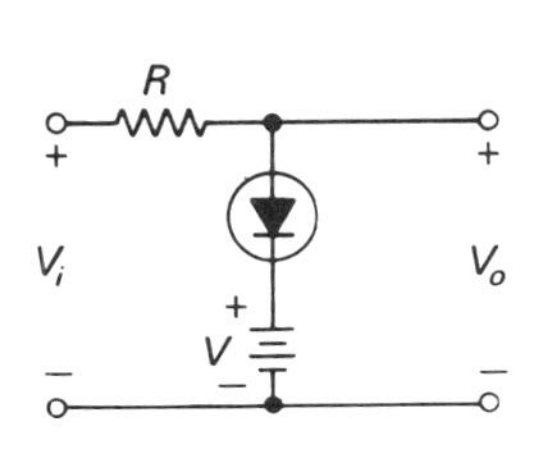

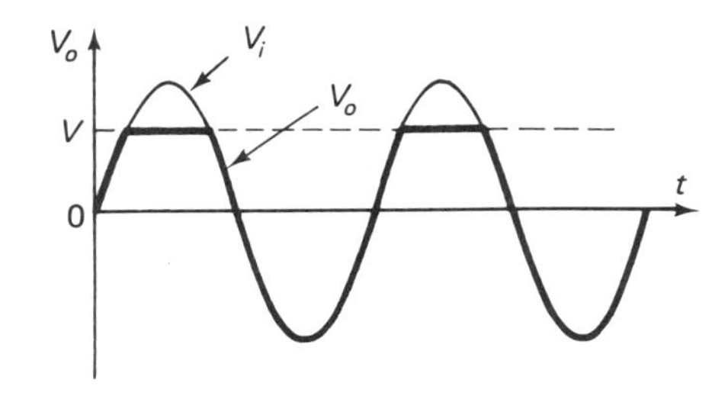

Figure 17-1. Positive clipping circuit.

biased); the output voltage is fixed at the reference voltage when the input voltage exceeds the reference voltage (diode, forward-biased).

An alternative for the positive clipping circuit is shown in Fig. 17-2. When the input waveshape is at a level lower than the reference voltage, the diode is forward biased (a short circuit) and the output waveshape follows the input. When the input waveshape is of a level higher than the reference voltage, the diode is reverse biased (an open circuit) and the output voltage is equal to the reference voltage (there being no current through R).

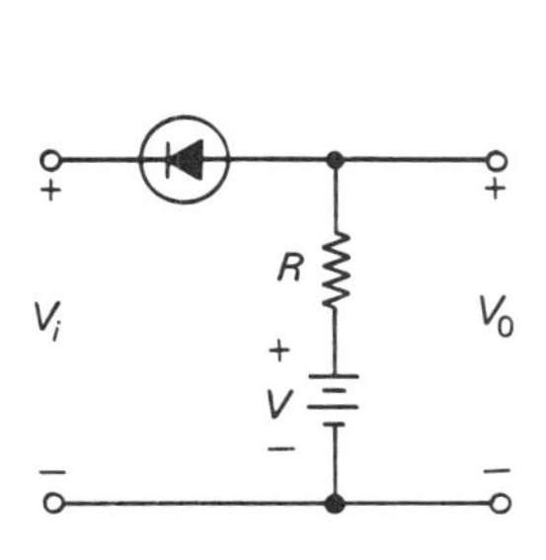

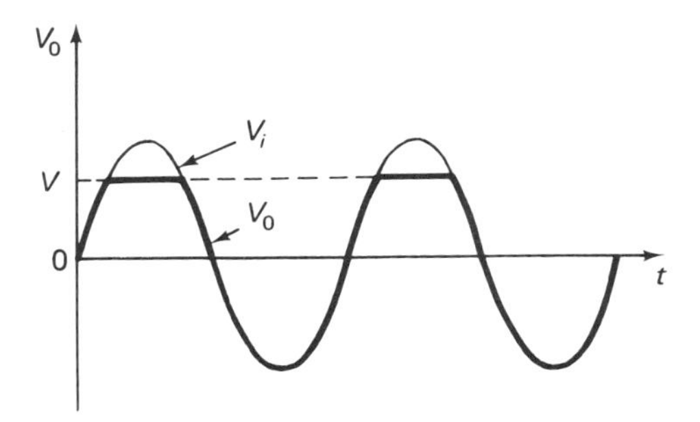

Figure 17-2. Alternate positive clipping circuit.

17.1.2 Negative Clipping Circuits

If both the diode and the polarity of the voltage reference in Fig. 17-1 are reversed, as shown in Fig. 17-3, the circuit clips the input waveshape at a negative voltage.

When the input is positive and less negative than the reference voltage V in Fig. 17-3, the diode is reverse biased and effectively open. There being no current through R, the output voltage is essentially the same as the input. When, however, the input becomes more negative than the reference voltage V, the diode becomes forward biased and effectively a short circuit. The output is then equal to the reference voltage, as indicated in Fig. 17-3.

We can also obtain a negative clipping circuit by reversing the diode and the polarity of the reference voltage in Fig. 17-2. The resulting circuit is illustrated in Fig. 17-4. When the input is positive or less negative than

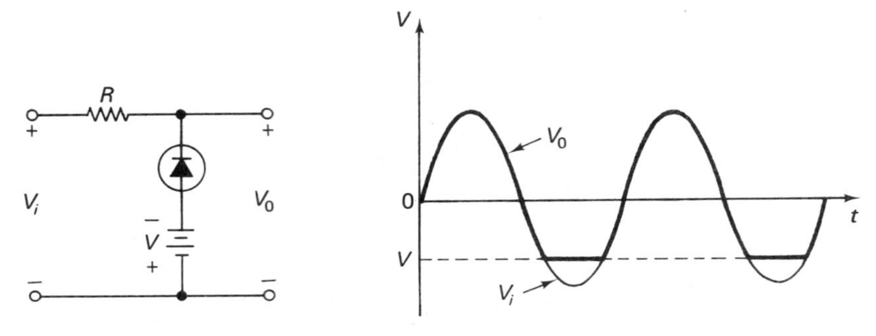

Figure 17-3. Negative clipping circuit.

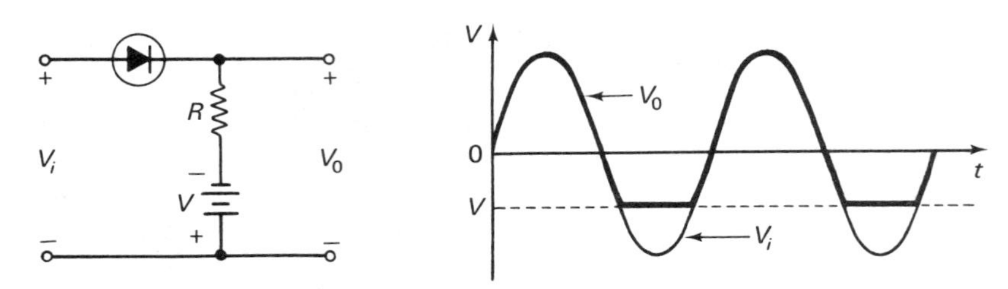

Figure 17-4. Alternate negative clipping circuit.

the reference voltage, the diode is forward biased and effectively a short circuit. The output voltage then follows the input voltage. However, when the input becomes more negative than the reference, the diode is reverse biased and effectively an open circuit. Therefore, we set the output voltage at the negative reference voltage, as shown in Fig. 17-4.

Consider the four clipping circuits when a load is placed across the output terminals. It should be obvious that the circuits in Figs. 17-1 and 17-3 will cause a *voltage* of the waveshape shown to exist across the load. At the same time, the circuits of Figs. 17-2 and 17-4 will cause a *current* of the given waveshape to flow through the load.

Figures 17-5, 17-6, 17-7 and 17-8 show the clipping circuits discussed above with the diode reversed. The resulting output waveshapes are shown. The verification of these waveshapes is left as an exercise.

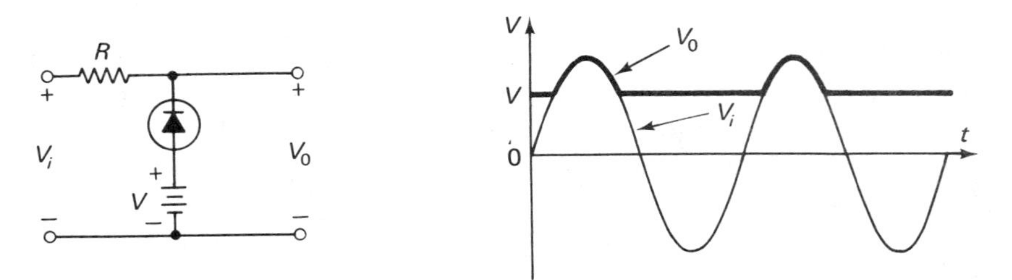

Figure 17-5. Positive comparator circuit.

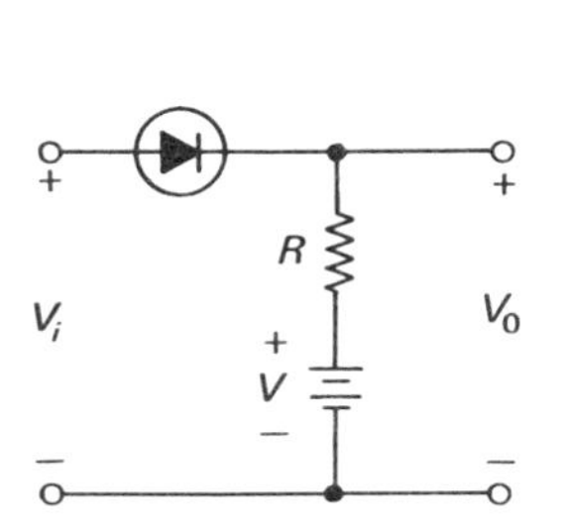

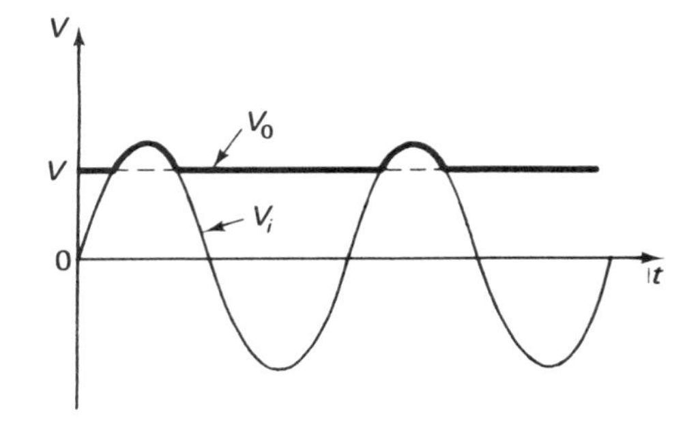

Figure 17-6. Alternate positive comparator circuit.

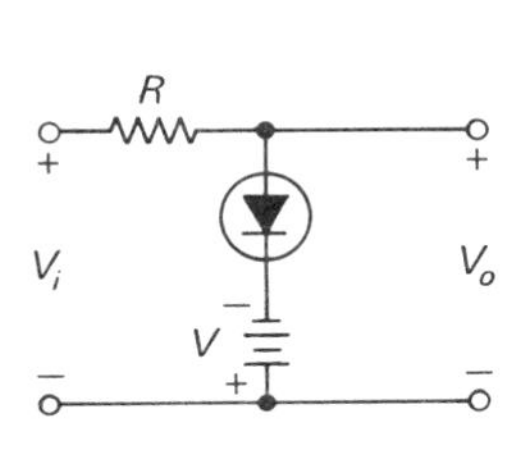

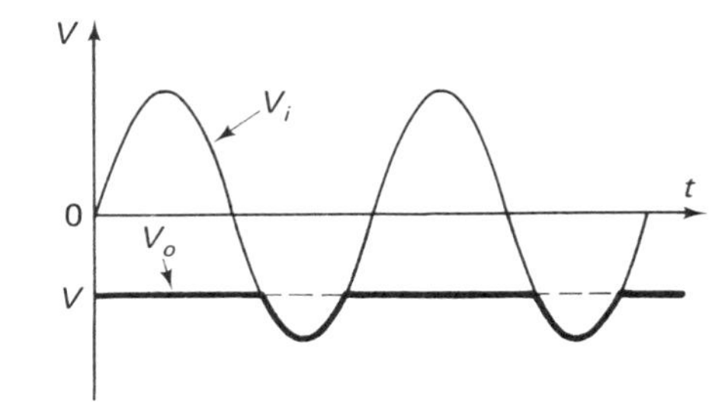

Figure 17-7. Negative comparator circuit.

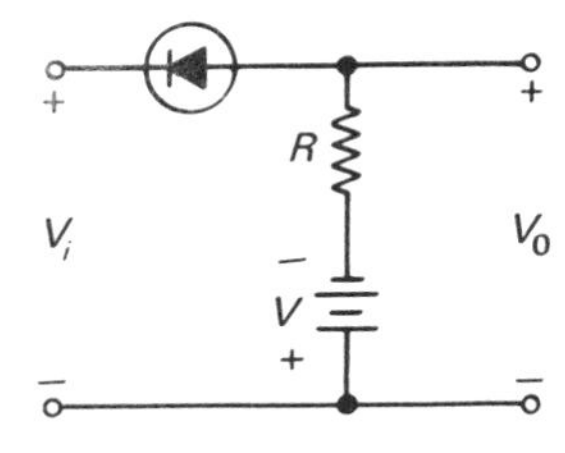

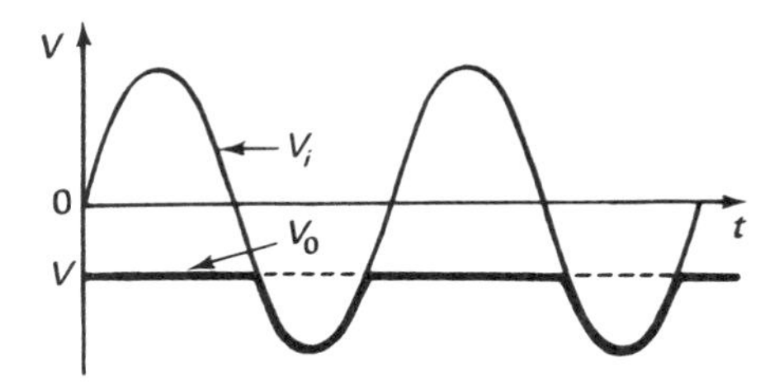

Figure 17-8. Alternate negative comparator circuit.

17.2 TWO-LEVEL CLIPPING CIRCUITS

In certain applications we have to clip the input waveshape at two preselected levels. The circuits shown in Figs. 17-9, 17-10, 17-11, and 17-12 perform this operation.

Consider the circuit in Fig. 17-9. If we assume that V_2 is greater than V_1, we can see that when the input is positive and less than V_1, diode $D1$ is forward biased, whereas diode $D2$ is reverse biased. The output voltage is then fixed at V_1. When the input is more positive than V_1 but less positive than V_2, $D1$ and $D2$ are both reverse biased. The output then follows the input. When the input exceeds both V_1 and V_2, $D1$ is still reverse biased and $D2$ is forward biased. The output voltage becomes fixed at V_2. When the input voltage goes negative, $D2$ is reverse biased with $D1$ forward

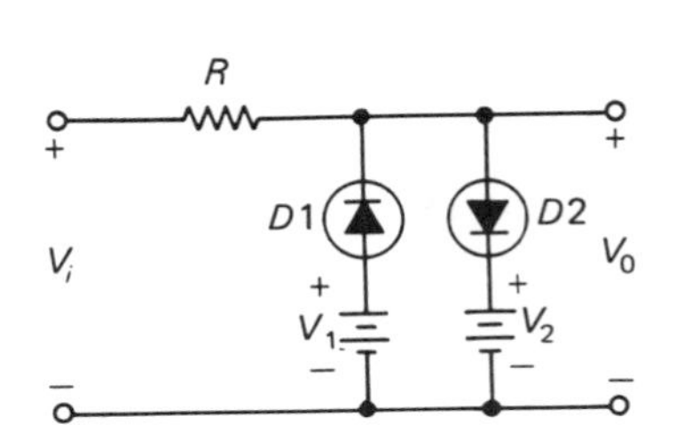

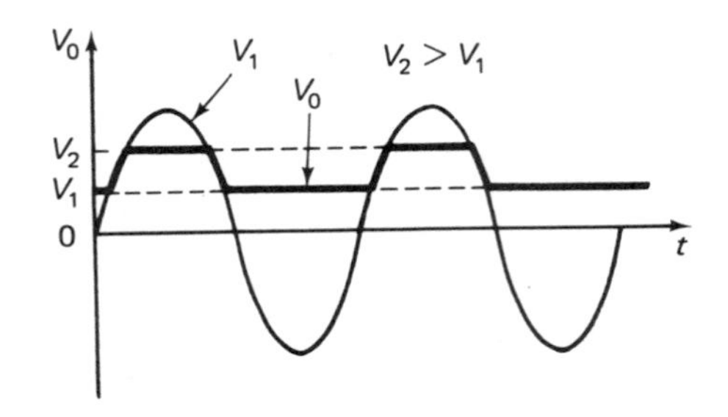

Figure 17-9. Circuit for clipping at two positive voltages.

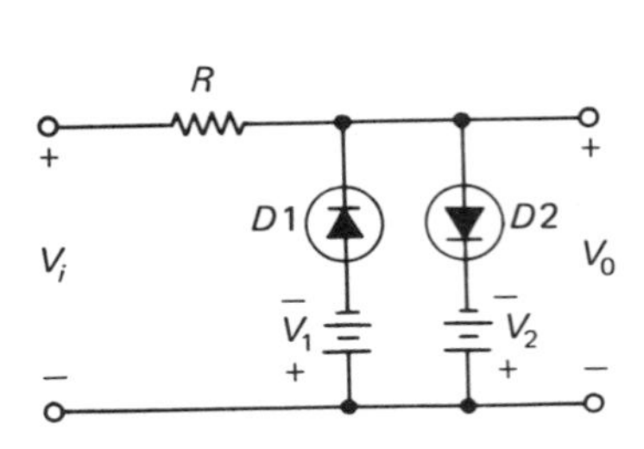

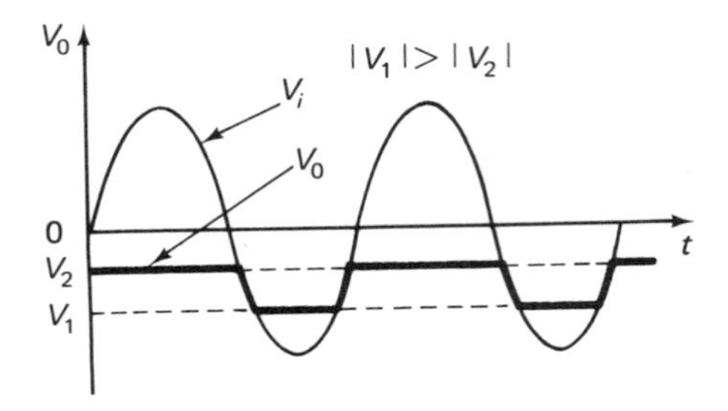

Figure 17-10. Circuit for clipping at two negative voltages.

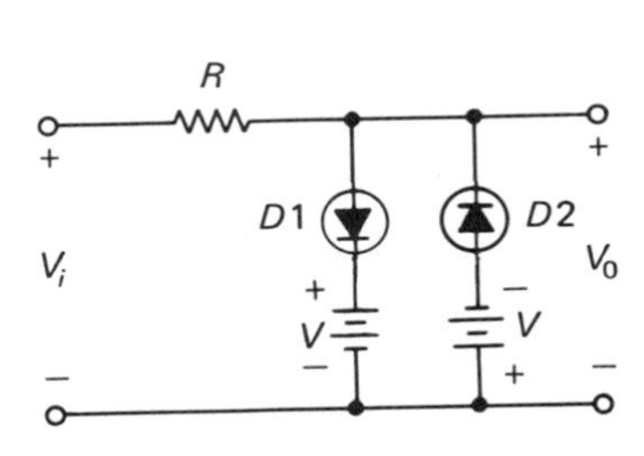

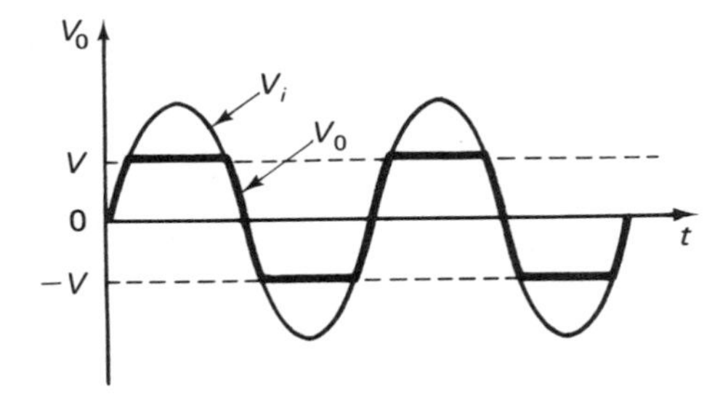

Figure 17-11. Circuit for clipping at one positive and one negative voltage.

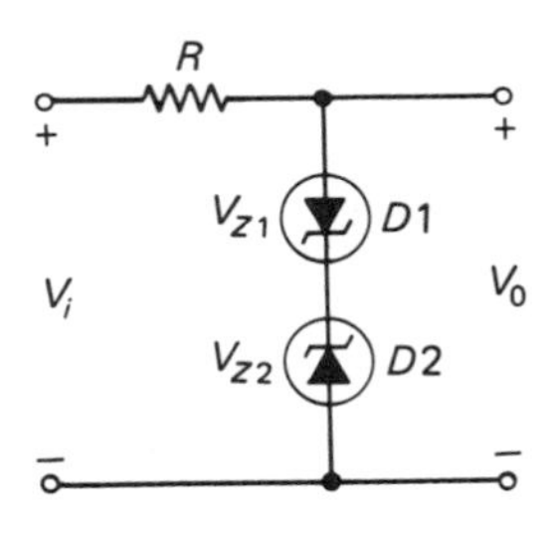

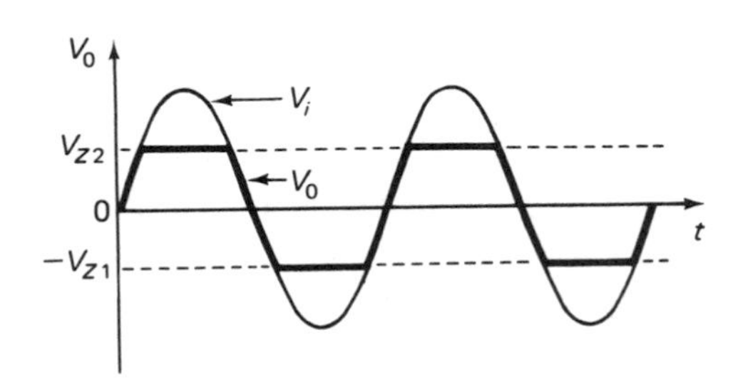

Figure 17-12. Clipping circuit using Zener diodes.

biased and conducting, so that the output is at V_1. The circuit in Fig. 17-9 is used when we want to clip the input voltage between two positive reference voltages V_1 and V_2.

If both $D1$ and $D2$ as well as the polarity of both reference voltages V_1 and V_2 in Fig. 17-9 are reversed, the circuit shown in Fig. 17-10 results. This circuit clips the input voltage between two predetermined negative reference levels and operates quite similarly to the circuit of Fig. 17-9.

When the input waveshape needs to be clipped between one positive and one negative level, we use the circuit of Fig. 17-11. (Although the reference voltages are shown to be the same, this need not be the case.) If the input voltage is less negative than the reference voltage and at the same time less positive than the other reference voltage, both $D1$ and $D2$ are reverse biased, and the output voltage follows the input. When the input becomes more positive than the $D1$ reference voltage, $D1$ conducts and the output is forced to the $D1$ reference voltage (V). When the input is more negative than the $D2$ reference voltage, $D2$ is forward biased (with $D1$ reverse biased), and the output is forced to the negative reference voltage of $D2$, as indicated.

An alternate way of achieving clipping at a negative and positive voltage is illustrated in Fig. 17-12. Here we use two Zener diodes. They need not be identical or have the same breakdown voltages. When the input is positive, $D1$ is forward biased, whereas $D2$ is reverse biased. When the input exceeds the breakdown voltage of $D2$ (this voltage is in the positive direction), the output is fixed at V_{Z2}. When the input is negative, $D2$ is forward biased, and $D1$ is reverse biased. When the input is more negative than the breakdown voltage of $D1$, $D1$ goes into breakdown and limits the output voltage to V_{Z1}. Note that when we deal with Zener diodes that have breakdown voltages lower than, say, 10 V, we cannot neglect the forward diode drop. The clipping levels then will be $V_Z + 0.7$ V for silicon diodes and $V_Z + 0.3$ V for germanium diodes.

17.3 CLAMPING CIRCUITS

One operation that we must carry out frequently is establishing the specific maximum or a minimum for the waveshape. These extremes specify the voltage to which the output is said to be *clamped*. The circuit accomplishing this function is called a *clamping circuit*.

17.3.1 DC Restorer Circuit

A circuit that clamps the output voltage to ground is often called a *dc restorer circuit*. It is shown in Fig. 17-13. With a sinusoidal input, circuit operation is as follows. As the input starts on its positive upswing, the diode is forward biased and conducts. The output voltage is essentially zero, and the diode current charges the capacitor (which is assumed to be uncharged initially). The capacitor voltage follows the input voltage until the positive peak is reached. At this time the voltage on the capacitor is V_p. When the input falls below V_p, the voltage across the capacitor exceeds the input voltage and the diode is reverse biased. Because of the reverse-biased

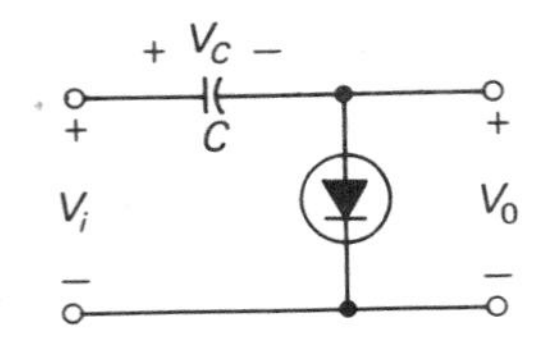

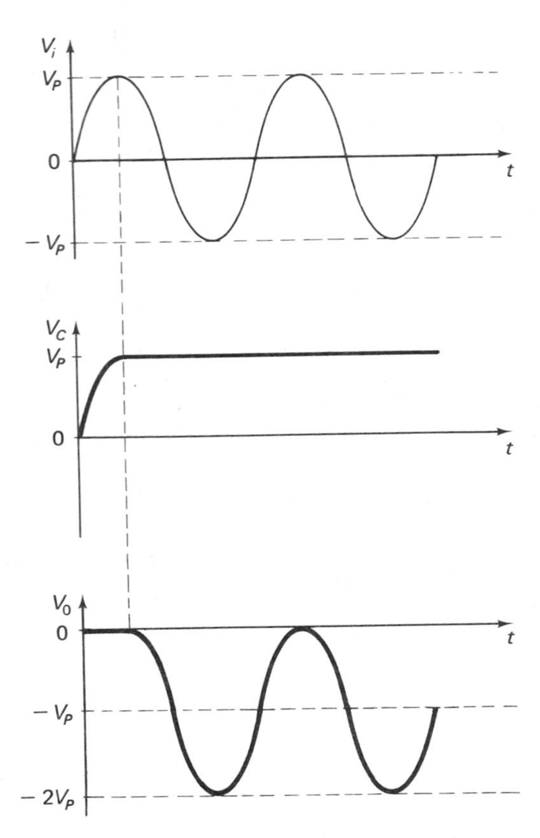

Figure 17-13. Negative dc restorer (clamping) circuit and waveshapes.

diode, no more current flows and the voltage across the capacitor (V_p) is maintained. Thus, the output now follows the input but starts at ground when the input is at V_p. From then on, the diode is always reverse biased (assuming ideal operation), the capacitor maintains its voltage (V_p) and the output is a sine wave with a negative dc level equal to the peak value of the input. The circuit thus has restored a dc level to the output. In other words, it has clamped the output at zero, not allowing it to go above zero.

If the diode in Fig. 17-13 is reversed, the circuit (shown in Fig. 17-14) restores a positive dc level to the output. Let us consider the circuit in Fig. Fig. 17-14, with the capacitor uncharged initially and a sinusoidal input as before. The input is positive, so the diode is reverse biased and the output

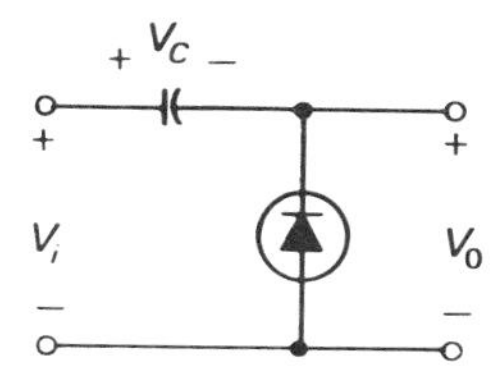

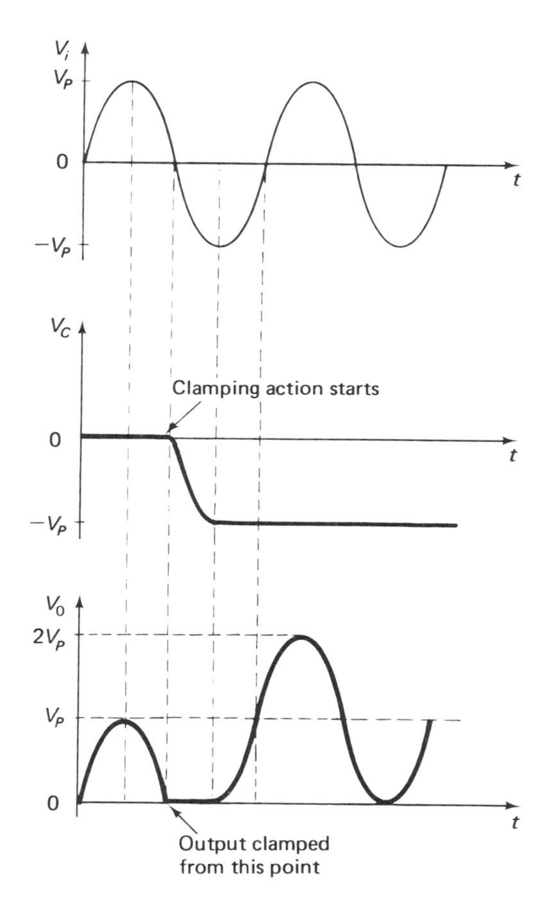

Figure 17-14. Positive dc restorer (clamping) circuit and waveshapes.

follows the input as shown. There is no clamping action to this point. Clamping action begins when the input starts on its negative downswing. During this time the diode is forward biased and allows the capacitor to charge up. Note that the capacitor charges to $-V_p$. Once the input becomes less negative than V_p, the diode is again reverse biased and the output is clamped at ground and follows the input. The output in this case has a positive dc level equal to the value of the input wave peak.

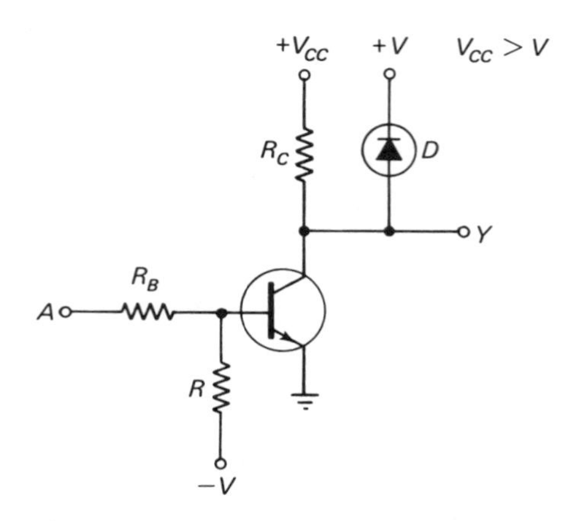

Figure 17-15. Diode-clamped gate.

17.3.2 Latching Circuits

In many applications we have to modify the performance of a circuit by causing the output to be clamped to a voltage other than the specific supply voltage being used for that circuit. An example is depicted in Fig. 17-15, where the diode (called a *latching* diode) and its reference supply V are added to the simple transistor inverter circuit (see section 16.2).

The addition of the latching diode does not change the basic circuit operation; the circuit is still an inverter. With a high input at A, the transistor saturates, point Y goes essentially to zero, and the diode is reverse biased. Thus, with a high input, the diode and its reference supply have no effect on the performance of the inverter. When the input to the inverter is low, the transistor is cut off and draws no current. We have redrawn the circuit in Fig. 17-16 to show the transistor disconnected. Because V_{CC} is larger than V, the diode is now forward biased and conducts. The current I is essentially given by:

$$I = \frac{V_{CC} - (V - 0.7)}{R_C} \tag{17-1}$$

where a forward diode drop of 0.7 V has been assumed. The output is now at $V - 0.7$ and not at V_{CC}, as would be the case were the latching circuit removed. We add the latching circuit in order to improve the fan-out of this type of gate.

To observe this improvement, let us consider the inverter circuit without the latching network. The logical 1 level at the output is V_{CC}.

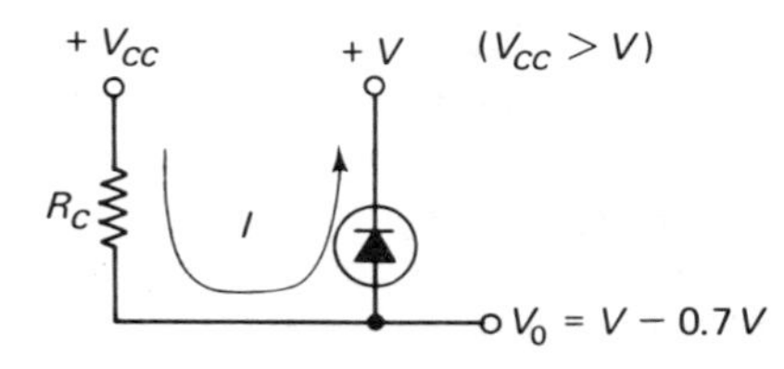

Figure 17-16. Equivalent circuit for Fig. 17-15 when the transistor is off with the output clamped to approximately V.

However, as soon as another gate is added to the output, it draws current and the logical 1 output voltage falls below V_{CC}. If enough gates are tied to the output, the voltage would fall below the specified minimum for a logical 1 level, and the gate would cease to function.

When the circuit of Fig. 17-15 is used, additional gates may be connected to the output without decreasing the logical 1 output voltage. This operation is illustrated in the following example.

Example 17-1. Determine the logical 1 output voltage $V_0(1)$ in the circuit of Fig. 17-17 with R_x of: (1) 22 kΩ and (2) 10 kΩ. Assume the forward diode drop to be 0.7 V.

Solution: Without R_x connected and with the transistor cut off, $V_0(1) = 5.7 - 0.7 = 5$ V. This figure will be the output voltage so long as I_x is smaller than I. We see that $I_x = I - I_D$. From Eq. (17-1):

$$I \cong \frac{10-(5.7-0.7)}{2.2} \text{ mA} \cong 2.27 \text{ mA}$$

For $R_x = 22$ kΩ:

$$I_x = \frac{V_o(1)}{R_x} \cong \frac{5}{22} \text{ mA} \cong 0.23 \text{ mA}$$

Thus, I_D is approximately 2.04 mA. For $R_x = 10$ kΩ:

$$I_x \cong \frac{5}{10} \cong 0.5 \text{ mA}$$

and I_D is approximately 1.77 mA. Note that the diode must be conducting for the output voltage to remain at 5 V. Thus, $I_D \geqslant 0$. As a result, we have a maximum I_x equal to I. For this worst-case condition, the value of R_x is:

$$R_{x(\min)} \cong \frac{5}{2.27} \text{ k}\Omega \cong 2.2 \text{ k}\Omega$$

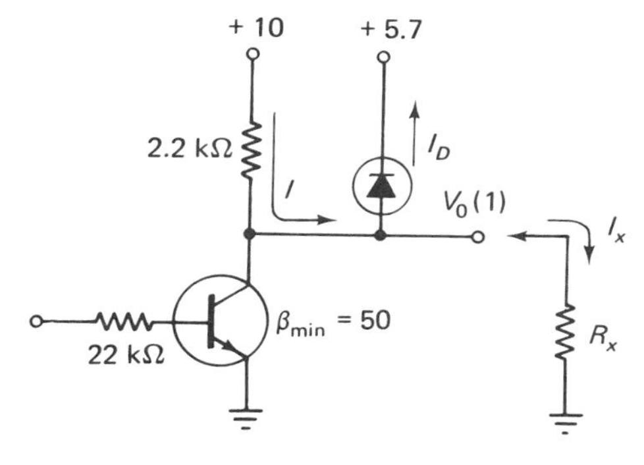

Figure 17-17. Latched inverter (see Example 17-1).

If R_x represents the net effect of connecting additional gates to the output and all the gates are simulated by an input resistance of R_B (22 kΩ), then we see that a maximum of 10 gates may be connected to the output of the gate in Fig. 17-17 and the logical 1 level of 5 V still be preserved. With the next gate added (the 11th), the output voltage would drop below 5 V. Therefore, fan-out for this gate is 10.

The performance of the latched inverter of the preceding example can be compared to an inverter without the latching circuit. The logical 1 output voltage for the inverter without the latching diode would vary from 10 V for no loading to 5 V with 10 identical gates connected to the output.

The diode in the latching circuit may be reversed and referenced to a voltage that is higher than V_{CC} if we wish to latch the output to a higher voltage.

The inverter example of a latching circuit shows only one of the many uses for diodes as latches.

17.4 WAVE-SHAPING CIRCUITS

In this section we consider some of the more common wave-shaping circuits using RC combinations. In many applications it is necessary (1) to form a "spike" at the output with a square-wave input or (2) to detect the average or peak value of the input waveshape. We can perform both of these functions with RC networks.

17.4.1 Spike-Forming Circuits

Consider the circuit shown in Fig. 17-18. If the input is a square wave as shown, the output may contain both positive and negative going spikes. To see how these spikes are obtained and what conditions must be

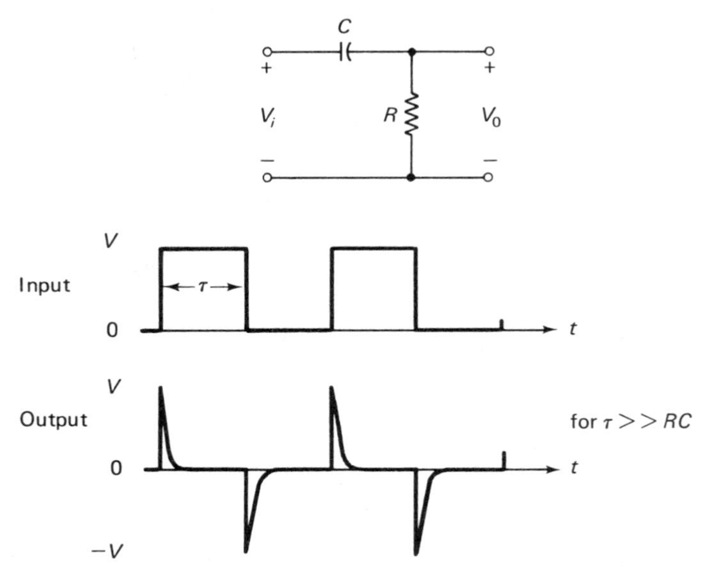

Figure 17-18. RC spike-forming (differentiator) circuit and waveshapes.

satisfied, we can analyze the circuit. With the input at zero and no initial charge on the capacitor, the output is also at zero. When the input switches to $+V$, the capacitor voltage cannot change instantaneously, so the change is transmitted to the output. If the RC time constant is very short in comparison to the duration of the input pulse, τ (typically $\frac{1}{5}$ of τ or shorter), the capacitor quickly charges to the input voltage V and the output falls to zero.

When the input switches to zero, the capacitor voltage is still $+V$, so the output switches from zero to $-V$, as shown. The capacitor now quickly discharges to zero through the resistor, and the output voltage goes to zero, as shown. When the input pulse is repeated, the operation just described is also repeated. Note that before the next pulse arrives, the capacitor is completely discharged. Consequently, the initial assumption of no charge on the capacitor is valid.

In order to achieve spikes at the output with a square wave at the input of the RC circuit illustrated in Fig. 17-18, the time constant must be much smaller than the pulse duration; that is, RC must be greater than 5τ.

In some applications we wish to generate only positive or only negative going spikes. In such cases, the RC circuit is coupled to a diode clipper circuit, as shown in Fig. 17-19. The operation of the two circuits (the RC and clipping circuits) is the same as it was when they were used

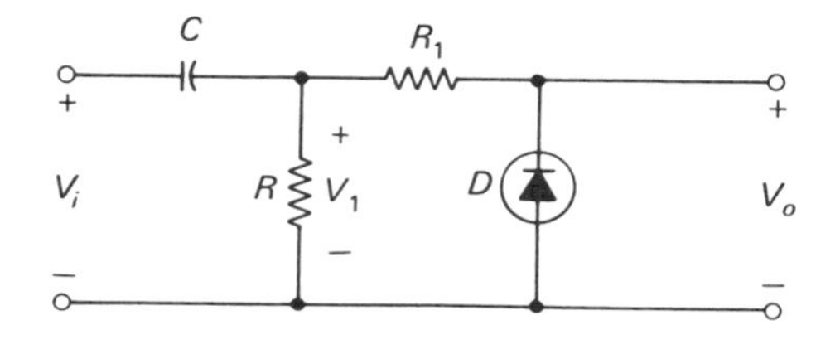

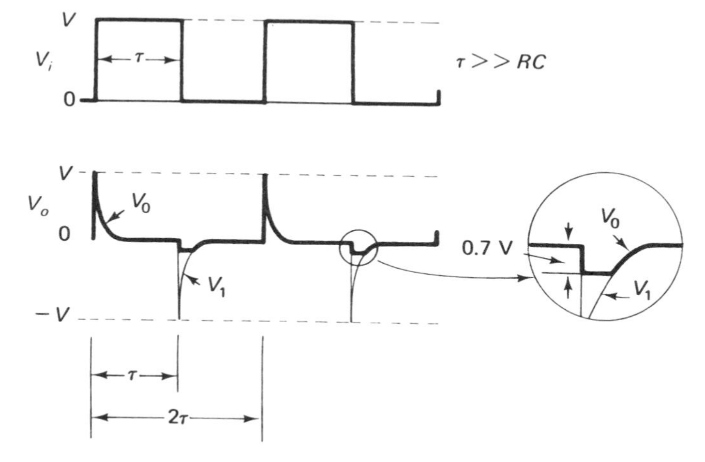

Figure 17-19. Circuit for forming positive going spikes.

individually. With the diode connected as indicated, the negative going spikes are clipped and the output contains only positive going spikes. If the diode is reversed, the output will contain only negative spikes. Note that the output has a positive going spike at every point where the input voltage makes a transition from zero to V.

If the voltage levels are low(say, less than 10 V) as they would be in most digital circuits, the output waveshape contains a small part of the negative spike, as shown in the detail in Fig. 17-19. This waveshape results because the forward drop across the diode is not zero but is 0.7 V for a silicon diode. This effect becomes very pronounced when the input pulse level is just a few volts. We can eliminate the undesired part of the waveshape if the diode clipper is replaced with a precision rectifier.

Precision Rectifier. Precision rectifier circuits are shown in Figs. 17-20 and 17-21. These circuits utilize an OP AMP to effectively reduce the diode forward drop. The two circuits in Fig. 17-20 give a positive output; the two in Fig. 17-21 give a negative output.

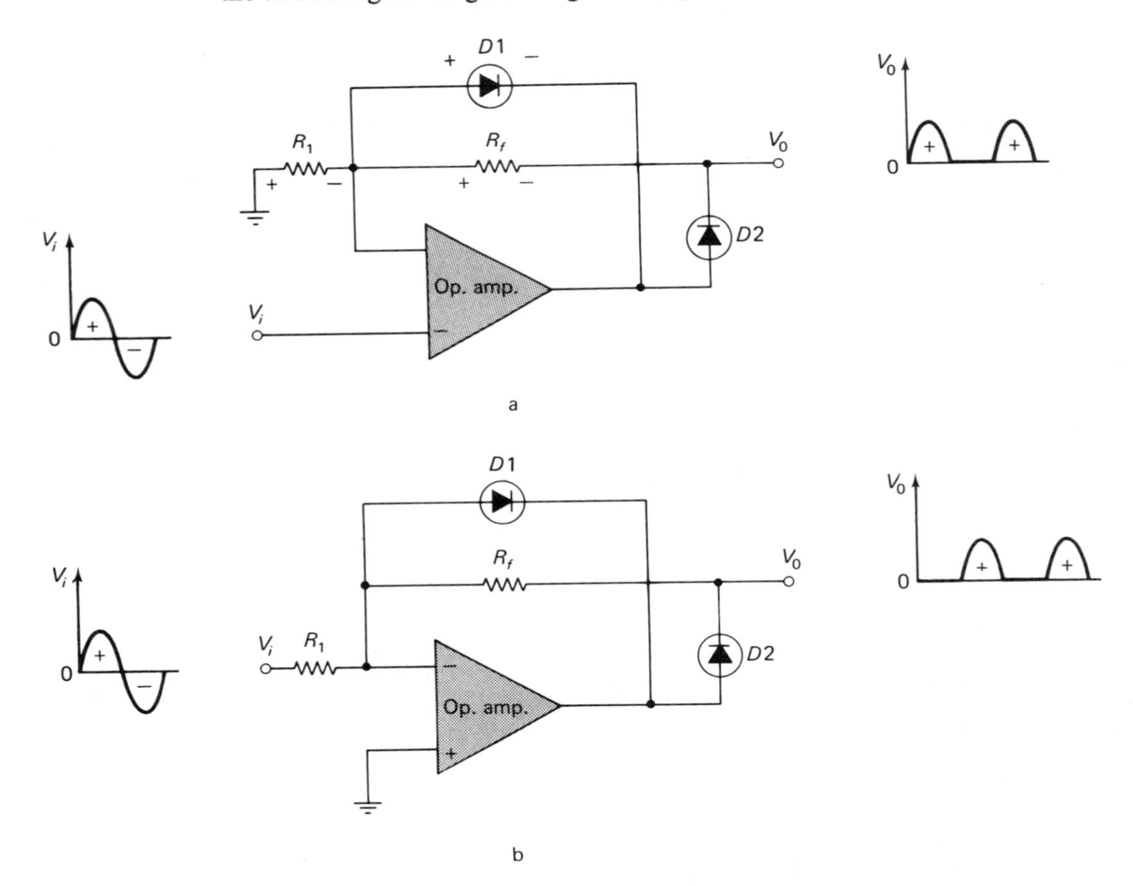

Figure 17-20. Positive-output precision half-wave rectifier using an OP AMP: (a) noninverting mode and (b) inverting mode.

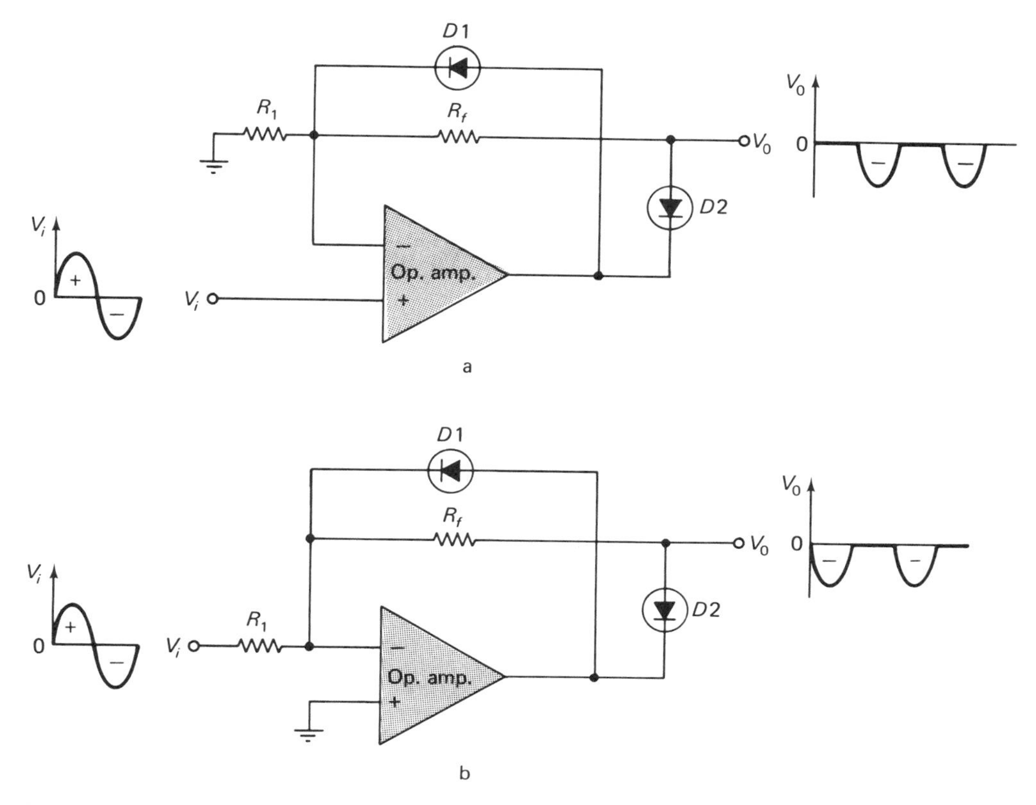

Figure 17-21. Negative-output precision half-wave rectifier using an OP AMP: (a) noninverting mode and (b) inverting mode.

Consider first the circuit in Fig. 17-20(a). The input is applied to the noninverting input-terminal of the OP AMP. When the input is positive, $D1$ is reverse biased and off; $D2$ is forward biased and conducting. The gain of the OP AMP in this case is given by $(R_1+R_f)/R_1$; the output follows the input with the given gain. When the input is negative, $D1$ is forward biased and conducting; $D2$ is reverse biased and off. The gain, with $D1$ effectively shorting the inverting input to the output, is essentially zero. Thus, the output is zero when the input is negative.

In the circuit of Fig. 17-20(b), the input is positive. Then $D1$ is forward biased and conducting; $D2$ is reverse biased and off. With $D1$ effectively shorting the inverting input to the output, the gain is essentially zero and the output is at zero. When the input is negative, $D1$ is off and $D2$ is on. As a result, the gain is given by $-R_f/R_1$. The output then is positive and equal to $-R_fV_i/R_1$, with V_i negative.

Operation of the two precision rectifiers with negative outputs, illustrated in Figs. 17-21(a) and 17-21(b), may be examined in the same manner. Note that the only difference between the circuits of Figs. 17-20 and 17-21 is that both $D1$ and $D2$ are reversed, thus giving a positive output in one case and a negative one in the other.

The main advantage of these precision rectifier circuits is their close approximation to an ideal diode. Many times we have to rectify signals whose amplitude is lower than a volt, and it becomes necessary to use a precision rectifier in place of a diode.

As an example, suppose that the input pulse has an amplitude V of less than 1 V. The precision rectifier can be used together with the RC wave-shaping network, as shown in Fig. 17-22, to give an output which

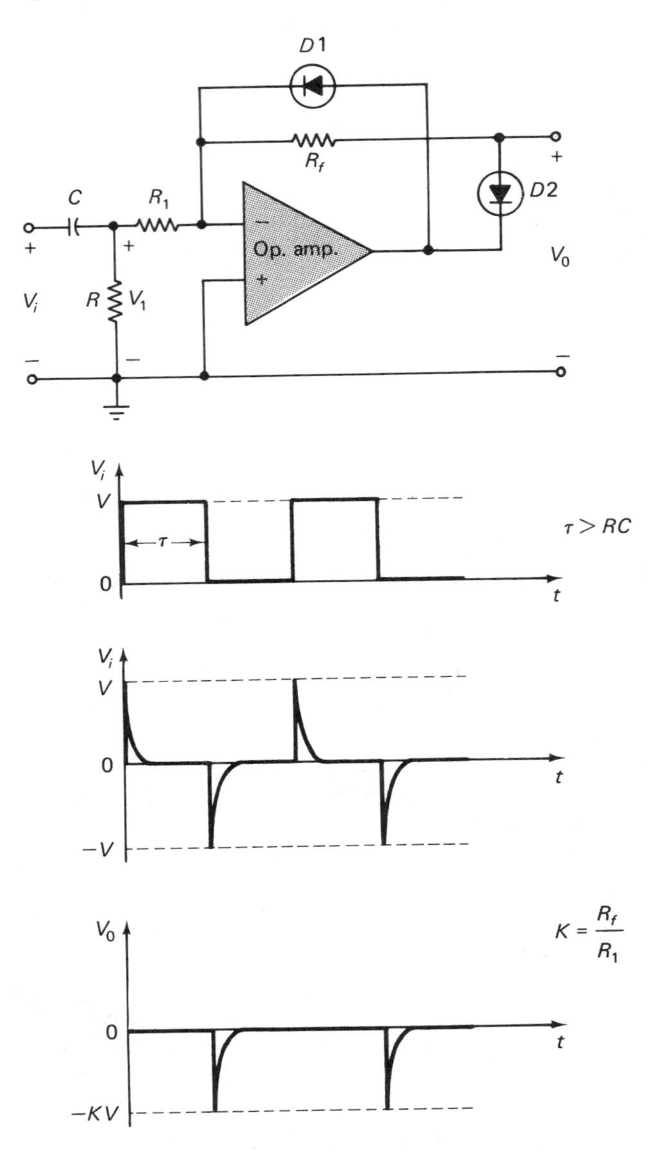

Figure 17-22. Improved spike-forming circuit for generating negative going spikes using a precision half-wave rectifier, and wave-shapes.

contains only a negative going spike. (For a positive going spike at the output, one of the two precision rectifiers shown in Fig. 17-20 can be used with the RC network.) When $\tau \gg RC$, voltage V_1 contains the indicated positive and negative spikes. The precision rectifier then eliminates the positive postion of V_1 and gives just negative going spikes. The amplitude is dependent on the gain set by R_f and R_1.

17.4.2 Peak and Average Detecting Circuits

In some cases, we must detect either the average or peak value of a time-varying signal. If the signal is sufficiently larger than the forward diode voltage, the simple diode detectors shown in Fig. 17-23 may be used. If very low signals are involved, the circuits using precision rectifiers, depicted in Fig. 17-24, should be used.

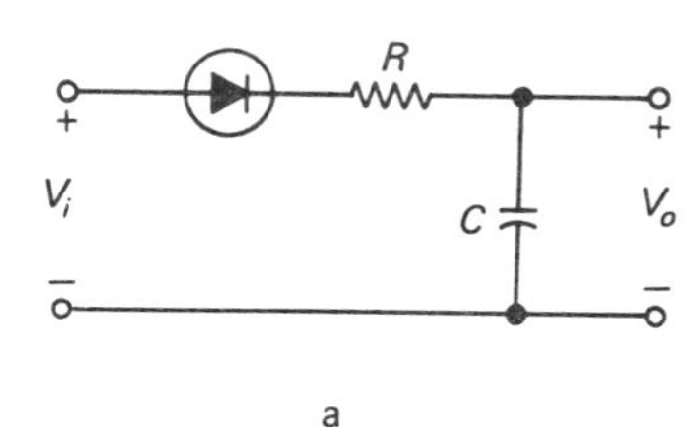

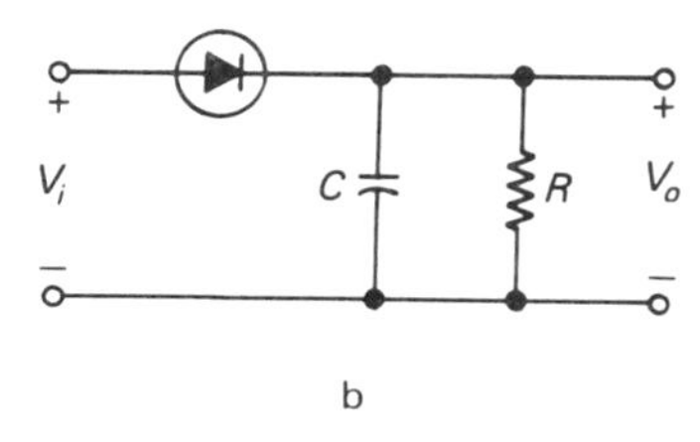

Figure 17-23 Diode detectors: (a) average and (b) peak.

Consider the circuit in Fig. 17-23(a) first. In the beginning there is no charge on the capacitor. When the input is positive, the diode is forward biased and conducts. The capacitor charges up to somewhat below the input voltage. If the input voltage falls below the capacitor voltage (or becomes negative), the diode turns off. The output, therefore, is proportional to the average value of the input signal.

In Fig. 17-23(b), the resistor is placed in parallel with the capacitor, and the circuit acts as a peak detector. Again the capacitor is uncharged at the start. When the input goes positive, the diode conducts and the capacitor charges up quickly through the low diode forward resistance. The output voltage then follows the input. When the input falls below the voltage stored across the capacitor, the diode turns off. The capacitor may discharge somewhat through the parallel resistance (which should be high). In essence, the capacitor voltage, and therefore the output, follows the peak value of the input voltage.

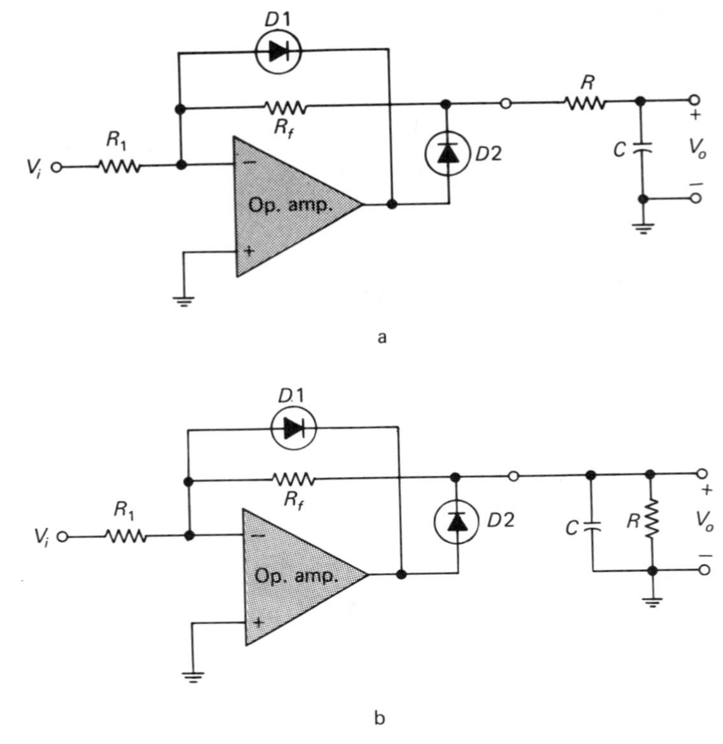

Figure 17-24. Detectors using precision rectifiers: (a) average detector and (b) peak detector.

The operation of the precision detector circuits in Fig. 17-24 is identical to that discussed for the diode circuits. The only exception is that the diodes are replaced with precision rectifiers.

Review Questions

1. What is a clipping circuit? What function does it perform?
2. What is the output of a single-level positive clipping circuit? Discuss in terms of the reference voltage.
3. What is the output of a single-level negative clipping circuit? Discuss in terms of the reference voltage.
4. What is the output of either the positive or negative single-level clipping circuit if the reference voltage is zero?
5. What is the operation of the two-level clipping circuit in Fig. 17-9? Discuss in terms of diode conduction and nonconduction.
6. What is the difference in the operation of the two-level clipping circuits shown in Figs. 17-9 and 17-10?
7. What is the operation of the two-level clipping circuit in Fig. 17-11? What is the output if the two reference voltages are not the same?

8. What is the difference in operation between the two-level clipping circuits shown in Figs. 17-11 and 17-12?
9. What is a clamping circuit? What function does it perform?
10. Why is the name of "dc restorer" given to the circuits in Figs. 17-13 and 17-14?
11. What is the role of the capacitor in the dc restorer circuits of Figs. 17-13 and 17-14?
12. What is a latching circuit? What function does it perform?
13. Explain the function of the latching diode used in the transistor inverter circuit of Fig. 17-15.
14. What function does the spike-forming circuit in Fig. 17-18 perform?
15. What is meant by a *precision* rectifier? How is it different from a regular diode rectifier?
16. Under what conditions must a precision rectifier be used (that is, when is a simple diode rectifier insufficient)?
17. Explain the operation of the peak (an average) detecting circuit.

Problems

1. The input to the clipping circuit of Fig. 17-11 is a sinusoid with peak amplitude of 10 V. With $V=2$ V and $R=1$ kΩ determine and sketch the diode current and output voltage.
2. The same input signal as in Problem 1 is applied to the circuit in Fig. 17-2. If the circuit values are also the same as in Problem 1 determine and sketch the diode current and output voltage.
3. A square wave with a positive peak of 5 V and a negative peak of -10 V is applied to the circuit of Fig. 17-3. The circuit values are: $V=5$ V and $R=100Ω$. Determine and sketch the diode current and the output voltage.
4. A 115 V-rms signal is applied to the circuit in Fig. 17-9. Determine the values of V_1 and V_2 to give an output centered around 60 volts, with a peak-to-peak amplitude of 30 V.
5. Determine the value of R needed in Problem 4 if the maximum diode forward current of 50 mA is not tc be exceeded.
6. We wish to obtain a symmetrical wave with a 5 V peak-to-peak amplitude from the circuit in Fig. 17-11. The input is a 10 V peak-amplitude sine wave. Silicon diodes ($V_{on}=0.6$ V) are used with $R=2$ kΩ. Determine the reference voltages needed.
7. In Problem 6 determine the two diode currents and sketch their waveshapes.
8. Silicon Zener diodes with $V_Z=5.6$ V and $V_{f(\text{on})}=0.5$ V are used in the circuit shown in Fig. 17-12. The input signal is a 20 V peak-amplitude sine wave. Sketch the output waveshape.

9. If $R=50\Omega$ in Problem 8, what is the peak power dissipated in the Zener diode?
10. Sketch the output waveshape if a 115 V-rms sine wave is applied to the circuit in Fig. 17-13.
11. Repeat Problem 10 for the circuit in Fig. 17-14.
12. Determine the logic voltages in the inverter circuit of Fig. 17-15 if: $R_B=10$ kΩ, $R=50$ kΩ, $R_C=200\Omega$, $V_{CC}=20$ V, $V=10$ V, and $\beta>100$. Both the diode and the transistor are silicon.
13. How many inverter circuits identical to that in Problem 12 could be tied to the collector of the inverter in Problem 12 without destroying the logic levels (that is, what is the fan-out of the inverter in Problem 12)?
14. A square wave amplitude of 0.5 V is applied to the input of the circuit shown in Fig. 17-22. If the on and off times of the square wave both equal 1 millisecond, determine the circuit values needed to provide a negative going spike with an amplitude of 5 V at the output.

18

Multivibrators

In this chapter we shall study three basic kinds of multivibrators: bistable, astable, and monostable. In addition, we shall discuss the RS, T, and JK flip-flops. All of these circuits are *sequential* circuits; that is, the operation (or state) of the circuit at any instant of time depends on the prior state of the circuit.

18.1 BISTABLE MULTI-VIBRATORS

The basic memory storage element is called a *bistable multivibrator*, or simply a *binary*. It is shown in Fig. 18-1. This circuit has two possible stable states. In one, the transistor $T1$ is saturated and transistor $T2$ is cut off; in the other, $T2$ is saturated and $T1$ is cut off. (In this chapter the abbreviations for a transistor are T, although the conventional method of labeling the outputs of a binary is with the letter Q.) The circuit is designed so that it is not possible for both $T1$ and $T2$ to be saturated or cut off at the same time.

Let us examine the circuit in detail. If $T1$ is saturated, its base is at approximately 0.7 V and its collector at 0.3 V. The equivalent circuit for $T2$ under these conditions is illustrated in Fig. 18-2. The base of $T2$ sees 0.7 V through R_B and $-V_{BB}$ through R. Thus, the base of $T2$ is reverse biased by $-V_{BB}$, and $T2$ is cut off; that is, it draws no collector current. Were it not for the fact that the R_B of $T1$ is connected to the collector of $T2$, it would be at essentially $+V_{CC}$. However, as shown in Fig. 18-2, $T1$ draws its base current through the collector resistor of $T2$. Thus, the collector of $T2$ is at:

$$V_{C2} = V_{CC} - I_{B1}R_C \tag{18-1}$$

This is the logical 1 voltage. The collector of $T1$ is at its saturation voltage (0.3 V), which corresponds to the logical 0 voltage. If one of the transistors is saturated, therefore, its low collector voltage keeps the other transistor cut off. In turn, the high collector voltage of the cutoff transistor maintains the first transistor in saturation. This state is stable and does not change unless the circuit is somehow disturbed. Note that either transistor

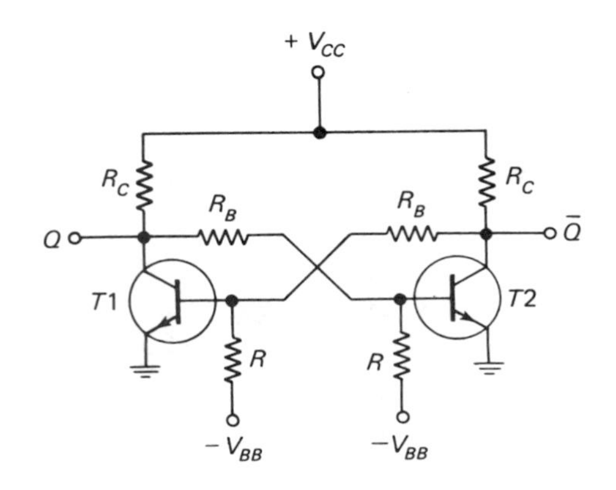

Figure 18-1. Bistable multivibrator (binary).

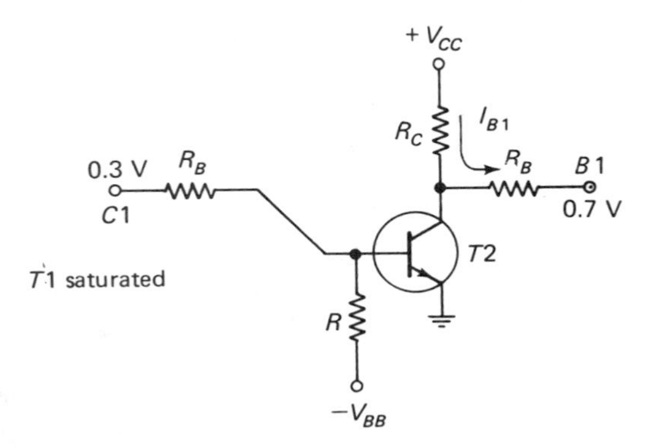

Figure 18-2. Binary with $T1$ saturated.

may be saturated when the other is cut off; a stable state always results. The two collectors are labeled Q and $\bar{Q}$ (not Q), because one is always logically 1 while the other is logically 0.

The ability of the circuit to *store* either a logical 1 or a logical 0 at one of its collectors makes it a basic memory element, or *binary*.

Example 18-1. The binary depicted in Fig. 18-3 uses silicon transistors with $\beta_{\min}=50$. The circuit values are given. We want to show that a stable state is set up with one transistor saturated and one cut off. We also need to determine the logic voltages.

Solution: We start by assuming that $T1$ is saturated and $T2$ cut off. But we have to justify this assumption.

If $T2$ is cut off, its base and collector currents must be zero, as shown. The current I_1 is determined:

$$I_1 = \frac{V_{CC} - V_{BES1}}{R_C + R_B} = \frac{10-0.7}{1+22}\text{ mA} = 0.4\text{ mA}$$

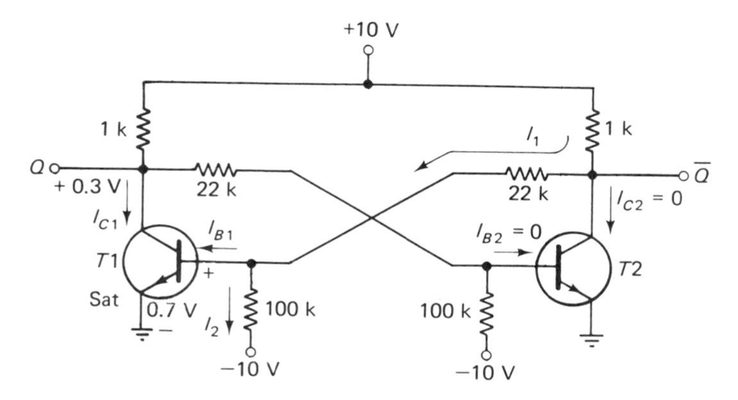

Figure 18-3. Example of a binary (see Example 18-1).

We next determine I_2:

$$I_2 = \frac{V_{BB} + V_{BES1}}{R} = \frac{10+0.7}{100}\text{ mA} = 0.1\text{ mA}$$

The base current of T_1 is found as the difference between I_1 and I_2:

$$I_{B1} = I_1 - I_2 \cong 0.4 - 0.1 \cong 0.3\text{ mA}$$

The collector current is calculated next:

$$I_{C1} = \frac{V_{CC} - V_{CES1}}{R_C} \cong \frac{10-0.3}{1}\text{ mA} \cong 9.7\text{ mA}$$

The fact that T_1 is indeed saturated can be established if we verify that $\beta_{\min} I_{B1}$ is greater than I_{C1}. Because $\beta_{\min} I_{B1} \cong 50(0.3\text{ mA}) \cong 15$ mA and I_{C1} is 9.7 mA, the saturation of $T1$ is assured.

The collector of $T1$ is at the logical 0 voltage. Thus,

$$V(0) \cong 0.3\text{ V}$$

The collector of $T2$ is at the logical 1 voltage. So,

$$V(1) \cong V_{C2} \cong V_{CC} - I_1 R_C \cong 10 - (0.4)1 \cong 9.6\text{ V}.$$

We still need to verify that $T2$ is indeed cut off. The circuit for determining V_{B2} is shown in Fig. 18-4. We first determine the current I:

$$I \cong \frac{0.3-(-10)}{22+100}\text{ mA} \cong 0.084\text{ mA}$$

V_{B2} may now be calculated:

$$V_{B2} = -V_{BB} + IR \cong -10 + (0.084)100 \cong -1.6\text{ V}$$

Because the base of $T2$ is negative, it is cut off.

In the previous example the state $Q=0$ and $\bar{Q}=1$ was established as a stable state. Obviously the symmetry of the circuit allows the other possible state ($Q=1, \bar{Q}=0$) to be established as a stable state in an

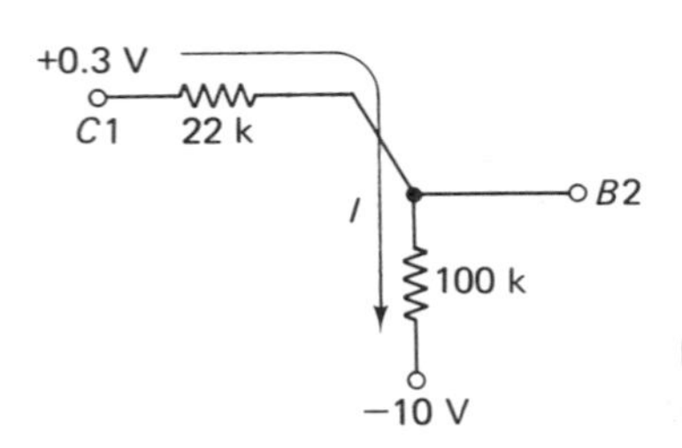

Figure 18-4. Circuit for determining V_{B2} (see Example 18-1).

identical manner. We simply start with the assumption that $T2$ is saturated and $T1$ is cut off.

Any multivibrator circuit discussed here may include compensating capacitors called *speed-up capacitors*, as shown in Fig. 18-5. The purpose of these capacitors is to compensate for the base-emitter capacitance of the transistors. If the time constant of the speed-up capacitor and R_B is adjusted to match that of the base-emitter junction, when a high is applied, the transistor turns on faster than it would without the speed-up capacitors.

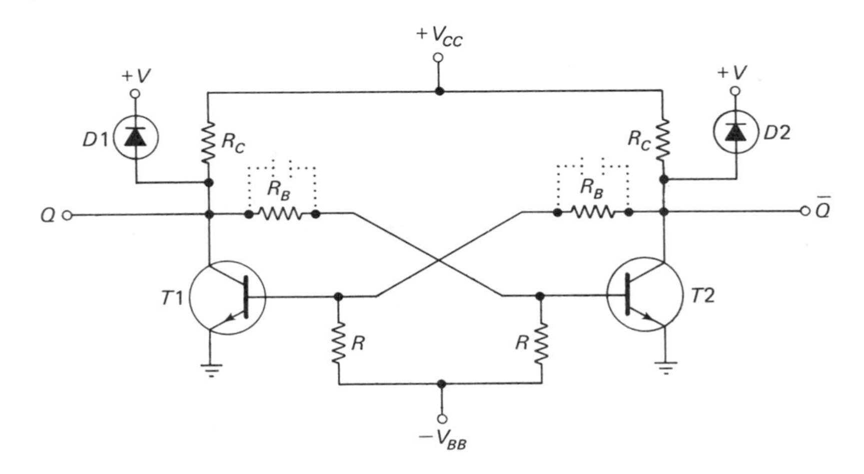

Figure 18-5. Binary with latching diodes.

The binary or bistable multivibrator (MV) circuit derives its name from the fact that it has two distinct stable states. An alternate bistable MV circuit with latching diodes (see section 17.3.2) is shown in Fig. 18-5. The operation of this circuit is the same as that discussed for the circuit of Fig. 18-1, except that the logical 1 voltage for the circuit with latching diodes is $V = 0.7$ (assuming that V is less than V_{CC}). The advantage of this circuit is that the logical 1 voltage is maintained even when other gates are connected to the output which is at the logical 1 level.

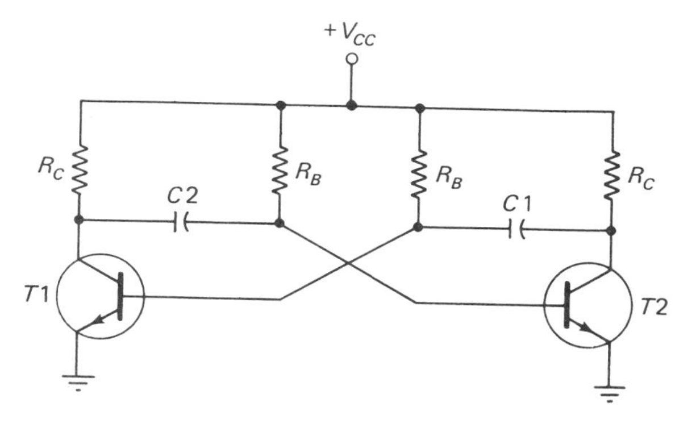

Figure 18-6. Astable multivibrator.

18.2 ASTABLE MULTI-VIBRATORS

The *astable* MV circuit (illustrated in Fig. 18-6), as the name indicates, has *no* stable states. It oscillates between the two states at a rate determined by the capacitor values.

Let us start the analysis at the point in time when $T1$ is cut off and $T2$, saturated. Capacitor $C2$ must have almost the full supply voltage across it in the direction indicated in Fig. 18-7. $T2$ is conducting and drawing a large collector current, made up of the sum of I_x and I_y. Current I_X is trying to charge up $C1$ to $+V_{CC}$. However, when the voltage (V_1) across $C1$ reaches about 0.4 V, the base of $T1$ is sufficiently positive $(0.4+V_{CES2})$ to turn on. With $T1$ now saturated, the collector of $T1$ is at 0.3 V, and the net voltage being applied to the base of $T2$ is negative because of the voltage across $C2$. Thus, $T2$ turns off. The conditions in the circuit at this instant ($T1$ turning on, $T2$ turning off) are: $V_1=0.4$ V; $V_2=V_{CC}-0.7$; $V_{B1}=0.7$ V; $V_{C1}=0.3$ V; and $V_{B2}=-(V_2-0.3)=-(V_{CC}-0.4\text{ V})$. Although $T2$ is cut off, its collector does not rise to V_{CC} immediately.

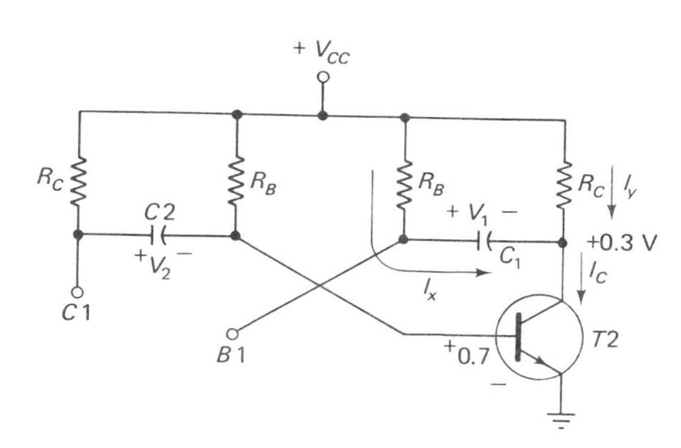

Figure 18-7. Astable multivibrator with $T1$ off and $T2$ on.

With $T1$ drawing a large collector current, part of it through the collector resistor and part of it through $C2$, the collector of $T2$ rises to $+V_{CC}$ as capacitor $C1$ discharges to 0 and charges up in the opposite direction. This charging of $C1$ is through R_C. Because R_C is relatively small, charging takes very little time. Capacitor $C2$ is discharging through R_B. R_B is relatively large, so this discharge takes longer than the charging of $C1$. V_2 eventually reaches 0 and starts becoming negative. (Polarity is indicated in Fig. 18-7.) As soon as V_2 becomes negative enough (at about 0.4 V), the base of $T2$ begins to conduct because its voltage has reached about 0.7 V. As soon as $T2$ turns on, its collector falls to 0.3 V. Because V_1 is negative and almost equal to the supply voltage, the base of $T1$ is now negative, and $T1$ therefore turns off. The conditions in the circuit at this time are: $V_1 = -(V_{CC} - 0.4)$; $V_2 = -0.4$ V; $V_{B1} = 0.7$ V; $V_{C1} = 0.3$ V; and $V_{B2} = -(V_{CC} - 0.4)$. The collector of $T1$ rises rapidly to $+V_{CC}$ as $C2$ charges up through R_C to $V_2 = V_{CC} - 0.7$. These conditions are the same as when the analysis was started, so we have completed one full cycle. The collector and base waveshapes for both transistors are shown in Fig. 18-8.

The timing of the circuit may be determined from the charging of $C1$ and $C2$. The length of time that transistor $T1$ is off is determined by the charging of $C1$ through R_B. The length of time that $T2$ is off is determined by the charging of $C2$ through R_B. The *period* is the sum of the times that $T1$ and $T2$ are off.

In each case the capacitor voltage is trying to change by almost $2V_{CC}$ – from $-V_{CC}$ to $+V_{CC}$. The time constant is given by the product of R_B and the respective capacitors.

Assuming $C1$ and $C2$ are equal and labeled by C, we can write the equation for the charging of the capacitor:

$$V_{\text{capacitor}} = (V_{\text{final}} - V_{\text{initial}})(1 - e^{-t/\tau}) + V_{\text{initial}} \tag{18-2}$$

In this case, the initial voltage is $-V_{CC}$; the final voltage is $+V_{CC}$; and the voltage attained during the actual charging is approximately zero. We can then find the length of time required to change the voltage across the capacitor by V_{CC}.

$$0 \cong [V_{CC} - (-V_{CC})](1 - e^{-t_1/R_B C}) + (-V_{CC}) \tag{18-3}$$

Solving for t_1 in the preceding equation, we obtain:

$$t_1 = (R_B C)\ln 2 = 0.69\ R_B C \tag{18-4}$$

With equal capacitors, the period T is just twice t_1. Thus, the frequency of the output may be calculated as the inverse of the period:

$$f = \frac{1}{T} \cong \frac{1}{2(0.69)R_B C} \cong \frac{0.721}{R_B C} \tag{18-5}$$

The output may be taken at either collector and is a symmetrical square wave with both capacitors the same. The circuit is very useful as a square-wave generator for digital circuits. In such applications it is also called a *clock*.

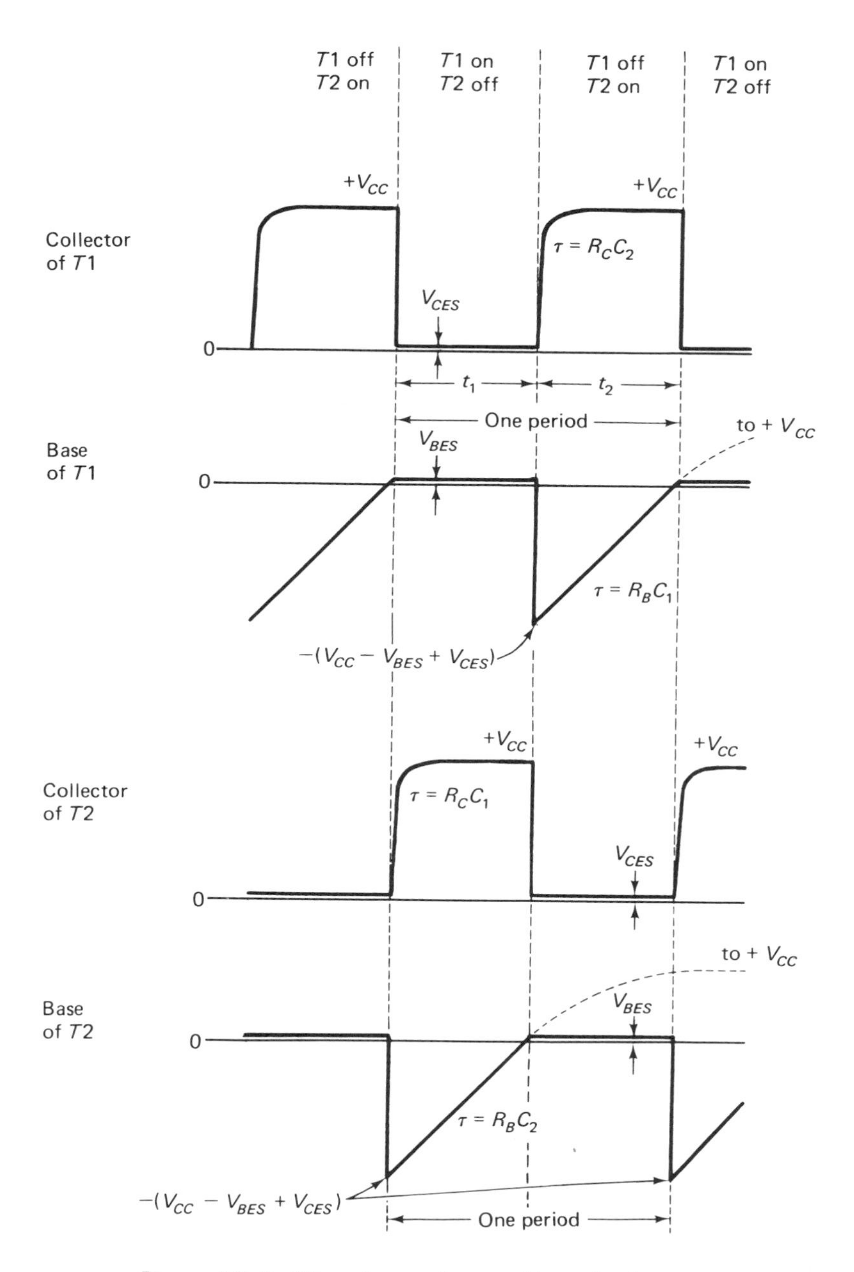

Figure 18-8. Waveshapes for astable multivibrator.

Note two important points. First, the clock frequency is independent of the supply voltage. Secondly, the output is not a perfect square wave but has a specific *rise time* which is not zero. The output frequency is independent of the supply voltage to the extent that transistor saturation voltages (V_{CES} and V_{BES}) are negligibly small compared to the supply voltage. We can calculate the rise time from the time constant for either collector to return from saturation to $+V_{CC}$. The charging path during this time is through R_C, so the time constant which gives the rise time is $R_C C$. The rise time is approximately given by:

$$t_r \cong 5R_C C \tag{18-6}$$

We can illustrate the operation of the circuit with the following example.

Example 18-2. The astable multivibrator illustrated in Fig. 18-6 has the following circuit components: $R_B = 33$ kΩ; $R_C = 2.2$ kΩ; and $C_1 = C_2 = C =$ 1000 pF. Determine the frequency and sketch the waveshapes if $V_{CC} =$ 10 V.

Solution: The frequency is calculated from Eq. (18-5):

$$f \cong \frac{0.721}{(33 \times 10^3)(10^{-9})} \text{ Hz} \cong 22 \text{ kHz}$$

The rise time for the collector waveshape is calculated from Eq. (18-6):

$$t_r \cong 5(2.2 \times 10^3)(10^{-9}) \text{ sec} \cong 11\ \mu\text{s}$$

The collector waveshape for these results is shown in Fig. 18-9.

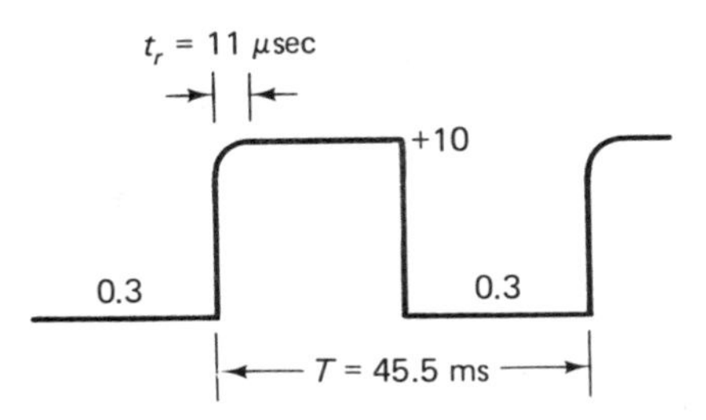

Figure 18-9. Collector waveshape for Example 18-2.

In some applications we may wish to have a clock which is not symmetrical; that is, we may require it to have on and off times which are unequal for the transistor where the output is being taken. We can provide for this uneven function either by using different capacitors $C1$ and $C2$ or by using different base resistors. Note the following example.

Example 18-3. We want to design a clock which has the waveshape shown in Fig. 18-10.

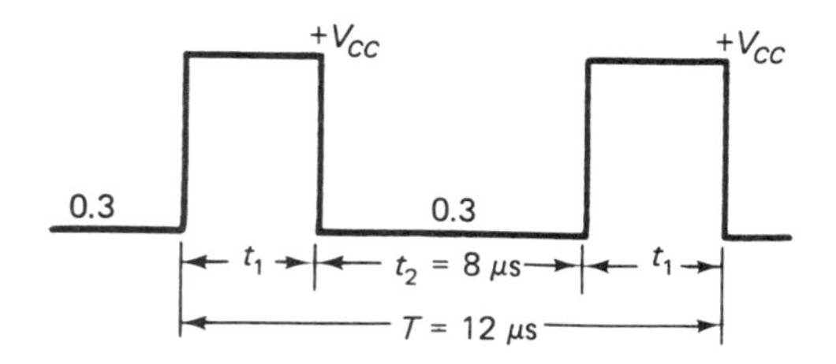

Figure 18-10. Waveshape for Example 18-3.

Solution: For a period of 12 μs, the frequency is approximately 83 kHz. If we take the output at the collector of $T2$ in Fig. 18-6, then:

$$t_1 = 0.69 R_B C_2 \quad \text{and} \quad t_2 = 0.69 R_B C_1$$

Let us assume R_B of, say, 22 kΩ. Thus,

$$C_2 = \frac{4 \times 10^{-6}}{0.69(22 \times 10^3)} \text{F} \cong 260 \text{ pF}$$

$$C_1 = \frac{8 \times 10^{-6}}{0.69(22 \times 10^3)} \text{F} \cong 520 \text{ pF}$$

The design is finished by choosing R_C to insure that the transistors will saturate when they should.

Gated Astable MV. In many applications it is necessary to have some logic signal control the clock, so that the clock will give an output

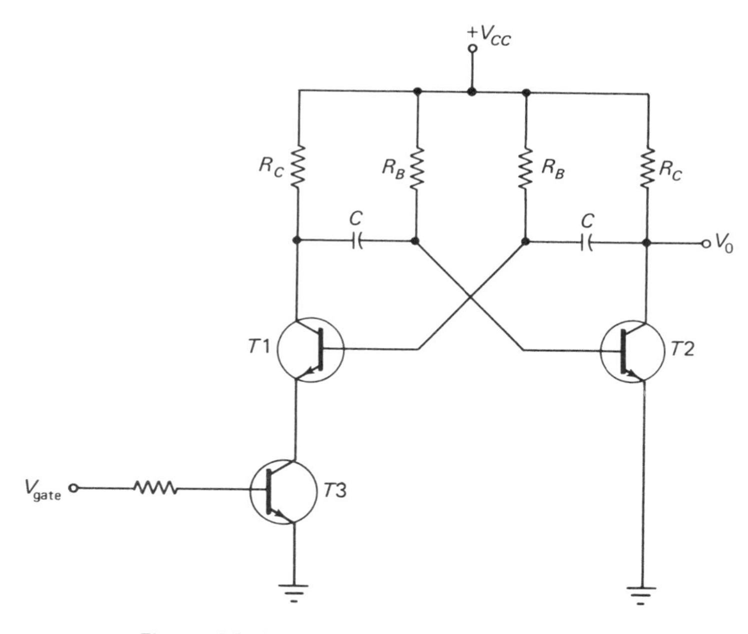

Figure 18-11. Gated astable multivibrator.

only when it is desired. We add, therefore, another transistor to the astable MV, as shown in Fig. 18-11. Circuit operation is identical to that of the astable MV so long as the gating signal is high and $T3$ is on. When the gating signal is low, $T3$ is off and prevents $T1$ from conducting when it would otherwise. The waveshape for the gated astable MV is depicted in Fig. 18-12. We determine the timing for this circuit the same way as for the astable MV circuit without the gate.

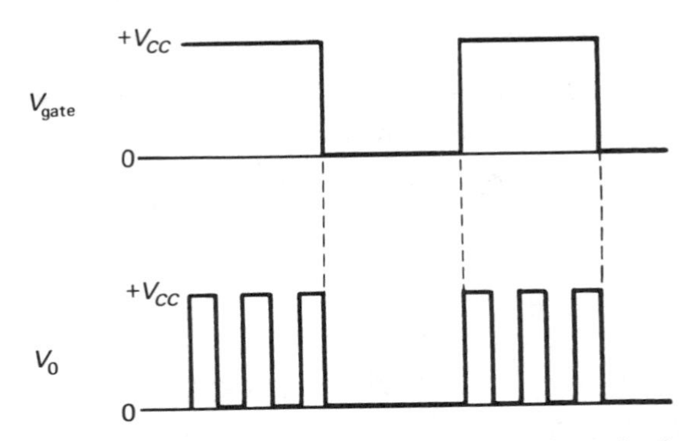

Figure 18-12. Waveshapes for gated astable multivibrator.

18.3 MONOSTABLE MULTIVIBRATORS (ONE-SHOT)

The third type of MV is shown in Fig. 18-13. It is a cross between the bistable and astable MVs and uses a single capacitor. It has only one stable state: $T2$ saturated and $T1$ cut off. Under these conditions, the capacitor voltage is almost V_{CC}. $T1$ is maintained in cutoff by the negative supply to its base, while $T2$ is maintained in saturation by the current through R_{B2}.

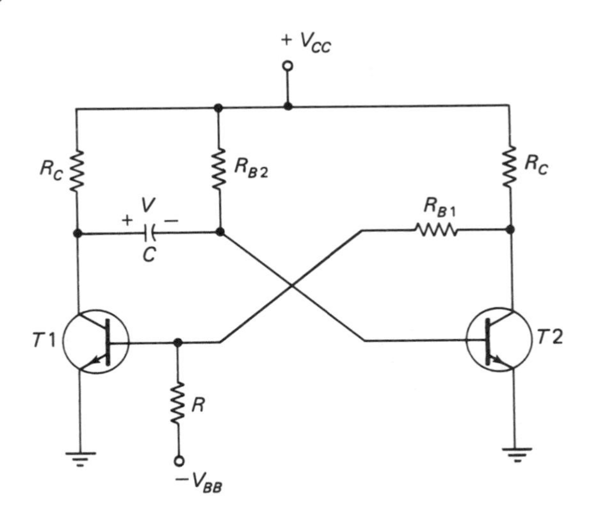

Figure 18-13. Monostable multivibrator (ONE-SHOT).

We can make the circuit change state (i.e., *trigger*) if $T2$ is turned off momentarily by applying a negative going spike to the base. As soon as this voltage turns off $T2$, its collector rises to V_{CC}. The base of $T1$, therefore, becomes positive, turning on $T1$. With the collector of $T1$ now at 0.3 V and the voltage across the capacitor as shown, the base of $T2$ is negative. Thus, a negative pulse of short duration (a spike) causes the circuit to change states. $T1$ now draws collector current through R_C and the $R_{B2}C$ branch. The voltage across C is then falling and trying to reach $-V_{CC}$. When it becomes just slightly negative (0.4 V), the base of $T2$ is made positive enough for $T2$ to saturate, turning off $T1$. Once again the circuit is in its stable state. It will remain stable until another triggering signal is applied. The waveshapes are shown in Fig. 18-14.

The timing of the ONE-SHOT may be determined in the same manner as that of the astable multivibrator. The delay, defined as the time that $T2$ is off, is the same as the half-period of the astable. Thus,

$$t_d = 0.69 R_{B2} C \tag{18-7}$$

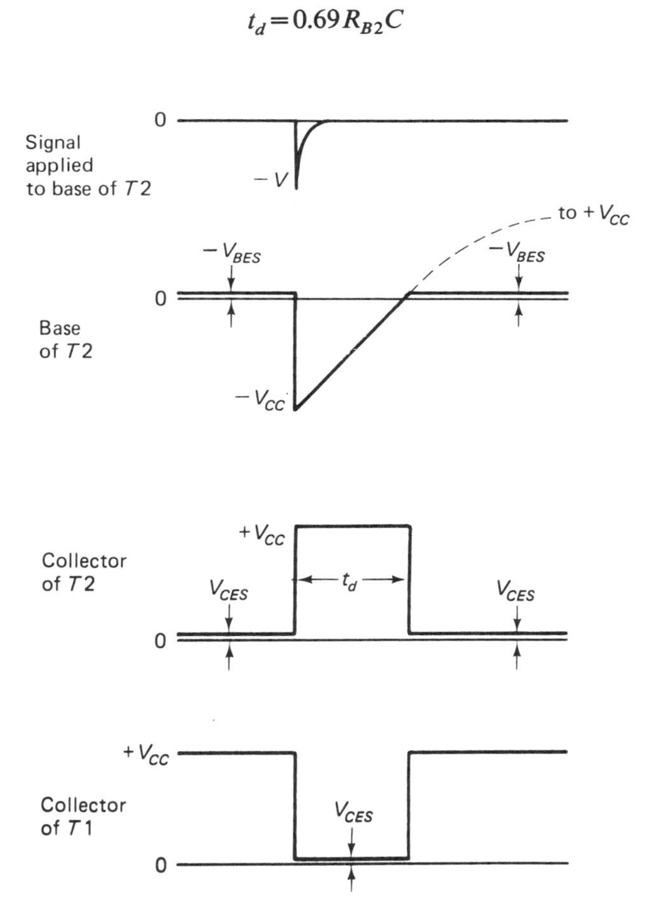

Figure 18-14. Waveshapes for monostable multivibrator.

Triggering of the ONE-SHOT when the input is a square wave may be accomplished in many different ways. One method is illustrated in Fig. 18-15, where the square-wave input is shaped into a negative spike by the RC network and the diode (see section 17.4.1). The circuit waveshapes are shown in Fig. 18-16.

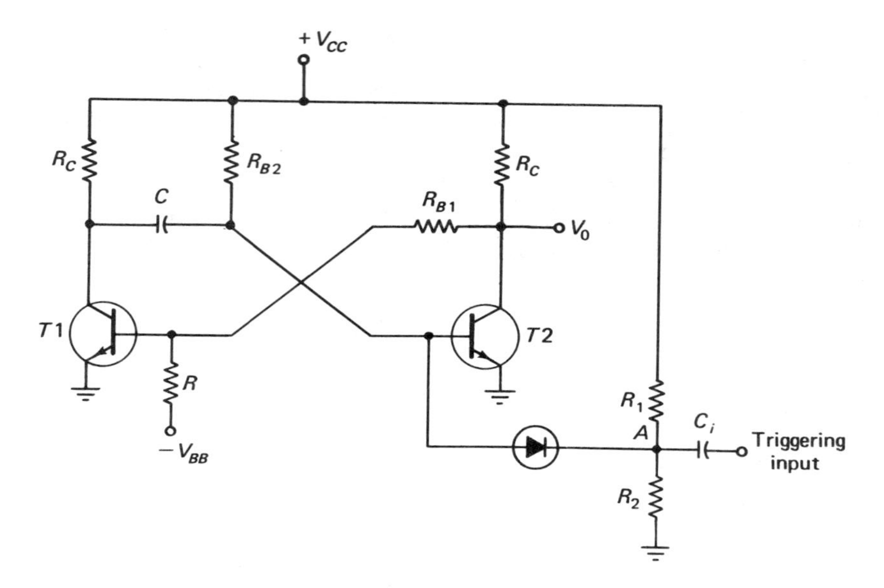

Figure 18-15. Circuit for triggering a ONE-SHOT.

In Fig. 18-15, if the input is indicated as low, the voltage divider, made up of R_1 and R_2, establishes a voltage at point A that is somewhat higher than the on voltage of the $T2$ base. In this way, we make certain that the diode is reverse biased and off. When the input goes high, the diode is still off. But when the input makes the transistion from high to low, a negative going pulse is impressed on the base of $T2$, thus turning it off. The circuit then takes over and generates the waveshapes shown in Fig. 18-16.

The monostable MV, or ONE-SHOT, generates a delay in the waveshape which is controlled by the timing of the device. The delay is given by Eq. (18-7) and can be seen in Fig. 18-16 as the difference in time from when the input waveshape goes to zero to the time when the output waveshape (collector of $T2$) goes to zero. Observe that the input waveshape must have a repetition rate that is sufficiently lower than the delay of the ONE-SHOT in order for the circuit to recover (i.e., reach its stable state) before the next pulse occurs.

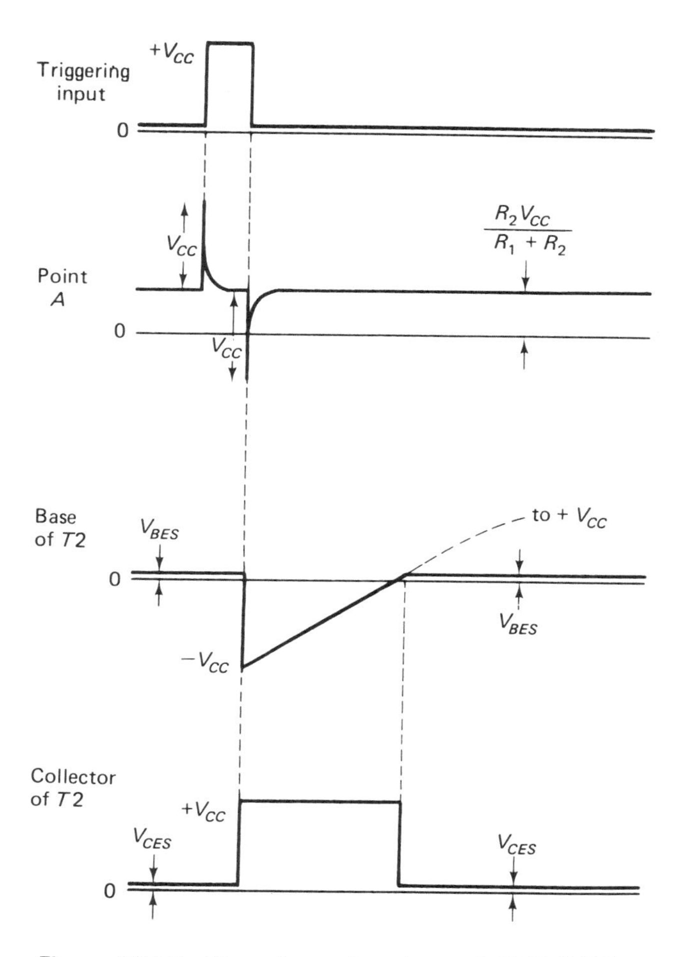

Figure 18-16. Waveshapes for triggered ONE-SHOT.

18.4 FLIP-FLOPS (FFS)

The basic storage element is a bistable MV because it can store either a logical 1 or 0. However, the bistable MV circuit has no mechanism by which the content (either 0 or 1) can be changed. The general name for circuits that can store binary information (either 0 or 1) is *flip-flop* (FF). The name is derived from the action of these circuits which can change their state or flip flop.

An example of an FF is shown in Fig. 18-17. We can consider this circuit in a number of ways: (1) It is an interconnection of four inverter circuits. (2) We can say that it is an interconnection of two NOR gates (RTL). We can look upon it as a bistable binary with two inverters added to provide a means of input. All of these views are acceptable.

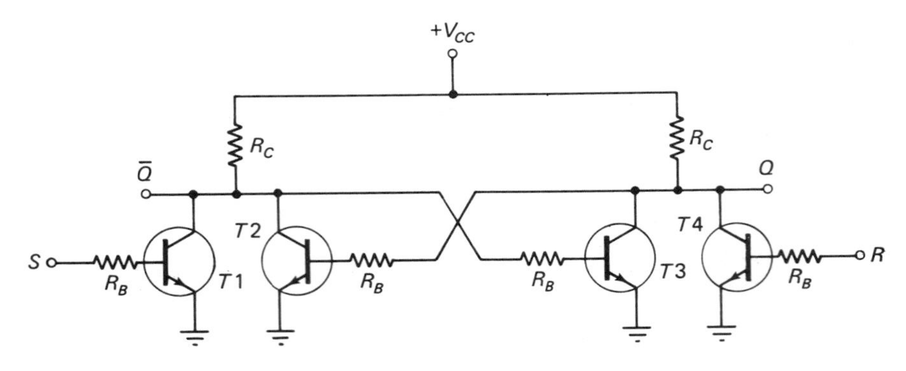

Figure 18-17. Discrete component RTL RS flip-flop.

RS Flip-Flop. You can understand the operation of the FF illustrated in Fig. 18-17 by considering the circuit as an interconnection of two NOR gates, as shown in Fig. 18-18. Transistors $T2$ and $T3$ comprise an astable MV, or binary. As such, one transistor must be saturated while the other one must be cut off. Let us assume for the time being that $T2$ is saturated, with $T1$ cut off. Consider the action of the circuit if one of the two inputs is high. (Both inputs cannot be high at the same time.) For example, if the input marked S is high, $T1$ saturates. However, the state of the outputs is not changed, because $\overline{Q}$ was already 0 and Q was 1. Thus, when the state of the FF is $Q=1$ and $\overline{Q}=0$, the S input has no effect on the output.

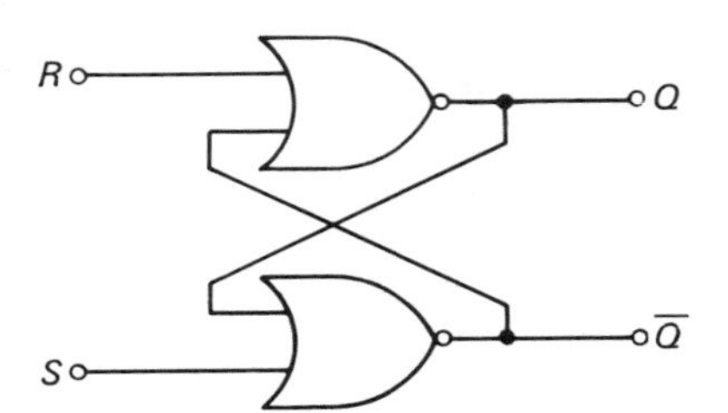

Figure 18-18. Two interconnected NOR gates form an RS flip-flop.

Next suppose that a high input is applied at the terminal marked R with the same state of $Q=1$ and $\overline{Q}=0$. As a result, $T4$ saturates and pulls the Q output to zero. This change is coupled to $T2$ removing the high input, and so $T2$ cuts off. The state of the FF has thus been changed. We say that a high input on the R terminal *resets* the Q output to zero; at the same time, it forces the $\overline{Q}$ output to 1. Another way of saying the same thing is that the R input enters a logical 0 into the FF at the Q output. Note that this change is permanent. It remains even after the high input has been removed, because the high output at $\overline{Q}$ is forcing $T3$ to be saturated even when the R input is removed.

If the same inputs are applied to the circuit starting in the opposite state, the results are the same. Suppose that the Q output is low and the $\overline{Q}$ output high. When a high input is applied at the S terminal, $T1$ saturates and forces the $\overline{Q}$ output to be low. This change is coupled to $T3$, and it now cuts off. As a result, the Q output becomes high because the inputs to both $T3$ and $T4$ are now low. A high input to the S terminal, therefore, *sets* the Q output to 1. This change of state is also permanent; even after the high input to S is removed, the high output at Q is maintained by the high at $\overline{Q}$ coupled to $T3$. If, in the same state, $(Q=0, \overline{Q}=1)$ we apply a high input to the R terminal, no change of state results; saturating $T4$ makes no difference when $T3$ is already saturated.

The circuit is called an *RS flip-flop*. The two input terminals set the FF to 1 or reset it to 0, depending on which input is applied. The operation of the RS-FF is summarized as follows. Regardless of the prior state of the FF, when a 1 is applied to the set terminal (S), the output at Q becomes a 1; when a 1 is applied to the reset terminal (R), the output at Q becomes a 0. This fact is summarized also in the truth table given in Table 18-1 where the original state of the output is labeled Q and the state after an input has been applied is labeled Q_+. Note that the condition of applying both a set and reset is not allowed because we could not be setting and resetting the FF at the same time.

Table 18-1. Truth Table for an RS-FF

Q	R	S	Q_+	
0	0	0	Q	no change
0	0	1	1	←SET
0	1	0	Q	no change
1	0	0	Q	no change
1	0	1	Q	no change
1	1	0	0	←RESET

T Flip-Flop. Let us examine the circuit shown in Fig. 18-19. This is another form of FF called a *T*, or *toggle, flip-flop*. The output of this FF changes state (i.e., toggles) every time the input makes a transition from high to low. Suppose that the Q output is high ($\overline{Q}$ low). Transistor $T1$ is saturated; $T2$ is cut off. With the input maintained high, capacitor $C1$ charges up to almost V_{CC} but $C2$ remains essentially uncharged. When the input goes low, the negative going pulse is applied to the base of $T1$ through the conducting diode $D1$. At the same time, $T2$ does not sense the input transition because its diode $D2$ is reverse biased. The heavily conducting diode $D1$ draws current through the collector and base resistors, thus starving the base current from $T1$. As a result, $T1$ turns off. Its collector, having gone high in turn, forces $T1$ into saturation. While $D1$ was conducting, it discharged capacitor $C1$; once $T2$ saturates, $C2$ charges up to essentially $-V_{CC}$. After the input goes negative, the situation is the reverse of what it was before.

The diodes in the circuit are called *steering diodes*. In effect, they steer the input negative going transition of the input waveshape to the

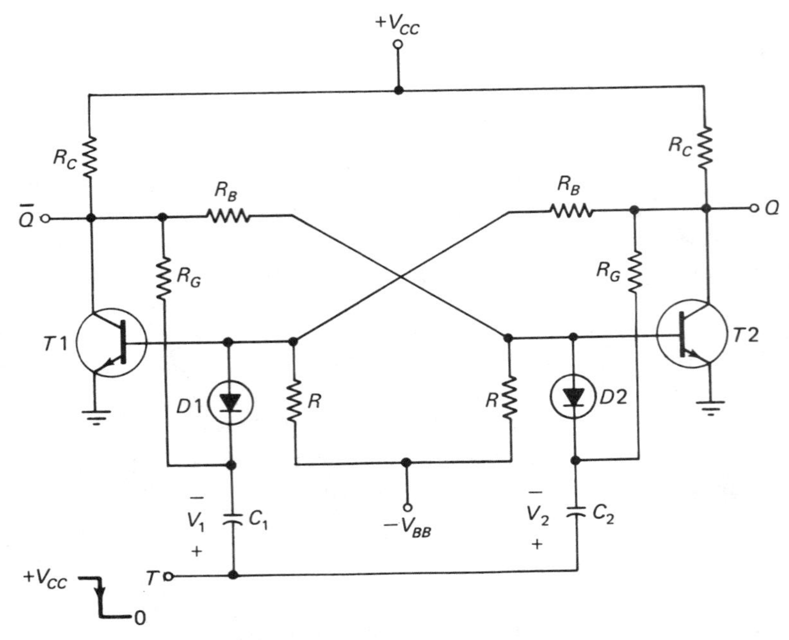

Figure 18-19. Toggle flip-flop.

transistor which is conducting so that it may be turned off. The truth table for the T-FF is given in Table 18-2.

Table 18-2. Truth Table for a T-FF

Q	T	Q_+
0	0	Q
0	1	$\bar{Q}$
1	0	Q
1	1	$\bar{Q}$

The T-FF can be used as a frequency divider. This application is illustrated in Fig. 18-20, where the input is a square wave and the output is also a square wave at one-half the frequency of the input. This waveshape results because the T-FF changes state only when the input makes a transition from a 1 to a 0. If the output of one T-FF is connected to the input of another T-FF, the eventual output is at a frequency of one-quarter of the original input. In this manner, we can divide a given frequency by any multiple of 2 through cascading the appropriate number of T-FFs.

These two filp-flops are both asynchronous; there is no control provided for external timing of the inputs. A flip-flop which has an additional input for timing pulses (clock) is called a *synchronous flip-flop*. An example of such a flip-flop is given in Fig. 18-21.

The flip-flop containing four interconnected NAND gates, as illustrated in Fig. 18-21, is called a *JK-FF*. It contains the capabilities of

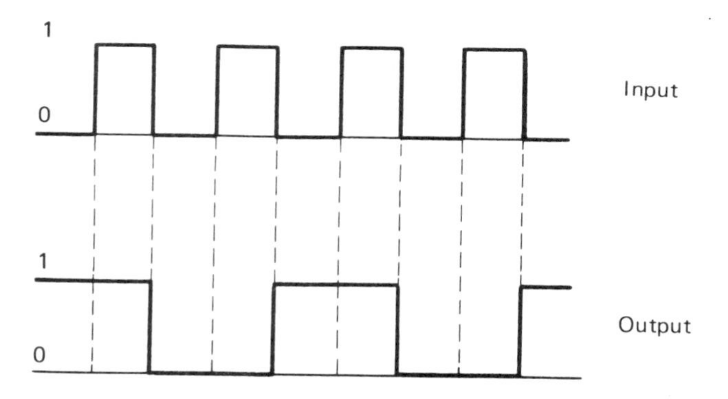

Figure 18-20. Waveshapes for a T-FF.

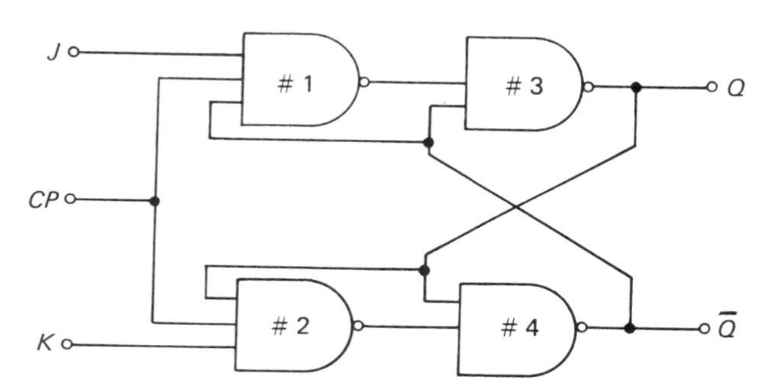

Figure 18-21. Basic JK-FF.

both the RS-FFs and T-FFs, with additional feature of synchronous or clocked operation.

The function of the clock (CP) is to either enable or disable the inputs. As we saw in Chap. 16, the output of a NAND gate is low only when all of its inputs are high. Let us then consider the JK-FF with outputs $Q=1$ and $\overline{Q}=0$. Because $\overline{Q}$ is one of the inputs to gate 1, the output of gate 1 cannot go high no matter what J or CP is. One of the inputs to gate 3, therefore, is always high when the other input ($\overline{Q}$) is low. The Q output will remain 1 no matter what is applied to J or CP. However, when both CP and K are high, all three inputs to gate 2 are high, and its output goes low. This low output is applied to the input of gate 4, forcing its output ($\overline{Q}$) to be high. This high output is coupled to the input of gate 3, which has both inputs high, and the Q output is forced to be low. Thus, the $K=1$ condition, *together* with $CP=1$, causes the FF to reset (i.e., go to $Q=0$).

If we make a similar analysis with the FF in the starting state of $Q=0$, we find that the $J=1$ input, *together* with $CP=1$, causes the FF to set to 1 (forces $Q=1$). In either of these cases, the J or K input has no effect *unless* the clock is also high.

However, consider the case when both J and K are high together, with CP high as well. Gates 1 and 2 then take the place of steering diodes and cause the output of the gate that has all 1 inputs to go low. As a result, the output Q is forced to change state. If both J and K inputs are applied at the same time, *together* with $CP=1$, the FF has to change state. In summary: The J input (together with $CP=1$) has the same effect as the S input of an RS-FF; the K input (together with $CP=1$) has the same effect as the R input of an RS-FF; and the J and K inputs at the same time (together with $CP=1$) have the same effect as the T input of a T-FF.

Our analysis of the JK-FF, although basically accurate, neglects one very important factor: the delay in the signal as it passes through a gate. Because the output in the JK-FF is connected to an input (Q to the input of gate 2 and $\overline{Q}$ to the input of gate 1), there is a problem in the operation known as *race-around*; it refers to the fact that an output may try to change twice or even more for a single input. A way of eliminating this problem is to use a *dual-rank* or *master-slave* JK-FF.

The functional block diagram for a master-slave JK-FF is depicted in Fig. 18-22. Operation of this FF is very similar to that of the JK-FF described earlier. However, when $CP1$ is high and enables the J and K inputs, $CP2$ is maintained low to disable the inputs of the slave FF (gates 5 and 6). The information is entered into the master FF. Then $CP1$ disables the J and K inputs while $CP2$ goes high to enable the information to be entered into the slave FF. Typically, there may be an inverter connected between $CP1$ and $CP2$ to insure that both cannot be high at the same time.

You can understand the operation from a single clock if you study the detail of the clocking waveshape shown in Fig. 18-23. When $CP1$ is at

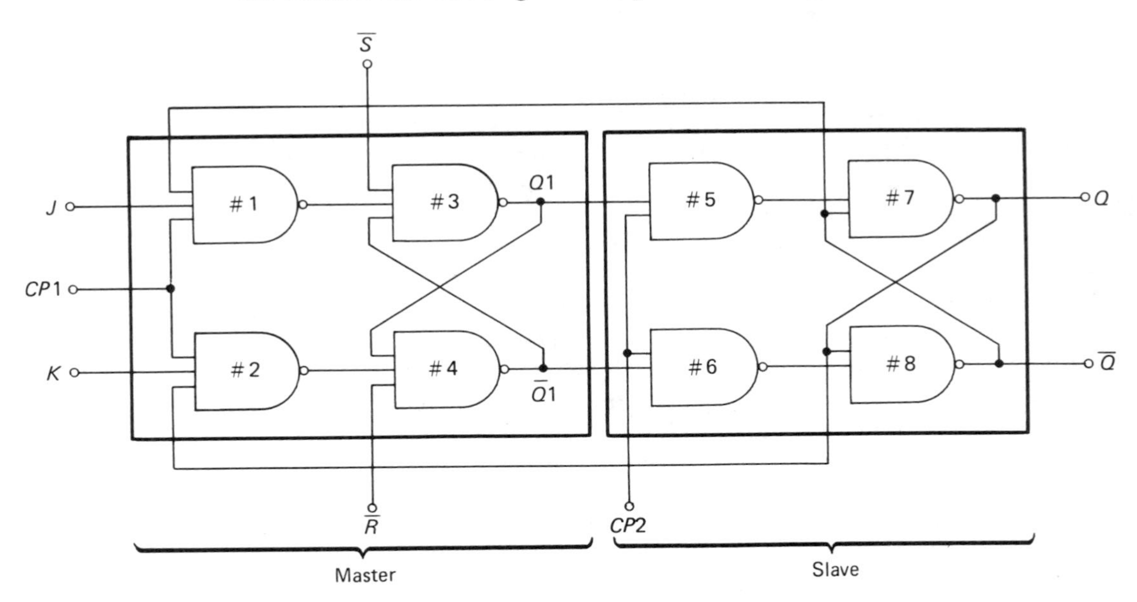

Figure 18-22. Master-slave clocked JK-FF.

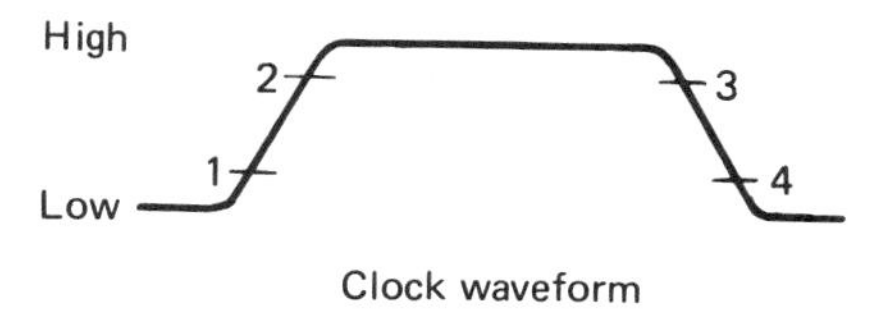

Figure 18-23. Detail of clock waveform ($CP1$) for a JK-FF.

point 1, the slave FF is disabled. When $CP1$ reaches point 2, that master FF is enabled and its inputs take effect. When $CP1$ gets to point 3, the inputs to the master FF are disabled. Finally when it gets to point 4, the slave FF inputs are enabled and the information in the master FF is transferrred into the slave FF. This dual clocking or master-slave principle prevents the race-around condition.

Let us consider now the master-slave JK-FF of Fig. 18-22 with an inverter connected between $CP1$ and $CP2$ and in the starting state of $Q1=0$. These features are shown in Fig. 18-24. Prior to interval 1, $CP1$ is low, so $CP2$ is high and Q is the same as $Q1$. Also prior to interval 1, because $CP1$ is low, no change in the FF results if either or both J and K are high. During time 1, $CP1$ and J are high, causing $Q1$ to set to 1. $CP2$ is low; therefore, this change is not transmitted to Q. During time 2, $CP1$ is low, thus disabling the J and K inputs. $CP2$ is now high, so that the information in $Q1$ is transferred to Q. During time 3, although $CP1$ is high, all the inputs (J and K) are low, so no change in either $Q1$ or Q results. During time 4, although K is high, its effect is disabled by the fact that $CP1$ is low. Again no change in either $Q1$ or Q occurs. During time 5, $CP1$ and K are both high, thus causing $Q1$ to reset to 0. Note that this change is not transmitted to Q because $CP2$ is low. During time 6, $CP1$ is low, thus preventing any change in $Q1$, but now $CP2$ is high, causing the information in $Q1$ to be transmitted to Q. During time 7, $CP1$ and both J and K are high, causing the state of $Q1$ to toggle (in this case, go to 1 because it was in 0). At the same time, $CP2$ is low and the change is not transmitted to Q. During time 8, $CP1$ is low and disables the J and K inputs. However, $CP2$ is high and so the information in $Q1$ is once again transmitted to Q. After time 8, if no further inputs are applied while $CP1$ is high, $Q1$ and Q are in the same state. We can now see the effect of this dual-clock system. The waveshapes at $Q1$ (the master FF) and at Q (the slave FF) are the same, except that the Q waveshape is *delayed* by one duration of the clock pulse.

Typical IC JK-FFs have additional inputs which are *asynchronous*; that is, they do not depend on the clock being high in order to perform their function. Two such inputs are illustrated in Fig. 18-22. When the input at the $\overline{S}$ terminal goes *low*, the FF is set to 1, no matter what the state of the clock. This 1 is transferred to the Q output when the clock goes low ($CP2$ high). Because the normal state for this input is a 1, it has no effect on the FF. The input labeled $\overline{R}$ is also normally high. When it goes

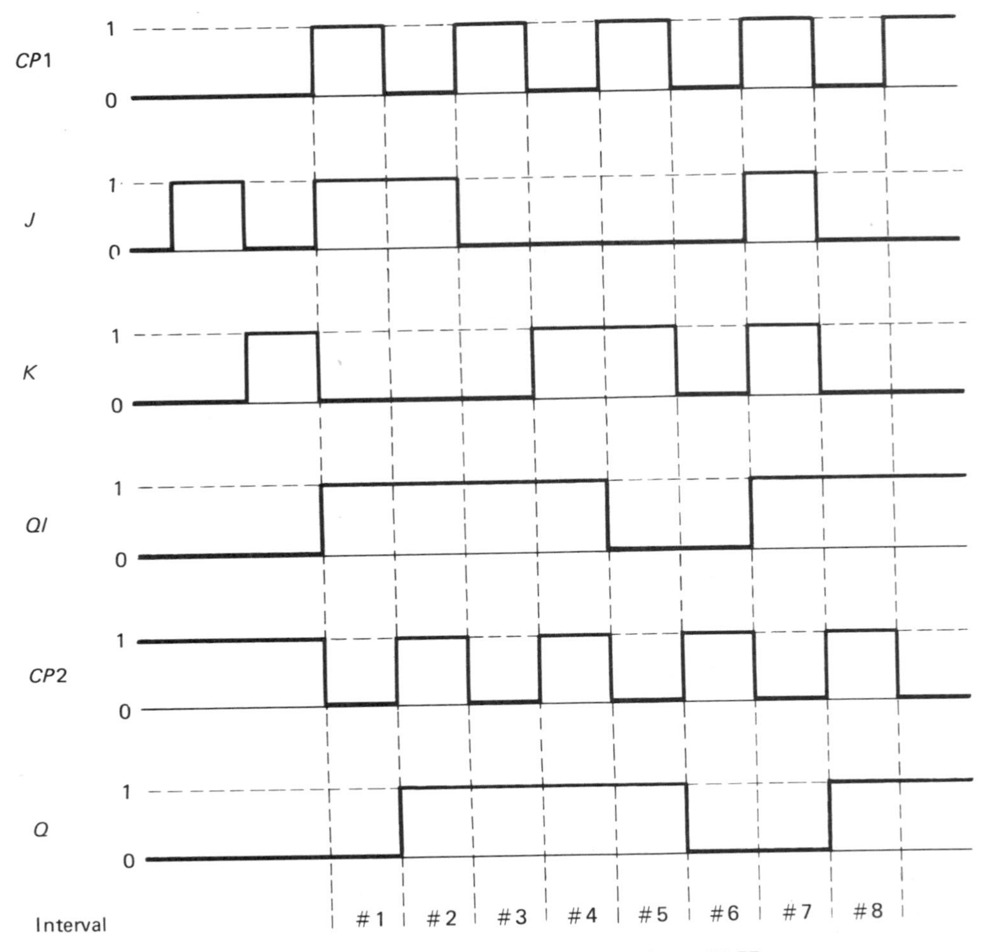

Figure 18-24. Timing in a master-slave JK-FF.

low, the master FF ($Q1$) is reset to 0, regardless of the state of the clock. The 0 is transferred on the next negative transition of the clock (when $CP2$ goes high).

The performance of all the FFs is summarized in Fig. 18-25. In particular, you should realize the flexibility of the JK-FF. For example, the J input performs the same operation as the S input (so long as the clock is high) of the RS-FF; the K input performs the same operation as the R input of the RS-FF. Thus, the JK-FF may be used instead of an RS-FF. We also saw that if J and K are both high, the FF toggles. The JK-FF, therefore, may be made into a T-FF by interconnecting the J and K inputs, as indicated in Fig. 18-26.

If an inverter gate is connected between the J and K inputs, as shown in Fig 18-27, a D-FF is formed. This FF has an output which is the same as the input. Because of the clocking, however, the output is *delayed* by one

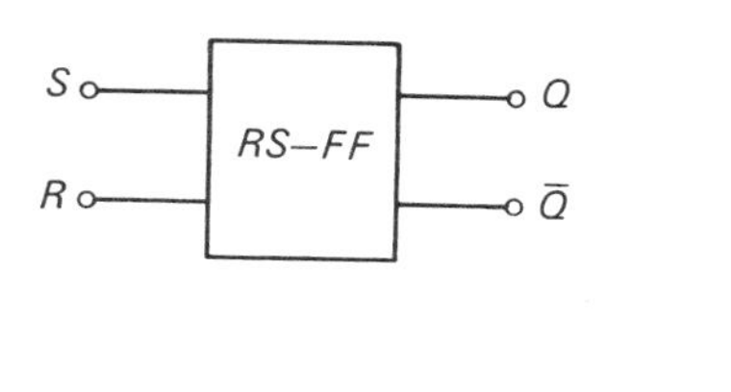

S	R	Q+
0	0	Q
0	1	0
1	0	1

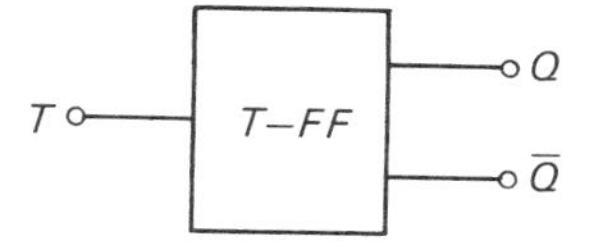

T	Q+
0	Q
1	$\bar{Q}$

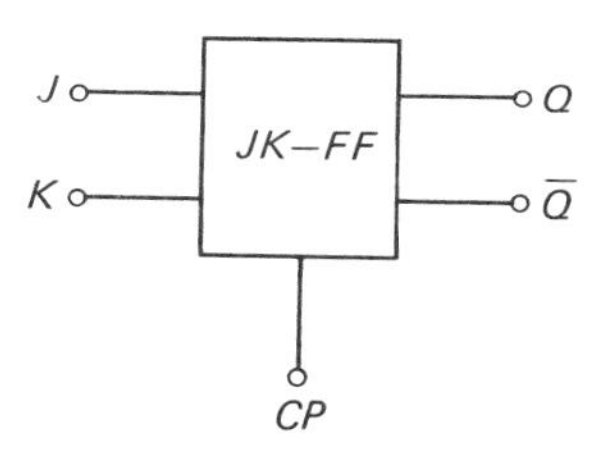

J	K	CP	Q+
0	0	0	Q
0	1	1	0
1	0	1	1
1	1	1	$\bar{Q}$

Figure 18-25. Summary of FF performance.

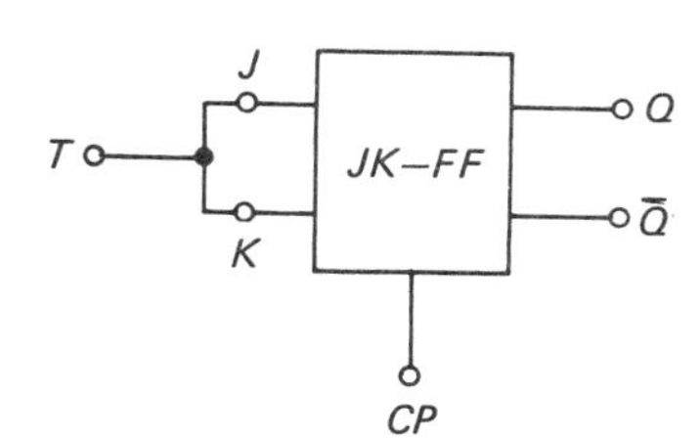

Figure 18-26. Interconnection of a JK-FF to form a clocked T-FF.

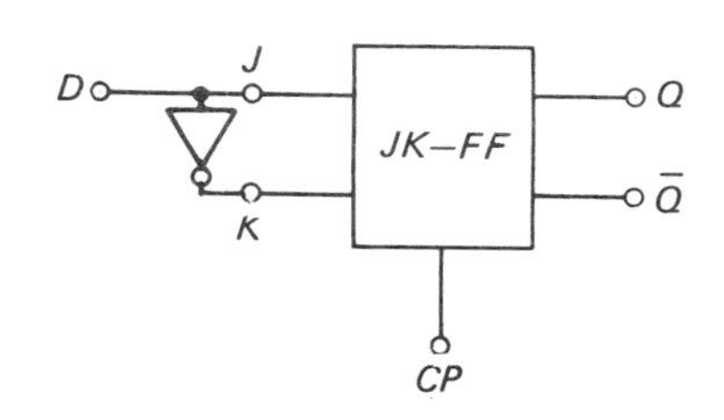

D	(J)	(K)	Q	
0	0	1	0	CP = 1
1	1	0	1	

Figure 18-27. Connection of a JK-FF to form a D-FF.

duration of the clock. It is available in IC form with two D-FFs in a single package (SN7474) and has important applications as a delay element and latch in digital circuits and systems.

18.5 IC ONE-SHOTS

The ONE-SHOT, or monostable MV, is an important digital building block. There are two types, both with IC form: a single, either positively or negatively triggered, ONE-SHOT (SN74121) and a dual ONE-SHOT (SN74123). The block diagrams for these devices are shown in Figs. 18-28 and 18-29, respectively, together with the external timing components that allow the user to obtain the desired delay.

In the circuit shown in Fig. 18-28, the output may be taken at either Q or $\overline{Q}$. The input possibilities are as follows. (1) With the B input high, the ONE-SHOT is triggered when either or both of the A inputs make a transition from high to low. (2) With either or both of the A inputs low, the ONE-SHOT will be triggered when the B input makes a transition from low to high. The timing of the circuit is accomplished with the external resistor and capacitor:

$$t_d = 0.69\ R_T C_T \qquad (18\text{-}8)$$

Thus, the delay time may be adjusted by changing either the resistance or the capacitance.

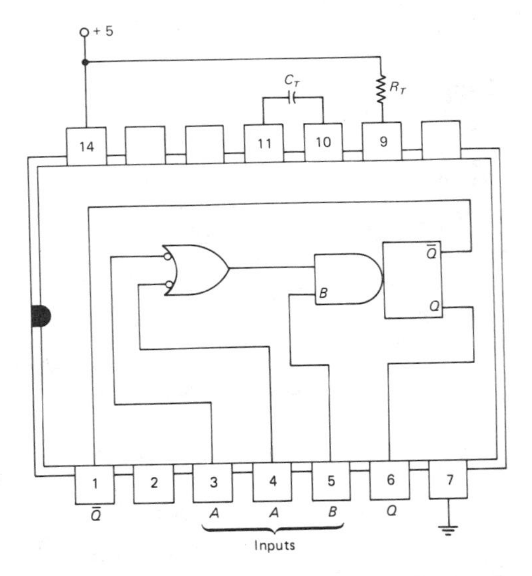

Figure 18-28. An IC ONE-SHOT (SN74121).

The dual ONE-SHOT depicted in Fig. 18-29 operates in much the same way as the single ONE-SHOT. The major difference is the lack of two A inputs and the addition of a clear input. The delay is given by:

$$t_d = 0.32\ R_T C_T \qquad (18\text{-}9)$$

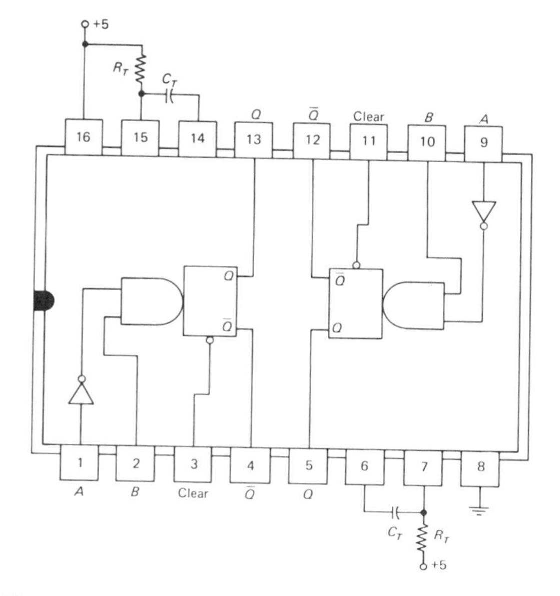

Figure 18-29. A dual IC ONE-SHOT (SN74123).

Two ONE-SHOTs may be interconnected together to provide a square-wave output. Such a circuit using a single IC is shown in Fig. 18-30. It uses a single timing resistor with two capacitors. The period of the square-wave output (which may be taken at either Q or $\bar{Q}$ terminal) is the sum of the two delays provided by the two ONE-SHOTs. Thus, the square wave may be adjusted to have unequal on and off times ($C_{T1} \neq C_{T2}$), or it may be adjusted to have a symmetrical output ($C_{T1} = C_{T2}$).

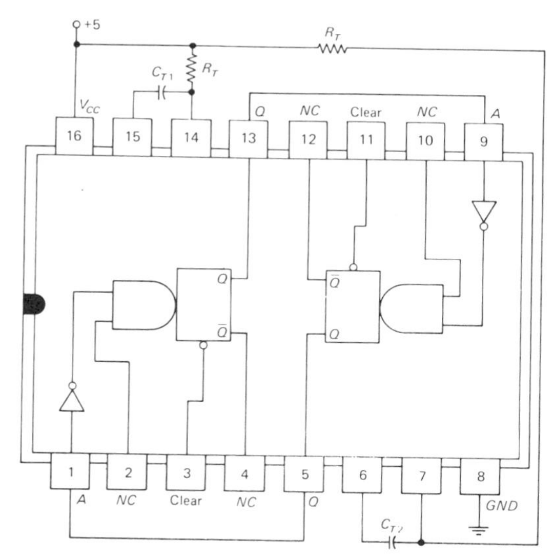

Figure 18-30. Dual ONE-SHOT (SN74123) interconnected as a clock (astable multivibrator).

18.6 THE TIMER

The 555 IC, which may be numbered differently depending on the manufacturer (SN72555, LM555, etc.), is an extremely versatile timing building-block. It incorporates the functions shown in the block-diagram of Fig. 18-31 in an eight-pin dual-in-line package. The schematic is shown in Fig. 18-32. Comparator 1 is formed by transistors $Q1$ through $Q4$; comparator 2 by *PNP* transistors $Q7$ through $Q10$ (transistors $Q5$, $Q6$ and $Q12$, $Q13$ are active loads for the respective comparators). The FF is formed by transistors $Q17$ through $Q20$—the reset input is to the base of $Q17$, the set to the base of $Q18$. Transistor $Q23$ is a split-load phase inverter which provides the necessary signal splitting to drive the totempole output stage comprised of $Q24$, $Q26$, $Q27$, and $Q28$. Output current (with the output high) is provided by the Darlington connection of $Q27$ and $Q28$; with the output low, output current is sunk by $Q24$ and $Q26$. Either sourcing or sinking, the maximum output current is 200 mA. External reset is provided by $Q25$. When the output is low, the discharge transistor, $Q14$, is on.

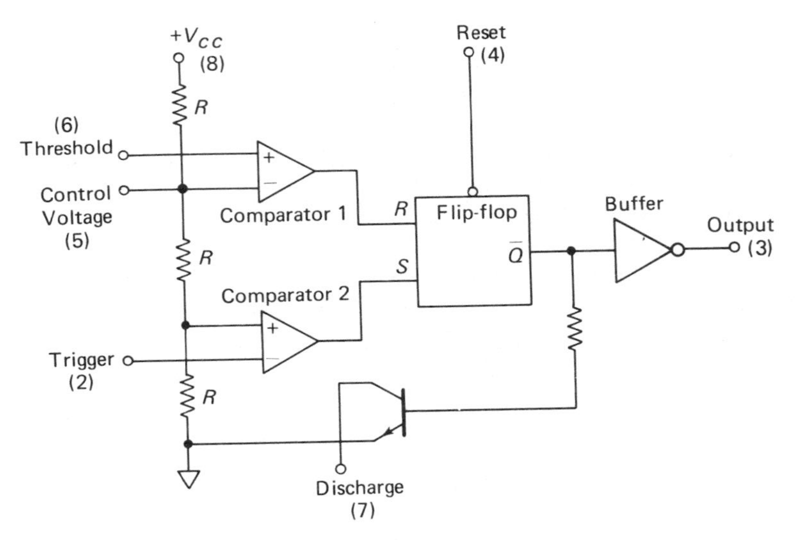

Figure 18-31. 555 Timer block diagram

The comparator trip-points are established by the voltage divider of R_3, R_4, and R_5. Although these three resistors may not be exactly 5 k, they are fabricated at the same time and will be matched (probably to better than 2%).

The 555 may be operated in many different modes. Consider first the monostable configuration shown in Fig. 18-33. Timing is provided by the external *RC* combination; the undisturbed state is output low, $\overline{Q}$ high, the discharge transistor on, and the external capacitor at approximately ground. Pin 5, the modulation input, is bypassed to ground with a capacitor to prevent pick-up of stray signals. Triggering is accomplished by bringing the trigger input (pin 2) low. Normally, pin 2 is maintained high

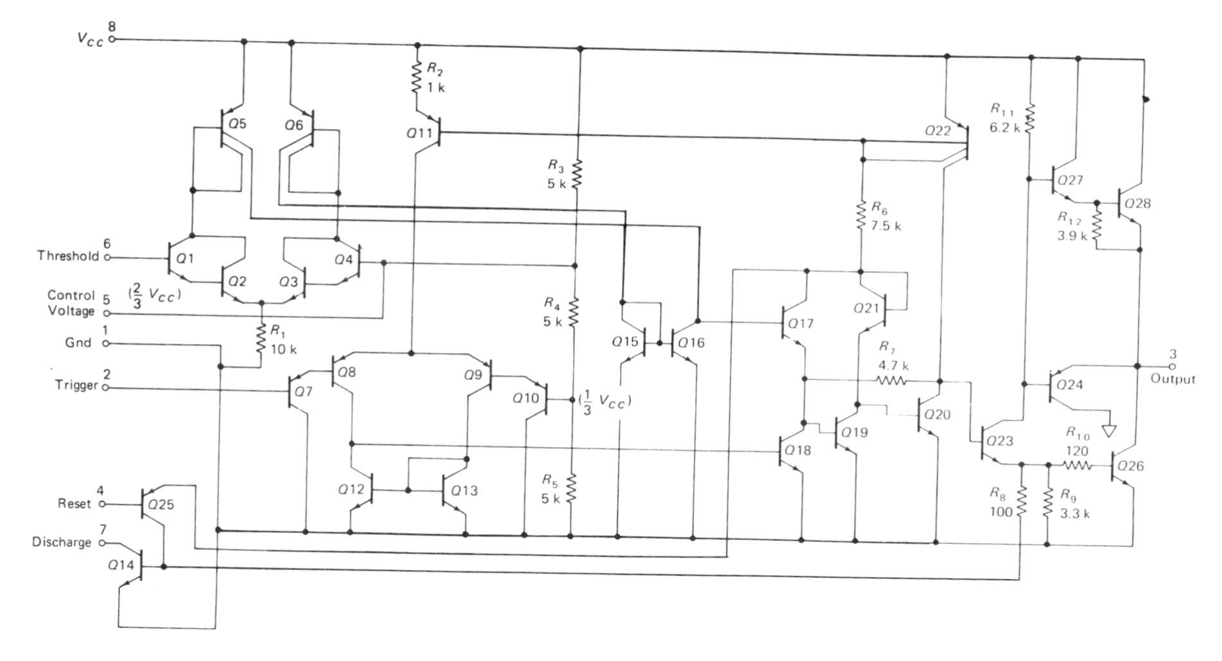

Figure 18-32. 555 Timer schematic

(near V_{CC}). When the trip-point for comparator 2 (see Fig. 18-31) is exceeded, that is, when pin 2 becomes more negative than $1/3\ V_{CC}$, the FF becomes set, $\overline{Q}$ goes low, the discharge transistor turns off, and the output goes high. At this point, the capacitor begins to charge through R toward V_{CC}. T, the duration of the output pulse, is determined by how long it takes the capacitor voltage to reach the trip-point of comparator 1. With an initial voltage of zero charging towards V_{CC} and a voltage of $2/3\,V_{CC}$ at

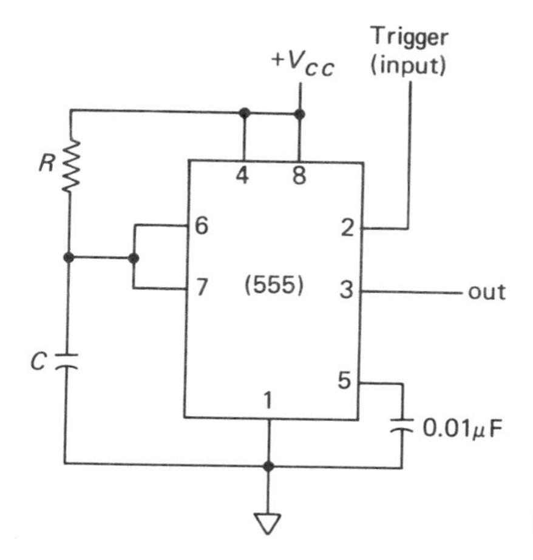

Figure 18-33. 555 One-shot

time T, we have the following expression for the voltage across the capacitor:

$$\frac{2}{3} V_{CC} = V_{CC}(1 - e^{-T/RC}) + \frac{1}{3} V_{CC} \tag{18-10}$$

Note that V_{CC} drops out of the timing equation to give a pulse duration which is not a function of the supply voltage.

Solving for T, we obtain the pulse duration:

$$T = RC \ln 2 = 0.693 RC \tag{18-11}$$

A time T after the one-shot has been triggered, the capacitor voltage has reached $2/3\ V_{CC}$, the FF is reset and the output is again low, with $\overline{Q}$ high. The discharge transistor turns on, and abruptly discharges the capacitor. The circuit has then returned to its original state.

The timing waveshapes for the 555 used as a one-shot are shown in Fig. 18-34. Note that the Darlington pair in the ouput pull-up circuit provides approximately 1.7 V reduction below V_{CC} in the output high-state voltage.

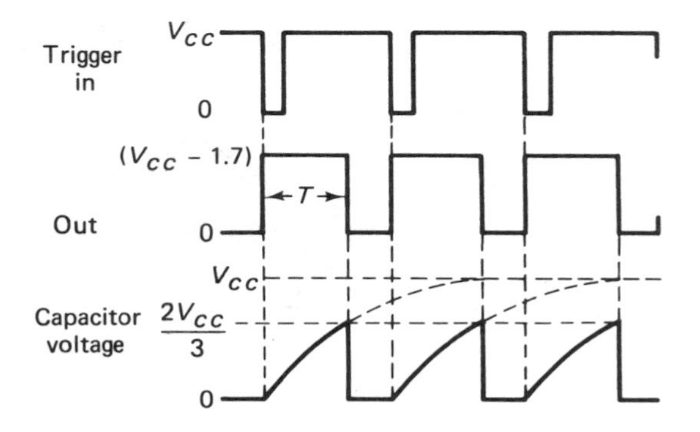

Figure 18-34. One-shot waveshapes

The trigger signal should be short, that is, pin 2 must be brought below $1/3\,V_{CC}$ only long enough to cause the FF to be set (100 ns should be sufficient). The minimum reliable pulse width attainable is limited by the delays of the comparators, the FF, and the output buffer, and should be at least 10 μs. The maximum pulse width is dictated by the quality of the capacitor, that is, the leakage resistance of the capacitor. As a general rule, the timing resistance should be smaller than the capacitor leakage resistance by a factor of at least 100. In addition, the timing resistor cannot be so large as to provide a charging current which is comparable to the threshold (or for the astable, the trigger) current, which is typically 0.5 μA. Therefore, the maximum pulse width is practically limited to 10 to 100 s. For extremely large delays (pulse widths), the output can be extended by the use of counters. Such an arrangement is provided by the XR2240, which, in addition to having a timer similar to the 555, has an eight-bit counter on one chip.

For normal one-shot applications, the external reset line, pin 4, is maintained high, as shown in Fig. 18-33. However, it is possible to terminate the pulse at any time prior to T simply by strobing pin 4 low. This external reset adds flexibility and in some applications eliminates the need for additional logic circuits.

The basic astable connection of the 555 is shown in Fig. 18-35. Here both the trigger input and the threshold input are tied together to sense the capacitor voltage. Operation is as follows. Capacitor charging occurs through both resistors R_1 and R_2; the capacitor is discharged through R_2. The lower limit of the capacitor voltage is $1/3\,V_{CC}$, established by the trip-point of comparator 2; the upper limit of the capacitor voltage is $2/3\,V_{CC}$, established by the trip-point of comparator 1. If we label the time the output is high as T_1 with a time constant equal to $(R_1+R_2)C$ (capacitor charging) and label by T_2 the time the output is low (capacitor discharging) with time constant R_2C, the timing is determined by noting that the capacitor charges from $1/3\,V_{CC}$ toward V_{CC}, and at T_1 its voltage is equal to $2/3\,V_{CC}$. Thus,

$$\frac{2}{3}V_{CC}=\left(V_{CC}-\frac{1}{3}V_{CC}\right)\left(1-e^{-T_1/(R_1+R_2)C}\right)+\frac{1}{3}V_{CC} \qquad (18\text{-}12)$$

Again, V_{CC} cancels, providing timing independent of the supply voltage:

$$T_1=(R_1+R_2)C\,ln2=0.693\,(R_1+R_2)C \qquad (18\text{-}13)$$

Similarly, while the output is low, the capacitor discharges from $2/3\ V_{CC}$ toward zero, and its voltage is equal to $1/3\ V_{CC}$ at time T_2 later. The time constant is now R_2C; otherwise operation is symmetrical to the charging. The time T_1 is then:

$$T_2=R_2C\,ln2=0.693\,R_2C \qquad (18\text{-}14)$$

The period of the output square wave is then the sum of T_1 and T_2:

$$\text{Period}=0.693\,(R_1+2R_2)C=1/\text{frequency} \qquad (18\text{-}15)$$

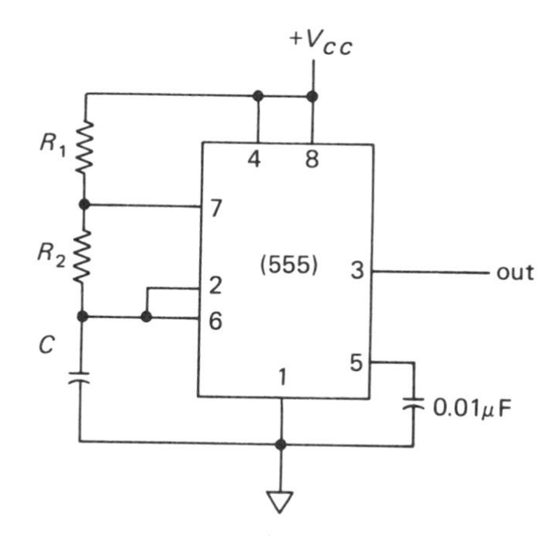

Figure 18-35. 555 Astable configuration

The duty cycle is the ratio of time in the high state to the period, and is given by:

$$\text{Duty Cycle} = \frac{T_1}{T_1 + T_2} = \frac{R_1 + R_2}{R_1 + 2R_2} \qquad (18\text{-}16)$$

Note that although the duty cycle is easily adjustable by varying either R_1 or R_2, it is always more than 50%. It is obvious from the equation that for a 50% duty cycle, R_1 would have to be zero. This cannot be implemented, since when the discharge transistor (collector at pin 7) would turn on and try to saturate, it would be trying to pull the supply voltage to ground. This cannot occur because excessive current would flow into the discharge transistor, causing permanent damage to it.

The timing waveshapes for the astable configuration are shown in Fig. 18-36. Note that the output in the high state does not go all the way to V_{CC}, but typically is 1.7 V lower. This is still sufficient to drive TTL directly, assuming a 5 V supply is used.

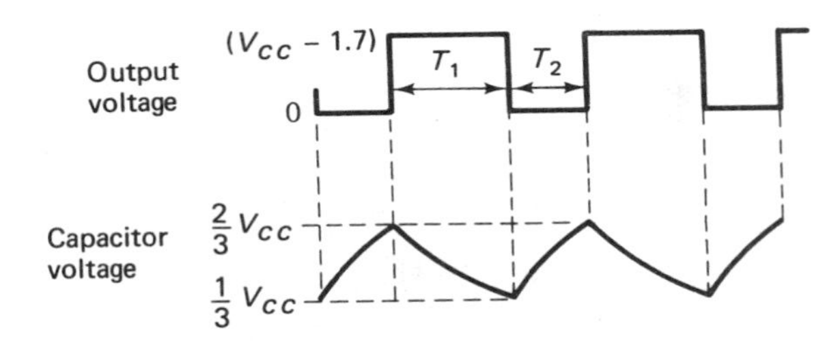

Figure 18-36. Astable waveshapes

The basic astable circuit must be modified if exactly 50% duty cycle is desired. A number of different schemes are possible; one is shown in Fig. 18-37. Here, charging is through R_1; discharging is through the parallel combination of R_1 and R_2. Since the equivalent voltage seen by the capacitor during discharge (pin 7 at ground) is $V_{CC}R_2/(R_1 + R_2)$, and since this voltage must be lower than $1/3\,V_{CC}$, R_1 must be greater than at least $2R_2$ or else the circuit will not oscillate. (If the equivalent voltage is not lower than $1/3\,V_{CC}$, the lower comparator trip-point is never reached and the output will latch in the low state.) The specific ratio between R_1 and R_2 for 50% duty cycle is calculated by equating the charging and discharging times. The resulting equation cannot be solved in closed form; iterative means yield the result that $R_1 = 2.362R_2$. In practice, this ratio is set by a potentiometer adjustment. The output frequency for this configuration is $0.72/R_1C$.

The circuits shown for astable operation have the external reset terminal tied to V_{CC}. It is possible to disable the square-wave output, that is, turn it off, simply by bringing the external reset line (pin 4) low. This causes the output to be low for as long as the reset line is maintained low. As soon as pin 4 is brought high, the output goes high and stays high for a

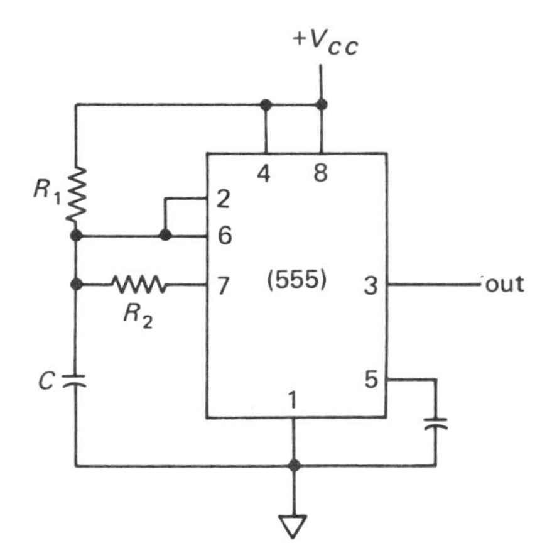

Figure 18-37. 555 Astable configuration for 0.5 duty cycle

time $1.1(R_1 + R_2)C$ for Fig. 18-35 and $1.1R_1C$ for Fig. 18-37; thereafter, normal timing resumes. (The assumption here is that the reset line is maintained low long enough for the capacitor to discharge fully; if that is not the case, the time in the high state just after the reset is made high is between the normal time and that specified above.)

This covers the basic operation of one-shot and astable configurations. In either application, the timing can be modified by using the modulation input, pin 5. The application of a voltage (from a low source impedance like the output of an OP AMP) to pin 5 causes the trip-point for both comparators to change, modifying the timing. When the modulation input (pin 5) is increased above its normal value of $2/3\,V_{CC}$, the timing is extended (lengthened); conversely, when the modulation input is made lower than $2/3\,V_{CC}$, the timing is shortened. In the astable mode, the modulation input acts to lengthen or shorten both T_1 and T_2. When the timing is modified through the modulation input, a well-regulated supply should be used, since the timing then is not independent of the supply. In addition, the relationship between the modulating voltage and the output pulse width or frequency is not a linear one. When not used, the modulation input should be bypassed to ground with a capacitor to prevent pickup of unwanted signals which would interfere with the proper timing of the circuit.

The 555 timer IC has a multitude of applications. The application notes available from the manufacturers contain many useful circuits. In addition, it is hoped that the understanding gained from the operating description above will allow the user an even wider range of applications through his own ingenuity and resourcefulness.

Review Questions

1. What is a multivibrator? Why is it also called a binary?
2. What must be the conditions in the two transistors of a binary?
3. What is an astable MV?
4. What function does an astable MV circuit serve?

5. Why is astable MV also referred to as a clock?
6. What circuit values determine the frequency of the output from an astable MV?
7. What is the rise time? How is it defined?
8. What is the operation of the gated astable MV shown in Fig. 18-11?
9. What is a monostable MV or ONE-SHOT?
10. What function does a ONE-SHOT serve?
11. What is a flip-flop?
12. What is an RS-FF? Describe its operation in terms of a truth table.
13. What is a T-FF? Describe its operation in terms of a truth table.
14. What are steering diodes and how are they used in a T-FF?
15. In what ways is a clocked FF different from an asynchronous FF?
16. What is the JK-FF? Describe in terms of a truth table.
17. How is a JK-FF different from an RS-FF? From a T-FF?
18. How can a JK-FF be used as a clocked RS-FF?
19. How can a JK-FF be used as a clocked T-FF?
20. What is the master-slave prinicple in a JK-FF? Why is the master-slave FF combination necessary?
21. What is a D-FF? Describe in terms of its truth table.
22. What basic purposes do all FFs serve?
23. How can a ONE-SHOT be connected to form a clock?
24. When the 555 timer is used as a one-shot (Fig. 18-33), the trigger (pin 2) is normally held high. If a 10V supply is used, how low must pin 2 be made to trigger the one-shot?
25. When using the 555 timer in fixed timing applications, pin 5 has a small capacitor connected to ground. What is the purpose of this capacitor?
26. When the 555 timer is used in the astable mode, with a 5V supply, specify the voltages at the output as well as across the timing capacitor.

Problems

1. Assuming $Q=1$ and $\bar{Q}=0$ in Fig. 18-1, and $R_C=200\ \Omega$, $R_B=5\ \text{k}\Omega$, $R=50\ \text{k}\Omega$, $V_{BB}=V_{CC}=8$ V, with transistor parameters: $\beta>50$, $V_{BES}=0.7$ V, and $V_{CES}=0.3$ V, determine the circuit currents and voltages.
2. Repeat Problem 1, assuming the saturation voltages to be zero.
3. Repeat Problem 1 for the circuit in Fig. 18-5, using silicon diodes and $V=5$ V.
4. Using silicon transistors with $\beta>60$ and $V_{CC}=10$ V, design the astable MV circuit in Fig. 18-6 for an output waveshape that has equal on and off times and a frequency of 10 kHz.

5. The astable MV circuit in Fig. 18-6 is constructed with: $R_C = 1$ kΩ, $R_B = 22$ kΩ, $V_{CC} = 20$ V using identical silicon transistors with $V_{BES} =$ 0.6 V and $V_{CES} = 0.2$ V. Determine the minimum value of β for the circuit to produce square waves.
6. In Problem 5, determine the frequency of the output waveshape if $C_1 = C_2 = 0.01$ μF.
7. Determine the rise time in Problem 6.
8. Sketch the output waveshape for Problem 5 if $C_1 = 0.1$ μF and $C_2 =$ 0.2 μF. What is the frequency of the output waveshape?
9. The ONE-SHOT in Fig. 18-15 is triggered with the input waveshape shown in Fig. 18-10. An output with equal on and off times is required. With $V_{CC} = 10$ V and $V_{BB} = 5$ V, determine the circuit values needed. (Neglect all saturation voltages and assume the transistor β is high enough to cause saturation.)
10. Sketch the waveshapes for Problem 9.
11. A 555 timer is to be used as a one-shot to provide a timing interval of 1 ms. Specify the resistor needed if a 0.1 μF capacitor is available. The input is triggered every (a) 2 ms, (b) 3 ms, (c) 4 ms. For each case, sketch the output waveshape (at pin 3).
12. The 555 timer is used in the astable mode as shown in Fig. 18-33. The desired frequency of operation is 10 kHz with a 70% duty cycle. Use a 0.05 μF capacitor and specify the resistor values needed. Sketch the output square wave if the supply voltage is 10 V.

3

Systems

In these days of space exploration, the importance of large-scale electronic systems is obvious. Therefore, we shall describe representative electronic systems in this part as a logical extension of the circuits covered in Part 2. From rectifiers and feedback amplifiers, the natural next step in discussion is regulated power supplies. In addition, power control systems and communication systems (AM and FM), using many of the relatively simple circuits developed in Part 2, are covered. Similarly, from our study of differential and operational amplifiers, we go on to analog systems; from basic digital circuits, we continue with digital systems, which are so important to all aspects of life today. The bridge between analog and digital systems is provided by A/D and D/A conversion systems.

19 Regulated Power Supplies

One of the most basic and necessary systems, or more specifically subsystems, in electronics is the dc power supply. Whether we are dealing with communication, instrumentation, computers, or any electronic system, small or extremely large, it invariably needs a source of dc power, which is furnished by a power supply. Basically, the function of a power supply is to convert the readily available 60 Hz 115 V-rms into a specific dc voltage. The power supply, therefore, usually contains a number of circuits: (1) the transformer, which either steps up or steps down the available line voltage, depending on the need; (2) the rectifier circuit, which converts the alternating current into unidirectional or pulsating direct current; (3) a filter circuit, which removes or minimizes the ripple; and (4) some sort of regulator circuit, which maintains the dc level at the output constant with varying loads. We have already discussed all the basic components of a power supply, with the exception of the regulator circuit. Therefore, we begin by studying regulators.

19.1 REGULATORS

A regulator is any circuit that maintains a rated output voltage under all conditions: either no load (open circuit) or full load supplying an output current, as shown in Fig. 19-1. No circuit provides perfect regulation; that is, maintains the output voltage at V_{OC} while supplying any current I_L. A practical regulator may have characteristics such as indicated in Fig. 19-2, where the output voltage under load V_L is somewhat lower than the no-load output voltage V_{OC}.

We measure the amount of regulation provided by a circuit as the ratio between (1) the difference in the output voltage with and without a load and (2) the output voltage under load conditions. Obviously, for a perfect regulator, this ratio is zero. So the smaller the ratio, the better will be the regulation for any given circuit. The amount of regulation in percentage is given by:

$$\%\ \text{regulation} = \frac{V_{OC} - V_L}{V_L} \times 100\% \qquad (19\text{-}1)$$

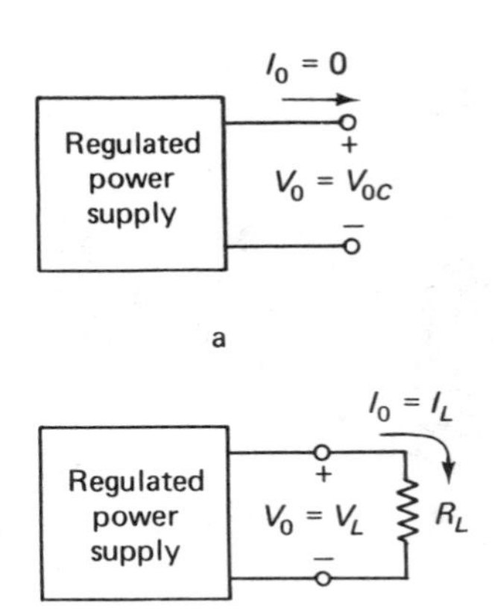

Figure 19-1. Voltage regulation: (a) no load (open circuit) and (b) full load.

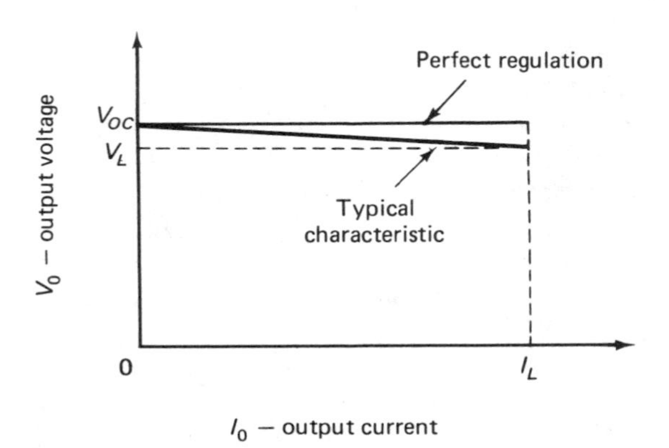

Figure 19-2. Voltage characteristics of a power supply.

Note that this ratio is defined with respect to a specific load condition or rated current.

Figure 19-1(b) tells us that $V_L = I_L R_L$, so that Eq. (19-1) may be rewritten:

$$\% \text{ regulation} = \frac{V_{OC} - V_L}{I_L R_L} \times 100\% \tag{19-2}$$

We can see the output resistance of the regulator as the ratio of the difference between the open circuit voltage and the voltage under load divided by the amount of current drawn. Thus,

$$R_o = \frac{V_{OC} - V_L}{I_L} \tag{19-3}$$

The percentage regulation then may be rewritten:

$$\% \text{ regulation} = \frac{R_o}{R_L} \times 100\% \tag{19-4}$$

For a given load, therefore, regulation improves (i.e., percentage regulation becomes lower) as the output resistance of the regulator decreases. One of the primary characteristics of a good regulator is low-output resistance. A perfect regulator has zero output resistance.

19.1.1 Zener Regulator Circuits

The most basic and inexpensive form of voltage regulator uses a Zener diode, as shown in Fig. 19-3. The raw or unregulated direct current, labeled by V, is applied to the series current limiting resistor R, and the regulated output is taken across the Zener diode. Note that the unregulated dc voltage at the input reverse biases the Zener diode and must be *larger* than the Zener voltage of the diode.

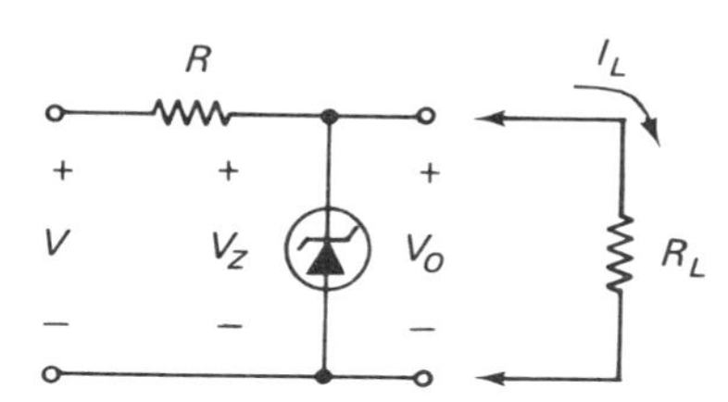

Figure 19-3. Basic Zener regulator.

The output voltage of this regulator is essentially equal to the Zener voltage. It does, however, change somewhat when a load is connected because of the nonzero resistance of the diode. We can see this change if we replace the Zener diode by its equivalent circuit consisting of a voltage V_Z and equivalent resistance R_Z, as indicated in Fig. 19-4. The output resistance of this regulator is the parallel combination of R and R_Z. Typically, the Zener resistance is 10 to 30 Ω. Therefore, the Zener regulator provides good regulation as long as the load resistance is sufficiently higher than R_Z.

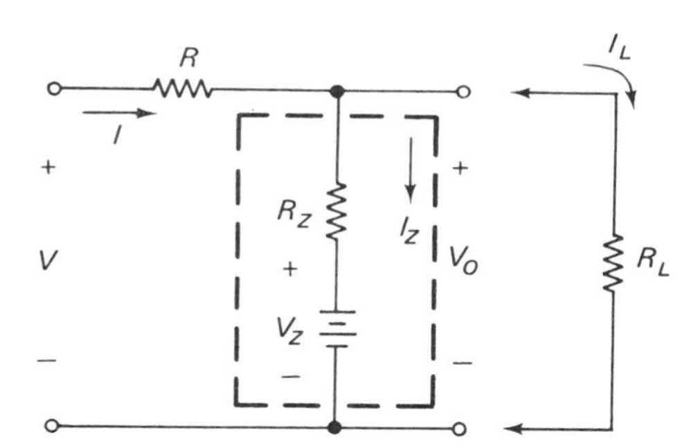

Figure 19-4. Equivalent circuit for Zener regulator.

Example 19-1. The Zener regulator circuit illustrated in Fig. 19-3 is constructed to provide a nominal regulated dc voltage of 10 V, with $V = 20$ V, $R = 100$ Ω, using a 1N758 (10 V Zener diode). The load is to be 200 Ω. We want to determine the percentage regulation and the actual voltage under full load.

Solution: The 1N758 diode has a maximum resistance of 17 Ω at a current of 20 mA (see Appendix 3). We may assume a typical resistance of, say, 10 Ω. The output resistance of the regulator is then:

$$R_o = \frac{(10)(100)}{10+100} \cong 9\ \Omega$$

The open-circuit voltage is determined from the equivalent circuit in Fig. 19-4 with $I_L = 0$, or $I = I_Z = (20-10)/(100+10) \cong 90$ mA. Thus,

$$V_{OC} = V_Z + I_Z R_Z \cong 10 + (0.09)(10) \cong 10.9\ \text{V}$$

The equivalent circuit for the regulator, therefore, is an open-circuit voltage of 10.9 V and a resistance of 9 Ω, as shown in Fig. 19-5. With the load connected, the output voltage is given by:

$$V_L = \frac{R_L V_{OC}}{R_L + R_o} = \frac{(200)(10.9)}{200+9} \cong 10.45\ \text{V}$$

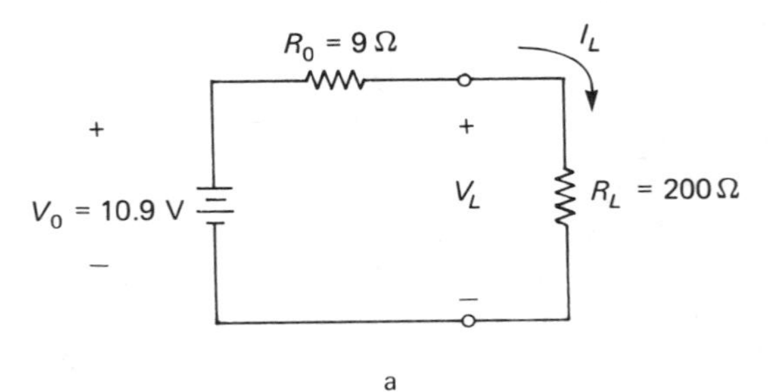

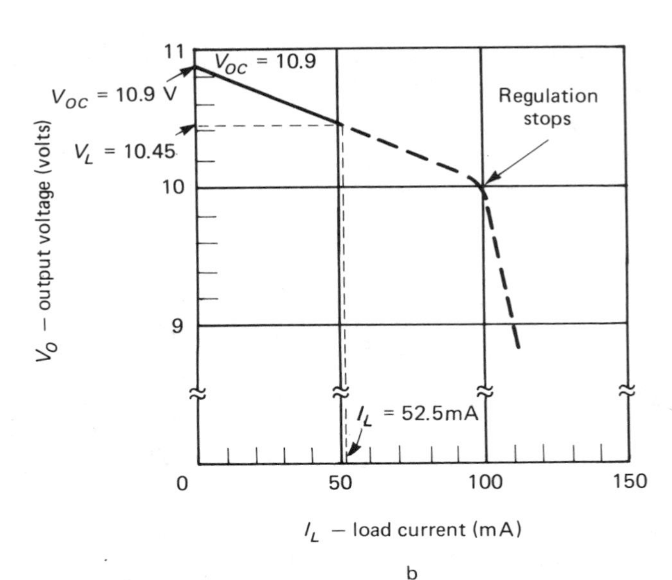

Figure 19-5. (a) Zener regulator equivalent circuit and (b) regulation characteristics.

The percentage regulation is:

$$\% \text{ regulation} \cong \frac{(9)(100\%)}{200} \cong 4.5\%$$

The regulation characteristics for this circuit are also given in Fig. 19-5.

In the design of Zener regulators, we must make certain of two facts. First, the power rating of the Zener should not be exceeded. Secondly, under the worst-case load (lowest R_L) the Zener must still draw some minimum current. In the previous example, the maximum current for the circuit occurs when the Zener draws zero current and its voltage is barely 10 V. Under these conditions the load is being supplied with the absolute maximum current and the Zener is about to stop regulating. Thus, if the load in the example were 100 Ω or less, the Zener would not regulate; the load voltage would rapidly decrease from 10 V as the load was decreased in size. This decrease is shown in Fig. 19-5(b). Furthermore, the circuit could not safely be operated without a maximum load of 200 Ω in order to keep the dissipation in the Zener at or below its rated maximum of 400 mW.

19.1.2 Basic Series Regulator

Regulation over a wider range of loads is possible with the *series regulator* circuit, illustrated in Fig. 19-6. The output voltage of this circuit is approximately the Zener voltage minus the base-emitter voltage of the transistor, or $V_Z - 0.6$.

The unregulated dc input voltage must exceed the desired output voltage by at least 1 V. The transistor ceases to provide regulation once it saturates. Its collector-emitter voltage is the difference between the unregulated input and regulated output voltages.

The series regulator operates in the following way. The current supplied to the load is essentially the same as the collector current of the transistor. So long as the transistor is operating in its active region, the collector current is essentially βI_B. Therefore, when the load resistance changes, the transistor base current changes. As a result, the collector

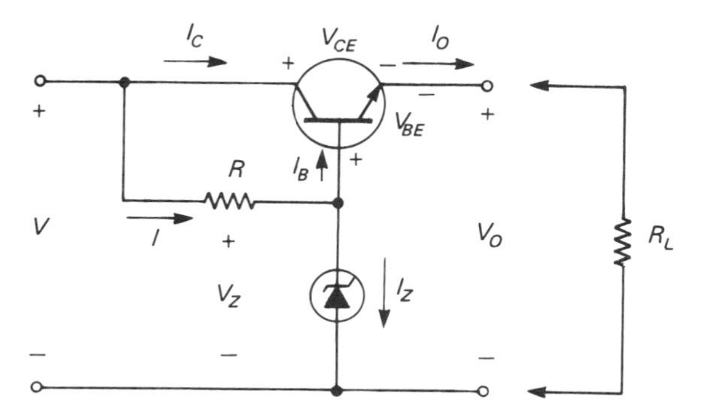

Figure 19-6. Basic series regulator.

current also changes, thus providing the proper output current in order to maintain the output voltage constant. The actual amount of change in the output voltage will be very small; it is governed by the amount of change in V_{BE} to cause the desired change in I_B.

Example 19-2. Consider the series regulator depicted in Fig. 19-6, with a silicon transistor $\beta = 100$, $V_{BE} = 0.6$ V, a 1N751 Zener diode ($V_Z = 5.6$ V, $R_Z < 11\ \Omega$ at 20 mA), $V = 10$ V, and $R = 220\ \Omega$. We want to determine the circuit conditions with R_L of 100 Ω and R_L of 50 Ω.

Solution: The output voltage with 100 Ω load is:

$$V_o = 5.6 - 0.6 \cong 5.0 \text{ V}$$

The output current is essentially the same as the collector current and must be:

$$I_C = \frac{V_o}{R_L} \cong \frac{5}{100} \cong 50 \text{ mA}$$

The current through R is given by:

$$I = \frac{V - V_Z}{R} \cong \frac{10 - 5.6}{220} \text{A} \cong 20 \text{ mA}$$

For the collector current calculated, the base current must be:

$$I_B = \frac{I_C}{\beta} \cong \frac{50}{100} \text{mA} \cong 0.5 \text{ mA}$$

Thus, the Zener current is:

$$I_Z = I - I_B \cong 20 - 0.5 \cong 19.5 \text{ mA}$$

With a 50 Ω load, the circuit values become: $I_C \cong 100$ mA, $I_B \cong 1$ mA, and $I_Z \cong 19$ mA. For this collector current, a typical value of h_{ie} may be roughly 500 Ω. With the increase in base current, the approximate increase in V_{BE} may be about 0.25 V. Thus, the output voltage decreases by the same amount; it becomes approximately 4.75 volts.

19.1.3 Shunt Regulators

The shunt regulator circuit is shown in Fig. 19-7. The regulated output voltage in this circuit is the sum of the Zener and base-emitter voltages. To understand the operation of this regulator, suppose that we decrease the load resistance, thus attempting to decrease the output voltage. A decrease in the output voltage must be reflected as a decrease in the base-emitter voltage because the Zener voltage is constant. A decrease in base-emitter voltage, therefore, causes both the base and collector currents to decrease. (The decrease in I_C is the larger and more important of the two.) If we assume that the input current I is essentially fixed, the current I' is also fixed. (Typically I_B is negligibly small when compared to either I or I'.) Because I_C has decreased, the output current I_o must increase if I and I' are constant. Thus, the cycle of cause and effect is complete. A

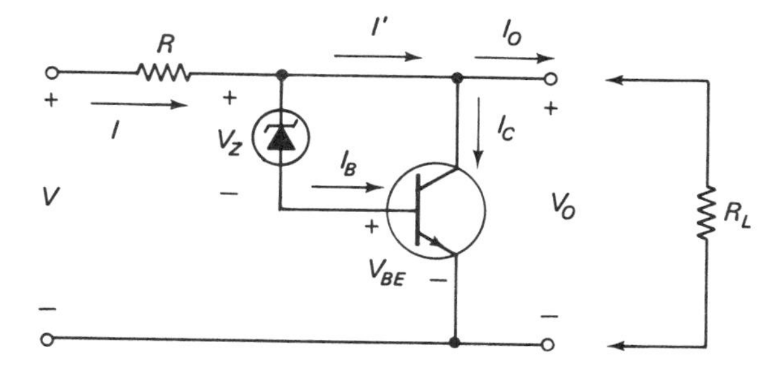

Figure 19-7. Basic shunt regulator.

tendency in the output voltage to decrease as a result of additional loading is eventually followed by an increase in the output current that tends to keep the output voltage constant.

The limit in the regulation for this circuit occurs when the load tries to draw all the current supplied by the unregulated supply (I). As a result, the transistor and Zener become cut off.

19.1.4 Series Regulator with Transistor Feedback

We can improve the performance of the basic series regulator if the output voltage is sensed and the series transistor is forced to adjust to the load. This procedure is the basic concept of feedback. One form of a circuit that uses feedback to regulate the output voltage is depicted in Fig. 19-8.

The output voltage in the circuit of Fig. 19-8 is always given by:

$$V_o = V - V_{CE1} \tag{19-5}$$

Regulation is achieved by forcing V_{CE1} to decrease by the same amount as the unregulated input, thus maintaining the desired constant output.

Suppose that the output voltage has decreased as a result of either a decrease in the unregulated input voltage (which might be caused by line voltage decrease) or an increase in the current drawn by the load. This

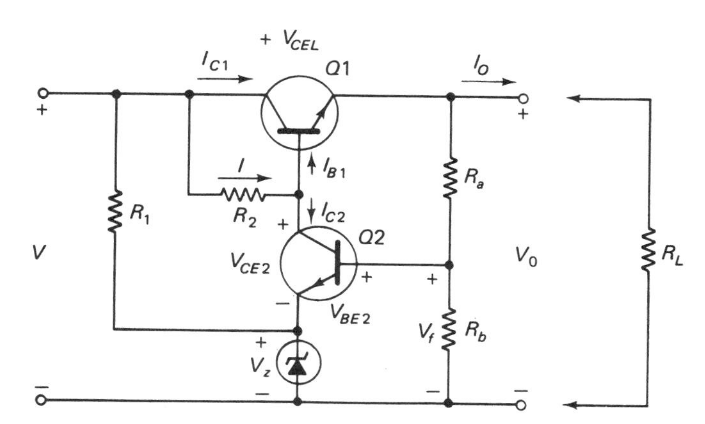

Figure 19-8. Series regulator with transistor feedback.

decrease in the output voltage is sensed across resistor R_b in terms of a decrease in the feedback voltage V_f. We can see that this voltage is given by:

$$V_f = V_{BE2} + V_Z \tag{19-6}$$

Because the Zener voltage is fixed, any decrease in the feedback voltage is reflected as a decrease in V_{BE2}. Consequently, I_{C2} decreases. I_{B1} is given by:

$$I_{B1} = I - 1_{C2} \tag{19-7}$$

Therefore, the decrease in I_{C2} causes I_{B1} to increase, which means that I_{C1} also increases (with V_{CE1} decreasing). In this way, it supplies additional current to the load to cancel out the decrease in output voltage.

Example 19-3. To illustrate the operation of the regulator circuit of Fig. 19-8, assume that the unregulated input is 30 V, the load voltage is 15 V, with $R_L = 30\ \Omega$, and the series transistor has a β of 20. Its output characteristics are shown in Fig. 19-9. We want to determine the increase in base current needed to maintain the specified output voltage when the load is 15 Ω. (Assume that the input voltage remains constant.)

Solution: With the 30 Ω load, the load line labeled case 1 on Fig. 19-9 applies. The conditions for transistor 1 are: $I_C \cong I_o \cong 0.5$ A, $I_B = 25$ mA, and $V_{CE} = 15$ V. (The current through R_a is negligible when compared to either I_C or I_o.) When the load is decreased to 15 Ω, load line labeled case 2 applies. If the output voltage is to remain at 15 V, then the collector

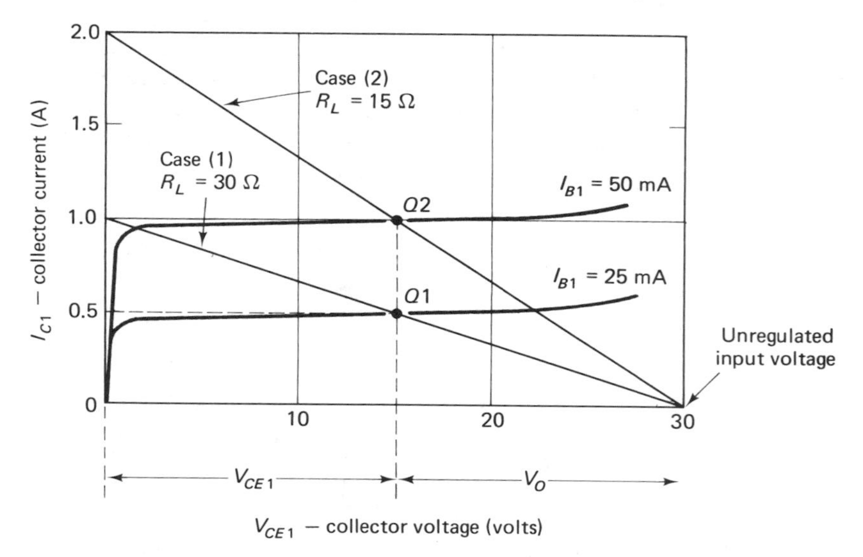

Figure 19-9. Series regulator: change in the *Q*-point of the series transistor due to increased load current requirement.

current must increase to 1 A, as indicated by the new operating point $Q2$. The increase in base current needed to maintain the output voltage is 25 mA. In this case, V_{CE1} remains essentially fixed.

In the preceding example we assumed that the unregulated input voltage remains fixed to illustrate how the circuit can adjust for increased current demands and maintain the regulated output voltage. Let us now consider the case when the unregulated input voltage decreases because of a decrease in the line (60 Hz) voltage.

Example 19-4. Consider the same situation as in Example 19-3, except that the load is constant at 30 Ω and that the input voltage drops from 30 to 25 V as a result of a decrease in line voltage. We want to determine the needed change in the operating point of the series transistor to maintain the output voltage.

Solution: With the input voltage at 30 V, we have the same load line and operating point as in the previous example. This result is shown in Fig. 19-10, labeled case 1. When the input drops to 25 V, the load line labeled case 2 applies. The slope is the same because the load is still 30 Ω. The base and collector currents are essentially unchanged, but there is a change in V_{CE1} as shown: from 15 to 10 V. Thus, even though the input has decreased by 5 V, the output voltage remains essentially unchanged at 15 V.

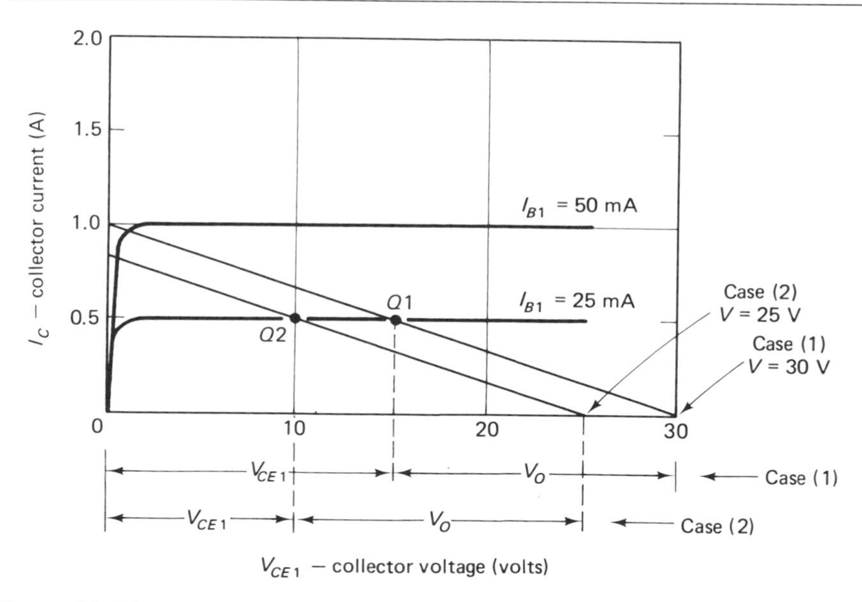

Figure 19-10. Series regulator: change in the Q-point of the series transistor due to a decrease in the unregulated input voltage.

Let us now consider the *worst case*, when the load demands additional current at the same time that the unregulated input voltage decreases. Note that this is usually the case; that is, when additional current is drawn from the unregulated dc supply, its terminal voltage decreases.

Example 19-5. With a 30 Ω load and a 30 V input, the output is at 15 V. When a 15 Ω load is connected, the additional current requirement forces the input voltage to fall to 25 V. We want to determine the change in the Q-point of the series transistor.

Solution: Under the original conditions, the load line is the same as case 1 in the two previous examples. When the load is changed to 15 Ω, both the starting point and the slope of the load line shift. The starting point moves from 30 V to 25 V; the slope now has to correspond to a 15 Ω resistor instead of a 30 Ω resistor. This change is indicated in Fig. 19-11, with the new conditions labeled by case 2. Note that the new operating point, $Q2$, gives: $V_{CE1} \cong 10$ V, $I_{C1} \cong 1$ A, and $I_{B1} \cong 50$ mA. The series transistor can adjust to maintain the output voltage even when an increase in output current is required at the same time that the input voltage decreases.

We can modify the series regulator circuit shown in Fig. 19-8 by replacing the series-pass transistor $Q1$ with a Darlington pair if the change in the collector current of $Q2$ is not sufficient to produce the desired change in the output current. The operation of the circuit will be essentially un-

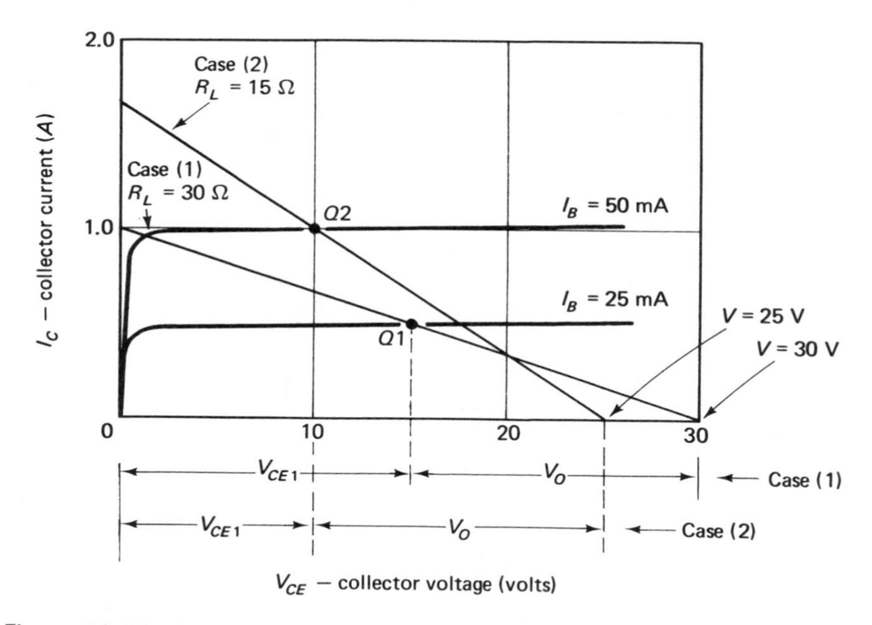

Figure 19-11. Series regulator: change in the Q-point due to increased load current requirement as well as a decrease in unregulated input voltage.

changed. However, the same change in the collector current of $Q2$ causes a larger change in I_{C1} and thus in I_o.

Example 19-6. The series regulator circuit as shown in Fig. 19-12 is constructed with the series transistor having a β of 40. The regulated output is to be 10 V at a maximum current of 1 A ($R_L \geqslant 10\ \Omega$). The unregulated input voltage is between 25 and 35 V. Determine the parameters for the transistors and Zener diode.

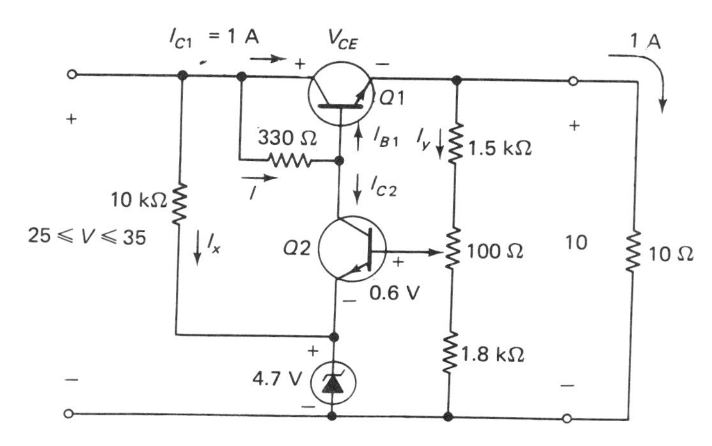

Figure 19-12. Series regulator example.

Solution: The worst-case conditions for the circuit occur when the input voltage is a maximum (35 V) and the output current is also a maximum (1 A). Assuming the output voltage to be approximately 10 V under these worst-case conditions, we calculate that $V_{CE1}=25$ V. Thus, the series-pass transistor is going to have to withstand 25 V and 1 A. The minimum power handling capability of this transistor should be 30 W.

The base of $Q1$ is at approximately 10 V, so the current I is:

$$I \cong \frac{35-10}{330} A \cong 76 \text{ mA}$$

I_{B1} is $I_{C1}/\beta \cong 1A/40 \cong 25$ mA; therefore, I_{C2} must be $76-25 \cong 51$ mA. The net current through the Zener is the sum of I_{C2} and the current through the 10 kΩ resistor, I_x. $I_x=(35-4.7)/10 \cong 3$ mA. Thus, the maximum current in the Zener is approximately $51+3 \cong 54$ mA.

The Zener must be able to dissipate (54 mA) (4.7 V) = 260 mW of power. A safe choice would be a 400 mW Zener. Transistor $Q2$ has a worst-case collector-emitter voltage of approximately $V_o - V_Z = 6$ V. The maximum collector current is slightly in excess of 50 mA, so that a 300 mW transistor would be acceptable for $Q2$.

19.1.5 Series Regulator with DIFF AMP Feedback

Figure 19-13 illustrates an improved series regulator circuit utilizing a basic DIFF AMP in the feedback loop. The DIFF AMP consists of transistors $Q3$ and $Q4$, together with their bias resistors and reference Zener $ZD2$. The circuit values are adjusted as follows. Resistors R_a and R_b are chosen to provide a reference voltage V_f that is the same as the Zener voltage of $ZD2$ when the output is at the desired value. The output of the DIFF AMP (at the collector of $Q3$) should be at about 0.6 V above V_{Z1} to insure that $Q2$ is conducting. Resistors R_1 and R_3 are chosen to limit the respective Zener currents to safe limits.

The DIFF AMP provides additional sensitivity to changes in the output voltage. Otherwise, the operation of the circuit is the same as that described in the previous section.

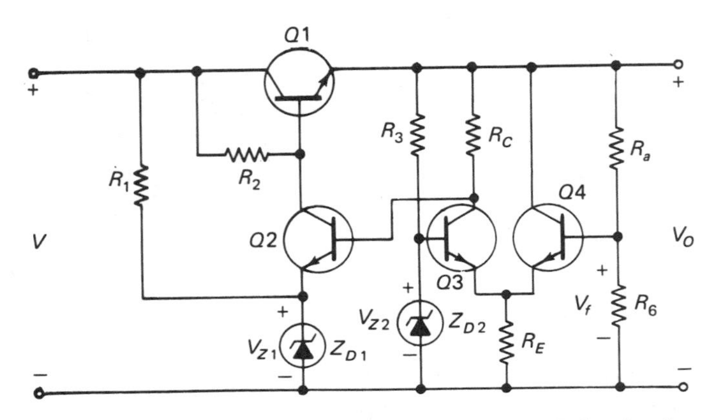

Figure 19-13. Series regulator with DIFF AMP feedback.

19.1.6 Series Regulator with OP AMP Feedback

An OP AMP can be used to give the highest sensitivity in the feedback loop. An example of a series regulator using an OP AMP in the feedback loop is depicted in Fig. 19-14. You can understand the circuit operation by considering that V_f and V_Z are approximately equal when the output voltage is at its desired value. The large open-loop gain of the OP AMP causes even the slightest difference between the two to be significant enough to change the Q-point of the series-pass transistor.

Consider the action when the output voltage tries to decrease: V_f becomes slightly smaller than V_Z; the output of the OP AMP becomes more positive and causes a larger base current in the series transistor, which, in turn, causes the collector current to increase. The output voltage is, therefore, maintained at its original level.

Should the output voltage try to increase (because of a lower output current demand), the reference voltage momentarily exceeds the Zener voltage; the output of the OP AMP is driven in the negative direction, thus reducing the base current drive for the transistor. This action, in turn, decreases the collector current, and the output voltage is again adjusted to its original value.

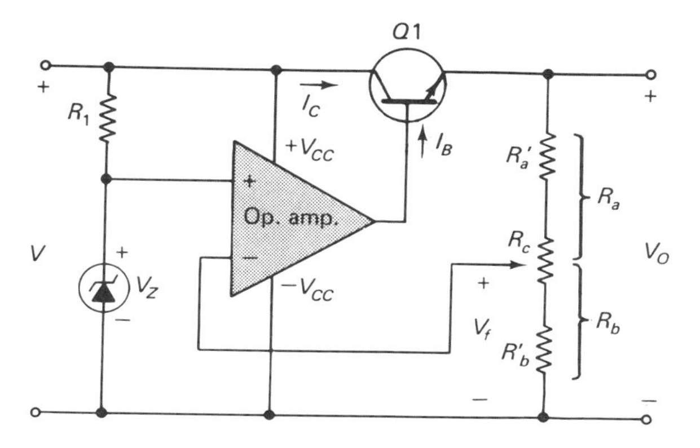

Figure 19-14. Series regulator with OP AMP feedback.

The circuit of Fig. 19-14 may be used for a variety of output voltages. However, the output current is limited because the OP AMP cannot provide a base current in excess of, typically, 10 to 20 mA. If additional output current is needed, the series transistor may be replaced with a Darlington circuit, where the OP AMP output current is amplified by two transistors instead of just one.

The minimum supply voltage for which the OP AMP will operate also limits the use of the circuit. The OP AMP takes its supply from the unregulated input voltage; therefore, the minimum value of the input voltage should be sufficient to bias the OP AMP. For a 741 OP AMP, this minimum supply voltage is about ± 5 volts. The absolute minimum unregulated input voltage is approximately 10 volts. If a potentiometer (R_c) is used, the output voltage is adjustable. Resistor values (R_a', R_b', and R_c) are chosen so that the arm of the potentiometer (pot) can be adjusted to a voltage at or near V_Z.

19.2 CURRENT-LIMITING CIRCUITS

All power supplies need some form of protection from overcurrent conditions caused by a component failure in the circuit that is being supplied or by accidental short circuits.

19.2.1 Diode Overcurrent Protection

We can modify the series regulator circuit as indicated in Fig. 19-15 to include overcurrent protection. As long as the load current is below the desired limit, the regulator behaves in the way already described. Diodes $D1$ and $D2$ will be essentially off (not conducting) and $I_D = 0$. When the load tries to draw an excessive current, the voltage drop across R_{limit} becomes sufficient to forward bias both diodes, which now conduct and limit the series-transistor emitter current. With both diodes conducting, the maximum voltage drop across the limiting resistor is $V_D - V_{BE1}$, where V_D is the sum of the diode voltage drops. If silicon diodes and transistors are used, $V_{BE1} = 0.7$ V and $V_D = 2(0.7)$ V. In this case, the emitter current is limited to a value of approximately $0.7\ V/R_{\text{limit}}$. For example, suppose

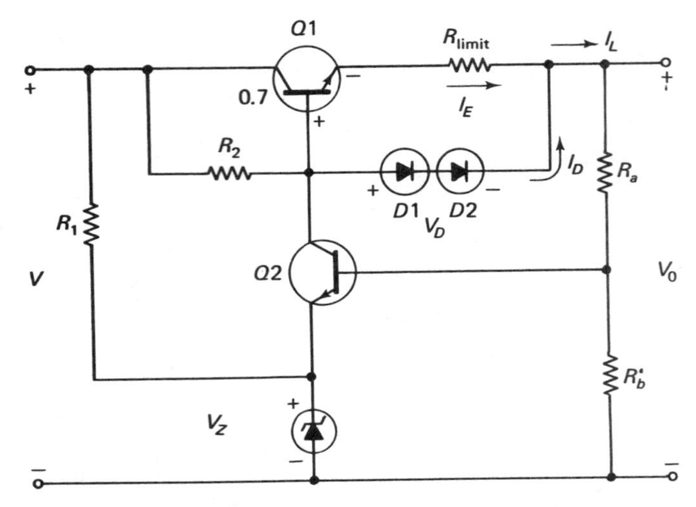

Figure 19-15. Current limiting in a series regulator using diodes ($D1$,$D2$).

that we need to limit the output current to 1 A. Using silicon devices, we would need a limiting resistor of approximately 0.7 Ω with a power rating of 1 W.

If, for some reason, a short-circuit condition existed at the output, the current would be limited to $0.7\ V/R_{limit} + I_D$, where I_D is determined by R_2 and is approximately V/R_2.

19.2.2 Transistor Overcurrent Protection

The same principle is used to limit the output current in the circuit shown in Fig. 19-16. Here the current is limited by a transistor. While the output current is below the desired limit, circuit operation is unaffected by the presence of $Q3$, because its base-emitter voltage is not high enough to

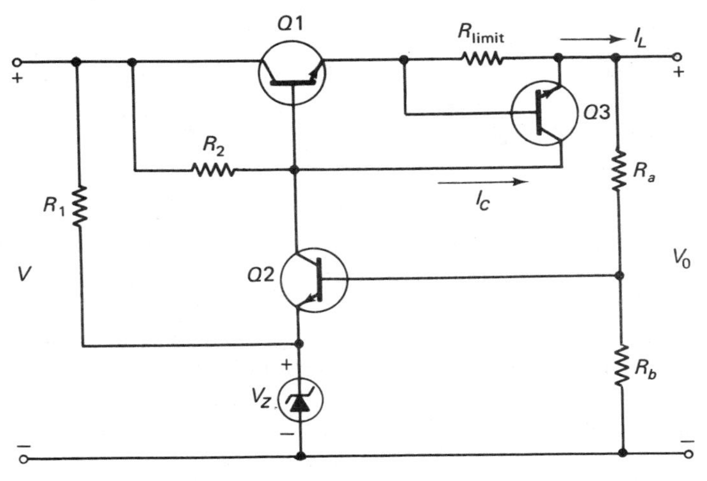

Figure 19-16. Current limiting in a series regulator using a transistor ($Q3$).

cause it to conduct. However, when the output current increases to the point that the voltage drop across R_{limit} nears 0.7 V, $Q3$ begins to turn on and draws collector current. This collector current of $Q3$ is supplied through R_2, which also supplies the base current for the series-pass transistor $Q1$. Thus, when an excessive current is demanded by the load, $Q3$ begins to conduct. It diverts the additional current that would otherwise become an increase in the base current of $Q1$. The load current is effectively limited to approximately $0.7\ V/R_{limit}$ if $Q3$ is a silicon transistor.

Under short-circuit conditions, the output current will not exceed $0.7\ V/R_{limit} + V/R_2$.

Either the diode or transistor current-limiting scheme can be used with any of the various series regulator circuits discussed in the previous sections.

Since most Zener diodes exhibit noise, they are usually bypassed by placing a capacitor, typically 0.01 to 0.1 μF, in parallel with the Zener diode.

19.3 IC REGULATORS

Low-cost fabricating techniques have made a large number of IC regulators available commercially. These devicies range from fairly simple, fixed-voltage types to high-quality precision regulators.

The IC regulators have a multitude of features built into them. In discrete component form, implementing these features would require a lot of extra space and significantly increase the cost of the regulator. Among these features are: current limiting (either variable or fixed), self-protection against overtemperature, remote control, remote shutdown, operation over a wide range of input voltages, and foldback current limiting.

As examples of IC regulators, we shall briefly discuss the LM309 and μA723, both of which are available from a number of different manufacturers. The LM309 is a completely self-contained 5 V, 1 A voltage regulator. It comes in a TO3 case (one of the standard power transistor packages). It has only three terminals: input, output, and case as a common terminal to both the input and output. For an input voltage of between 7 and 25 V, it provides an output voltage that is compatible under worst-case conditions with TTL circuits (between 4.75 and 5.25 V) at a current in excess of 1 A. The regulator is virtually fail-safe; that is, the output current is automatically limited to a safe value (the actual value depends on the input voltage). Thermal shutdown is also provided. If the internal dissipation becomes too great, the regulator shuts itself off to prevent burnout. Using the LM 309 requires only three connections and *no* external components, other than capacitors for stability.

The μA723 is a precision voltage regulator that can accommodate input voltages between 9.5 and 40 V and can provide output (or regulated) voltages between 2 and 37 V. It can supply an output current up to 65 mA without an external series-pass transistor (limited to a programmable

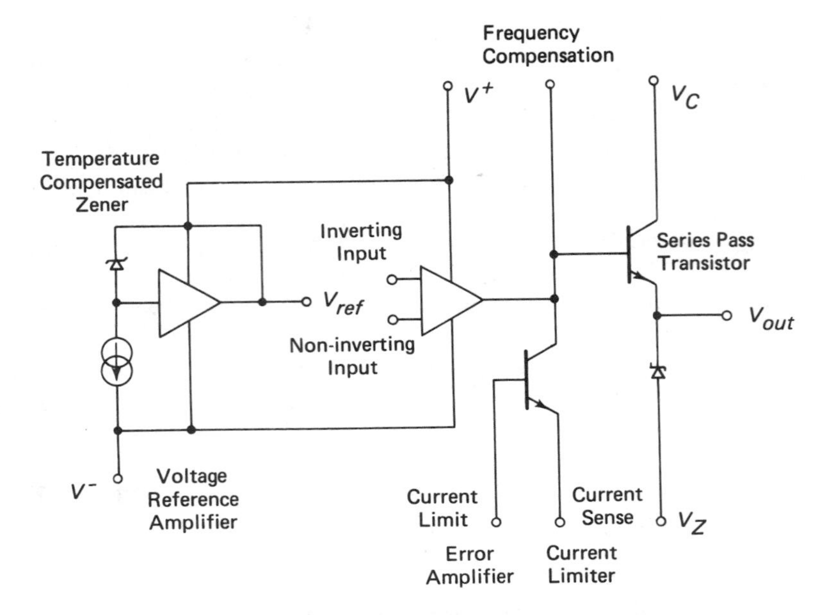

Figure 19-17. 723 regulator block diagram

value). The output current may be increased with the addition of an external power transistor. We can employ an IC regulator as a negative as well as a positive voltage regulator. The μA723 contains a temperature-compensated reference amplifier, error amplifier, series-pass transistor, and current limiter with remote shutdown access. The block diagram is shown in Fig. 19-17.

Typical applications of the μA723 regulator are illustrated in Figs. 19-18, 19-19, and 19-20. Note the graph of output voltage versus current. It illustrates the principle of *foldback* current limiting.

The output current is limited to a certain value under load conditions but is forced to an even lower value when the resistance of the load is too low or a short-circuit condition exists. Observe how useful foldback current limiting is. When the output is short-circuited as a worst case, the series transistor must dissipate the highest power because the voltage across it is the highest possible. To keep dissipation down, the current is "folded back" or decreased from the normal limit.

You do not need to be completely familiar with the circuit inside the IC regulator in order to use it. However, it is necessary to understand the terminology used in the specification sheets.

Input Voltage Range: The upper and lower limit on the input voltage (unregulated or *raw* direct current) that may safely be applied.

Output Voltage Range: The range of possible regulated output voltages obtainable from the regulator. (Note: The regulated dc output voltage is always lower than the unregulated or raw dc input voltage. In order

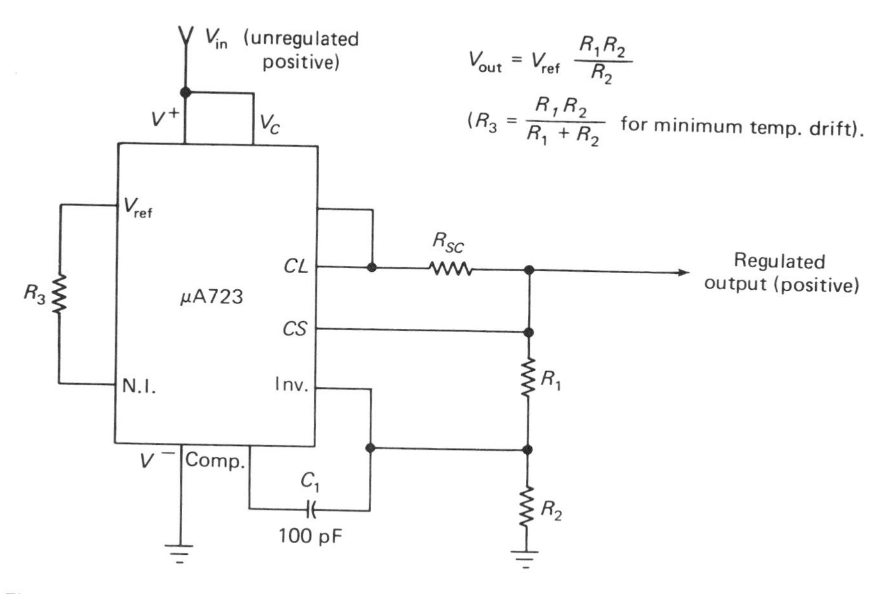

Figure 19-18. μA723 as a positive voltage regulator; $7\ \text{V} \leqslant V_{out} \leqslant 37\ \text{V}$. (Courtesy of Signetics Corporation)

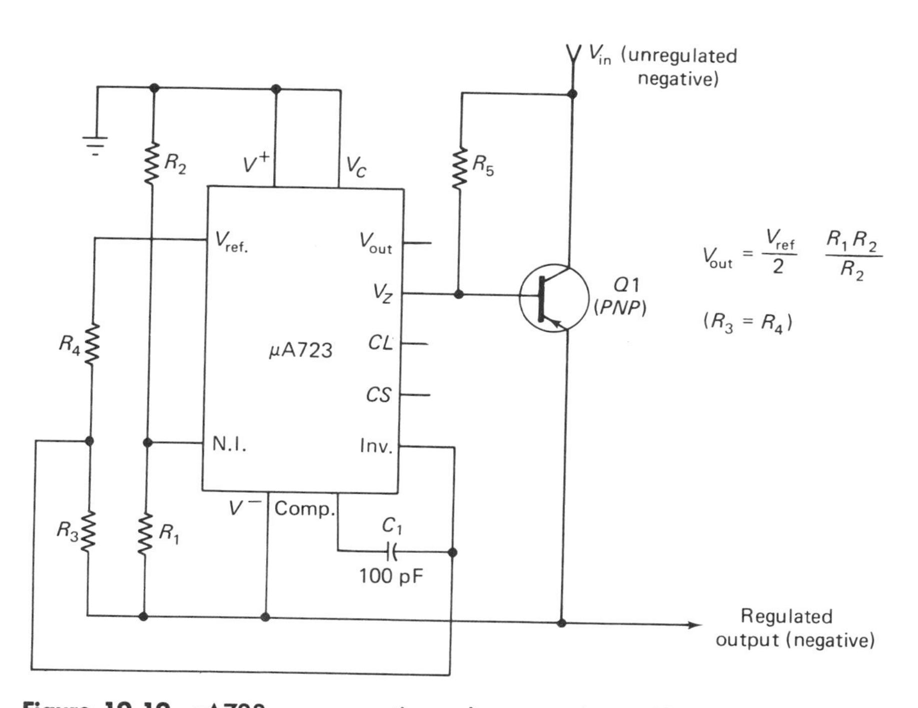

Figure 19-19. μA723 as a negative voltage regulator. (Courtesy of Signetics Corporation)

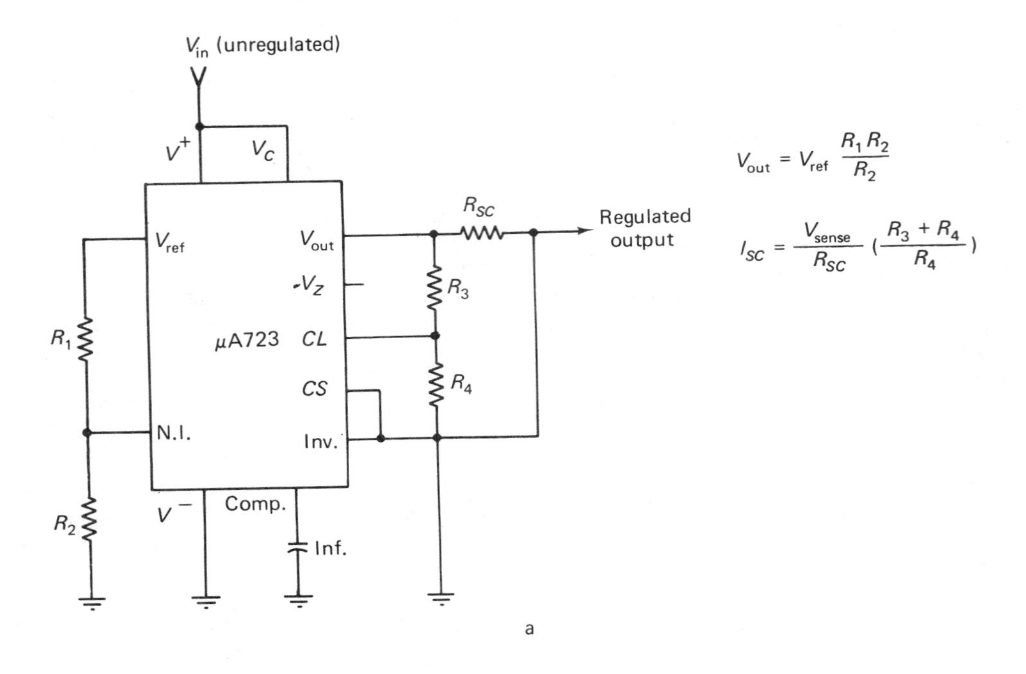

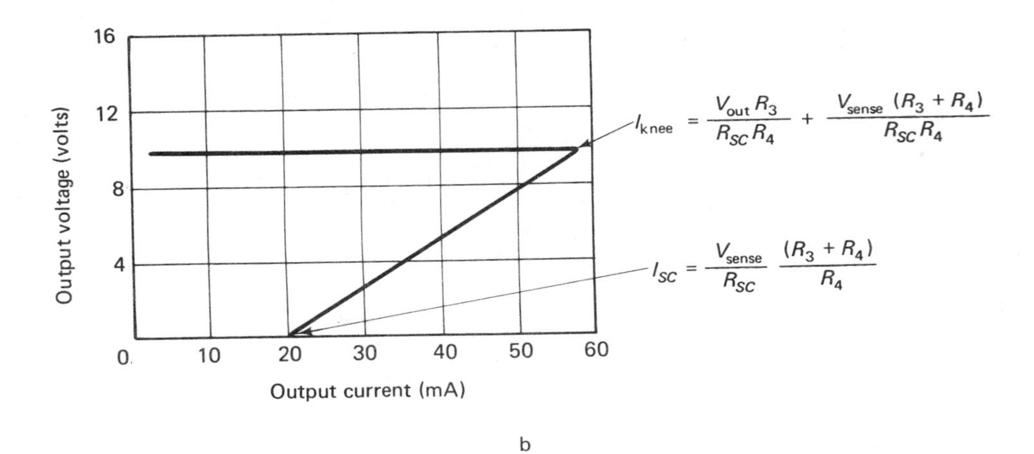

Figure 19-20. μA723 as a positive voltage regulator with foldback current limiting: (a) circuit and (b) output *V-I* characteristics. (Courtesy of Signetics Corporation)

for the regulator to work, there must be some voltage dropped across the series transistor or transistors.)

Line Regulation: The percentage change in regulated output voltage for a specified change in unregulated input voltage.

Load Regulation: The percentage change in regulated output voltage for a change in load current from zero to the specified or rated maximum load current.

Ripple Rejection: The amount of decrease (specified as a percentage or in dB) in the ac component from the input to the output.

Temperature Stability: An indication of the change in the output voltage for a change in operating temperature.

Standby Current: The amount of current drawn off by the regulator when no load is connected.

Output Noise Voltage: The rms value of the ac voltage at the output under load with no ripple at the input.

19.4 COMPLETE POWER SUPPLY

We have discussed the operation of all the components in a regulated dc power supply separately. In this section let us examine the interfacing of the components to form a complete system.

In a dc power supply, the desired output voltage (or range of output voltages) and the maximum current needed are known. Typically, the input is derived from the line, either 115 V-rms at 60 Hz or 230 V-rms at 50 Hz.

The unregulated voltage available from the filter is always larger than the final regulated output voltage. The transformer, rectifier, and filter should be chosen so that they can safely handle the maximum anticipated current and be able to supply a dc voltage *under full load* that is sufficiently higher than the desired regulated output voltage within the rated input voltage range for the regulator. It is important to specify "under load," because the dc (or average) level at the output of the filter decreases under load from its no-load value. This decrease is especially true of capacitive filters.

As an example of a minimum number of components in a well-regulated supply, examine the circuit of Fig. 19-21. There are only five components in the complete supply: (1) a transformer, which should be

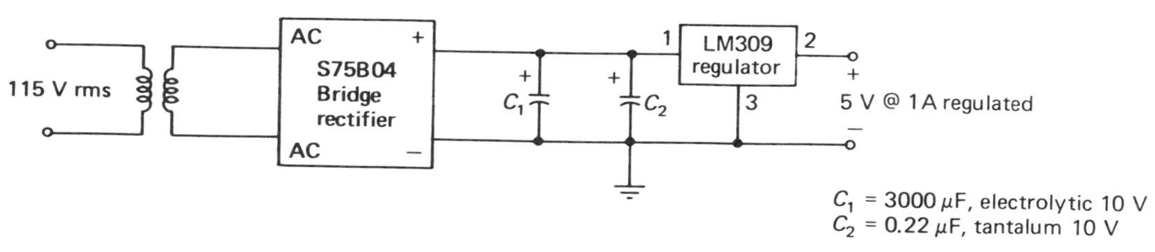

Figure 19-21. 5 V (1 A) regulated power supply.

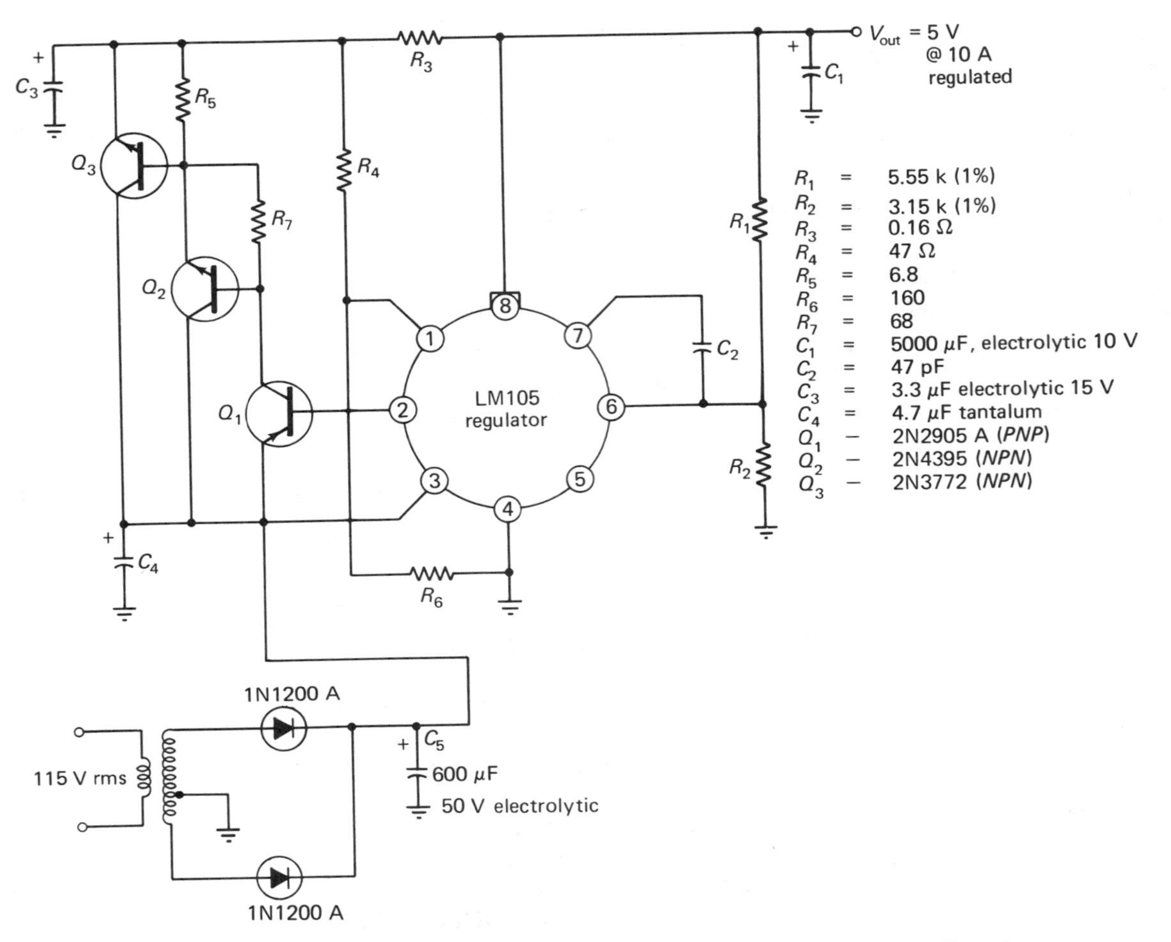

Figure 19-22. 5V (10 A) regulated power supply with foldback current limiting. (Regulator courtesy of National Semiconductor Corp.)

capable of providing about 10 V-rms at 1 A; (2) a full-wave bridge rectifier assembly with a peak-inverse volts rating in excess of 50 and a current rating in excess of 1 A; (3) two capacitors, $C1$ for ripple filtering and $C2$ for noise suppression; and (4) a hybrid IC voltage regulator with a heat sink.

Figure 19-22 depicts a high-current version of a 5 V regulated supply with foldback current limiting. The transformer should be capable of providing 20 to 24 V center-tapped at 10 A. The two 1N1200A diodes provide full-wave rectification. Filtering is done by $C1$ and $C5$. For regulation, the LM105 IC regulator acts in conjunction with additional Darlington-connected series-pass transistors $Q1$, $Q2$, and $Q3$.

Figure 19-23 shows a dual (positive and negative) low-current supply suitable for OP AMP or other IC circuits. The transformer windings should produce + and -41 V at the output of the filter, as indicated.

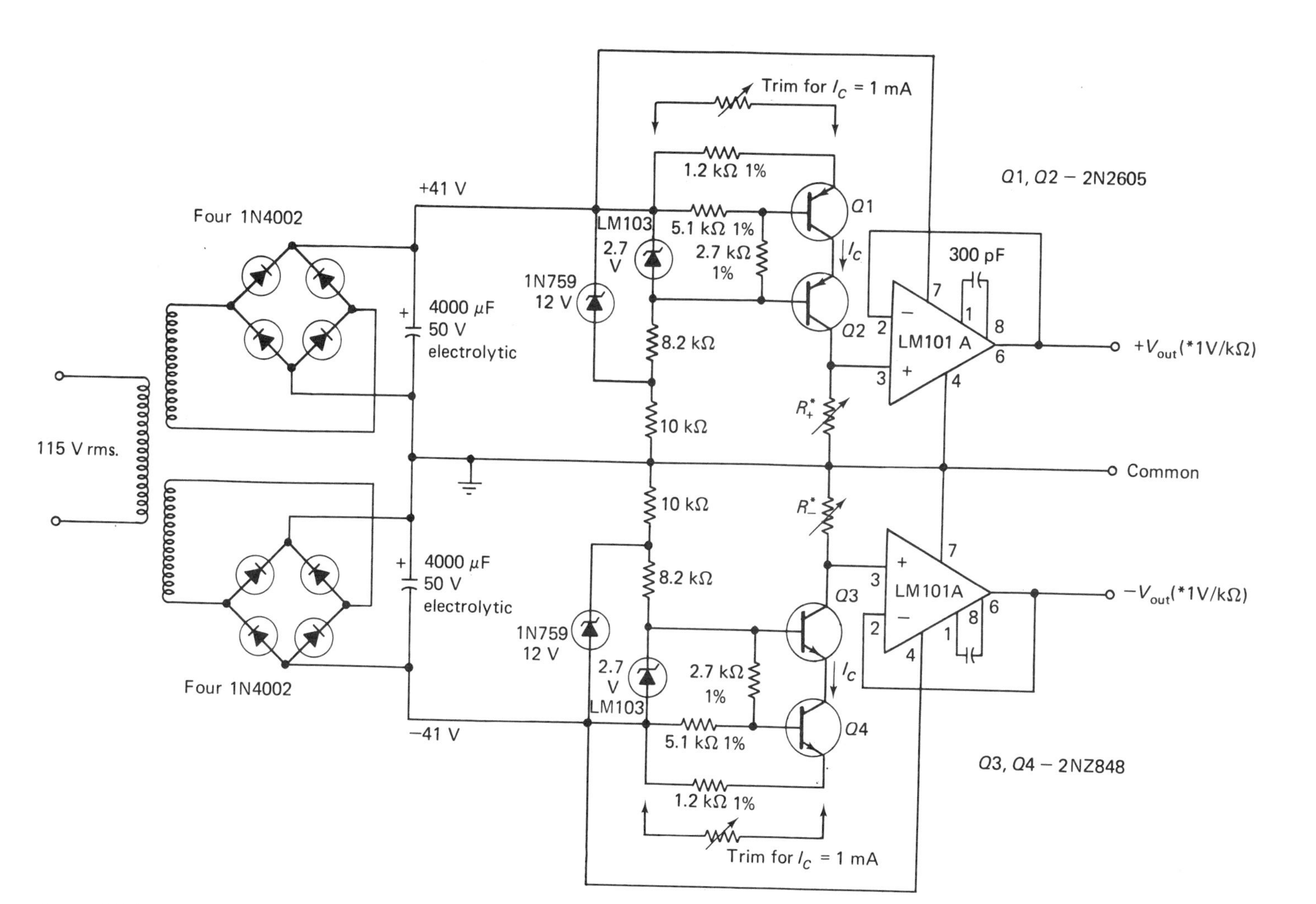

Figure 19-23. Programmable split (+ and −) low power supply for OP AMP circuits. (Regulator circuits courtesy of National Semiconductor Corp.)

Rectification is performed by means of two bridge circuits, with a capacitive filter in each side. Two OP AMPs, each connected in a voltage-follower configuration, provide regulation. Because the inverting input of each OP AMP is connected to the output, whatever voltage is applied to the noninverting input appears at the output. The input reference voltage is developed across variable resistors R_+ and R_-, which allow the output voltage to be varied at the rate of 1 V for each 1000 Ω of resistance. We achieve this variation by using constant current generators comprised of two Zener diodes and two transistors in each half of the supply. In each case, the 1.2 kΩ resistor should be trimmed for exactly 1 mA collector current through the two transistors. In this way, a constant current of 1 mA is provided through the programming resistors R_+ and R_-.

Review Questions

1. What are the components of a regulated power supply?
2. What is a regulator?
3. What distinguishes an unregulated power supply from a regulated one?
4. For best regulation, what should be the output resistance of a voltage supply?
5. How may a Zener diode together with a resistor be used to provide voltage regulation?
6. In the basic series regulator shown in Fig. 19-6, how is voltage regulation obtained?
7. How is regulation provided in the shunt regulator illustrated in Fig. 19-7?
8. How does the addition of transistor feedback improve the regulation in a series regulator?
9. How does the use of DIFF AMP feedback improve the performance of a series regulator?
10. How does the use of an OP AMP in the feedback of a series regulator improve its performance?
11. What are the advantages of the different feedback schemes used with a series regulator? The disadvantages?
12. Make a comparison of the different regulator circuits in terms of (a) performance, (b) circuit complexity (cost), and (c) operation.
13. What is meant by current limiting in a voltage supply?
14. How does the simple diode circuit of Fig. 19-15 provide current limiting?
15. How does the addition of $Q3$ in Fig. 19-16 provide current limiting?
16. What is meant by foldback current limiting?
17. What is the line regulation in a power supply?
18. What is load regulation in a power supply?

19. What is meant by the ripple rejection of a regulator?
20. What is the function of each part of a complete power supply?

Problems

1. In Fig. 19-1 the power supply voltage is 25 V with no load. When a 100 Ω load is connected, the voltage is 24 V. What is the percentage regulation and what is the output resistance of the supply?
2. The simple Zener diode regulator shown in Fig. 19-3 is connected using a 12 V Zener diode with $R=100\ \Omega$ and $V=15$ V. Determine (a) the power rating of the Zener necessary for operation without a load: (b) the highest current that can be supplied with the diode still regulating, assuming a Zener resistance of 10 Ω and $I_{Z\min}$ of 5 mA.
3. Determine the equivalent resistance and voltage for the power supply in Problem 2. Make a plot of its regulation curve (see Fig. 19-5).
4. The series regulator in Fig. 19-6 is constructed with: $R=180\ \Omega$, a TIP29 transistor and a 1N751A Zener diode. Make a plot of the output voltage as a function of output current when the load is varied from 1 kΩ to 10 Ω. Assume the unregulated input voltage to be 10 V with a source resistance of 25 Ω.
5. Circuit components for the shunt regulator in Fig. 19-7 are: $R=50\ \Omega$, $V_Z=15$ V, and $R_Z=10\ \Omega$; a silicon transistor with β of 50 is used. For an input voltage of 20 V, determine the no-load power dissipation in the transistor. Also determine the output voltage with no load and again with a 250 Ω load.
6. In the series regulator circuit in Fig. 19-8, circuit values are: $V=40$ V, $R_1=3.3$ kΩ, $R_2=220\ \Omega$, with both silicon transistors have a β of 50. The desired output voltage is to be nominally 20 V. If we use a 6.8 V Zener together with $R_b=6.8$ kΩ, determine the circuit conditions (voltages and currents) as well as the value of R_a needed.
7. When an output current of 100 mA is drawn, the unregulated input voltage drops to 35 V in Problem 6. Determine the new circuit conditions and the output voltage.
8. The OP AMP in the regulator circuit in Fig. 19-14 is capable of supplying a maximum of 20 mA output current. The circuit values are: $R_1=5.6$ kΩ, $V=35$ V, $V_Z=6.8$ V, and the desired output voltage is 15 V. Determine the values of R_a and R_b needed. (This pair of resistors should not draw more than 5 mA.)
9. If the supply in Problem 8 is to be able to provide the rated output voltage into a 5 Ω load, what must be the β of $Q1$? Also determine the worst-case power dissipation in $Q1$. (Assume that at the highest output current V falls to 20 V.)

Power Control Systems

There are many industrial and consumer applications where the net amount of power delivered must be controlled. These systems range from the simple light dimmer to very sophisticated lighting-control installations or motor-speed controls.

20.1 PRINCIPLES OF POWER CONTROL

The basic control element is a *thyristor*. Depending on the specific application, this device may be an SCR, Triac, SCS, or any other member of the thyristor family. Most often, the source of power is the line, either 115 V-rms or 230 V-rms, at 60 Hz (or sometimes at 50 Hz). The most common form of power control is *phase* control. In this mode of operation, the thyristor is held in an off condition, in which it blocks all current flow in the circuit except for a very small leakage current. It stays off for a portion of the positive (and/or negative) half-cycle; then it is *triggered* or *fired* into conduction at a time in the half-cycle determined by the control circuitry.

A single SCR in series with the load, as shown in Fig. 20-1, can be used to control the amount of power delivered to the load. The control signal keeps the SCR in its off state for the first part of the positive half-cycle of the ac input and then fires it at the desired point to allow current to flow. When the SCR is conducting, almost all the applied ac voltage appears across the load (with the exception of approximately 1 V, which is across the SCR). The current in the circuit is only a function of the applied voltage and the load. Thus, the SCR controls only the voltage. Once the control circuit has fired the SCR by applying a sufficiently high gate signal, it loses control, but the SCR continues to conduct so long as its anode is positive with respect to the cathode. When the positive half cycle is terminated and the input voltage starts to go negative, the SCR reverts to its off or nonconducting state. (Actually, the SCR turns off when the anode current falls below the holding current; see section 6.3.1. This decrease occurs while the input is still slightly positive.) The SCR cannot be fired again until the input becomes positive; thus, a single SCR can

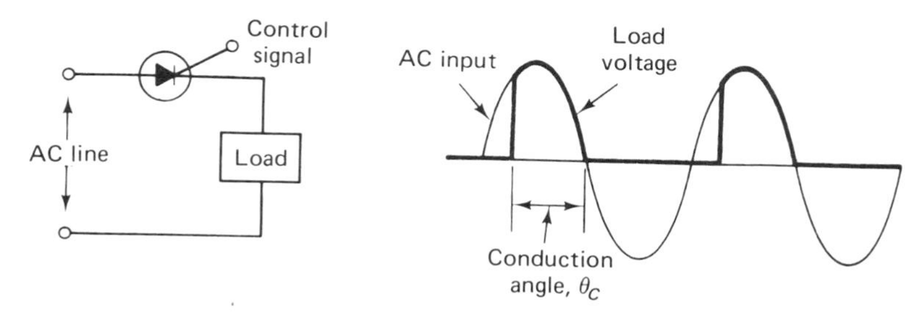

Figure 20-1. Single SCR power control.

control only one half-cycle. The amount of power delivered to the load is proportional to the length of time during which the SCR is conducting. This duration is called the *conduction angle*; it can be varied from almost 50% of the available input power (SCR on all the time; conduction angle = 180°) to almost 0% (SCR off all the time; conduction angle = 0°).

The circuit shown in Fig. 20-2 uses two SCRs connected in inverse parallel (anode of one to the cathode of the other) and can control both the negative as well as the positive half-cycle of the ac input. During the positive half-cycle, the anode of *SCR*2 is negative and *SCR*2 is off, no matter what the control signal. *SCR*1 is fired into conduction during the positive half-cycle in the same manner as in the single SCR circuit of Fig. 20-1. When the input goes from positive to negative, *SCR*1 turns off and *SCR*2 can be turned on by an appropriate signal to its gate. Thus, *SCR*1 controls the positive half-cycle, whereas *SCR*2 controls the negative half-cycle to achieve full-wave power control.

The two SCRs in the circuit of Fig. 20-2 can be replaced with a single Triac as shown in Fig. 20-3. Circuit operation is unchanged because the Triac is nothing more than a bilateral SCR; it can conduct in either direction (assuming that the proper gating signal is applied). In a full-wave control system, the power delivered to the load can be varied from 100% of the available power (conduction angle 180°) to almost 0% (conduction angle 0°).

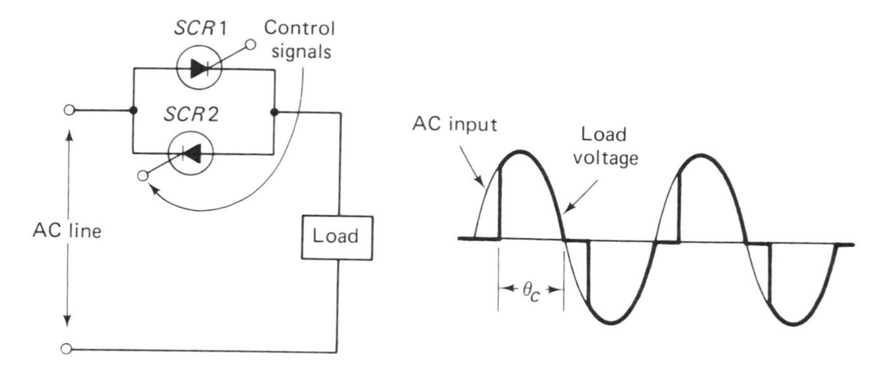

Figure 20-2. Two SCR power control.

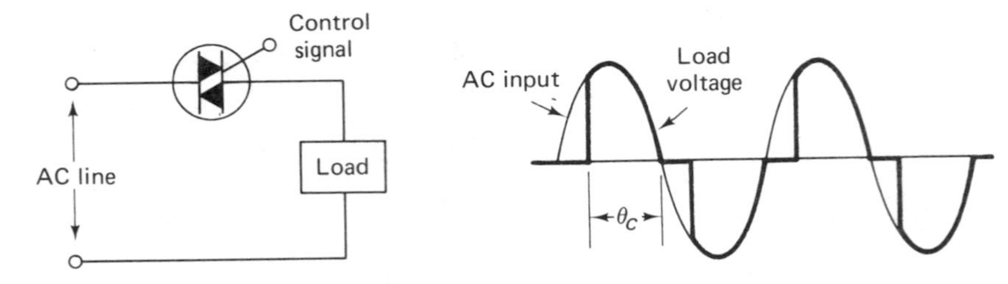

Figure 20-3. Triac power control.

An alternate means of full-wave power control is shown in Fig. 20-4. The ac input is full-wave rectified by the bridge consisting of diodes $D1$ through $D4$. It is applied to the load through an SCR. We can vary the amount of power delivered to the load from 0% and to essentially 100% of the available input power by controlling the conduction angle through the gating signal to the SCR. Similar operation can be achieved with the circuit depicted in Fig. 20-5, where diodes $D1$ and $D2$ have been replaced with SCRs. During the positive half-cycle of the input, power is applied to the load only when $SCR1$ is gated and load current flows through $SCR1$ and $D3$. During the negative half-cycle of the input, power is applied to the load only when $SCR2$ is gated and load current flows through $SCR2$ and $D4$. Note that the output is a controlled full-wave rectified waveshape.

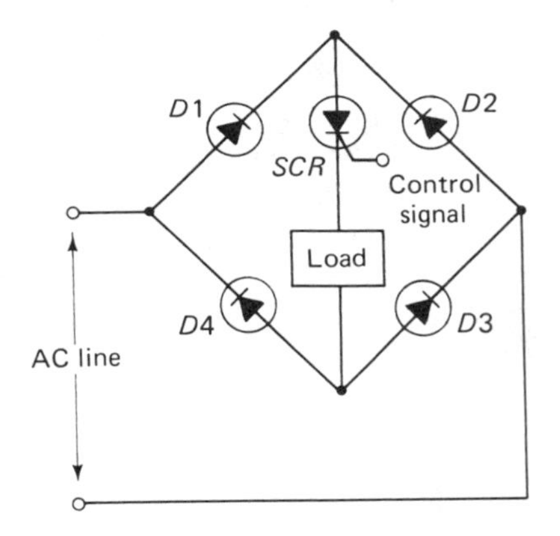

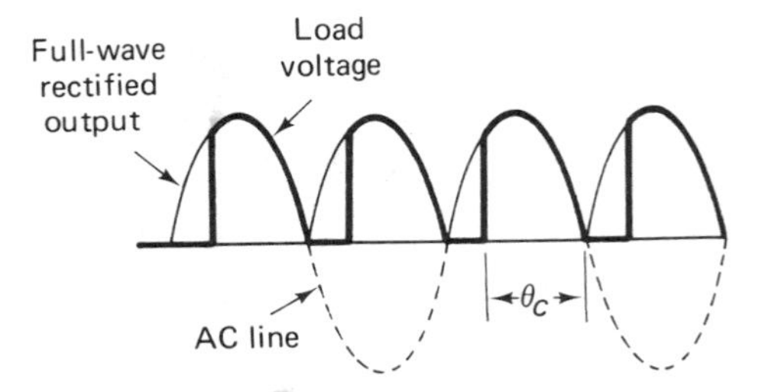

Figure 20-4. Full-wave bridge and one SCR power control.

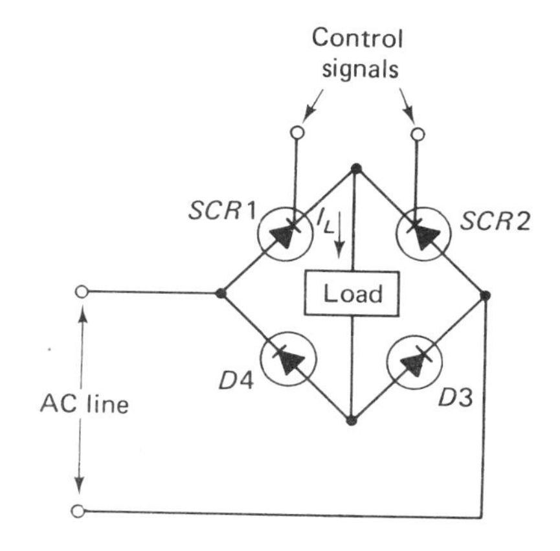

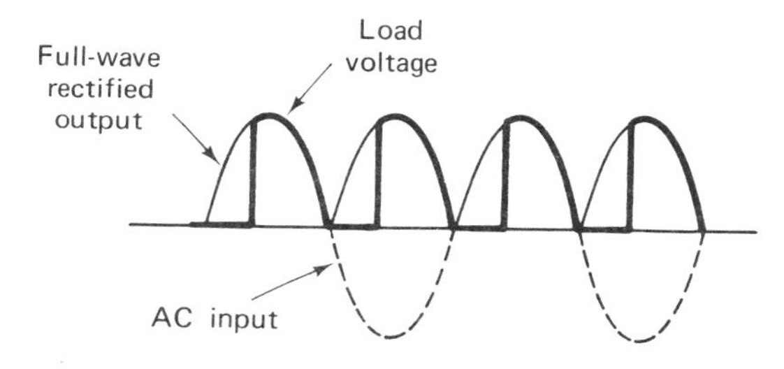

Figure 20-5. Two SCR full-wave bridge power control.

Figure 20-6 indicates the percentage of the available input power that is delivered to the load in a half-wave power control circuit as a function of the conduction angle. The maximum power delivered to the load is 50% for a conduction angle of 180°. However, approximately 45% is delivered at a conduction angle of 150°. It is not practical, therefore, to try to achieve conduction angles in excess of 150°. Similarly, only about 2% of the available input power is delivered to the load at a conduction angle of 30°, so it becomes pointless to try to achieve conduction angles of less than 30°.

The full-wave power control characteristics as a function of the conduction angle are given in Fig. 20-7. As is the case with half-wave control, conduction angles between 30° and 150° provide almost the entire range (from about 3% to 97%) of power delivered to the load in the full-wave control circuits. Note that the full-wave control circuits can deliver 100% of the available input power to the load, whereas the half-wave control circuits can deliver a maximum of 50%.

A simple means of firing an SCR is depicted in Fig. 20-8. During the positive half-cycle, the capacitor charges up through the adjustable resistor *R*. When the capacitor voltage reaches the gate firing potential for the SCR, the SCR is turned on. It conducts until the input voltage goes to zero. When the SCR fires, its gate current discharges the capacitor, so that

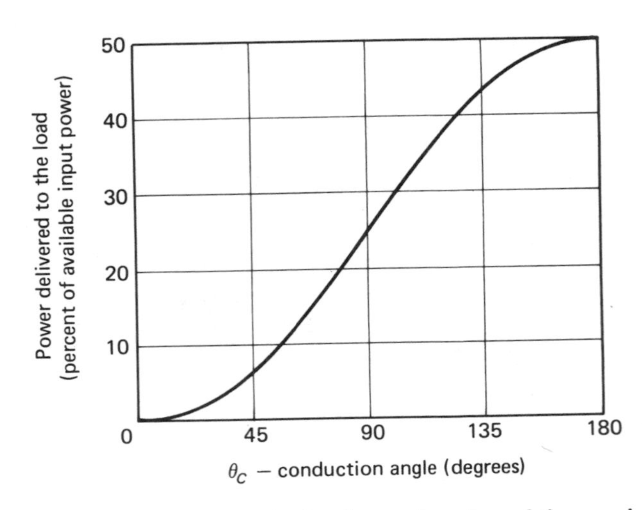

Figure 20-6. Power delivered to the load as a function of the conduction angle in a half-wave circuit.

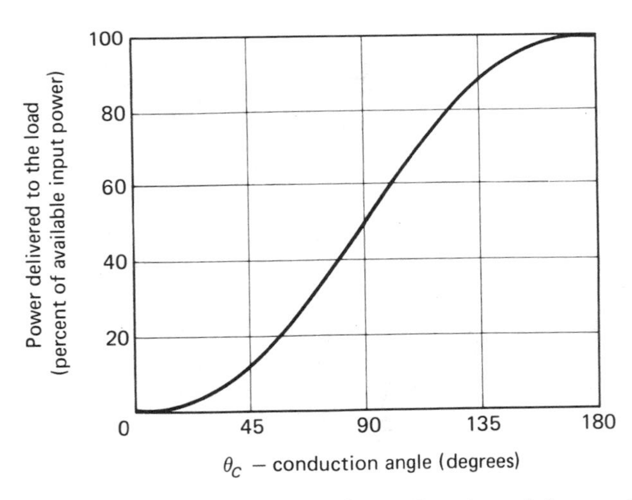

Figure 20-7. Power delivered to the load as a function of the conduction angle in a full-wave circuit.

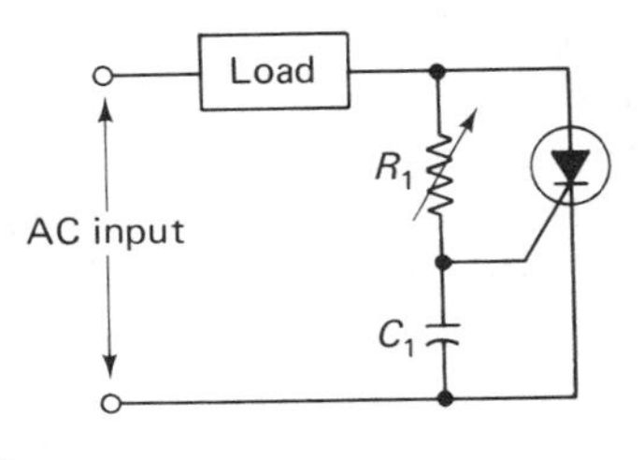

Figure 20-8. Simple means of controlling the conduction angle of an SCR.

the cycle can be repeated when the input becomes positive again. The rate at which C charges determines how quickly its voltage becomes high enough to fire the SCR; therefore, it controls the conduction angle. The charging rate is a function of both R and C. By adjusting R, the conduction angle may be controlled. For conduction angles below 90°, the voltage across the capacitor is affected by the decreasing input voltage. It does not increase at the same rate above 90° that it did below 90°. This problem can be minimized by the addition of a series resistor and a parallel capacitor between $C1$ and the SCR gate. Such a circuit offers superior performance for conduction angles below 90°. It also can be used to control the firing of an SCR (or Triac) to give conduction angles between essentially 0° and 170°.

We can further refine the SCR by inserting a trigger diode in series with the SCR (or Triac) gate, as indicated in Fig. 20-9. This circuit operates as follows. During the positive half-cycle capacitor $C1$ charges through $R1$; the voltage across $C1$ causes $C2$ to charge through $R2$. The SCR remains off as long as the voltage across $C2$ is below the breakover voltage of the trigger diode. When the voltage across $C2$ reaches the breakover voltage of the trigger diode, the diode fires (i.e., conducts), and supplies a fast-rising, high-current gate pulse that in turn fires the SCR. The two capacitors are discharged through the trigger diode and gate of the SCR. The point at which the voltage across $C2$ reaches the firing potential for the trigger diode controls the conduction angle. It can be adjusted by the variable resistor $R1$: the smaller $R1$, the faster the charging of the capacitors and the sooner the firing of both the trigger diode and SCR. Thus, a decreasing $R1$ will increase the conduction angle. A unilateral (Shockley or four-layer) diode, which fires in one direction only, may be used if the control element is an SCR (providing half-wave control). A bilateral diode (Diac or trigger diode) together with a Triac can be used if full-wave control is desired.

The UJT (see section 6.1) is another device that we frequently use to fire an SCR. A typical UJT control circuit is shown in Fig. 20-10. The UJT is connected as a relaxation oscillator. The ac input is full-wave rectified by the bridge consisting of $D1$-$D4$. This rectified signal is then clipped by the Zener diode $D5$ to give the waveshape shown in Fig. 20-11. When the UJT circuit has a voltage applied, capacitor C_T charges exponentially

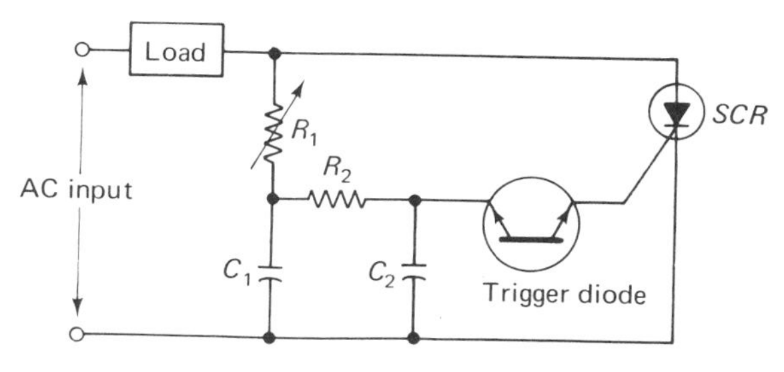

Figure 20-9. Improved circuit for firing an SCR.

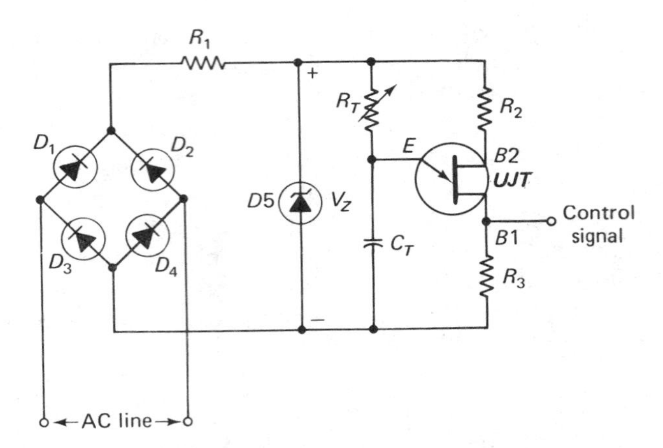

Figure 20-10. UJT line-synchronized relaxation oscillator which provides the signal for firing an SCR or Triac.

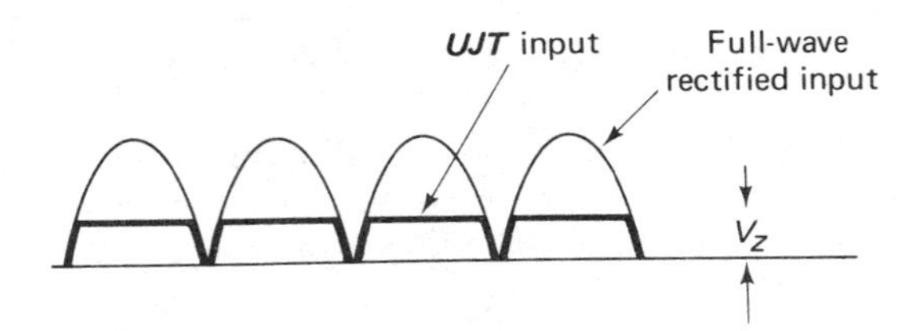

Figure 20-11. Input waveshapes for UJT circuit of Fig. 20-10.

through R_T. When the capacitor voltage reaches the UJT peakpoint voltage V_p, the UJT fires and the emitter-base-1 resistance becomes very low. Capacitor C_T is then discharged through this low emitter-base-1 resistance and $R3$. The discharge current causes a voltage spike across $R3$, which is used as a control signal to fire an SCR. The capacitor and base-1 waveshapes are depicted in Fig. 20-12. The timing of the charging and discharging of C_T can be controlled by R_T. Increasing R_T increases the charging time of C_T and thus increases the length of time between control signal pulses. The UJT relaxation oscillator is synchronized to the line because at the end of every half-cycle of the line input, the input to the

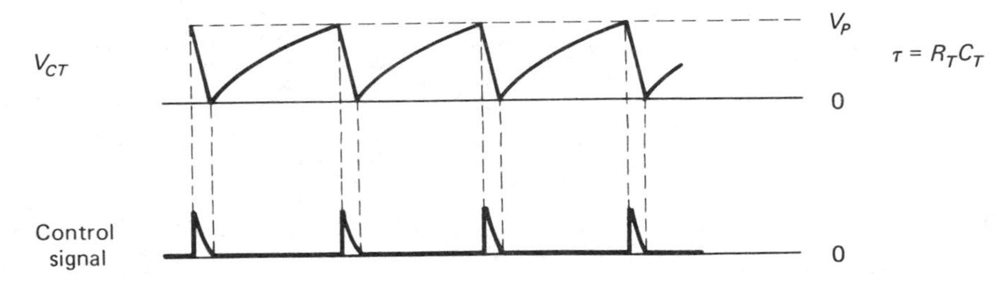

Figure 20-12. UJT relaxation oscillator waveshapes.

relaxation oscillator (shown in Fig. 20-11) goes to zero. Thus, we can make certain that the timing capacitor is discharged at the beginning of the next half-cycle.

20.2 TRIAC LIGHT-INTENSITY CONTROL

We can utilize the triggering circuit shown in Fig. 20-9 in a simple light-intensity control (light dimmer) circuit, as shown in Fig. 20-13. On each half-cycle (both positive and negative) capacitor $C2$ charges through the phase-shift network of $R1$, $C1$, and $R2$, until it reaches the firing potential of diode $D1$ (about 20 V). Once the trigger diode has fired, its voltage drops and the Triac is turned on. For the rest of the half-cycle, the timing network is effectively shorted through the trigger diode and the Triac gate. No further pulsing can occur until the next half-cycle. During the positive half-cycle, the discharging current flows into the Triac gate; on the negative half-cycle, the Triac gate supplies the discharge current. Because the Triac and the trigger diode are symmetrical, the conduction angle is the same during both the negative and positive half-cycles. It is controlled by the setting of $R1$. Increasing $R1$ has the effect of reducing the conduction angle and thus reducing the amount of power to the load–and thus dimming the light.

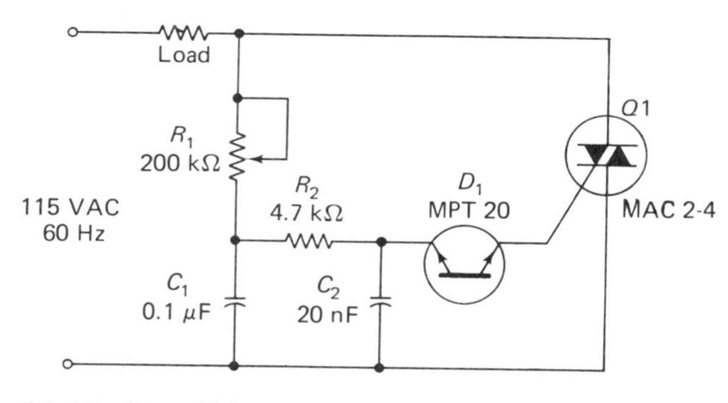

Figure 20-13. Triac 800 W light dimmer. (Courtesy of Motorola, Inc.)

20.3 SCS ALARM CIRCUIT

There are many applications in which we want an audible or a visible indication of an event. In such cases, the SCS alarm circuit (illustrated in Fig. 20-14) may be used. We have incorporated it here because it can indicate certain conditions that may be useful in power control applications. The circuit may be expanded to accommodate as many inputs as desired.

The figure gives one possible form for an input. It does not matter whether the sensor resistor R_s is sensitive to temperature or to light or to radiation so long as it is normally in its high-resistance state. Then when it experiences a decrease in its resistance, the condition to turn on the alarm occurs. Resistor R_1 is set to a value that allows the input to the SCS to be below the triggering level with the sensor resistance high. When the sensor

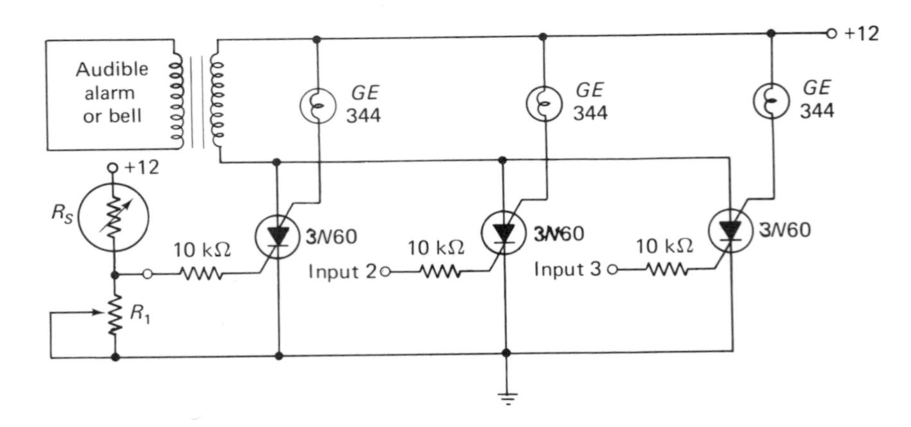

Figure 20-14. SCS alarm circuit.

resistance decreases because of a change in its temperature, light, or radiation, a positive pulse hits the gate of the SCS, causing the SCS to conduct. The audible alarm is triggered and at the same time gives a visual indication by the lighting of the appropriate lamp. Any of the inputs thus triggered will cause an audible alarm. But only a specific input will cause the lighting of the lamp.

20.4 SCR UNIVERSAL MOTOR SPEED AND DIRECTION CONTROL

In a series-wound dc motor, the speed is governed by the voltage and current in the armature, whereas direction is determined by the direction of current through the field winding. A circuit for controlling both the direction and speed of such a motor is depicted in Fig. 20-15. The ac input is full-wave rectified by the bridge comprised of diodes $D1$-$D4$. As the voltage V_x increases, capacitor C_1 charges through R_1, R_2, and the primary of either $T1$ or $T2$, depending on the position of switch $S1$. Let us assume that for the time being $S1$ is directing current through $T1$. The Zener diode $D5$ is off until the capacitor charges up to and a little beyond the Zener voltage (about 51 V). At this point, the Zener diode conducts, and a positive voltage is developed across R_3. This voltage eventually fires $SCR5$, which now discharges the capacitor through the winding of transformer $T1$. The discharge current causes a positive pulse to appear at the secondary of transformer $T1$; the pulse, in turn, causes $SCR1$ and $SCR4$ to conduct. As a result, power is applied to the motor and a current flows downward through the field winding. The speed of the motor is determined by the conduction angle of $SCR1$ and $SCR4$. This conduction angle can be varied by R_1. Increasing R_1 decreases the conduction angle and also decreases the speed of the motor.

The direction of the motor can be reversed by connecting $S1$ to the primary winding of $T2$. Thus, when $SCR5$ fires, it discharges the capacitor through the primary of $T2$, causing a positive pulse on the gates of $SCR2$ and $SCR3$. These two SCRs are then fired and power is once again applied

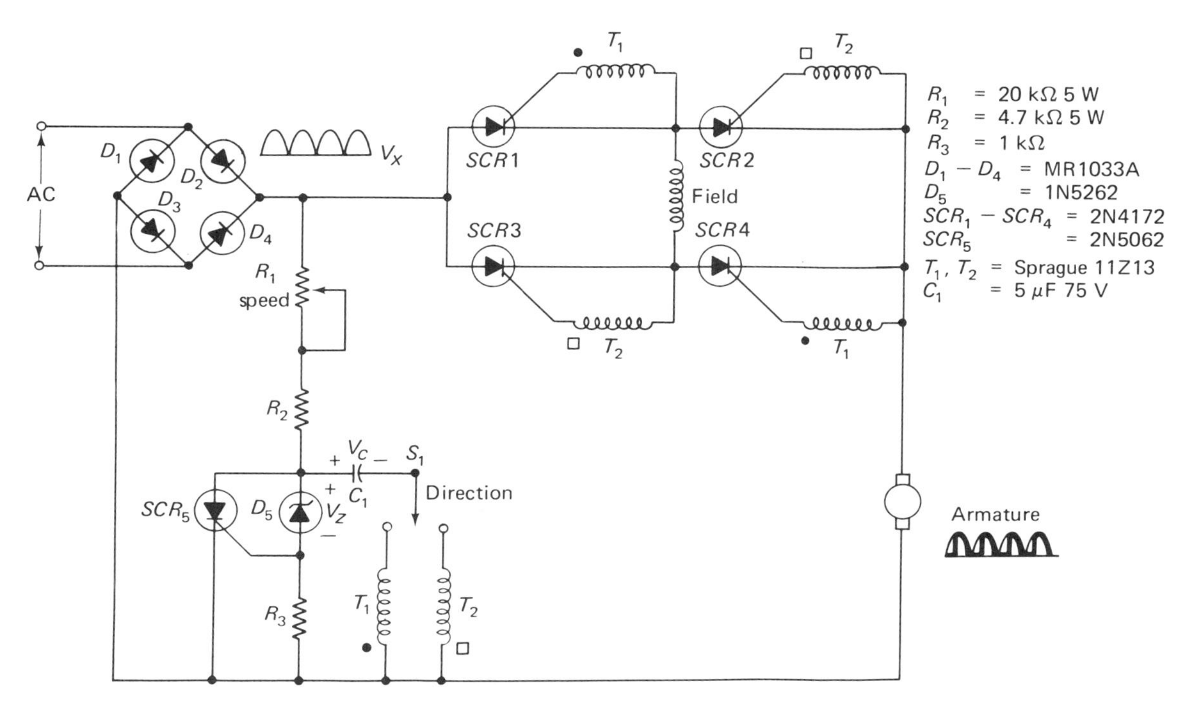

Figure 20-15. Direction and speed control for a series-wound universal motor. (Courtesy of Motorola, Inc.)

to the motor. However, in this case, the current through the armature is in the upward direction, causing the motor to turn in a direction opposite to that when $S1$ is energizing $T1$. The timing is still determined by R_1, R_2, C_1, and the Zener diode.

We can use the same circuit to control the speed and direction of a shunt-wound motor if the field and armature windings are reversed. In a shunt motor the speed is governed by the power applied and the direction is determined by the direction of the current through the armature.

20.5 12-VOLT BATTERY CHARGER

The 12-volt battery charger circuit shown in Fig. 20-16, which is capable of supplying 8 A, uses a programmable UJT (PUT) in a relaxation oscillator circuit. The circuit utilizes the programmable peak-point voltage characteristic of the PUT. When power is applied, the battery to be charged supplies current to charge capacitor $C1$ through $R1$. When the capacitor voltage reaches the peak-point voltage of the PUT, the PUT fires and, in turn, causes a positive voltage pulse in the secondary of transformer $T2$. This voltage pulse fires the SCR and charging begins. So long as the battery voltage is low, the PUT relaxation oscillator supplies the firing pulses to maintain charging current through the battery. As the battery charges up, its voltage increases. As the battery voltage increases,

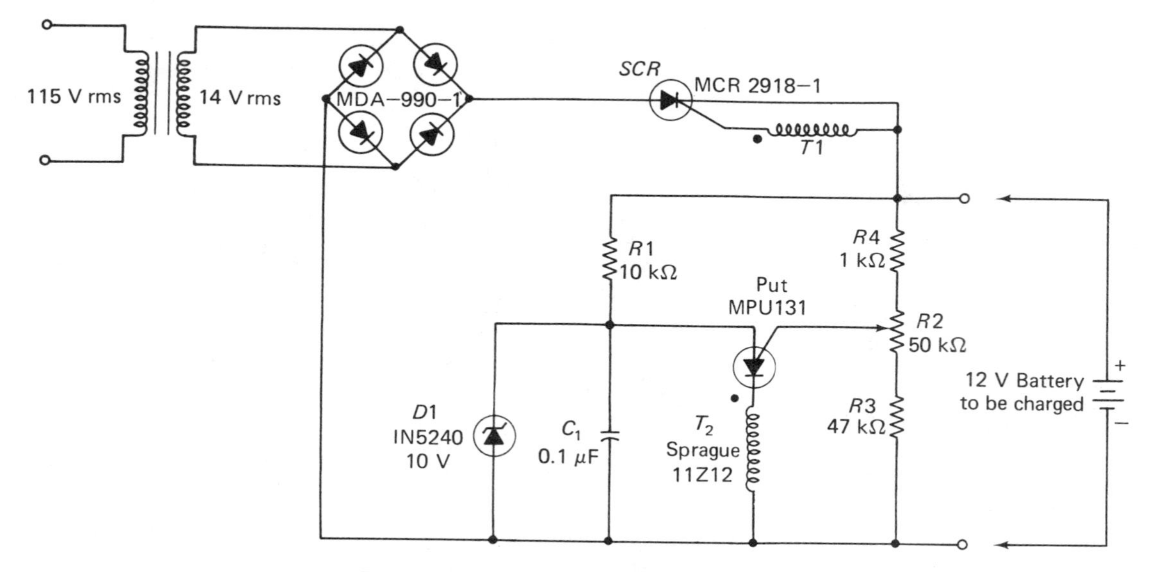

Figure 20-16. 12-volt battery charger.

the peak-point voltage of the PUT also increases, so that the capacitor must charge to a slightly higher voltage before the PUT (and, in turn, the SCR) can fire. The maximum capacitor voltage (in this case, 10 V) is set by the Zener diode, and the capacitor cannot charge above this voltage. Thus, when the battery is charged and its voltage is above a certain limit (set by *R*2), the PUT relaxation oscillator ceases to function. The capacitor cannot reach the PUT peak-point voltage; therefore, the SCR firing pulses are no longer supplied. The charger turns itself off once the battery is charged.

Note that the charger will not function unless the battery is connected properly. In order to start, the relaxation oscillator takes its voltage and current from the battery; if the polarity of the battery is inadvertently reversed, the oscillator will not start. The charged voltage of the battery is variable (with *R*2) between the lower limit set by the Zener diode (10 V) and the voltage available from the bridge rectifier (14 V).

In these sections, we have examined only a small representative sample of the many power control circuits. If you wish information beyond this introduction, consult the many manufacturers' application notes and manuals.

Review Questions

1. What is power control?
2. What is the most common method of power control?
3. What is phase control?
4. What is the conduction angle? What determines it?
5. In a half-wave control circuit, what is the maximum percentage of the available power that can be applied to the load?

6. What is the relationship between the conduction angle and the percentage of the available power that reaches the load?
7. In the half-wave control circuit of Fig. 20-1, what determines the conduction angle?
8. In the half-wave control circuit of Fig. 20-1, once the SCR is fired, how is it turned off?
9. Explain how two SCRs, as shown in Fig. 20-2, may be used for full-wave control.
10. Directly below the load voltage for Fig. 20-2 sketch the waveshapes for the two control signals.
11. How is a Triac used for full-wave control? (Refer to Fig. 20-3.)
12. Explain the operation of the bridge control circuits in Figs. 20-4 and 20-5.
13. What are the advantages and disadvantages of the two bridge control circuits in Figs. 20-4 and 20-5?
14. Why is the firing circuit of Fig. 20-9 superior to that of Fig. 20-8?
15. How does varying R_1 in Fig. 20-9 control the conduction angle for the SCR?
16. What function does the trigger diode in Fig. 20-9 perform?
17. Explain the operation of the UJT relaxation oscillator.
18. How is the UJT relaxation oscillator in Fig. 20-10 synchronized to the line frequency?
19. The light dimmer circuit of Fig. 20-13 provides for full-wave control. How is this control accomplished and what determines the conduction angle?
20. How should the motor speed and direction circuit shown in Fig. 20-15 be modified to provide control for a shunt-wound motor?
21. How is the battery charger in Fig. 20-16 turned off once the battery has been fully charged?
22. What determines when the battery charger (Fig. 20-16) turns off?
23. If the battery is connected with the polarity reversed, the battery charger in Fig. 20-16 will not operate. Why?

21 AM Receivers

In this chapter we take up the basic principles of amplitude modulation (AM) and describe some typical circuits in amplitude modulated receivers. Probably the first significant commercial use of semiconductor devices (specifically, transistors) was in small portable AM receivers. In fact, most laymen understand the word "transistor" to mean a small portable radio, because the first really portable radios used transistors.

21.1 PRINCIPLES OF AMPLITUDE MODULATION

Before we examine what is meant by amplitude modulation, let us first consider the need for any modulating scheme. Basically, modulation is changing or modifying one waveshape in accordance with another waveshape. Thus, in any modulating scheme we are dealing with two waveshapes: (1) the *carrier*, which is the signal being modulated, and (2) the information, or modulating, signal. If we want to send some sort of signal, either voice or music, from one point to another, the air waves would get unbearably noisy if all the signals were transmitted in audible form. Another problem is that audio signals are attenuated very quickly, so we would have to be within a few hundred feet of the originating source in order to hear the signal. Obviously, this is undesirable. Different modulating schemes have been devised to accomplish long-range transmission. Any such system must have multiple station capabilities so that the listener can select the specific station (i.e., signal) that he wishes to receive.

In amplitude modulation, we make the amplitude of the carrier signal vary in accordance with the amplitude of the modulating signal. This operation is shown in Fig. 21-1. The modulating signal is assumed to be sinusoidal with peak amplitude E_m. The carrier is of a much higher frequency (f_c) and has peak amplitude E_c. The AM signal has the frequency of the carrier and has an amplitude whose *envelope* follows the amplitude of the modulating signal. In reality, the carrier frequency can be from 540 to 1620 kHz, whereas the modulating signal is typically less than 10 kHz. The relative frequencies indicated in Fig. 21-1, therefore, are not to scale.

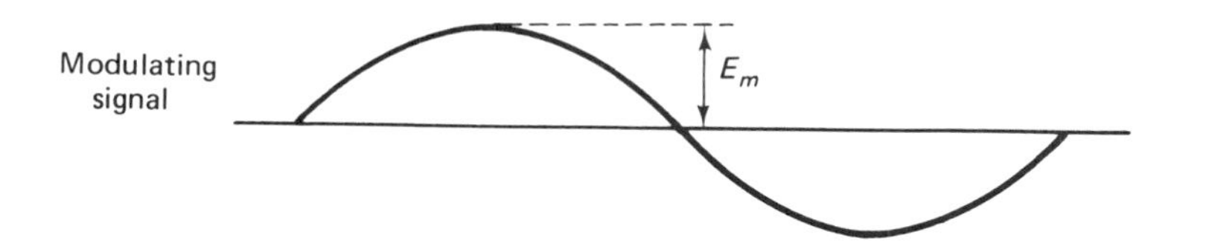

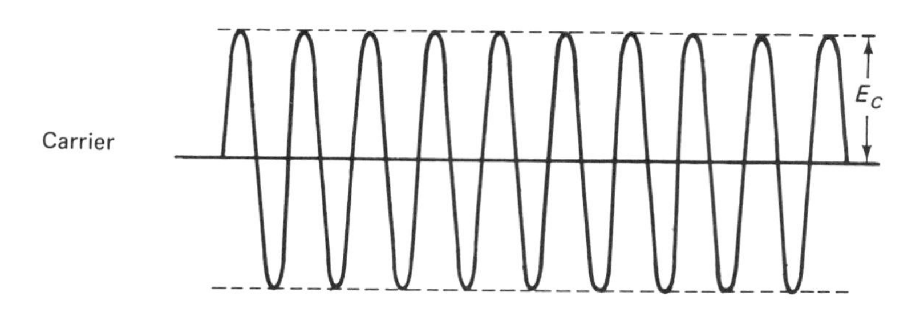

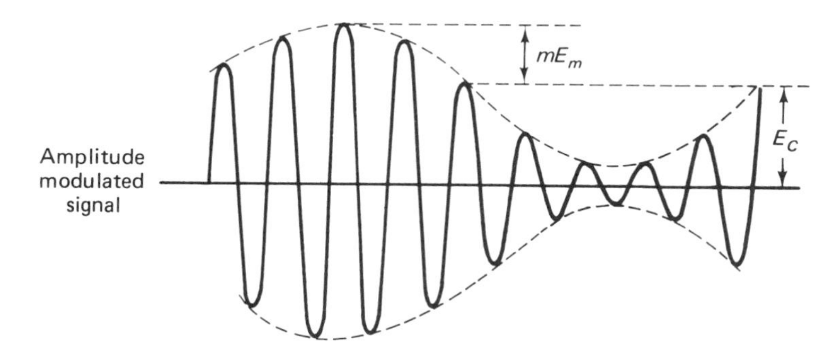

Figure 21-1. Waveshapes showing the components of the AM signal.

In terms of equations, we can write:

$$\text{modulating signal} = E_m \cos 2\pi f_m t \tag{21-1}$$

$$\text{carrier} = E_c \cos 2\pi f_c t \tag{21-2}$$

$$\text{AM signal} = E_c(1 + m\cos 2\pi f_m t)\cos 2\pi f_c t \tag{21-3}$$

where m is the modulation index (percentage). It is defined as the ratio of the peak modulating-signal amplitude to the peak carrier amplitude:

$$m = \frac{E_m}{E_c} \tag{21-4}$$

The amplitude of the AM signal is: $E_c(1 + m\cos 2\pi f_m t)$ and varies at the modulating-signal frequency f_m. So Eq. (21-3) can be rewritten as

$$\text{AM signal} = E_c \cos 2\pi f_c t + E_c m \cos 2\pi f_m t \cos 2\pi f_c t$$

The second term here contains the product of two cosine functions. We use the following trigonometric expansion:

$$\cos x \cos y = \frac{1}{2}\cos(x+y) + \frac{1}{2}\cos(x-y)$$

Therefore,

$$\text{AM signal} = E_c \cos 2\pi f_c t + \frac{mE_c}{2}\cos 2\pi(f_c + f_m)t + \frac{mE_c}{2}\cos 2\pi(f_c - f_m)t \quad (21\text{-}5)$$

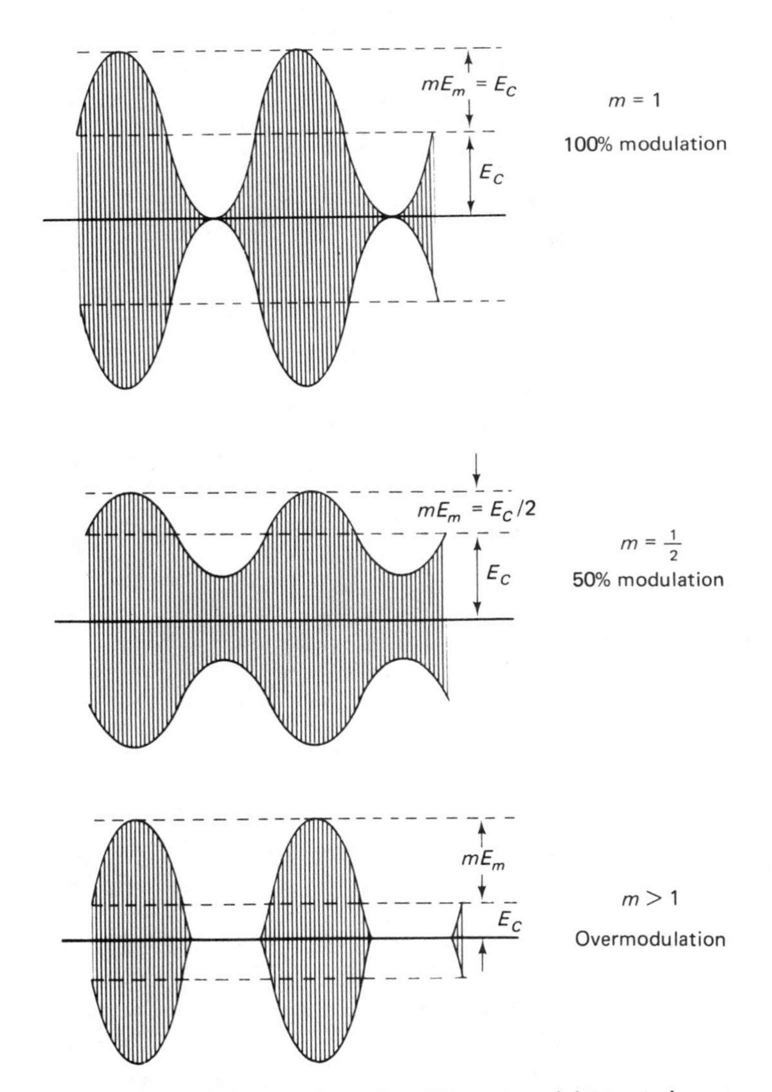

Figure 21-2. AM waveshape for different modulation indexes.

The amplitude modulated signal, rewritten as in Eq. (21-5), obviously contains three components: one at the carrier frequency f_c, one above the carrier frequency (at $f_o + f_m$), and one below the carrier frequency (at $f_c - f_m$). The complete signal is said to contain the carrier and the *upper* and *lower sidebands*. The upper sideband is at a frequency $f_c + f_m$, while the lower sideband is at a frequency $f_c - f_m$. We shall discuss the significance of the sidebands after considering the modulation index.

Figure 21-2 illustrates the AM waveshape for different values of the modulation index m. Note that the maximum modulation percentage without envelope distortion is 100%. The normal modulation range is then up to 100%. When the modulation index m exceeds 1, the waveshape is said to be *overmodulated*.

If we allow for the highest modulation index ($m = 1$) in Eq. (21-5), we see that each sideband peak amplitude is half the peak amplitude of the carrier component. The component amplitudes for different modulation index values are shown in Fig. 21-3. For $m = 1$, of the total power contained in the AM signal, one-half is at the carrier frequency and one-quarter in each of the two sidebands. (The power is proportional to the square of the voltage amplitude.) The two sidebands actually contain the same signal information, and the carrier component has no (modulating) signal information. Consequently, only one-quarter of the total power in the complete AM signal carries the necessary information.

We can employ different schemes to improve efficiency. For example, in the *suppressed-carrier* system only the two sidebands are transmitted, with a 50% saving in the power needed. In the *single-sideband* (SSB)

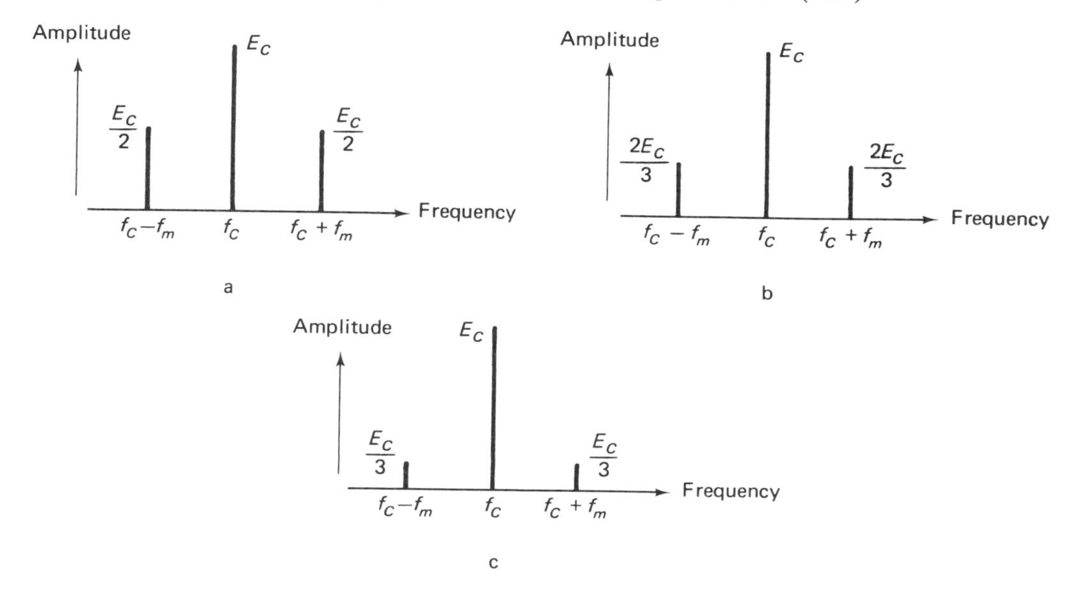

Figure 21-3. AM frequency spectrum for different modulation indexes: (a) $m = 1$, (b) $m = 0.65$, and (c) $m = 0.33$.

system, the carrier and one of the two sidebands are suppressed; only the single remaining sideband is transmitted. Although these systems offer more efficient operation, their detection requires quite complex receivers and they are not used in the broadcast band, although they are used in the amateur bands.

21.2 SUPERHETERODYNE AM RECEIVERS

We now turn our attention to the AM receiver whose role it is to take the AM signal, which is available in the form of electromagnetic radiation, process it, decode the modulating signal, and eventually reproduce the original sound (voice or music). We call this type a *superheterodyne* AM receiver. A block diagram is shown in Fig. 21-4. We shall discuss the complete receiver in general and then focus on each of the parts separately, examining typical circuits.

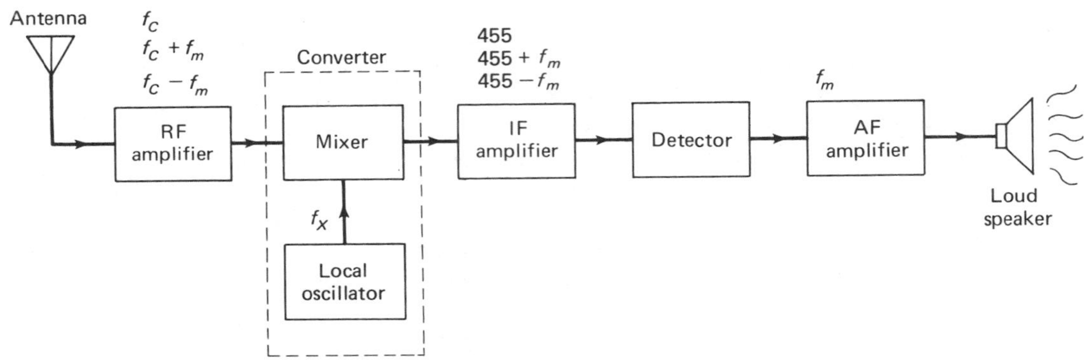

Figure 21-4. Block diagram of a superheterodyne AM receiver.

The AM signal is first picked up by an *antenna*, which is a loop-stick inductor. The antenna coil, together with a tuning capacitor, forms the input tuned circuit for a *radio-frequency* (RF) amplifier. This amplifier is tuned to the carrier frequency of the station to be received. The bandwidth of the RF amplifier is wide enough to pass both sideband signals as well as the carrier signal. At the same time, the selectivity of this amplifier should be high enough to reject (i.e., attenuate) all other signals of adjacent stations. The Federal Communications Commission (FCC) allocates carrier frequencies so that there is a 10 kHz minimum separation between adjacent stations; however, in any one locality, there is usually an even larger separation between adjacent stations. The primary role of the RF amplifier is to select and amplify the desired AM signal.

The next stage is the *converter* stage. The converter changes the AM signal centered at the carrier frequency to an AM signal centered at the *intermediate frequency* (IF). This procedure is accomplished with the aid of the *local oscillator* signal, which is tuned in such a way that it is always a fixed frequency away from the carrier. The tuned circuits in the RF amplifier and the local oscillator utilize capacitors mounted on a common shaft. Therefore, once the local oscillator is adjusted for the proper frequency difference between itself and the center of the passband of the RF amplifier, the frequency difference between the two remains the same even when tuned to a different station.

The output of the converter is an AM signal centered around 455 kHz (the intermediate frequency). This signal is then further amplified in the *intermediate frequency* (IF) amplifier stage (or stages).

Next the signal is decoded in the *detector*, as depicted in Fig. 21-4. The detector removes the carrier and provides an output which contains only the original modulating signal frequencies. (These are in the audio range.)

The audio signal from the detector is further amplified in the AF amplifier stage (or stages) and finally applied to the loudspeaker for reproduction.

Before we proceed to the actual circuits, we should familiarize ourselves with some of the more common terms used to specify receiver performance.

Fidelity: Refers to how well and truthfully the receiver has reproduced the original sound. You should realize that the overall fidelity is only as good as the fidelity of any and all the intermediate processes. For example, suppose that the source of music—an orchestra or band—is first recorded on tape; then a master record is made from this tape; the record is produced in quantity, played, and transmitted by the radio station; then the signal is received and played back on an AM receiver. The eventual quality of reproduction is a function of *all* the processes; even the best receiver could not make up for shortcomings in the processes leading up to transmission.

Selectivity: The ability of the receiver to reject unwanted stations (also see Chap. 11).

Sensitivity: Specifies the minimum signal strength that the receiver can respond to. It is usually specified in mV/meter.

21.2.1 RF Amplifier

As was stated earlier, the role of the RF amplifier is twofold. First, it must be selective; second, it must provide gain. A typical RF amplifier is illustrated in Fig. 21-5. Transformer $T1$ is the antenna, tuned to the desired frequency by capacitor $C1$, which is one gang of a multigang variable capacitor. Capacitors $C3$, $C4$, and $C5$ are dc blocking and bypass capacitors, all of which are essentially short circuits at the carrier frequency. The second tuned circuit is made up of the primary of $T2$ and variable capacitor $C2$, which shares a common shaft with $C1$. Base bias is provided by the AGC (automatic gain control) input, which will be discussed later.

The circuit shown is a double-tuned circuit; as such, it may be either synchronously or stagger tuned. The output is taken across the secondary

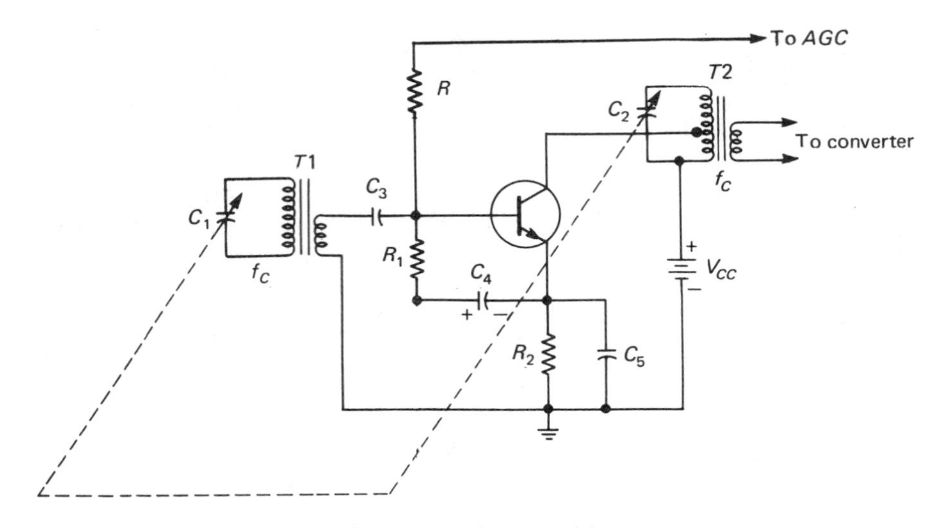

Figure 21-5. Tuned RF amplifier stage.

of $T2$. It contains only signals close to the carrier frequency selected by the tuned circuits. Certain receivers dispense with the RF amplifier and apply the antenna output directly to the converter. However, noise can be significantly reduced if we use one or more stages of RF amplifiers. Furthermore, there is an obvious increase in signal amplitude caused by the gain of the RF amplifier.

21.2.2 Converter

The converter circuit changes the frequency of the signal. Frequency conversion, or *heterodyning*, produces a *beat*, or difference frequency, between two signals. In the case of the superheterodyne receiver, the input to the converter is at the carrier frequency; the local oscillator provides the second signal, which is heterodyned with the incoming signal to produce the desired fixed-frequency IF signal.

Any nonlinear device (a transistor or FET) may be used as a converter. We can express the nonlinear relationship between the input voltage of a device and its output current as a power series:

$$i_0 = a + bv_i + cv_i^2 + \cdots \tag{21-6}$$

where the coefficients a, b, and c have the proper dimensions and are characteristic of the particular device being used. The input voltage is also the sum of two signals of different frequencies:

$$v_i = \cos 2\pi f_c t + \cos 2\pi f_x t \tag{21-7}$$

The amplitudes have been left out, because the frequency is the item of importance at this time. For simplicity, we may neglect the higher order terms in Eq. (21-6) and consider the square-law converter:

$$\begin{aligned} i_0 = a &+ b(\cos 2\pi f_c t + \cos 2\pi f_x t) \\ &+ c(\cos 2\pi f_c t + \cos 2\pi f_x t)^2 \end{aligned} \tag{21-8}$$

Expanding the last term above, we obtain:

$$i_0 = a + b(\cos 2\pi f_c t + \cos 2\pi f_x t) \\ + c[\cos^2 2\pi f_c t + \cos^2 2\pi f_x t + (\cos 2\pi f_c t)(\cos 2\pi f_x t)] \quad (21\text{-}9)$$

To simplify Eq. (21-9), we can use the following trigonometric identities:

$$\cos^2 x = \frac{1}{2} + \frac{1}{2}\cos 2x$$

$$(\cos x)(\cos y) = \frac{1}{2}\cos(x+y) + \frac{1}{2}\cos(x-y)$$

The output current is, therefore,

$$i_0 = \cos 2\pi f_c t + \cos 2\pi f_x t + \cos 2\pi 2 f_c t + \cos 2\pi 2 f_x t \\ \times \cos 2\pi(f_c + f_x)t + \cos 2\pi(f_c - f_x)t \quad (21\text{-}10)$$

where the direct current terms and coefficients have been omitted for simplicity. The output current contains frequency components at: f_c, $2f_c$, f_x, $2f_x$, $(f_c + f_x)$, and $(f_c - f_x)$, besides the dc terms. If frequency conversion is desired, the output circuit should be tuned to the difference frequency, $f_c - f_x$, thus rejecting all other signal components.

Had we started with the complete AM signal (the carrier together with the two sidebands) and the local oscillator signal for v_i, we would have obtained additional frequency components. Specifically, we are interested in the components originating in the crossproduct of the two sidebands with the local oscillator signal. As a result, we have these frequency components: $f_c + f_m + f_x$, $f_c + f_m - f_x$, $f_c - f_m + f_x$, *and* $f_c - f_m - f_x$. If the output is still tuned to the difference frequency $f_c - f_x$, and if the tuning is such that it keeps $f_c - f_x = \text{constant} = 455$ kHz, then the converter output will be given by:

$$v_o = \cos 2\pi(455 + f_m)t + \cos 2\pi(455 - f_m)t \quad (21\text{-}11)$$

where all other frequency components have been filtered out by the output tuned circuit. The output is still an AM signal; however it has been translated in frequency from being centered around the carrier frequency to being centered around the IF frequency of 455 kHz.

To understand the desirability of this frequency conversion, consider additional signal processing. The signal still needs to be further amplified. We perform this amplification in the succeeding IF amplifier stages, which are now of *fixed* tuning. Regardless of the carrier frequency, the signal coming out of the converter is always at the IF frequency. For example, if the carrier is at 700 kHz, the local oscillator is automatically tuned to 245 kHz because of a common shaft on the variable tuning capacitors in the RF section and the converter output. The output is centered around the IF (455 kHz) frequency. If the RF section were tuned to receive, for example, 900 kHz, the local oscillator would be tuned to 445 kHz, once again insuring that the output of the converter would be centered around 455 kHz. Thus, any stages after the converter can be pretuned by the manufacturer, and need not be tunable by the user. This feature simplifies the design and reduces the cost of the IF amplifiers.

Basically, there are two ways of achieving frequency conversion. (1) A single transistor can function as both the local oscillator and mixer; this type of circuit is called an *autodyne converter*. (2) Separate devices can be used, one for the local, or beat frequency oscillator (BFO), and one for the mixer. The first type of circuit is shown in Fig. 21-6. The second will be taken up in the next chapter, which deals with FM receivers.

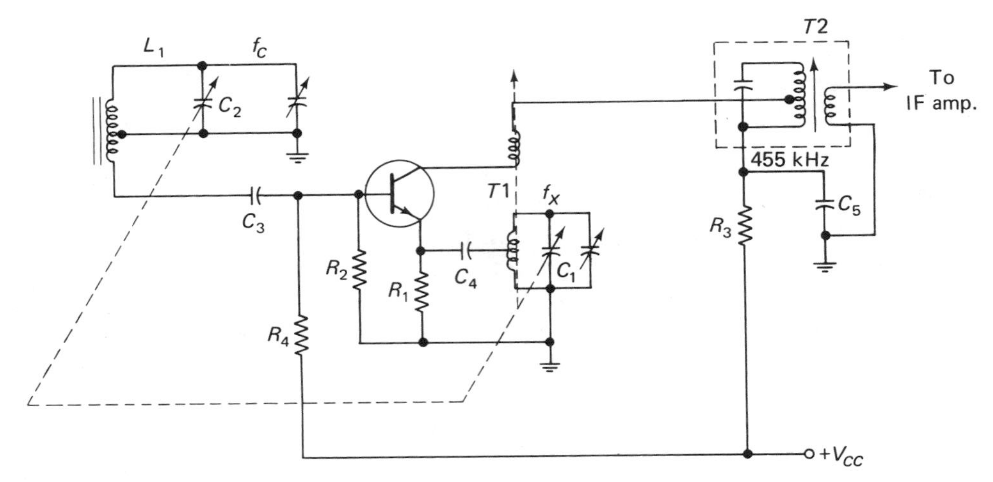

Figure 21-6. Autodyne converter.

The converter circuit illustrated in Fig. 21-6 may derive its input from an RF amplifier stage or directly from an antenna, as shown ($L1$). The oscillator is formed by the tuned circuit of transformer $T1$. The primary of $T1$ is in the collector circuit, with a tuned secondary (tuned by $C1$ and its trimmer capacitor) that feeds back into the emitter circuit through $C4$. Any noise present in the base circuit becomes amplified and fed into the primary of $T1$; with the proper phasing of the secondary, this noise is fed back to the emitter in phase to cause regenerative action. The tuned circuit of $T1$ and $C1$ determines the oscillator frequency and is tuned to f_x.

The mixer circuit is comprised of the RF and oscillator signals adding in the base circuit; they are amplified by the transistor, which is operating in a nonlinear fashion to give the difference (and sum) frequencies at the output transformer $T2$. The input circuits ($L1$ and $C2$) are tuned to the carrier frequency f_c. The output transformer is tuned to the IF frequency (455 kHz). It is available with a fixed-tuning capacitor inside the case. (Touch-up tuning is accomplished by tuning the ferrite transformer core.)

Resistors $R1$, $R2$, $R3$, and $R4$ provide bias for the transistor. $R1$ also provides the regenerative feedback for the oscillator. $C5$ is a bypass capacitor to prevent signal loss across $R3$. $C3$ is a coupling capacitor, chosen to be a virtual short at the carrier frequencies. $C4$ is also a coupling capacitor, chosen to be a short at the oscillator frequencies. The additional capacitors in parallel with $C1$ and $C2$ are small trimmer capacitors; they are available on most air-gap variable capacitors. They can provide fine tuning for exactly the difference frequency of 455 kHz at the output.

21.2.3 IF Amplifier

The IF amplifier circuit is shown in Fig. 21-7. The amplifier takes the output of the converter circuit, which is centered around 455 kHz, and provides the additional gain needed. A receiver may have a single-stage IF amplifier, or it may have as many as four stages of the type illustrated in Fig. 21-7.

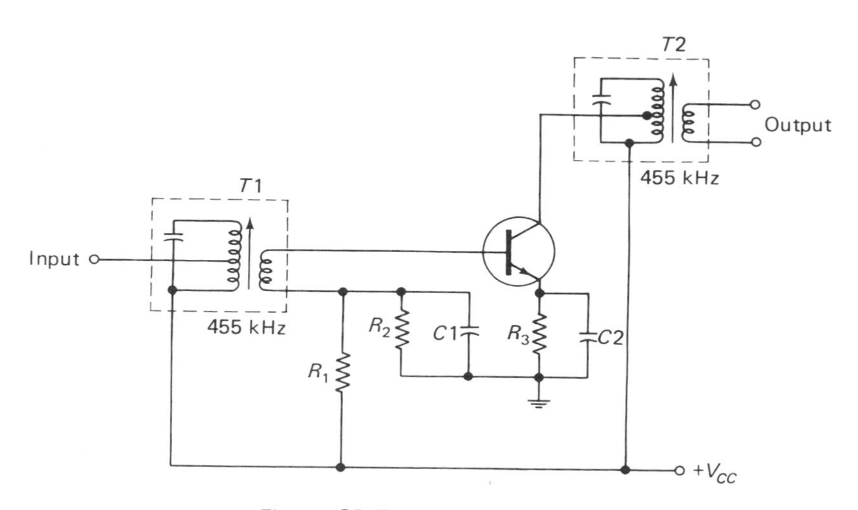

Figure 21-7. IF amplifier stage.

The IF amplifier is a tuned amplifier, but it need not be tuned to a different frequency to receive a different station (for the reasons already mentioned). Tuning is accomplished in the IF transformer $T2$. ($T1$ is the coupling transformer of either the previous IF amplifier or the converter (if this is the first stage in the IF strip). Adjustments in the tuning of the IF amplifier are made by turning the transformer core. For tuning of multiple IF amplifiers, see Chap. 11.

21.2.4 Detector

The output of the IF amplifier is still an AM signal, although it is raised significantly in amplitude from the level at the antenna and shifted in frequency. This signal is applied to the *detector* circuit, which decodes or demodulates the AM signal.

The basic detector circuit, shown in Fig. 21-8, is a diode envelope detector. Capacitors $C1$ and $C2$, together with $R1$, form a low-pass filter, which essentially rejects any signals outside the audio range. Capacitor $C3$ removes any direct current from the output. $R3$ is a volume control.

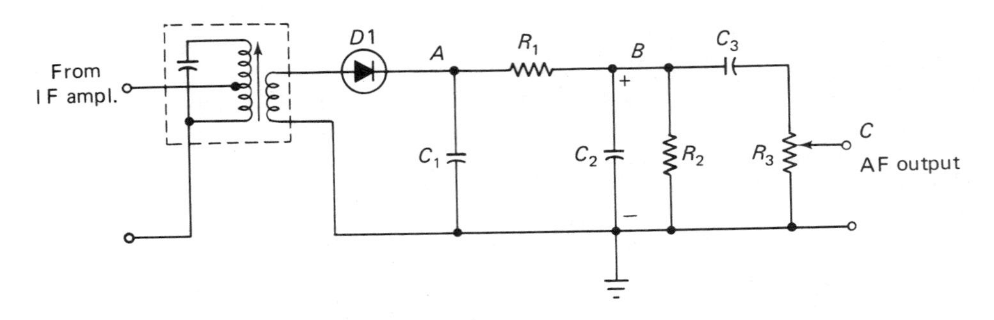

Figure 21-8. Diode detector circuit.

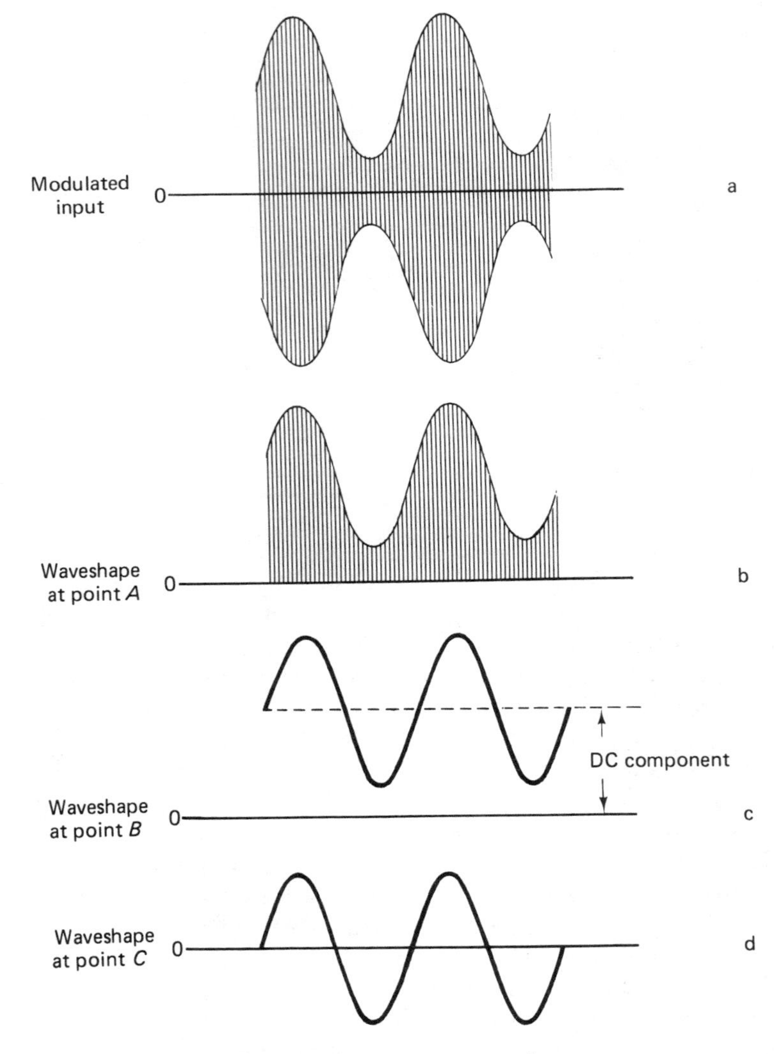

Figure 21-9. Detector waveshapes.

You can understand circuit operation by studying the circuit waveshapes in Fig. 21-9. The AM input signal from the last IF amplifier is depicted in Fig. 21-9(a). This signal is rectified by diode $D1$ to provide the waveshape shown in Fig. 21-9(b) at point A in the detector circuit. The low-pass filters ($C1$, $C2$, and $R1$) remove the high-frequency components from the waveshape to provide the original modulating signal waveshape at point B in the detector circuit, as indicated in Fig. 21-9(c). This waveshape has a certain dc component, as shown, which is eliminated by coupling capacitor $C3$ to give the output waveshape illustrated in Fig. 21-9(d). Resistor $R3$ acts as a volume (loudness) control by determining how much of the available output is applied to the audio amplifier stages.

21.2.5 Automatic Gain Control

Suppose that we want to tune to a distant station (with relatively low signal strength) and adjust the volume control for a normal listening level. Now what would happen if we left the volume control as set and retuned to a nearby station with a very strong signal? We would be blasted by the sound. Moreover, had we first tuned to the nearby station and adjusted the volume control for a normal listening level, chances are that at the same volume setting we might not even be able to "find" the distant station. To eliminate, or at least minimize, this problem, some form of *automatic volume control* (AVC) or *automatic gain control* (AGC) is necessary.

Automatic gain control is achieved by sensing the output of the detector and feeding a dc voltage to the input of a transistor to reduce its gain. Figure 21-10 depicts one such scheme. The detector diode $D1$ is connected in such way that it passes only the negative portion of the AM signal arriving from the IF amplifier. The negative dc voltage (V) developed across $C4$ is fed back to the base of the IF amplifier transistor $Q1$.

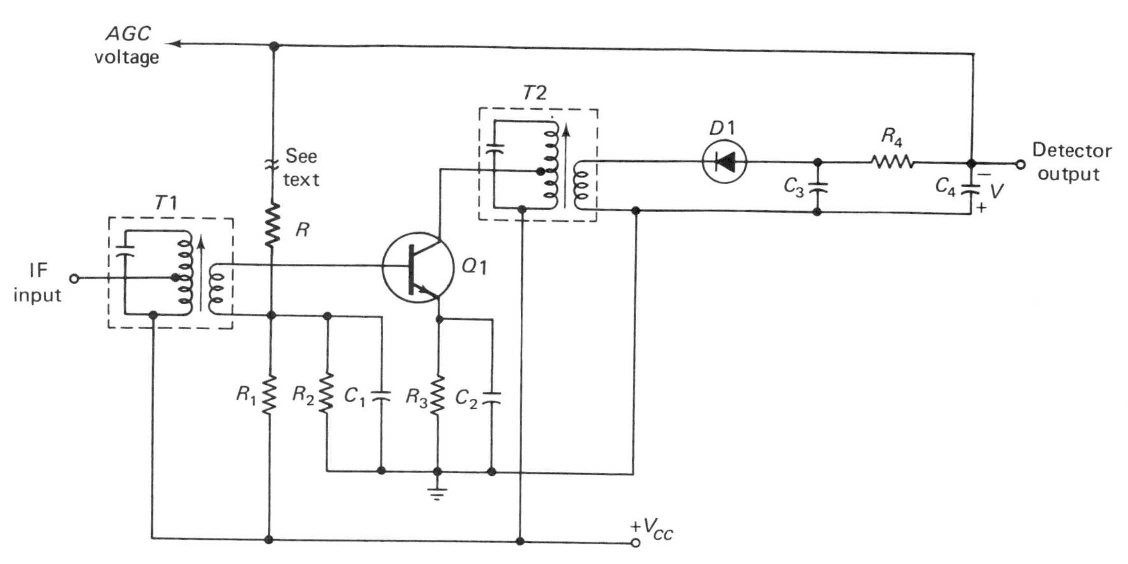

Figure 21-10. AGC circuit.

When the output of the detector increases because of a stronger signal at the antenna, V becomes more negative, and so the bias point for $Q1$ is moved nearer to cutoff. In effect, the amplification of the transistor (its h_{fe}) is reduced, thus keeping the output within certain limits. If the output of the detector should decrease, the transistor is biased to conduct harder. Therefore, its amplification is increased but the output remains essentially fixed. (For a plot of the variation of transistor h_{fe} as a function of its quiescent current, see Fig. 4-13.)

In the case of multiple IF amplifiers, the AGC is usually applied to the first IF amplifier. If the receiver has an RF amplifier, the AGC would be applied to the RF amplifier, as shown in Fig. 21-5. Again the AGC would tend to move the transistor Q-point in such a way that it maintains a constant level at the output, independent of the strength of the signal being received.

The AGC scheme depicted in Fig. 21-10 works relatively well so long as the signals are not extremely weak or extremely strong. Improved performance for very low signals can be achieved by increasing the system gain; however, this gain will be at the loss of performance for strong signals.

An auxiliary diode AGC system can significantly improve the performance. Such a circuit is illustrated in Fig. 21-11. Diode $D2$ is normally reverse biased and not conducting. When the signal level becomes high, the AGC signal from the detector reduces the direct current in the transistor. This reduced current causes the voltage drop across $R5$ to decrease, thus bringing the anode end of $D2$ toward V_{CC}. Eventually $D2$ becomes forward biased and draws off to ground some of the signal from $T2$. In this way it adds to the AGC action.

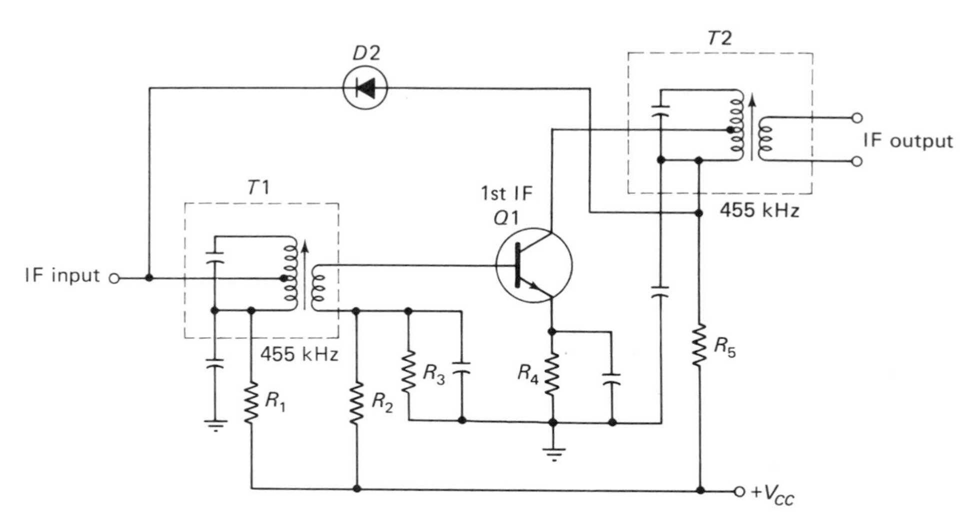

Figure 21-11. Auxiliary AGC (provided by $D2$).

21.2.6 AF Amplifier

The decoded signal from the detector is amplified for the final time in the audio-frequency (AF) amplifier section before it is applied to the loudspeaker. The AF amplifier complexity depends greatly on the size and efficiency of the speaker. For small low-power speakers, the complete AF amplifier might be a single class-A transistor amplifier. For larger and less efficient speakers, we may need some form of a driver plus push-pull or complementary-symmetry configuration (see Chap. 12).

21.3 EXAMPLES OF AM RECEIVERS

Figures 21-12 and 21-13 give examples of transistor AM receivers: One is battery powered and one is line powered.

The circuit shown in Fig. 21-12 uses low-voltage transistors and operates from a 3 V battery. It contains no RF amplifier. Instead, it relies on the two IF amplifiers ($Q2$ and $Q3$) and the AF amplifier for sufficient gain. It uses the autodyne converter $Q1$. $D1$ acts as the detector. AGC feedback is accomplished via $R6$ and $C11$ to the base of the first IF amplifier $Q2$. Note that the AGC voltage is positive in this circuit, because *PNP* transistors are used. Capacitor $C3$ acts as a low-pass filter, with $C10$ blocking the direct current from entering the AF amplifier. The AF amplifier consists of (1) a transformer coupled driver amplifier $Q4$ and (2) a push-pull power amplifier output stage, operating in class AB, that is also transformer-coupled to the speaker.

The circuit illustrated in Fig. 21-13 is a line-operated receiver. The dc power is supplied by the half-wave rectifier $D3$ and associated filter capacitors. This circuit also features the autodyne detector circuit, $Q1$ and its tuned circuits, with no RF amplifier. Two IF amplifiers, $Q2$ and $Q3$, are used with AGC from the detector $D1$ to the base of $Q2$ through the 15 kΩ resistor. It also has an auxiliary AGC circuit consisting of $D2$ and the 180 Ω resistor. Transistor $Q4$ is the driver in the AF amplifier. It is transformer-coupled to the push-pull output stage ($Q5$ and $Q6$), which is operating in class AB. The speaker is transformer-coupled at the output.

Review Questions

1. What does the term "modulation" mean?
2. What is amplitude modulation?
3. What are the components of the complete AM signal?
4. Which of the components of an AM signal carry information?
5. Over what frequency band are commercial AM broadcasts made?
6. What are the features of a suppressed carrier system?
7. What are the features of a single-sideband system?
8. What is heterodyning?
9 What are the features of a superheterodyne receiver?
10. What is the role of an RF amplifier?
11. What is the role of a converter?

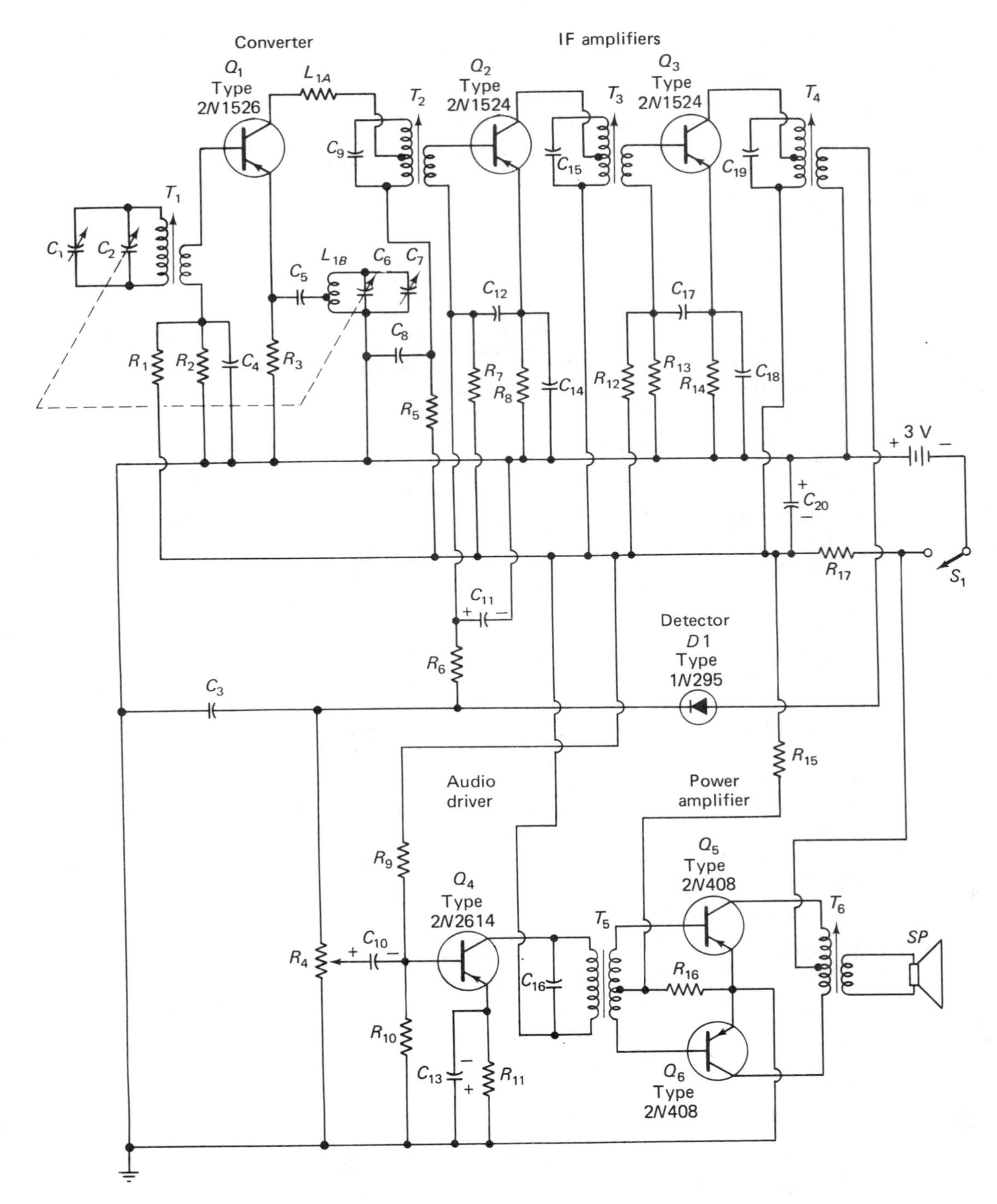

Figure 21-12. Portable 3 V AM receiver. (Courtesy of RCA)

Parts List

C_1 = trimmer, 3 to 15 pF
C_2, C_6 = ganged tuning capacitor, C_2 = 9.5 to 141 pF; C_6 = 7.2 to 109 pF
C_3, C_4 = 0.02 μF, ceramic
C_5 = 0.005 μF, ceramic
C_7 = trimmer, 3 to 20 pF
C_8, C_{12}, C_{14}, C_{17}, C_{18} = 0.05 μF, ceramic
C_9 = 128 pF (part of T_2)
C_{10} = 2 μF, electrolytic, 3 V
C_{11} = 10 μF, electrolytic, 3 V
C_{13}, C_{20} = 100 μF, electrolytic, 3 V
C_{15} = 125 pF, (part of T_3)
C_{16} = 0.005 μF, ceramic
C_{19} = 125 pF, (part of T_4)
L_1 = oscillator coil; wound from No. 3/44 Litz wire on coil form suitable for a No. 10-32 slug; L_{1A}, 19 turns; L_{1B}, 155 turns, tapped at 8 turns from ground end, tunes with 100 pF at 990 kHz

R_1, R_9 = 10000 ohms, 0.5 watt
R_2 = 3900 ohms, 0.5 watt
R_3, R_{15} = 1500 ohms, 0.5 watt
R_4 = volume-control potentiometer, 5000 ohms, audio taper (part of assembly with ON-OFF switch S_1)
R_5 = 470 ohms, 0.5 watt
R_6 = 6800 ohms, 0.5 watt
R_7 = 39000 ohms, 0.5 watt
R_8 = 330 ohms, 0.5 watt
R_{10} = 2700 ohms, 0.5 watt
R_{11} = 270 ohms, 0.5 watt
R_{12} = 10000 ohms, 0.5 watt
R_{13} = 2200 ohms, 0.5 watt
R_{14} = 240 ohms, 0.5 watt
R_{16} = 100 ohms, 0.5 watt
R_{17} = 47 ohms, 0.5 watt
S_1 = ON-OFF switch (part of assembly with potentiometer R_4)
SP = speaker; voice-coil impedance, 12 to 15 ohms
T_1 = antenna transformer; primary, 110 turns of No. 10/41 Litz wire wound on a 3/4″-by-1/8″-by-4″ ferrite rod (pitch, 50 turns per inch); secondary, 6 turns of No. 10/41 Litz wire wound at the start of the primary; Q = 100 with transformer mounted on chassis; transformer should tune with 135 pF at 535 kHz
T_2 = 1st if transformer; Thompson-Ramo-Wooldridge EO-13550, or equivalent
T_3 = 2nd if transformer; Thompson-Ramo-Wooldridge EO-13551, or equivalent
T_4 = 3rd if transformer; Thompson-Ramo-Wooldridge EO-13552, or equivalent
T_5 = driver transformer; primary impedance, 10000 ohms; secondary impedance, 2000 ohms, center-tapped
T_6 = output transformer; primary impedance, 100 ohms, center-tapped; secondary impedance, 15 ohms (to match voice-coil impedance of 12 to 15 ohms)

Figure 21-12. (Cont.)

12. What is the role of an IF amplifier?
13. What is the role of the detector?
14. What is the role of the AF amplifier?
15. What is meant by fidelity?
16. What is meant by selectivity?
17. What is meant by sensitivity?
18. What is AGC (or AVC)?
19. Why is AGC necessary?
20. What is the difference between a converter containing a mixer and local oscillator and one using the autodyne principle?
21. What is the action of the auxiliary AGC circuit in Fig. 21-11?
22. What is an envelope detector? Explain its operation.
23. What is the intermediate frequency used in the broadcast AM system?
24. Why is it advantageous to amplify the AM signal after it has been converted to the IF frequencies as opposed to before it has been converted?
25. How and why is the common tuning of the RF amplifier and local oscillator accomplished?

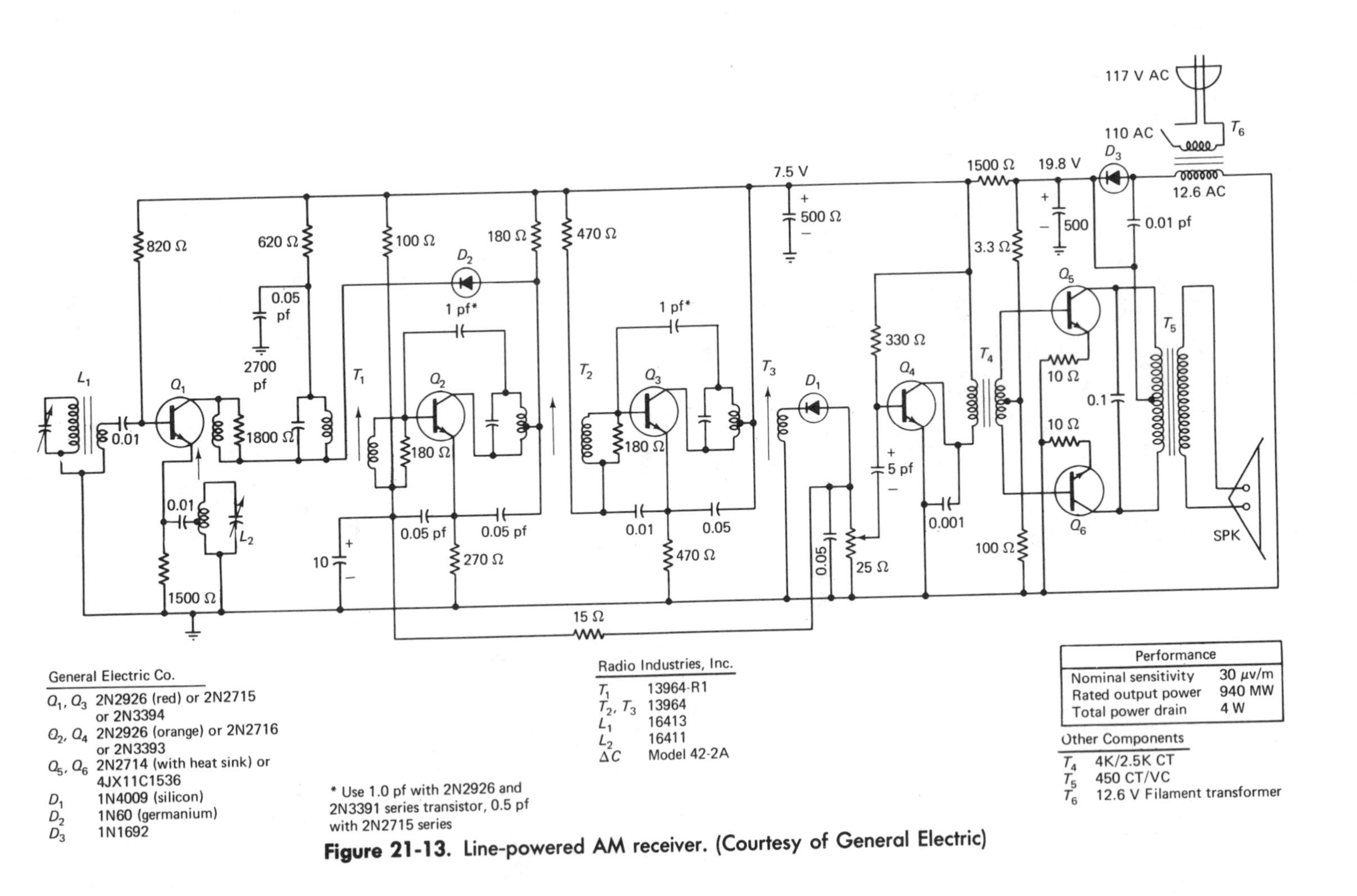

Figure 21-13. Line-powered AM receiver. (Courtesy of General Electric)

26. What is the mechanism by which a single station is selected among all the others in a superheterodyne receiver?

Problems

1. Make a plot of the AM signal if the carrier frequency is 10 kHz and the modulating signal frequency is 1 kHz. (Plot at least 20 cycles of the carrier signal.)
2. An AM signal with carrier frequency of 900 kHz is modulated by audio signals of 400, 800, and 1500 Hz. Sketch the frequency spectrum. What is the minimum bandwidth of the RF amplifier required to receive the signal?
3. An AM signal with frequency components at 1, 1.001, and 0.999 MHz is received by an AM receiver. Specify the tuning of (a) the RF amplifier, (b) the local oscillator, (c) the output of the converter, and (d) the IF amplifiers.
4. For the signal in Problem 3, specify the frequencies at each of the blocks in Fig. 20-4. What is the frequency of the signal that will eventually be reproduced by the loudspeaker?
5. In terms of the operation of an AM receiver, explain why the tuning dial is not linear, that is, why frequencies near the high end of the AM band are compressed.
6. The receiver in Fig. 21-12 is tuned to receive a broadcast at a center frequency of 770 kHz. Specify the frequencies to which each of the tuned circuits must be tuned.
7. Draw a block diagram of the receiver in Fig. 21-12 placing the transistor numbers inside the appropriate blocks. Also specify the frequencies present in each block.

22

FM Receivers

In this chapter we shall examine the basic principles of *frequency modulation* (FM), together with the basic receiver circuits. Because of FM, high-quality (hi-fi) audio programs can be transmitted over commercial stations. Stereo FM broadcasts have now become commonplace. Indeed, with modern FM-multiplex stereo receivers, listening to an FM broadcast is the next best thing to hearing the program in person.

22.1 PRINCIPLES OF FREQUENCY MODULATION

As we saw in the previous chapter, one way of encoding an audio signal onto a carrier is by modulating the carrier *amplitude* in accordance with the audio signal. Another, and in some respects a better, scheme is to modulate the carrier *frequency* in accordance with the audio signal. One of the drawbacks to AM is the basic lack of immunity to *noise*, i.e., random unwanted signals. Noise in an AM system causes undesirable amplitude changes whose appearance in the decoded signal destroys the fidelity of reproduction. These problems are eliminated in FM systems, because the carrier amplitude does not convey any signal information.

Study Fig. 22-1. If the carrier signal remains at a frequency f_c and the modulating signal is also as shown, the FM waveshape will have constant amplitude and a frequency that is varied according to the amplitude of the modulating signal. This variation is not easily indicated graphically although an attempt has been made in Fig. 22-1. When the modulating signal is zero, the FM wave is at the nominal carrier frequency f_c. When the modulating signal is positive, the FM wave has a frequency higher than f_c by an amount proportional to the modulating signal amplitude. When the modulating signal is negative, the FM wave has a frequency below f_c by an amount proportional to the amplitude of the modulating signal. Information regarding the frequency of the modulating signal (the rate at which the modulating signal amplitude is varied) is encoded into the FM wave as the *rate* at which the frequency varies around the carrier frequency. It should be apparent that, in an FM system, noise signals that vary the amplitude of the FM signal do not present a problem and are not propagated to the output in a receiver.

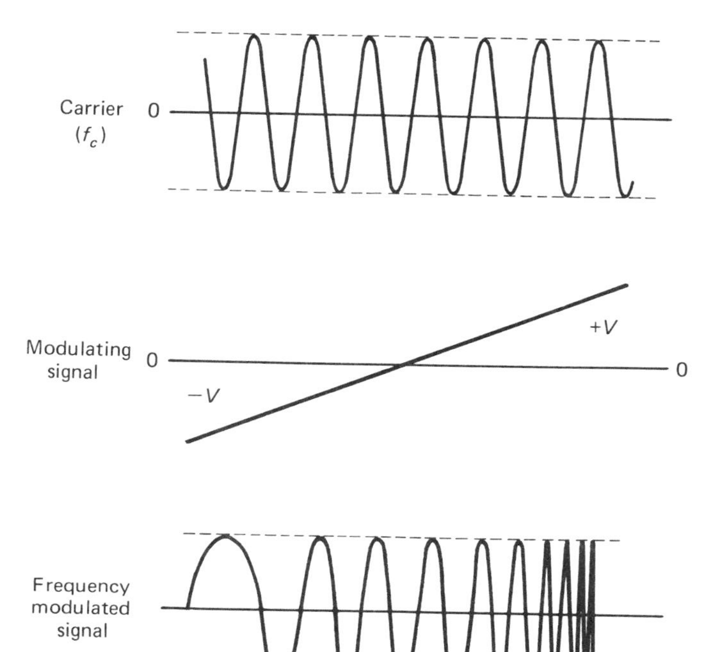

Figure 22-1. Waveshapes for FM.

The basic block diagram for an FM receiver is shown in Fig. 22-2. There is a marked similarity between the AM and FM receivers in the front end. Both employ an RF amplifier, a frequency converter (mixer and oscillator), and IF amplifiers. However, they differ in the signal conditioning after the IF amplifier as well as in the detection. Although the RF amplifier operates in much the same manner as described previously, it is tuneable only over the FM band, from 88 to 108 MHz. Frequency conversion is accomplished by beating the incoming FM signal from the RF amplifier with the signal provided by the local oscillator in the mixer stage. The local oscillator is adjusted to track the incoming signal so as to

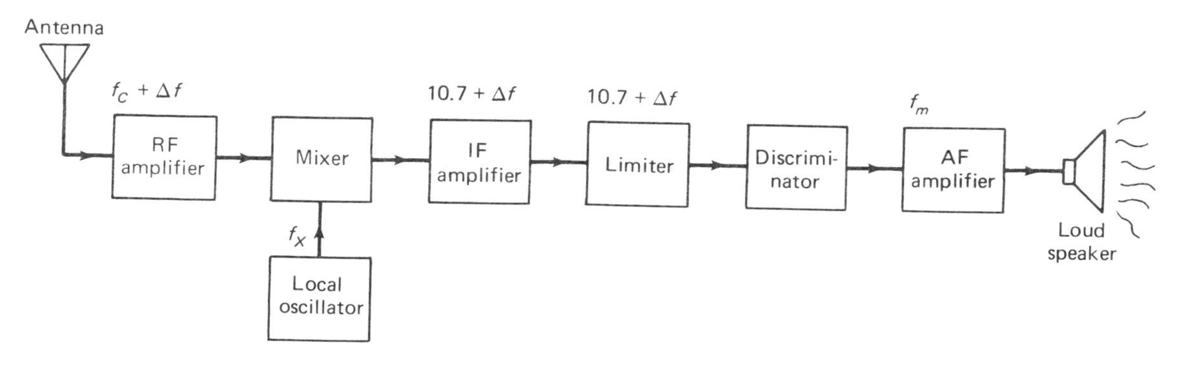

Figure 22-2. Block diagram for a basic FM receiver.

provide a fixed difference frequency, the IF. In the case of FM, the IF is 10.7 MHz.

Another major difference between AM and FM RF and IF amplifiers is their bandwidth. In AM the bandwidth need be only 10 kHz because that is the width of the sidebands. In FM, the bandwidth requirement is more extensive, and a typical stage might have a bandwidth of 200 kHz. We need much wider bandwidth in FM because of the higher upper limit in the modulating signal frequency (typically 15 kHz) as well as the modulation index used. With a modulation index of 5 and the highest modulating frequency of 15 kHz, the required bandwidth is slightly under 200 kHz.

As in AM, the IF amplifier in an FM receiver may contain a number of stages, usually no less than two. The signal coming from the IF amplifier is basically at 10.7 MHz (±the frequency deviation of the modulating signal) and is fed into the *limiter* stage. The limiter makes certain that the amplitude of the FM signal provided to the detector is of constant amplitude. The amplitude of the FM signals carries no information; therefore, it is important to have all the frequency components at the same amplitude. Detection is accomplished in the frequency *discriminator*, which provides the audio output proportional to the frequency of the incoming signal. Processing of the audio signal is carried out in the AF amplifier section. Usually, the AF amplifier in an FM receiver is quite a bit more sophisticated and at a higher power than its AM counterpart because of the better fidelity available from an FM system. Its audio signal is usually amplified in low-distortion, high-fidelity (hi-fi) amplifiers to drive high-quality speakers. Frequently, the first half of an FM receiver, called the *tuner*, containing everything except the AF amplifiers and speakers is manufactured separately to allow the purchaser to use the audio reproducing system of his choice.

If we compare the costs of FM and AM receivers, we quickly see that the FM tuner is more sophisticated and so more costly. Both the RF and IF sections operate at a much higher frequency. However, this sophistication alone does not justify the price difference between the two. The cost differential stems mainly from the fact that in the FM system, more care is taken in the processing of the signal. For example, a high-quality FM tuner might have two or three RF amplifiers, whereas its AM counterpart might have one or even none. The FM set might have two to four IF stages; the AM set might have one or two.

22.2 FM TUNER

The basic FM tuner, as described earlier, contains the RF amplifier, mixer and local oscillator, IF amplifiers, a limiter, and a detector or discriminator. High-quality tuners also incorporate an AGC circuit (which functions almost the same as discussed for the AM system) and an *automatic frequency control* (AFC) for the local oscillator. The AFC circuit senses the accuracy of the IF frequency. If it is not 10.7 MHz, the AFC circuit must force the local oscillator to adjust its frequency until the IF frequency is exactly 10.7 MHz. The need for this frequency is critical; any error in the IF frequency would be decoded by the detector as an audio signal.

22.2.1 FM Front End

The FM signal, which is available in the form of electromagnetic radiation, is picked up by the antenna. The input circuit of the RF amplifier tunes the antenna to the specific station to be received. A typical front end consists of an FET RF amplifier and mixer ($Q1$ and $Q2$) and a BJT local oscillator ($Q3$). It is depicted in Fig. 22-3. The RF amplifier is a double-tuned circuit sharing a commonly tuned capacitor shaft with the local oscillator. The RF amplifier is tuneable over the entire FM band, from 88 to 108 MHz. The local oscillator is a tuned-collector type with a feedback path through $C7$ and the variable-capacitance (*varactor*) diode to the base circuit. The net capacitance between collector and base of $Q3$, together with the tuned circuit ($L3C3$), determines the frequency of oscillation for the local oscillator. This frequency is adjusted with the tuning of the RF amplifier to provide a constant difference frequency between the incoming RF and the oscillator output of 10.7 MHz. The oscillator signal is coupled to the source of mixer FET $Q2$ by $C6$, while the RF signal is applied to the gate of the FET. The transfer characteristics of an FET have a square term; that is, the output current is proportional to the square of the input voltages. The FET, therefore, is ideal for mixing (see section 21.2.2). Resistors $R2$ and $R6$ provide bias for $Q1$; $R5$, $R7$, and $R8$ provide bias for $Q2$; while $R3$, $R4$, $R9$, and $R10$ provide bias for the oscillator transistor $Q3$.

The AGC signal is a dc voltage that biases the RF amplifier FET Q1 for more gain or less gain in order to keep the amplitude of the signal constant, regardless of the weakness or strength of the signal transmitted by the station. The AFC signal is also averaged by $C8$ so as to provide dc bias for the varactor diode $D1$. $D1$ is always reverse biased; depending on the exact bias, if offers the appropriate capacitance to correct the local oscillator. In this way, the oscillator provides the proper frequency for mixing with the radio frequency to give the desired 10.7 MHz IF. The source for the AGC and AFC signals will be discussed later.

Integrated circuits are widely used in FM systems. The circuit diagram for an IC RF amplifier is depicted in Fig. 22-4. It features a differential pair, $Q1$ and $Q2$, together with the current source transistor $Q3$. Figure 22-5 shows the RCA CA3005 RF amplifier connected to provide the same functions as the discrete component front-end circuit of Fig. 22-3. The RF input is applied to the base of $Q3$, which functions as the RF amplifier. We tune the RF amplifier by using the single-tuned circuit consisting of $L1$ and $C1$. Transistors $Q1$ and $Q2$ are connected in a positive feedback configuration to form the local oscillator. The frequency of this oscillator is determined by the tuned circuit formed by $L2$ and $C2$. Mixing the RF and the local oscillator signals also takes place in the

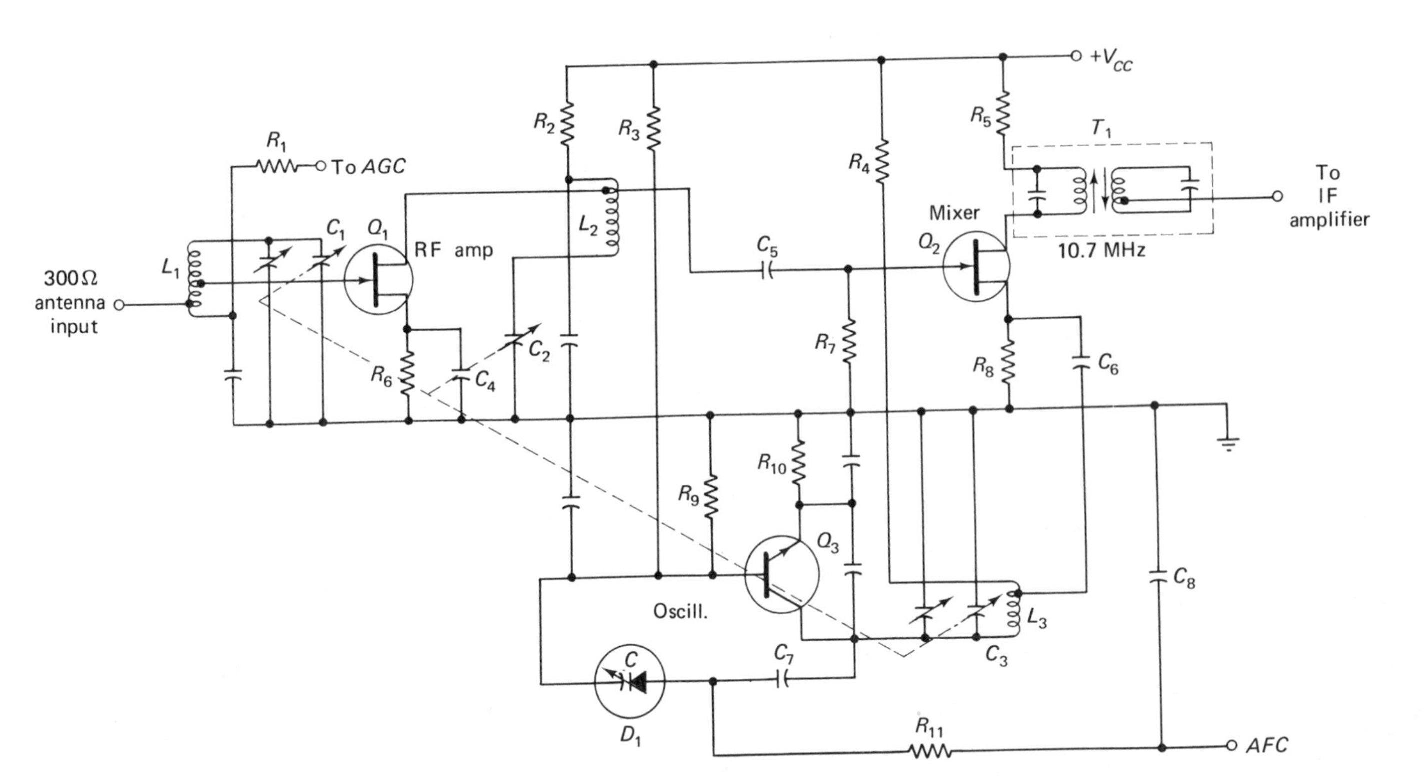

Figure 22-3. FM front end showing AGC and AFC connections.

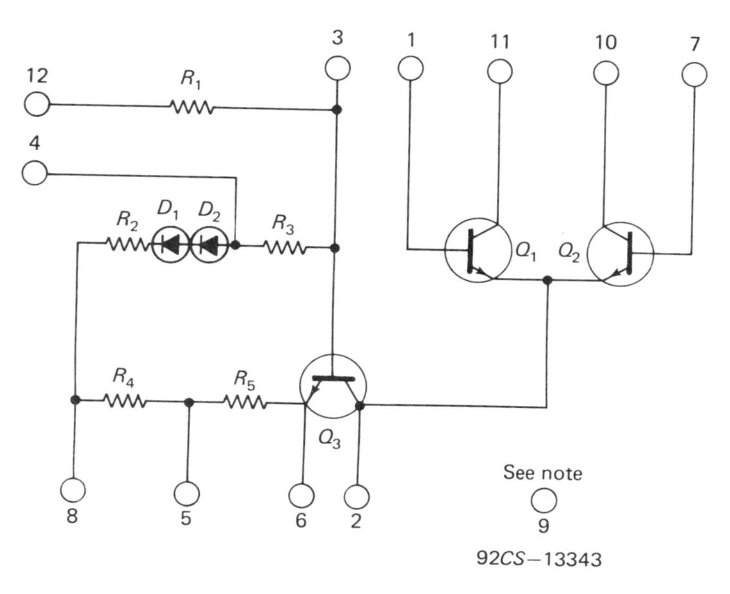

Note: Connect terminal 9 to most positive dc supply voltage

Figure 22-4. Schematic diagram of RCA CA3005 integrated circuit RF amplifier. (Courtesy of RCA)

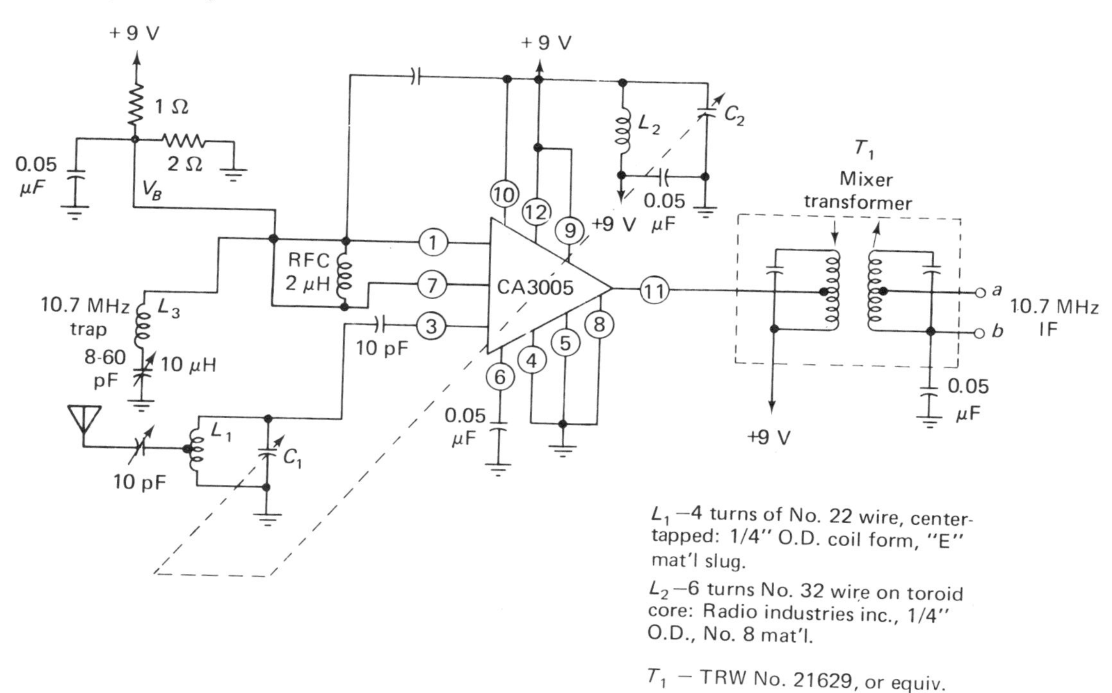

Figure 22-5 Single IC FM front end RF amplifier, local oscillator, and mixer. (Courtesy of RCA)

differential transistors $Q1$ and $Q2$, because the RF signal is injected directly into the emitters of $Q1$ and $Q2$. The output is taken at the collector of $Q1$ and filtered by the tuned transformer $T1$ to remove all signals not near the IF frequency of 10.7 MHz.

A dual-gate MOSFET suitably performs the function of a mixer. Both gates control the amount of current flowing in the drain-source circuit, so that the RF signal is applied to one gate and the oscillator signal to the other. The difference-frequency signal is available at the drain. Such a circuit, using a dual-gate MOSFET as a mixer, is shown in Fig. 22-6. The RF (FM) signal is developed across $R2$ and applied to gate 1; the oscillator signal is developed across $R1$ and applied to gate 2. Transformer $T1$ is tuned to the IF frequency; it rejects all other signals not within its pass-band centered at 10.7 MHz. The output of $T1$ shows the use of "tapped" capacitors, instead of tapped inductors, to provide impedance matching.

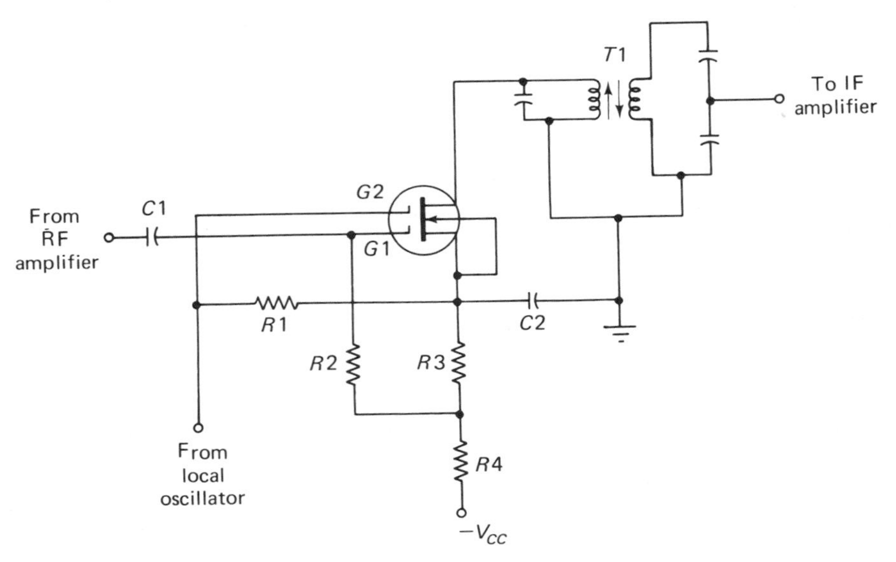

Figure 22-6. Example of a dual-gate MOSFET mixer.

22.2.2 FM Limiter

The limiter circuit maintains constant amplitude for all signals within the passband of the IF amplifiers. Even though the receiver may have an AGC circuit, the limiter is still needed. The AGC circuit keeps the system amplification such that the detector output amplitude is essentially constant for all different signal levels at the antenna.

However, the AGC circuit can do nothing about a certain signal that may have different amplitudes at different frequencies. This condition may exist as a result of *electromagnetic interference* (EMI), or distortion of the signal between the transmitting and receiving antennas. The limiter clips the incoming signal, so that its output is at a constant amplitude over the

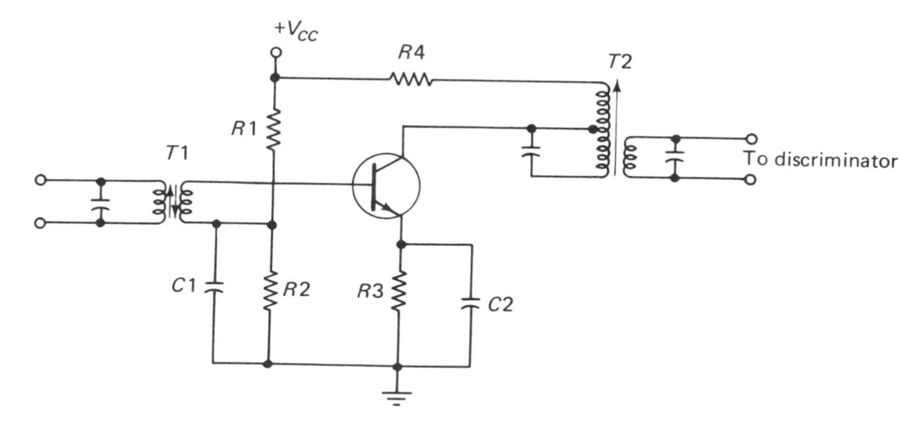

Figure 22-7. BJT limiter circuit.

entire range of frequencies supplied to it by the IF amplifier. The wave-shape distortion introduced by the clipping action is removed by the tuned circuit in the output of the limiter.

Figure 22-7 depicts a typical BJT limiter circuit. The transistor is biased in such a way that it saturates when the positive peaks of even the weakest signal occur. The input transformer tuned circuit is the last IF transformer. The output circuit is also tuned to provide the detector (or discriminator) with an IF FM signal having constant amplitude.

The transfer characteristics for a limiter circuit are shown in Fig. 22-8. For very low input levels, the limiter acts as a linear amplifier. For any input levels above the limiter threshold, V_{Lim}, limiter action occurs and the output is clamped to a fixed value V_{OL}.

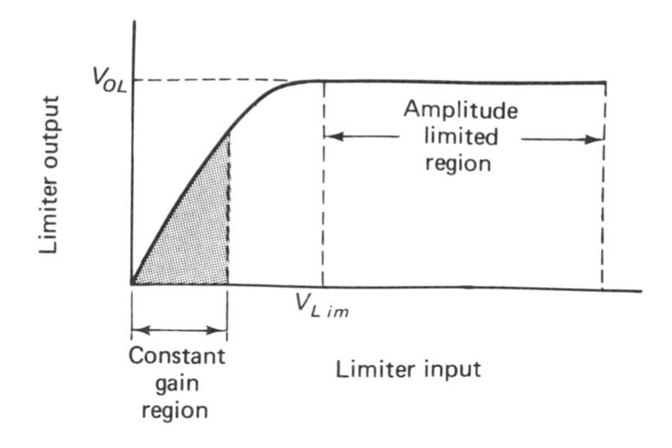

Figure 22-8 Limiter transfer characteristics.

The limiter circuit is not essential with all detectors (namely, the ratio detector). However, it can improve the performance of any detector.

22.2.3 FM Detection

To detect, i.e., decode, an FM signal we need a circuit that gives an output voltage whose amplitude is proportional to the frequency of the input signal. Such circuits are generally termed *frequency discriminators*. Perhaps the simplest means of FM detection uses a tuned amplifier circuit that is detuned either above or below resonance. For example, suppose that we tune the amplifier to have a resonant frequency f_o above the center of the IF band (10.7 MHz), as indicated in Fig. 22-9. The output of the tuned amplifier will have an amplitude that is proportional to the frequency of the incoming signal but over a limited range.

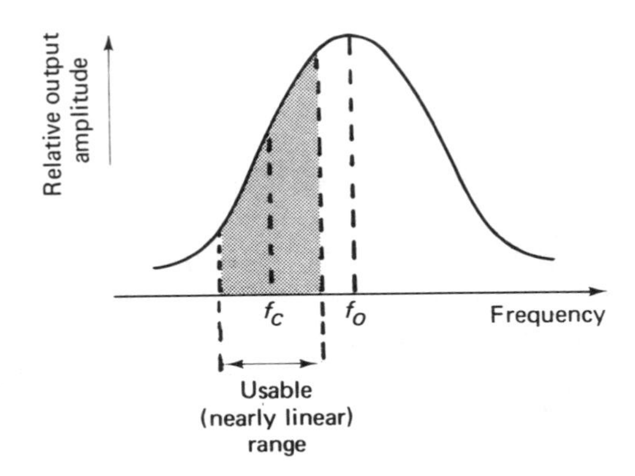

Figure 22-9. Using the skirt of a tuned circuit for FM detection.

Figure 22-10 illustrates another method, which is somewhat more efficient. It uses the *Foster-Seeley discriminator*. Operation of this circuit hinges on the phase relationship between the primary and secondary voltages of the doubly tuned transformer, consisting of $L1$ and $L2$. We choose capacitor $C2$ in order to provide a short at the signal frequencies (around 10 MHz) and to furnish isolation at dc. At resonance, the primary

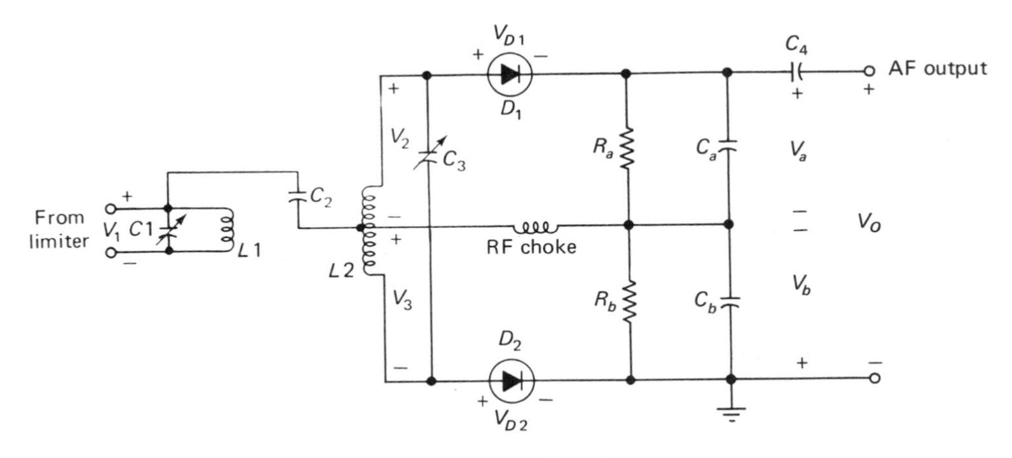

Figure 22-10. Foster-Seeley discriminator.

voltage $V1$ is 90° out of phase with the total secondary voltage $V2+V3$. The voltages applied to the two diodes are exactly equal because of the center tap on $L2$. The output voltages V_a and V_b are also equal.

When the input signal deviates from the resonant frequency, the phase between the primary and secondary voltages is either above or below 90°. When the input signal frequency is below the resonant frequency, the phase shift between the primary and secondary is larger than 90°, as shown in Fig. 22-11. The phasors for V_a and V_b correspond to the diodes conducting. The output voltage is essentially direct current and is given by the difference between the magnitudes of V_a and V_b. Obviously, as long as the frequency of the input signal is equal to the resonant frequency, the output voltage is zero. When the input frequency goes below the resonant frequency, the magnitude of V_a decreases, whereas the magnitude of V_b increases. The output voltage is then negative. The larger the frequency deviation of the input signal below resonance, the larger V_b and the smaller V_a become. As a result, an output voltage is below zero (negative) by an amount proportional to the frequency deviation below resonance.

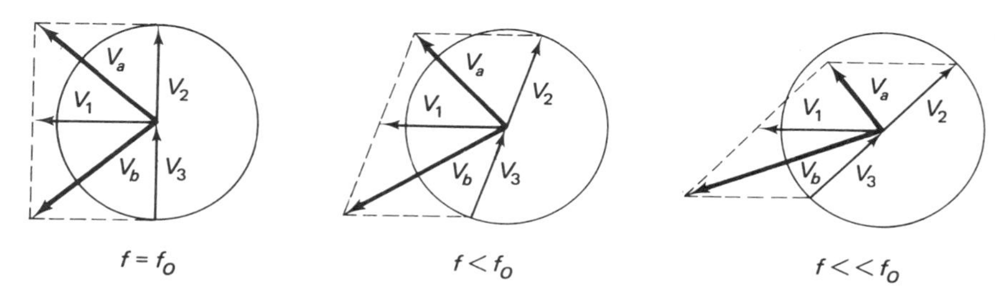

Figure 22-11. Discriminator phasor diagrams: input at or below resonance.

When the input signal deviates above resonance, the phase shift between the primary and secondary of the transformer becomes less than 90°, as indicated in Fig. 22-12. The magnitude of V_a now exceeds the magnitude of V_b, and the output becomes positive. The more the input frequency exceeds the resonant frequency, the larger V_a and the smaller V_b become. The magnitude of the output voltage in the positive direction is proportional to the amount by which the input frequency exceeds the

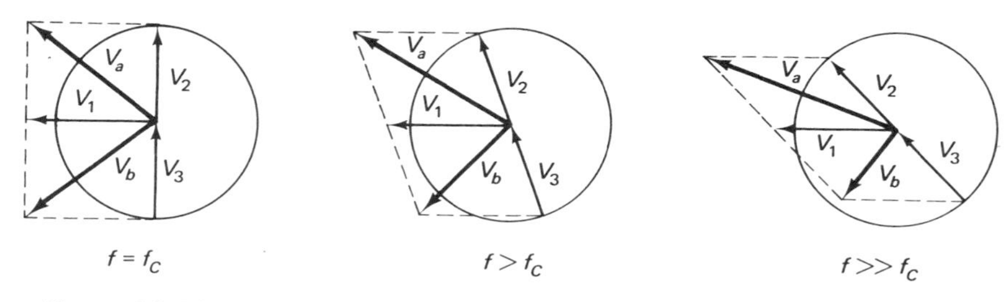

Figure 22-12. Discriminator phasor diagrams: input at or above resonance.

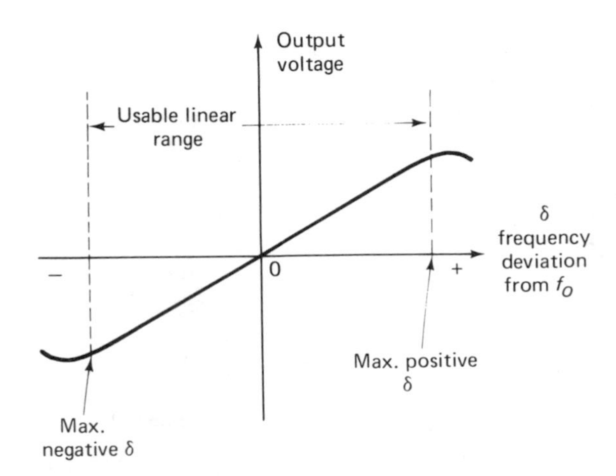

Figure 22-13. Typical discriminator characteristics.

resonant frequency. Figure 22-13 supplies a plot of the output amplitude as a function of the frequency deviation away from resonance for a typical discriminator.

Operation of the discriminator hinges on the limiting action of the previous stage. If a signal of an amplitude lower than all the other signals is allowed to reach the discriminator, it will be decoded as a lower amplitude signal rather than the higher amplitude signal at the same frequency. This decoding corresponds to amplitude distortion.

The discriminator circuit may be modified as shown in Fig. 22-14 by including resistor $R1$ and capacitor C and, at the same time, reversing the direction of $D2$. Circuit operation is similar to that of the discriminator. The output voltage across C is essentially constant because of the large time constant offered by the relatively high capacitance of C. At resonance, the voltages across C_a and C_b are equal and no current flows through $R1$, giving an output voltage of zero.

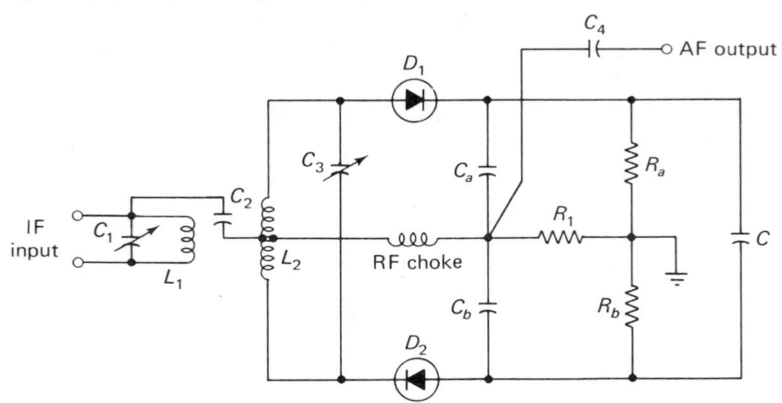

Figure 22-14. Ratio detector for FM.

When the input signal deviates above the resonant frequency, the circuit is unbalanced and voltage V_a exceeds voltage V_b. The output voltage becomes negative. Because the total voltage across C is essentially constant and equal to the sum of V_a and V_b, the output voltage is now proportional to the *ratio* of V_a and V_b. Thus, the circuit is called a *ratio detector*. To understand this process fully, let us assume that at resonance, the undeviated dc voltage across C is 10 V, meaning that both V_a and V_b are 5 V. When the input signal frequency exceeds the resonant frequency, let us assume that the voltage across C is still 10 V and V_a is, say, 7 V. At this time, V_b must be $10-7=3$ V. The output voltage across $R1$ is proportional to the ratio of V_a to V_b. We further assume that the input signal amplitude changes (as long as no limiter is used) so that the net voltage across C changes accordingly to, say, 5 V. The voltage across C_a would now be 3.5 V and that across C_b, 1.5 V. The output voltage, however, would be essentially unchanged because the ratio of V_a and V_b would still be the same. Thus, with the ratio detector circuit, the amplitude limiter is not essential, although its inclusion does enhance circuit performance.

For frequency deviations below resonance, V_b exceeds V_a and the output voltage is negative and still proportional to the ratio of V_a and V_b, because the total voltage across C is maintained.

22.2.4 IC IF Amplifiers

The circuit diagram for a wide-band IC amplifier (RCA CA3012) is shown in Fig. 22-15. It consists of three differential pairs $Q1$-$Q2$, $Q4$-$Q5$, and $Q7$-$Q8$. It also contains its own voltage regulator circuit: $Q9$, which is the series-pass transistor with the reference voltage provided by the series diodes $D1$ and $D2$. Emitter-follower transistors, $Q3$ and $Q6$, perform interstage coupling. They also act as level translators.

A typical IC IF strip containing two CA3012 amplifiers, together with a ratio detector stage, is illustrated in Fig. 22-16. The first IC offers 65 dB of gain; the second IC provides 57 dB of gain. The coupling IF transformer tuned circuit causes a loss of 8 dB, providing for a total IF gain of 114 dB (an input of 3.5 μV causes the output to the detector of 1.75 V).

22.2.5 AGC and AFC

As mentioned earlier, an FM tuner (i.e., receiver) may have both automatic gain control and automatic frequency control. The AGC signal is obtained in the form of direct current from the detector output and fed back to change the bias, and therefore the gain, of the first RF amplifier (see Fig. 22-3). It functions in the same manner as the AGC circuits that we described for AM.

The AFC signal is also a dc signal obtained from the detector output. It is usually switched, providing either manual frequency control or automatic frequency control (at the operator's option). In the automatic mode, the dc signal proportional to the audio output is fed back into a voltage variable capacitor (varactor). The dc level determines the amount of

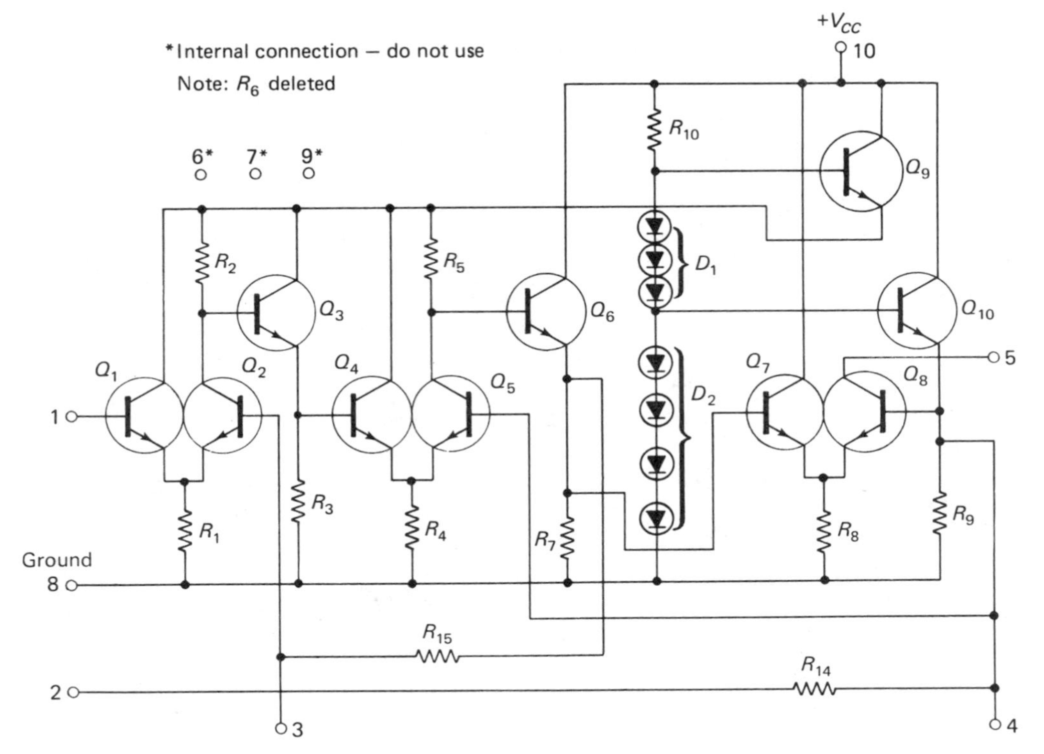

Figure 22-15. Schematic diagram of RCA CA3012 wide-band amplifier. (Courtesy of RCA)

reverse bias on the varactor (see Fig. 22-3) and therefore its capacitance. As the level of direct current varies, the capacitance also varies, causing the oscillator frequency to adjust to the optimum value. A difference signal is provided at the output of the mixer, which is always centered at 10.7 MHz, the IF frequency.

22.3 COMPLETE FM TUNER

We can study an example of a complete high-quality FM tuner in Fig. 22-17. The antenna transformer $T1$ has common tuning with both the oscillator circuit ($C17$) and the output of the RF amplifier ($C7$). Transistor $Q1$ acts as the single RF amplifier; $Q3$ serves as the local oscillator. We couple the RF and the local oscillator signals into the base circuit of mixer transistor $Q2$ through capacitors $C10$ and $C18$, respectively. Antenna transformer $T1$ and transformer $T2$ are both tuned to the desired RF frequency of the incoming signal. Oscillator transformer $T4$ is tuned to provide the proper (10.7 MHz) difference frequency between the incoming RF signal and the local oscillator. The mixer output transformer $T3$ and all the interstage IF transformers are tuned for the IF frequency. Three IF stages of amplification are provided by transistors $Q4$, $Q5$, and $Q6$. $T7$

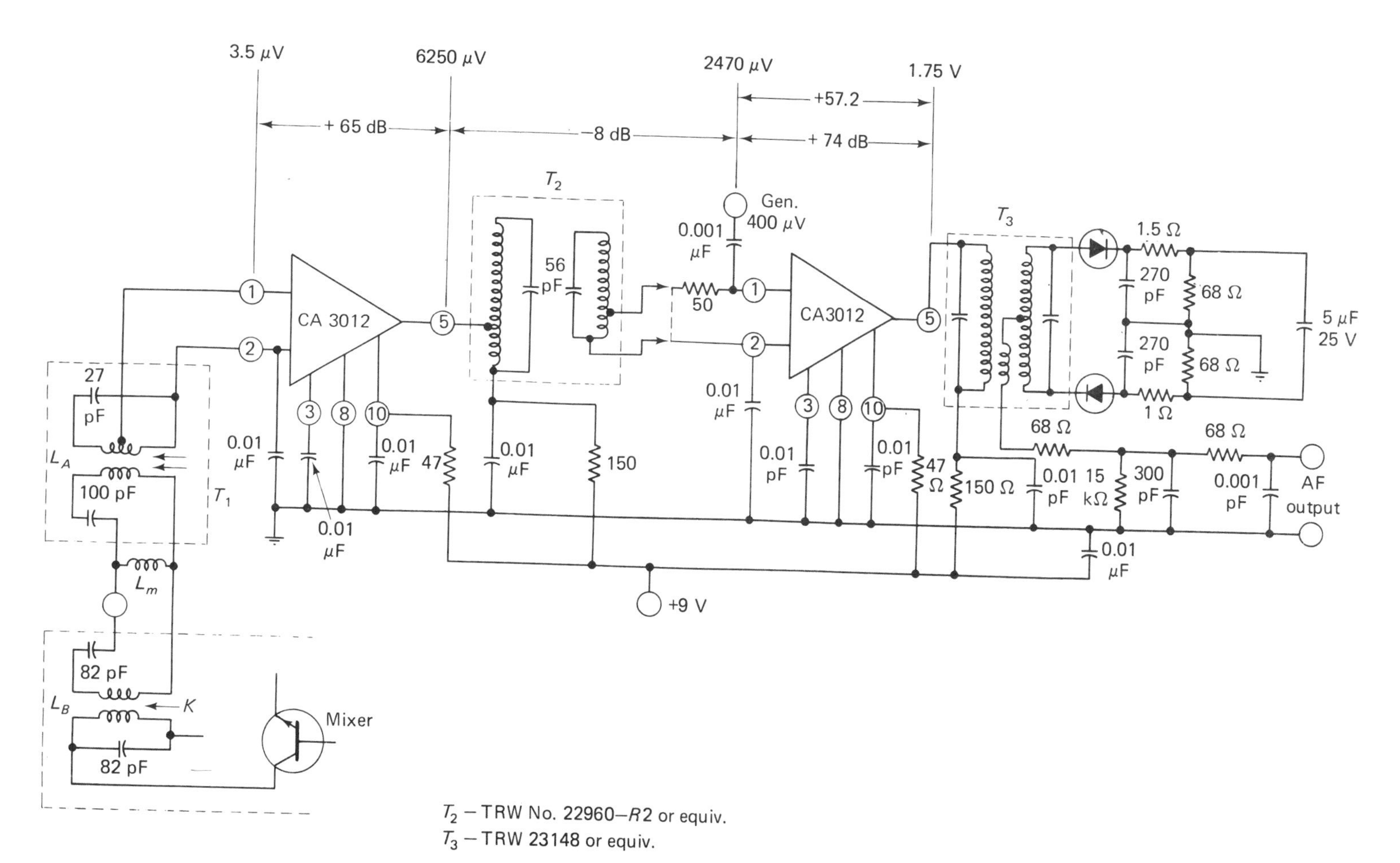

T_2 − TRW No. 22960−$R2$ or equiv.
T_3 − TRW 23148 or equiv.

Figure 22-16. IC IF amplifier strip with detector. (Courtesy of RCA)

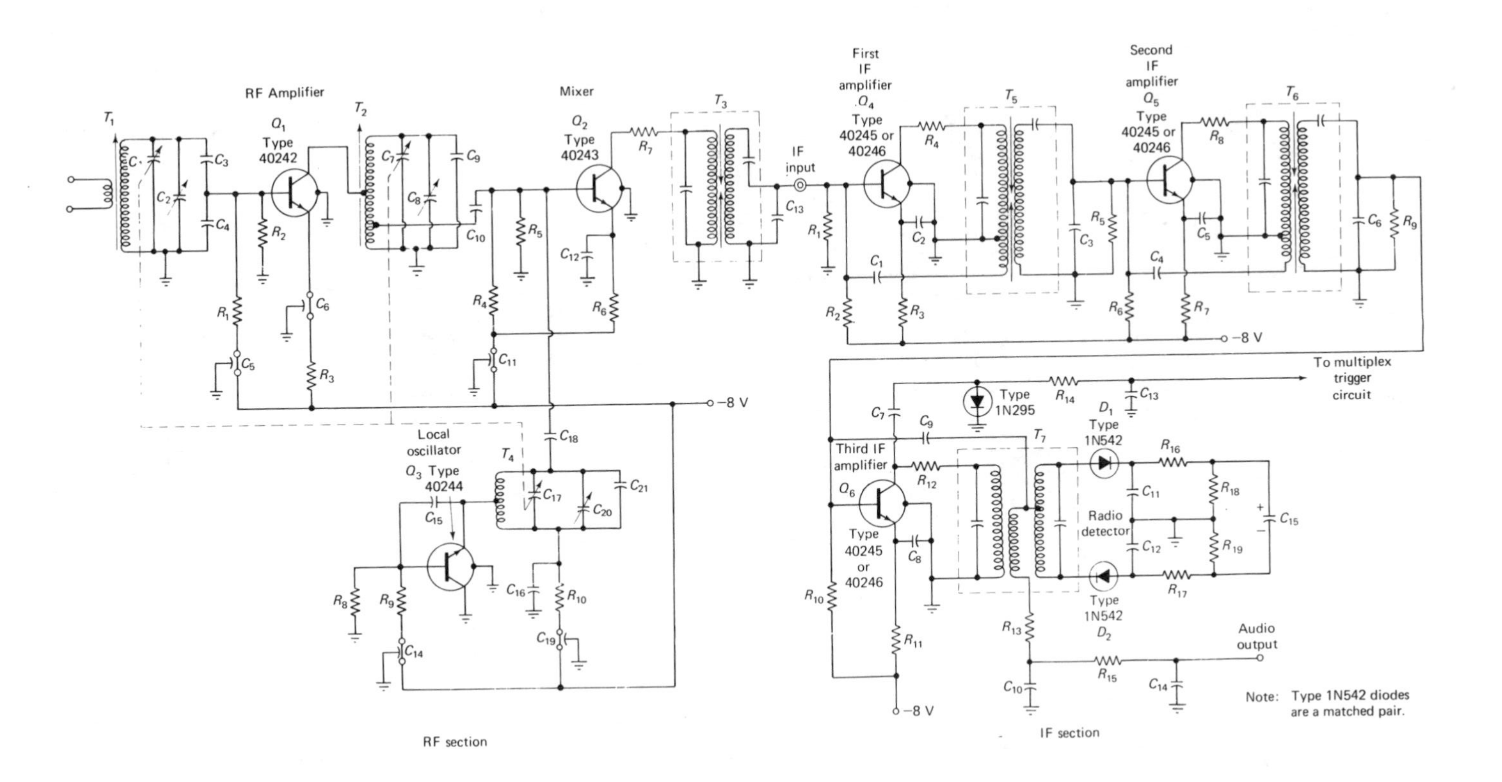

Figure 22-17. Discrete component FM tuner. (Courtesy of RCA)

Parts List for RF Section

C_1, C_7, C_{17} = ganged tuning capacitors, C_1, C_7 = 7.25 to 19 pF; C_{17} = 6 to 21 pF

C_2, C_8 = trimmer capacitor (part of ganged tuning capacitor assembly), approximately 17 pF maximum

C_3, C_9 = 5.6 pF, minature ceramic

C_4 = 27 pF, ceramic disc

C_5, C_6, C_{11}, C_{14}, C_{19} = feedthrough capacitor, 1000 pF

C_{10} = 2000 pF, ceramic disc, 1000 V

C_{12} = 0.01 μF, ceramic disc

C_{13}, C_{16} = 1000 pF, ceramic disc, 1000 V

C_{15} = 3.3 pF, NPO ceramic

C_{18} = 0.22 pF to 3.3 pF (value determines oscillator injection voltage and is dependent upon factors such as circuit layout and placement of components)

C_{20} = tubular trimmer capacitor, 1.5 to 10 pF

C_{21} = 12 pF, ceramic disc

R_1, R_4 = 3300 ohms, 0.5 watt

R_2, R_5 = 18000 ohms, 0.5 watt

R_3, R_6 = 330 ohms, 0.5 watt

R_7 = 100 ohms, 0.5 watt

R_8 = 8200 ohms, 0.5 watt

R_9 = 4700 ohms, 0.5 watt

R_{10} = 1500 ohms, 0.5 watt

T_1 = FM antenna transformer; slug-tuned; slug, 0.250 inch long, 0.181 inch in diameter, Arnold Type 1RN9 or equivalent; secondary, 4 turns of No. 22 bare-tinned copper wire wound with 1 wire-diameter spacing between adjacent turns or 7/32-inch outer-diameter coil form, resonates with 27-pF capacitance at 100 MHz, impedance = 6100 ohms; primary, 2 turns of No. 30 Gripeze wire close wound below cold end of secondary and in same direction, impedance (includes shunting effect of rf amplifier biasing network) = 460 ohms

T_2 = rf interstage coil; 4 turns of No. 18 bare-tinned copper wire wound with approximately 1/8-inch spacing between turns on 5/16-inch diameter coil form (coil form is removed after coil is wound); resonates with 27-pF capacitance at 100 MHz; impedance of full winding, 6100 ohms; input tap located so that impedance at tap = 590 ohms; output tap located so that impedance at input tap is 540 ohms with the transformer properly loaded

T_3 = oscillator coil; 3½ turns of No. 18 bare-tinned copper wire wound with 3/32-inch spacing between turns on 7/32-inch-diameter coil form (coil form is removed after coil is wound), center tapped

T_4 = first if (10.7-MHz) transformer, Thompson-Ramo-Wooldridge No. E019309-R4 or equivalent

Parts List for IF Section

C_1, C_4 = 4.7 pF, ceramic disc

C_2, C_5, C_8 = 0.01 μF, ceramic disc

C_3, C_6 = 1000 pF, ceramic disc, 1000 V

C_7 = 5 pF, ceramic disc

C_9 = 1.0 pF, ceramic disc

C_{10}, C_{11}, C_{12} = 330 pF, ceramic

C_{13} = 0.05 μF, ceramic disc

C_{14} = 0.02 μF, ceramic disc

C_{15} = 5 μF, electrolytic, 10 V

R_1, R_5, R_9 = 12000 ohms, 0.5 watt

R_2, R_6, R_{10} = 2700 ohms, 0.5 watt

R_3, R_4, R_7, R_8, R_{11} = 220 ohms, 0.5 watt

R_{12} = 470 ohms, 0.5 watt

R_{13} = 68 ohms, 0.5 watt

R_{14} = 22000 ohms, 0.5 watt

R_{15} = 3900 ohms, 0.5 watt

R_{16} = 1000 ohms, 0.5 watt

R_{17} = 1500 ohms, 0.5 watt

R_{18}, R_{19} = 6800 ohms, 0.5 watt

T_5 = second if (10.7-MHz) transformer, Thompson-Ramo-Wooldridge No. E019310-R2 or equivalent.

T_6 = third if (10.7 MHz) transformer, Thompson-Ramo-Wooldridge No. E019311-R1 or equivalent

T_7 = ratio-detector transformer, Thompson-Ramo-Wooldridge No. 3019312-R3 or equivalent

Figure 22-17. (Cont.)

serves as the tuned input transformer for the ratio detector circuit comprised of matched diodes $D1$ and $D2$. The audio output is available across $C14$. We have also provided for an output for the multiplex trigger circuit across $C13$. A multiplexed FM signal contains the information for two sources of audio (or two channels of a stereo signal).

Review Questions

1. What is frequency modulation?
2. How is frequency modulation different from amplitude modulation?
3. What is the range of frequencies allocated for FM commercial transmission?
4. Why must the bandwidth allocation for FM be much larger than for AM?
5. Why is the fidelity (quality) of an FM-derived signal usually superior to that of an AM-derived signal?
6. What blocks in an FM tuner are the same as in an AM receiver?
7. What circuits comprise the FM front end, and what functions do they perform?
8. Why is it commonly found that a FET is used in the front end (specifically, in the RF amplifier stage)?
9. Which parts of an FM tuner have common variable tuning and why?
10. Which parts of an FM tuner have fixed tuning and why?
11. What is the role of a limiter?
12. Why is the limiter necessary? What are the undesirable consequences of leaving out the limiter stage?
13. What are the different methods used for the detection of FM?
14. What does the discriminator stage in an FM tuner discriminate?
15. What is the intermediate frequency used in commercial FM receivers?
16. How is AGC accomplished in an FM tuner?
17. What is AFC?
18. Why is AFC desirable in an FM tuner?
19. What would be the characteristics of an ideal limiter?
20. What are the characteristics of a practical limiter? Discuss in terms of the difference between the ideal and that realizable.
21. Explain how a tuned amplifier that has been tuned off resonance may be used to decode an FM signal.
22. How does the Foster-Seely discriminator decode an FM signal?
23. How does the ratio detector decode an FM signal?
24. Which functional blocks in an FM tuner are essential for operation and which may be left out? Why?

1. The modulating signal is a square wave having equal on and off times at a frequency of 1Hz. The carrier frequency is 10 Hz. Sketch the corresponding frequency-modulated signal with the amplitude-modulated signal for comparison.
2. We wish to receive an FM signal with an undeviated frequency of 100 MHz. The maximum frequency deviation is 75 kHz. In the block diagram for an FM tuner (like the one shown in Fig.22-2), specify the tuning and bandwidth required in the stages where applicable, and specify the frequency ranges for the other stages.
3. If the drain current for a dual-gate MOSFET is given by:

$$i_d = K(v_{g1s} + v_{g2s})^2$$

and the MOSFET is used as a mixer (Fig. 22-6), what are the frequency components in the output for an input signal of undeviated frequency of 104 MHz? What must be the frequency of the local oscillator signal?

23 Analog Systems

In this chapter we introduce the functions performed in an analog computer. We shall also show how programs can be specifically implemented to solve certain problems. Analog circuits have a wide range of applications outside analog computers. Most are extremely easy to implement, requiring only an OP AMP and a few discrete components.

23.1 PRINCIPLES OF ANALOG COMPUTATION

Some experts maintain that analog computers have become obsolete since the advent of the high-speed digital computer. Although the use of analog computers has definitely declined with the increased availability of sophisticated programming subroutines for high-speed digital computers, the analog computer can still be useful, mainly as a simulator. It is able to simulate the physical world in real time. Because both analog and digital computers have applications for which each is best suited, they will continue to exist side by side. This is evidenced by the rise in hybrid computers, using both digital and analog signal processing.

An analog system is distinguished by the fact that the input and output waveshapes are linearly related: The output waveshape is proportional to or at least a linear function of the input waveshape.

Let us discuss the chief building block in an analog computer: the OP AMP. The basic inverting configuration, using generalized impedances, is shown in Fig. 23-1. As described in Chap. 14, the output voltage for this configuration is given by:

$$V_o = -\frac{Z_f}{Z_1} V_1 \qquad (23\text{-}1)$$

The function performed by the circuit is determined by the specific elements used for the feedback and input impedances. To minimize the error in the output voltage caused by offset current, the noninverting input should be returned to ground through a resistance equal to the parallel combination of Z_f and Z_1 (if these are resistive).

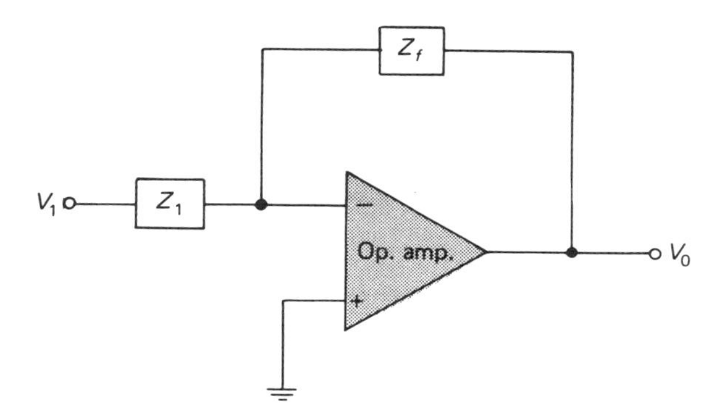

Figure 23-1. General negative feedback circuit using the inverting input.

Figure 23-2 depicts the same basic connection, with the input applied to the noninverting terminal. In this case, the output voltage is given by:

$$V_o = \frac{Z_f + Z_1}{Z_1} V_1 \qquad (23\text{-}2)$$

These two basic general circuits may be used to implement a variety of analog functions.

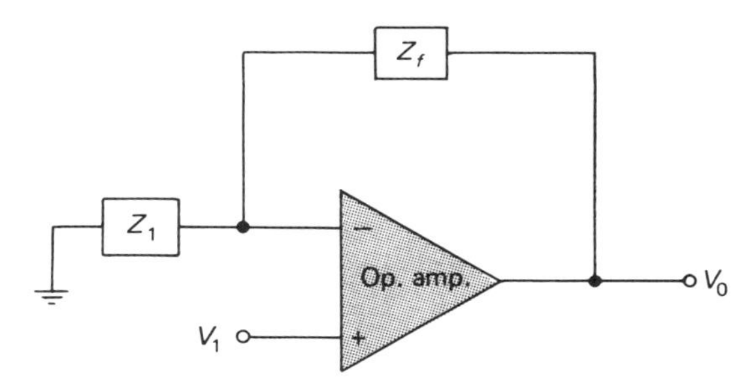

Figure 23-2. General negative feedback circuit using the noninverting input.

23.1.1 Sign Changer (Inverter)

The most basic analog function needing implementation is that of sign changing. Although we call it inversion, it should not be confused with the function of the digital inverter. The analog inverter is implemented by making the two impedances in Fig. 23-1 equal; precision resistors are commonly used, as shown in Fig. 23-3. The output voltage then is the negative of the input voltage.

23.1.2 Scaler

In many applications we have to scale or change the amplitude of a signal. This process may involve either increasing or decreasing the amplitude. For the scaler circuit shown in Fig. 23-4, the scale factor K is negative, given by $-R_f/R_1$, and may be larger or smaller than 1.

The noninverting scaler circuit is illustrated in Fig. 23-5. Note that this circuit can only scale up; that is, the output voltage is always larger

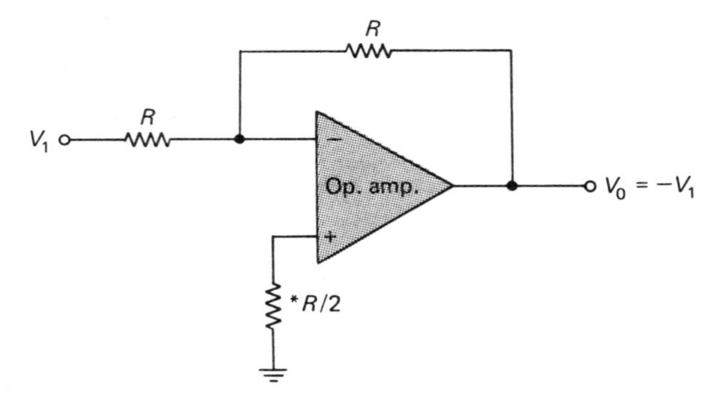

Figure 23-3. Sign charger (inverter). (*$R/2$ is included to minimize the effect of offset current.)

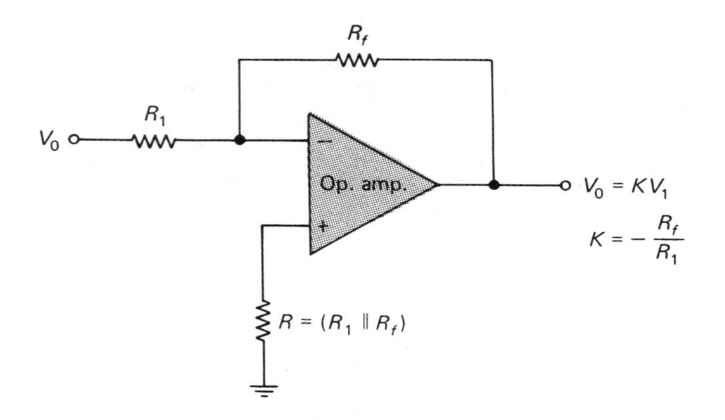

Figure 23-4. Inverting scaler.

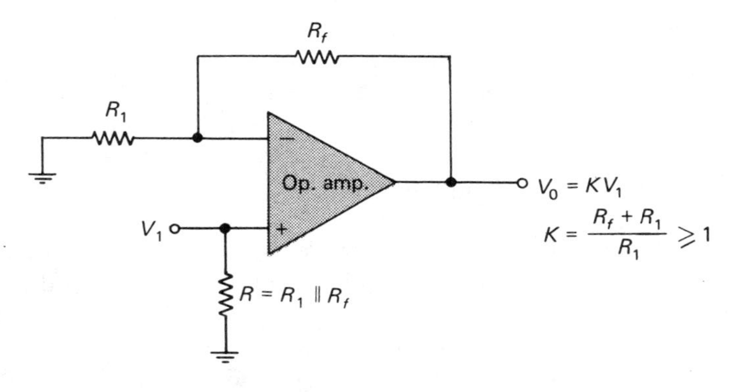

Figure 23-5. Noninverting scaler.

than the input voltage. In both scaler circuits, we use precision resistors to improve the accuracy of the scale factor K. A variable scale factor can be achieved by using a calibrated (usually 5 or 10 turn) potentiometer for either R_1 or R_f.

23.1.3 Summing Amplifier (Adder)

Figure 23-6 shows the OP AMP connected as a summing amplifier. The output voltage is proportional to the sum of the input voltages. Although three inputs are shown, any number may be used by connecting additional input resistors. To see how the addition is performed, remember that the input of the OP AMP is a virtual ground and draws no input current. The current through R_1 is labeled I_1; through R_2, I_2; and through R_3, I_3. These currents are given by:

$$I_1 = \frac{V_1}{R_1} \quad I_2 = \frac{V_2}{R_2} \quad I_3 = \frac{V_3}{R_3} \tag{23-3}$$

The current through the feedback resistor, labeled I_f, must be the sum of I_1, I_2, and I_3 because the input of the OP AMP draws no current. Thus,

$$I_f = I_1 + I_2 + I_3 = -\frac{V_o}{R_f} \tag{23-4}$$

Making use of Eq. (23-3), we obtain:

$$V_o = -(K_1V_1 + K_2V_2 + K_3V_3) \tag{23-5}$$

where we have defined scale factors:

$$K_1 = \frac{R_f}{R_1} \quad K_2 = \frac{R_f}{R_2} \quad K_3 = \frac{R_f}{R_3} \tag{23-6}$$

Note that the output voltage given in Eq. (23-5) is proportional to the negative of the sum of the input voltages. If $R_1 = R_2 = R_3 = R_f$, it is obvious that $V_o = -(V_1 + V_2 + V_3)$.

The circuit offers flexibility by allowing us to use a different scale factor for each of the voltages to be added. The following example, although not very practical, illustrates the use of different scale factors.

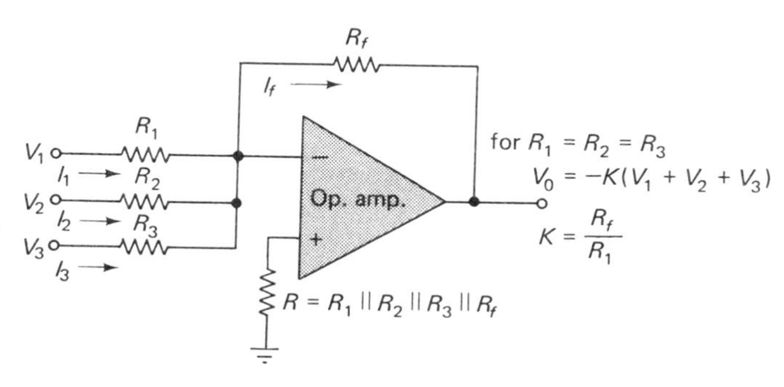

Figure 23-6. Summing amplifier (adder).

Suppose that the three input voltages represent the number of coins; that is, the number of millivolts for V_1 corresponds to the number of quarters, the number of millivolts for V_2 corresponds to the number of dimes, and the number of millivolts for V_3 corresponds to the number of nickels. We can have the output voltage correspond to the number of total coins (irrespective of worth) by making all the resistors in Fig. 23-6 equal. The total number of coins would then be given by the number of millivolts at the output. (Note that this number would be negative.) If, on the other hand, we wanted to know the net amount of money represented by the coins, we might choose $K_1=25$, $K_2=10$, and $K_3=5$. Thus, the output would give us an amount (in cents) totaling the worth of all the coins represented by the number of millivolts for V_o. If we wanted an output where each millivolt represents a dollar, we would use scale factors $K_1=0.25$, $K_2=0.10$, and $K_3=0.05$.

The noninverting summing amplifier circuit is shown in Fig. 23-7. With all the input resistors equal, the scale factor K is set by the values of R_1 and R_f, as given by Eq. (23-2). The output voltage is:

$$V_o=K(V_1+V_2+V_3) \tag{23-7}$$

In this case, the output voltage is proportional to the sum of the input voltages and is not negative (as happened with the circuit of Fig. 23-6.)

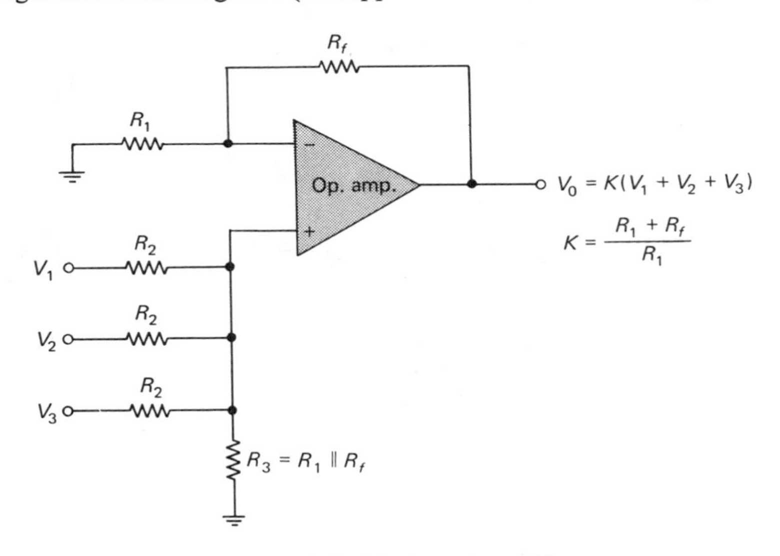

Figure 23-7. Noninverting adder.

23.1.4 Difference Amplifier (Subtractor)

The circuit whose output voltage is proportional to the difference of the two input voltages is depicted in Fig. 23-8. The scale factor K is once again set by R_1 and R_f, and the output voltage is given by:

$$V_o=K(V_2-V_1) \tag{23-8}$$

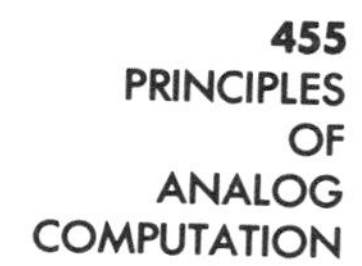

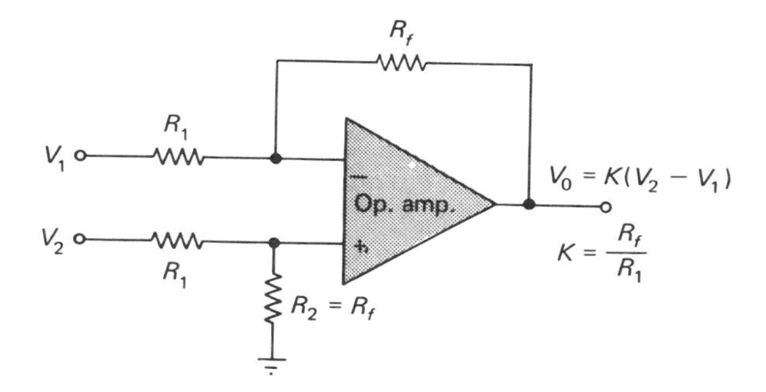

Figure 23-8. Difference amplifier (subtractor).

This equation is the basic differential configuration for the OP AMP and is used often in a large number of instrumentation applications. For example, we may use it to replace a micorammeter in a bridge circuit, as shown in Fig. 23-9. Any imbalance in the bridge circuit is reflected as a voltage between points A and B and is amplified by the difference amplifier. When the bridge is balanced, $R = R_b R_x / R_a$, and the output of the difference amplifier goes to zero. As the bridge is brought closer to balance, R_f should be increased to provide higher gain and therefore better sensitivity.

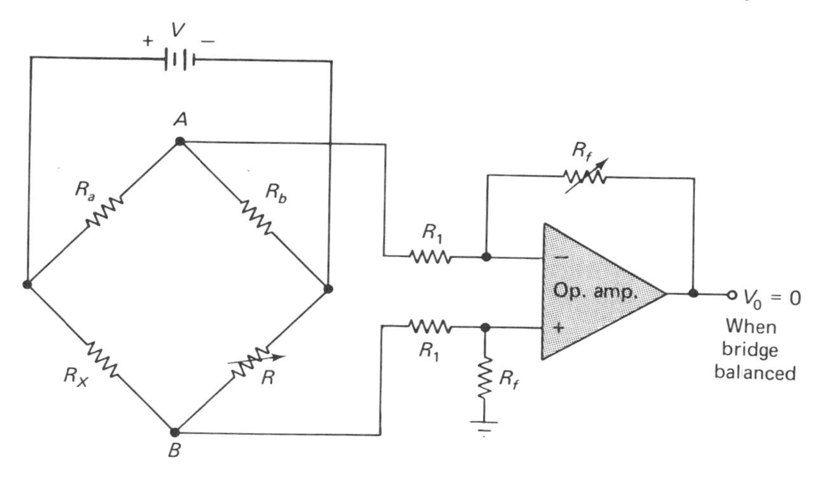

Figure 23-9. Difference amplifier used to indicate balance in the bridge.

23.1.5 Integrator

The basic OP AMP integrator circuit is illustrated in Fig. 23-10. In this case, the feedback impedance is a capacitor whose impedance in operational form is $1/sC$. The output voltage is given by:

$$V_o = -\frac{1}{RCs} V_1 \tag{23-9}$$

where both V_1 and V_o are the functions of s, the complex frequency variable. Network theory tells us that dividing by s in the frequency

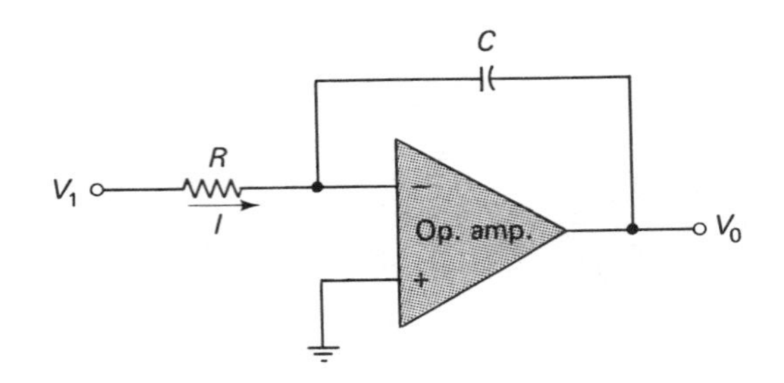

Figure 23-10. Integrator circuit.

domain corresponds to integrating in the time domain. Thus, the output voltage is proportional to the input voltage divided by s in the frequency domain, and so the circuit performs integration on the input voltage. To fully understand this point, assume that the input voltage is zero and then abruptly rises to a voltage V, as shown in Fig. 23-11. While the input is at zero, the output is also at zero. We assume there is no initial charge on the capacitor. When the input becomes V, the voltage across the capacitor cannot change instantaneously, so that the output voltage builds up as the capacitor charges up. The current through R is the current charging the capacitor. One end of R is at V; the other end is connected to the inverting input of the OP AMP, which is essentially at ground. Therefore, the current through R is constant throughout the charging period and given by V/R. For a capacitor, the rate at which the terminal voltage changes is directly proportional to the charging current. Thus, if the charging current is constant, the rate of change in the capacitor terminal voltage is also constant. The capacitor is charging to provide a *linear* increase in output voltage inversely proportional to the circuit time constant RC.

There is another way in which we can approach the operation of the integrator circuit of Fig. 23-10. One end of the capacitor is permanently

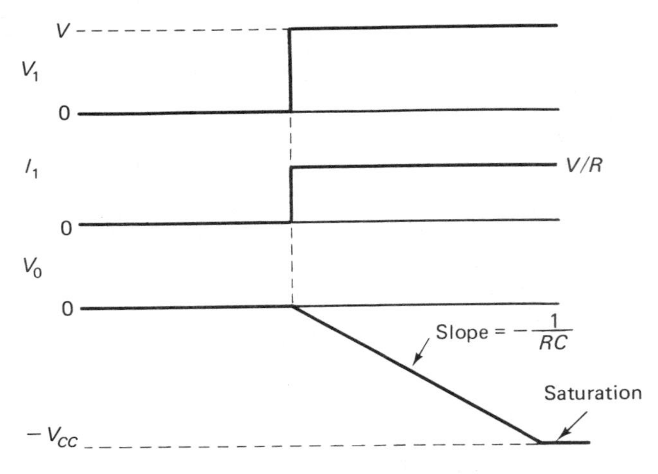

Figure 23-11. Waveshapes for the integrator circuit.

held at ground by the inverting input of the OP AMP. If it were not for the capacitor, once the input is applied the output would try to reach a voltage that is essentially the input voltage multiplied by the difference gain of the amplifier. Obviously, this voltage is extremely large. For example, a 1 V input with a typical difference gain of 200,000 would mean that the output is trying to reach −200,000 V. A practical OP AMP is limited to operating between its negative and positive supply voltages. However, as far as the capacitor is concerned, it is trying to charge up to this extremely large voltage. The fact that it can never reach it is unimportant. The important fact is that during the first fractions of time constant, the charging of a capacitor is linear. Thus, the OP AMP has the effect of providing a linear charging characteristic.

When we integrate mathematically, we are finding, in effect, the area under the curve of the function we are integrating between specified limits. The circuit shown in Fig. 23-10 can be used to integrate a function of time over a specified interval. The output voltage provides a plot of the area under the curve of the input waveshape in Fig. 23-11. (Note: The output is actually the *negative* of the integral of the input.) When the output voltage reaches the supply voltage V_{CC}, the OP AMP is said to be saturated, and the output voltage cannot increase further. At this point, the circuit ceases to function as an integrator.

If the input voltage is a succession of voltage pulses, the output waveshape becomes a linear triangle wave. If we wish to integrate a repetitive function, the capacitor must be discharged to zero at the end of each cycle. This discharge is accomplished by placing some form of a transistor (usually an FET) switch in parallel with the capacitor, as shown in Fig. 23-12. The switch is normally open during the integration interval. At the end of the integration, the gate of the switch is pulsed, causing the transistor to saturate and discharge the capacitor through its very low resistance. A typical set of waveshapes is given in Fig. 23-13. The circuit operates as an integrator so long as the FET is off; that is, its gate-source

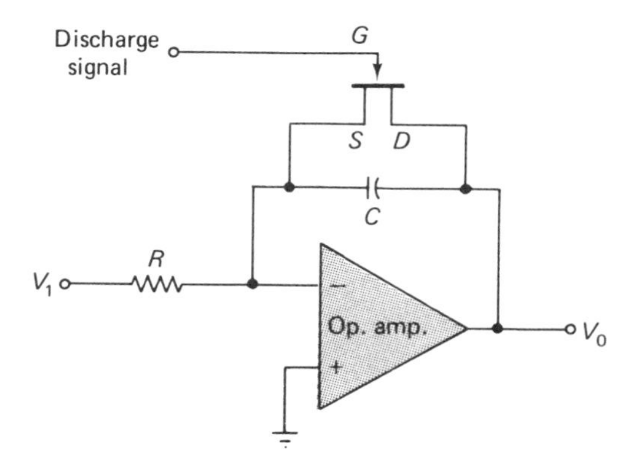

Figure 23-12. FET provides discharge in an integrator circuit.

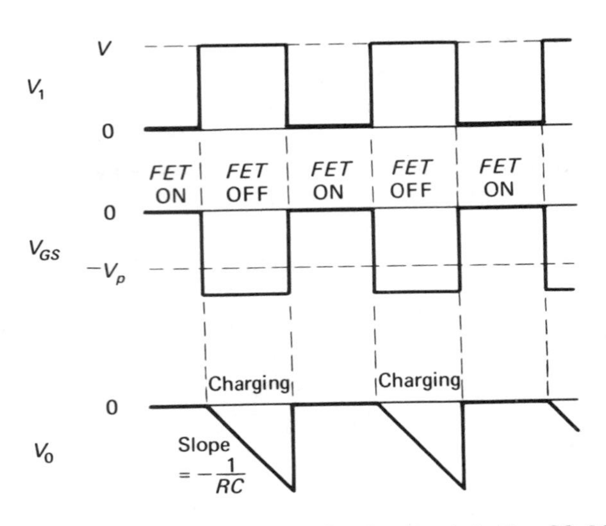

Figure 23-13. Waveshapes for the circuit in Fig. 23-12.

junction is biased beyond pinchoff. The capacitor is discharged when the gate-source junction is biased above pinchoff.

In certain cases, it is necessary to start the integration with certain initial values, which are introduced as an initial charge (supplied by a dc power source) on the capacitor.

23.1.6 Differentiator

A useful but less common circuit is the *differentiator* circuit, illustrated in Fig. 23-14. The resistor and capacitor of the integrator circuit has been interchanged. The input impedance is now $1/Cs$, so that the output voltage is given by:

$$V_o = -RCsV_1 \tag{23-10}$$

where both V_1 and V_o are functions of s. Multiplication by s in the frequency domain corresponds to differentiation in the time domain. Consequently, the output voltage (as a function of time) is the negative of the differentiated input voltage.

Figure 23-15 shows that the output waveshape for a simple case of a ramp input voltage is a step function. Circuit operation is as follows. While

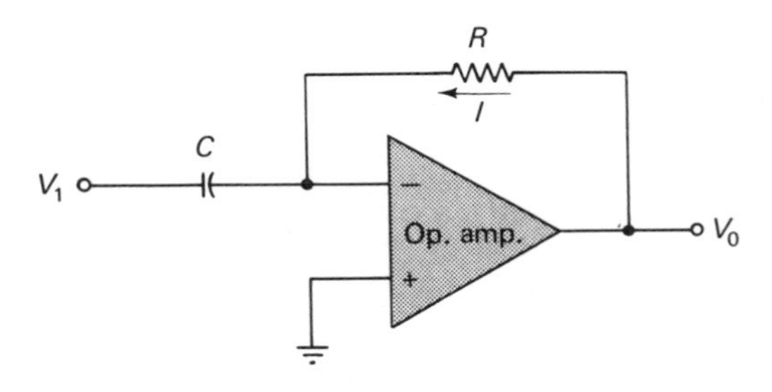

Figure 23-14. Differentiator circuit.

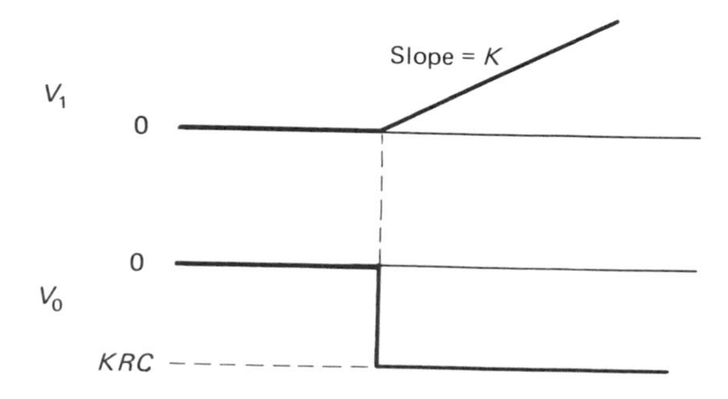

Figure 23-15. Differentiator waveshapes.

no input is applied, the capacitor is uncharged and the output is zero. When the input begins to increase at a linear rate (i.e., slope = constant), the current through the capacitor must be constant. It is supplied through R, one end of which is at virtual ground (the inverting input of the OP AMP) and the other end at the output. Because the current through R is constant, the voltage drop across R must also be constant. This voltage drop is equal to the output voltage.

If repetitive waveshapes are to be differentiated, the capacitor must be discharged at the end of each cycle. Discharge may be accomplished with the use of an FET switch. We would follow a procedure similar to that shown in Fig. 23-12 for the integrator; the source of the FET would be connected to the end of the capacitor at the invering input.

23.2 ASSORTED ANALOG CIRCUITS

In this section we shall take up a few assorted analog circuits which do not rightfully belong under the classification of analog computer circuits but which, nevertheless, are useful enough to warrant our attention.

23.2.1 Voltage-to-Current Converter

In many cases a voltage signal needs to be converted into a proportional current signal. We can perform this conversion by using the circuit illustrated in Fig. 23-16. The load is placed in the feedback slot. This converter is especially useful when it is necessary to have both ends of the load isolated from ground.

The circuit is nothing more than a noninverting amplifier. Because the OP AMP draws no input current, I_o and I_1 must be equal. Moreover, the differential input voltage for the OP AMP must be zero, so $V_1 = I_1 R_1 = I_o R_1$. Thus,

$$I_o = \frac{V_1}{R_1} \qquad (23\text{-}11)$$

The output current is directly proportional to the input voltage. Note that for a given V_1 the output current is independent of the load resistance R_f; therefore, the magnitude of the output current can be adjusted by changing R_1.

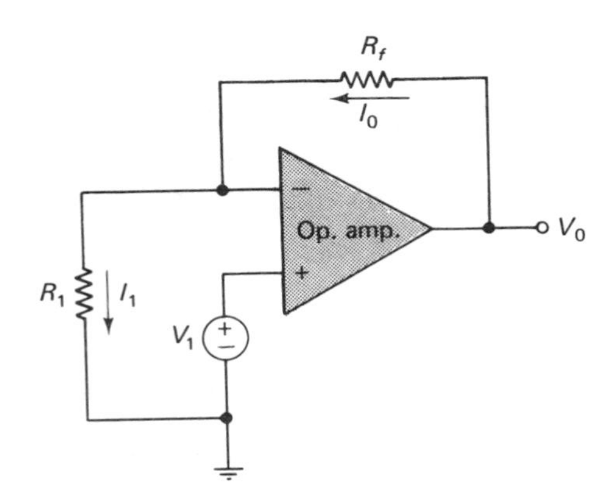

Figure 23-16. Voltage-to-current converter.

23.2.2 Current-to-Voltage Converter

The circuit shown in Fig. 23-17 provides an output voltage which is proportional to the input current. It is useful in applications where the source provides an extremely low voltage under load conditions, for example, to amplify the output of a photocell.

Because of the virtual ground at the input of the OP AMP, V_1 is zero, and the source resistance R_1 draws no current. All of the input current flows through R_f, causing an output voltage. Note that the output voltage is not a function of the source resistance R_1; and, for a given I_1, the output voltage may be varied by simply changing R_f.

We can use this circuit to amplify very low currents. However, for extremely low currents, the OP AMP input bias current causes an error; that is, the actual circuit would have a part of the input current flowing in R_1 and into the input of the OP AMP. Both of these currents are negligible for input currents sufficiently larger than the OP AMP input bias current.

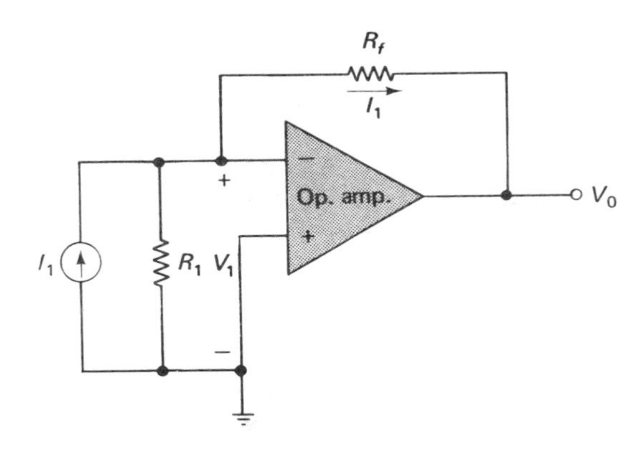

Figure 23-17. Current-to-voltage converter.

23.2.3 Voltage Follower (Buffer)

In some instances it is necessary to amplify a voltage from a source which has a very high source resistance. The simple circuit shown in Fig. 23-18 may be used in such cases.

Because the output is tied to the inverting input, which is *virtually* tied to the noninverting input, the output voltage follows (or is equal to) the input voltage. OP AMPs are specifically designed for such applications. They have a very high input impedance and an extremely low input current drain. An example is the Signetics μA740 FET input OP AMP, which has an input bias current of less than 0.1 nA (10^{-10} A) and a typical input resistance of 1 million MΩ (10^{12} Ω).

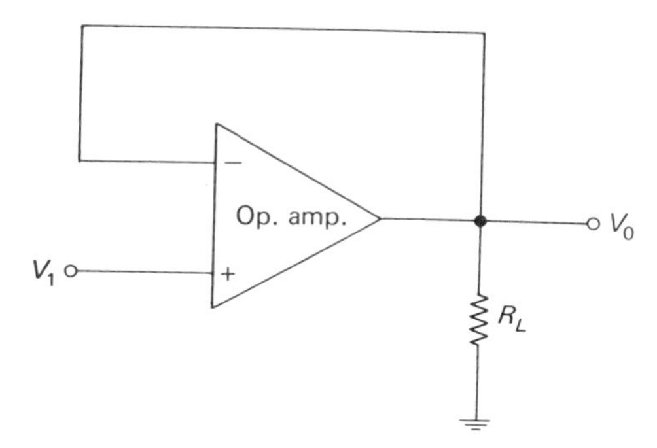

Figure 23-18. Voltage follower (buffer).

23.3 APPLICATION OF ANALOG SYSTEMS

Analog computation is useful in problems where we need to evaluate (1) the effect of varying one or more parameters and (2) the solution to specific values. For example, let us again determine the net worth of a given number of quarters, dimes, and nickels. The circuit for solving this problem is given in Fig. 23-19. Each coin to be counted corresponds to (i.e., is analogous to), say, 1 millivolt at one of the input potentiometers. The number of quarters is represented by the number of millivolts at the input to $R_1(V_a)$. The number of dimes is represented by the number of millivolts at the input to $R_2(V_b)$. The number of nickels is represented by the number of millivolts applied to the input of $R_3(V_c)$. For the resistor values shown, voltage V' is given by:

$$V' = -(25V_a + 10V_b + 5V_c) \tag{23-12}$$

The final output voltage is:

$$V_o = 250V_a + 100V_b + 50V_c \tag{23-13}$$

Assume that the input voltages are set to represent 16 quarters, 21 dimes, and 7 nickels; that is, $V_a = 16$ mV, $V_b = 21$ mV, and $V_c = 7$ mV. The output voltage for this case would be:

$$V_o = (250)(16) + (100)(21) + (50)(7) = 6450 \text{ mV} = 6.45 \text{ V}$$

The answer in volts corresponds to the sum of the coins in dollars, so that 16 quarters, 21 dimes, and 7 nickels equal \$6.45.

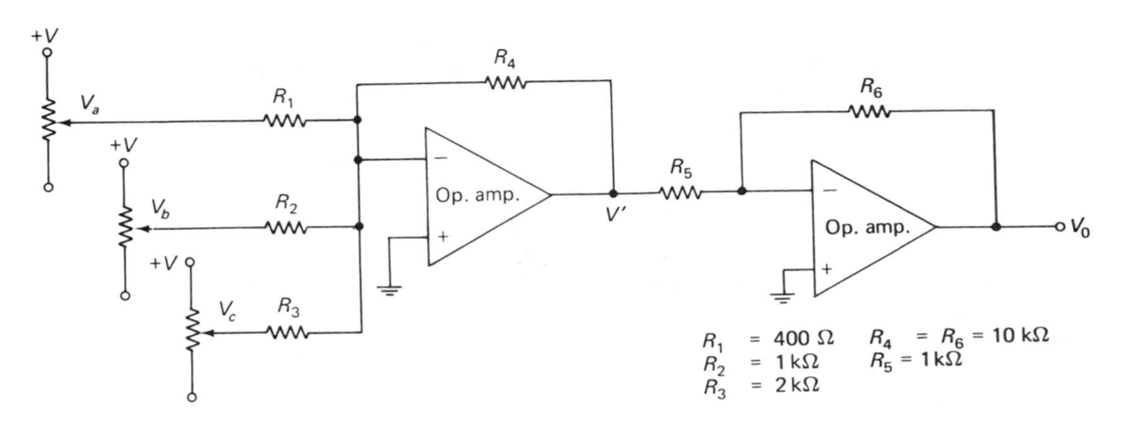

Figure 23-19. Example for computing the worth of a given number of coins.

You can see that the real power of this method is the ease of changing the numbers of the coins involved by simply resetting the input voltages. Although we are not suggesting that this problem is best solved by analog computer methods, it is a simple problem that easily illustrates these methods.

As another example, consider the solution of simultaneous equations. Assume that two items costing X and Y dollars are purchased. The first time, three of one item and four of the other are purchased for a net cost of \$0.95. The second time, one of each item is purchased for a net cost of \$0.30. This is expressed in equation form as:

$$3X+4Y=0.95$$
$$X+Y=0.30 \tag{23-14}$$

Let us use the analogy that each dollar is equivalent to a volt. We can rearrange the equations, solving one for X and the other for Y:

$$Y=0.2375-0.75X$$
$$S=0.30-Y \tag{23-15}$$

These two equations are implemented as indicated in Fig. 23-20. The output voltage of OP AMP 1 ($OA1$) is analogous to X and the output voltage of $OA2$ is analogous to Y. Verification of the answers $X=0.25$ and $Y=0.05$ is left as an exercise.

As a more practical example, suppose that we want to examine the response (current waveshape) of the series RLC circuit shown in Fig. 23-21, for (1) different excitations (input voltages), (2) different initial conditions, and (3) different circuit parameters. One obvious way would be to connect the circuit in the laboratory and perform the tests on the actual circuit. The testing is easily done with simple circuits, but the problems involved with performing experiments on actual complicated circuits should be obvious. With an analog computer, however, we can easily simulate any such operation and many other more complicated circuits.

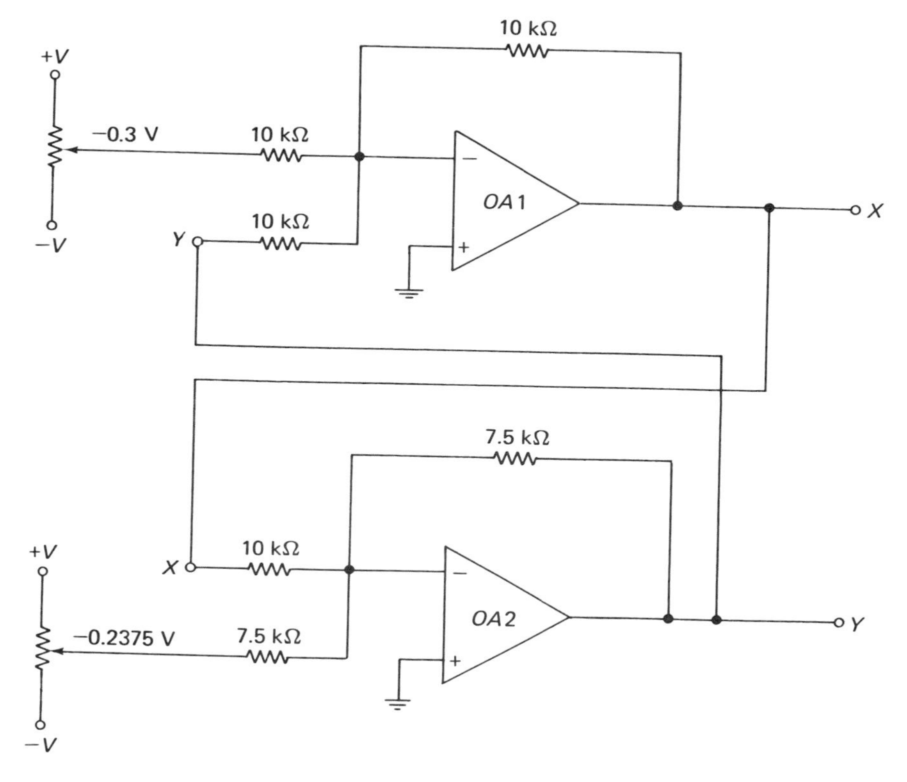

Figure 23-20. Solution of simultaneous equations.

The applied voltage is given by:

$$V = IR + \frac{1}{C} I\,dt + L\frac{dI}{dt} \tag{23-16}$$

To simplify matters, we can use a variable other than the current, namely, the charge in the circuit. Remember that current is the rate of flow of charge or that the charge is the integral of the current. Thus,

$$V = R\frac{dQ}{dt} + \frac{Q}{C} + L\frac{d^2Q}{dt^2} \tag{23-17}$$

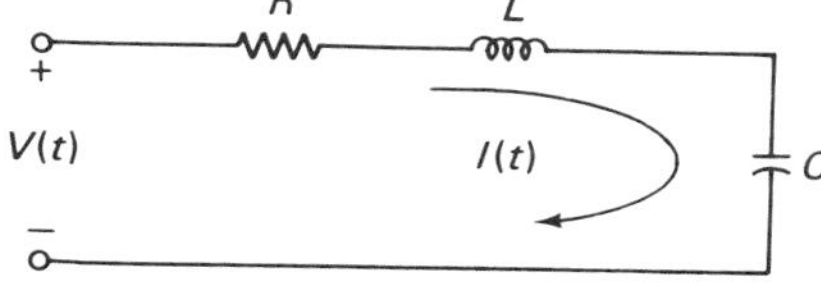

Figure 23-21. Series RLC circuit.

To implement this second order differential equation on an analog computer, we first solve for the highest derivative of the variable Q.

$$-\frac{d^2Q}{dt^2} = -\frac{V}{L} + \frac{Q}{LC} + \frac{R}{L}\frac{dQ}{dt} \tag{23-18}$$

The implementation of this equation on an analog computer is shown in Fig. 23-22. The inputs to the inverting or summing amplifier $OA1$ are the functions on the righthand side of Eq. (23-18). The proper scaling factors are indicated. The output of $OA1$ then must be equal to the lefthand side of Eq. (23-18), namely, $-d^2Q/dt^2$. This signal is integrated by $OA2$ to provide an output dQ/dt. This signal is again integrated by $OA3$ to provide $-Q$ and also multiplied by a -1 in inverter $OA4$.

We can now obtain the circuit response for any excitation voltage V, which is derived from a function generator. It may be obtained for any given set of initial conditions—provided by voltages V_1 and V_2—and for any set of circuit values—provided by the scaling resistors R_1, R_2, R_3, and R_4. The current is the circuit response. Its analog is obtained at the output

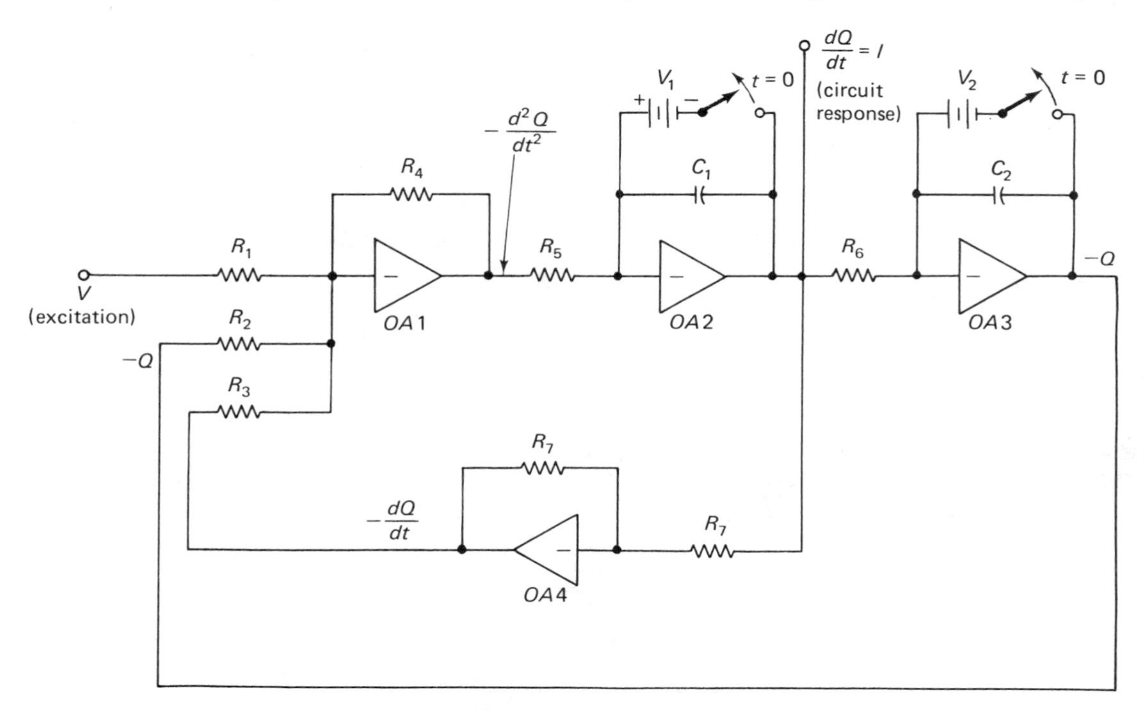

Analogies $\frac{R_4}{R_1} = \frac{1}{L}$; $\frac{R_4}{R_2} = \frac{1}{LC}$; $\frac{R_4}{R_3} = \frac{R}{L}$; $\frac{1}{R_5C_1} = \frac{1}{R_6C_2} = 1$;

$V_1 = I_0$ (current in L at $t = 0$) ; $V_2 = Q_0$ (charge on C at $t = 0$)

Figure 23-22. Analog computer simulation for a series RLC circuit.

of $OA2$ (because $I = dQ/dt$). The analog may be displayed on an oscilloscope or recorded on an $x-y$ plotter or strip-chart recorder.

The three examples just discussed are by no means the only problems which can be handled on an analog computer. They simply illustrate the basic operation of analog systems, which are used to simulate very complex problems. If you are interested in analog computer programming, consult one of the many books on the subject.

Review Questions

1. What is the basic building block in an analog system?
2. What must be the conditions in the circuit of Fig. 23-3 for the circuit to change the sign of the input signal?
3. What is a scaler?
4. What are the uses of a scaler? Illustrate.
5. What is an adder? What are the requirements if the output is to be truly equal to the sum of the inputs?
6. What is a subtractor?
7. What is an integrator?
8. If the input to an integrator is a dc voltage, what is the output?
9. What are the limitations of a practical OP AMP integrator?
10. What is a differentiator?
11. If the input to a differentiator is a voltage increasing linearly with time, what is the output?
12. How is a voltage signal converted into a proportional current signal?
13. Why is it not enough to simply connect a resistor in series with the voltage signal to produce a current signal?
14. How is a current signal converted into a proportional voltage signal?
15. Why is it not enough to simply connect a resistor in parallel with the current signal to convert to a proportional voltage signal?
16. What is a buffer?
17. What are some applications for a voltage follower?

19\. Give some concrete examples of the application of analog systems.

19\. What type of problems is the analog computer best suited for?

20\. If the OP AMPs used in the example of Fig. 23-19 have (output) saturation voltages of +20 V and −20 V, what is the highest count that the system can accomplish? Why?

24 Digital Systems

In this chapter we discuss the principles of digital systems. To most people, digital systems mean digital computers. But there are many other uses for digital circuits: instrumentation, communication, industrial applications, as well as computation. As the cost decreases, the availability of high-quality digital ICs spreads. Even quite complex functions on a single chip (for example, the calculator on a single chip) and other applications of digital circuits are being used in consumer and industrial products. With larger scale integration comes increased reliability because of fewer components, a tremendous decrease in size, and eventually a cost saving (in either the actual cost of components or the cost of assembly or both).

24.1 BINARY ADDERS

One of the basic building blocks of digital systems—of the computer, in particular—is a binary *adder*. It uses the digital, or binary, system, which has only two digits: 1 and 0. The addition of two digits, therefore, results simply in a carry (a 0 or a 1) and the sum, which is 0 or 1.

To add binary numbers, we usually use the digital circuit depicted in Fig. 24-1. This circuit is given a special name: an *exclusive-OR* gate. Suppose that the two inputs A and B are the two binary digits to be added. The output corresponds to the binary sum of these digits. We want the output to be a 1 only if either of the two inputs is a 1 with the other input a 0. This is the definition of the exclusive-OR gate. It provides a 0 output if both inputs are a 1 and also if both inputs are a 0. Figure 24-2 gives the symbol for an exclusive-OR gate.

Although the exclusive-OR gate provides the proper sum, we need to add an AND gate, as shown in Fig. 24-3, to provide the carry. When both A and B are 1s, the output of the exclusive-OR gate indicates the proper sum (0), while the output of the AND gate goes to 1, indicating the proper carry. The circuit of Fig. 24-3 is called a *half adder*, because it can be used to add two single digit numbers only. The symbol for a half adder, together with its truth table, is given in Fig. 24-4. The truth table summarizes what we have said here.

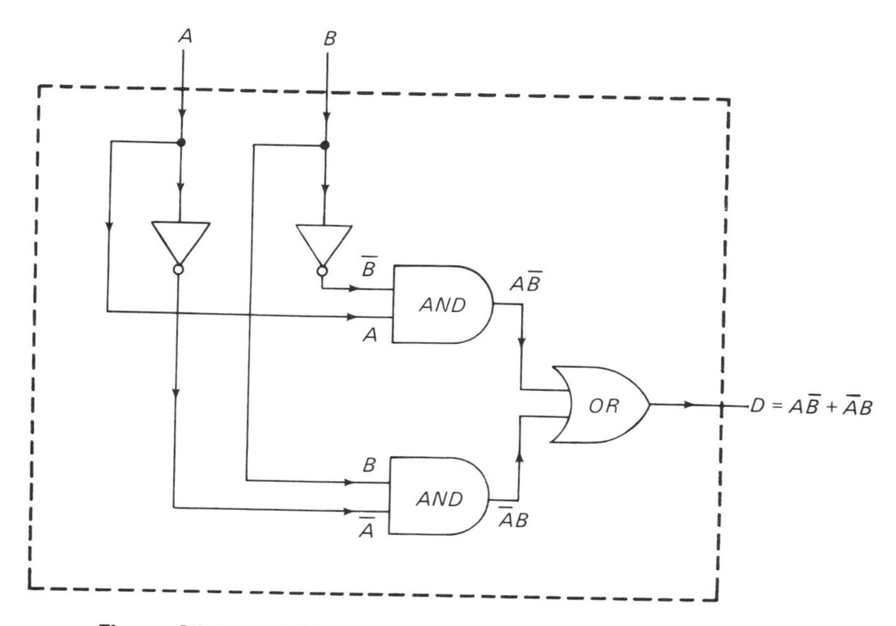

Figure 24-1. Individual gates to make up an exclusive-OR gate.

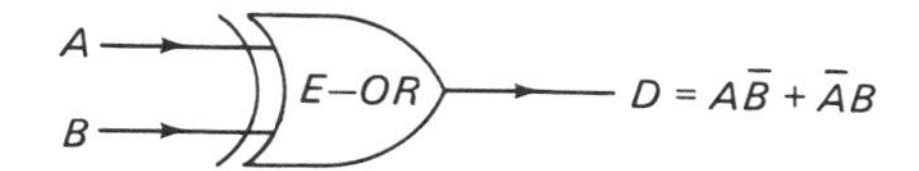

Figure 24-2. Symbol for an exclusive-OR gate.

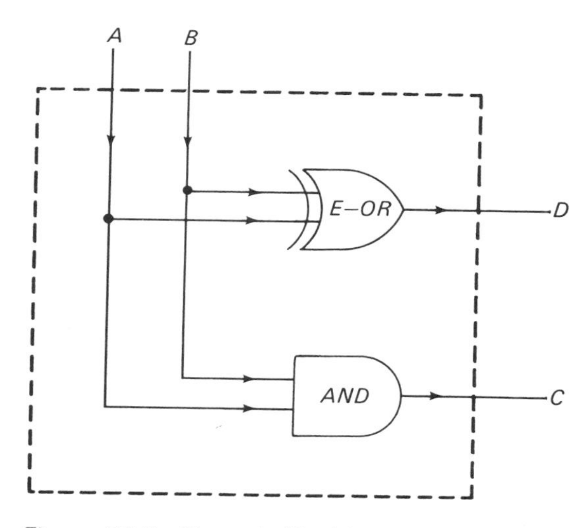

Figure 24-3. Binary half adder implementation.

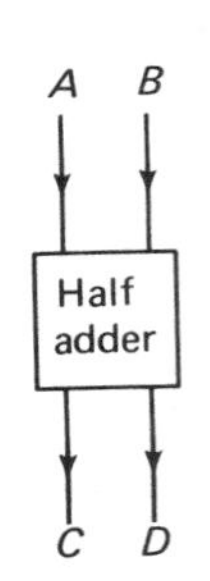

Input		Output		
A	B	C	D	Sum
0	0	0	0	00
0	1	1	0	01
1	0	1	0	01
1	1	1	1	10

Figure 24-4. Binary half adder: symbol and truth table.

When multidigit numbers are to be added, we use the half adder to provide the sum and carry for the least significant digits. The addition of the next significant digits has three inputs: the two digits to be added and the carry from the previous addition. The addition of these three digits is carried out with the use of two half adders and an OR gate, as indicated in Fig. 24-5. The digits A and B are added in the first half adder. The result of this addition is then added with the carry from the previous stage in the second half adder. We provide the carry for this stage by feeding the carry outputs of the two half adders through the OR gate, as shown. The circuit thus obtained is called a *full adder*, because it can add two digits and the carry from the previous stage.

The difference between a half and a full adder becomes obvious when their symbols and truth tables are compared (Fig. 24-6). The full adder symbol shows three inputs, whereas the half adder only has two. The input possibilities for the full adder total 8; for the half adder, they total 4.

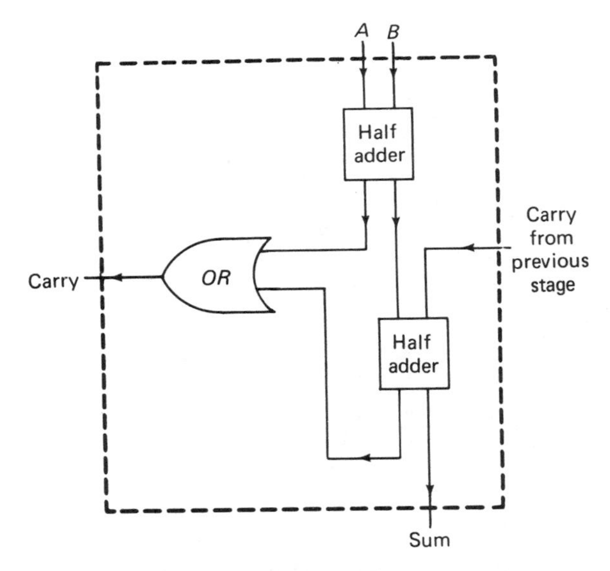

Figure 24-5. Full adder.

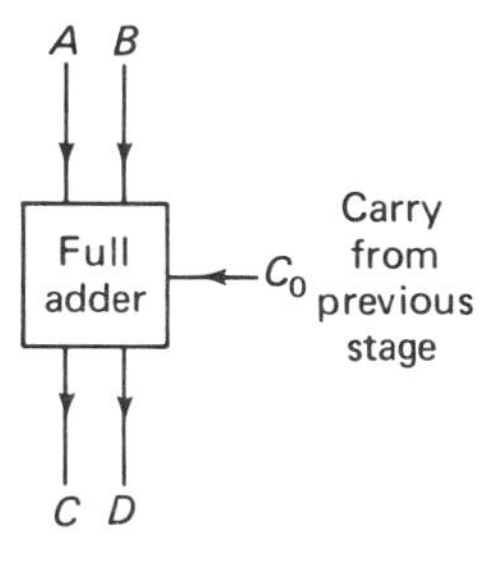

Inputs			Outputs		
A	B	C_0	D	C	Sum
0	0	0	0	0	00
0	1	0	1	0	01
1	0	0	1	0	01
1	1	0	0	1	10
0	0	1	1	0	01
0	1	1	0	1	10
1	0	1	0	1	10
1	1	1	1	1	11

Figure 24-6. Full adder: symbol and truth table.

To add a complete set of digits comprised of two binary numbers, we use either serial or parallel operation. Figure 24-7 illustrates the parallel addition of two five-digit binary numbers: number A—11010 (26 in decimal) and number B—11011 (27 in decimal). The addition of the two least significant digits is carried out using a half adder, because for the first sum there is no carry from the previous stage. For the example numbers,

$$A = \begin{cases} A_4 & A_3 & A_2 & A_1 & A_0 \\ 1 & 1 & 0 & 1 & 0 \end{cases} = 26$$

Plus

$$B = \begin{cases} 1 & 1 & 0 & 1 & 1 \\ B_4 & B_3 & B_2 & B_1 & B_0 \end{cases} = 27$$

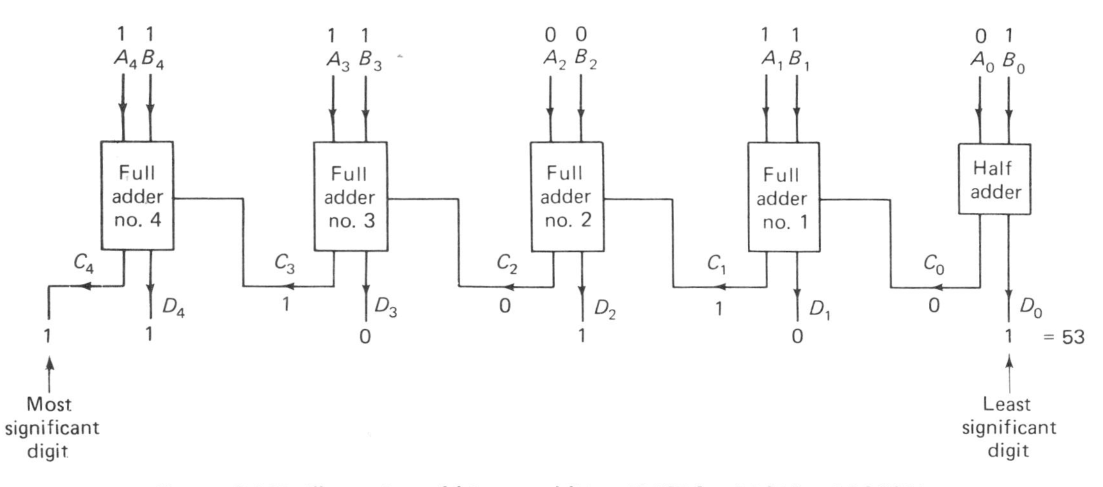

Figure 24-7. Illustration of binary addition (11010 + 11011 = 110101).

when A_0 is 0 and B_0 is 1, the sum is 1 with the carry 0. The carry of this calculation is applied to the full adder, which is to add the next two digits A_1 and B_1. This process is then repeated as many times as there are digits (in this case, three more times). Note that the full adder adding the two most significant digits has two outputs: (1) the normal sum output D_4 and (2) the carry C_4. C_4 becomes the most significant digit in the answer, because there is no adder to the left of this one. This operation is termed *parallel*. In effect, all the adders are working at the same time; that is, the digits are being added simultaneously (or in parallel).

Serial addition is accomplished by using only a single full adder, shown in Fig. 24-8. The digits of the numbers are added one at a time. First, we apply the least significant digits to the adder. The sum is obtained and stored. The carry from the first addition is delayed or temporarily stored in a type-D flip-flop until the input is changed and the next two digits to be added are applied. When this step occurs, the carry is transferred from the D-FF into the carry input of the full adder. The sum and carry for this addition are then formed. The sum is stored and the carry delayed. This operation is repeated as many times as there are digits in the numbers to be added. It may seem that serial addition offers a saving in the number of adders used; it uses only a single adder as opposed to many adders for parallel operation. However, the parallel adder does not require the use of timing, storing, and shifting circuits that the serial adder must have.

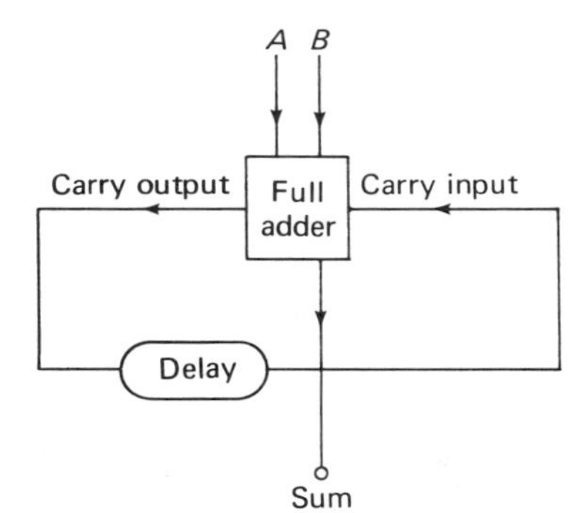

Figure 24-8. Serial addition.

The arithmetic operation of addition is the heart of a computer's calculating capability. Multiplication of two numbers is accomplished through repeated addition; subtraction is also accomplished through addition; and division is done through repeated subtraction, which also relies on addition.

Binary subtraction is based on the fact that the binary number system has only two digits. As a result, when the complement of any binary number is added to the binary number, the sum is always a series of 1s with a 1 carry for the most significant digit. For example, if we consider the binary number 10101, its complement is 01010 (obtained by changing all 0s to 1s and all 1s to 0s). The sum of this number and its complement is

111111: five 1s and a 1 carry. If the carry is ignored and the number 1 (00001) is added to the sum of the number and its complement, a 0 will always result. Thus, the original number may be written as:

$$\text{number} = -(\text{complement number}) - 1$$

with the most significant digit ignored. With the example number 10101, this statement can be verified. The negative of the number is then just the addition of the complement of the number and 1 with the most significant digit ignored. Thus, when we want to subtract the number B from a number A, we merely add to A the complement of B and 1, ignoring the most significant digit in the answer.

Suppose that $A = 11010$ and $B = 10101$. We want to subtract B from A. First, we form the complement of $B(\overline{B})$, which is $\overline{B} = 01010$. Next, we add 1 to the complement of B:

$$\overline{B} + 1 = 01010 + 00001 = 01011$$

The subtraction is completed by adding this result to A and ignoring the most significant digit:

$$\begin{aligned} A - B &= A + \overline{B} + 1(-\text{MSD}) \\ &= 11010 \text{ plus } 01011\ (-\text{MSD}) \\ &= 100101\ (-\text{MSD}) \\ &= 00101 \end{aligned}$$

The answer may be verified by noting that in decimal notation $A = 26$ and $B = 21$ so that $A - B = 5$, which is indeed the binary number 00101.

In terms of circuit implementation of this scheme for subtraction, note that each digit can be complemented by just feeding it through an inverter gate. We perform the addition by using adders. We can show that the carry will always be a 1 if the answer is positive; therefore, the carry can be reintroduced as the carry for the least significant digit adder. We call this process the *end-around* carry. The block diagram for the subtraction of two five-digit numbers A and B is shown in Fig. 24-9. The values correspond to the numbers used in the previous example.

The same scheme for subtraction can be used even if the result is negative. Under these circumstances, the most significant digit carry is always 0, and the answer is the complement of that obtained. For example, consider forming B minus A for the same numbers as above. This procedure may be done on Fig. 24-9 by interchanging the A and B inputs on top. The end-around carry in this case is 0, and the answer on the bottom will be 11010. This is the complement of the actual negative answer -5 (00101).

As mentioned earlier, division is accomplished through repeated subtraction. Obviously, the basic building block of the arithmetic unit in a digital computer is an adder. We do not mean, however, that the associated circuitry may be deleted. An arithmetic unit consists of adders, shift registers, multiplexers, and other circuitry.

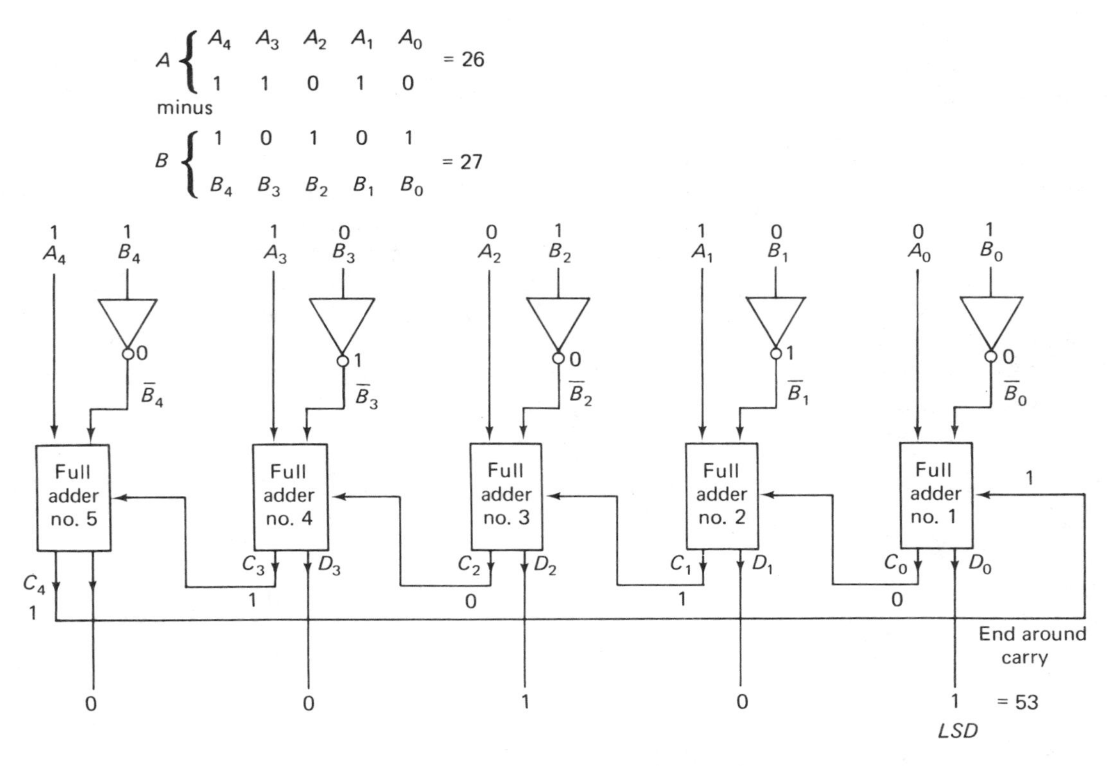

Figure 24-9. Illustration of binary subtraction (11010 − 10101 = 00101).

24.2 DECODERS

Another important building block used in digital computers, as well as in many other digital systems, is the *decoder*. The basic function of any decoder is to convert information from one number system into another. For example, the result of a calculation on a digital computer is probably in binary form. To be more usable, this binary form is decoded (i.e., translated) into decimal form before it is displayed or typed.

Besides the pure binary number system, it is often advantageous to use the *binary coded decimal* (BCD) notation. In this scheme, each digit in the decimal number is allocated four places in the BCD number. The units of the decimal number are represented by the first four binary digits on the right; the 10s of the decimal number are represented by the next four binary digits, and so on. This scheme is illustrated in Fig. 24-10. The

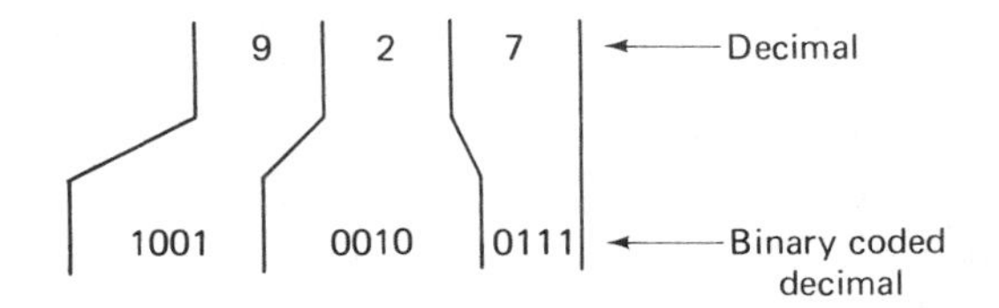

Figure 24-10. Example of a BCD number and its decimal equivalent.

decimal number 927 is represented by the binary number 1001 (for the 900), 0010 (for the 20), and 0111 (for the 7). Thus, the BCD number is a form of binary notation where the digits are grouped in fours. The four binary digits are necessary to represent the decimals 0 through 9.

24.2.1 BCD-to-Decimal Decoder

As a result of calculations or some processing, we have a numerical answer available in BCD form. But the advantage of having it displayed in decimal form rather than in BCD can be readily seen by comparing the two forms of the simple number 927 in Fig. 24-10.

To decode a single BCD number (consisting of four binary digits) into its decimal equivalent requires the use of a *BCD-to-decimal decoder*. This circuit is available in a single 16-lead package (SN7445). It has four inputs: the four binary digits of the BCD number. It has ten outputs, which indicate the specific decimal number (0 through 9) corresponding to the BCD input. The truth table for the BCD-to-decimal decoder is given in Fig. 24-11. For example, suppose that the BCD input is 0111. The output labeled 7 will be high, with all other outputs low, indicating that the BCD input corresponds to the decimal number 7. We can display the decoded number in any of a number of ways. One method is to have 10 indicator lights with the decimal numerals 0 through 9 printed on each of the lights. These lights can be electrically connected to the 10 outputs of the decoder. We indicate the decoded decimal number by lighting the indicator lamp having the appropriate numeral on it.

Another scheme uses a single vacuum tube, called a *nixie tube*, which has 10 filaments inside it. Each filament is shaped in the form of one of the numerals 0 through 9. When the signal is applied to one of the filaments, it glows, giving an indication of the decimal number corresponding to the BCD input to the decoder. Because nixie tubes require a high voltage, an

Input				Output low*
D	*C*	*B*	*A*	
0	0	0	0	0
0	0	0	1	1
0	0	1	0	2
0	0	1	1	3
0	1	0	0	4
0	1	0	1	5
0	1	1	0	6
0	1	1	1	7
1	0	0	0	8
1	0	0	1	9

*All other outputs high

Figure 24-11. Truth table for a BCD-to-decimal decoder (SN7445).

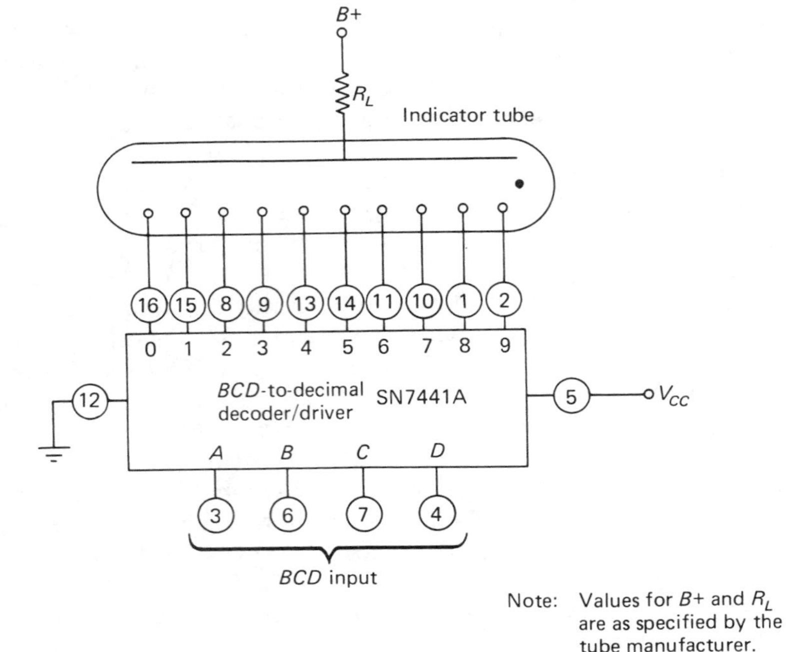

Figure 24-12. BCD-to-decimal decoder with nixie tube display.

external supply (of between 150 and 170 V) is connected in series with a current-limiting resistor (typically 10 kΩ) to the anode of the display tube. The 10 electrodes are connected to the 10 decoder outputs. This design is shown in Fig. 24-12. The output of the decoder used with the tube type display (SN7441 A) indicates that the decoded number is high. All other outputs are low, which is just the opposite of the truth table in Fig. 24-11.

24.2.2 Seven-Segment Decoder

Another popular way to display decimal numerals 0 through 9 is the seven-segment readout, so called because it uses 7 lines (segments) to display the proper numeral. This decoder is shown in Fig. 24-13, together with examples of possible displays. The display unit may be an LED type with the seven segments made up of rows of LEDs, or it may be an incandescent display with the seven segments made up of filaments somewhat similar to the ones used in a light bulb. These filaments light up when a current is passed through them. Other forms of seven-segment displays use liquid crystals or cold-cathode (glow) segments.

The decoder, in any case, must take the four-digit binary number in BCD form and decode it to provide the proper drive for the seven-segment display. A single IC package (SN7447) is available to perform this function. It has four inputs corresponding to the BCD information to be decoded and seven outputs capable of driving the seven-segment readout. Zero-suppressing inputs and coded outputs are included. It can also test

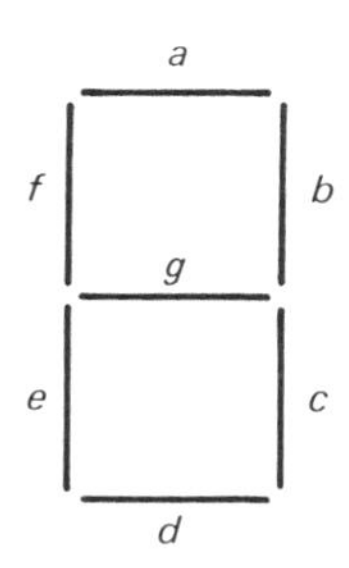

Segment identification

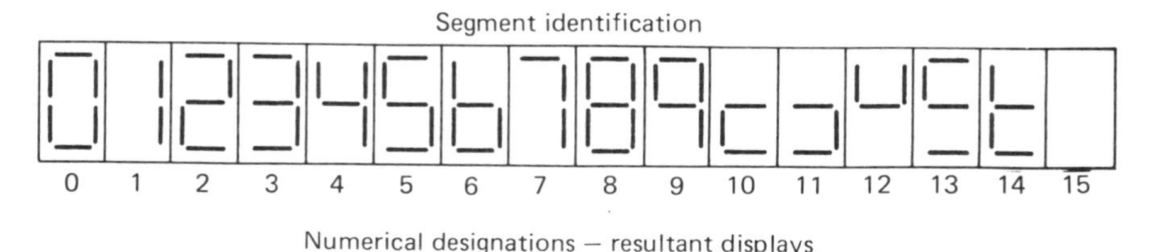

Numerical designations — resultant displays

Figure 24-13. Seven-segment readout.

the segments, that is, light them all at the same time. These functions will be discussed in more detail under applications. The truth table for the seven-segment decoder is given in Fig. 24-14. If we wish to display the numerals from 0 to 9, only the first 10 lines of the truth table are necessary, and BCD inputs in excess of 1001 are not allowed. In any case, they are not legitimate BCD numbers.

24.3 COUNTERS (FREQUENCY DIVIDERS)

Most digital applications require that some signal consisting of a series of 0s and 1s be counted. There are many possible counting schemes; we shall discuss the most basic of these: ripple and decade counters.

Suppose that the signal is a series of 0s and 1s equally spaced in time with frequency f_x. Such a signal might be derived from an astable MV or other square-wave generator. If we apply this signal to the clock input of a JK-FF, as shown in Fig. 24-15, with the JK terminals permanently tied to $+V_{CC}$ (logical 1), the FF will toggle on every transition of the input. Recall that the master FF changes state, or toggles, on every transition from 0 to 1. This information is transferred into the slave FF on the transition from 1 to 0 (see Fig. 18-23). The Q output of the FF is indicated in Fig. 24-16 for the input shown. Note that the frequency of the output is exactly one-half of the input waveshape frequency; thus, the single FF circuit of Fig. 24-15 is a divide-by-2 counter.

By simply connecting additional FFs to the output of the preceding one, we divide the frequency of the input waveshape by any multiple of 2. For example, two FFs provide an output that has one-fourth the frequency of the input; three FFs divide the frequency by 8; four FFs, by 16, and so on.

Inputs								Outputs							
Decimal or function	*LT*	*RBI*	*D*	*C*	*B*	*A*	*BI/RBO*	*a*	*b*	*c*	*d*	*e*	*f*	*g*	Note
0	1	1	0	0	0	0	1	0	0	0	0	0	0	1	1
1	1	X	0	0	0	1	1	1	0	0	1	1	1	1	1
2	1	X	0	0	1	0	1	0	0	1	0	0	1	0	
3	1	X	0	0	1	1	1	0	0	0	0	1	1	0	
4	1	X	0	1	0	0	1	1	0	0	1	1	0	0	
5	1	X	0	1	0	1	1	0	1	0	0	1	0	0	
6	1	X	0	1	1	0	1	1	1	0	0	0	0	0	
7	1	X	0	1	1	1	1	0	0	0	1	1	1	1	
8	1	X	1	0	0	0	1	0	0	0	0	0	0	0	
9	1	X	1	0	0	1	1	0	0	0	1	1	0	0	
10	1	X	1	0	1	0	1	1	1	1	0	0	1	0	
11	1	X	1	0	1	1	1	1	1	0	0	1	1	0	
12	1	X	1	1	0	0	1	1	0	1	1	1	0	0	
13	1	X	1	1	0	1	1	0	1	1	0	1	0	0	
14	1	X	1	1	1	0	1	1	1	1	0	0	0	0	
15	1	X	1	1	1	1	1	1	1	1	1	1	1	1	
BI	X	X	X	X	X	X	0	1	1	1	1	1	1	1	2
RBI	1	0	0	0	0	0	0	1	1	1	1	1	1	1	3
LT	1	X	X	X	X	X	1	0	0	0	0	0	0	0	4

Notes: 1. *B1/RBO* is wire—*OR* logic serving as blanking input (*BI*) and/or ripple-blanking output (*RBO*). The blanking input must be open or held at a logical 1 when output functions 0 through 15 are desired and ripple-blanking input (*RBI*) must be open or at a logical 1 during the decimal 0 input. *X* = input may be high or low.

2. When a logical 0 is applied to the blanking input (forced condition) all segment outputs go to a logical 1 regardless of the state of any other input condition.

3. When ripple-blanking input (*RBI*) is at a logical 0 and *A* = *B* = *C* = *D* = logical 0, all segment outputs go to a logical 1 and the ripple-blanking output goes to a logical 0 (response condition).

4. When blanking input/ripple-blanking output is open or held at a logical 1, and a logical 0 is applied to lamp-rest input, all segment outputs go to a logical 0.

Figure 24-14. Seven-segment decoder (SN7447) truth table.

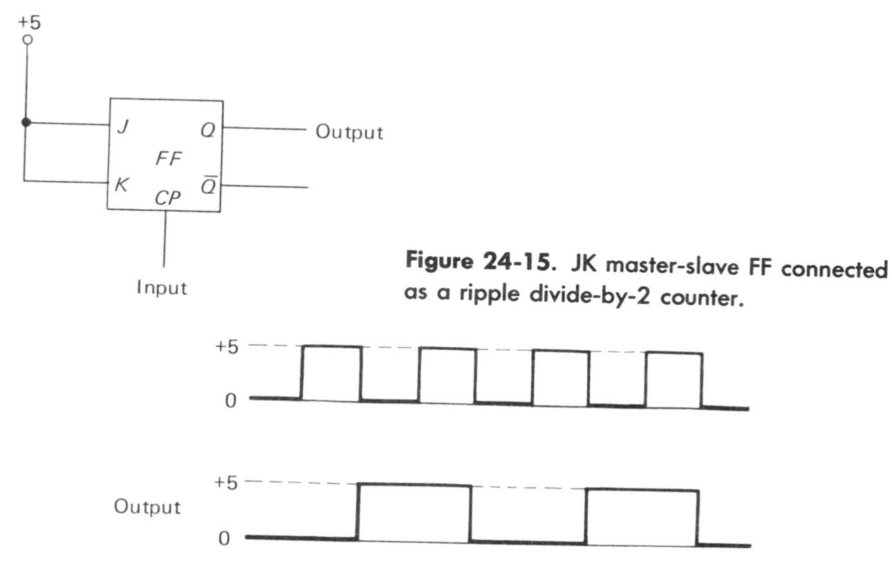

Figure 24-15. JK master-slave FF connected as a ripple divide-by-2 counter.

Figure 24-16. Waveshapes for the ripple counter of Fig. 24-15.

In addition, this scheme can count the number of pulses that have been applied at the input. Consider the case of interconnecting four JK-FFs, as shown in Fig. 24-17. This circuit divides the frequency of the input signal by 16. However, assume that the input contains a certain number of pulses between 0 and 16. The FFs are all in the $Q=0$ state; that is, they have been reset prior to the initiation of the count. The input pulse train is now initiated (t_0) and will terminate a certain time later (at t_1). If, for example, the input waveshape contains 11 pulses, the ripple counter waveshapes are as depicted in Fig. 24-18. Note that at the end of the pulse train $A=B=D=1$ and $C=0$. This state corresponds to the binary number 1011 (with A being the least significant digit), which is exactly the count of the pulses (11) in binary form. Thus, the circuit may be used as a counter. It is called a *ripple counter* because the negative transitions (from 1 to 0) ripple through the counter.

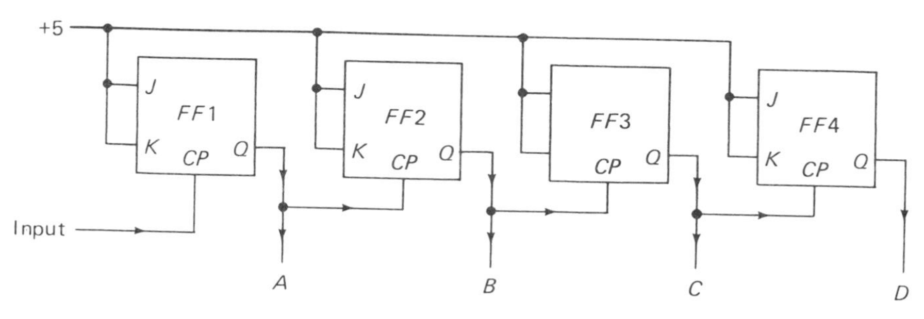

Figure 24-17. Four-stage ripple counter.

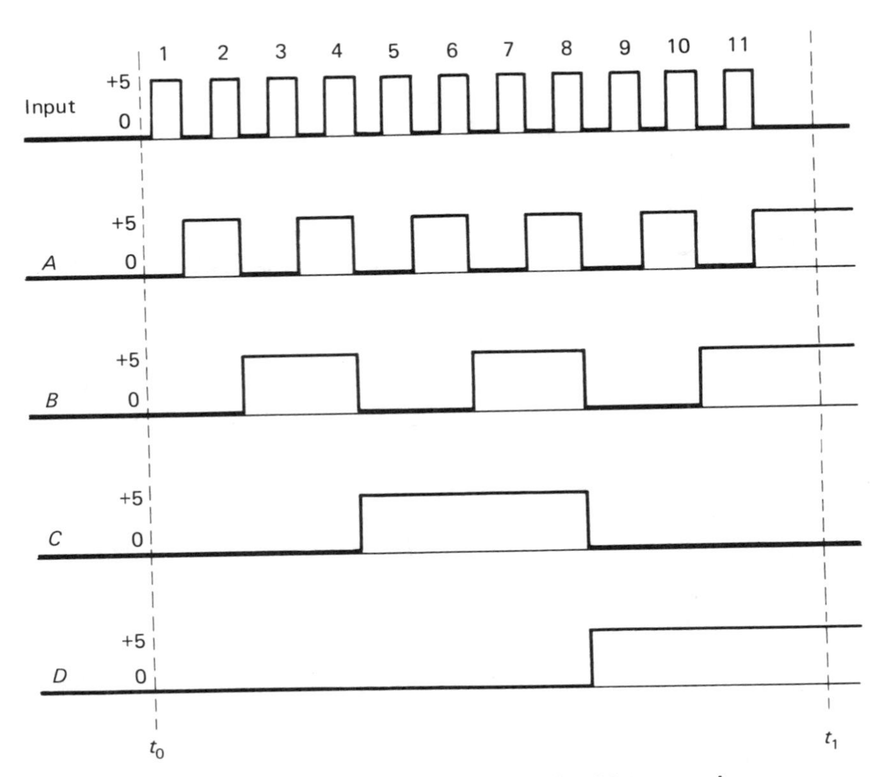

Figure 24-18. Counter waveshapes for 11 input pulses.

We can use the basic ripple counter principle to count to almost any base. However, the problem with a ripple counter is the delay that is propagated (ripples) through the counter. The worst case is when all the FFs are in the 1 state, and the input makes a transition from 1 to 0. Each of the FFs in turn makes the same transition from 1 to 0. There is a time lag between the first FF making its transition and the last FF making its transition because the transition must pass through all the FFs in between. This problem is avoided if the number of FFs in the counter is kept low.

The *decade counter* is used in a multitude of digital systems and so deserves special attention. The input to a decade counter consists of a series of pulses. The output is a BCD number corresponding to the number of pulses counted. When the count reaches 9, the counter resets itself, having provided the initiating pulse for the next decade counter.

The decade counter is available on a single IC chip. Its schematic is shown in Fig. 24-19. In order to use the counter illustrated (SN7490) as a decade counter, the output A of the first FF (pin 12) must be connected to the input of the second FF (pin 1). The truth table for the decade counter is given in Fig. 24-20. Note the availability of manual reset inputs R_0. When high (+5 V), they cause the counter to reset to 0000. There are also reset inputs R_9, which when energized (+5 V) cause the counter to go to a 9 state: 1001.

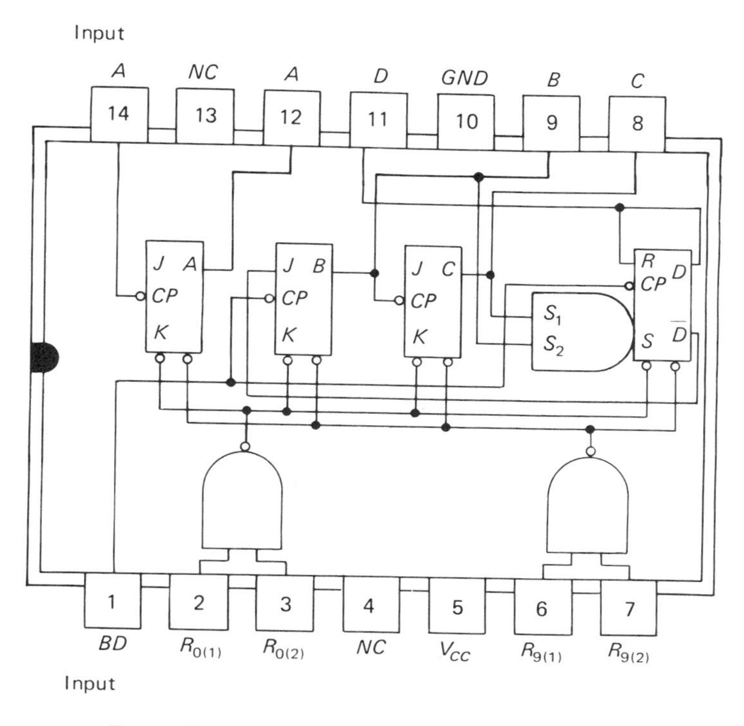

Figure 24-19. An IC decade counter (SN7490).

Count	Output D	Output C	Output B	Output A
0	0	0	0	0
1	0	0	0	1
2	0	0	1	0
3	0	0	1	1
4	0	1	0	0
5	0	1	0	1
6	0	1	1	0
7	0	1	1	1
8	1	0	0	0
9	1	0	0	1

Reset inputs				Output
$R_{0(1)}$	$R_{0(2)}$	$R_{9(1)}$	$R_{9(2)}$	D C B A
1	1	0	X	0 0 0 0
1	1	X	0	0 0 0 0
X	X	1	1	1 0 0 1
X	0	X	0	Count
0	X	0	X	Count
0	X	X	0	Count
X	0	0	X	Count

Figure 24-20. Truth table for decade counter (SN7490).

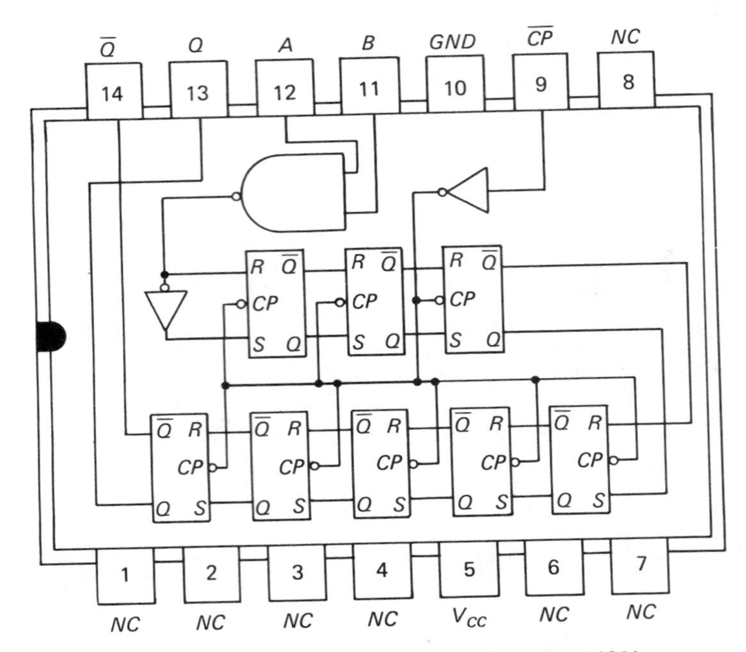

Figure 24-21. Eight-bit shift register (SN7491).

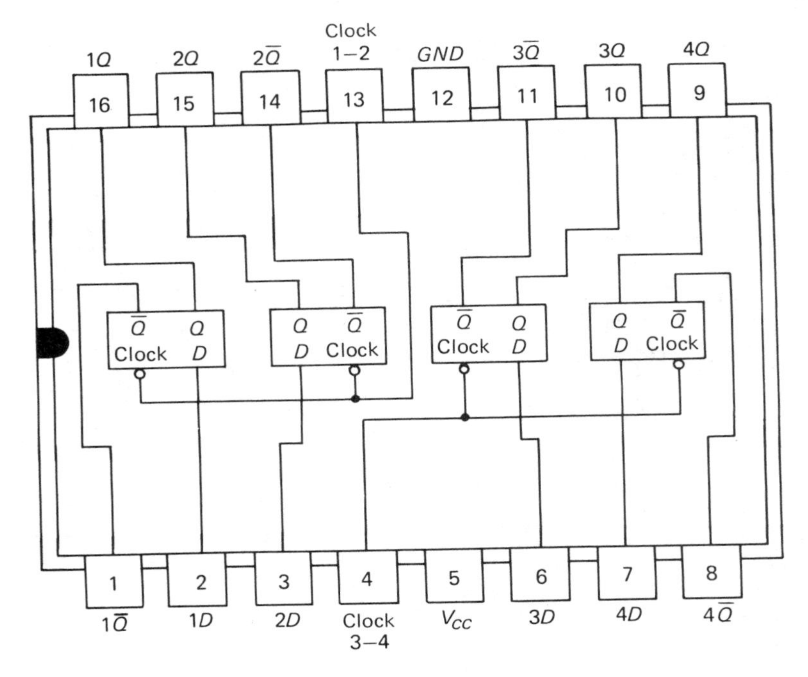

Figure 24-22. Four-bit latch (SN7475).

24.4 SHIFT REGISTERS AND LATCHES

As we have already mentioned in dealing with arithmetic functions, it is often necessary to transfer digital data from one place to another. This function is performed by circuits called *shift registers*. In addition, we frequently need to store data temporarily and then release it upon the proper signal. Circuits called *latches* perform this operation.

An example of an IC shift register (SN7491) is shown in Fig. 24-21. The circuit shifts the digital information through its eight flip-flops at the command of the clock-pulse CP. After each clock-pulse, the input information is shifted through one FF. Thus, after eight clock-pulses, the information is available at the output terminals. RS-FFs have both Q and $\overline{Q}$ outputs, with the input gated; that is, information is transferred only if both inputs are high.

An example of a four-bit latch is given in Fig. 24-22. The circuit stores the input information after the clock pulse and rejects any inputs when the clock pulse is low. It can store four digital numbers (0 or 1) in its four D-FFs, thus its name: four-bit latch. It is used in counters, among other applications, to store the result of a particular count after it is completed (see the section on applications).

24.5 MULTIPLEXERS

A *multiplexer* places a number of digital signals on one channel. For example, we may have several different inputs, all of which are to be handled on one line. These inputs are fed into a multiplexer circuit, which has only one output selected by the code applied to it. For one code, the output provides one of the inputs; for a different code, it can provide one of the other inputs at the output terminals.

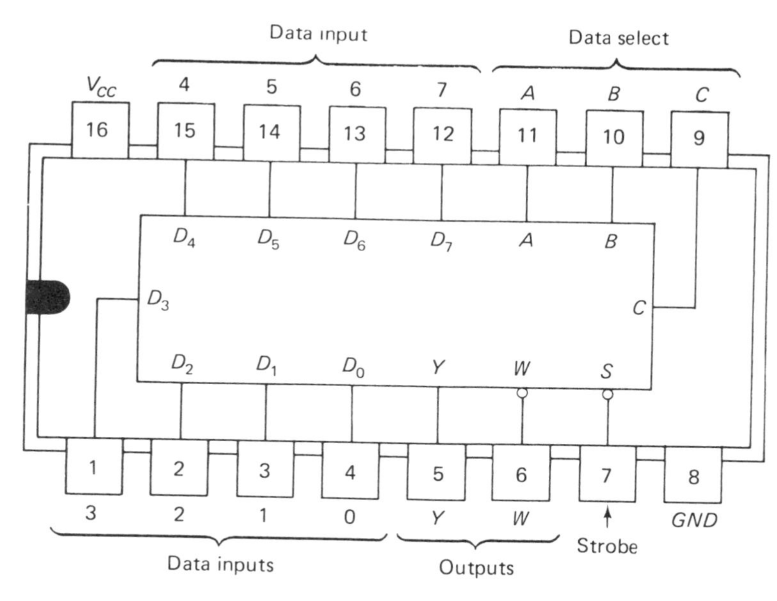

Figure 24-23. Multiplexer (SN74151).

An example of an IC multiplexer that can accommodate eight inputs is shown in Fig. 24-23. The data selection code is applied to terminals 9, 10, and 11 in accordance with the truth table in Fig. 24-24. Note the additional flexibility of the "strobe" input. When high (+5 V), it has the effect of disabling the multiplexer. The operation of this circuit is as follows. Up to eight digital inputs can be accommodated at the input terminals labeled D_0 through D_7. The data selection code applied to terminals A, B, and C determines which of the eight input signals will be available at the output terminal Y. Note that the inverted output is available at terminal W. For example, suppose that the data select code is $A=0$, $B=1$, and $C=1$. Then the output signal will follow the input signal that is applied to the D_3 terminal.

24.6 EXAMPLES OF DIGITAL SYSTEMS

As a simple digital system, let us consider a *decimal counting unit* (DCU). The input to the unit is a series of pulses to be counted; the output is a visual display of the count. Such a unit is shown in Fig. 24-25. It consists of (1) a decade counter (SN7490), (2) a four-bit latch (SN7475), (3) a decoder—in this case, a seven-segment decoder (SN7447), and (4) a seven-segment readout. To initiate the count, a logical 1 (+3.6 to +5 V) is applied to pins 2 and 3, the reset input of the decade counter. As a result, all of the BCD output lines (at pins 12, 7, 8, and 11) become zero. During the input cycle, the pulses to be counted are applied to the decade counter input (pin 14). In addition, the latch clock is kept low, thus preventing the BCD inputs from being transferred to the seven-segment decoder. The BCD outputs (pins 16, 10, 9, and 15) of the latch are at zero, thus causing all of the decoder outputs to be high. Under this condition, none of the segments is lit up.

While the input pulses are being counted, the latch clock is maintained low to prevent the count from being displayed. In this way, the readout display will not flicker from one number to the next as the count increases. Once the count sequence is terminated, the latch clock is made high (+3.6 to +5 V), and the BCD output of the counter is allowed to transfer into the decoder. As an example, suppose that the counter outputs at this time are $A=B=C=1$ and $D=0$. When the latch clock goes high, this information is stored in the latch and applied to the decoder inputs. As a result, the decoder outputs will be $a=b=c=0$, with all other segments maintained high (logical 1). The readout segments that are connected to ground (at a logical 0 level) are the ones that are lit up. Thus, in this example, the a, b, and c segments (see Fig. 24-13) are lit up, indicating that the decimal number 7 corresponds to the BCD count at the output of the counter 0111. (Note that A represents the least significant digit in the BCD number.)

As soon as the count has been transferred, the latch clock is made low, thus preserving this count and preventing the output of the counter from being transmitted into the decoder. The decimal counting unit is now ready for the next counting sequence. Note that the action of the latch

Inputs												Outputs	
C	B	A	Strobe	D_0	D_1	D_2	D_3	D_4	D_5	D_6	D_7	Y	W
X	X	X	1	X	X	X	X	X	X	X	X	0	1
0	0	0	0	0	X	X	X	X	X	X	X	0	1
0	0	0	0	1	X	X	X	X	X	X	X	1	0
0	0	1	0	X	0	X	X	X	X	X	X	0	1
0	0	1	0	X	1	X	X	X	X	X	X	1	0
0	1	0	0	X	X	0	X	X	X	X	X	0	1
0	1	0	0	X	X	1	X	X	X	X	X	1	0
0	1	1	0	X	X	X	0	X	X	X	X	0	1
0	1	1	0	X	X	X	1	X	X	X	X	1	0
1	0	0	0	X	X	X	X	0	X	X	X	0	1
1	0	0	0	X	X	X	X	1	X	X	X	1	0
1	0	1	0	X	X	X	X	X	0	X	X	0	1
1	0	1	0	X	X	X	X	X	1	X	X	1	0
1	1	0	0	X	X	X	X	X	X	0	X	0	1
1	1	0	0	X	X	X	X	X	X	1	X	1	0
1	1	1	0	X	X	X	X	X	X	X	0	0	1
1	1	1	0	X	X	X	X	X	X	X	1	1	0

Figure 24-24. Truth table for SN74151 multiplexer.

maintains the previous count on the display while the counter is counting the next sequence of pulses. Consequently, the display does not flicker. The timing of the latch clock determines the rate at which the output of the counter is sampled and displayed.

The "light test" input to the decoder is normally kept at a logical 1. When brought to a logical 0 (essentially ground), it allows current to pass through all the segments, causing them all to light up. This method of testing the readout makes certain that all the segments are operational.

The readout may have a light to indicate a decimal point (shown as a "DP" input to the readout). When this point is at a logical 1, the decimal point is not lit up. When it is brought to a logical 0, the decimal point is lit up.

We must also consider the operation of a *digital frequency counter*, which represents a more complex digital system. The basic principles involved in a frequency counter may be extended to a *digital voltmeter* (see Chap. 25), an event counter, an interval indicator, and an indicator of speed and distance.

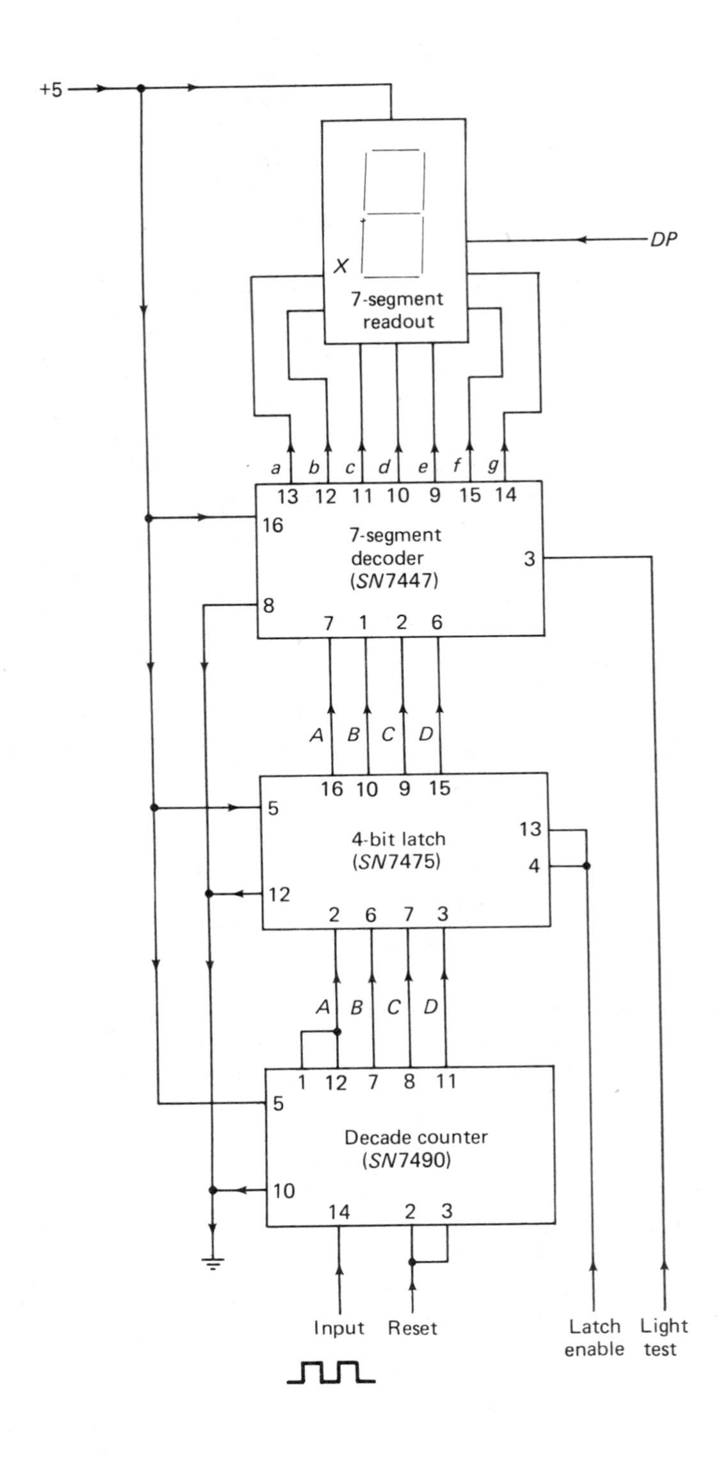

Figure 24-25. Decimal counting unit with 7-segment readout.

The frequency of a signal is measured in the following way. The signal whose frequency is to be measured is first transformed into a series of pulses having the same frequency. These pulses are fed through a gate into a series of decimal counting units that count the number of pulses. The frequency of the signal is determined from the number of pulses passing through the gate in a known length of time. For example, suppose that the pulses are allowed to pass through the gate for exactly 1 second. The number of pulses counted by the circuit will then be proportional to the signal's frequency in Hertz. If the gate is enabled for only 1 millisecond, then the count would correspond to the signal's frequency in kHz.

The block diagram for a frequency counter is depicted in Fig. 24-26. The input signal, which may be sinusoidal, is first conditioned by the squaring circuit. The output of this circuit is a series of pulses having exactly the same frequency as the input signal. These pulses are counted to provide the display of the frequency of the input signal. The squared signal is fed through the NAND gate, $G1$, whose other (enabling) input is the precise time-reference signal.

We derive the time-reference signal by counting down a precise clock signal. The diagram shows that this clock is a 1 MHz crystal controlled oscillator. The clock frequency is divided down from 1 MHz to 1 Hz in a series of six decade counters, as indicated. The 1 Hz signal is then divided by 2 to provide a 1-second pulse used to gate the input signal into the DCUs. The individual DCUs are connected in such a way that when the count reaches 9 in the first one, the next pulse resets the first DCU and initiates the count in the second DCU. Thus, DCU 1 is the units counter; DCU 2 is the 10s counter; DCU 3 is the 100s counter; and so on. The output of each DCU is displayed on its readout. The last DCU is connected to an overflow indicator that lights up when the last DCU makes a transition from 9 to 0. At this point, the frequency of the input would be in excess of the highest count possible. (In this diagram the highest count is 999.999 kHz.)

Latch timing is also provided by the crystal controlled gating signal. When the gate is enabled, inverter $I1$ keeps the outputs of gates $G2$ and $G3$ low, so that the count is disabled. In addition, when the gate is low, the output of $I1$ is high. In this way, the transfer pulse is provided by $G2$ when the count in the decoder (a BCD-to-decimal decoder SN7441) reaches 3. The transfer gate is once again closed when the count in the decoder reaches 4. The output of $G3$ is maintained low until the count in the decoder reaches 5. At this time, $G3$ goes high, providing the reset pulse to all DCUs. At count 6 in the decoder, the reset pulse is removed, and $G3$ becomes low again. From this point until the gate signal goes high, the display reads the previous count of the DCUs, which is stored in the latches. All decade counters inside the DCUs are reset to zero.

The gate signal remains low for 1 second and then again goes high to enable the gate and allow the signal to be applied to the DCUs. During this counting sequence, the latch is kept low because the output of inverter

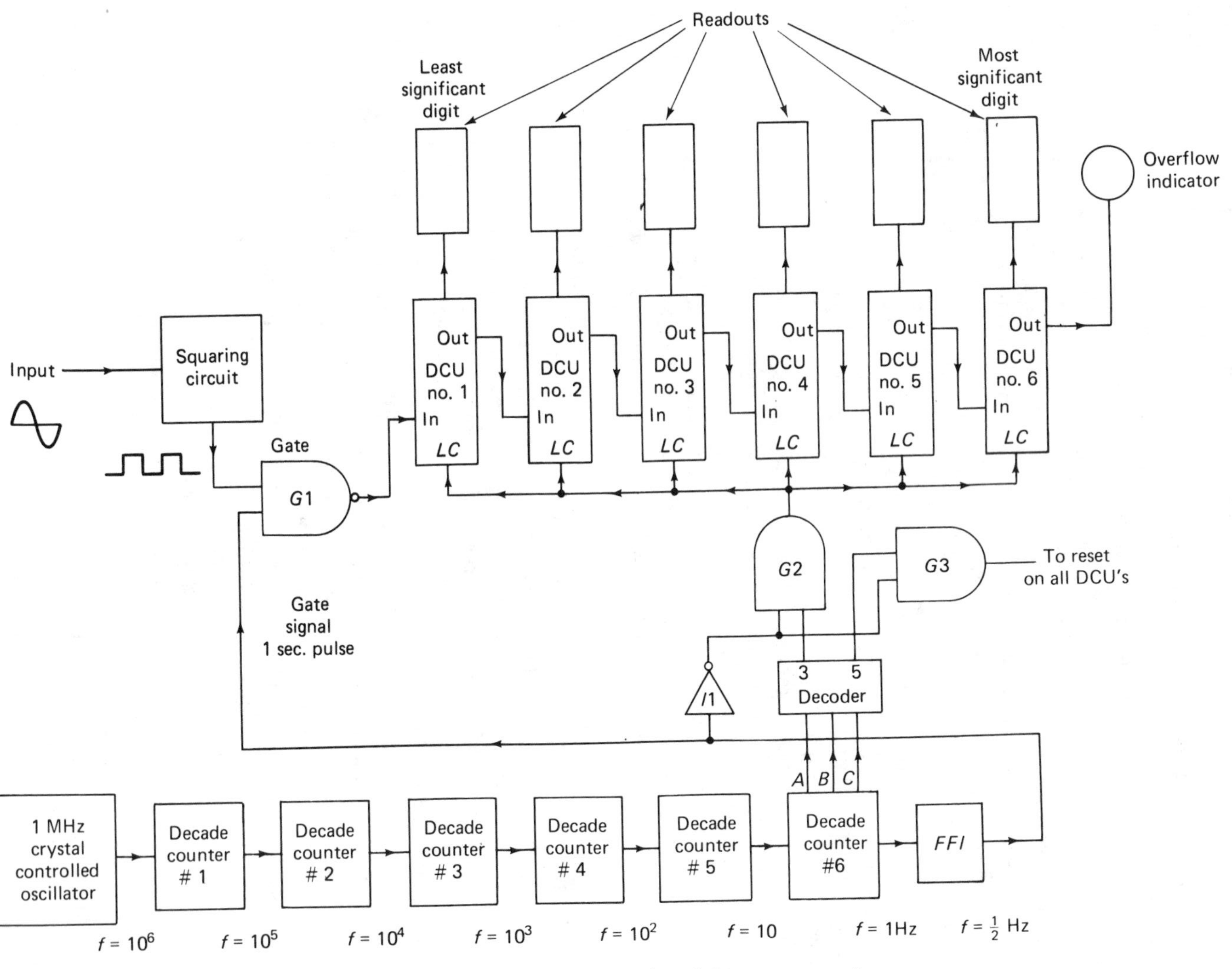

Figure 24-26. Block diagram for a digital frequency counter.

$I1$ is low, thus maintaining the readouts in the previous count. The readouts obtain a new count every 2 seconds.

The accuracy of the count obtained is a function of the accuracy and stability of the 1 MHz clock. In addition, there is always a 1 count inaccuracy in the least significant digit because the gate may have been opened in between input pulses.

Observe that additional DCUs may be added to the right of DCU 6 if additional counting capability is required. For example, with one additional DCU, the highest count would correspond to 9.999999 MHz. This count could be increased by connecting the output of the last DCU to an indicator light with a number 1 on it to provide $7\frac{1}{2}$ decades (count to 19.999999 MHz).

We can employ fewer readouts together with a multiplexer to achieve the same measurements. The outputs of each of the decade counters would be applied as inputs to the multiplexer, with the code inputs provided by a rotary switch appropriately connected to +5 V and ground. The time base could then be switch selected to provide a gating pulse of any duration between 1 second and 1 μs. The output of the multiplexer would be connected to FF1, with appropriate modifications to the transfer and reset circuits.

Review Questions

1. What is an exclusive-OR gate? (Discuss in terms of its truth table.)
2. What is a half adder?
3. How can a half adder be used to add two signal-digit binary numbers?
4. What is a full adder?
5. When must a full adder be used? When can a half adder be used?
6. What is the serial method of binary addition?
7. What is the parallel method of binary addition?
8. How are two binary numbers multiplied?
9. How can binary addition be used to *subtract* two binary numbers?
10. What is the complement of a binary number? What is the complement of: 110011, 101100, 1000001?
11. How is one binary number divided by another binary number?
12. What is a decoder?
13. What is a decoder used for?
14. What are some common types of decoders?
15. What form is the input (or the output) for a BCD-to-decimal decoder?
16. What is a seven-segment decoder?
17. Give some examples for the application of BCD-to-decimal and seven-segment decoders.
18. What is a digital counter?
19. What is a ripple counter?

20. What is a decade counter?
21. What is a shift register?
22. What is a binary latch?
23. What is a multiplexer?
24. What is a decimal counting unit?
25. What are some typical applications of digital systems (besides in computers)?

Problems

1. If the input to an exclusive-OR gate is: $A=1$ and $B=0$, what is the output? Repeat with $A=B=0$.
2. If the inputs to a half adder are: $A=1$ and $B=1$, what are the two outputs? Repeat if $A=0$ and $B=1$.
3. The inputs to a full adder are: $A=1$ and $B=1$. What are the two outputs if the carry input is 1? Repeat if the carry input is 0.
4. Sketch a diagram of the type shown in Fig. 24-7 that can be used to add two six-digit binary numbers.
5. If the numbers in Problem 4 are: $A=110011$ and $B=011111$, fill in the values in the diagram and also show the answer (sum).
6. Repeat Problems 4 and 5 if $A=111111$ and $B=111111$.
7. Sketch a diagram (of the type in Fig. 24-9) for the subtraction of two six-digit binary numbers.
8. If the numbers in Problem 7 are $A=111111$ and $B=111110$, fill in the values in the diagram and give the answer. (Note: Take A minus B.)
9. Repeat Problems 7 and 8 if the numbers are $A=001100$ and $B=101011$.
10. What are the following decimal numbers in BCD notation: (a) 14, (b) 301, (c) 999, (d) 5674?
11. What are the outputs of a BCD-to-decimal decoder for the following inputs: (a) 0100, (b) 0011, (c) 0110, (d) 1001?
12. Which of the following numbers are legitimate BCD numbers: 1000, 0111, 0100, 1100, 0011, 1001, 1110?
13. The input to a four-stage ripple counter (Fig. 24-17) is a series of pulses. What is the count (outputs A, B, C, and D) after: (a) 2 pulses, (b) 3 pulses, (c) 6 pulses, (d) 7 pulses?
14. The conditions in the DCU shown in Fig. 24-25 are: latch clock low; latch count $A=0$, $B=1$, $C=0$, and $D=1$, decade counter reset to 0000. Specify the outputs.
15. After a 9 input pulse (to the counter) in Problem 14, the latch clock goes high and is returned to low; the counter R input is made high. Specify the conditions throughout the circuit. (Also show the output.)

25

Digital-To-Analog and Analog-To-Digital Conversion

In many applications an analog input signal needs to be converted into digital form for processing. After it is processed, it has to be converted back into analog form for control or display. Such systems require the use of both *digital-to-analog (D/A)* and *analog-to-digital (A/D) converters*. In this chapter we shall examine the principles involved in such conversions and discuss some typical applications.

25.1 D/A CONVERSION

The input to a D/A converter is in digital form. It may be pure binary, *BCD*, or any binary form. The output of a D/A converter is an analog voltage (or current), which represents the digital input. As a simple example, suppose that the input is a binary coded number; the output would be a voltage corresponding to the number.

25.1.1 Binary-to-Analog Conversion

If the digital signal consists of a pure binary *word* (represented by a series of ones and zeros), the simplest D/A conversion is provided by the ladder network illustrated in Fig. 25-1. The digital word is coded into the switches with a switch connecting to V, corresponding to a logical 1, and a ground representing a logical 0. For example, the digital word 10101 would be represented by switches $S1$, $S3$, and $S5$ connecting to V and switches $S2$ and $S4$ connecting to ground. Switch $S1$ regulates the most significant digit. Obviously a digital word containing any number of digits, or *bits*, could be accommodated by simply extending the ladder network to the left. The last switch on the left ($S5$ in Fig. 25-1) would control the least significant bit.

The current supplied by any of the closed switches is met by a resistance of $2R$ to either side, so that it splits equally: half goes to the left and half to the right. To see how the circuit operates, assume that the binary word is 10000. Only $S1$ is connecting V; all other switches are grounded. Under these conditions, the equivalent circuit is shown in Fig. 25-2. The equivalent circuit taken at the output terminals has a resistance equal to R and an open circuit voltage equal to $V/2$, as indicated in Fig. 25-3. The output voltage is $V/3$.

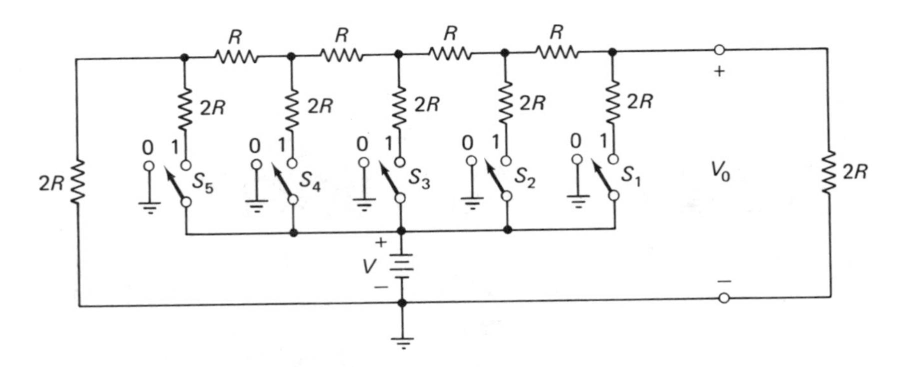

Figure 25-1. Ladder-type D/A converter.

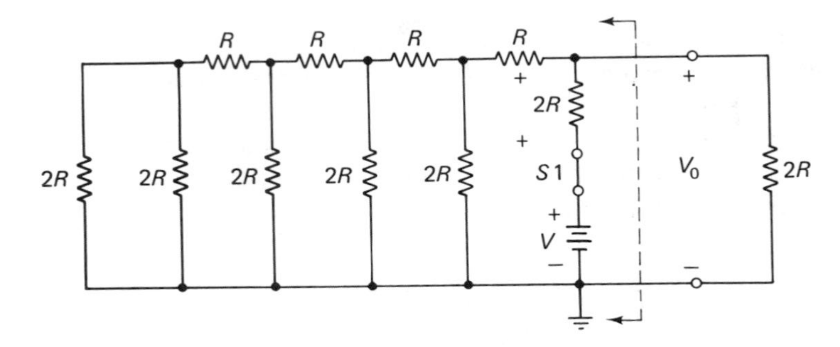

Figure 25-2. Example of ladder network with $S1 = 1$ and all other switches at 0.

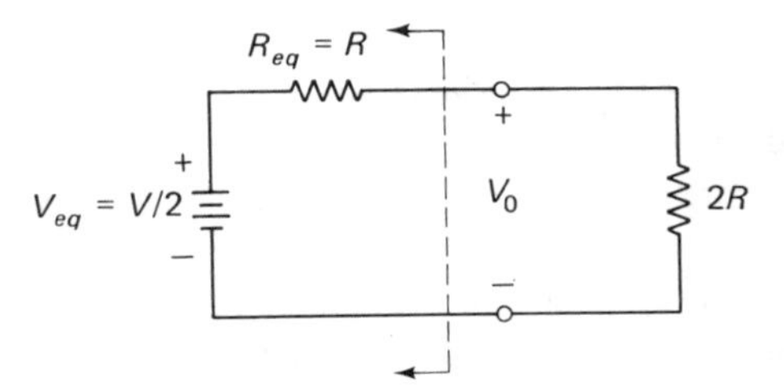

Figure 25-3. Thevenin equivalent of circuit in Fig. 25-2.

By using the principle of superposition, we can see that if all the switches are in logical 1 position, the output voltage would be given by:

$$V_o = \frac{2}{3}\left(\frac{V}{2} + \frac{V}{4} + \frac{V}{8} + \frac{V}{16} + \frac{V}{32}\right) \tag{25-1}$$

To remove the 2/3 scale factor, we can use an OP AMP set to have a gain of 3/2. The OP AMP provides a negative output for a positive input; therefore, we reverse the polarity of the reference voltage V to obtain a positive output voltage V_o. The complete circuit is shown in Fig. 25-4, where the voltage applied to any of the switches is either 0 (ground) or

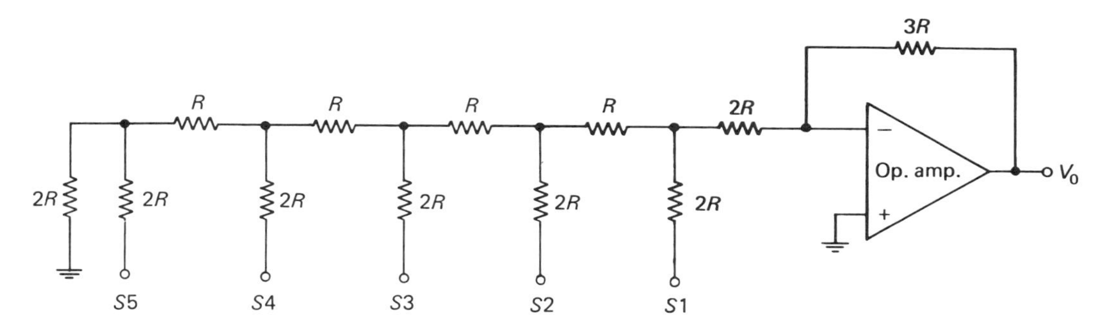

Figure 25-4. D/A decoder.

$-V$, depending on the input information. The output voltage now becomes:

$$V_o = V\left(\frac{S_1}{2} + \frac{S_2}{4} + \frac{S_3}{8} + \frac{S_4}{16} + \frac{S_5}{32}\right) \tag{25-2}$$

For example, if the binary word is 10101, the output voltage is obtained by setting $S1 = S3 = S5 = 1$ and $S2 = S4 = 0$ in Eq. (25-2) to obtain $V_o = V(0.5 + 0.125 + 0.03125) = V(0.65625)$.

Note that with a five-bit word, there are 32 possible words (starting with 00000 and ending with 11111). The output voltage corresponding to the word 11111 (all switches in the logical 1 position) is $V(0.96875)$. If the binary input words are considered to be numbers, the largest number is 31 with an analog output voltage of $V(0.96875)$. For the binary number 10101 (decimal number 21), the analog output voltage is calculated to be $V(0.65625)$. The ratio of 31/21 is the same as the ratio of the two analog output voltages $V(0.96875)$ and $V(0.65625)$. Scaling the output by a factor of 32 yields the analog voltages corresponding to the digital input; i.e., $32(0.96875) = 31$ and $32(0.65625) = 21$. Thus, the output voltage magnitude is directly proportional to the binary number input.

The mechanical switches are usually replaced by FET or other solid-state switches, available in integrated circuit form. For example, Fig. 25-5 shows the AH0146 J-FET analog switch (National Semiconductor), which

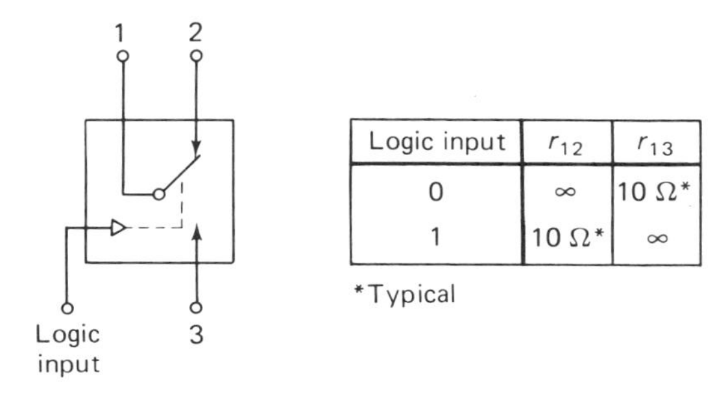

Logic input	r_{12}	r_{13}
0	∞	10 Ω*
1	10 Ω*	∞

*Typical

Figure 25-5. Analog STDP switch (National Semiconductor AH0146).

accomplishes the single-throw-double-pole (STDP) switching function necessary in this application. When the logic input is 0, the switch connects terminals 1 and 3, with terminal 2 disconnected. When the logic input is 1, the switch connects terminals 1 and 2, with terminal 3 disconnected.

Figure 25-6 depicts a complete five-bit ladder-type D/A converter using analog switches. Terminals 3 of all the switches are tied to ground, so that when a logical 0 input is applied to any of the switches, the appropriate 20 kΩ resistor is connected to ground. All terminals 2 are tied to the reference voltage, which is negative in order to provide a positive output. Therefore, when a logical 1 is applied to a switch, its 20 kΩ resistor (terminal 1) is connected to $-V$, thus decoding the digital input into an analog signal at the output (V_o).

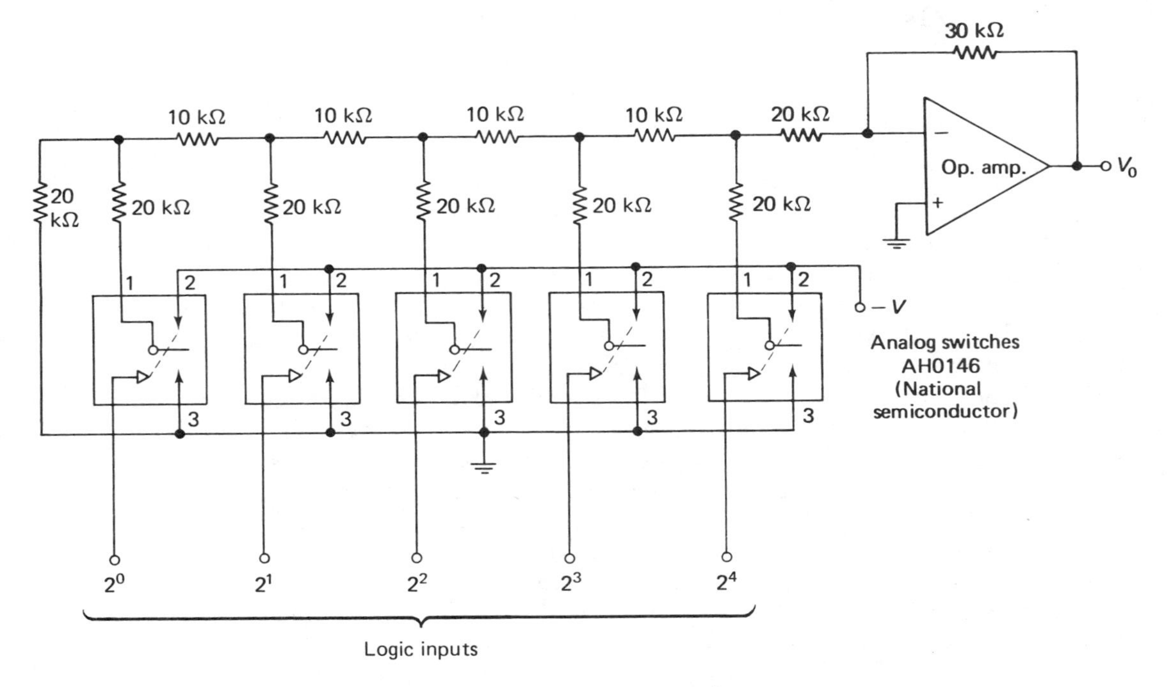

Figure 25-6. Five-bit ladder-type D/A converter.

Another type of D/A converter, shown in Fig. 25-7, uses weighted resistors. Switches $S1$ through $S4$ represent ground for a logical 0, and $-V$ for a logical 1 at any of the inputs. Note that we convert from binary by providing different gains for the different input digits. The output voltage is given by:

$$V_o = V[(8)S1 + (4)S2 + (2)S3 + (1)S4] \tag{25-3}$$

For example, suppose that the input is 1011 ($S1 = S3 = S4 = 1, S2 = 0$). The output voltage is then $V_o = V(8+2+1) = 11$ V: a direct conversion

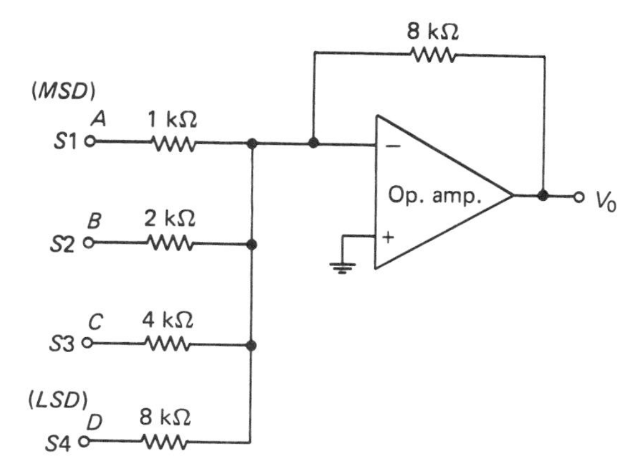

Figure 25-7. Four-bit D/A converter.

from the binary input number 11. The switches can also be replaced here by solid-state digitally controlled switches of the type shown in Fig. 25-5.

Note the disadvantage of this method of D/A conversion over the ladder-type. Here we need many different sized resistors, whereas the ladder converter needs just two sizes of resistors. This type of converter is normally used when the number of bits is low, typically when decoding from BCD (four bits).

The accuracy of the D/A conversion in both schemes depends on the precision of the resistors used. More exactly, the actual resistance is not of great importance, but the *ratio* of resistors must be controlled precisely if accuracy is desired.

25.1.2 BCD-to-Analog Conversion

Any BCD word consists of four binary bits and ranges from 0000 to 1001 (decimal equivalents 0 through 9). Any binary input higher than 1001 is not an allowed BCD word. Thus, to convert from BCD to analog, we need implement only a four-bit D/A converter. An example is given in Fig. 25-8. The BCD inputs control the analog switches in such a way that $-V$ is applied if the input is at a logical 1 and the input is grounded if the logical signal is 0. The output is an analog voltage corresponding to the BCD input.

25.2 A/D CONVERSION

An example of a system employing A/D conversion is a digital voltmeter. The output is a numerical readout of the voltage being measured. A digital readout implies accuracy far beyond that of a regular analog (meter-type) voltmeter, so the conversion between the analog voltage and the digital signal must be very accurate. Such accuracy is especially critical in digital voltmeters (DVMs) with more than a three-digit readout. There are many different gradations of accuracy in A/D conversion; these range from the simplest on-off indication to eight- or even nine-digit DVMs.

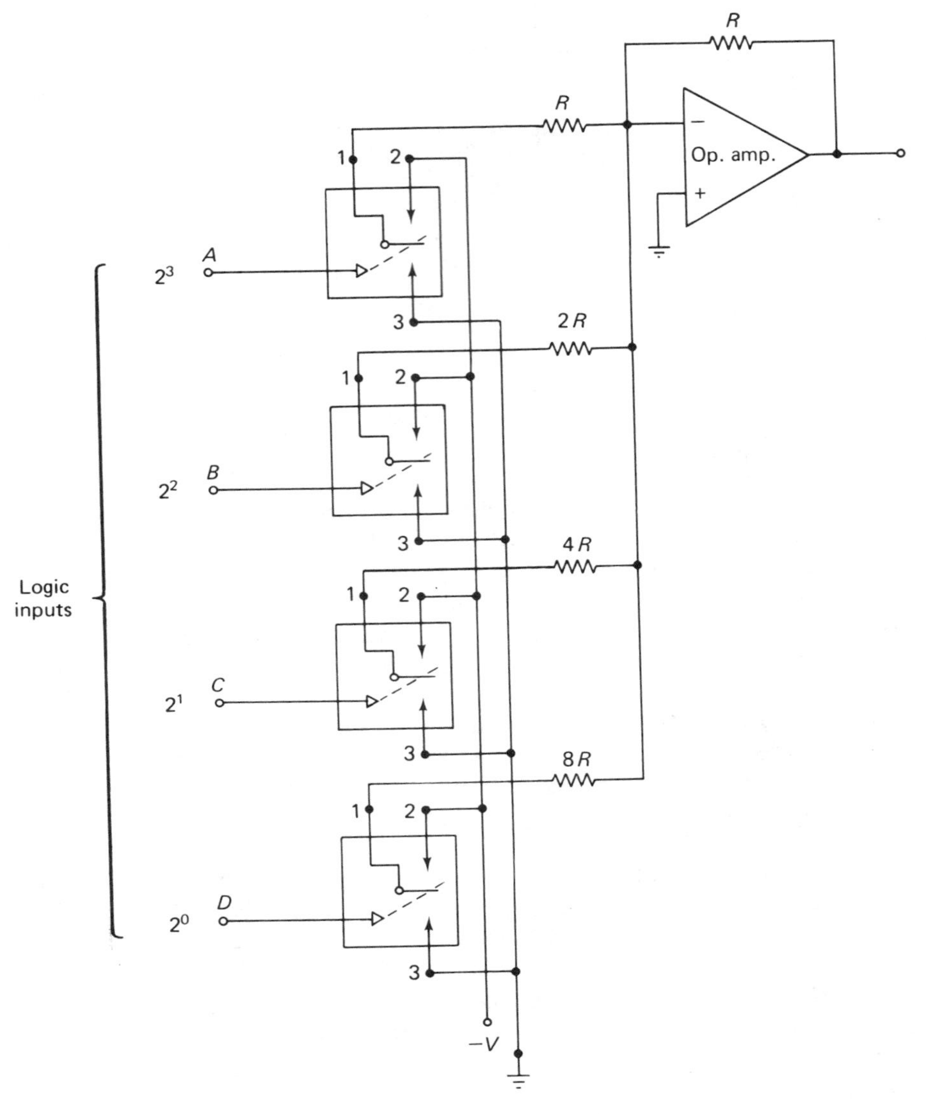

Figure 25-8. BCD-to-analog converter with weighted resistors.

25.2.1 Comparators

One of the very common circuits used in converters, both D/A and A/D, is the *voltage comparator*. Basically, a comparator has two possible output voltages (or states), corresponding to input conditions that are below or above certain voltage limits. The primary use of comparators is as level detectors that provide a change in the output voltage when the input has exceeded or fallen below a certain level preset by the reference voltage.

One example of a comparator is a DIFF AMP without any feedback, with one input tied to a reference voltage. The voltage to be compared with the reference is applied to the other input, as indicated in Fig. 25-9. With a

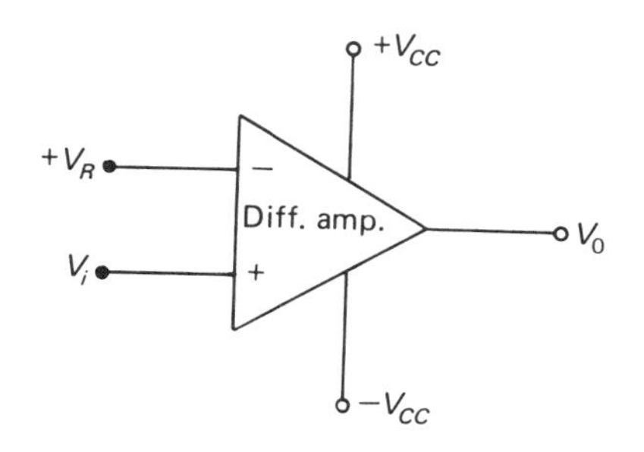

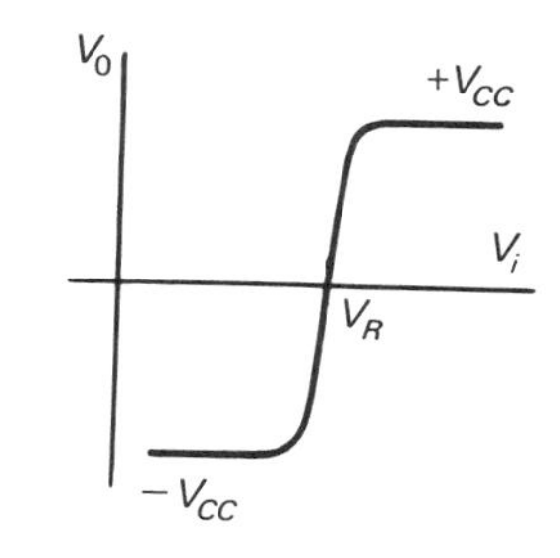

Figure 25-9. DIFF AMP comparator (positive).

reference voltage applied to the inverting input, the output is saturated at the negative supply voltage as long as the input voltage applied to the noninverting input is lower than the reference voltage. As soon as the input exceeds the reference voltage, the output is driven to and saturates at the positive supply voltage.

The input and the reference voltage may be interchanged, as shown in Fig. 25-10, with circuit operation unchanged. However, the output switches from the positive to the negative supply voltage as the input exceeds the reference voltage.

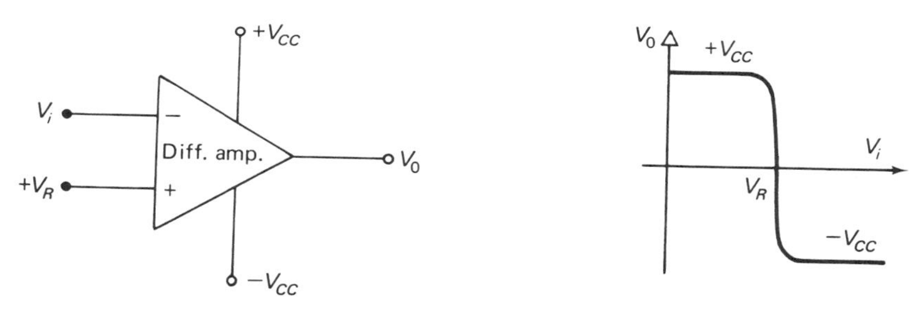

Figure 25-10. DIFF AMP comparator (negative).

We may use the DIFF AMP comparator circuit in cases where a comparison within a few millivolts between the input voltage and the reference voltage is sufficient.

In addition to the DIFF AMP comparator, many different integrated circuits are available for use a specific comparators. These units feature improved switching characteristics and output levels that are compatible for directly driving logic circuits.

Schmitt Trigger. One form of comparator that offers added flexibility is the *Schmitt trigger* circuit. It is illustrated, using an OP AMP or comparator, in Fig. 25-11. This circuit is also called a *regenerative comparator*, because it utilizes positive or regenerative feedback.

The feedback factor β for the circuit is given by:

$$\beta = \frac{R_2}{R_1 + R_2} \tag{25-4}$$

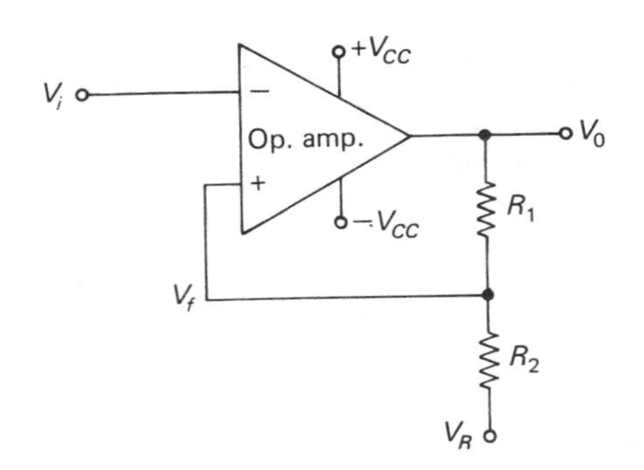

Figure 25-11. Schmitt trigger.

If the feedback factor is adjusted so that the loop gain is exactly 1, the circuit acts as a comparator. The output is at essentially $+V_{CC}$ when the input voltage is below the reference voltage. It switches to $-V_{CC}$ when the input just exceeds the reference voltage. However, to adjust the two resistors for a unity loop gain is not easy because the OP AMP gain is not constant over the entire output swing. The circuit is usually designed to have a certain amount of *hysteresis*. In this case, we adjust the two resistors for a loop gain in excess of 1. Note that the feedback voltage V_f is given by:

$$V_f = V_o \frac{R_2}{R_1 + R_2} + V_R \frac{R_1}{R_1 + R_2} \tag{25-5}$$

Let us assume that the output voltage is at the positive extreme, somewhat lower than $+V_{CC}$. When the input just slightly exceeds the feedback voltage given in Eq. (25-5), the difference input to the OP AMP is positive. This positive input is amplified by the OP AMP and forces the output to its negative extreme voltage, somewhat below V_{CC}. If we now start decreasing the input voltage, it needs to fall just slightly below the reference voltage given in Eq. (25-5). However, the output voltage is now different. Consequently, the input voltage must decrease *beyond* the voltage that caused the transition to $+V$ in order to cause a transition to $-V$.

Consider the following example: $R_1 = 100\ \text{k}\Omega$; $R_2 = 1\ \text{k}\Omega$; $V_R = 0$; the positive extreme output swing is 10 V ($+V$); the negative extreme output swing is -10 V($-V$). With the output at $+10$ V, the input voltage must exceed:

$$V_{f1} = 10 \frac{(1)}{1+100} + 0 \cong 0.1\ \text{V}$$

When it does, the output switches to $-V(-10\ \text{V})$. In order to switch the output back to $+V$, the input must be decreased beyond the 0.1 V that caused the first transition. It must be decreased beyond:

$$V_{f2} = -10 \frac{1}{1+100} \cong -0.1\ \text{V}$$

This process is repetitive; that is, every time the input goes above 0.1 V, the output goes to −10 V. Every time the input goes below −0.1 V, the output switches to + 10 V. The voltage difference between V_{f1} and V_{f2} is the hysteresis voltage.

The input-output characteristics for the preceding example are shown in Fig. 25-12. Note that any reference voltage could be used with the same results: V_{f1} is always slightly above V_R, with V_{f2} slightly below V_R. This will always be the case so long as the loop gain is larger than 1. In other words, if R_1/R_2 is less than the OP AMP open loop gain, hysteresis will be present.

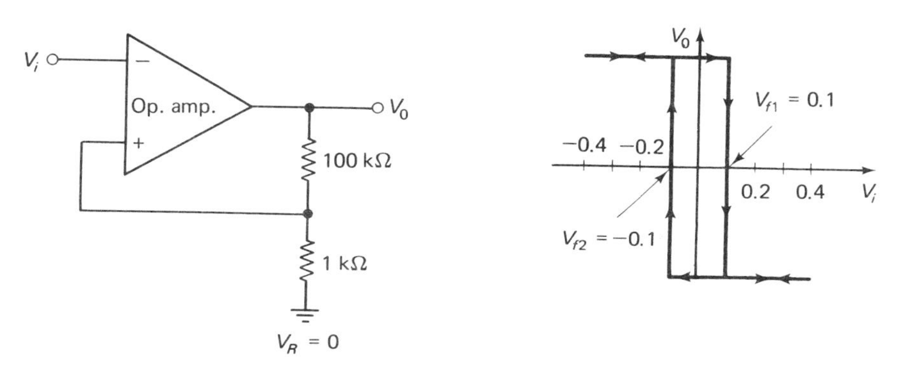

Figure 25-12. Schmitt trigger: zero-crossing detector.

The circuit discussed in the example with $V_R = 0$ is called a *zero-crossing detector*. It indicates when the input voltage becomes positive or negative (i.e., crosses 0 V).

One of the important uses of a Schmitt trigger circuit is in "squaring" a slowly varying input voltage. When any periodic waveshape is applied to the input of a Schmitt trigger, the output waveshape is a square wave with almost vertical sides. Figure 25-13 illustrates this waveshape for a sine

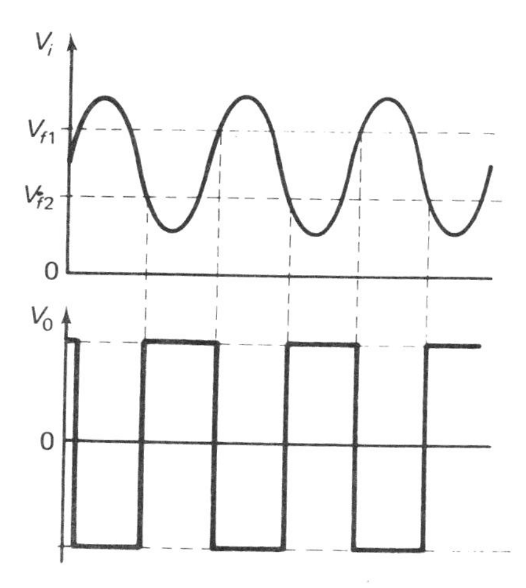

Figure 25-13. Schmitt trigger used as a squaring circuit.

wave input. We can now fill in the block in the frequency counter circuit shown in Fig. 24-26. The squaring circuit consists of a Schmitt trigger and protective diodes.

25.2.2 Voltage-to-Frequency Conversion

In one of the most basic A/D conversions, we change a voltage signal to a square wave signal whose frequency is proportional to the voltage amplitude. This conversion is the heart of a DVM. There are numerous ways of obtaining a signal whose frequency is controlled by a voltage. The generic name applied to such circuits is *voltage controlled oscillator* (VCO).

A simple VCO can be constructed by interconnecting two ONE-SHOTs to form a clock, as shown in Fig. 18-30. But instead of connecting the $+V_{CC}$ terminal to a fixed 5-V supply, we would connect the input signal to the $+V_{CC}$ terminal. Such a circuit provides a fairly linear relationship between the input voltage and the clock frequency for input voltages between 4 and 6 V.

In many applications, conversion over a wider range and with better linearity (i.e., accuracy) is needed. Another method is often used in A/D conversion. It is illustrated in Fig. 25-14. A capacitor may be charged linearly if fed from a constant current source. If the analog input voltage (a constant for a single conversion) is applied to the input of an integrator, as shown in Fig. 23-12, the charging current is directly proportional to the analog input voltage (given by V/R). The output voltage is a ramp whose rate of increase (or slope) is given by I/C. Thus, the length of time

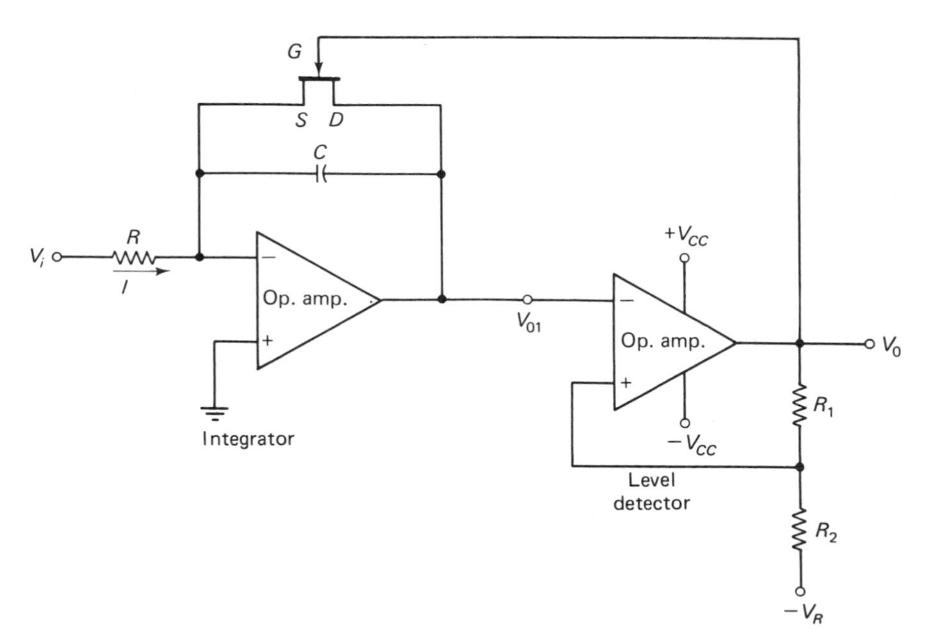

Figure 25-14. Voltage-to-frequency converter.

required for the output voltage to reach a predetermined voltage level is directly related to the analog input voltage.

A circuit utilizing this principle is depicted in Fig. 25-14. It provides a square wave output whose frequency is proportional to the amplitude of the (dc) input voltage. Circuit operation is as follows. A positive dc voltage at the input is converted into a negative-going ramp at the output of the integrator. The level detector is a Schmitt trigger adjusted to fire between two negative voltage levels, V_1 and V_2, determined by resistors R_1 and R_2, the negative reference voltage V_R, and the positive and negative output saturation voltages of the Schmitt trigger OP AMP. As the voltage V_{o1} becomes more negative, the output of the Schmitt trigger is at its negative saturation value (somewhat below the $-V_{CC}$ level). When V_{o1} reaches $-V_2$, the Schmitt trigger fires and its output goes to the positive saturation level (slightly below the $+V_{CC}$ level). While V_o was at the negative extreme, the FET was pinched off and not conducting. When V_o becomes positive, the FET is turned on and begins to discharge the capacitor. The drain-source resistance is low while the FET is conducting, and the capacitor is discharging toward zero. However, as soon as V_{o1} falls to $-V_1$, the Schmitt trigger fires. Once again its output is at the negative saturation value, thus causing the FET to turn off. The capacitor again begins to ramp down toward $-V_2$, and the cycle is repeated.

The waveforms for the circuit are shown in Fig. 25-15. If the input voltage is increased, the charging current also increases proportionally. Then the length of time required for the capacitor voltage to fall from $-V_1$ to $-V_2$ is lowered, and so the frequency of the output wave is increased. For similar reasons, when the input voltage is decreased, the period of the output waveshape also increases. Thus, the frequency decreases.

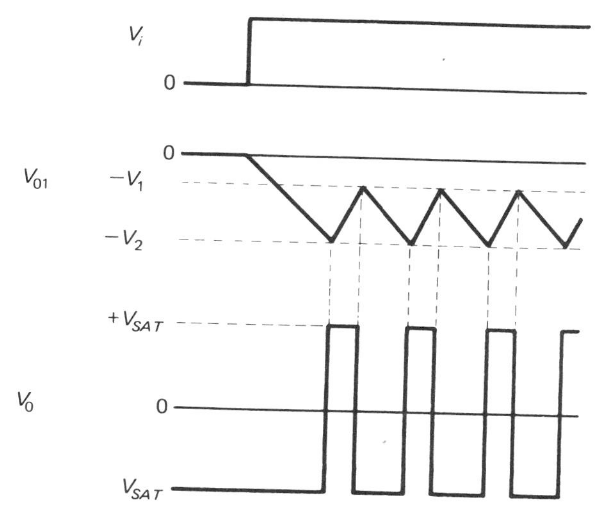

Figure 25-15. Waveshapes for the circuit in Fig. 25-14.

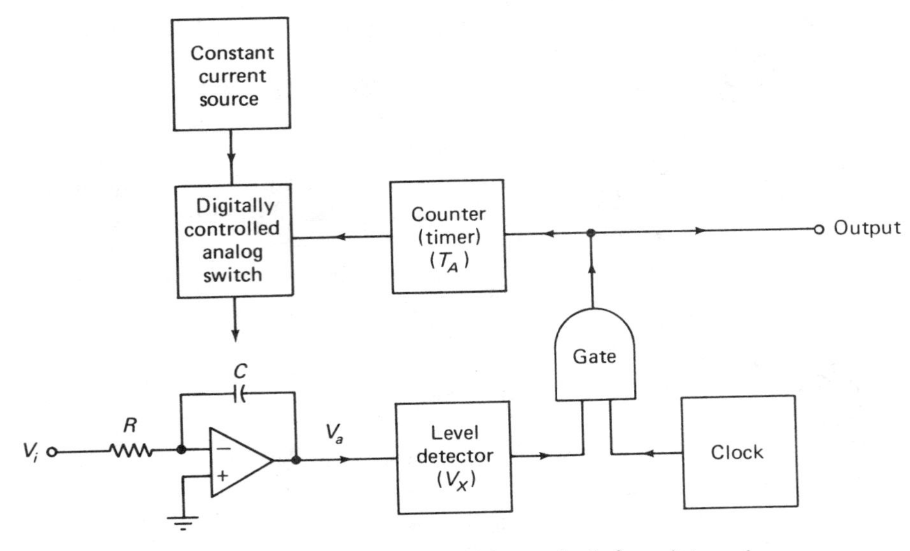

Figure 25-16. Block diagram for a dual-slope integrator.

A still more accurate method of voltage-to-frequency conversion uses the so-called *dual-slope integration* principle. The basic block diagram for a dual-slope integrator is given in Fig. 25-16, with the waveforms illustrated in Fig. 25-17. The dc analog input V_i causes the capacitor to start a negative ramp at the output of the integrator. When this level exceeds the threshold voltage (V_x) of the level detector, the level detector output goes

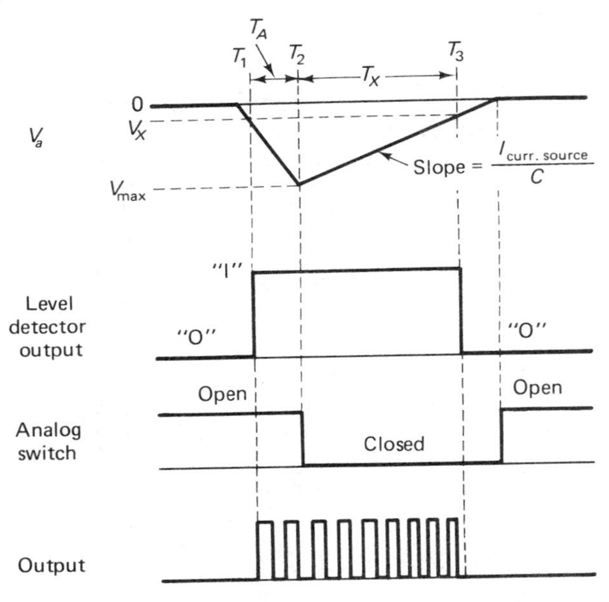

Figure 25-17. Dual-slope integrator waveshapes.

to a logical 1 state. As a result, the clock signal can pass through the gate. The counter is used to measure the length of time the capacitor is allowed to charge up. This time T_A is kept a constant. When it is reached, the output of the counter causes the digitally controlled analog switch to close. The constant current from the current generator is then applied to the capacitor, thus causing the capacitor to begin discharging. This operation occurs at time T_2. The slope of the capacitor discharge is controlled by the constant current source. Therefore, it is the same for all input voltages. When the capacitor discharges to the threshold voltage of the level detector, the output of the level detector switches and goes to a logical 0 state. In this way, the gate is disabled; that is, it does not allow the clock signal to pass. This operation occurs at time T_3. The total time that the clock is available at the output is $T_A + T_x$. The time is proportional to the input voltage. Note that T_A is fixed; that is, it is controlled by the counter. Thus, depending on the input voltage, the capacitor will charge to a different maximum voltage V_{MAX} in time T_A. This voltage is obtained from the slope relationship during T_A:

$$\frac{V_{MAX} - V_x}{T_A} = \text{slope} = -\frac{V_i}{RC} \qquad (25\text{-}6)$$

All the quantities in Eq. (25-6) are fixed with the exception of V_i and V_{MAX}. Thus, V_{MAX} is directly related to the input voltage. The discharge slope (during T_x) is given by:

$$\text{slope} = \frac{I_{\text{current source}}}{C} = \frac{V_x - V_{MAX}}{T_x} \qquad (25\text{-}7)$$

With V_x, C, and $I_{\text{current source}}$ all fixed, the time T_x is seen as directly proportional to the input voltage through its proportionality to V_{MAX}. Eliminating V_{MAX} from Eqs. (25-6) and (25-7), we get:

$$T_x = \frac{V_i T_A}{RI_{\text{current source}}} \qquad (25\text{-}8)$$

This equation tells us that the length of time that clock pulses are available at the output is directly porportional to the amplitude of the analog input voltage. To get a digital "count" corresponding to the amplitude of the input signal, all we need do is apply the output of Fig. 25-16 to a digital counter. The count of the output pulses will then give a direct indication of the amplitude of the input voltage, as seen in Fig. 25-18. In Fig. 25-18(a) a larger input voltage than in Fig. 25-18(b) is applied. As a result, V_{M1} is greater than V_{M2}, and T_{x1} is greater than T_{x2}. Thus, for the higher input voltage, the gate is "open" (logical 1) for a longer period of time, allowing the counter connected to the output to receive a larger number of pulses.

A complete DVM contains the additional counter and decoder sections with a readout. It must also have logic circuits that provide the proper reset pulses to both counters. In this way, the count can be started over, the whole process repeated, and a new display count provided.

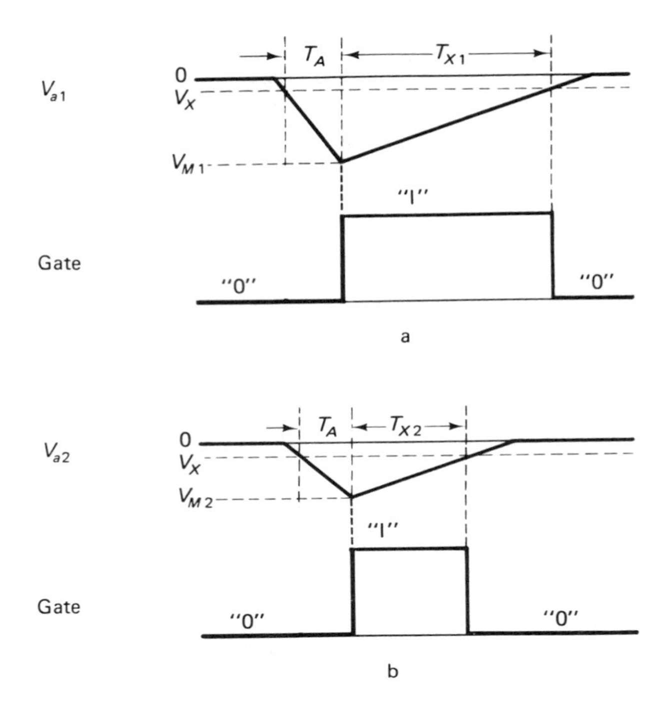

Figure 25-18. Dual-slope integrator waveforms for two different input voltages $V_i(a) > V_i(b)$.

The final accuracy of the instrument will depend on the stability of the clock; in most cases, a crystal-controlled high-frequency clock is used. The stability of the constant current generator is also important. Some form of a temperature-compensated circuit is used for the current generator in critical applications.

Some IC manufacturers offer digital panel meter (DPM) chips or subsystems based on the dual-slope conversion or some modification thereof. For additional information, the reader is encouraged to contact the manufacturer for complete data and application notes. (Motorola, National Semiconductor, Intersil, and Siliconix are some of the manufacturers with such products.)

25.2.3 Other A/D Converters

Another method of A/D conversion utilizes a digital counter whose input is controlled by a comparator, as depicted in Fig. 25-19. When the analog input voltage V_i is applied, the output of the comparator is high, enabling the clock to be counted in the digital counter. As the count increases, so does the output voltage V_R from the D/A converter. This converter may be any of the types discussed in section 25.1. When the output of the D/A converter exceeds the input voltage, the comparator switches to a logical 0 state and disables the gate. The count in the counter cannot increase therefore. At this time the count stored in the counter is the digital equivalent of the input voltage. We can see that for a lower

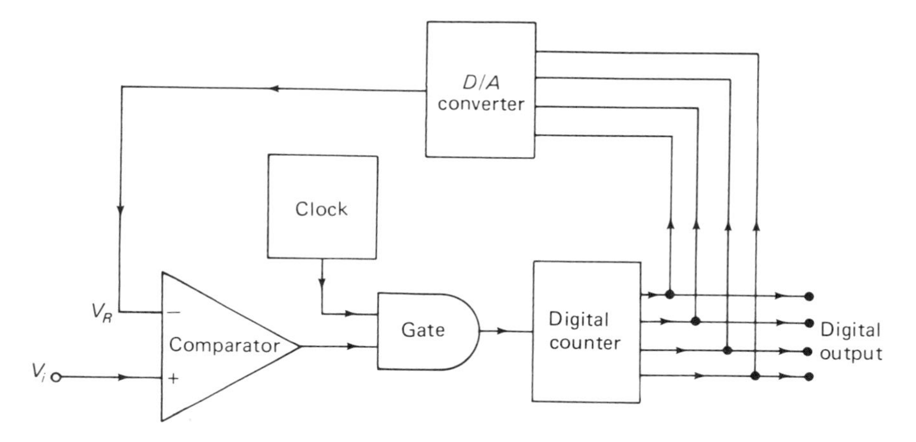

Figure 25-19. A/D converter.

input voltage, a lower count will have been reached when $V_R = V_i$ and the comparator closes the gate. The waveshapes for this A/D converter are shown in Fig. 25-20.

Another method, called *successive approximation*, is also used in A/D conversion. This method requires a comparator and a D/A converter but no counter. Instead, a programmable register is used. The most significant bit in the register is set to 1; all other bits are at 0. This bit is then decoded by the D/A converter and applied to the comparator for a comparison with the input signal. If the decoded signal is greater than the input signal, the *MSB* (most significant bit) is set to 0. If the decoded signal is smaller than the input, the *MSB* is left at 1. In either case, the next operation is to set the second *MSB* to 1 (leaving all other lower order bits at 0). Once again, we convert the binary number to an analog signal and compare it

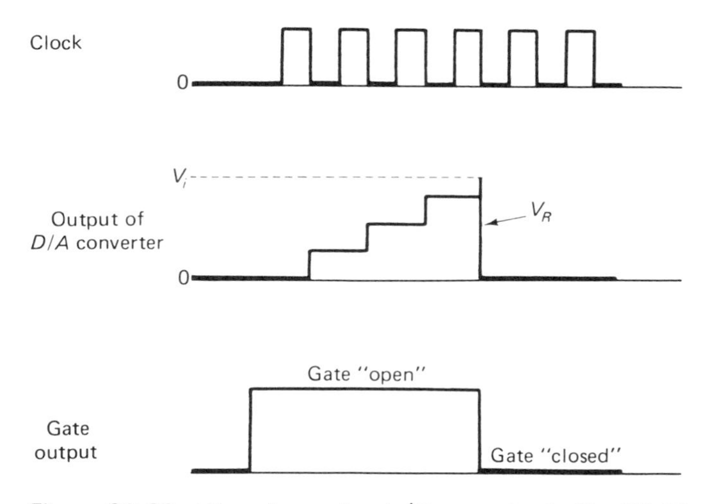

Figure 25-20. Waveshapes for A/D converter in Fig. 25-19.

with the input signal. In this manner, each of the binary bits is set to a proper value, one at a time, until the analog input signal is correctly represented by the binary word contained in the register. Such A/D converters are available in single large-scale integration (LSI) packages.

25.3 SUMMARY

As we have described in Chaps. 23 and 24, both analog and digital systems have definite applications. Actually many more exist than were discussed. There are times when a signal available in one form (analog or digital) must be converted into the other form. A large complex system may involve (1) signal sensing, (2) processing in both analog and digital form, (3) conversions between analog and digital, (4) conversions from digital to analog, (5) human and machine (computer) interfaces, and so on. Most of these interfaces use some form of conversion; therefore, A/D and D/A converters comprise a very inportant subsystem and will continue to do so.

Review Questions

1. What are the input and output for a D/A converter?
2. How is a binary "word" converted into its analog form?
3. What are the differences between the ladder converter and the converter in Fig. 25-7? What are the limitations on each?
4. What is an analog switch?
5. What is a digitally controlled switch?
6. How is a BCD "word" converted into its analog equivalent?
7. What are the input and output for an A/D converter?
8. What is a voltage comparator?
9. What are the terminal characteristics of a voltage comparator?
10. How may a DIFF AMP be used as a voltage comparator?
11. What is a Schmitt trigger?
12. What determines the characteristics of a Schmitt trigger?
13. What is the significance of hysteresis as applied to a Schmitt trigger?
14. What is a zero-crossing detector?
15. What is a VCO?
16. How may a VCO be used as an A/D converter?
17. How can an integrator be used in an A/D converter?
18. What is a dual-slope integrator?
19. How is the dual-slope principle used to obtain accurate A/D conversion?
20. What is the successive approximation method of A/D conversion?
21. What method of A/D conversion offers the most accuracy?

22. What are the basic components of a digital readout voltmeter? Explain in some detail, using the concepts developed in the previous two chapters as well as this one.

Problems

1. In the D/A converter in Fig. 25-8, determine the output voltage if the digital inputs are: (a) $A=C=1$ and $B=D=0$; (b) $A=0$ and $B=C=D=1$; (c) $A=B=C=D=1$.
2. Assume that each of the inputs in Fig. 25-7 is tied to the collector of an inverter-transistor. A logical 1 is represented by the transistor being cut off; a logical 0 corresponds to the transistor being saturated ($V_{CES}=0.3$ V). The collector voltage when the transistor is off is +5 V. What are the output voltages for the following inputs: (a) $A=C=1$ and $B=D=0$, (b) $A=0$ and $B=C=D=1$, (c) $A=B=C=D=1$, (d) $A=B=C=D=0$?
3. What is the "error" in the conversions in Problem 2? What should the analog output voltage be as compared to what it is actually?
4. An OP AMP with an open-loop gain of 200,000 is used in the Schmitt trigger circuit of Fig. 25-11. The output saturation voltages are ±10 V, with $R_1=10$ kΩ and $R_2=100$ Ω. The reference voltage is 1 V. What is the output if the input is a linear ramp from −10 to +10 V? Sketch both input and output waveshapes.
5. Repeat Problem 4 for reference voltages of: 0, −5, +5 V.
6. Determine the amount of hysteresis (in volts) in Problem 4.
7. We wish to design a Schmitt trigger circuit to give an indication every time the input voltage goes above +5 V and falls below 0.4 V. The OP AMP parameters are the same as in Problem 4. Specify the reference voltage and the values of resistors needed.

4

Appendices

Measurement of Transistor Hybrid Parameters

As discussed in Chap. 4, the transistor curve tracer can be used to determine transistor hybrid parameters. In some cases, this piece of equipment may not be available or more accurate results may be necessary. Under these conditions, the test circuits shown in Figs. A1-1 and A1-2 may be used to determine by actual circuit measurement the values of the four transistor parameters.

Measurement of h_{ie} and h_{fe}. The circuit shown in Fig. A1-1 is used for experimentally determining h_{ie} and h_{fe}. First, we set up the proper dc conditions for the BJT under test by adjusting the dc power supplies V_{CC} and V_{BB}. With V_{CC} set to a value just slightly higher than the desired V_{CEQ} (collector-emitter operating voltage), V_{BB} is adjusted until the desired quiescent base current flows in the circuit. The dc value of I_B is monitored by a dc meter at the $M1$ position. (Note: Meters $M1$ and $M2$ must be "floating"; that is, they cannot have a ground on either terminal.)

$$I_{BQ} = \frac{V_{M1}}{1} \text{ (in } \mu\text{A)}$$

where V_{M1} is in volts. With I_{BQ} properly set, V_{CC} is adjusted until V_{CEQ} is its desired value. (This voltage can be monitored with an additional dc voltmeter from collector to emitter.)

With ac meters in positions $M1$ and $M3$, readings are obtained to calculate h_{ie}:*

$$h_{ie} = \frac{v_{be}}{i_b} \cong \frac{V_{M3}}{V_{M1}} \times 1 \ M\Omega$$

where both V_{M3} and V_{M1} are rms voltages.

*The oscillator is set to 1 kHz and an amplitude convenient for measurement. It is good practice to monitor the collector with a scope to insure that the waveshape is undistorted. If distorted, reduce the amplitude of the oscillator.

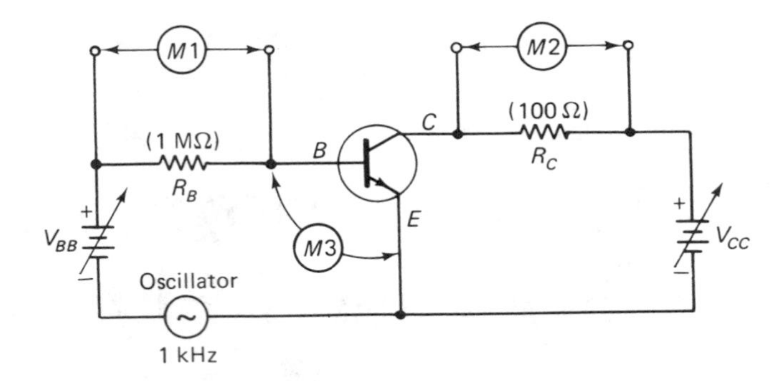

Figure A1-1. Test setup for determining h_{ie} and h_{fe}. (All resistors ±1% tolerance or smaller; all meters high impedance–10 M or larger).

By connecting an ac voltmeter at position $M2$ and taking readings again, h_{fe} can be determined:

$$h_{fe} = \frac{i_c}{i_b} \cong \frac{V_{M2}}{V_{M1}} \times 10^4$$

where both V_{M1} and V_{M2} are rms voltages.

Measurement of h_{re} and h_{oe}. With the same dc conditions as were set up previously for the circuit in Fig. A1-2, the oscillator is adjusted to produce a measurable ac voltage on meter $M3$. We then calculate h_{re} from:

$$h_{re} = \frac{v_{be}}{v_{ce}} = \frac{V_{M3}}{V_{M1}}$$

where h_{re} may be expected to be small (10^{-4}). Convenient levels are v_{ce} of a few volts rms, producing a v_{be} of a few hundred microvolts rms. We calculate h_{oe} from:

$$h_{oe} = \frac{i_c}{v_{ce}} \cong \frac{V_{M2}}{V_{M1}} \times 10^{-2} \text{ (in mhos)}$$

where both V_{M1} and V_{M2} are rms voltages. Typically, h_{oe} may be approximately 10 to 200 micromhos.

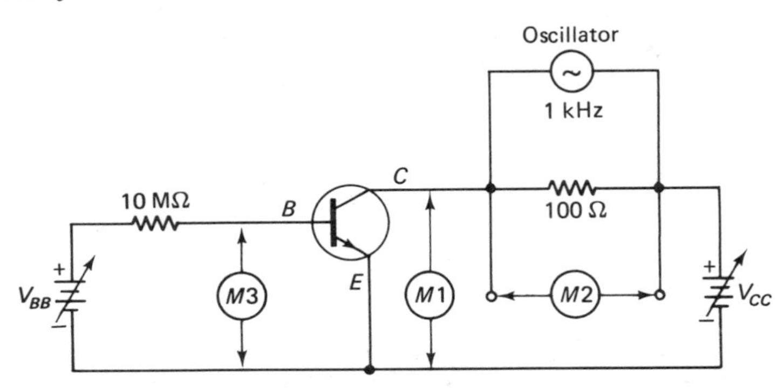

Figure A1-2. Test setup for determining h_{re} and h_{oe}.

A2

Vacuum Tubes

Modern electronics relies heavily on solid-state devices and systems. However, there are a number of applications, specifically those demanding very high power and high frequencies, where there is no solid-state replacement for the vacuum tube. In addition, there are many cases where equipment with vacuum tubes is still in use, although more up-to-date solid-state equipment is available.

Since the advent of transistors and other solid-state devices, the words "solid-state" or "transistorized" have come to mean automatically "better." The qualitative difference between vacuum-tube equipment and the solid-state counterpart is relatively small. The essential "improvement" resulting from the use of solid-state devices is in their longer (theoretically infinite) life. Solid-state devices also have other advantages: smaller size, lower bias voltages, lower operating temperatures, no need for applying "heater" power, and finally lower cost. Consequently, the vacuum tube has become obsolete in many applications.

Although the future does not seem to hold many new and important applications for the vacuum tube, the technologist of the future will undoubtedly have to know something of the basic operation and terminal characteristics of vacuum tubes.

A2.1 VACUUM-TUBE DIODES

The basic vacuum diode, like its solid-state counterpart, consists of two electrodes, the *cathode* and *anode*. These two electrodes are encased in an evacuated envelope, usually made of glass. For proper operation, two additional terminals are incorporated; these are the *heater* terminals.

The vacuum diode is constructed in a way that enhances its operation. The heater is either in close proximity to the cathode or is actually a part of the cathode. The cathode is fabricated of a metal or other material that readily emits electrons when heated. Such materials are tungsten (the same material used in incandescent lights), thoriated tungsten, barium, and strontium oxides. The anode, also termed the *plate*, is a metal electrode that surrounds the cathode in order to collect as many of the electrons

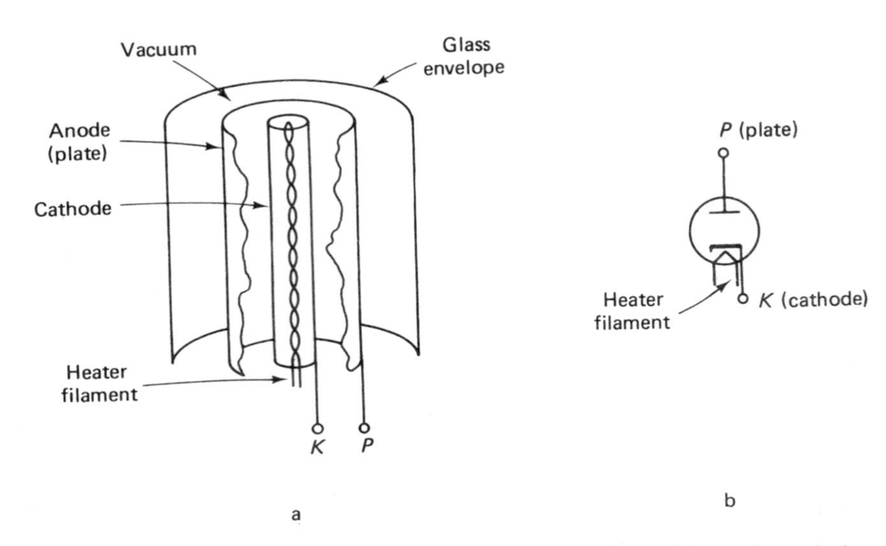

Figure A2-1. Vacuum tube diode: (a) construction and (b) circuit symbol.

emitted from the cathode as is possible. Thus, the operation of a vacuum diode is based on the heating of the cathode so that it emits electrons into the evacuated space between the cathode and the plate. Then a voltage is applied between the plate and cathode to cause collection of the electrons at the plate. This flow of electrons from cathode to plate constitutes the *plate current*. The construction and the circuit symbol for the vacuum diode are illustrated in Fig. A2-1.

The heater element may be an actual part of the cathode; this is called a *directly heated* cathode. Or the heater element may be separate, forming an *indirectly heated* cathode. To achieve the proper heating of the cathode, a relatively low ac potential is applied to the heater elements. This voltage is developed across a separate winding of the power transformer (called the *filament winding*). It may be from 1 to 12 V-rms, depending on the tube. The most common voltage is either 6.3 or 12.6 V-rms. The heater or filament is a wire that gets heated as a result of the I^2R power applied by the filament winding in the transformer.

Once power is applied to the heater, the cathode temperature is elevated. Typical operating temperatures vary for different tubes; they are between 1000°K and 2500°K. The cathode material, once heated, readily emits relatively large numbers of electrons. We have to provide the vacuum so the cathode will not burn. A positive potential at the plate (positive with respect to the cathode) causes the electrons that have left the cathode because of *thermionic emission* to be accelerated toward and eventually captured by the plate.

Let us consider the vacuum diode terminal characteristics. First, we apply heater voltage with no external potential to the plate. As the cathode is heated, a certain number of electrons will be emitted into the region between the cathode and plate. Eventually, an equilibrium will be reached;

exactly the same number of electrons are emitted by the cathode as fall back to the cathode. Note that the liberated electrons do not move toward the plate, because no external force (in the form of a voltage) is present. Thus, there is a *space-charge region* of free electrons near the surface of the cathode. Under these conditions, the plate current is essentially zero. A few electrons that are emitted from the cathode have sufficient kinetic energy to reach the plate, but they constitute a net current that is at most in the microampere range and is negligible.

As external plate potential (plate positive with respect to cathode) is applied, some of the emitted electrons are attracted toward the plate; they constitute a net transport of charge (electrons) from the cathode to the plate. We observe this phenomenon in the external circuit as a plate current entering the plate and leaving the cathode. The higher the applied voltage at the plate, the larger the observed current. The plate current increases exponentially with increases in plate voltage, as shown in Fig. A2-2, until all of the emitted electrons are set in motion toward the plate and take part in contributing to the plate current. Further increases in plate voltage do not cause additional increases in plate current; all of the available electrons are already contributing to the plate current. This condition is known as *saturation*, also indicated in Fig. A2-2.

If the plate-to-cathode potential is negative (+at the cathode and − at the plate), the field set up between the two electrons prevents electrons from reaching the plate. In the reverse direction the vacuum diode is essentially an open circuit; that is, no plate current flows.

It should be obvious that the circuit operation of the vacuum diode is essentially the same as that of the solid-state diode. The major difference is the relative magnitude of the forward voltage needed to establish conduction. Thus, the uses of vacuum diodes parallel those of solid-state diodes.

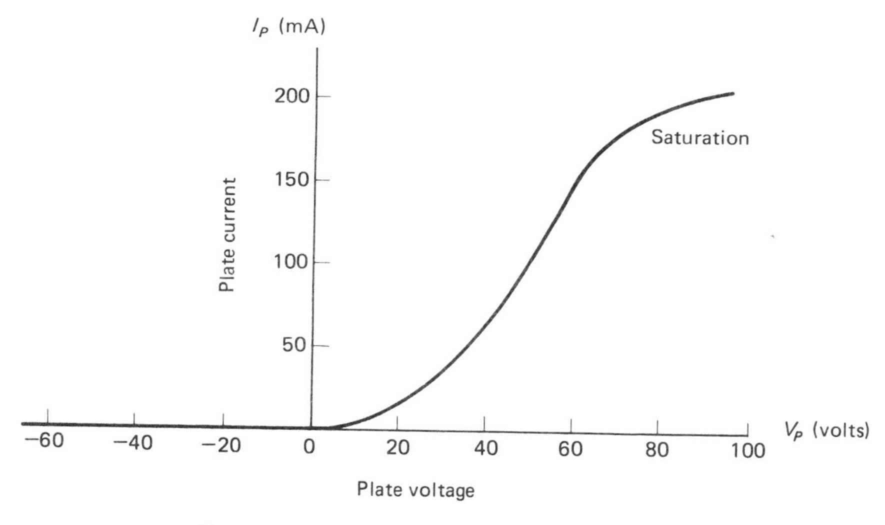

Figure A2-2. Vacuum diode characteristics.

In fact, the same procedures used for determining the static and dynamic resistances and conductances of the solid-state diode can be and are used for vacuum diodes.

A2.2 VACUUM-TUBE TRIODES

The triode is a three-terminal vacuum device that has amplifying capabilities. These capabilities result from the addition of a third electrode, called the *grid*, between the plate and cathode of a vacuum diode.

The physical construction and the circuit symbol for a triode are illustrated in Fig. A2-3. You can understand the operation of the triode by considering the zero reference potential to be at the cathode. A current flows from the plate to the cathode when the plate is made positive. With no bias on the control electrode, the grid operation is quite similar to that of a diode. Heating of the cathode results in thermionic emission at the cathode. These free electrons are attracted to the plate, with a very small number colliding with the structure of the grid. However, if a slight negative potential is applied to the grid, the flow of electrons away from the cathode is impeded. If the negative potential at the grid is small compared to the positive potential at the plate, the number of electrons reaching the plate will be smaller than there would be had there not been any voltage impressed on the grid. Thus, for the same plate voltage, a smaller plate current flows when the grid is biased negatively than when there is no bias on the grid.

With a fixed negative bias on the grid, applying successively larger plate voltage causes the plate current to increase, much in the same manner as it does in the vacuum diode. If the grid bias is made slightly more negative and the plate voltage is again made successively larger, the plate current again increases. However, the grid is more negative; therefore, it causes a lower plate current (for the same plate voltage). The bias on the grid, therefore, *controls* the amount of plate current for a given plate

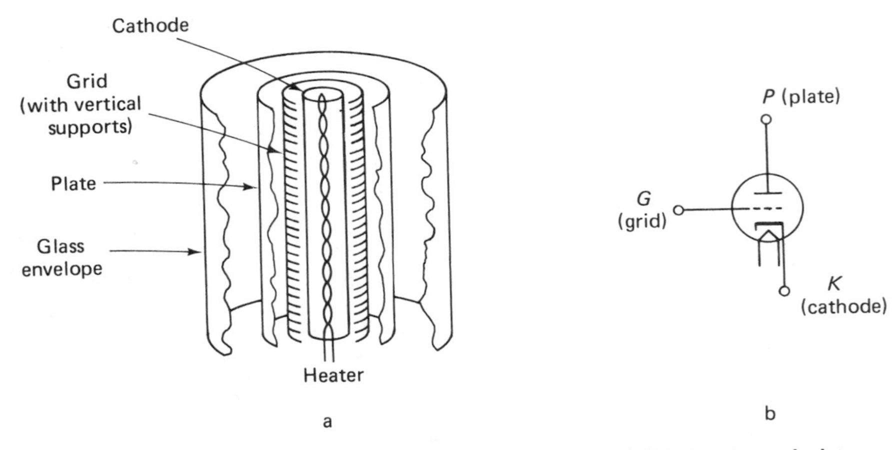

Figure A2-3. Vacuum triode: (a) construction and (b) circuit symbol.

voltage. It is this controlling effect of the grid that is the key to the amplification possible in a triode. For example, if the grid bias is caused to vary (as in the case of a small ac signal applied to the grid), the plate current varies accordingly. This variation of the plate current is used to vary the voltage across a resistor in the plate circuit. The triode amplifies an input voltage signal when the variation of voltage in the plate circuit is larger than the variation of voltage in the grid circuit.

The output characteristics (i.e., the plate characteristics) for a typical triode are given in Fig. A2-4. One possible biasing scheme for a triode is shown in Fig. A2-5. Analysis and design of the bias circuit is similar to that for any of the solid-state devices. The summation of voltages in the plate loop is zero. The load line equation is:

$$V_{PP} = V_{PK} + I_P R_L \tag{A2-1}$$

The quiescent, or operating, point for the triode is established by plotting the load line on the plate characteristics and noting where the load line intersects the characteristic curve for $V_{GK} = -V_{GG}$.

The need for two supply voltages (V_{PP} and V_{GG}) is eliminated in the more practical bias circuit shown in Fig. A2-6. The negative grid bias voltage is developed across the cathode resistor R_K and is given by:

$$V_{GK} = -I_P R_K \tag{A2-2}$$

The load line equation is modified by the additional voltage drop across R_K:

$$V_{PP} = V_{PK} + I_P(R_L + R_K) \tag{A2-3}$$

For this bias scheme, the Q-point is established by plotting both the load line [Eq. (A2-3)] and the bias curve [Eq. (A2-2)] on the plate characteristics and noting the intersection of the two.

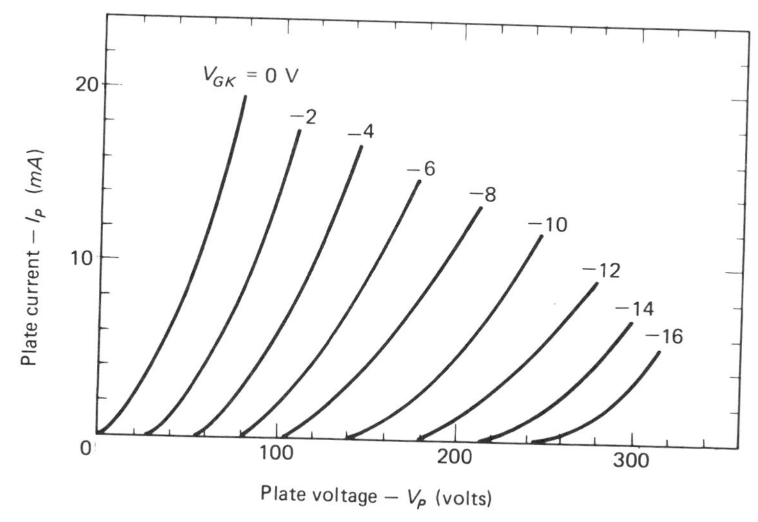

Figure A2-4. Triode characteristics.

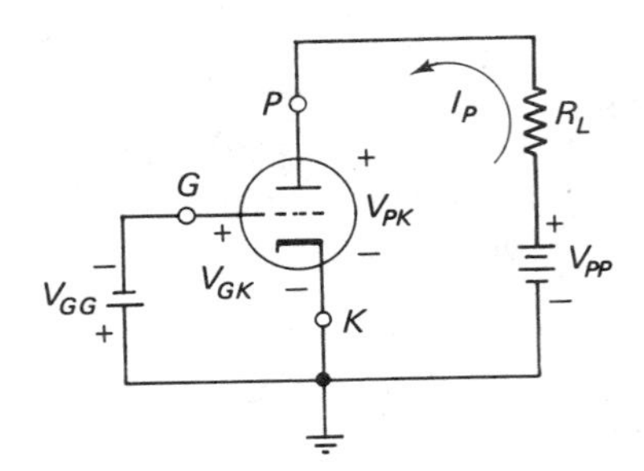

Figure A2-5. Basic triode biasing circuit.

Because the grid is maintained negative with respect to the cathode, essentially no grid current flows. Thus, there is no dc voltage drop across R_G. The ac signal is impressed across R_G (usually coupled through a capacitor), and the ac output signal is taken across R_L (usually through a capacitor). To eliminate the feedback across R_K for ac signals, this resistor is normally bypassed with a capacitor, so that the ac performance of the circuits illustrated in Figs. A2-5 and A2-6 is the same.

We can duplicate the ac performance of a triode amplifier graphically or through the use of a small-signal model. The vacuum triode is a voltage-controlled device. As such, it is quite similar to the FET. The ac small-signal models of the FET and the triode are also similar. The major difference is in the magnitudes of their individual parameters. The triode parameters are: *plate resistance* r_p, *transconductance* g_m, and *amplification* μ. These are defined as:

$$r_p = \left. \frac{v_{pk}}{i_p} \right|_{V_{GKQ}} \tag{A2-4}$$

$$g_m = \left. \frac{i_p}{v_{gk}} \right|_{V_{PKQ}} \tag{A2-5}$$

$$\mu = g_m r_p \tag{A2-6}$$

The circuit models for the triode are illustrated in Fig. A2-7. Typical values for the triode parameters are: r_p between 10 and 50 kΩ, g_m between 2 and 10 millimho, and μ less than 100.

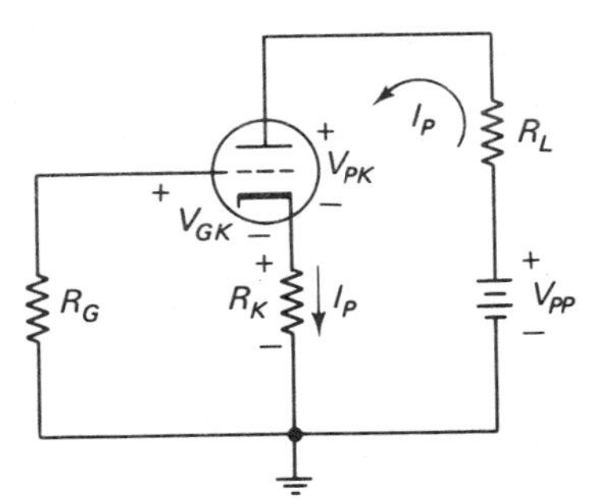

Figure A2-6. Single-supply triode bias circuit.

At high frequencies, the model must include the effects of interelectrode capacitances. This function is accomplished by placing capacitors C_{gk}, C_{pk}, and C_{pg} between the terminals indicated by the subscripts. The voltage between the plate and grid of a triode is rather large, so the grid-to-plate capacitance is also large. This large capacitance degrades the amplification of triodes at high frequencies. This net capacitance can be broken up and thus reduced by the insertion of an additional electrode, the *screen grid* (or simply *screen*), between the control grid and plate. This configuration is called a *tetrode*. It has a much lower net capacitance between the control grid and plate. The screen is biased positive with respect to the cathode, but it is usually biased somewhat lower than the plate.

The presence of the screen has another less desirable effect on the characteristics. Because the screen is positive, it draws current by trapping some of the electrons on their way to the plate. In addition, the electrons that reach the plate liberate some electrons from the surface of the plate. This phenomenon is termed *secondary emission*. The screen also collects these secondary electrons from the plate, causing the plate characteristics of tetrodes to exhibit a negative (or dynamic) resistance region. For this reason, the applications for tetrodes are limited to oscillator circuits.

A2.3 PENTODES

To take advantage of the desirable characteristics of tetrodes, and at the same time eliminate their undesirable ones, we can insert an additional grid between the screen and plate. Such a five electrode tube is called a pentode. This is the *suppressor grid*, so called because it acts to suppress screen-grid current caused by secondary emission at the plate. The suppressor is usually biased at 0 V, i.e., the same potential as the cathode. In some pentodes, the cathode and the suppressor grids are tied together internally; in others, this connection is external.

The small-signal model of the pentode is the same as that of the triode, which is shown in Fig. A2-7. The major difference between triodes and pentodes lies in the magnitudes of the respective parameters. All three parameters, g_m, r_p, and μ, for the pentode are higher numerically than for

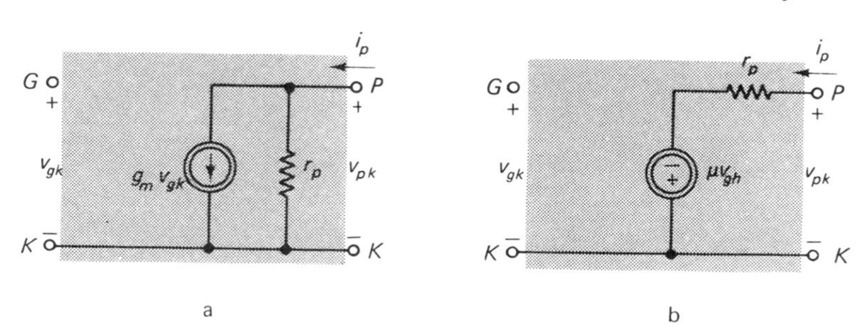

Figure A2-7. Triode model: (a) current-source equivalent and (b) voltage-source equivalent.

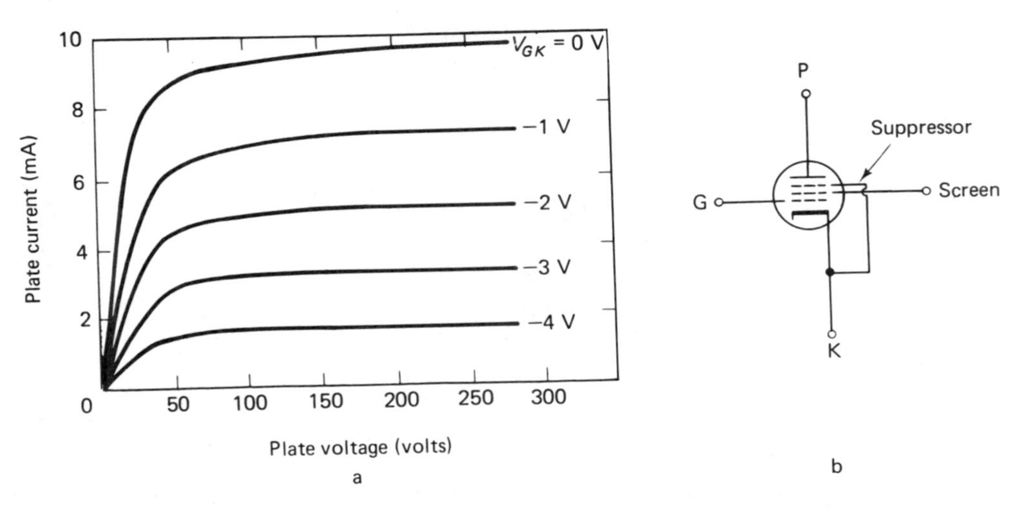

Figure A2-8. Pentode: (a) characteristics (suppressor and screen grid potentials constant) and (b) circuit symbol.

the typical triode. Consequently, a typical pentode amplifier will provide higher voltage gain and higher output resistance than its triode counterpart. The output or plate characteristics for a typical pentode are depicted in Fig. A2-8.

The obvious similarities between the terminal characteristics of pentodes and FETs (with the exception in the magnitudes of operating and biasing voltages) cannot be overemphasized. At high frequencies, we can observe the interelectrode capacitances of the pentode to be at least a few orders of magnitude lower than for the triode: Specifically, C_{gp}, which is the determining factor in the frequency response, for a pentode may be as small as 0.002 pF. The dual-gate MOSFET has parameter values similar to that of the pentode, and the two are used in much the same way. In fact, the vacuum tube–either diode, triode, or pentode–may be used to perform similar functions in many circuits utilizing solid-state devices.

A3

Manufacturers' Data Sheets

This appendix contains a compilation of manufacturers' data sheets for most of the devices discussed in the text. These were chosen as being representative of the types of devices readily available, and are not chosen due to any superior performance characteristics. The author wishes to acknowledge his gratitude to the manufacturers who made these data sheets available.

TYPES 1N456, 1N457, 1N458, 1N459
SILICON GENERAL PURPOSE DIODES

TYPES 1N456, 1N457, 1N458, 1N459
BULLETIN NO. DL-S 688612, MARCH 1966
REVISED MAY 1968

$V_{RM(wkg)}$ . . . 25 to 175 Volts

- Rugged Whiskerless Construction
- Small Size
- Low Reverse Current

mechanical data

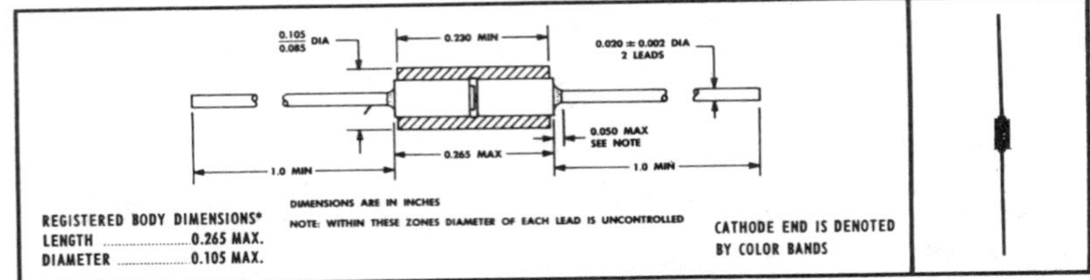

REGISTERED BODY DIMENSIONS*
LENGTH .... 0.265 MAX.
DIAMETER .... 0.105 MAX.

CATHODE END IS DENOTED BY COLOR BANDS

***absolute maximum ratings at 25°C free-air temperature (unless otherwise noted)**

		1N456	1N457	1N458	1N459	UNIT
V_{RM}	Peak Reverse Voltage	30	70	150	200	V
$V_{RM(wkg)}$	Working Peak Reverse Voltage	25	60	125	175	V
I_O	Average Rectified Forward Current at (or below) 25°C Free-Air Temperature (See Notes 1 and 2)	90	75	55	40	mA
I_F	Steady State Forward Current at (or below) 25°C Free-Air Temperature (See Note 2)	135	110	80	60	mA
$I_{FM(surge)}$	Peak Surge Current, One Second (See Note 3)	0.7	0.6	0.5	0.4	A
$I_{FM(surge)}$	Peak Surge Current, Two Microseconds (See Note 4)	1.2	1	0.8	0.7	A
P	Continuous Power Dissipation at (or below) 25°C Free-Air Temperature (See Note 5)	200				mW
T_{stg}	Storage Temperature Range	−80 to 200				°C
	Altitude	Any				

***electrical characteristics at 25°C free-air temperature (unless otherwise noted)**

PARAMETER		TEST CONDITIONS	1N456		1N457		1N458		1N459		UNIT
			MIN	MAX	MIN	MAX	MIN	MAX	MIN	MAX	
$V_{(BR)}$	Reverse Breakdown Voltage	$I_R = 100\ \mu A$	30		70		150		200		V
I_R	Static Reverse Current	V_R = Rated $V_{RM(wkg)}$		25		25		25		25	nA
		V_R = Rated $V_{RM(wkg)}$, T_A = 150°C		5		5		5		5	μA
V_F	Static Forward Voltage	I_F = 40 mA		1							V
		I_F = 20 mA				1					V
		I_F = 7 mA						1			V
		I_F = 3 mA								1	V

NOTES: 1. These values may be applied continuously under single-phase 60-c/s half-sine-wave operation with resistive load.
2. Derate linearly to 0 at 200°C free-air temperature.
3. These values apply for a one-second square-wave pulse with the device at nonoperating thermal equilibrium immediately prior to the surge.
4. These values apply for 2-μs pulses, duty cycle ≤ 1%, with the device at nonoperating thermal equilibrium immediately prior to the surge.
5. Derate linearly to 200°C free-air temperature at the rate of 1.14 mW/deg.

†Trademark of Texas Instruments
*Indicates JEDEC registered data

PRINTED IN U.S.A.
568

TEXAS INSTRUMENTS INCORPORATED
POST OFFICE BOX 5012 • DALLAS, TEXAS 75222

MZ1000-1 thru MZ1000-37

1 WATT
ZENER DIODES

SILICON
OXIDE PASSIVATED

3.3-100 VOLTS

JUNE 1966 — DS 7031 R1
(Replaces DS 7031)

MINIATURE PLASTIC ENCAPSULATED ZENER DIODES

. . . for regulated power supply circuits, surge protection, arc suppression and other functions in television, automotive and other consumer product applications.

- No larger than conventional 250 mW case yet conservatively rated at 1 watt (to 3 watts dissipation possible).
- 100% oscilloscope tested to assure sharp breakdown and long-term, reliable operation.

MAXIMUM RATINGS

Characteristic	Rating	Unit
DC Power Dissipation @ T_L = 50°C Derate above 50°C	1.5 8.33	Watts mW/°C
Lead Temperature*	-65 to +175	°C

*Maximum lead temperature for 10 seconds at 1/16'' from case = 230°C

FIGURE 1 — POWER-TEMPERATURE DERATING CURVE

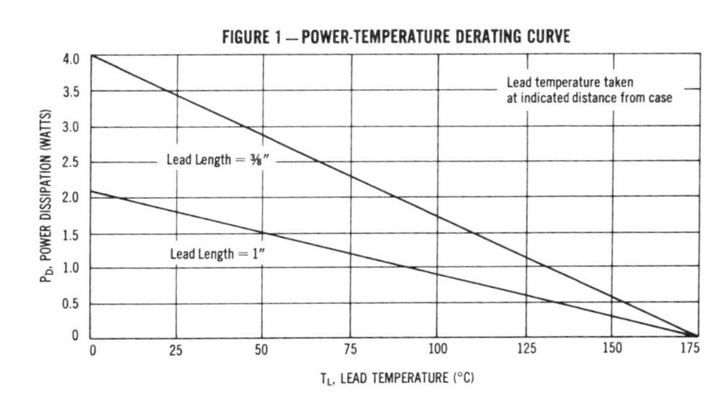

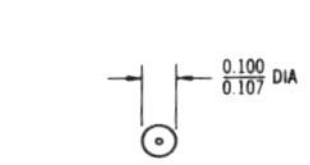

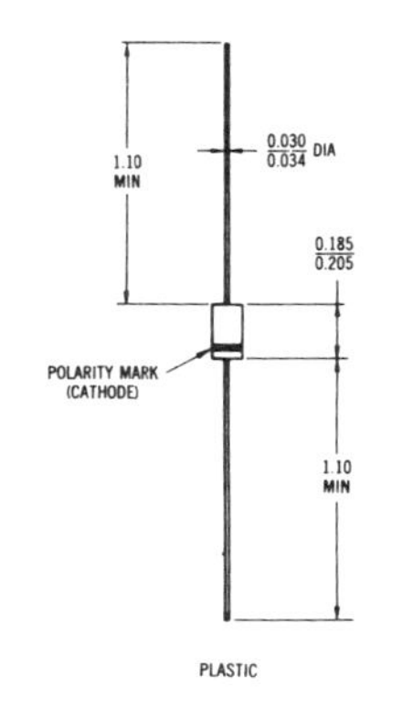

PLASTIC

MECHANICAL CHARACTERISTICS

CASE: Void free, transfer molded.

FINISH: All external surfaces are corrosion resistant. Leads are readily solderable.

POLARITY: Cathode indicated by color band. When operated in zener mode, cathode will be positive with respect to anode.

MOUNTING POSITION: Any.

WEIGHT: 0.42 gram (approximately).

SILICON ZENER DIODES
MZ1000-1 thru MZ1000-37
DS 7031 R1

MZ1000-1 thru MZ1000-37

ELECTRICAL CHARACTERISTICS (T_A = 25°C unless otherwise noted) V_F = 1.5 V max @ 200 mA on all types

Motorola Type No.	Zener Voltage V_Z @ I_{ZT} Volts *			Test Current I_{ZT} mA	Typical Z_{ZT} @ I_{ZT} Ohms	Max DC Zener Current I_{ZM} mA	Maximum Reverse Leakage Current		Temperature Coefficient %/°C
	Min	Nom	Max				I_R µA Max	@ V_R Volts	
MZ1000-1	2.97	3.3	3.63	76	15	276	150	1	-.070
MZ1000-2	3.24	3.6	3.96	69	15	252	150	1	-.065
MZ1000-3	3.51	3.9	4.29	64	13.5	234	75	1	-.060
MZ1000-4	3.87	4.3	4.73	58	13.5	217	20	1	-.050
MZ1000-5	4.23	4.7	5.17	53	12	193	20	1	-.043
MZ1000-6	4.59	5.1	5.61	49	10.5	178	20	1	±.030
MZ1000-7	5.04	5.6	6.16	45	7.5	162	20	2	±.028
MZ1000-8	5.58	6.2	6.82	41	3	146	20	3	+.045
MZ1000-9	6.12	6.8	7.48	37	5.25	133	20	4	.050
MZ1000-10	6.75	7.5	8.25	34	6	121	20	5	.058
MZ1000-11	7.38	8.2	9.02	31	6.75	110	20	5.9	.062
MZ1000-12	8.19	9.1	10.01	28	7.5	100	20	6.6	.068
MZ1000-13	9	10	11	25	10.5	91	20	7.2	.075
MZ1000-14	9.9	11	12.1	23	12	83	10	8.0	.076
MZ1000-15	10.8	12	13.2	21	13.5	76	10	8.6	.077
MZ1000-16	11.7	13	14.3	19	15	69	10	9.4	.079
MZ1000-17	13.5	15	16.5	17	21	61	10	10.8	.082
MZ1000-18	14.4	16	17.6	15.5	24	57	10	11.5	.083
MZ1000-19	16.2	18	19.8	14	30	50	10	13.0	.085
MZ1000-20	18	20	22	12.5	33	45	10	14.4	.086
MZ1000-21	19.8	22	24.2	11.5	34.5	41	10	15.8	.087
MZ1000-22	21.6	24	26.4	10.5	37.5	38	10	17.3	.088
MZ1000-23	24.3	27	29.7	9.5	52.5	34	10	19.4	.090
MZ1000-24	27	30	33	8.5	60	30	10	21.6	.091
MZ1000-25	29.7	33	36.3	7.5	67.5	27	10	23.8	.092
MZ1000-26	32.4	36	39.6	7	75	25	10	25.9	.093
MZ1000-27	35.1	39	42.9	6.5	90	23	10	28.1	.094
MZ1000-28	38.7	43	47.3	6	105	22	10	31.0	.095
MZ1000-29	42.3	47	51.7	5.5	120	19	10	33.8	.095
MZ1000-30	45.9	51	56.1	5	142.5	18	10	36.7	.096
MZ1000-31	50.4	56	61.6	4.5	165	16	10	40.3	.096
MZ1000-32	55.8	62	68.2	4	177.5	14	10	44.6	.097
MZ1000-33	61.2	68	74.8	3.7	225	13	10	49.0	.097
MZ1000-34	67.5	75	82.5	3.3	262.5	12	10	54.0	.098
MZ1000-35	73.8	86	90.2	3	300	11	10	59.0	.098
MZ1000-36	81.9	91	100.1	2.8	375	10	10	65.5	.099
MZ1000-37	90	100	110	2.5	525	9	10	72.0	.100

*1. Nominal voltages other than those stated above, matched sets, and tighter voltage tolerances are available as listed on DS 7030 R1 (available from your local Motorola sales office or distributor) . . . Motorola 1N4728 thru 1N4764 series (1M3.3ZS10 thru 1M100ZS10).

2. Voltages to 200 volts are available in other package configurations on request.

FIGURE 2 — TYPICAL ZENER DIODE CHARACTERISTICS and SYMBOL IDENTIFICATION

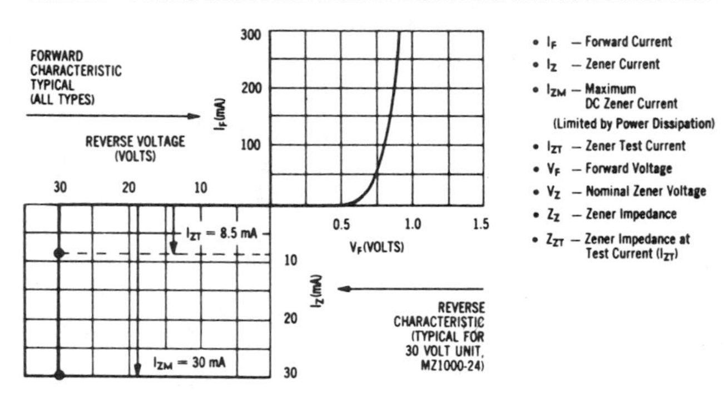

- I_F — Forward Current
- I_Z — Zener Current
- I_{ZM} — Maximum DC Zener Current (Limited by Power Dissipation)
- I_{ZT} — Zener Test Current
- V_F — Forward Voltage
- V_Z — Nominal Zener Voltage
- Z_Z — Zener Impedance
- Z_{ZT} — Zener Impedance at Test Current (I_{ZT})

MOTOROLA Semiconductor Products Inc.

BOX 955 • PHOENIX, ARIZONA 85001 • A SUBSIDIARY OF MOTOROLA INC.

DS 7031 R1

MV2101 thru MV2115

VVC

SILICON EPICAP▲ DIODES

... designed in the popular PLASTIC PACKAGE for high volume requirements of FM Radio and TV tuning and AFC, general frequency control and tuning applications; providing solid-state reliability in replacement of mechanical tuning methods.

- High Q with Guaranteed Minimum Values
- Controlled and Uniform Tuning Ratio
- Standard Capacitance Tolerance – 10%
- Complete Typical Design Curves
- Case TO-92 with Two Leads

VOLTAGE-VARIABLE CAPACITANCE DIODES

6.8–100 pF
30 VOLTS

FEBRUARY 1969 – DS 8528

CASE 182

MAXIMUM RATINGS

Rating	Symbol	Value	Unit
Reverse Voltage	V_R	30	Volts
Forward Current	I_F	200	mA
Device Dissipation @ $T_A = 25^oC$ Derate above 25^oC	P_D	280 2.8	mW mW/^oC
Junction Temperature	T_J	+125	^oC
Storage Temperature Range	T_{stg}	-65 to +150	^oC

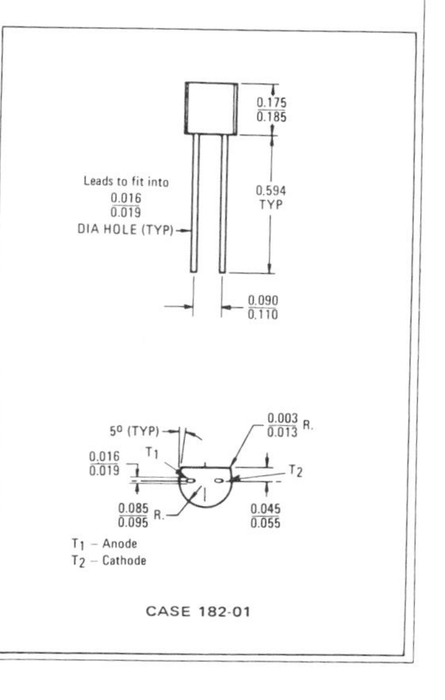

CASE 182-01

▲Trademark of Motorola Inc.

ELECTRICAL CHARACTERISTICS (T_A = 25°C unless otherwise noted)

Characteristic—All Types	Symbol	Min	Typ	Max	Unit
Reverse Breakdown Voltage (I_R = 10 μAdc)	BV_R	30	–	–	Vdc
Reverse Voltage Leakage Current (V_R = 25 Vdc, T_A = 25°C)	I_R	–	–	0.10	μAdc
Series Inductance (f = 250 MHz, Lead Length ≈ 1/16")	L_S	–	6.0	–	nH
Case Capacitance (f = 1.0 MHz, Lead Length ≈ 1/16")	C_C	–	0.18	–	pF
Diode Capacitance Temperature Coefficient (V_R = 4.0 Vdc, f = 1.0 MHz)	TC_C	–	280	400	ppm/°C

	C_T, Diode Capacitance V_R = 4.0 Vdc, f = 1.0 MHz pF			Q, Figure of Merit V_R = 4.0 Vdc, f = 50 MHz	TR, Tuning Ratio C_2/C_{30} f = 1.0 MHz		
Device	Min	Nom	Max	Min	Min	Typ	Max
MV2101	6.1	**6.8**	7.5	450	2.5	2.7	3.2
MV2102	7.4	**8.2**	9.0	450	2.5	2.8	3.2
MV2103	9.0	**10.0**	11.0	400	2.5	2.9	3.2
MV2104	10.8	**12.0**	13.2	400	2.5	2.9	3.2
MV2105	13.5	**15.0**	16.5	400	2.5	2.9	3.2
MV2106	16.2	**18.0**	19.8	350	2.5	2.9	3.2
MV2107	19.8	**22.0**	24.2	350	2.5	2.9	3.2
MV2108	24.3	**27.0**	29.7	300	2.5	3.0	3.2
MV2109	29.7	**33.0**	36.3	200	2.5	3.0	3.2
MV2110	35.1	**39.0**	42.9	150	2.5	3.0	3.2
MV2111	42.3	**47.0**	51.7	150	2.5	3.0	3.2
MV2112	50.4	**56.0**	61.6	150	2.6	3.0	3.3
MV2113	61.2	**68.0**	74.8	150	2.6	3.0	3.3
MV2114	73.8	**82.0**	90.2	100	2.6	3.0	3.3
MV2115	90.0	**100.0**	110.0	100	2.6	3.0	3.3

PARAMETER TEST METHODS

1. L_S, SERIES INDUCTANCE

L_S is measured on a shorted package at 250 MHz using an impedance bridge (Boonton Radio Model 250A RX Meter).

2. C_C, CASE CAPACITANCE

C_C is measured on an open package at 1.0 MHz using a capacitance bridge (Boonton Electronics Model 75A or equivalent).

3. C_T, DIODE CAPACITANCE

($C_T = C_C + C_J$). C_T is measured at 1.0 MHz using a capacitance bridge (Boonton Electronics Model 75A or equivalent).

4. TR, TUNING RATIO

TR is the ratio of C_T measured at 2.0 Vdc divided by C_T measured at 30 Vdc.

5. Q, FIGURE OF MERIT

Q is calculated by taking the G and C readings of an admittance bridge at the specified frequency and substituting in the following equations:

$$Q = \frac{2\pi fC}{G}$$

(Boonton Electronics Model 33AS8). Use Lead Length ≈1/16".

6. TC_C, DIODE CAPACITANCE TEMPERATURE COEFFICIENT

TC_C is guaranteed by comparing C_T at V_R = 4.0 Vdc, f = 1.0 MHz, T_A = -65°C with C_T at V_R = 4.0 Vdc, f = 1.0 MHz, T_A = +85°C in the following equation which defines TC_C:

$$TC_C = \frac{C_T(+85^\circ C) - C_T(-65^\circ C)}{85 + 65} \cdot \frac{10^6}{C_R(25^\circ C)}$$

Accuracy limited by measurement of C_T to ± 0.1 pF.

TYPES 2N5449, 2N5450, 2N5451
BULLETIN NO. DL-S 6810924, MAY 1968

SILECT† TRANSISTORS

Encapsulated in Plastic for Such Applications as Medium-Power Amplifiers, Class B Audio Outputs, and Hi-Fi Drivers

- **Electrically Equivalent to 2N3704, 2N3705, and 2N3706**
- **For Complementary Use with 2N5447 and 2N5448**
- **Rugged, One-Piece Construction Features Standard 100-mil TO-18 Pin Circle**

mechanical data

These transistors are encapsulated in a plastic compound specifically designed for this purpose, using a highly mechanized process‡ developed by Texas Instruments. The case will withstand soldering temperatures without deformation. These devices exhibit stable characteristics under high-humidity conditions and are capable of meeting MIL-STD-202C method 106B. The transistors are insensitive to light.

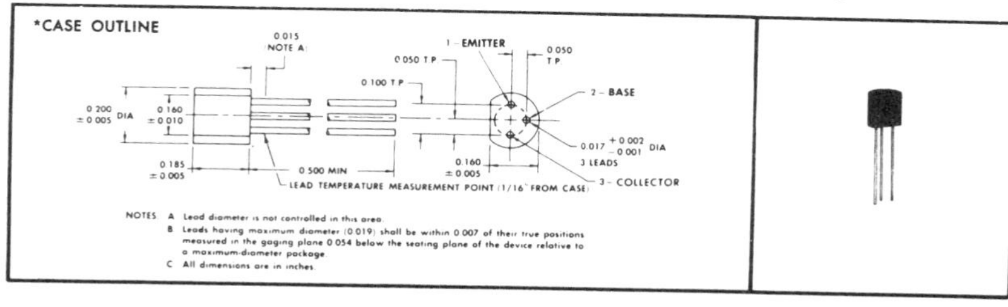

*absolute maximum ratings at 25°C free-air temperature (unless otherwise noted)

	2N5449 2N5450	2N5451
Collector-Base Voltage	50 V	40 V
Collector-Emitter Voltage (See Note 1)	30 V	20 V
Emitter-Base Voltage	5 V	5 V
Continuous Collector Current	←800 mA→	
Continuous Device Dissipation at (or below) 25°C Free-Air Temperature (See Note 2)	←360 mW→	
Continuous Device Dissipation at (or below) 25°C Lead Temperature (See Note 3)	←500 mW→	
Storage Temperature Range	−65°C to 150°C	
Lead Temperature 1/16 Inch from Case for 10 Seconds	←260°C→	

NOTES: 1. These values apply when the base-emitter diode is open-circuited.
2. Derate linearly to 150°C free-air temperature at the rate of 2.88 mW/deg.
3. Derate linearly to 150°C lead temperature at the rate of 4 mW/deg. Lead temperature is measured on the collector lead 1/16 inch from the case.

†Trademark of Texas Instruments
‡Patent pending
*Indicates JEDEC registered data

TYPES 2N5449, 2N5450, 2N5451
N-P-N EPITAXIAL PLANAR SILICON TRANSISTORS

***electrical characteristics at 25°C free-air temperature**

PARAMETER		TEST CONDITIONS	2N5449 MIN	2N5449 MAX	2N5450 MIN	2N5450 MAX	2N5451 MIN	2N5451 MAX	UNIT
$V_{(BR)CBO}$	Collector-Base Breakdown Voltage	$I_C = 100\ \mu A, I_E = 0$	50		50		40		V
$V_{(BR)CEO}$	Collector-Emitter Breakdown Voltage	$I_C = 10\ mA$, $I_B = 0$, See Note 4	30		30		20		V
$V_{(BR)EBO}$	Emitter-Base Breakdown Voltage	$I_E = 100\ \mu A$, $I_C = 0$	5		5		5		V
I_{CBO}	Collector Cutoff Current	$V_{CB} = 20\ V$, $I_E = 0$		100		100		100	nA
I_{EBO}	Emitter Cutoff Current	$V_{EB} = 3\ V$, $I_C = 0$		100		100		100	nA
h_{FE}	Static Forward Current Transfer Ratio	$V_{CE} = 2\ V$, $I_C = 50\ mA$, See Note 4	100	300	50	150	30	600	
V_{BE}	Base-Emitter Voltage	$V_{CE} = 2\ V$, $I_C = 100\ mA$, See Note 4	0.5	1	0.5	1	0.5	1	V
$V_{CE(sat)}$	Collector-Emitter Saturation Voltage	$I_B = 5\ mA$, $I_C = 100\ mA$, See Note 4		0.6		0.8		1	V
$\|h_{fe}\|$	Small-Signal Common-Emitter Forward Current Transfer Ratio	$V_{CE} = 2\ V$, $I_C = 50\ mA$, $f = 20\ MHz$	5		5		5		
C_{cb}	Collector-Base Capacitance	$V_{CB} = 10\ V$, $I_E = 0$, $f = 1\ MHz$, See Note 5		12		12		12	pF

NOTES: 4. These parameters must be measured using pulse techniques. $t_p = 300\ \mu s$, duty cycle $\leq 2\%$.
5. C_{cb} is measured using three-terminal measurement techniques with the emitter guarded.

*Indicates JEDEC registered data

TYPICAL CHARACTERISTICS

2N5450

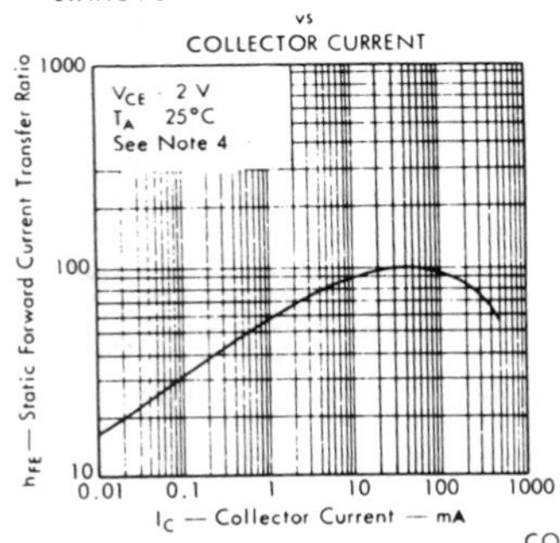

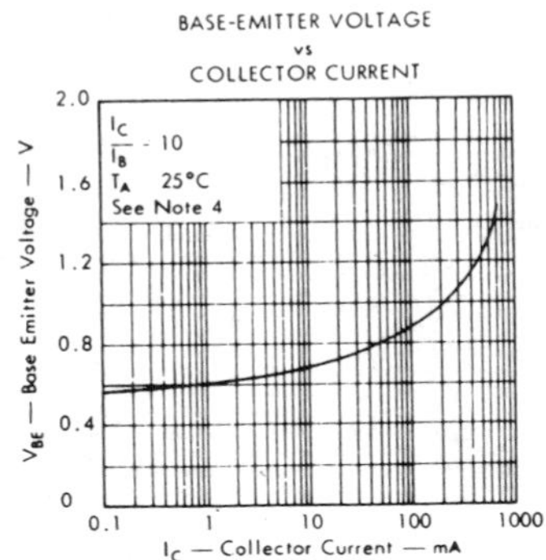

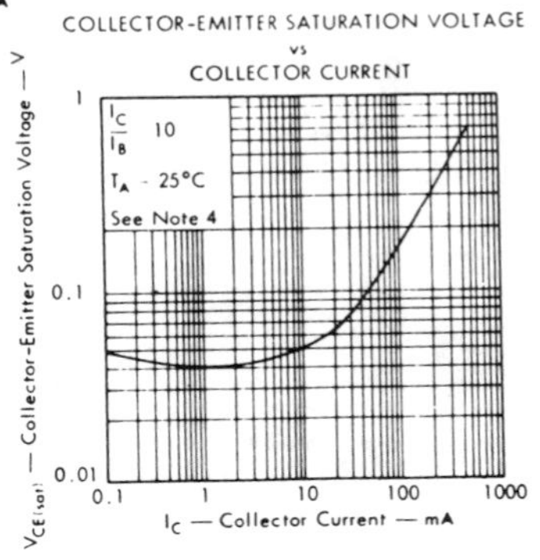

PRINTED IN U.S.A.

668

TEXAS INSTRUMENTS
INCORPORATED
POST OFFICE BOX 5012 • DALLAS, TEXAS 75222

PRELIMINARY SPECIFICATION

MPF109

JUNCTION FIELD-EFFECT TRANSISTOR

SYMMETRICAL SILICON N-CHANNEL

DECEMBER 1968 – PS 107

SILICON N-CHANNEL JUNCTION FIELD-EFFECT TRANSISTOR

Depletion mode (Type A) transistor designed for general-purpose audio and switching applications.

- Devices are Classified and Identified in 2:1 Zero-Gate Voltage Drain Current Ranges (2:1 I_{DSS} Ranges)
- Drain and Source Interchangeable
- High AC Input Impedance
- High DC Input Resistance
- Low Transfer and Input Capacitance
- Low Cross-Modulation and Intermodulation Distortion
- Unibloc* Plastic Encapsulated Package

MAXIMUM RATINGS

Rating	Symbol	Value	Unit
Drain-Source Voltage	V_{DS}	25	Vdc
Drain-Gate Voltage	V_{DG}	25	Vdc
Gate-Source Voltage	V_{GS}	-25	Vdc
Forward Gate Current	$I_{G(f)}$	10	mAdc
Total Device Dissipation @ T_A = 25°C Derate above 25°C	P_D	310 2.82	mW mW/°C
Operating Junction Temperature Range	T_J	-65 to +135	°C
Storage Temperature Range	T_{stg}	-65 to +150	°C

*Trademark of Motorola Inc.

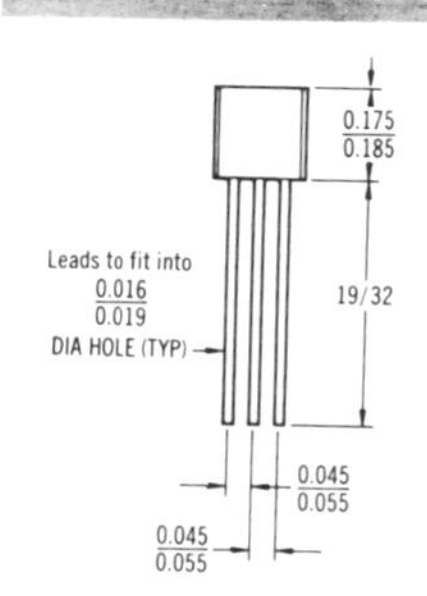

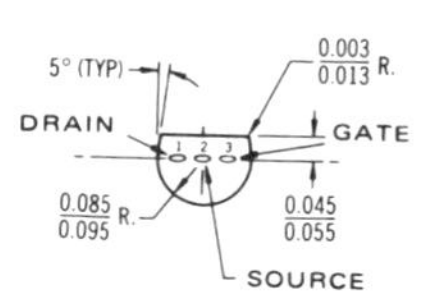

CASE 29(5)
TO-92

Drain and Source may be Interchanged

ELECTRICAL CHARACTERISTICS (T_A = 25°C unless otherwise noted)

Characteristic	Symbol	Min	Max	Unit
OFF CHARACTERISTICS				
Gate-Source Breakdown Voltage (I_G = 10 μAdc, V_{DS} = 0)	$V_{(BR)GSS}$	-25	-	Vdc
Gate-Source Cutoff Voltage* (V_{DS} = 15 Vdc, I_D = 10 μAdc)	$V_{GS(off)}$*	0.2	8.0	Vdc
Gate Reverse Current (V_{GS} = -15 Vdc, V_{DS} = 0)	I_{GSS}	-	-1.0	nAdc
ON CHARACTERISTICS				
Zero-Gate Voltage Drain Current* (V_{DS} = 15 Vdc, V_{GS} = 0)	I_{DSS}*	0.5	24	mAdc
SMALL-SIGNAL CHARACTERISTICS				
Forward Transadmittance* (V_{DS} = 15 Vdc, V_{GS} = 0, f = 1.0 kHz)	y_{fs}*	800	6000	μmhos
Output Admittance (V_{DS} = 15 Vdc, V_{GS} = 0, f = 1.0 kHz)	y_{os}	-	75	μmhos
Input Capacitance (V_{DS} = 15 Vdc, V_{GS} = 0, f = 1.0 MHz)	C_{iss}	-	7.0	pF
Reverse Transfer Capacitance (V_{DS} = 15 Vdc, V_{GS} = 0, f = 1.0 MHz)	C_{rss}	-	3.0	pF
Common-Source Noise Figure (V_{DS} = 15 Vdc, V_{GS} = 0, R_G = 1.0 Megohm, f = 1.0 kHz)	NF	-	2.5	dB

*To characterize these devices to narrower limits, regarding I_{DSS}, $V_{GS(off)}$ and y_{fs}, the entire production lot is tested and divided into color-coded groups, with each color dot representing a relatively small range compared with the total min-max limit of the whole distribution. The color codes and their associated limits are given in the following table.

When packaged for shipment, the colors are randomly selected and no specific color distribution is implied or guaranteed.

Color	I_{DSS}	$V_{GS(off)}$	y_{fs}
White	0.5 mAdc Min, 1.0 mAdc Max	0.2 Vdc Min, 2.0 Vdc Max	800 to 3200 μmhos
Red	0.8 mAdc Min, 1.6 mAdc Max	0.4 Vdc Min, 4.0 Vdc Max	1000 to 4000 μmhos
Orange	1.5 mAdc Min, 3.0 mAdc Max	0.4 Vdc Min, 4.0 Vdc Max	1000 to 4000 μmhos
Yellow	2.5 mAdc Min, 5.0 mAdc Max	1.0 Vdc Min, 6.0 Vdc Max	1500 to 5000 μmhos
Green	4.0 mAdc Min, 8.0 mAdc Max	1.0 Vdc Min, 6.0 Vdc Max	1500 to 5000 μmhos
Blue	7.0 mAdc Min, 14 mAdc Max	2.0 Vdc Min, 8.0 Vdc Max	2000 to 6000 μmhos
Violet	12 mAdc Min, 24 mAdc Max	2.0 Vdc Min, 8.0 Vdc Max	2000 to 6000 μmhos

MOTOROLA Semiconductor Products Inc.
BOX 20912 • PHOENIX, ARIZONA 85036 • A SUBSIDIARY OF MOTOROLA INC.

4279 PRINTED IN USA 12-68 IMPERIAL LITHO

3N169
3N170
3N171

MOS FIELD-EFFECT TRANSISTORS

N-CHANNEL

TYPE C

MAY 1969 – DS 5320

SILICON N-CHANNEL MOS FIELD-EFFECT TRANSISTORS

Enhancement Mode (Type C) transistors designed for low-power switching applications.

- Low Switching Voltages – $V_{GS(th)} \leqslant 3.0$ Vdc
- Fast Switching Times $t_r \leqslant 10$ ns
- Low Drain-Source Resistance $r_{ds(on)} = 200$ Ohms (Max)
- Low Reverse Transfer Capacitance $C_{rss} = 1.3$ pF (Max)
- Manufactured Using the New Silicon Nitride Process Resulting in a Stable $V_{GS(th)}$ and Gate Oxide Breakdown Protection to Typical Transients of ±150 Volts Peak

MAXIMUM RATINGS (T_A = 25°C unless otherwise noted)

Rating	Symbol	Value	Unit
*Drain-Source Voltage	V_{DS}	25	Vdc
*Drain-Gate Voltage	V_{DG}	±35	Vdc
*Gate-Source Voltage	V_{GS}	±35	Vdc
*Drain Current	I_D	30	mAdc
Power Dissipation @ T_A = 25°C Derate above 25°C	P_D	300 1.7	mW mW/°C
*Power Dissipation @ T_C = 25°C *Derate above 25°C	P_D	800 4.56	mW mW/°C
Operating Junction Temperature	T_J	175	°C
*Storage Temperature Range	T_{stg}	-65 to +200	°C

*Indicates JEDEC Registered Data.

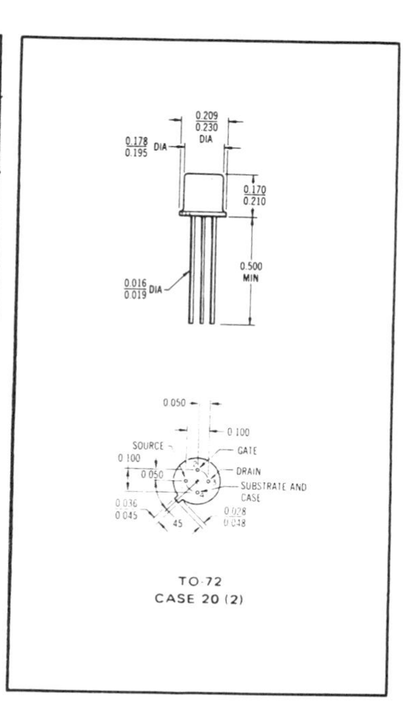

TO-72
CASE 20 (2)

HANDLING PRECAUTIONS:

MOS field-effect transistors have extremely high input resistance. They can be damaged by the accumulation of excess static charge. Avoid possible damage to the devices while handling, testing, or in actual operation, by following the procedures outlined below:

1. To avoid the build-up of static charge, the leads of the devices should remain shorted together with a metal ring except when being tested or used.
2. Avoid unnecessary handling. Pick up devices by the case instead of the leads.
3. Do not insert or remove devices from circuits with the power on because transient voltages may cause permanent damage to the devices.

ELECTRICAL CHARACTERISTICS (T_A = 25°C unless otherwise noted)

Substrate connected to source.

Characteristic	Figure No.	Symbol	Min	Max	Unit
OFF CHARACTERISTICS					
Drain-Source Breakdown Voltage (I_D = 10 μAdc, V_{GS} = 0)	–	$V_{(BR)DSS}$	25	–	Vdc
*Gate Leakage Current	–	I_{GSS}			pAdc
(V_{GS} = –35 Vdc, V_{DS} = 0)			–	10	
(V_{GS} = –35 Vdc, V_{DS} = 0, T_A = 125°C)			–	100	
*Zero-Gate-Voltage Drain Current	–	I_{DSS}			
(V_{DS} = 10 Vdc, V_{GS} = 0)			–	10	nAdc
(V_{DS} = 10 Vdc, V_{GS} = 0, T_A = 125°C)			–	1.0	μAdc
***ON CHARACTERISTICS**					
Gate-Source Threshold Voltage (V_{DS} = 10 Vdc, I_D = 10 μAdc) 3N169	–	$V_{GS(th)}$	0.5	1.5	Vdc
3N170			1.0	2.0	
3N171			1.5	3.0	
"ON" Drain Current (V_{GS} = 10 Vdc, V_{DS} = 10 Vdc)	3	$I_{D(on)}$	10	–	mAdc
Drain-Source "ON" Voltage (I_D = 10 mAdc, V_{GS} = 10 Vdc)	–	$V_{DS(on)}$	–	2.0	Vdc
SMALL SIGNAL CHARACTERISTICS					
*Drain-Source Resistance (V_{GS} = 10 Vdc, I_D = 0, f = 1.0 kHz)	4	$r_{ds(on)}$	–	200	Ohms
Forward Transfer Admittance (V_{DS} = 10 Vdc, I_D = 2.0 mAdc, f = 1.0 kHz)	1	$\|y_{fs}\|$	1000	–	μmhos
*Reverse Transfer Capacitance (V_{DS} = 0, V_{GS} = 0, f = 1.0 MHz)	2	C_{rss}	–	1.3	pF
*Input Capacitance (V_{DS} = 10 Vdc, V_{GS} = 0, f = 1.0 MHz)	2	C_{iss}	–	5.0	pF
*Drain-Substrate Capacitance ($V_{D(SUB)}$ = 10 Vdc, f = 1.0 MHz)	–	$C_{d(sub)}$	–	5.0	pF
***SWITCHING CHARACTERISTICS** (V_{DD} = 10 Vdc, $I_{D(on)}$ = 10 mAdc, $V_{GS(on)}$ = 10 Vdc, $V_{GS(off)}$ = 0, R_G' = 50 Ohms)					
Turn-On Delay Time	6,10	$t_{d(on)}$	–	3.0	ns
Rise Time	7,10	t_r	–	10	ns
Turn-Off Delay Time	8,10	$t_{d(off)}$	–	3.0	ns
Fall Time	9,10	t_f	–	15	ns

*Indicates JEDEC Registered Data.

FIGURE 1 – FORWARD TRANSFER ADMITTANCE

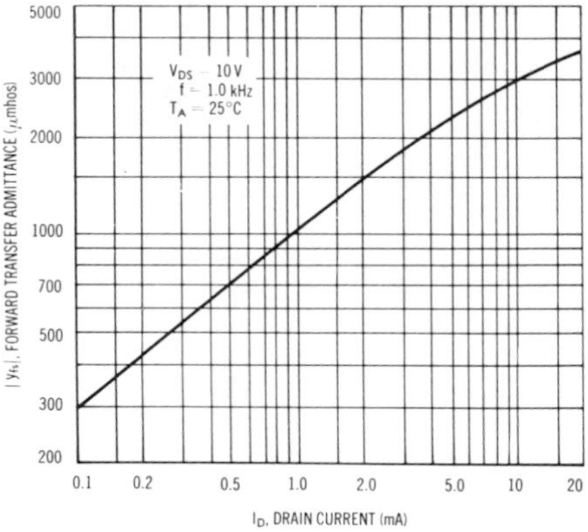

FIGURE 2 – CAPACITANCE

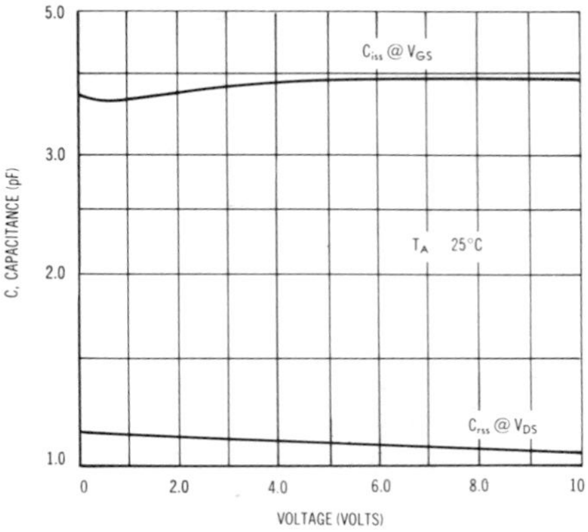

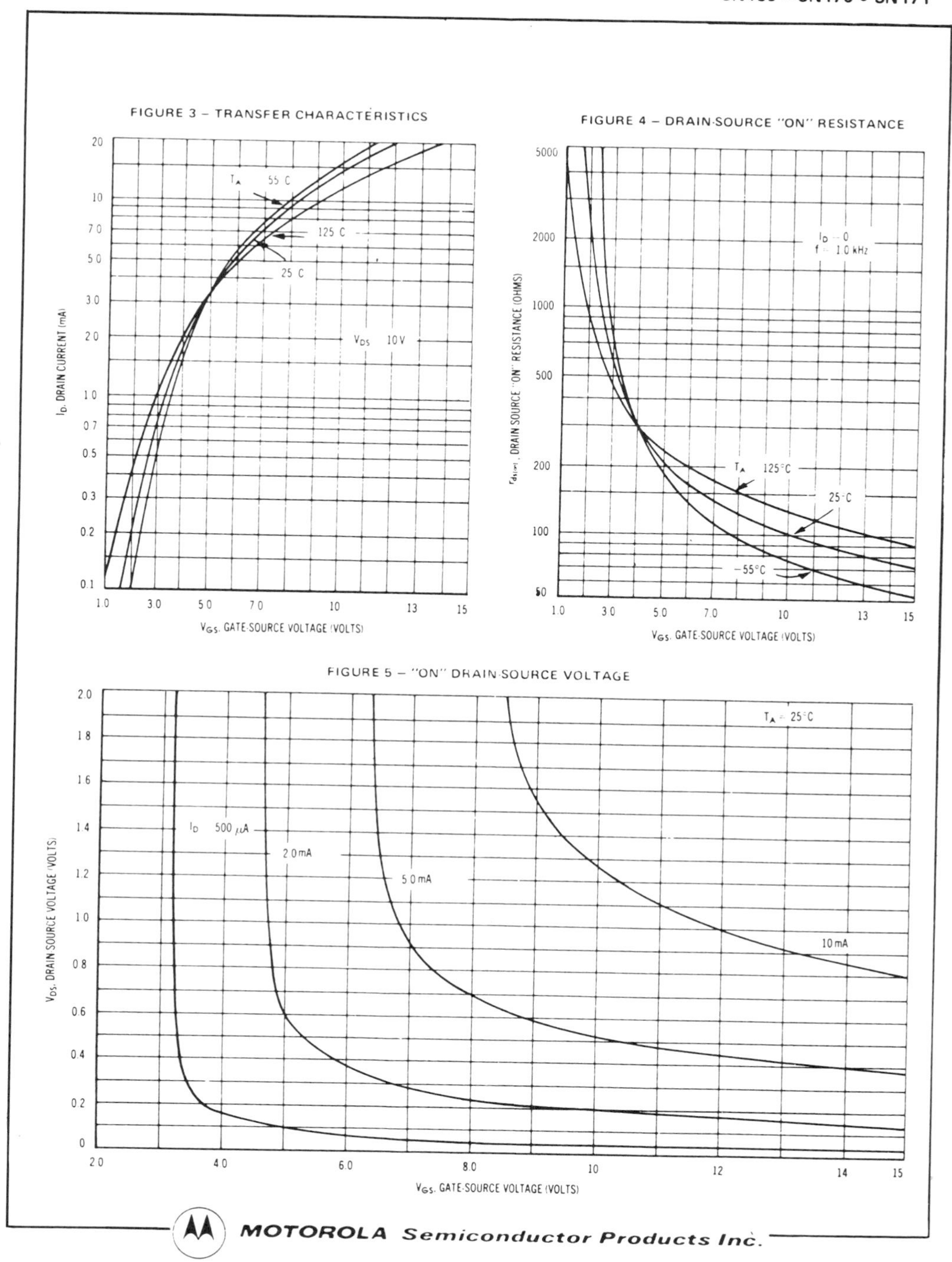
FIGURE 3 – TRANSFER CHARACTERISTICS
T_A 55 C
125 C
25 C
V_{DS} 10 V
I_D, DRAIN CURRENT (mA)
V_{GS}, GATE-SOURCE VOLTAGE (VOLTS)
FIGURE 4 – DRAIN-SOURCE "ON" RESISTANCE
I_D = 0
f = 1.0 kHz
T_A 125°C
25°C
−55°C
$r_{ds(on)}$, DRAIN SOURCE "ON" RESISTANCE (OHMS)
V_{GS}, GATE-SOURCE VOLTAGE (VOLTS)
FIGURE 5 – "ON" DRAIN-SOURCE VOLTAGE
T_A = 25°C
I_D 500 μA
2.0 mA
5.0 mA
10 mA
V_{DS}, DRAIN SOURCE VOLTAGE (VOLTS)
V_{GS}, GATE-SOURCE VOLTAGE (VOLTS)

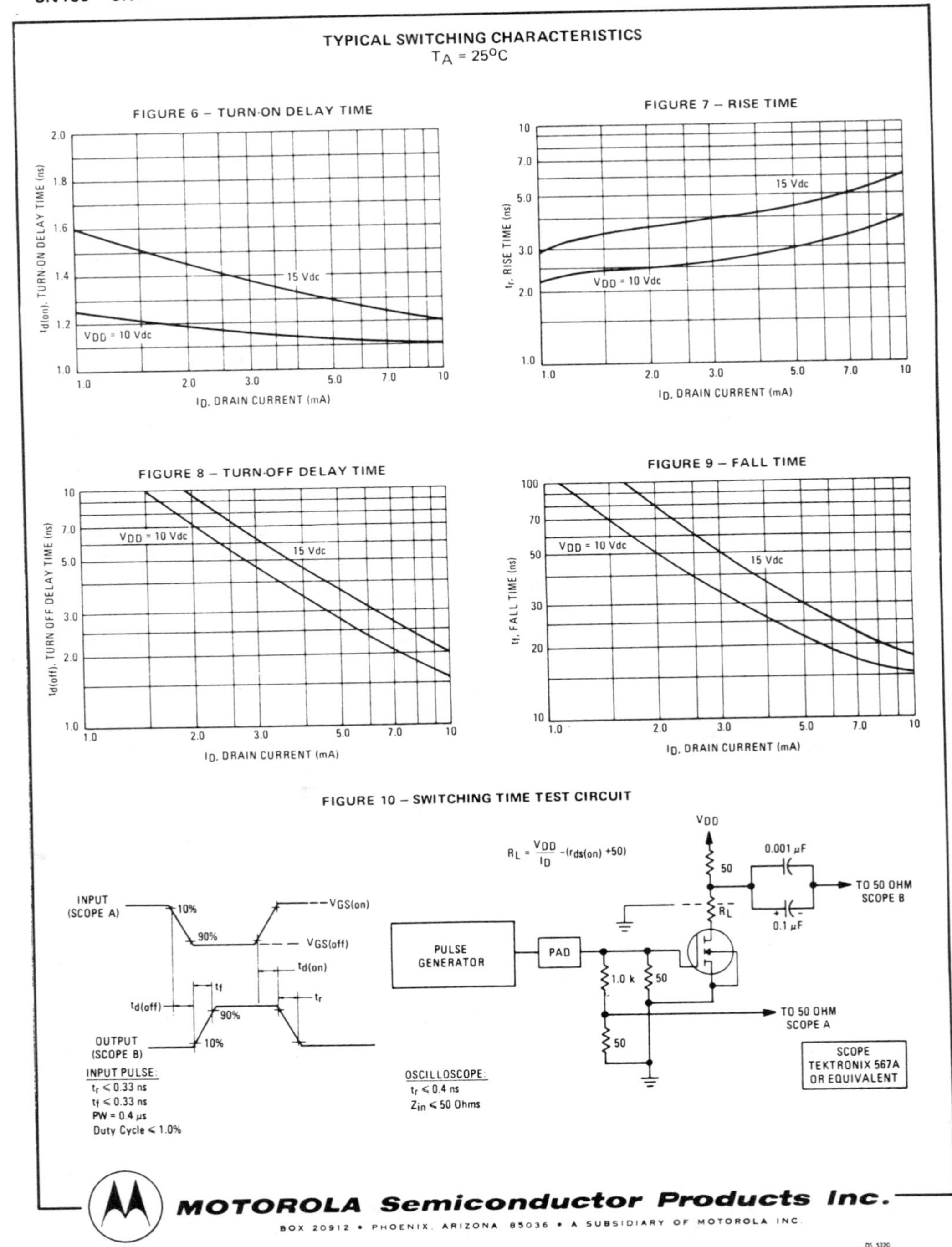

MOTOROLA Semiconductor Products Inc.
BOX 20912 • PHOENIX, ARIZONA 85036 • A SUBSIDIARY OF MOTOROLA INC.
4527 PRINTED IN USA 5-69 IMPERIAL LITHO B11152 15M
DS 5320

N-CHANNEL EPITAXIAL PLANAR SILICON FIELD-EFFECT TRANSISTOR

TYPE 2N3823
BULLETIN NO. DL-S 657816, JULY 1965

SYMMETRICAL N-CHANNEL FIELD-EFFECT TRANSISTOR FOR VHF AMPLIFIER AND MIXER APPLICATIONS

- Low Noise Figure: ≤ 2.5 db at 100 Mc
- Low C_{rss}: ≤ 2 pf
- High y_{fs}/C_{iss} Ratio (High-Frequency Figure-of-Merit)
- Cross Modulation Minimized by Square-Law Transfer Characteristic

*mechanical data

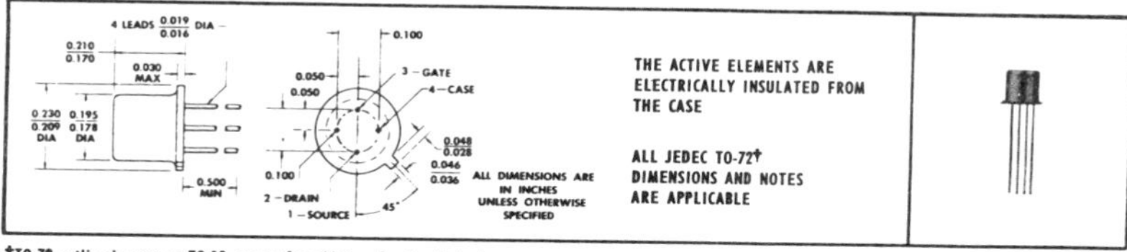

†TO-72 outline is same as TO-18 except for addition of a fourth lead.

* absolute maximum ratings at 25°C free-air temperature (unless otherwise noted)

Drain-Gate Voltage	30 v
Drain-Source Voltage	30 v
Reverse Gate-Source Voltage	−30 v
Gate Current	10 ma
Continuous Device Dissipation at (or below) 25°C Free-Air Temperature (See Note 1)	300 mw
Storage Temperature Range	−65°C to + 200°C
Lead Temperature 1/16 Inch from Case for 10 Seconds	300°C

*electrical characteristics at 25°C free-air temperature (unless otherwise noted)

	PARAMETER	TEST CONDITIONS‡	MIN	MAX	UNIT
$V_{(BR)GSS}$	Gate-Source Breakdown Voltage	$I_G = -1\ \mu a$, $V_{DS} = 0$	−30		v
I_{GSS}	Gate Cutoff Current	$V_{GS} = -20$ v, $V_{DS} = 0$		−0.5	na
		$V_{GS} = -20$ v, $V_{DS} = 0$, $T_A = 150°C$		−0.5	μa
I_{DSS}	Zero-Gate-Voltage Drain Current	$V_{DS} = 15$ v, $V_{GS} = 0$, See Note 2	4	20	ma
V_{GS}	Gate-Source Voltage	$V_{DS} = 15$ v, $I_D = 400\ \mu a$	−1	−7.5	v
$V_{GS(off)}$	Gate-Source Cutoff Voltage	$V_{DS} = 15$ v, $I_D = 0.5$ na		−8	v
$\|y_{fs}\|$	Small-Signal Common-Source Forward Transfer Admittance	$V_{DS} = 15$ v, $V_{GS} = 0$, f = 1 kc, See Note 2	3500	6500	μmho
$\|y_{os}\|$	Small-Signal Common-Source Output Admittance	$V_{DS} = 15$ v, $V_{GS} = 0$, f = 1 kc, See Note 2		35	μmho
C_{iss}	Common-Source Short-Circuit Input Capacitance	$V_{DS} = 15$ v, $V_{GS} = 0$, f = 1 Mc		6	pf
C_{rss}	Common-Source Short-Circuit Reverse Transfer Capacitance			2	pf
$\|y_{fs}\|$	Small-Signal Common-Source Forward Transfer Admittance	$V_{DS} = 15$ v, $V_{GS} = 0$, f = 200 Mc	3200		μmho
$Re(y_{is})$	Small-Signal Common-Source Input Conductance			800	μmho
$Re(y_{os})$	Small-Signal Common-Source Output Conductance			200	μmho

NOTES: 1. Derate linearly to 175°C free-air temperature at the rate of 2 mw/C°.
2. These parameters must be measured using pulse techniques. PW = 100 msec, Duty Cycle ≤ 10%.

*Indicates JEDEC registered data.
‡The fourth lead (case) is connected to the source for all measurements.

TEXAS INSTRUMENTS
INCORPORATED
SEMICONDUCTOR-COMPONENTS DIVISION
POST OFFICE BOX 5012 • DALLAS 22, TEXAS

*** operating characteristics at 25°C free-air temperature**

PARAMETER		TEST CONDITIONS‡	MAX	UNIT
NF	Common-Source Spot Noise Figure	$V_{DS} = 15$ v, $V_{GS} = 0$, f = 100 Mc, $R_G = 1$ kΩ	2.5	db

TYPICAL CHARACTERISTICS‡

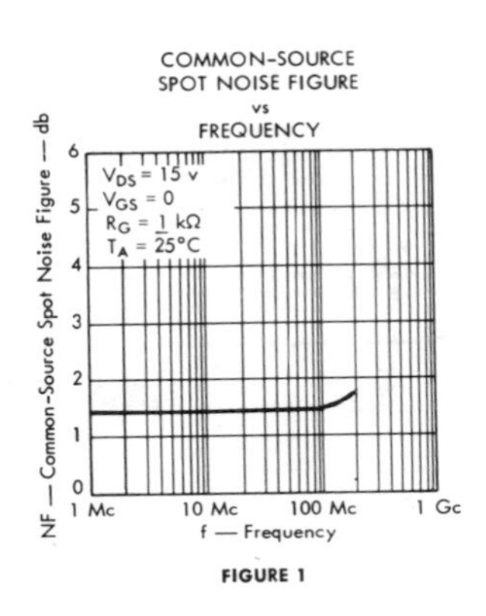

FIGURE 1

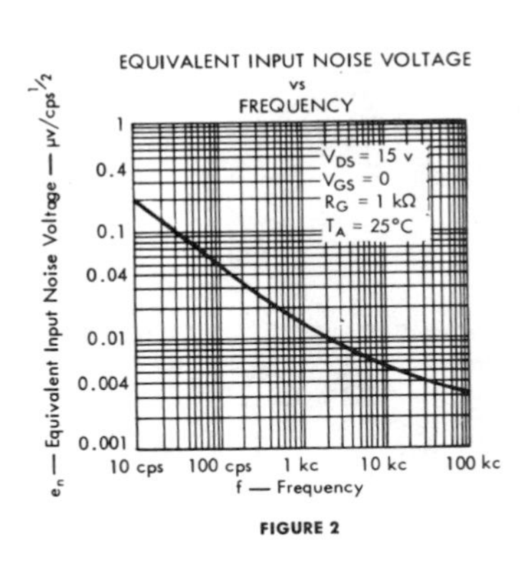

FIGURE 2

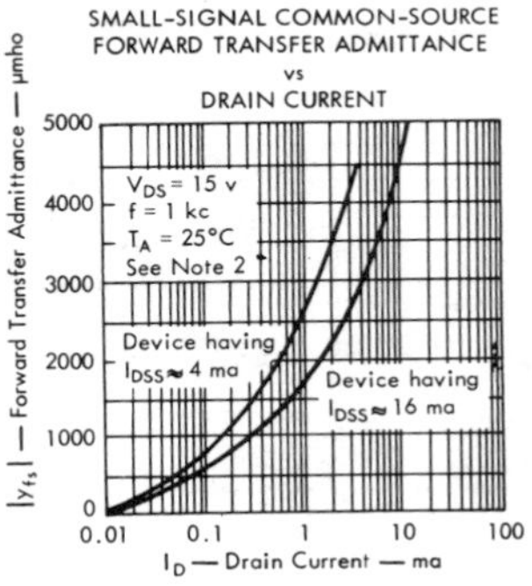

SMALL-SIGNAL COMMON-SOURCE FORWARD TRANSFER ADMITTANCE vs DRAIN CURRENT

FIGURE 3

NOTE 2: These parameters must be measured using pulse techniques. PW = 100 msec, Duty Cycle ≤ 10%.

*Indicates JEDEC registered data.
‡The fourth lead (case) is connected to the source for all measurements.

TYPICAL CHARACTERISTICS‡

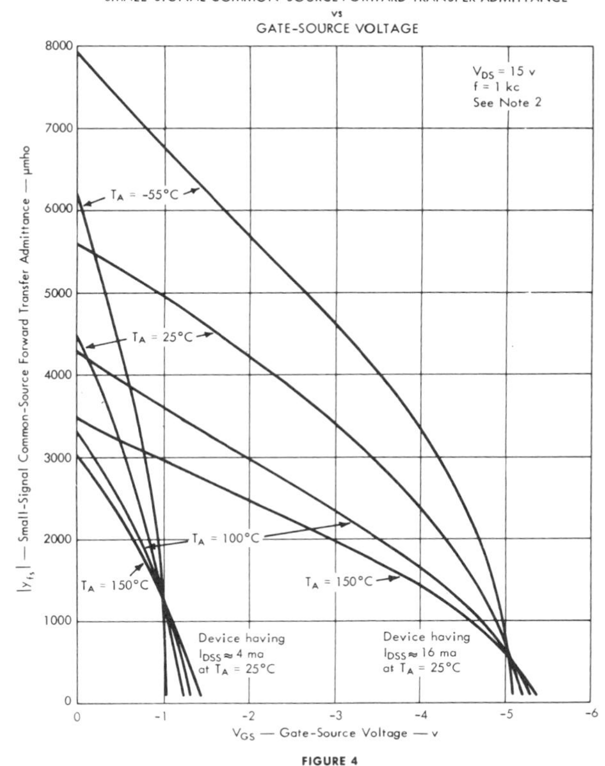

FIGURE 4

NOTE 2: These parameters must be measured using pulse techniques. PW = 100 msec, Duty Cycle ≤ 10%.

‡The fourth lead (case) is connected to the source for all measurements.

TYPE 2N3823
N-CHANNEL EPITAXIAL PLANAR SILICON FIELD-EFFECT TRANSISTOR

TYPICAL CHARACTERISTICS‡

GATE CUTOFF CURRENT vs FREE-AIR TEMPERATURE

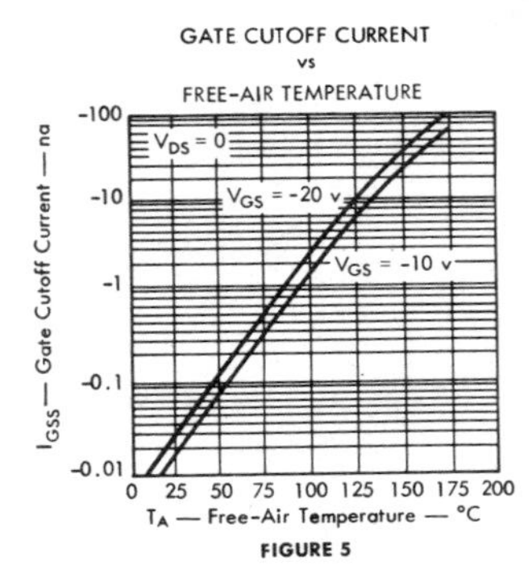

FIGURE 5

SMALL-SIGNAL COMMON-SOURCE INPUT ADMITTANCE vs FREQUENCY

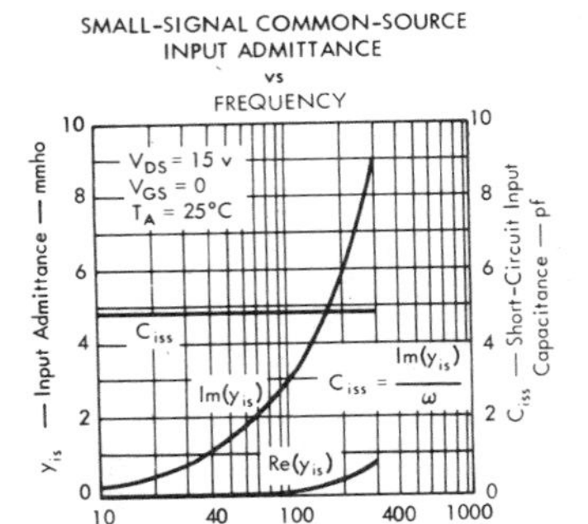

FIGURE 6

SMALL-SIGNAL COMMON SOURCE FORWARD TRANSFER ADMITTANCE vs FREQUENCY

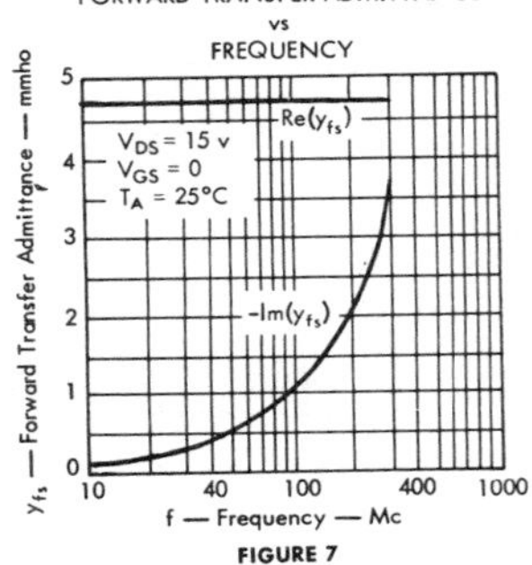

FIGURE 7

SMALL-SIGNAL COMMON-SOURCE REVERSE TRANSFER ADMITTANCE vs FREQUENCY

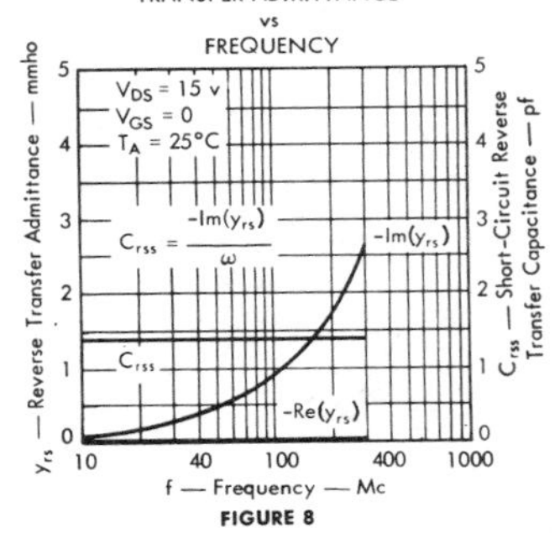

FIGURE 8

SMALL-SIGNAL COMMON-SOURCE OUTPUT ADMITTANCE vs FREQUENCY

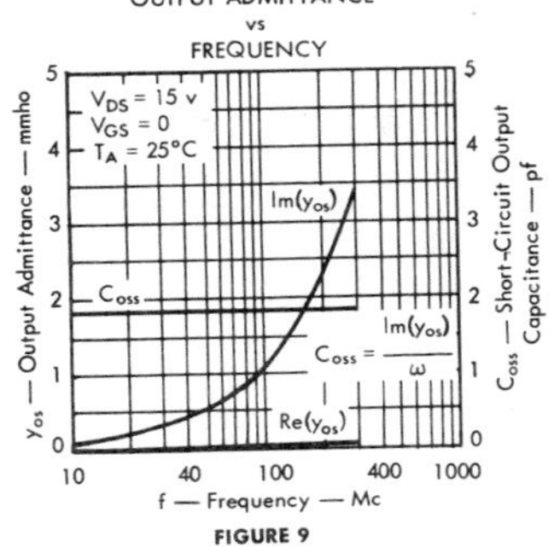

FIGURE 9

‡The fourth lead (case) is connected to the source for all measurements.

COMMON-SOURCE SHORT-CIRCUIT INPUT AND REVERSE-TRANSFER CAPACITANCES vs GATE-SOURCE VOLTAGE

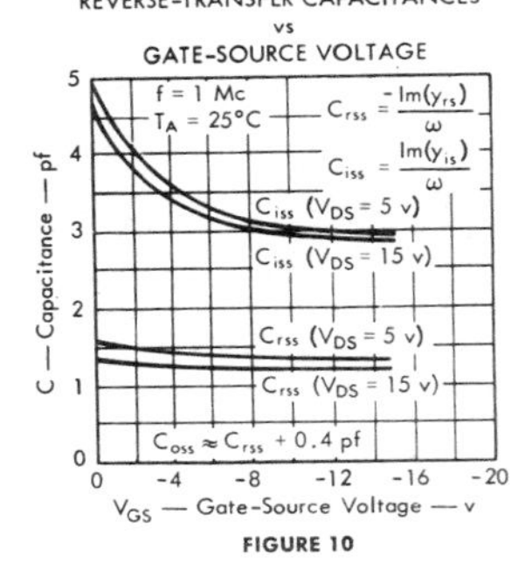

FIGURE 10

TYPE 3N160
P-CHANNEL INSULATED-GATE PLANAR SILICON FIELD-EFFECT TRANSISTOR

TYPE 3N160
BULLETIN NO. DL-S 7011149, MARCH 1970
REPLACES BULLETIN NO. DL-S 6811149, DECEMBER 1968

ENHANCEMENT-TYPE† METAL-OXIDE-SEMICONDUCTOR TRANSISTOR

For Applications Requiring Very High Input Impedance, Such as Series and Shunt Choppers, Multiplexers, and Commutators

- **Channel Cut Off with Zero Gate Voltage**
- **Square-Law Transfer Characteristic Reduces Distortion**
- **Independent Substrate Connection Provides Flexibility in Biasing**

*mechanical data

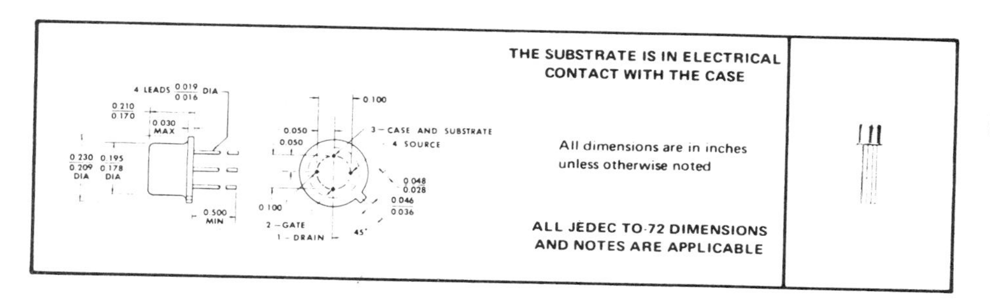

handling precautions

Curve-tracer testing and static-charge buildup are common causes of damage to insulated-gate devices. Permanent damage may result if either gate-voltage rating is exceeded even for extremely short time periods. Each transistor is protected during shipment by a gate-shorting device, which should be removed only during testing and after permanent mounting of the transistor. Personnel and equipment, including soldering irons, should be grounded.

*absolute maximum ratings at 25°C free-air temperature (unless otherwise noted)

Drain-Gate Voltage	−25 V
Drain-Source Voltage	−25 V
Forward Gate-Source Voltage	−25 V
Reverse Gate-Source Voltage	25 V
Continuous Drain Current	−125 mA
Continuous Device Dissipation at (or below) 25°C Free-Air Temperature (See Note 1)	360 mW
Continuous Device Dissipation at (or below) 25°C Case Temperature (See Note 2)	1.8 W
Storage Temperature Range	−65°C to 200°C
Lead Temperature 1/16 Inch from Case for 10 Seconds	300°C

NOTES: 1. Derate linearly to 175°C free-air temperature at the rate of 2.4 mW/°C.
2. Derate linearly to 175°C case temperature at the rate of 12 mW/°C.

*JEDEC registered data. This data sheet contains all applicable registered data in effect at the time of publication.

†Enhancement-mode operation entails the use of a forward gate-source voltage to increase drain current from I_{DSS}, the drain current at $V_{GS} = 0$, as opposed to depletion-mode operation wherein a reverse gate-source voltage is used to decrease drain current. An enhancement-type transistor is in the "off" state at $V_{GS} = 0$ and hence will not operate normally in the depletion mode.

MFE3006 thru MFE3008

N-CHANNEL

DUAL GATE MOS FIELD-EFFECT TRANSISTORS

TYPE B

MAY 1969 – DS 5318
(Replaces DS 5275 and DS 5296)

N-CHANNEL DUAL-GATE SILICON-NITRIDE PASSIVATED MOS FIELD-EFFECT TRANSISTORS

. . . depletion mode (Type B) dual gate transistors designed for VHF amplifier and mixer applications. These types are specified as follows:

MFE3006 – RF Amplifier @ 100 MHz
MFE3007 – RF Amplifier @ 200 MHz
MFE3008 – Mixer @ 100 and 200 MHz

- Silicon Nitride Passivation for Excellent Long Term Stability
- High Common-Source Power Gain –
  MFE3006: G_{ps} = 20 dB (Min) @ f = 100 MHz
  MFE3007: G_{ps} = 18 dB (Min) @ f = 200 MHz
- High Common-Source Conversion Gain –
  MFE3008: G_{ps} = 14 dB (Min) @ f = 100 MHz
  G_{ps} = 10 dB (Min) @ f = 200 MHz
- Low Reverse Transfer Capacitance –
  C_{rss} = 0.02 pF (Typ) @ V_{DS} = 15 Vdc

MAXIMUM RATINGS

Rating	Symbol	Value	Unit
Drain-Source Voltage	V_{DS}	+25	Vdc
Gate 1 Source Voltage	V_{G1S}	±35	Vdc
Gate 2 Source Voltage	V_{G2S}	±35	Vdc
Drain Current	I_D	30	mAdc
Total Device Dissipation @ T_A = 25°C Derate above 25°C	P_D	300 1.7	mW mW/°C
Operating Junction Temperature Range	T_J	-65 to +175	°C
Storage Temperature Range	T_{stg}	-65 to +175	°C

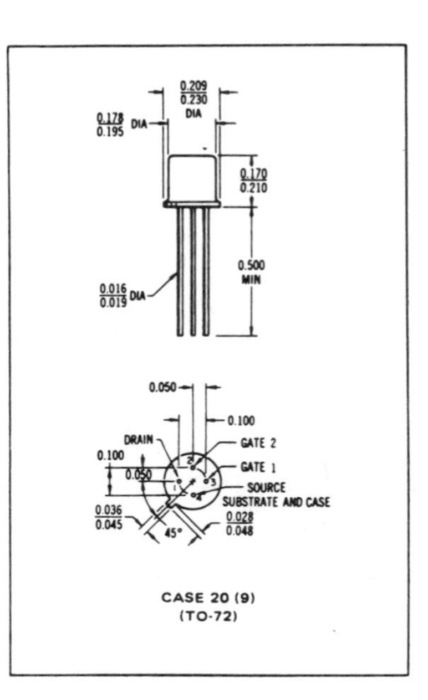

CASE 20 (9)
(TO-72)

HANDLING PRECAUTIONS:

MOS field-effect transistors have extremely high input resistance. They can be damaged by the accumulation of excess static charge. Avoid possible damage to the devices while handling, testing, or in actual operation, by following the procedures outlined below:

1. To avoid the build-up of static charge, the leads of the devices should remain shorted together with a metal ring except when being tested or used.
2. Avoid unnecessary handling. Pick up devices by the case instead of the leads.
3. Do not insert or remove devices from circuits with the power on because transient voltages may cause permanent damage to the devices.

ELECTRICAL CHARACTERISTICS (T_A = 25°C unless otherwise noted)
Substrate Connected to Source

Characteristic		Symbol	Min	Typ	Max	Unit
OFF CHARACTERISTICS						
Drain-Source Breakdown Voltage (I_D = 10 μAdc, V_S = 0, V_{G1} = V_{G2} = -4.0 Vdc)		$V_{(BR)DSX}$	25	–	–	Vdc
Gate 1 to Source Cutoff Voltage (V_{DS} = 15 Vdc, V_{G2S} = 4.0 Vdc, I_D = 200 μAdc)		$V_{G1S(off)}$	–	–	-3.0	Vdc
Gate 2 to Source Cutoff Voltage (V_{DS} = 15 Vdc, V_{G1S} = 0, I_D = 200 μAdc)		$V_{G2S(off)}$	–	–	-3.0	Vdc
Gate 1 Reverse Leakage Current (V_{G1S} = -10 Vdc, V_{G2S} = 0, V_{DS} = 0) (V_{G1S} = -35 Vdc, V_{G2S} = 0, V_{DS} = 0)		I_{G1SS}	– –	– –	1.0 10	nAdc
Gate 2 Reverse Leakage Current (V_{G2S} = -10 Vdc, V_{G1S} = 0, V_{DS} = 0) (V_{G2S} = -35 Vdc, V_{G1S} = 0, V_{DS} = 0)		I_{G2SS}	– –	– –	1.0 10	nAdc
ON CHARACTERISTICS						
Zero-Gate Voltage Drain Current (V_{DS} = 15 Vdc, V_{G1S} = 0, V_{G2S} = 4.0 Vdc)	MFE3006 MFE3007 MFE3008	I_{DSS}	2.0 5.0 2.0	7.0 10 9.0	18 20 20	mAdc
SMALL-SIGNAL CHARACTERISTICS						
Forward Transadmittance (Gate 1 to Drain) (V_{DS} = 15 Vdc, V_{G2S} = 4.0 Vdc, I_D = 10 mAdc, f = 1.0 kHz)	MFE3006/8 MFE3007	y_{fs}	8000 10,000	– –	18,000 18,000	μmhos
Input Capacitance (V_{DS} = 15 Vdc, V_{G2S} = 4.0 Vdc, I_D = 10 mAdc, f = 1.0 MHz)	MFE3006/8 MFE3007	C_{iss}	– –	4.5 4.5	6.0 5.5	pF
Output Capacitance (V_{DS} = 15 Vdc, V_{G2S} = 4.0 Vdc, I_D = 10 mAdc, f = 1.0 MHz)	MFE3006/8 MFE3007	C_{oss}	– –	2.5 2.5	4.0 3.5	pF
Reverse Transfer Capacitance (V_{DS} = 15 Vdc, V_{G2S} = 4.0 Vdc, I_D = 10 mAdc, f = 1.0 MHz)		C_{rss}	–	0.02	–	pF
Common-Source Noise Figure (V_{DS} = 15 Vdc, V_{G2S} = 4.0 Vdc, I_D = 10 mAdc, R_S = 1000 Ohms) f = 100 MHz, Figure 1 f = 200 MHz, Figure 4	MFE3006 MFE3007	NF	– –	2.5 3.0	4.0 4.0	dB
Common-Source Power Gain (V_{DS} = 15 Vdc, V_{G2S} = 4.0 Vdc, I_D = 10 mAdc) f = 100 MHz, Figure 1 f = 200 MHz, Figure 4	MFE3006 MFE3007	G_{ps}	20 18	25 21	– –	dB
Level of Unwanted Signal for 1.0% Cross Modulation (V_{DS} = 15 Vdc, V_{G2S} = 4.0 Vdc, I_D = 10 mAdc)		–	–	100	–	mV
Common-Source Conversion Power Gain (V_{DS} = 15 Vdc, V_{G2S} = 0.5 Vdc, Local Oscillator Voltage = 3.0 Vrms) Signal Frequency = 100 MHz, Local Oscillator Frequency = 130 MHz, Figure 3 Signal Frequency = 200 MHz, Local Oscillator Frequency = 230 MHz, Figure 6	MFE3008 MFE3008	G_{ps}	14 10	17 13	– –	dB

TYPES TIP29, TIP29A, TIP29B, TIP29C
N-P-N SINGLE-DIFFUSED MESA SILICON POWER TRANSISTORS

TYPES TIP29, TIP29A, TIP29B, TIP29C
BULLETIN NO. DL-S 7011371, OCTOBER 1970
REPLACES BULLETIN NO. DL-S 6810954, JULY 1968

FOR POWER-AMPLIFIER AND HIGH-SPEED-SWITCHING APPLICATIONS
DESIGNED FOR COMPLEMENTARY USE WITH TIP30, TIP30A, TIP30B, TIP30C

- 30 W at 25°C Case Temperature
- 1 A Rated Collector Current
- Min f_T of 3 MHz at 10 V, 200 mA

mechanical data

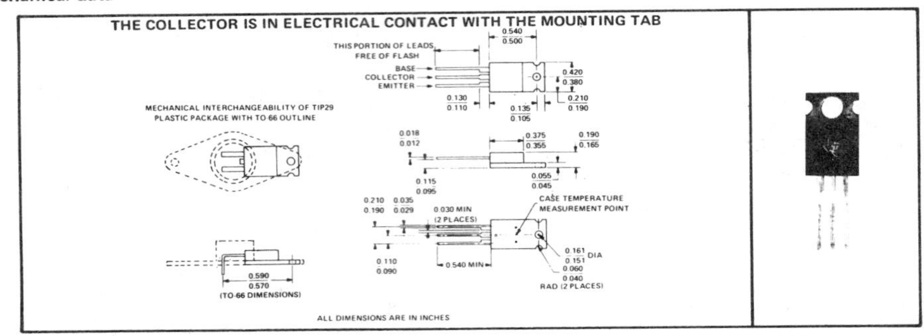

absolute maximum ratings at 25°C case temperature (unless otherwise noted)

	TIP29	TIP29A	TIP29B	TIP29C
Collector-Base Voltage	40 V	60 V	80 V	100 V
Collector-Emitter Voltage (See Note 1)	40 V	60 V	80 V	100 V
Emitter-Base Voltage	← 5 V →			
Continuous Collector Current	← 1 A →			
Peak Collector Current (See Note 2)	← 3 A →			
Continuous Base Current	← 0.4 A →			
Safe Operating Region at (or below) 25°C Case Temperature	← See Figure 5 →			
Continuous Device Dissipation at (or below) 25°C Case Temperature (See Note 3)	← 30 W →			
Continuous Device Dissipation at (or below) 25°C Free-Air Temperature (See Note 4)	← 2 W →			
Unclamped Inductive Load Energy (See Note 5)	← 32 mJ →			
Operating Collector Junction Temperature Range	← −65°C to 150°C →			
Storage Temperature Range	← −65°C to 150°C →			
Lead Temperature 1/8 Inch from Case for 10 Seconds	← 260°C →			

NOTES: 1. This value applies when the base-emitter diode is open-circuited.
2. This value applies for $t_w \leq 0.3$ ms, duty cycle $\leq 10\%$.
3. Derate linearly to 150°C case temperature at the rate of 0.24 W/°C.
4. Derate linearly to 150°C free-air temperature at the rate of 16 mW/°C.
5. This rating is based on the capability of the transistor to operate safely in the circuit of Figure 2. $L = 20$ mH, $R_{BB1} = 100\ \Omega$, $V_{BB2} = 0$ V, $R_S = 0.1\ \Omega$, $V_{CC} = 10$ V. Energy $\approx I_C{}^2L/2$.

TEXAS INSTRUMENTS
INCORPORATED
POST OFFICE BOX 5012 • DALLAS, TEXAS 75222

electrical characteristics at 25°C case temperature

PARAMETER		TEST CONDITIONS		TIP29 MIN	TIP29 MAX	TIP29A MIN	TIP29A MAX	TIP29B MIN	TIP29B MAX	TIP29C MIN	TIP29C MAX	UNIT
$V_{(BR)CEO}$	Collector-Emitter Breakdown Voltage	I_C = 30 mA, See Note 6	I_B = 0,	40		60		80		100		V
I_{CEO}	Collector Cutoff Current	V_{CE} = 30 V,	I_B = 0		0.3		0.3					mA
		V_{CE} = 60 V,	I_B = 0						0.3		0.3	
I_{CES}	Collector Cutoff Current	V_{CE} = 40 V,	V_{BE} = 0		0.2							mA
		V_{CE} = 60 V,	V_{BE} = 0				0.2					
		V_{CE} = 80 V,	V_{BE} = 0						0.2			
		V_{CE} = 100 V,	V_{BE} = 0								0.2	
I_{EBO}	Emitter Cutoff Current	V_{EB} = 5 V,	I_C = 0		1		1		1		1	mA
h_{FE}	Static Forward Current Transfer Ratio	V_{CE} = 4 V, See Notes 6 and 7	I_C = 0.2 A,	40		40		40		40		
		V_{CE} = 4 V, See Notes 6 and 7	I_C = 1 A,	15	75	15	75	15	75	15	75	
V_{BE}	Base-Emitter Voltage	V_{CE} = 4 V, See Notes 6 and 7	I_C = 1 A,		1.3		1.3		1.3		1.3	V
$V_{CE(sat)}$	Collector-Emitter Saturation Voltage	I_B = 125 mA, See Notes 6 and 7	I_C = 1 A,		0.7		0.7		0.7		0.7	V
h_{fe}	Small-Signal Common-Emitter Forward Current Transfer Ratio	V_{CE} = 10 V, f = 1 kHz	I_C = 0.2 A,	20		20		20		20		
$\|h_{fe}\|$	Small-Signal Common-Emitter Forward Current Transfer Ratio	V_{CE} = 10 V, f = 1 MHz	I_C = 0.2 A,	3		3		3		3		

NOTES: 6. These parameters must be measured using pulse techniques. t_w = 300 μs, duty cycle ≤ 2%.
7. These parameters are measured with voltage-sensing contacts separate from the current-carrying contacts.

thermal characteristics

PARAMETER		MAX	UNIT
$R_{\theta JC}$	Junction-to-Case Thermal Resistance	4.17	°C/W
$R_{\theta JA}$	Junction-to-Free-Air Thermal Resistance	62.5	

switching characteristics at 25°C case temperature

PARAMETER		TEST CONDITIONS†	TYP	UNIT
t_{on}	Turn-On Time	I_C = 1 A, $I_{B(1)}$ = 100 mA, $I_{B(2)}$ = −100 mA,	0.5	μs
t_{off}	Turn-Off Time	$V_{BE(off)}$ = −4.3 V, R_L = 30 Ω, See Figure 1	2	

† Voltage and current values shown are nominal; exact values vary slightly with transistor parameters.

TYPES TIP30, TIP30A, TIP30B, TIP30C
BULLETIN NO. DL-S 7011401, DECEMBER 1970
REPLACES BULLETIN NO. DL-S 6810956, JULY 1968

FOR POWER-AMPLIFIER AND HIGH-SPEED-SWITCHING APPLICATIONS
DESIGNED FOR COMPLEMENTARY USE WITH TIP29, TIP29A, TIP29B, TIP29C

- 30 W at 25°C Case Temperature
- 1 A Rated Collector Current
- Min f_T of 3 MHz at 10 V, 200 mA

mechanical data

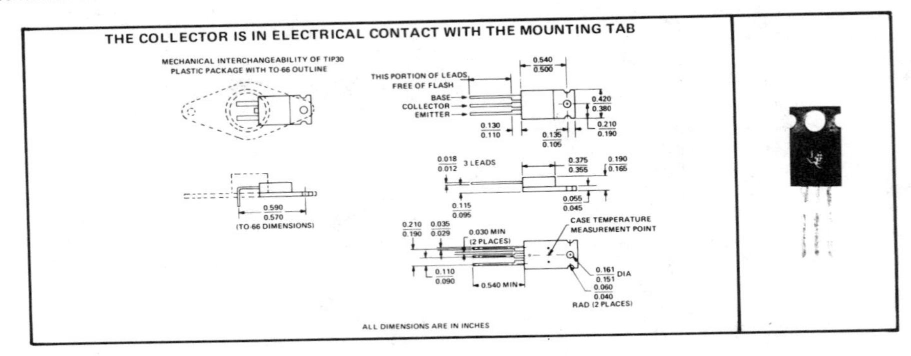

absolute maximum ratings at 25°C case temperature (unless otherwise noted)

	TIP30	TIP30A	TIP30B	TIP30C
Collector-Base Voltage	−40 V	−60 V	−80 V	−100 V
Collector-Emitter Voltage (See Note 1)	−40 V	−60 V	−80 V	−100 V
Emitter-Base Voltage	← −5 V →			
Continuous Collector Current	← −1 A →			
Peak Collector Current (See Note 2)	← −3 A →			
Continuous Base Current	← −0.4 A →			
Safe Operating Region at (or below) 25°C Case Temperature	← See Figure 5 →			
Continuous Device Dissipation at (or below) 25°C Case Temperature (See Note 3)	← 30 W →			
Continuous Device Dissipation at (or below) 25°C Free-Air Temperature (See Note 4)	← 2 W →			
Unclamped Inductive Load Energy (See Note 5)	← 32 mJ →			
Operating Collector Junction Temperature Range	← −65°C to 150°C →			
Storage Temperature Range	← −65°C to 150°C →			
Lead Temperature 1/8 Inch from Case for 10 Seconds	← 260°C →			

NOTES:
1. This value applies when the base-emitter diode is open-circuited.
2. This value applies for $t_w \leq 0.3$ ms, duty cycle $\leq 10\%$.
3. Derate linearly to 150°C case temperature at the rate of 0.24 W/°C.
4. Derate linearly to 150°C free-air temperature at the rate of 16 mW/°C.
5. This rating is based on the capability of the transistor to operate safely in the circuit of Figure 2. L = 20 mH, $R_{BB1} = 100\ \Omega$, $V_{BB2} = 0$ V, $R_S = 0.1\ \Omega$, $V_{CC} = 10$ V. Energy $\approx I_C{}^2L/2$.

electrical characteristics at 25°C case temperature

PARAMETER		TEST CONDITIONS	TIP30		TIP30A		TIP30B		TIP30C		UNIT
			MIN	MAX	MIN	MAX	MIN	MAX	MIN	MAX	
$V_{(BR)CEO}$	Collector-Emitter Breakdown Voltage	$I_C = -30$ mA, $I_B = 0$, See Note 6	−40		−60		−80		−100		V
I_{CEO}	Collector Cutoff Current	$V_{CE} = -30$ V, $I_B = 0$		−0.3		−0.3					mA
		$V_{CE} = -60$ V, $I_B = 0$						−0.3		−0.3	
I_{CES}	Collector Cutoff Current	$V_{CE} = -40$ V, $V_{BE} = 0$		−0.2							mA
		$V_{CE} = -60$ V, $V_{BE} = 0$				−0.2					
		$V_{CE} = -80$ V, $V_{BE} = 0$						−0.2			
		$V_{CE} = -100$ V, $V_{BE} = 0$								−0.2	
I_{EBO}	Emitter Cutoff Current	$V_{EB} = -5$ V, $I_C = 0$		−1		−1		−1		−1	mA
h_{FE}	Static Forward Current Transfer Ratio	$V_{CE} = -4$ V, $I_C = -0.2$ A, See Notes 6 and 7	40		40		40		40		
		$V_{CE} = -4$ V, $I_C = -1$ A, See Notes 6 and 7	15	75	15	75	15	75	15	75	
V_{BE}	Base-Emitter Voltage	$V_{CE} = -4$ V, $I_C = -1$ A, See Notes 6 and 7		−1.3		−1.3		−1.3		−1.3	V
$V_{CE(sat)}$	Collector-Emitter Saturation Voltage	$I_B = -125$ mA, $I_C = -1$ A, See Notes 6 and 7		−0.7		−0.7		−0.7		−0.7	V
h_{fe}	Small-Signal Common-Emitter Forward Current Transfer Ratio	$V_{CE} = -10$ V, $I_C = -0.2$ A, f = 1 kHz	20		20		20		20		
$\|h_{fe}\|$	Small-Signal Common-Emitter Forward Current Transfer Ratio	$V_{CE} = -10$ V, $I_C = -0.2$ A, f = 1 MHz	3		3		3		3		

NOTES: 6. These parameters must be measured using pulse techniques. $t_w = 300$ μs, duty cycle ≤ 2%.
7. These parameters are measured with voltage-sensing contacts separate from the current-carrying contacts.

thermal characteristics

PARAMETER		MAX	UNIT
$R_{\theta JC}$	Junction-to-Case Thermal Resistance	4.17	°C/W
$R_{\theta JA}$	Junction-to-Free-Air Thermal Resistance	62.5	

switching characteristics at 25°C case temperature

PARAMETER		TEST CONDITIONS†	TYP	UNIT
t_{on}	Turn-On Time	$I_C = -1$ A, $I_{B(1)} = -100$ mA, $I_{B(2)} = 100$ mA, $V_{BE(off)} = 4.3$ V, $R_L = 30\ \Omega$, See Figure 1	0.3	μs
t_{off}	Turn-Off Time		1.0	

† Voltage and current values shown are nominal; exact values vary slightly with transistor parameters.

2N6027
2N6028

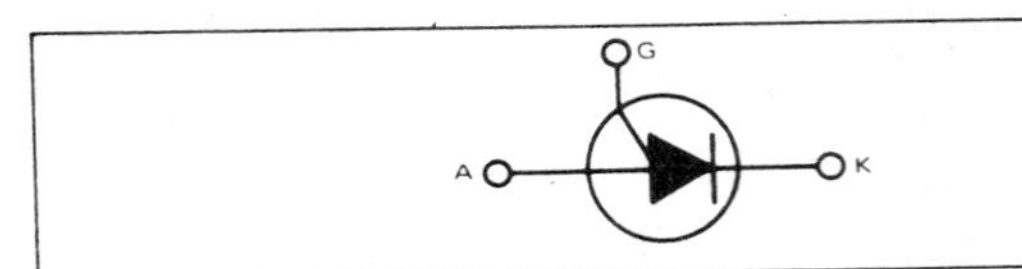

SILICON PROGRAMMABLE UNIJUNCTION TRANSISTORS

40 VOLTS
375 mW

MAY 1971 – DS 2520

SILICON PROGRAMMABLE UNIJUNCTION TRANSISTORS

. . . designed to enable the engineer to "program" unijunction characteristics such as R_{BB}, η, I_V, and I_P by merely selecting two resistor values. Application includes thyristor-trigger, oscillator, pulse and timing circuits. These devices may also be used in special thyristor applications due to the availability of an anode gate. Supplied in an inexpensive TO-92 plastic package for high-volume requirements, this package is readily adaptable for use in automatic insertion equipment.

- Programmable – R_{BB}, η, I_V and I_P.
- Low On-State Voltage – 1.5 Volts Maximum @ I_F = 50 mA
- Low Gate to Anode Leakage Current – 10 nA Maximum
- High Peak Output Voltage – 11 Volts Typical
- Low Offset Voltage – 0.35 Volt Typical (R_G = 10 k ohms)

A
G
K

MAXIMUM RATINGS

Rating	Symbol	Value	Unit
Power Dissipation (1) Derate Above 25°C	P_F $1/\theta_{JA}$	375 5.0	mW mW/°C
DC Forward Anode Current (2) Derate Above 25°C	I_T	200 2.67	mA mA/°C
*DC Gate Current	I_G	±50	mA
Repetitive Peak Forward Current 100 μs Pulse Width, 1.0% Duty Cycle *20 μs Pulse Width, 1.0% Duty Cycle	I_{TRM}	 1.0 2.0	 Amp Amp
Non-Repetitive Peak Forward Current 10 μs Pulse Width	I_{TSM}	5.0	Amp
*Gate to Cathode Forward Voltage	V_{GKF}	40	Volt
*Gate to Cathode Reverse Voltage	V_{GKR}	-5.0	Volt
*Gate to Anode Reverse Voltage	V_{GAR}	40	Volt
*Anode to Cathode Voltage	V_{AK}	±40	Volt
Operating Junction Temperature Range	T_J	-50 to +100	°C
*Storage Temperature Range	T_{stg}	-55 to +150	°C

*Indicates JEDEC Registered Data
(1) JEDEC Registered Data is 300 mW, derating at 4.0 mW/°C.
(2) JEDEC Registered Data is 150 mA.

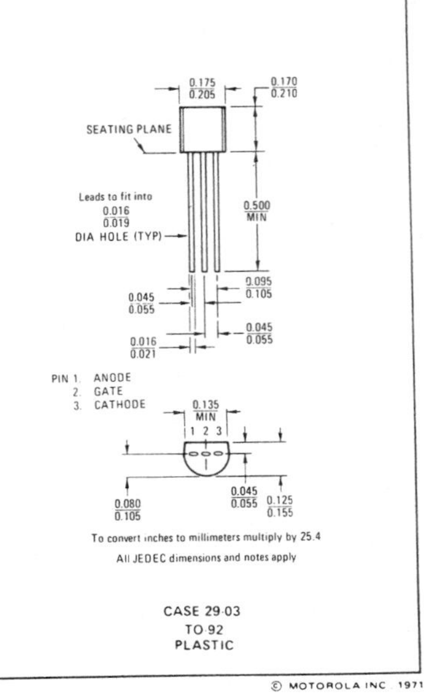

CASE 29-03
TO-92
PLASTIC

ELECTRICAL CHARACTERISTICS (T_A = 25°C unless otherwise noted)

Characteristic		Figure	Symbol	Min	Typ	Max	Unit
*Peak Current		2,9,11	I_P				μA
(V_S = 10 Vdc, R_G = 1.0 MΩ)	2N6027			–	1.25	2.0	
	2N6028			–	0.08	0.15	
(V_S = 10 Vdc, R_G = 10 k ohms)	2N6027			–	4.0	5.0	
	2N6028			–	0.70	1.0	
*Offset Voltage		1	V_T				Volts
(V_S = 10 Vdc, R_G = 1.0 MΩ)	2N6027			0.2	0.70	1.6	
	2N6028			0.2	0.50	0.6	
(V_S = 10 Vdc, R_G = 10 k ohms)	(Both Types)			0.2	0.35	0.6	
*Valley Current		1,4,5,	I_V				μA
(V_S = 10 Vdc, R_G = 1.0 MΩ)	2N6027			–	18	50	
	2N6028			–	18	25	
(V_S = 10 Vdc, R_G = 10 k ohms)	2N6027			70	270	–	
	2N6028			25	270	–	
(V_S = 10 Vdc, R_G = 200 Ohms)	2N6027			1.5	–	–	mA
	2N6028			1.0	–	–	
*Gate to Anode Leakage Current		–	I_{GAO}				nAdc
(V_S = 40 Vdc, T_A = 25°C, Cathode Open)				–	1.0	10	
(V_S = 40 Vdc, T_A = 75°C, Cathode Open)				–	3.0	–	
Gate to Cathode Leakage Current (V_S = 40 Vdc, Anode to Cathode Shorted)		–	I_{GKS}	–	5.0	50	nAdc
*Forward Voltage (I_F = 50 mA Peak)		1,6	V_F	–	0.8	1.5	Volts
*Peak Output Voltage (V_B = 20 Vdc, C_C = 0.2 μF)		3,7	V_o	6.0	11	–	Volts
Pulse Voltage Rise Time (V_B = 20 Vdc, C_C = 0.2 μF)		3	t_r	–	40	80	ns

*Indicates JEDEC Registered Data

FIGURE 1 – ELECTRICAL CHARACTERIZATION

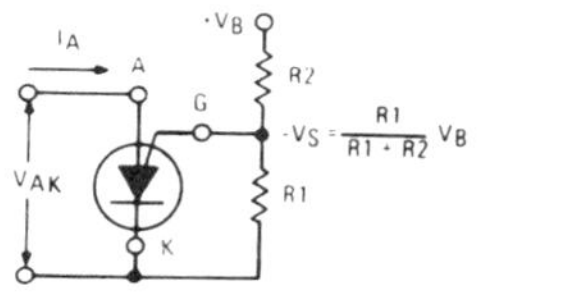

1A PROGRAMMABLE UNIJUNCTION WITH "PROGRAM" RESISTORS R1 and R2

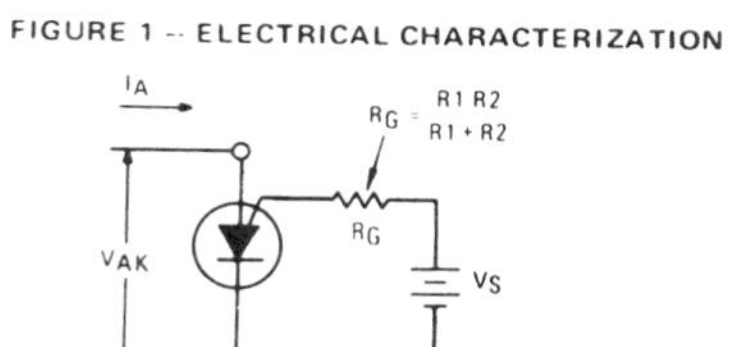

1B EQUIVALENT TEST CIRCUIT FOR FIGURE 1A USED FOR ELECTRICAL CHARACTERISTICS TESTING (ALSO SEE FIGURE 2)

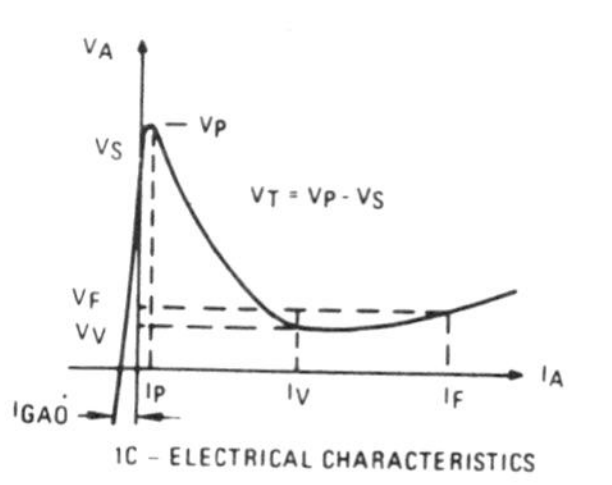

1C – ELECTRICAL CHARACTERISTICS

FIGURE 2 – PEAK CURRENT (I_P) TEST CIRCUIT

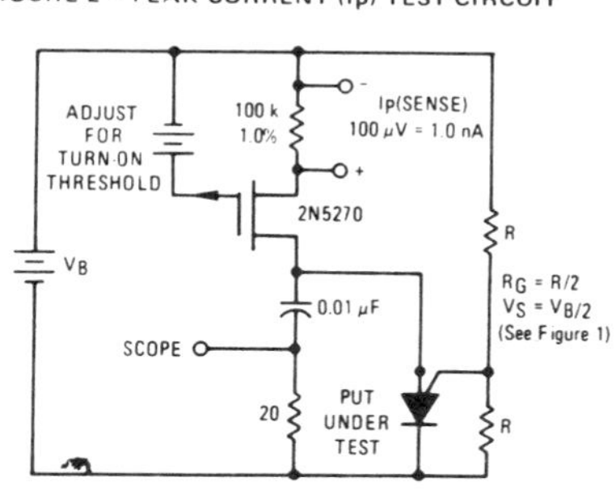

FIGURE 3 – V_o AND t_r TEST CIRCUIT

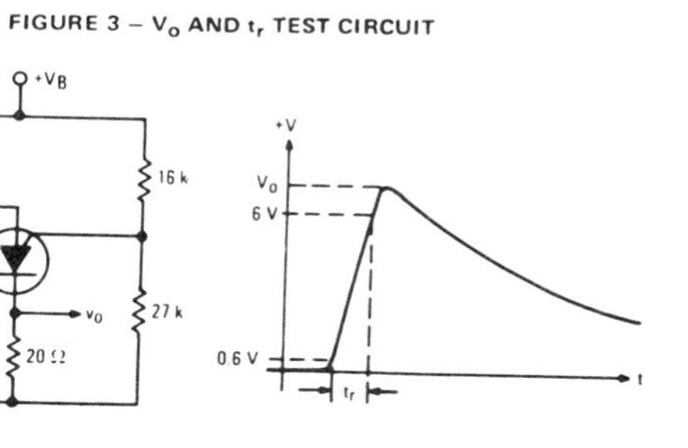

MOTOROLA
Semiconductors
BOX 20912 • PHOENIX, ARIZONA 85036

MPT20

PLASTIC SILICON
ANNULAR
BILATERAL TRIGGER

APRIL 1968 – DS 6518

PLASTIC SILICON 3-LAYER BILATERAL TRIGGER

. . . annular*, two-terminal devices that exhibit bi-directional negative resistance switching characteristics. These economical, durable devices have been developed for use in thyristor triggering circuits for lamp drivers and universal motor speed controls.

- Low Switching Voltage – $V_{(BR)12}$ and $V_{(BR)21}$ = 20 V (Typ)
- Large Switchback Voltages – ΔV_{12} and ΔV_{21} = 7.0 V (Typ)
- Passivated Surface for Reliability and Uniformity

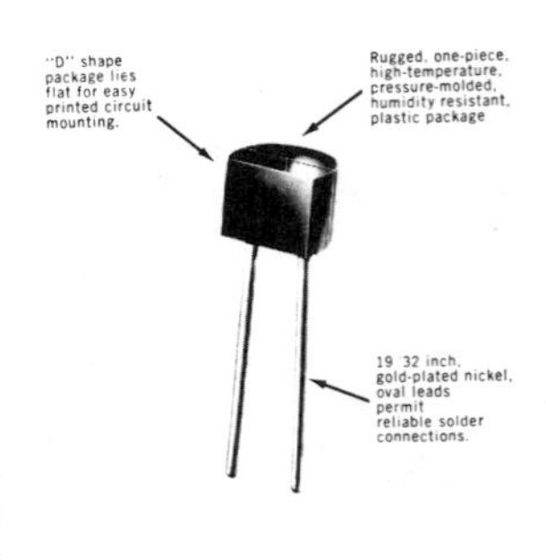

MAXIMUM RATINGS (T_A = 25°C unless otherwise noted)

Rating	Symbol	Value	Unit
Peak Pulse Current (30 μs duration, 120 Hz repetition rate)	I_{pulse}	2.0	Amp
Power Dissipation @ T_A = -40 to +25°C Derate above 25°C	P_D	300 4.0	mW mW/°C
Operating Junction Temperature Range	T_J	-40 to +100	°C
Storage Temperature Range	T_{stg}	-40 to +150	°C

ELECTRICAL CHARACTERISTICS (T_A = 25°C unless otherwise noted)

Characteristic	Symbol	Min	Typ	Max	Unit
Breakover (Switching) Voltage - both directions	$V_{(BR)12}$ & $V_{(BR)21}$	16	20	24	Volt
Breakover (Switching) Current - both directions	$I_{(BR)12}$ & $I_{(BR)21}$	-	35	100	μAmp
Switchback (Delta) Voltage - both directions ($I_{12} = I_{21}$ = 10 mAdc)	ΔV_{12} & ΔV_{21}	5.0	7.0	-	Volt
Peak Blocking Current - both directions Voltage Applied = 14 V	$I_{(BL)12}$ & $I_{(BL)21}$	-	0.5	10	μA

† Motorola Plastic Trigger (MPT) devices have bi-directional characteristics and as such the terminal leads are interchangeable. For purposes of symbol clarification, the leads have arbitrarily been designated 1 and 2. A 12 designation indicates that terminal 1 is positive with respect to terminal 2, vice versa for a 21 designation. (See Figure 1)

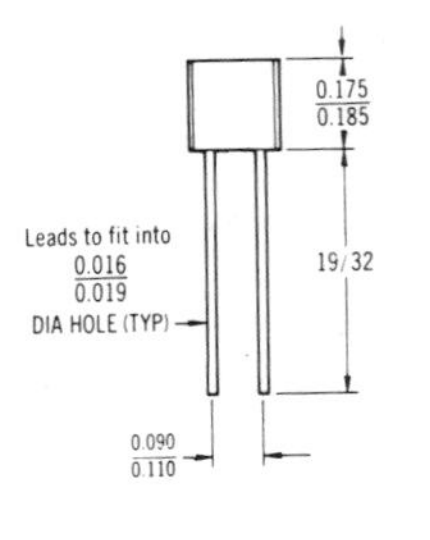

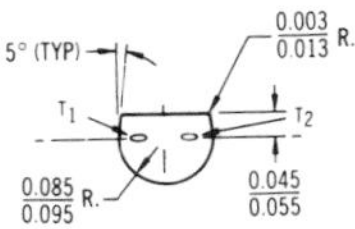

CASE 29B
(TO-92 Outline)

BILATERAL TRIGGER
MPT20
DS 6518

*Annular Semiconductors Patented by Motorola, Inc.

TYPICAL ELECTRICAL CHARACTERISTICS

FIGURE 1 – VOLT-AMPERE CHARACTERISTICS †

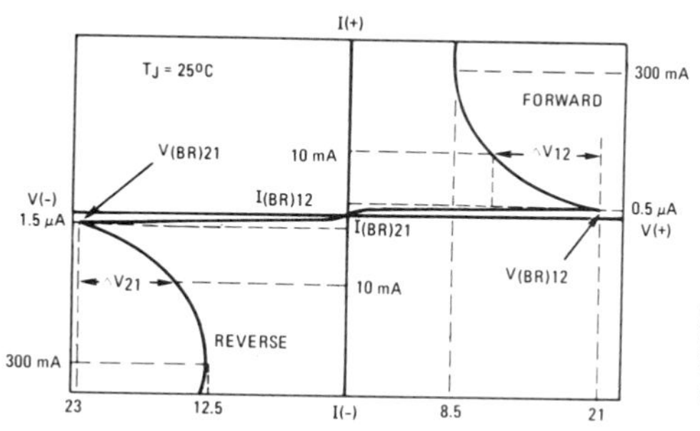

FIGURE 2 – INSTANTANEOUS "ON" VOLTAGE

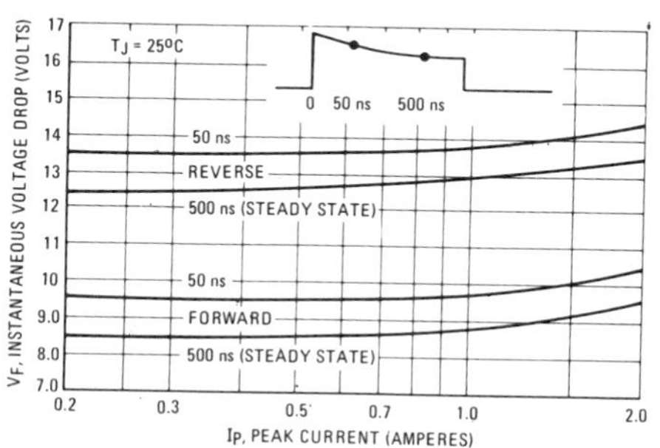

FIGURE 3 – BREAKOVER VOLTAGE BEHAVIOR

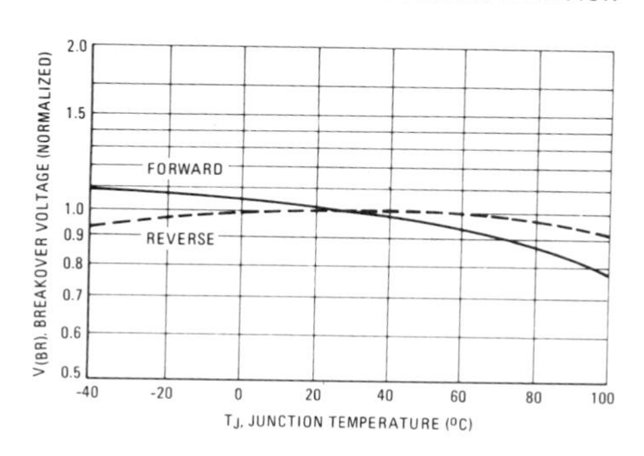

FIGURE 4 – NORMALIZED OUTPUT VOLTAGE BEHAVIOR

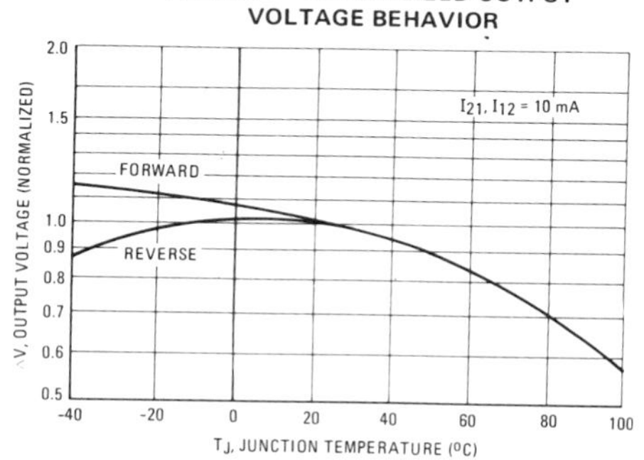

FIGURE 5 – SWITCHING TIMES

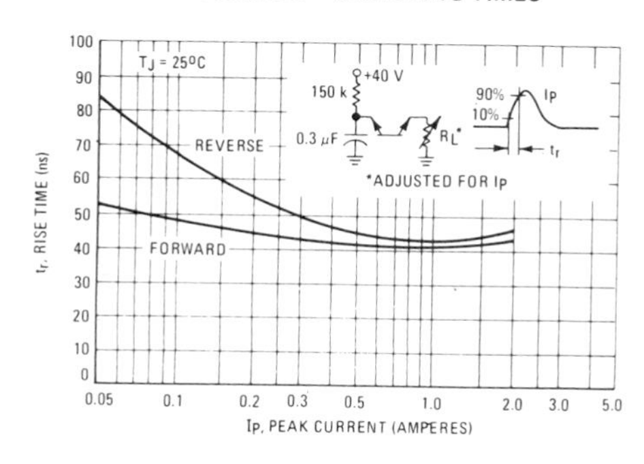

FIGURE 6 – CONTROL CIRCUIT

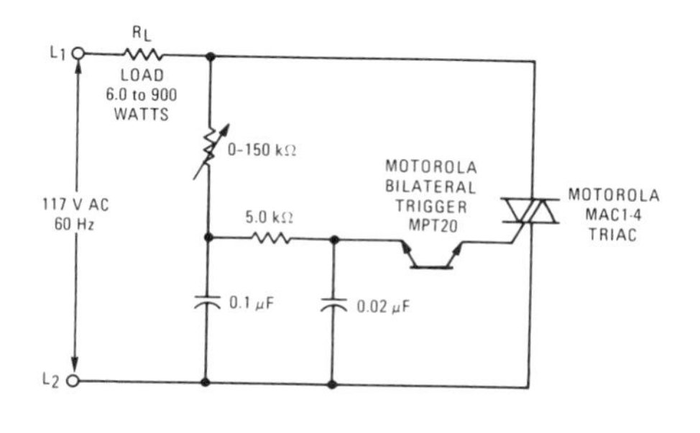

MOTOROLA Semiconductor Products Inc.

BOX 955 • PHOENIX, ARIZONA 85001 • A SUBSIDIARY OF MOTOROLA INC.

1N5158 thru 1N5160
(Formerly M4L3052 thru M4L3054)
1N5779 thru 1N5793

PNPN 4-LAYER DIODES

. . . two terminal, fast-switching devices specifically designed for low voltage applications such as logic circuits, pulse generators, memory and relay drivers, relay replacements, alarm circuits, multivibrators, ring counters, and telephone switching circuits. These devices feature:

- Low Breakover (Switching) Voltage — 10 to 15-Volt Ratings
- Fast Switching Speeds — t_{on} = 75 ns (Typ)
  t_{off} = 250 ns (Typ)
- Low Junction Capacitance — 45 pF (Typ)
- Low Breakover Currents
- Subminiature Glass Package

EPITAXIAL 4-LAYER DIODES
10-15 VOLTS
150 mW

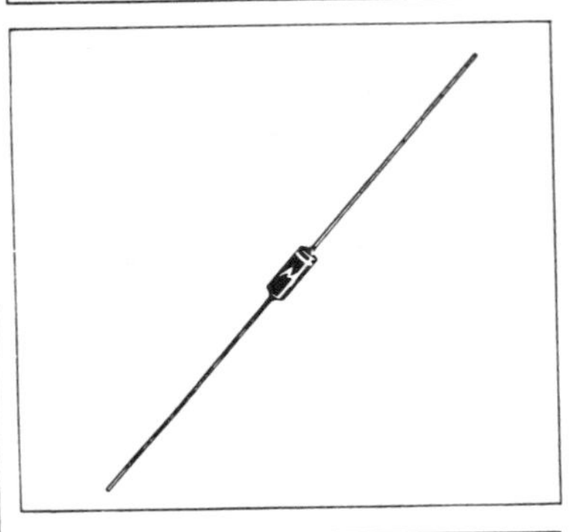

MAXIMUM RATINGS (T_A = 25°C unless otherwise noted)

Rating		Symbol	Value	Unit
•Reverse Voltage	1N5158, 1N5782, 1N5788	V_{RM}	10	Volts
	1N5159, 1N5783, 1N5789		11	
	1N5160, 1N5784, 1N5790		12	
	1N5779, 1N5785, 1N5791		13	
	1N5780, 1N5786, 1N5792		14	
	1N5781, 1N5787, 1N5793		15	
•Continuous Forward Current		I_F	150	mA
•Steady State Power Dissipation @ T_A = 50°C Derate above 50°C		P_D	150 1.5	mW mW/°C
•Peak Pulse Current (50 μs maximum pulse width)		I_{pulse}	10	Amps
•Operating Junction Temperature Range		T_J	-65 to +150	°C
Storage Temperature Range		T_{stg}	-65 to +175	°C

•Indicates JEDEC Registered Data.

FIGURE 1 — POWER-TEMPERATURE DERATING CURVE

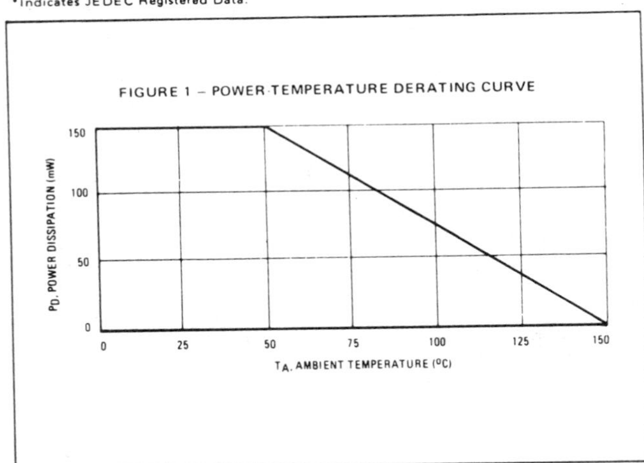

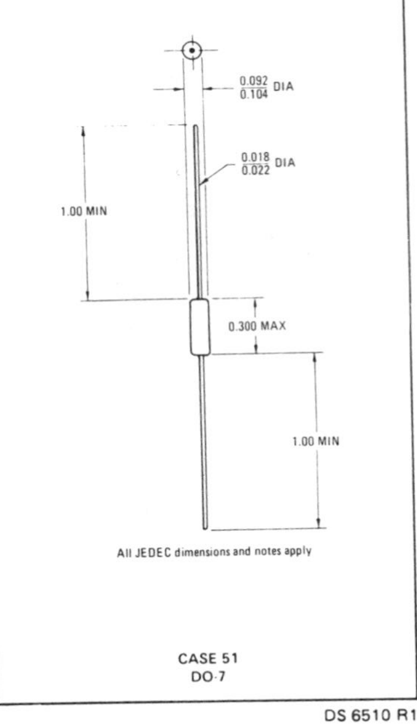

CASE 51
DO-7

DS 6510 R1

BOX 20912 • PHOENIX, ARIZONA 85036

2N5060 thru 2N5064

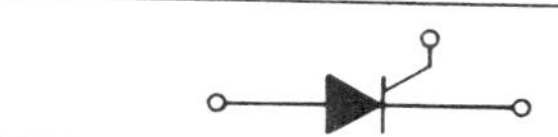

PLASTIC THYRISTORS

. . . Annular♦ PNPN devices designed for high volume consumer applications such as relay and lamp drivers, small motor controls, gate drivers for larger thyristors, and sensing and detection circuits. Supplied in an inexpensive plastic TO-92 package which is readily adaptable for use in automatic insertion equipment.

- Sensitive Gate Trigger Current – 200 μA Maximum
- Low Reverse and Forward Blocking Current – 50 μA Maximum, T_C = 125°C
- Low Holding Current – 5.0 mA Maximum
- Passivated Surface for Reliability and Uniformity

PLASTIC SILICON CONTROLLED RECTIFIERS

0.8 AMPERE RMS
30 thru 200 VOLTS

JUNE 1970 – DS 6525
(Replaces PS 41)

MAXIMUM RATINGS(1)

Rating	Symbol	Value	Unit
Peak Reverse Blocking Voltage	V_{RRM}		Volts
2N5060		30*	
2N5061		60*	
2N5062		100*	
2N5063		150*	
2N5064		200*	
Forward Current RMS (See Figures 4 & 5) (All Conduction Angles)	$I_{T(RMS)}$	0.8	Amp
Peak Forward Surge Current, T_A = 25°C (1/2 cycle, Sine Wave, 60 Hz)	I_{TSM}	6.0*	Amp
Circuit Fusing Considerations, T_A = 25°C (t = 1.0 to 8.3 ms)	I^2t	0.15	A^2s
Peak Gate Power – Forward, T_A = 25°C	P_{GM}	0.1*	Watt
Average Gate Power – Forward, T_A = 25°C	$P_{GF(AV)}$	0.01*	Watt
Peak Gate Current – Forward, T_A = 25°C (300 μs, 120 PPS)	I_{GFM}	1.0*	Amp
Peak Gate Voltage – Reverse	V_{GRM}	5.0*	Volts
Operating Junction Temperature Range @ Rated V_{RRM} and V_{DRM}	T_J	-65 to +125*	°C
Storage Temperature Range	T_{stg}	-65 to +150*	°C
Lead Solder Temperature (<1/16" from case, 10 s max)	–	+230*	°C

THERMAL CHARACTERISTICS

Characteristic	Symbol	Max	Unit
Thermal Resistance, Junction to Case	θ_{JC}	75	°C/W
Thermal Resistance, Junction to Ambient	θ_{JA}	200	°C/W

♦Annular Semiconductor Patented by Motorola Inc.
*Indicates JEDEC Registered Data.
(1) Temperature reference point for all case temperatures in center of flat portion of package. (T_C = +125°C unless otherwise noted.)

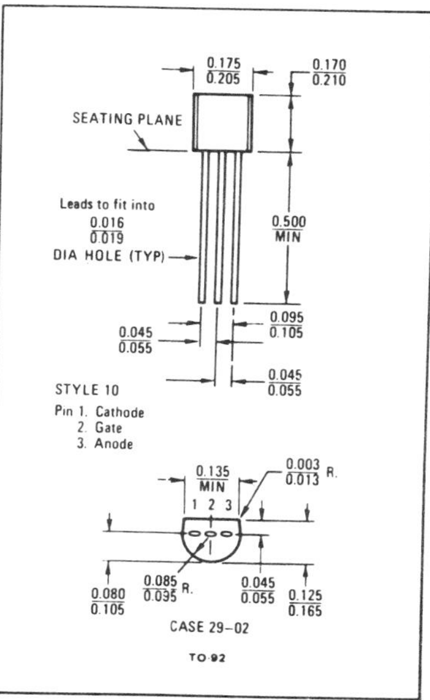

CASE 29-02

TO-92

ELECTRICAL CHARACTERISTICS (R_{GK} = 1000 Ohms)

Characteristic		Symbol	Min	Max	Unit
Peak Forward Blocking Voltage (Note 1) (T_C = 125°C)	2N5060 2N5061 2N5062 2N5063 2N5064	V_{DRM}	30* 60* 100* 150* 200*	– – – – –	Volts
Peak Forward Blocking Current (Rated V_{DRM} @ T_C = 125°C)		I_{DRM}	–	50*	μA
Peak Reverse Blocking Current (Rated V_{RRM} @ T_C = 125°C)		I_{RRM}	–	50*	μA
Forward "On" Voltage (Note 2) (I_{TM} = 1.2 A peak @ T_A = 25°C)		V_{TM}	–	1.7*	Volts
Gate Trigger Current (Continuous dc) (Note 3) (Anode Voltage = 7.0 Vdc, R_L = 100 Ohms)	T_C = 25°C T_C = -65°C	I_{GT}	– –	200 350*	μA
Gate Trigger Voltage (Continuous dc) (Anode Voltage = 7.0 Vdc, R_L = 100 Ohms)	T_C = 25°C T_C = -65°C	V_{GT}	– –	0.8 1.2*	Volts
(Anode Voltage = Rated V_{DRM}, R_L = 100 Ohms)	T_C = 125°C	V_{GD}	0.1	–	
Holding Current (Anode Voltage = 7.0 Vdc, initiating current = 20 mA)	T_C = 25°C T_C = -65°C	I_H	– –	5.0 10*	mA
Thermal Resistance, Junction to Case (Note 4)		θ_{JC}	–	75*	°C/W
Thermal Resistance, Junction to Ambient		θ_{JA}	–	200	°C/W

*Indicates JEDEC Registered Data.

1. V_{DRM} and V_{RRM} for all types can be applied on a continuous dc basis without incurring damage. Ratings apply for zero or negative gate voltage but positive gate voltage shall not be applied concurrently with a negative potential on the anode. When checking forward or reverse blocking capability, thyristor devices should not be tested with a constant current source in a manner that the voltage applied exceeds the rated blocking voltage.
2. Forward current applied for 1.0 ms maximum duration, duty cycle ≤ 1.0%.
3. R_{GK} current is not included in measurement.
4. This measurement is made with the case mounted "flat side down" on a heat sink and held in position by means of a metal clamp over the curved surface.

FIGURE 1 – SURGE RATINGS

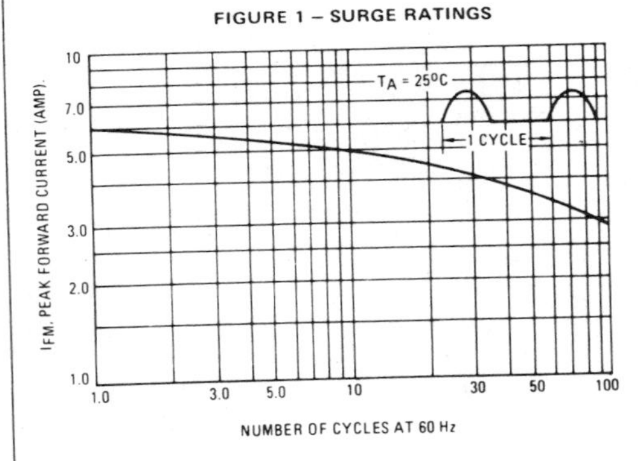

FIGURE 2 – POWER DISSIPATION

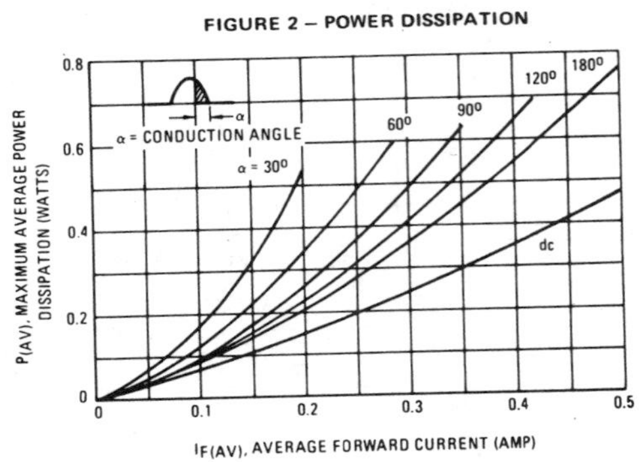

2N6068 thru 2N6075

(Formerly MAC77-1 thru MAC77-8)

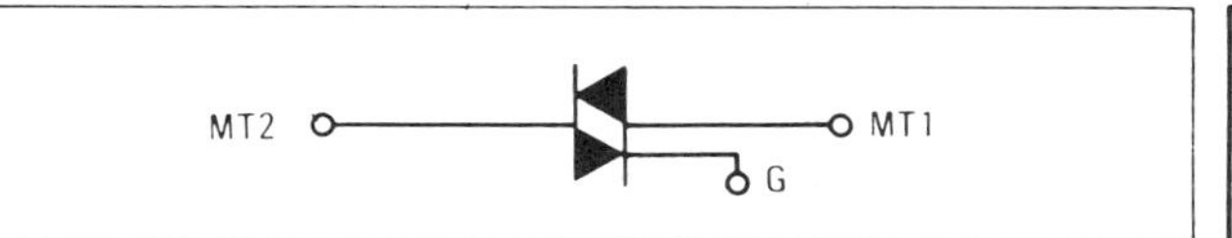

TRIACS
(THYRISTORS)
4 AMPERES RMS
25 THRU 600 VOLTS

DS 8544
(Replaces ADI-76)

SILICON BIDIRECTIONAL THYRISTORS

. . . designed primarily for full-wave ac control applications, such as light dimmers, motor controls, heating controls and power supplies; or wherever full-wave silicon gate controlled solid-state devices are needed. Triac type thyristors switch from a blocking to a conducting state for either polarity of applied anode voltage with positive or negative gate triggering. [MT2(+)G(+), MT2(-)G(-)]

- All Diffused and Passivated Junctions for Greater Parameter Uniformity and Stability
- Small, Rugged, Thermopad▲ Construction for Low Thermal Resistance, High Heat Dissipation and Durability

MAXIMUM RATINGS

Rating	Symbol	Value	Unit
*Repetitive Peak Off-State Voltage, Note 1 (T_J = 110°C)	V_{DRM}		Volts
2N6068		25	
2N6069		50	
2N6070		100	
2N6071		200	
2N6072		300	
2N6073		400	
2N6074		500	
2N6075		600	
*On-State Current RMS (T_C = 85°C)	$I_{T(RMS)}$	4.0	Amp
*Peak Surge Current (One Full cycle, 60 Hz, T_J = -40 to +110°C)	I_{TSM}	30	Amp
Circuit Fusing Considerations (T_J = -40 to +110°C, t = 1.0 to 8.3 ms)	I^2t	3.6	A^2s
*Peak Gate Power	P_{GM}	10	Watts
*Average Gate Power	$P_{G(AV)}$	0.5	Watt
*Peak Gate Voltage	V_{GM}	5.0	Volts
*Operating Junction Temperature Range	T_J	-40 to +110	°C
*Storage Temperature Range	T_{stg}	-40 to +150	°C
Mounting Torque (6-32 Screw) Note 2		8.0	in. lb.

THERMAL CHARACTERISTICS

Characteristic	Symbol	Max	Unit
*Thermal Resistance, Junction to Case	$R_{\theta JC}$	3.5	°C/W
Thermal Resistance, Case to Ambient	$R_{\theta CA}$	60	°C/W

*Indicates JEDEC Registered Data

NOTES:

1. Ratings apply for open gate conditions. Thyristor devices shall not be tested with a constant current source for blocking capability such that the voltage applied exceeds the rated blocking voltage.
2. Torque rating applies with use of torque washer (Shakeproof WD19523 or equivalent). Mounting torque in excess of 6 in. lb. does not appreciably lower case-to-sink thermal resistance. Main terminal 2 and heat-sink contact pad are common.

For soldering purposes (either terminal connection or device mounting), soldering temperatures shall not exceed +200°C.

▲Trademark of Motorola Inc.

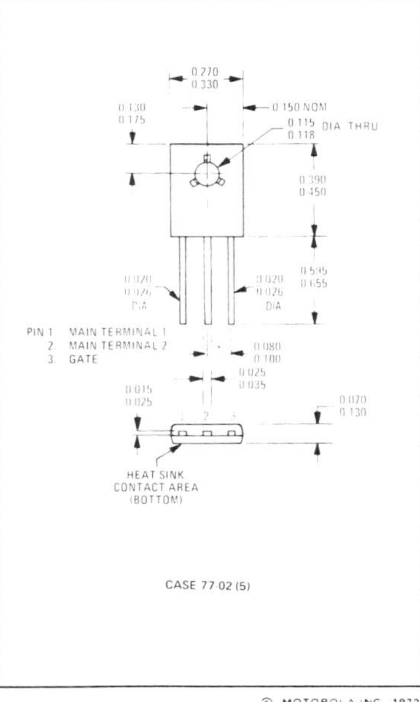

CASE 77-02 (5)

TYPE TIL01
P-N PLANAR GALLIUM ARSENIDE DIODE LIGHT SOURCE

TYPE TIL01
BULLETIN NO. DL-S 6810709, FEBRUARY 1968
REPLACES BULLETIN NO. DL-S 657272, FEBRUARY 1965

DESIGNED TO EMIT NEAR-INFRARED LIGHT WHEN FORWARD BIASED

- **Light Source Spectrally Matched to Silicon Sensors**
- **Recommended for Application in Character Recognition, Tape and Card Readers, and Encoders**
- **Unique Package Design Allows for Matrix Assembly Directly into Printed Circuit Boards**
- **Narrow Light Beam**

mechanical data

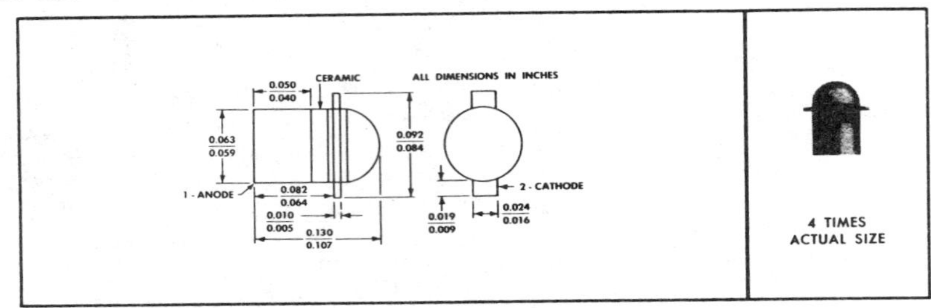

absolute maximum ratings

Reverse Voltage at 25°C Case Temperature . . . . 2 V
Continuous Forward Current at 25°C Case Temperature (See Note 1) . . . . 50 mA
Operating Case Temperature Range . . . . −65°C to 125°C
Storage Temperature Range . . . . −65°C to 150°C
Soldering Temperature (3 minutes) . . . . 240°C

operating characteristics at 25°C case temperature

	PARAMETER	TEST CONDITIONS	MIN	TYP	MAX	UNIT
λ_{peak}	Wavelength at Peak Emission	$I_F = 50$ mA		0.9		μm
P_O	Radiant Power Output Into 90° Circular Cone	$I_F = 50$ mA, See Note 2	50†			μW
P_O	Radiant Power Output Into 10° Circular Cone	$I_F = 50$ mA, See Note 3	3			μW
V_F	Static Forward Voltage	$I_F = 50$ mA			1.3	V

NOTES: 1. Derate linearly to 125°C case temperature at the rate of 0.5 mA/°C.
2. The radiant power output into a 90° right circular cone coaxial with the device axis of symmetry is measured by use of a calibrated silicon solar cell held in close proximity to the glass lens of the device.
3. The radiant power output into a 10° circular cone coaxial with the device axis of symmetry is measured by use of a calibrated silicon solar cell which is covered by a 0.175 cm aperture. The measurement is made with an aperture mounted coaxial with the device axis of symmetry and with the solar cell surface at a distance of 1.00 ± 0.05 cm from the device lens surface.

†This power is equivalent to approximately 2.25×10^{14} photons per second at 0.9 micron.

TEXAS INSTRUMENTS
INCORPORATED
POST OFFICE BOX 5012 • DALLAS, TEXAS 75222

TYPICAL CHARACTERISTICS

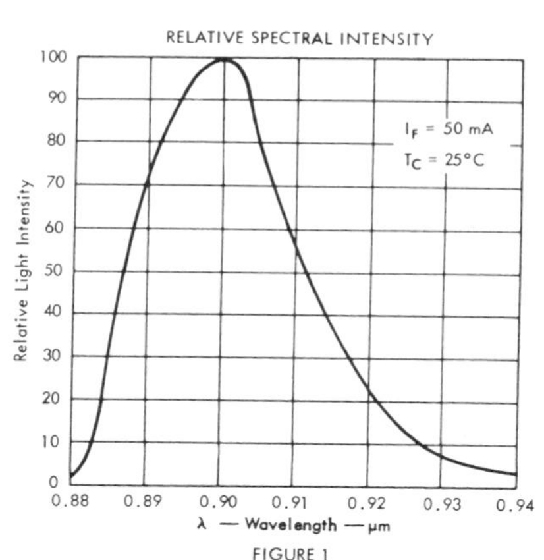

FIGURE 1

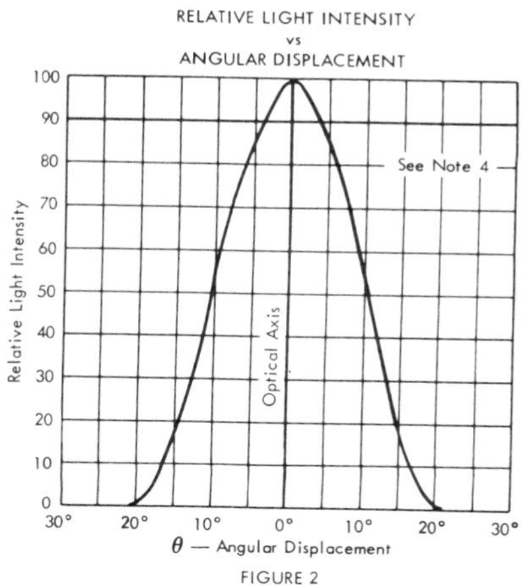

FIGURE 2

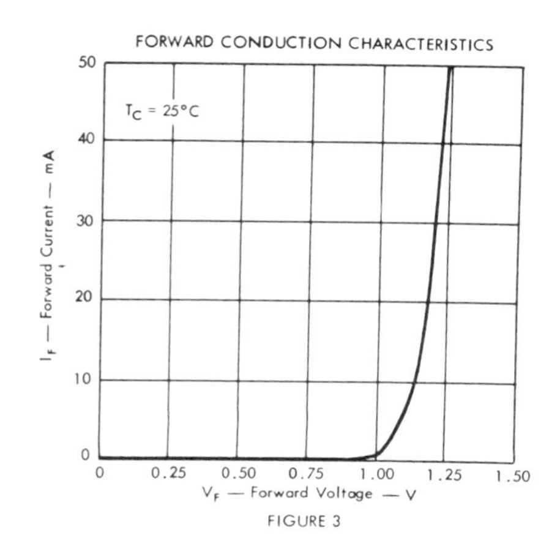

FIGURE 3

NOTE 4: The diagram for relative light intensity shows that most of the radiant output of the TIL01 is concentrated within a narrow circular cone. Radiant power output into various view angles can be measured by a small-diameter aperture and a calibrated solar cell. The solar cell, covered by the aperture, is held at a distance S from the light source, where S is related to the view angle θ and the aperture radius r by the following equation:

$$S = \frac{r}{\tan\left(\frac{\theta}{2}\right)}$$

PRINTED IN U.S.A.

TEXAS INSTRUMENTS
INCORPORATED
POST OFFICE BOX 5012 • DALLAS, TEXAS 75222

BOX 20912 • PHOENIX, ARIZONA 85036

MRD150

PLASTIC NPN SILICON PHOTO TRANSISTORS

. . . designed for application in punched card and tape readers, pattern and character recognition equipment, shaft encoders, industrial inspection processing and control, counters, sorters, switching and logic circuits, or any design requiring radiation sensitivity, stable characteristics and high-density mounting.

- Economical Plastic Package
- Sensitive Throughout Visible and Near Infra-Red Spectral Range **for Wide Application**
- Small Size for High-Density Mounting
- High Light Current Sensitivity (0.20 mA) for Design Flexibility
- Annular♦ Passivated Structure for Stability and Reliability

40 VOLT
MICRO-T
NPN SILICON
PHOTO TRANSISTOR

50 MILLIWATTS

CASE 173

MAXIMUM RATINGS

Rating	Symbol	Value	Unit
Collector-Emitter Voltage	V_{CEO}	40	Volts
Emitter-Collector Voltage	V_{ECO}	6.0	Volts
Total Device Dissipation @ $T_A = 25^oC$ Derate above 25^oC	P_D	50 0.67	mW $mW/^oC$
Operating and Storage Junction Temperature Range	$T_J(1), T_{stg}$	-40 to +100	^oC

(1) Heat Sink should be applied to leads during soldering to prevent Case Temperature from exceeding 85^oC.

FIGURE 1 – COLLECTOR-EMITTER SENSITIVITY

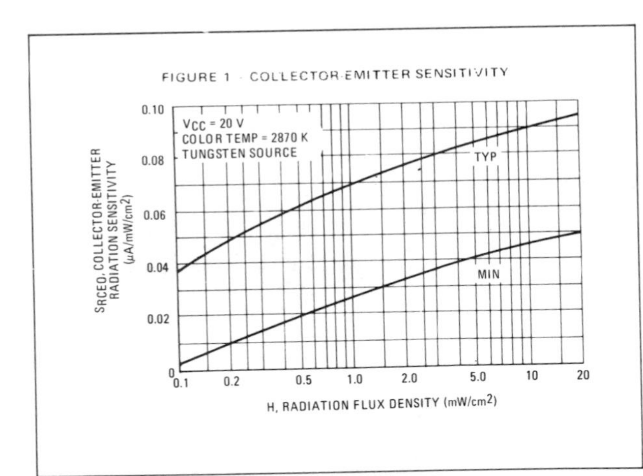

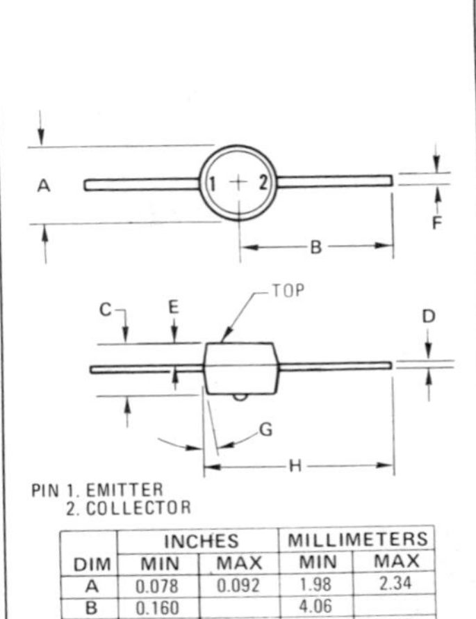

PIN 1. EMITTER
2. COLLECTOR

DIM	INCHES		MILLIMETERS	
	MIN	MAX	MIN	MAX
A	0.078	0.092	1.98	2.34
B	0.160		4.06	
C	0.048	0.058	1.22	1.47
D	0.004	0.006	0.102	0.152
E	0.020	0.030	0.508	0.762
F	0.010	0.016	0.254	0.406
G	3°	7°	3°	7°
H	0.213 REF		5.41 REF	

CASE 173

♦Annular Semiconductors Patented by Motorola Inc.

DS 2605 R2

μA740
FET INPUT OPERATIONAL AMPLIFIER
FAIRCHILD LINEAR INTEGRATED CIRCUITS

FEATURES
- **HIGH INPUT IMPEDANCE . . . 1,000,000 MΩ**
- **NO FREQUENCY COMPENSATION REQUIRED**
- **SHORT-CIRCUIT PROTECTION**
- **OFFSET VOLTAGE NULL CAPABILITY**
- **LARGE COMMON-MODE AND DIFFERENTIAL VOLTAGE RANGES**
- **NO LATCH UP**

GENERAL DESCRIPTION — The μA740 is a high performance FET input operational amplifier constructed on a single silicon chip, using the Fairchild Planar® epitaxial process. It is intended for a wide range of analog applications where very high input impedance is required and features very low input offset current and very low input bias current. High slew rate, high common mode voltage range and absence of "latch up" make the μA740 ideal for use as a voltage follower. The high gain and wide range of operating voltages provide superior performance in active filters, integrators, summing amplifiers, sample and holds, transducer amplifiers, and other general feedback applications. The μA740 is short circuit protected and has the same pin configuration as the popular μA741 operational amplifier. No external components for frequency compensation are required as the internal 6 dB/octave roll-off insures stability in closed loop applications. (For other Fairchild operational amplifiers, see listing on back page.)

ABSOLUTE MAXIMUM RATINGS

Supply Voltage	±22 V
Internal Power Dissipation (Note 1)	500 mW
Differential Input Voltage	±30 V
Input Voltage (Note 2)	±15 V
Voltage between Offset Null and V+	±0.5 V
Storage Temperature Range	−65°C to +150°C
Operating Temperature Range	−55°C to +125°C
Lead Temperature (Soldering, 60 seconds)	300°C
Output Short-Circuit Duration (Note 3)	Indefinite

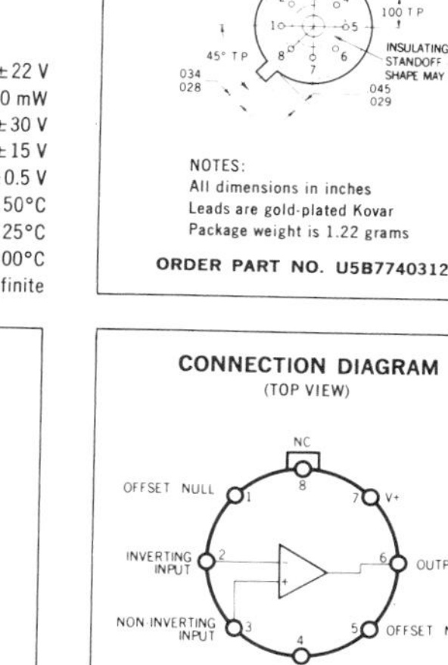

ORDER PART NO. U5B7740312

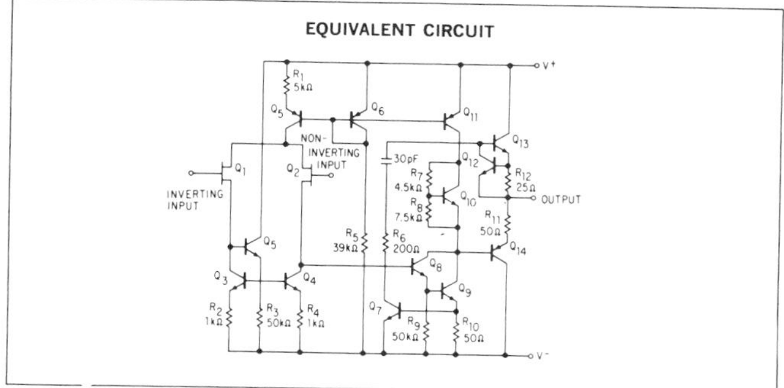

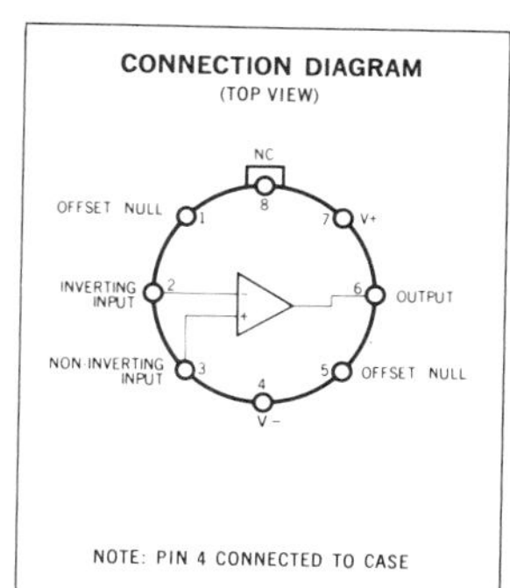

NOTES:
(1) Rating applies for case temperatures to 125°C; derate linearly at 6.5 mW/°C for ambient temperatures above +75°C.
(2) For supply voltages less than ±15 V, the absolute maximum input voltage is equal to the supply voltage.
(3) Short circuit may be to ground or either supply. Rating applies to +125°C case temperature or +75°C ambient temperature.

®Planar is a patented Fairchild process.

FAIRCHILD SEMICONDUCTOR
A DIVISION OF FAIRCHILD CAMERA AND INSTRUMENT CORPORATION

313 FAIRCHILD DRIVE, MOUNTAIN VIEW, CALIFORNIA ... TWX: 910-379-6435

ELECTRICAL CHARACTERISTICS ($V_S = \pm 15$ V, $T_C = 25°C$ unless otherwise specified)

PARAMETER	CONDITIONS	MIN.	TYP.	MAX.	UNITS
Input Offset Voltage	$R_S \leq 100$ kΩ		10	20	mV
Input Offset Current			40		pA
Input Current (either input)			100	200	pA
Input Resistance			1,000,000		MΩ
Large Signal Voltage Gain	$R_L \geq 2$ kΩ, $V_{out} = \pm 10$ V	50,000	1,000,000		
Output Resistance			75		Ω
Output Short-Circuit Current			20		mA
Common Mode Rejection Ratio		64	80		dB
Supply Voltage Rejection Ratio			70	300	μV/V
Supply Current			4.2	5.2	mA
Power Consumption			126	156	mW
Slew Rate			6.0		V/μs
Unity Gain Bandwidth			3.0		MHz
Transient Response (Unity Gain)	$C_L \leq 100$ pF, $R_L = 2$ kΩ, $V_{in} = 100$ mV				
Risetime			110		ns
Overshoot			10	20	%
The following specifications apply for $T_C = -55°C$ to $+85°C$:					
Input Voltage Range		±10		±12	V
Large Signal Voltage Gain		25,000			
Output Voltage Swing	$R_L \geq 10$ kΩ	±12	±14		V
	$R_L \geq 2$ kΩ	±10	±13		V
Input Offset Voltage	$R_S \leq 100$ kΩ		15	30	mV
Input Offset Current	$T_A = -55°C$		185		pA
	$T_A = +85°C$		30		pA
Input Current (either input)	$T_A = -55°C$			200	pA
	$T_A = +85°C$		2.5	4.0	nA

VOLTAGE OFFSET NULL CIRCUIT

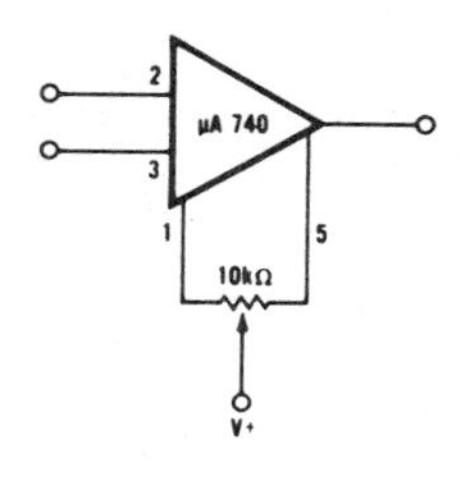

TRANSIENT RESPONSE TEST CIRCUIT

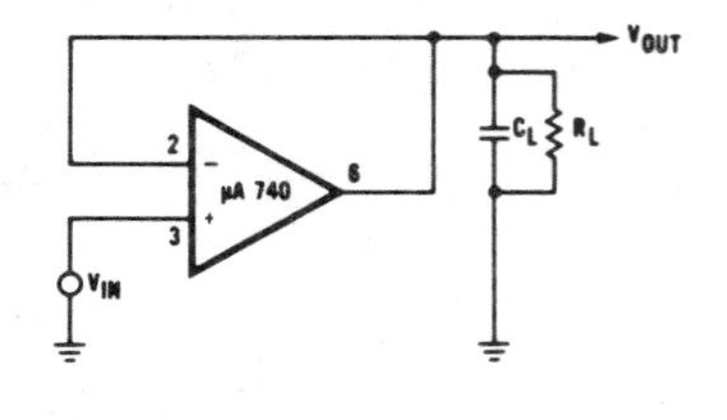

NOTES:

FOR OTHER FAIRCHILD OPERATIONAL AMPLIFIERS, SEE THE FOLLOWING DATA SHEETS:

HIGH SPEED — μA715, μA715C
INSTRUMENTATION — μA725, μA725B, μA725C
TEMPERATURE STABLE PREAMPS — μA727, μA727B
MICROPOWER — μA735, μA735B, μA735C
FREQUENCY COMPENSATED — μA741, μA741C
DUALS — μA739C, μA749, μA749C, μA747, μA747C
HIGH PERFORMANCE — μA748, μA748C
PRECISION — μA777, μA777B, μA777C

μA741

FREQUENCY-COMPENSATED OPERATIONAL AMPLIFIER

FAIRCHILD LINEAR INTEGRATED CIRCUITS

FEATURES:

- **NO FREQUENCY COMPENSATION REQUIRED**
- **SHORT-CIRCUIT PROTECTION**
- **OFFSET VOLTAGE NULL CAPABILITY**
- **LARGE COMMON-MODE AND DIFFERENTIAL VOLTAGE RANGES**
- **LOW POWER CONSUMPTION**
- **NO LATCH UP**

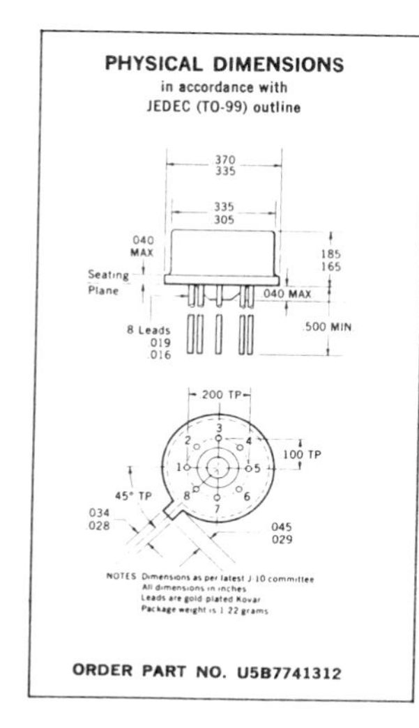

GENERAL DESCRIPTION — The μA741 is a high performance monolithic operational amplifier constructed on a single silicon chip, using the Fairchild Planar* epitaxial process. It is intended for a wide range of analog applications. High common mode voltage range and absence of "latch-up" tendencies make the μA741 ideal for use as a voltage follower. The high gain and wide range of operating voltage provides superior performance in integrator, summing amplifier, and general feedback applications. The μA741 is short-circuit protected, has the same pin configuration as the popular μA709 operational amplifier, but requires no external components for frequency compensation. The internal 6dB/octave roll-off insures stability in closed loop applications.

ABSOLUTE MAXIMUM RATINGS

Supply Voltage	±22 V
Internal Power Dissipation (Note 1)	500 mW
Differential Input Voltage	±30 V
Input Voltage (Note 2)	±15 V
Voltage between Offset Null and V−	±0.5 V
Storage Temperature Range	−65°C to +150°C
Operating Temperature Range	−55°C to +125°C
Lead Temperature (Soldering, 60 sec)	300°C
Output Short-Circuit Duration (Note 3)	Indefinite

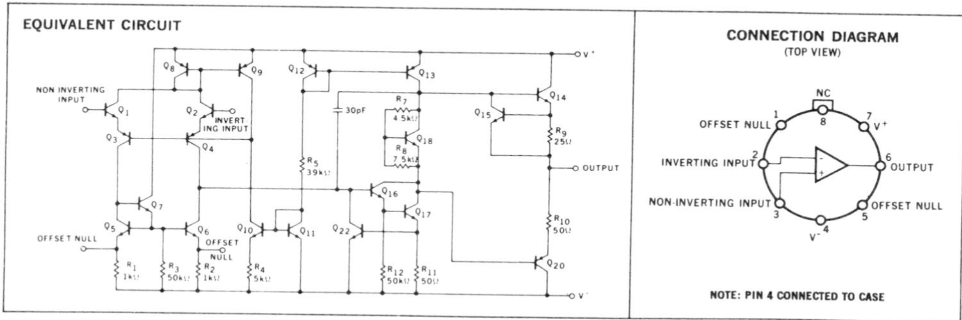

NOTES:

(1) Rating applies for case temperatures to 125°C; derate linearly at 6.5 mW/°C for ambient temperatures above +75°C.
(2) For supply voltages less than ±15 V, the absolute maximum input voltage is equal to the supply voltage.
(3) Short circuit may be to ground or either supply. Rating applies to +125°C case temperature or +75°C ambient temperature.

*Planar is a patented Fairchild process.

313 FAIRCHILD DRIVE, MOUNTAIN VIEW, CALIFORNIA, (415) 962-5011, TWX: 910-379-6435

FAIRCHILD
SEMICONDUCTOR
A DIVISION OF FAIRCHILD CAMERA AND INSTRUMENT CORPORATION

ELECTRICAL CHARACTERISTICS ($V_S = \pm 15$ V, $T_A = 25°C$ unless otherwise specified)

PARAMETERS (see definitions)	CONDITIONS	MIN.	TYP.	MAX.	UNITS
Input Offset Voltage	$R_S \leq 10$ kΩ		1.0	5.0	mV
Input Offset Current			20	200	nA
Input Bias Current			80	500	nA
Input Resistance		0.3	2.0		MΩ
Input Capacitance			1.4		pF
Offset Voltage Adjustment Range			±15		mV
Large-Signal Voltage Gain	$R_L \geq 2$ kΩ, $V_{out} = \pm 10$ V	50,000	200,000		
Output Resistance			75		Ω
Output Short-Circuit Current			25		mA
Supply Current			1.7	2.8	mA
Power Consumption			50	85	mW
Transient Response (unity gain)	$V_{in} = 20$ mV, $R_L = 2$ kΩ, $C_L \leq 100$ pF				
Risetime			0.3		μs
Overshoot			5.0		%
Slew Rate	$R_L \geq 2$ kΩ		0.5		V/μs
The following specifications apply for $-55°C \leq T_A \leq +125°C$:					
Input Offset Voltage	$R_S \leq 10$ kΩ		1.0	6.0	mV
Input Offset Current	$T_A = +125°C$		7.0	200	nA
	$T_A = -55°C$		85	500	nA
Input Bias Current	$T_A = +125°C$		0.03	0.5	μA
	$T_A = -55°C$		0.3	1.5	μA
Input Voltage Range		±12	±13		V
Common Mode Rejection Ratio	$R_S \leq 10$ kΩ	70	90		dB
Supply Voltage Rejection Ratio	$R_S \leq 10$ kΩ		30	150	μV/V
Large-Signal Voltage Gain	$R_L \geq 2$ kΩ, $V_{out} = \pm 10$ V	25,000			
Output Voltage Swing	$R_L \geq 10$ kΩ	±12	±14		V
	$R_L \geq 2$ kΩ	±10	±13		V
Supply Current	$T_A = +125°C$		1.5	2.5	mA
	$T_A = -55°C$		2.0	3.3	mA
Power Consumption	$T_A = +125°C$		45	75	mW
	$T_A = -55°C$		60	100	mW

TYPICAL PERFORMANCE CURVES

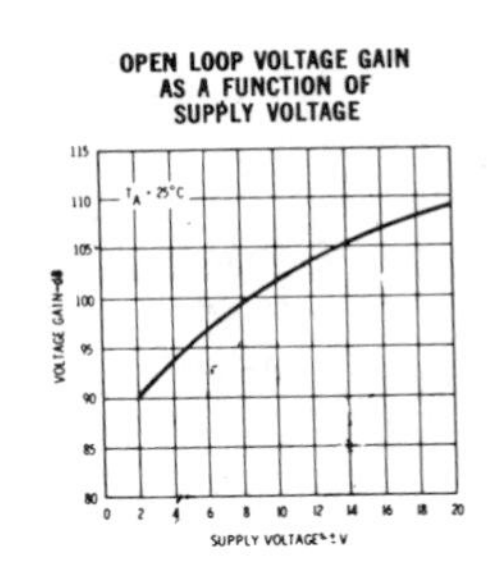

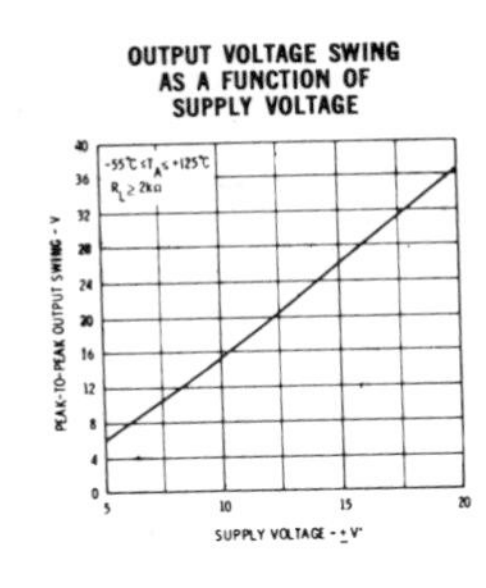

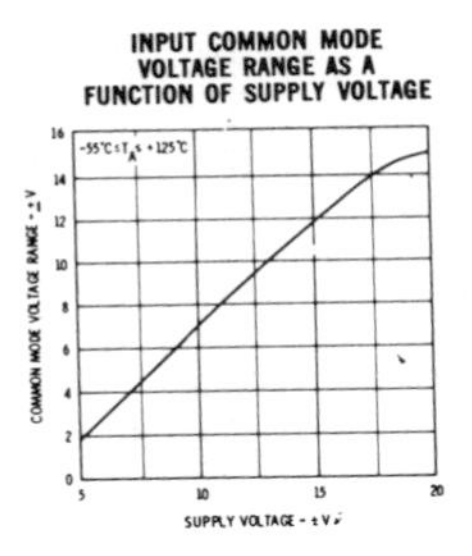

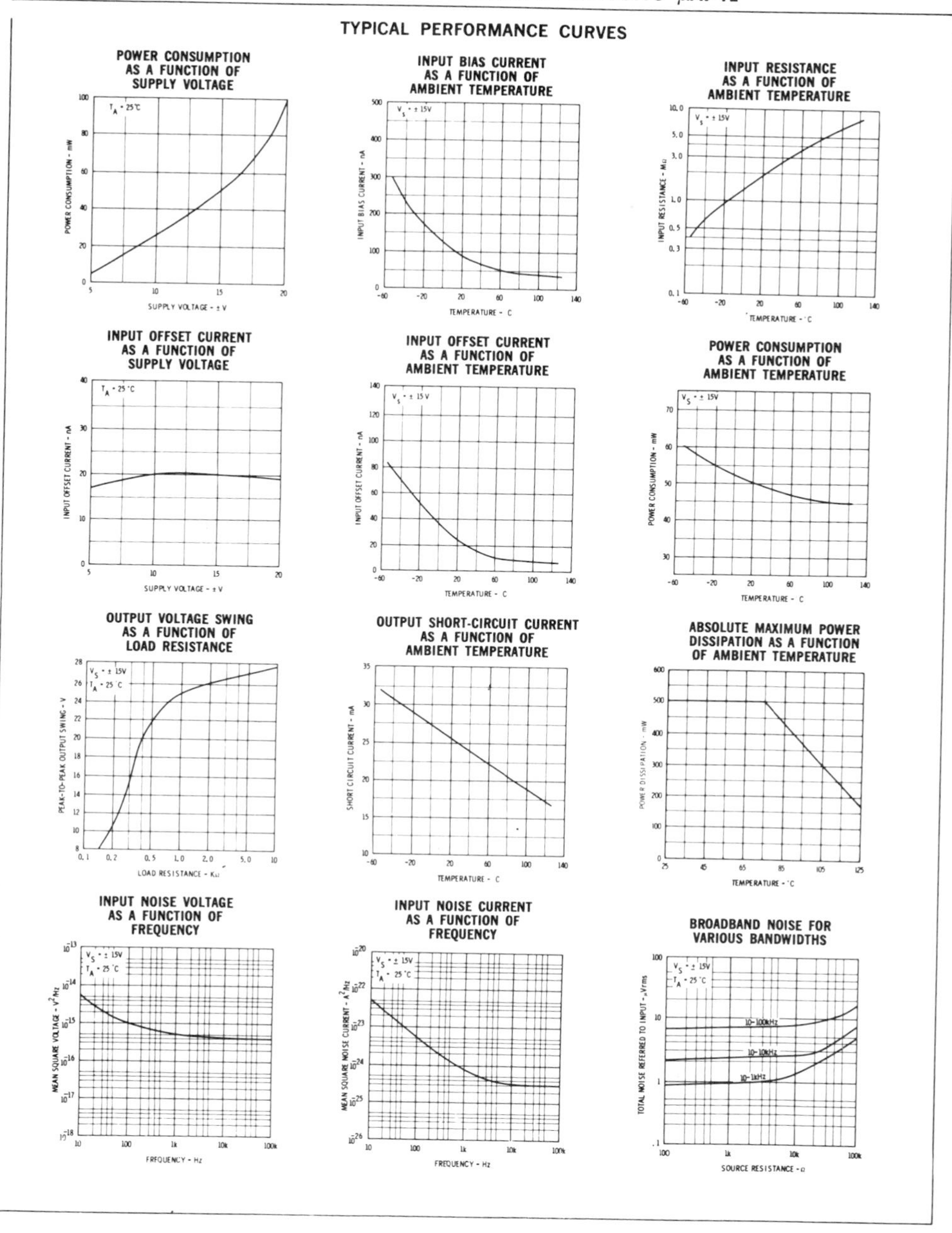
TYPICAL PERFORMANCE CURVES
POWER CONSUMPTION AS A FUNCTION OF SUPPLY VOLTAGE
INPUT BIAS CURRENT AS A FUNCTION OF AMBIENT TEMPERATURE
INPUT RESISTANCE AS A FUNCTION OF AMBIENT TEMPERATURE
INPUT OFFSET CURRENT AS A FUNCTION OF SUPPLY VOLTAGE
INPUT OFFSET CURRENT AS A FUNCTION OF AMBIENT TEMPERATURE
POWER CONSUMPTION AS A FUNCTION OF AMBIENT TEMPERATURE
OUTPUT VOLTAGE SWING AS A FUNCTION OF LOAD RESISTANCE
OUTPUT SHORT-CIRCUIT CURRENT AS A FUNCTION OF AMBIENT TEMPERATURE
ABSOLUTE MAXIMUM POWER DISSIPATION AS A FUNCTION OF AMBIENT TEMPERATURE
INPUT NOISE VOLTAGE AS A FUNCTION OF FREQUENCY
INPUT NOISE CURRENT AS A FUNCTION OF FREQUENCY
BROADBAND NOISE FOR VARIOUS BANDWIDTHS
SUPPLY VOLTAGE - ± V
TEMPERATURE - °C
LOAD RESISTANCE - KΩ
FREQUENCY - Hz
SOURCE RESISTANCE - Ω
10-100kHz
10-10kHz
10-1kHz

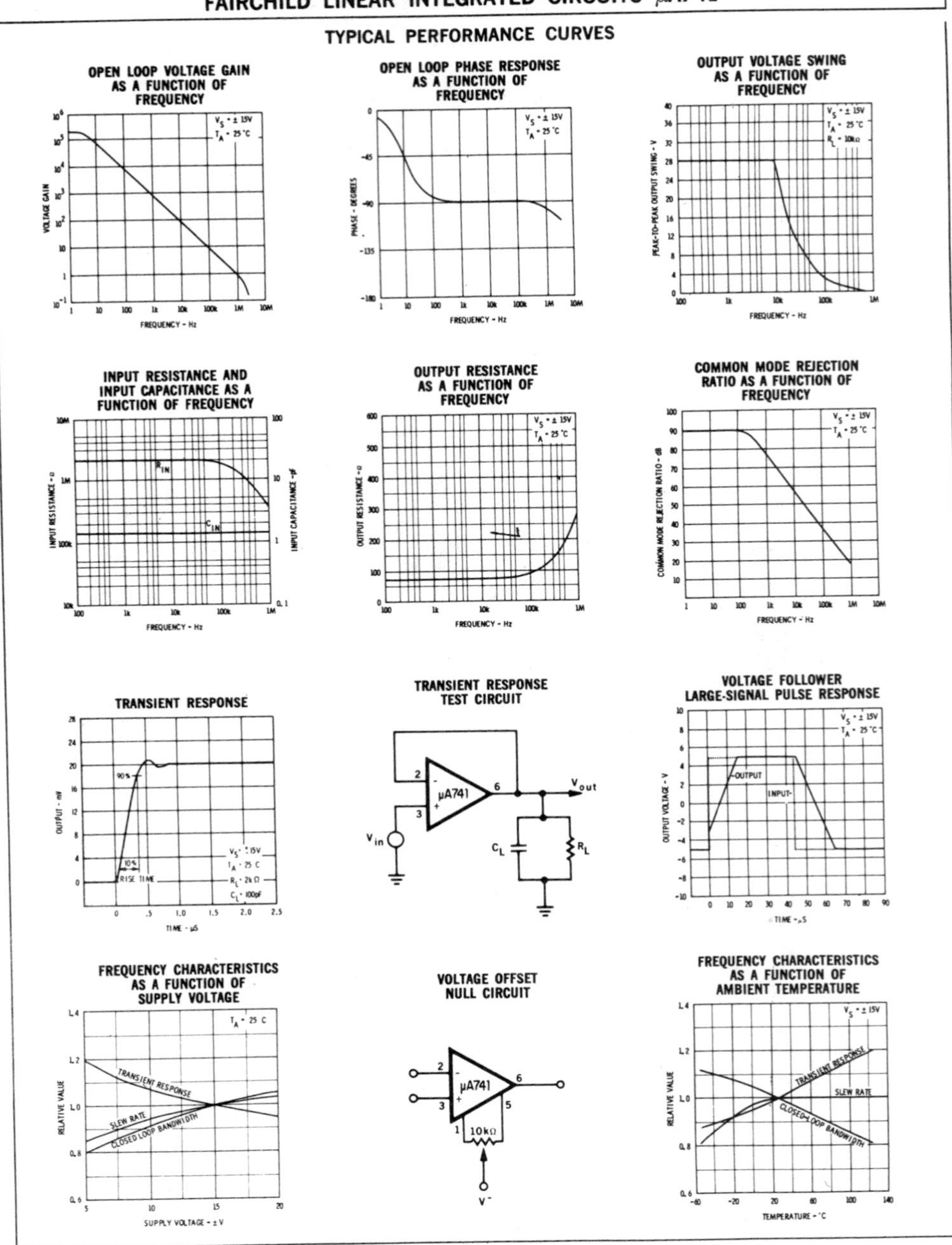
TYPICAL PERFORMANCE CURVES
OPEN LOOP VOLTAGE GAIN AS A FUNCTION OF FREQUENCY
VOLTAGE GAIN
FREQUENCY - Hz
OPEN LOOP PHASE RESPONSE AS A FUNCTION OF FREQUENCY
PHASE - DEGREES
OUTPUT VOLTAGE SWING AS A FUNCTION OF FREQUENCY
PEAK-TO-PEAK OUTPUT SWING - V
INPUT RESISTANCE AND INPUT CAPACITANCE AS A FUNCTION OF FREQUENCY
INPUT RESISTANCE - Ω
INPUT CAPACITANCE - pF
OUTPUT RESISTANCE AS A FUNCTION OF FREQUENCY
OUTPUT RESISTANCE - Ω
COMMON MODE REJECTION RATIO AS A FUNCTION OF FREQUENCY
COMMON MODE REJECTION RATIO - dB
TRANSIENT RESPONSE
OUTPUT - mV
TIME - μS
RISE TIME
TRANSIENT RESPONSE TEST CIRCUIT
μA741
VOLTAGE FOLLOWER LARGE-SIGNAL PULSE RESPONSE
OUTPUT VOLTAGE - V
OUTPUT
INPUT
TIME - μS
FREQUENCY CHARACTERISTICS AS A FUNCTION OF SUPPLY VOLTAGE
RELATIVE VALUE
TRANSIENT RESPONSE
SLEW RATE
CLOSED LOOP BANDWIDTH
SUPPLY VOLTAGE - ± V
VOLTAGE OFFSET NULL CIRCUIT
μA741
10kΩ
FREQUENCY CHARACTERISTICS AS A FUNCTION OF AMBIENT TEMPERATURE
RELATIVE VALUE
TRANSIENT RESPONSE
SLEW RATE
CLOSED-LOOP BANDWIDTH
TEMPERATURE - °C

DEFINITION OF TERMS

INPUT OFFSET VOLTAGE — That voltage which must be applied between the input terminals to obtain zero output voltage. The input offset voltage may also be defined for the case where two equal resistances are inserted in series with the input leads.
INPUT OFFSET CURRENT — The difference in the currents into the two input terminals with the output at zero volts.
INPUT BIAS CURRENT — The average of the two input currents.
INPUT RESISTANCE — The resistance looking into either input terminal with the other grounded.
INPUT CAPACITANCE — The capacitance looking into either input terminal with the other grounded.
LARGE-SIGNAL VOLTAGE GAIN — The ratio of the maximum output voltage swing with load to the change in input voltage required to drive the output from zero to this voltage.
OUTPUT RESISTANCE — The resistance seen looking into the output terminal with the output at null. This parameter is defined only under small signal conditions at frequencies above a few hundred cycles to eliminate the influence of drift and thermal feedback.
OUTPUT SHORT-CIRCUIT CURRENT — The maximum output current available from the amplifier with the output shorted to ground or to either supply.
SUPPLY CURRENT — The DC current from the supplies required to operate the amplifier with the output at zero and with no load current.
POWER CONSUMPTION — The DC power required to operate the amplifier with the output at zero and with no load current.
TRANSIENT RESPONSE — The closed-loop step-function response of the amplifier under small-signal conditions.
INPUT VOLTAGE RANGE — The range of voltage which, if exceeded on either input terminal, could cause the amplifier to cease functioning properly.
INPUT COMMON MODE REJECTION RATIO — The ratio of the input voltage range to the maximum change in input offset voltage over this range.
SUPPLY VOLTAGE REJECTION RATIO — The ratio of the change in input offset voltage to the change in supply voltage producing it.
OUTPUT VOLTAGE SWING — The peak output swing, referred to zero, that can be obtained without clipping.

TYPICAL APPLICATIONS

UNITY-GAIN VOLTAGE FOLLOWER

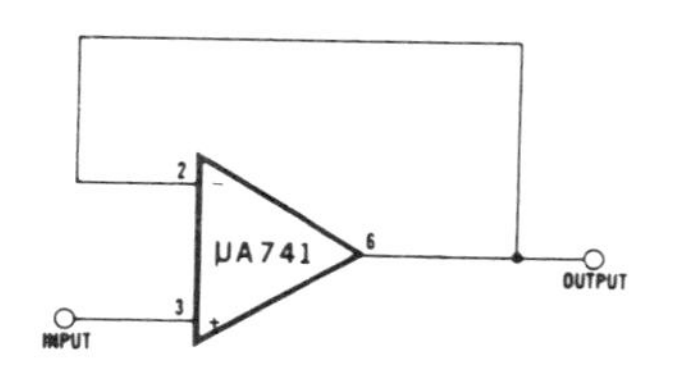

R_{IN} = 400 MΩ
C_{IN} = 1 pF
R_{out} << 1 Ω
B.W. = 1 MHz

NON-INVERTING AMPLIFIER

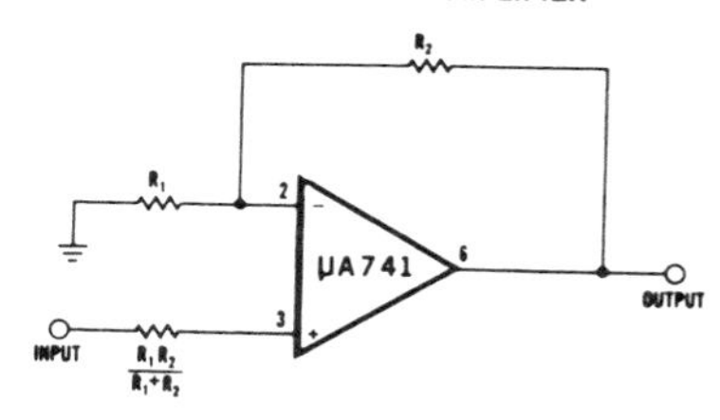

GAIN	R_1	R_2	B.W.	R_{IN}
10	1 kΩ	9 kΩ	100 kHz	400 MΩ
100	100 Ω	9.9 kΩ	10 kHz	280 MΩ
1000	100 Ω	99.9 kΩ	1 kHz	80 MΩ

INVERTING AMPLIFIER

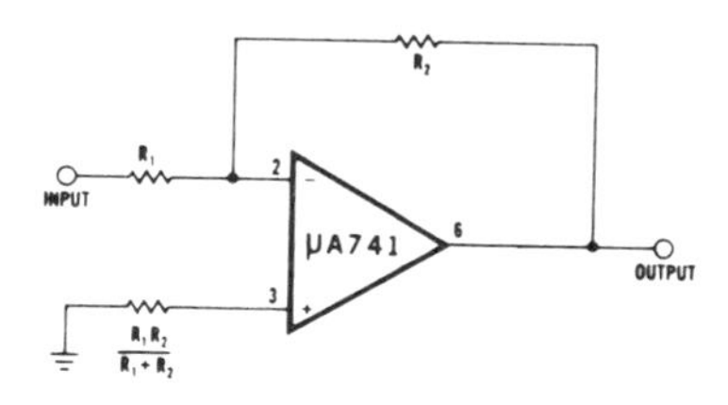

GAIN	R_1	R_2	B.W.	R_{IN}
1	10 kΩ	10 kΩ	1 MHz	10 kΩ
10	1 kΩ	10 kΩ	100 kHz	1 kΩ
100	1 kΩ	100 kΩ	10 kHz	1 kΩ
1000	100 Ω	100 kΩ	1 kHz	100 Ω

CLIPPING AMPLIFIER

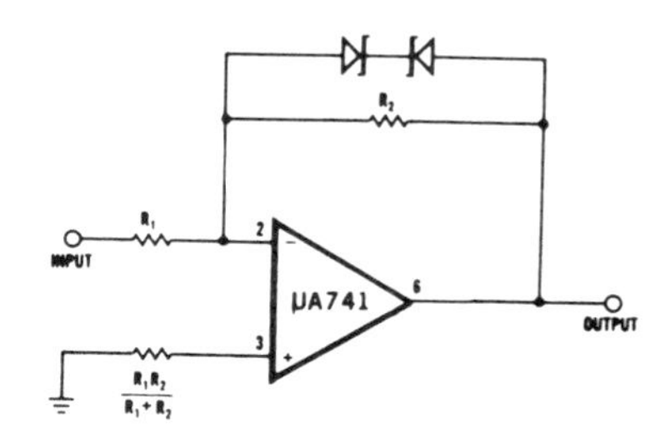

$$\frac{E_{out}}{E_{IN}} = \frac{R_2}{R_1} \text{ if } |E_{out}| \leq V_Z + 0.7\text{ V}$$

where V_Z = Zener breakdown voltage

TYPICAL APPLICATIONS

SIMPLE INTEGRATOR

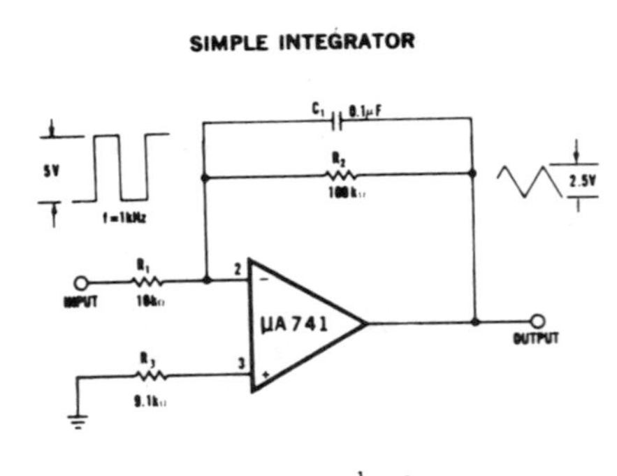

$$E_{out} = -\frac{1}{R_1C_1}\int E_{IN}dt$$

SIMPLE DIFFERENTIATOR

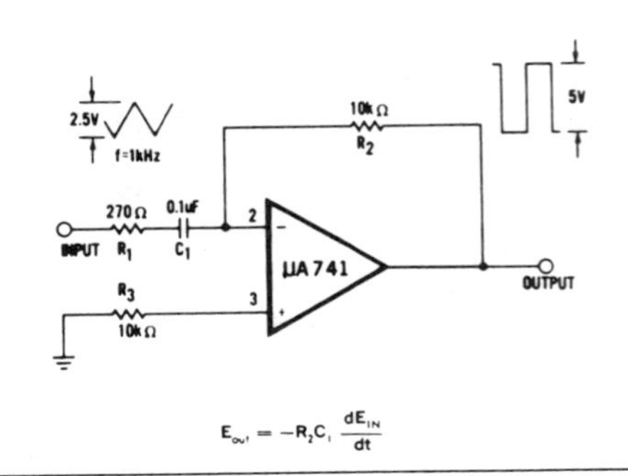

$$E_{out} = -R_2C_1\frac{dE_{IN}}{dt}$$

LOW DRIFT LOW NOISE AMPLIFIER

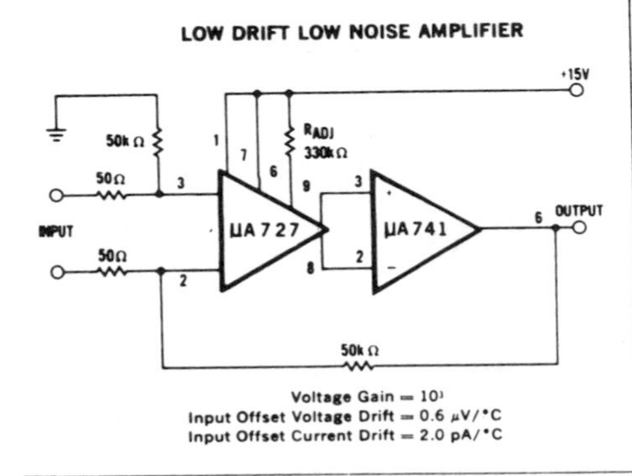

Voltage Gain = 10^3
Input Offset Voltage Drift = 0.6 μV/°C
Input Offset Current Drift = 2.0 pA/°C

HIGH SLEW RATE POWER AMPLIFIER

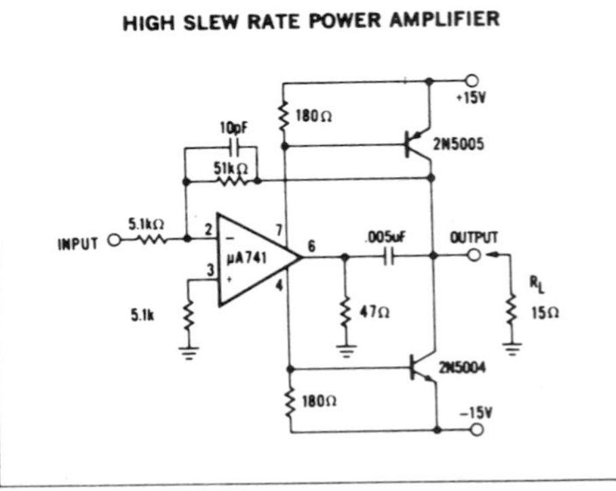

NOTCH FILTER USING THE μA741 AS A GYRATOR

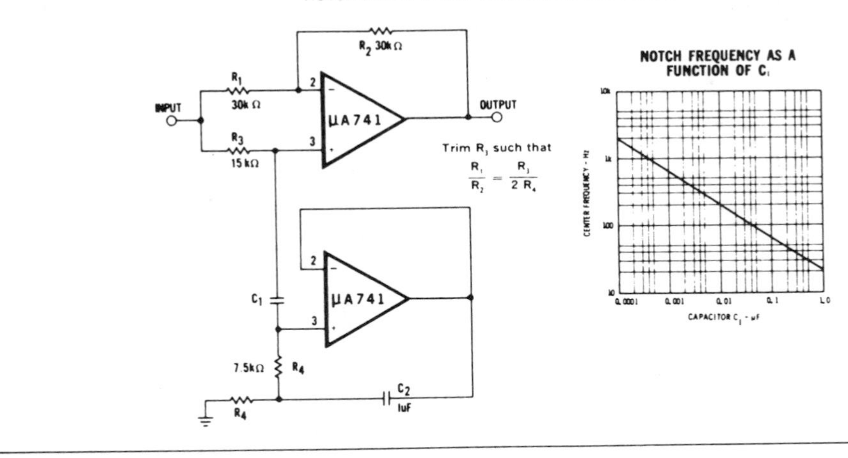

Index

C

D

L

M

N

O

P

R

S

T

U

V

W

Z